Encyclopedia of
Plant Physiology

New Series Volume 17

Editors

A. Pirson, Göttingen
M.H. Zimmermann, Harvard

Cellular Interactions

Edited by

H. F. Linskens and J. Heslop-Harrison

Contributors

A.A. Ager F.-W. Bentrup P. Bubrick J.A. Callow P.R. Day
S.C. Ducker K. Esser M. Galun J.L. Harley J. Heslop-Harrison
N.G. Hogenboom B.J. Howlett R.B. Knox T.M. Konijn
H.F. Linskens F. Meinhardt T. Nagata D. de Nettancourt
W. Reisser J.A.M. Schrauwen L. Sequeira L. Stange R.C. Starr
R.F. Stettler D.L. Taylor H. Van den Ende P.J.M. Van Haastert
F.R. Whatley J.M. Whatley L. Wiese W. Wiessner
N. Yanagishima M.M. Yeoman

With 198 Figures

Springer-Verlag
Berlin Heidelberg NewYork Tokyo 1984

Professor Dr. Hans F. Linskens
Botanisch Laboratorium
Faculteit der Wiskunde en Natuurwetenschappen
Katholieke Universiteit
Toernooiveld
NL-6525 ED Nijmegen, The Netherlands

Professor Dr. John Heslop-Harrison
University College of Wales
Welsh Plant Breeding Station
Plas Gogerddan Near Aberystwyth
SY23-3EB, Great Britain

ISBN 3-540-12738-0 Springer-Verlag Berlin Heidelberg New York Tokyo
ISBN 0-387-12738-0 Springer-Verlag New York Heidelberg Berlin Tokyo

Library of Congress Cataloging in Publication Data. Main entry under title: Cellular interactions. (Encyclopedia of plant physiology; new ser., v. 17). 1. Plant cells and tissues. 2. Cell interactions. I. Linskens, H.F. (Hans F.), 1921–. II. Heslop-Harrison, J. (John) III. Series. QK711.2.E5 vol. 17 [QK725] 581.1s [581.87] 83-27187
ISBN 0-387-12738-0 (U.S.).

Typesetting, printing and bookbinding: Universitätsdruckerei H. Stürtz AG, Würzburg
2131/3130-543210

Contents

4   Autotrophic Eukaryotic Freshwater Symbionts

W. REISSER and W. WIESSNER (With 4 Figures)

5   Autotrophic Eukaryotic Marine Symbionts

D.L. TAYLOR

6   Endosymbiotic Cyanobacteria and Cyanellae

W. REISSER (With 6 Figures)

7   Epiphytism at the Cellular Level with Special Reference to Algal Epiphytes

S.C. DUCKER and R.B. KNOX (With 4 Figures)

8   Genetics of Recognition Systems in Host-Parasite Interactions

P.R. DAY

12 Mating Systems in Unicellular Algae
L. WIESE

13 Colony Formation in Algae
R.C. STARR (With 67 Figures)

14 Cellular Interaction in Plasmodial Slime Moulds
J.A.M. SCHRAUWEN (With 5 Figures)

19 Mating Systems and Sexual Interactions in Yeast

N. YANAGISHIMA (With 13 Figures)

20 Cellular Interactions During Early Differentiation

L. STANGE (With 4 Figures)

21   Cellular Recognition Systems in Grafting

M.M. YEOMAN (With 5 Figures)

22   Cellular Polarity

F.-W. BENTRUP (With 7 Figures)

23   Fusion of Somatic Cells

T. NAGATA (With 5 Figures)

24   Pollen – Pistil Interactions

R.B. KNOX (With 21 Figures)

25 Mentor Effects in Pollen Interactions
R.F. STETTLER and A.A. AGER

26 Incompatibility
D. DE NETTANCOURT (With 2 Figures)

27 Incongruity: Non-Functioning of Intercellular and Intracellular Partner Relationships Through Non-Matching Information
N.G. HOGENBOOM

28 Allergic Interactions

B.J. Howlett and R.B. Knox (With 4 Figures)

List of Contributors

A.A. AGER
  College of Forest Resources
  University of Washington
  Seattle, WA 98195/USA

F.-W. BENTRUP
  Botanisches Institut I
  der Justus-Liebig-Universität
  Senckenbergstraße 17–21
  D-6300 Giessen/FRG

P. BUBRICK
  George S. Wise Faculty
  of Life Science
  Department of Botany
  Tel-Aviv University
  69978 Tel-Aviv/Israel

J.A. CALLOW
  Department of Plant Biology
  University of Birmingham
  P.O. Box 363
  Birmingham, B15 2TT/United Kingdom

P.R. DAY
  Plant Breeding Institute
  Maris Lane
  Trumpington
  Cambridge CB2 2LQ/United Kingdom

S.C. DUCKER
  School of Botany
  University of Melbourne
  Parkville
  Victoria 3052/Australia

K. ESSER
  Ruhr-Universität Bochum
  Lehrstuhl für Allgemeine Botanik
  Postfach 10 21 48
  D-4630 Bochum 1/FRG

M. GALUN
  George S. Wise Faculty of Life
  Science
  Department of Botany
  Tel-Aviv University
  69978 Tel-Aviv/Israel

J.L. HARLEY
  Commonwealth Forestry Institute
  South Parks Road
  Oxford, OX1 3RB/United Kingdom

J. HESLOP-HARRISON
  University College of Wales
  Welsh Plant Breeding Station
  Plas Gogerddan Near Aberystwyth
  SY23-3EB/United Kingdom

N.G. HOGENBOOM
  Institute for Horticultural Plant
  Breeding (IVT)
  P.O. Box 16
  Wageningen/The Netherlands

B.J. HOWLETT
  Plant Cell Biology Research Centre
  School of Botany
  University of Melbourne
  Parkville
  Victoria 3052/Australia

R.B. KNOX
  School of Botany
  University of Melbourne
  Parkville
  Victoria 3052/Australia

T.M. KONIJN
  Cell Biology and Morphogenesis Unit
  Zoological Laboratory
  University of Leiden
  Kaiserstraat 63
  2311 GP Leiden/The Netherlands

H.F. LINSKENS
  Botanisch Laboratorium
  Faculteit der Wiskunde en
  Natuurwetenschappen
  Katholieke Universiteit
  Toernooiveld
  6525 ED Nijmegen/The Netherlands

F. MEINHARDT
  Ruhr-Universität Bochum
  Lehrstuhl für Allgemeine Botanik
  Postfach 10 21 48
  D-4630 Bochum 1/FRG

T. NAGATA
  Department of Cell Biology
  National Institute for Basic Biology
  Myodaiji-cho, Okazaki 444/Japan

D. DE NETTANCOURT
  Commission of the European
  Communities
  Rue de la Loi 200
  DG XII (SDM 2/67)
  1049 Brussels/Belgium

W. REISSER
  Fachbereich Biologie der
  Universität
  Lahnberge
  D-3550 Marburg/FRG

J.A.M. SCHRAUWEIN
  Botanisch Laboratorium
  Universiteit Nijmegen
  Toernooiveld
  6525 ED Nijmegen/The Netherlands

L. SEQUEIRA
  Department of Plant Pathology
  University of Wisconsin
  Madison, WI 53706/USA

L. STANGE
  Arbeitsgruppe Pflanzenphysiologie
  Universität Gesamthochschule Kassel
  Heinrich-Plett-Str. 40
  D-3500 Kassel/FRG

R.C. STARR
  Department of Botany
  The University of Texas at Austin
  Austin, Texas 78712/USA

R.F. STETTLER
  College of Forest Resources
  University of Washington
  Seattle, WA 98195/USA

D.L. TAYLOR
  Centre for Environmental and
  Estuarine Studies
  University of Maryland
  Cambridge, MD 21613/USA

H. VAN DEN ENDE
  Department of Plant Physiology
  University of Amsterdam
  Kruislaan 318
  1098 SM Amsterdam/The Netherlands

P.J.M. VAN HAASTERT
  Cell Biology and Morphogenesis Unit
  Zoological Laboratory
  University of Leiden
  Kaiserstraat 63
  2311 GP Leiden/The Netherlands

F.R. WHATLEY
  Botany School
  South Parks Road
  Oxford OX1 3RA/United Kingdom

J.M. WHATLEY
  Botany School
  South Parks Road
  Oxford OX1 3RA/United Kingdom

L. WIESE
  Department of Biological Science
  Florida State University
  Tallahassee, FL 32306/USA

W. WIESSNER
  Abteilung für experimentelle
  Phykologie
  Pflanzenphysiologisches Institut
  der Universität
  Untere Karspüle 2
  D-3400 Göttingen/FRG

N. YANAGISHIMA
  Department of Biology
  Faculty of Science
  Nagoya University
  Chikusa-ku
  464 Nagoya/Japan

M.M. YEOMAN
  Botany Department
  University of Edinburgh
  King's Buildings
  Mayfield Road
  Edinburgh, Scotland/United Kingdom

1 Introduction

H.F. Linskens and J. Heslop-Harrison

The chapters of this volume deal with intercellular interaction phenomena in plants. Collectively they provide a broad conspectus of a highly active, if greatly fragmented, research field.

Certain limitations have been imposed on the subject matter, the most important being the exclusion of long-range interactions within the plant body. It is true that pervasive hormonal control systems cannot readily be demarcated from controls mediated by pheromones or information-carrying molecules with more limited spheres of action, but consideration is given in this volume to the main classes of plant hormones and their functions only incidentally, since these are treated adequately in other volumes of this Encyclopedia series (Volume 9–11) and in numerous other texts and reviews. Similarly, certain other effects, such as those associated with nutrients and ions, are not considered in any detail. Furthermore, we have excluded intracellular interactions, and also consideration of transport phenomena, which are treated in detail in Volume 3 of this Series. Other aspects of inter-cellular interaction, such as cell surface phenomena and implications of lectin-carbohydrate interactions, and plant-virus inter-relationships, are treated in other sections of this Encyclopedia (Volumes 13 B and 14 B, respectively). In the volume on physiological plant pathology (Volume 4 of this series) special attention has been given to host-pathogen interaction. These aspects of our subject will therefore be excluded in the present treatise.

On the other hand, the volume includes in its scope various genetic aspects of cellular interaction in plants, and also topics such as algal colony formation and short-range interaction during cell differentiation, which are particularly significant because of the opportunities they so clearly offer for further research.

In the introductory chapter we consider briefly aspects of recent animal cell biological work which have some pertinence for cellular interaction in plants, and offer an outline classification for the various phenomena as a background to the detailed treatments in the main body of the text.

We acknowledge with thanks the help of many people who have advised us – above all, we wish to thank Professor André Pirson, one of the series editors, who initiated the topic and made many valuable suggestions during the preparation of the volume.

2 Cellular Interaction: a Brief Conspectus

J. HESLOP-HARRISON and H.F. LINSKENS

2.1 Introduction

During the last few decades the investigation of cellular interactions has assumed a prominent place in both animal and plant biology in a remarkable range of contexts. These have included development and differentiation, sexual reproduction, parasitism, symbiosis, commensalism and various other associations between cells and tissues of the same or different genetic stocks. The scope has been extended also to embrace certain specific interactions between cells and non-living factors, such as substrates and toxins.

The topic is accordingly not one to which limits are readily set. If it can be said to be unified at all, it is scarcely by pre-occupation with a single set of problems, nor yet by the application of a particular armoury of techniques, but rather by being concerned with one seemingly widespread property, namely the capacity possessed by various eukaryotic cells for sensing constituents of their immediate environment, including other cells, and responding to the information so gained.

Many cellular interaction phenomena fall into the general category of "recognition" events. It so happens that the idea of cellular recognition was introduced into biology in a botanical context by DARWIN (1877), who, discussing an angiosperm heterostylous self-incompatibility system, commented that "... It may be said that the two pollens and the two stigmas mutually *recognise* each other." In the last two decades short-range cellular interactions in which a high degree of specificity exists have commonly been referred to as recognition systems, and the description has also been applied to interaction between the living cell and constituents of its non-living environment where high specificity or selectivity is expressed, whether or not an element of "learning" is involved.

In the context of animal cell biology the terminology is not always appropriate, and it is still less so in application to plants, where specific cellular interactions depend in the main upon responses attributable to pre-determined systems already present in or upon the partners, the presentation of which is not contingent upon previous intercourse. Such systems might be more appropriately termed cognitive. It may be noted nevertheless that it remains a matter of considerable scientific interest and practical importance to establish how widely "recognition" systems dependent on a pre-conditioning event – simple analogues, perhaps, of animal immune systems – do operate in plant tissues, as currently seems probable in certain forms of host response to parasite.

2.2 The Cell Surface

The present lively appreciation of the importance of short-range cellular interactions has originated in the main from intensive and wide-ranging work during the last two decades on animal cells and tissues, and it is to animal cell biologists that we owe most of the current ideas concerning the nature of the communication systems that might be involved (see, e.g. the various reviews in CURTIS 1978). A central conception has been that many functions of the cell are determined by reactions and responses centred in the first instance in the plasmalemma, now seen as a dynamic system, undergoing progressive changes in structure, chemistry and function in the course of growth, cell division, movement and differentiation. Most conceptions of membrane organisation accept the general basis formalized in the fluid-mosaic model of SINGER and NICOLSON (1972), according to which the continuous component of the membrane is the lipid bi-layer, within which are embedded the integral membrane proteins. A common view is that the stability of the proteins within the thickness of the membrane is determined by the asymmetric distribution of hydrophilic and hydrophobic groups in the molecule, which establishes that part will be held embedded in, or oriented across, the lipid bi-layer, with part protruding. Movement in the plane of the membrane is unconstrained by any inherent properties, the lateral distribution of the membrane proteins being governed by such forces as they may develop between themselves, or which may act upon them from outside the membrane. Virtually all the surface-mediated properties known in animal cells are satisfactorily accommodated in the fluid-mosaic model, and it is true to say that the essential proposition is now so thoroughly assimilated into the lore of animal cell biology as no longer to require justification, or indeed much further discussion.

The conception of membrane dynamism has been especially valuable in interpreting ontogenetic changes in the behaviour of the cell surface and in investigating interactions between cells and with the secular environment, as the briefest catalogue of contexts serves to show. Modulation of surface properties under genetical control may arise through the synthesis and insertion into the membrane of receptor or sensor molecules, with the possibility of great diversity according to their disposition and density (BODMER 1978). Chemical variation in lipid composition may affect permeability properties and reaction to stress, and the accretion of surface materials can control adhesivity, rigidity and shape. Motility effected by changes in membrane conformation may be controlled through linkage between the membrane proteins and microfilaments or other cytoskeletal components. Membrane properties might also be affected directly by interaction with other cells, or with constituents of the environment, for example through induced structural change, as in the phenomenon of capping (NICOLSON 1974).

Although the field is still fraught with many uncertainties and some of the interpretations of experimental observations remain highly speculative, the work on animal cells has undoubtedly thrown up results and ideas with far-reaching implications for plant cell biology, many yet to be followed up. Yet there are limits to the relevance of the experience with animal cells, resultant from the fact that some pathways of intercellular communication are not available to plants.

The membrane systems of plant cells are highly heterogeneous, and include one class, the chloroplast membranes, not represented at all in animal cells. But there is no reason to suppose that the outer cell membrane, the plasmalemma, is radically different in organization in the two kingdoms. Because of the severe technical difficulties associated with preparing the membranes in adequate quantity and purity for analysis, the chemical composition of plant plasmalemmas is not well-known, but from such evidence as is available, it appears that it is generally similar to that of animal cells in the major constituents, although there may be differences in the representation of certain classes of lipids (MORRÉ 1975). The physical organization, as expressed in permeability properties and observable structural features, is also broadly similar. It is particularly significant that recent work on the plasmalemmas of isolated plant protoplasts has shown that they share with those of animal cells the property of internal mobility, which is the central conception of the fluid mosaic model. This has been proved, for example, by the experiments of BURGESS and LINSTEAD (1977), who found that the receptor sites for gold-labelled concanavalin A on naked tobacco-leaf protoplasts undergo re-distribution following high-temperature treatment in a manner suggesting lateral movement in the plane of the plasmalemma.

2.3 The Plant Cell Wall

In respect to the properties of the cell membrane itself there is therefore nothing to suggest radical differences between plant and animal. The essential departure is, of course, in the possession by plant cells of an external wall. Animal cells are invested by sheaths of varying degrees of emphasis formed by the carbohydrate moieties of surface glycoproteins, and these are sometimes highly conspicuous in electron micrographs of suitably fixed and stained material. The universality – or at least near-universality – of this feature caused BENNETT (1969) to introduce the concept of the glycocalyx, and to suggest a direct homology between the animal cell carbohydrate investment and the plant cell wall. The usefulness of such a comparison is limited, however. The bulk of animal cells are not impeded in their interactions with other cells and with the outside world by their glycocalyces, and these may indeed form part of their means of communication. On the other hand, the plant cell wall can act as a very effective barrier isolating the plasmalemma from the direct impact of external factors.

The heterogeneous nature of plant cell walls needs no emphasis. The algae in particular show a spectacular chemical versatility in wall composition, the skeletal polysaccharides including xylans and mannans as well as glucans, associated with various matrix polysaccharides (PRESTON 1974). Moreover, it seems from the recent demonstrations that a glycoprotein constitutes a major part of the wall of the green alga *Chlorogonium* (ROBERTS and HILLS 1976), and that sporopollenin forms the outer sheath of the wall of *Chlorella* (GUNNING and ROBARDS 1976) indicates that there might be much still to be learned about the investments of many familiar algal groups. The skeletal polysaccharides of the primary walls of higher plants have long been known to be glucans with a predominance of β-1,4-linkages. However, with the accretion of new knowledge about the heterogeneity of the associated matrix polysaccharides resulting from the increasing sophistication of analytical techniques, the simpler views of the likely organisation of the wall of earlier years have given place to more complex formulations, and to models which essentially represent the primary wall as a form of macromolecule (KEEGSTRA et al. 1973). Simultaneously with the advance in knowledge of the carbohydrates of the primary wall, the role of protein as a wall constituent has been increasingly appreciated.

Covalently bound glycoprotein evidently forms an important structural element (LAMPORT 1970), and primary walls have commonly been found to be the repositories of enzymic and other proteins which seemingly have no structural role but are transferred into the wall by the secretory activity of the plasmalemma.

Viewing the complexity of wall structure and chemistry in the various plant groups, and especially the dynamic nature of wall metabolism in young cells, one may well agree with PRESTON's conclusion (1979): "The time has long gone since the wall of a living cell could be regarded merely as an outer, inactive, secreted envelope. It is now recognised instead as a delicately balanced entity with specific functions, open to the metabolic machinery of the plant, essential to the continuing existence of the cell, and therefore, in a tissue, to the continuation of the plant itself".

The point of the foregoing discussion is to emphasize that a major difference between plant and animal cells in respect to short-range communication must lie in the containment of most plant cells within the wall. At the same time one may note that there are circumstances in which wall-less plant cells do interact, and these are already providing valuable opportunities for exploring the applicability of some of the ideas and models for cellular interaction derived from animal systems. Mostly they are related to reproduction. They include gametic contact in algae and fungi, where the initial interaction may be between flagella or between naked gametes, or between specialized fertilization organs, the walls of which are eroded in contact zones. In archegoniates, the contact is between naked sperm and egg, or between naked sperm and an egg invested by a persistent wall or a special egg membrane; in the latter circumstances contact between the plasmalemmas first required the rupture of the barrier or the penetration of gelatinous sheaths derived from degenerating walls. In vascular groups with siphonogamous fertilization, direct gamete contact does not occur until the wall of the tube and the walls and membranes surrounding the female gametophyte, or the egg itself, are ruptured. Direct cellular contacts without the intervention of walls are otherwise rare, occurring mainly during brief phases of differentiation, as in cleavage stages associated with spore and gamete formation, or in certain instances of parasite invasion or in the formation of symbiotic associations. The last may be exemplified by intracellular organelles, if these are indeed symbionts.

2.4 Cell–Cell Communication: Plant and Animal Situations Compared

Contemplating the circumstances of the walled plant cell, the obvious question arising is, to what extent does the isolation preclude the functioning of the sensing mechanisms, centred in, or located upon, the plasmalemma, analogous to those which seemingly play so important a part in the life of the animal cell? We have noted that, whereas direct contact between cell membranes is a common circumstance in the animal body, it is not found in somatic plant tissues. The isolation of plant cells within polysaccharide or other envelopes may exclude certain types of electrical coupling, and certainly must prevent chemical signalling between cells dependent upon direct plasmalemma contacts. Except in the case of the slime moulds – dubious members of the plant kingdom in any event – specific communication between different individuals, whether of the same or different species, must initially be through, or be mediated by, the external walls, and must somehow circumvent their protective incrustations. Within the plant body, communication may similarly take place through the wall matrix, the so-called apoplastic system, but here the alternative route through the plasmodesmata, the symplastic system, is also available.

Notwithstanding Preston's dictum, it is improbable that the metabolic involvement of the wall extends to the possession of an active transport system. The passage of chemical signals through the apoplast must therefore depend upon diffusion. Since water is the continuous phase, the agents would seemingly need to be water-soluble, although the fact that lipidic secretions and the precursors of waxy wall incrustations undoubtedly do traverse watery polysaccharide walls in certain tissues indicates that mechanisms must be available for the transfer of hydrophobic materials, perhaps through the agency of protein carriers, as suggested in the model proposed by Hallam (1982). The slow rate of diffusion would seem to exclude the apoplastic pathway as the exclusive one for long-range intercellular communication, and the current view of the movement of auxin in the plant is that the major part of the pathway is within the symplast, movement by diffusion being restricted to the transfer from cell to cell through the intervening walls (Goldsmith 1977). However, short-range chemical communication can obviously be achieved through the secretion into the wall of diffusible factors.

A widely accepted view of the function of the plasmodesma is that it provides for communication between cells through the desmotubule, an extension of the endoplasmic reticulum which would allow the passage of materials contained within the lumen of the reticulum between cells, and also through the surrounding annulus, which would provide a cytosol link (Gunning and Robards 1976). This conception of the plasmodesma indicates that it might form a major channel for the transfer of signals between contiguous cells and throughout tissues. Some published dimensions for the two channels have suggested, further, that passage might be permitted for molecules of considerable size, perhaps even information-carrying peptides or proteins. This view is not supported by the most recent observations on the fine structure of plasmodesmata in *Azolla* (Overall et al. 1983), who conclude that the desmotubule has an extremely narrow lumen of cross-sectional area equivalent to only a few water molecules, and that the cytoplasmic annulus is partially occluded. These authors cautiously suggest that the passages between contiguous cells through the annulus may be of dimensions comparable to those of animal cell gap junctions, with a molecular weight exclusion limit in the range 800–1700. If it should prove that the results from *Azolla* have general validity, then it must be supposed that plasmodesmata would not normally provide for the transfer of large information-carrying protein molecules between cells of the plant body.

2.5 Specific Interactions: Models and Theories

Some of the principal types of close-range cellular interactions and "recognition" responses in animals where specificity may be shown are listed in Table 1. Investigation of these diverse systems has produced a wealth of data and led to the proposal of numerous explanatory hypotheses, many in conflict. Perhaps the fact that the field is still full of uncertainties and remains one for vigorous debate may offer some comfort for plant cell biologists in contemplating their own mass of unsolved problems.

Table 1. Examples of specific responses of cells to other cells and environmental factors mediated by the cell surface

Type of response	Expression and possible mechanisms	Examples
Cell–Cell		
Cell sorting	Chemotaxis; contact guidance; specific or differential adhesion	Colony formation; cell colony and tissue disaggregation experiments
Development and morphogenesis	Cell differentiation and directed growth and movement controlled by chemospecific recognition signals and/or contact guidance	Various examples in invertebrates and vertebrates
Fertilisation	Chemotaxis; contact recognition by surface receptors	All groups
Immune responses	Specific surface receptors, coupled to intracellular response mechanisms	General among vertebrates, with analogues of various kinds in other groups
Cell-Inanimate environment		
Food recognition	Cell surface and/or internal receptors	Unicells etc.
Antigen recognition	Specific surface receptors, coupled to intracellular response mechanisms	Vertebrates
Selective adhesion and locomotion	Surface receptors coupled to intracellular motility mechanisms	Fibroblasts, lymphocytes and other cells in culture

In summarizing the proposals that may have special significance in the study of plant systems, it is useful to begin by formalizing the possible modes of interaction in a simple classification, based in part upon the analysis given by LOEWENSTEIN (1968) of the pathways of communication between animal cells:

1. *Surface-Surface.* In the case of adhesion, specific ligands bound to the plasmalemma; in the case of communication, signal molecules and receptors on or in the plasmalemma.

2. *Surface Receptor, Mobile (Soluble) Signal.* The signal is synthesized in the transmitting cell, secreted into the external space, and accepted at the surface of the receiving cell.

3. *Internal receptor,* information transferred through gap junctions or cytoplasmic bridges.

The property of seemingly specific cell adhesion has been central to the discussion of animal cell recognition events since the work of WILSON (1907) on regeneration in sponges. WILSON's experiments showed that the formation of specific cell assemblages, a primitive form of morphogenesis, could arise through directed cell movement. The essential observation was that when the cells of the sponge were dissociated mechanically, they were capable of reassociating to form a sponge again, and that when the cells of different species were mixed, they sorted themselves out to re-form the original parent types. The behaviour of sponge cells has been adopted as a model for other types of cell-sorting, including those involved in animal embryogenesis. The experiment illustrates several processes. Evidently the response depends upon the capacity for movement, and ultimately upon contact and adhesion.

Adhesion is certainly a major factor in certain types of cellular interaction in plants, and its importance is exemplified especially be the work on the reproductive stages of yeasts and flagellate algae. The contrast between the situations in these two groups is highly instructive. The initial engagement between sexual *Chlamydomonas* cells is via the flagella; that between yeast cells, through wall contact, the flagellar membrane may be regarded as a differentiation of the plasmalemma, so the analogy with specific animal cell adhesion is close. The agglutination factors in the yeasts are carried on the outside of the wall (e.g. CRANDALL 1978), and the initial interaction is not between plasmalemmas, but between walls.

The yeasts provide a model for various types of cell interaction associated with reproductive stages in filamentous fungi, and parallel situations may be found also in algae, as in the male gamete–trichogyne interaction in red algae. Among vascular groups, adhesion has been stressed as a factor in the pollen–stigma interaction in angiosperms especially by KNOX and co-workers (CLARKE et al. 1979), and it is certainly significant in grafting. There is as yet no indication of high specificity in these cases, however.

Various molecular models have been proposed for specific cell adhesion, based both upon unipolar and bipolar complementarity (e.g. BURGER et al. 1975). A popular proposal for the lock-and-key relationship postulated for animal cells is that the specifity is based upon the presence of lectin-like molecules on the one partner and the appropriate complementary sugars or sugar sequences on the other (see e.g. the review by ROTH 1973). It is suggestive, therefore, that the work on *Chlamydomonas* (WIESE and WIESE 1975) and some of the evidence from yeasts (CRANDALL 1978) point towards carbohydrate–protein interaction. The attractiveness of this idea explains part of the intense current interest in – and speculation upon – the function of plant lectins in development, sexual interactions and in disease (MIRELMAN et al. 1975; CALLOW 1978; BOLWELL et al. 1979).

The ideas arising from the work on cell adhesion, agglutination and similar events have, of course, a wider significance in the context of cellular communication, for such phenomena form a subset of the whole class of interactions mediated by specific factors held at the cell surface. Indeed, it is common now to look upon the cell surface as a mediator in many developmental and differentiational processes in animals – the key to many crucial events, particularly in early ontogeny (SUBTELNY and WESSELLS 1980).

In plants it can hardly now be disputed that there must be some form of specific, short-range intercellular signalling during development and tissue differentiation. This has long been suspected in the differentiation of apical meristems (HESLOP-HARRISON 1963), but perhaps the most convincing indications at present come from cases of extremely local cell positioning, as in the well-known example of the differentiation of the grass stomatal complex. The precise disposition of the division planes – and of the microtubule assemblies foreshadowing them – which this illustrates can scarcely be explained except by appeal to extremely localized interaction between neighbouring cells. How, then, is this achieved?

It is in this connection that work such as that on the yeasts attains its special importance for the walled cells of higher plants. The findings illustrate without ambiguity that the factors responsible for the specific adhesion – the primary recognition, should one wish to use the terminology – are secreted from the cell into the wall and pass then to the surface, exercising their function at a distance from the plasmalemma. Pursuing the comparison with animal cells, it might be said that if the wall is not part of the cell membrane quite in the sense of BENNETT (1969), it acts here as agent of it. We have already noted that in addition to containing structural proteins, the walls of higher plant cells are often the repositories of other proteins, including glycoproteins and lipoproteins, at sites remote from the plasmalemma. The insertion of these from the parent cells been most fully investigated in pollen and stigmas, where granulocrine and exfoliative types of secretion are known. Vesiculation at the plasmalemma, including the formation of paramural bodies (MARCHANT and ROBARDS 1968), is a common feature of many developing plant cells in somatic tissues, and this may reflect active protein secretion; in other instances, mass-secretion systems are probably not involved, the transfer being by membrane perfusion, the so-called eccrine mode. The internal secretory systems are obvious candidates for the transfer of information-carrying molecules between somatic cells, and similarly those operating at the epidermis and its derivatives may be involved in external communication. The "dry" type of stigma provides an example from the higher plant to match the yeast cell; here the secretory products, including a complex spectrum of proteins, pass across the pectocellulosic wall and then through gaps in the cuticle to form an outer layer which is a prime candidate as a recognition site in the interaction with pollen (Mattsson et al. 1974; Heslop-Harrison 1982).

Referring, then, to the classification given earlier in this section, we see that plants have the potential for a form of cell–cell communication with something of the characteristics of both modes (1) and (2). Signalling may not be directly between contiguous plasmalemmas, but the transfer of information-carrying molecules into the wall and thence to the plasmalemmas of contiguous cells, or to outer surfaces for interaction with the cells of other individuals, provides the substitute (Heslop-Harrison 1978). An extension of this last principle provides the obvious model for short-range communication between cells not in contact, the characteristic of pheromonal or soluble-phase communication. This is now established in fungi (Duntze et al. 1970), and is certainly to be suspected in many other groups, as for example in conjugating filamentous algae, and in other circumstances such as in the sensing of host by parasite or symbiont.

As for the form of communication under (3) in the earlier classification, that mediated by the direct passage of information between neighbouring somatic cells through cytoplasmic links, it would seem from the findings of Overall et al. (1983) already mentioned that the potential for transfer of macromolecules via plasmodesmata in somatic tissues may be quite limited. The view that contiguous cells in the plant body do not freely exchange information-carrying molecules through this route either via cytosol links or desmotubules is supported by the mass of circumstantial evidence cited in the comprehensive review by Carr (1976), which indicates that cells in plasmodesmatal communication may pursue different pathways of differentiation.

There are plant tissues within which continuity between the cytoplasmic domains of different nuclei is established or maintained by links of a much more massive kind than those offered by plasmodesmata. In coenocytic algae and fungi, somatic nuclei effectively share a common cytoplasmic matrix; although it is indicative here that sexual differentiation evidently requires that the continuum be broken up by the formation of walls. In the angiosperm anther, the meiocytes are linked transiently by massive cytoplasmic connections, large enough even for the passage of organelles. During this period the cells act in close synchrony, implying that they share the informational molecules that are presumably concerned in controlling their activities. Later, the continuities are broken, and the meiotic products become ensheathed by uninterrupted callosic walls. Thenceforth their behaviour diverges, both in the timing of further development, and in their metabolic activities; they may, for example, show segregation for haplophase-active genes. This indicates, in turn, that insulation from interaction with neighbouring cells is a prerequisite for the expression of genetic independence in the closely packed environment of the anther, and that callosic ensheathment provides a protection that cellulosic walls do not (Heslop-Harrison 1964).

Direct cellular contacts may also be important in certain types of host parasite interaction. One of the most fruitful developments in this connection has been the gene-for-gene concept of FLOR (1962). According to this hypothesis, for each "resistance" gene in the host there is a specific and related gene for virulence in those parasites to which it is susceptible (PERSON et al. 1962). It has been suggested that temporary intercellular bridges permit the exchange of cellular inducer substances between parasite and host. After crossing the bridges, the inducer substances are envisaged as attaching themselves to repressor substances produced by regulator genes of the recipient. Accepting the Jacob-Monod model of gene regulation, it can be predicted that the protein constitution of the receiver will resemble that of the inducer. Therefore the transfer of inducers may evoke the production of altered proteins, resulting in one or more changes in the metabolic pathways, perhaps in this way triggering the formation of toxins, controlling the expression of other genes, or forming so-called immunological masks in the cells (JONES 1967). While these proposals remain largely hypothetical, they do suggest ways in which intercellular interactions may be mediated in the special case of host–parasite relationships.

2.6 Self and Non-Self

The immune systems of vertebrates provide for the tolerance of cells and cell products of an individual's own body, while enabling it to react against foreign antigens; in the jargon, to recognize and reject "non-self" while ignoring "self". In apparent contrast, some of the best-known recognition systems in plants operate in the converse mode, acting to reject "self" while accepting "non-self": these are, of course, the self-incompatibility systems which enforce outbreeding in various plant groups. Self-recognition and its implication for animal immune systems were discussed by BURNET (1971), who cited a remarkable self-rejection system, that of the colonial tunicate, *Botryllus*. As in many angiosperm self-incompatibility systems, the breeding system in *Botryllus* is controlled by a single locus with many alleles; any male gamete carrying an allele in common is rejected by diploid tissue surrounding the ovum, thus ensuring that all zygotes are heterozygous at the controlling locus. Thus far the formal similarity with plant self-incompatibility is obviously very great (MORGAN 1910, 1942); but the system reveals another foible. Fusion between the diploid colonies will take place if they possess a gene at the controlling locus in common; if both alleles are unlike, then a barrier is formed at the contact zone. The recognition system in *Botryllus,* therefore, acts to ensure both acceptance *and* rejection of "self" in the different contexts of somatic and sexual fusion.

BURNET concluded that in this system there must be three distinct sets of recognition phenomena, (1) recognition of the presence of a common allele between somatic cells in contact which ensures that the somatic rejection reaction is inhibited; (2) recognition of the absence of a common allele between somatic cells in contact, but at the same time of conspecificity ("self" in the wider sense) so that the rejection reaction is mounted; and (3) recognition of the presence of a common allele between a male gamete and somatic tissue surrounding the ovum which again leads to rejection. His further analysis of the implications of systems like this is of considerable interest in the context of plants. He noted that the common view of the nature of specific "recognition" reactions, whether they be in the context of reaction to antigens, drug-receptors or enzyme and substrate, is that they must be based upon steric complementarity; but observed that in the case of "self" recognition, acceptance of this axiom left the question of how the same allele in different cells could code for complementary recognition products, "a still totally unresolved problem".

The same problem is inherent in other examples of self-recognition in animals, and it has arisen in a striking way in recent work on the vertebrate immune system (Katz 1978). The precursors of antibody-forming cells, B cells, interact in the initiation of immune responses with T cells, which perform as yet undefined auxiliary functions. T cells from a particular animal are capable of co-operating with B cells of histocompatible, but not with those of histo-*in*compatible, donor. Such a response indicates mutual recognition between T and B cells, but in this instance recognition of common genetic origin, and thus of "self". Katz (1978) argues that "self-recognition is the critical mechanism by which cell–cell communication takes place, certainly in the immune system, and perhaps as well in other control processes of differentiation and organogenesis", and reaches the unorthodox conclusion that the "... immune system appears to have evolved with inherent mechanisms for self-recognition, rather than having developed mechanisms to prevent such recognition".

Like Burnet (1971), Katz (1978) observes that no decision can yet be made as to whether self-recognition in the higher animal involves complementary interactions or like–like interactions, and notes the difficulty of explaining complementarity through the action of one and the same allele in different cells. While this must still be accounted a major problem in both animal and plant systems, it is noteworthy that hypotheses have been proposed for mechanisms through which such an end might be achieved. A particularly interesting proposal is that cells may interact through surface-held glycosyl transferases on the one and complementary sugar groupings on the other, an idea that links specific cell–cell recognition phenomena with the enzyme–substrate relationship (see the review by Roth 1973). Such schemes may allow the reciprocal specificity to be governed through the action of the same allele on each side, and this may be the resolution of the dilemma. Convincing evidence for systems of this kind appears still to be lacking, however. It may be that they could be tested to good effect with recognition systems involving walled plant cells, where the glycosyl moieties may be integral components of the polysaccharide wall and the complementary receptors themselves located in sites remote from the plasmalemma.

2.7 Secondary Responses

This last proposition raises the question of the "second message", which assumes a special dimension in walled plant cells, as various chapters in this volume show. It is characteristic of animal immune systems that the initial interaction at the plasmalemma, the recognition event itself, is followed by a sequence of responses – mitosis and cell proliferation, or lysis and the release of antibody. With the possible exception of simple adhesion reactions, this is no doubt generally true for other cases of specific cell communication in the animal body, from those involving cell contact to long-range hormonal systems, such as in the action of glycoprotein gonadotrophic hormones, the surface binding sites of which have been identified (e.g. MENDELSON et al. 1975). The implication is that the plasmalemma acts as a transducer, generating after the appropriate stimulus secondary stimuli which provoke various kinds of intracellular activity. In several cases, the "second messenger" has been identified as cyclic-AMP.

What kind of transducing mechanism can be envisaged for walled plant cells when the primary recognition reaction is remote from the plasmalemma? One model derived from the animal cell example might be based upon the proposition that the binding reaction on or in the wall generates a second message, which then moves to a further set of receptors at the plasmalemma, which in turn activate the metabolic response. At present it is difficult to see how reactions such as that induced at the plasmalemma of a stigma cell by contact of the outer face of the wall with pollen-derived proteins from an incompatible partner (DICKINSON and LEWIS 1973) can be explained without invoking a relay system at least of this degree of elaboration, and this has been formalized in a model for the response (HESLOP-HARRISON et al. 1975). Numerous instances can be quoted from algae and fungi involving similar induced changes in activity at the plasmalemma after the initial recognition event, a common yet striking example being the wall-softening and subsequent directional outgrowth of conjugation tubes that follows upon the first cell contact in the sexual processes of many genera. It may be that these will provide the most suitable material for unravelling the stages of the dialogue and defining them in molecular terms.

In some instances, the cells responding as a result of the initial recognition event may be remote from the site at which it takes place; reported reactions include the induction of wilting and protein synthesis. It is not excluded here that the long-range hormonal systems of the plant may be involved. However, other possibilities are not to be overlooked, including electrical coupling. A relationship has been demonstrated between the presence of special cells and the registration of bioelectric potential changes, and we may glimpse here another type of transducing mechanism. Analogues are of course well known in the leaf motility systems of *Mimosa* and *Dionaea*.

2.8 Types of Cellular Interaction in Plants

The interactions considered in this volume fall into three principal categories, (1) between cells and tissues of the same physiological organism, (2) between those of the same species, but of different individuals, and (3) between different species.

Category (1) is concerned with interactions within the plant body, between cells which are, at least initially, genetically identical. The category subsumes the reactions involved in development and differentiation. As we have already noted, in the plant body no real distinction can be made between long- and short-distance signalling between cells and tissues, for range of action does not enter into the definition of a hormone. However, as we have tried to emphasize in this chapter, it seems that as the analysis of controlling systems operating within the plant body advances a distinction of a different kind will have to be made – between long-range, *low* specificity regulators of the character of the known plant hormones, and short-range communicators of *high* specificity, the principal feature of which is that they carry more information. The distinction has, of course, long been acknowledged in animal physiology.

Categories (2) and (3) share in common the fact that the communication occurs between physiologically independent individuals which normally will be genetically distinct. The signalling may be between cells in contact, or over a physical gap, in which case the communication may conveniently be described as pheromonal, following the animal model. The interactions in category (1) are mainly sexual (although grafting provides an instance of an artificially imposed somatic association between genetically and physiologically distinct individuals which may have parallels in the natural environment). In the normal circumstances of sexual interaction, the genetical differences between the partners will be mainly significant for the functioning of the reproductive processes, and they are often associated with sophisticated signalling systems by which the potential mates recognise each other and respond appropriately to exchanged stimuli. Category (3) includes a wide range of interactive associations where members of different groups are involved.

A general classification is provided in Table 2, which includes an assessment of the specificity that may be displayed in the different circumstances and lists a few conspicuous examples of particular relationships.

It is already evident that in the inter-individual communication systems of categories (2) and (3) models can be found for the kinds of controls that are likely to be concerned in development and differentiation within the individual. This is notably so for the types of interaction concerned in sexual processes, which provide analogues for long-range hormonal control, shorter-range pheromonal signalling, and control by short-range agencies which transfer information faithfully reflecting the genetic constitution of the interacting cells. The investigation of these systems in which the partners and their products can be investigated independently and in all stages of the interaction may well provide critical clues for interpreting events in the more confused environment of the soma.

Various other ways of classifying cellular interactions than that of Table 2 have been developed (MALCOLM 1966). One of the most convenient approaches is to distinguish between spatial and functional interactions, a distinction which can be applied to host–pathogen interactions, which constitute a special case of inter-relations between systems of different genetic background (LINSKENS 1968).

Table 2. A classification of associations between living plants, illustrating different degrees of intimacy and specificity

Type	Specificity	Examples
Hemiepiphytism Epiphyte is independently rooted, at least initially, but gains support from the host	Low: host choice usually related to mechanical factors such as bark texture	Many common lianes
Epiphytism Epiphyte gains support, protection and sometimes favourable microclimate	Low to moderate host specificity	Many vascular epiphytes, including ferns, bromeliads and orchids
	Moderate to high host specificity	Associations between unicellular and multicellular algae, and between algae, mosses and ferns and woody vascular plants
	High host specificity	Associations between various groups of marine algae, and between bryophytes and vascular plant hosts
Symbiosis Partners both gain benefit from the association, but the advantage may be very one-sided	Low to very high host specificity	Bacteria and fungi in the rhizosphere and phyllosphere; angiosperm root systems in association with mycorrhizal fungi; algae and fungi in the lichen symbiosis; legumes and other angiosperms in association with nitrogen fixing bacteria and blue-green algae
Hemiparasitism Parasite is autotrophic, but is partly dependent on the host for nutrients	Moderate to high host specificity	Root hemiparasites, including species of Scrophulariaceae and Santalaceae; stem hemiparasites, including species of Loranthaceae
Parasitism Ectoparasitism Parasite remains mainly on the exterior of the host	Moderate to high host specificity	Many vascular plant parasites, including *Cuscuta* and *Lathraea* spp., and species of Orobanchaceae, Balanophoraceae, Rafflesiaceae
Endoparasitism Parasite develops mainly within the tissues or cells of the host	High to very high host specificity	Many viral, bacterial and fungal pathogens attacking plant groups from fungi and algae to angiosperms

References

Bennett HS (1969) In: Lima de Faria A (ed) Handbook of molecular cytology. North Holland, Amsterdam, pp 1261–1293

Bodmer WF (ed) (1978) Genetics of the cell surface. A Royal Society Discussion. Proc R Soc Lond B 202:1–189

Bolwell GP, Callow JA, Callow ME, Evans LV (1979) Fertilisation in brown algae. II. Evidence for lectin-sensitive complementary receptors involved in gamete recognition in *Fucus serratus*. J Cell Sci 36:19–30

Burger MM, Turner RS, Kuhns WJ, Weinbaum G (1975) A possible model for cell–cell recognition via surface macromolecules. Phil Trans R Soc Lond B 271:379–393

Burgess J, Linstead PJ (1977) Membrane mobility and the concanavalin A binding system of the plasmalemma of higher plant protoplasts. Planta 136:253–259

Burnet FM (1971) "Self-recognition" in colonial marine forms and flowering plants in relation to the evolution of immunity. Nature 232:230–235

Callow JA (1978) Recognition, resistance and the role of plant lectins in host–parasite interactions. In: Preston RD, Woolhouse HW (eds) Advances in botanical research. Academic Press, London New York, pp 1–49

Carr DJ (1976) Plasmodesmata in growth and development. In: Gunning BES, Robards AW (eds) Intercellular communication in plants: studies on plasmodesmata. Springer, Berlin Heidelberg New York, pp 243–287

Clarke A, Gleeson P, Harrison S, Knox RB (1979) Pollen–stigma interactions: identification and characterisation of surface components with recognition potential. Proc Natl Acad Sci USA 76:3358–3362

Crandall M (1978) Mating-type interactions in yeasts. In: Curtis ASG (ed) Cell–cell recognition. Symp Soc Exp Biol Vol 32. Cambridge Univ Press, Cambridge London New York Melbourne, pp 105–120

Curtis ASG (1978) Cell–cell recognition. Symp Soc Exp Biol Vol 32. Cambridge Univ Press, Cambridge London New York Melbourne

Darwin C (1877) The different forms of flowers on plants of the same species. Murray, London

Dickinson HG, Lewis D (1973) Cytochemical and ultrastructural differences between intraspecific compatible and incompatible pollinations in *Raphanus*. Proc R Soc Lond B 183:21–28

Duntze W, Mackay V, Manney TR (1970) *Saccharomyces cerevisiae*: a diffusible sex factor. Science 168:1472–1473

Flor HH (1942) Inheritance of pathogenicity in *Melampsora lini*. Phytopathology 32:653–669

Goldsmith MHM (1977) The polar transport of auxin. Annu Rev Plant Physiol 28:439–478

Gunning BES, Robards AW (1976) Plasmodesmata and symplastic transport. In: Wardlaw IE, Passioura J (eds) Transport and transfer processes in plants. Academic Press, London New York, pp 15–41

Hallam ND (1982) Fine structure of the leaf cuticle and the origin of leaf waxes. In: Cutler DF, Alvin KL, Price CE (eds) The plant cuticle. Linn Soc Symp Ser Vol 10. Academic Press, London New York San Francisco, pp 197–214

Heslop-Harrison J (1963) Sex expression in flowering plants. Brookhaven Symp Biol 16:109–122

Heslop-Harrison J (1964) Cell walls, cell membranes and protoplasmic connections during meiosis and pollen development. In: Linskens HF (ed) Pollen physiology and fertilisation. North-Holland, Amsterdam, pp 39–47

Heslop-Harrison J (1978) Genetics and physiology of angiosperm incompatibility systems. In: Bodmer WF (ed) Genetics of the cell surface. Proc R Soc Lond, pp 73–92

Heslop-Harrison J (1982) Pollen–stigma interaction and cross-incompatibility in the grasses. Science 215:1358–1364

Heslop-Harrison J, Knox RB, Heslop-Harrison Y, Mattsson O (1975) Pollen-wall proteins: emission and role in incompatibility responses. In: Duckett JG, Racey PA (eds) The biology of the male gamete Suppl Vol 1. Biol J Linn Soc Vol 7, pp 189–202

Jones BM (1967) How cells interact. Sci J 3:73–78

Katz DH (1978) Self-recognition as a means of communication in the immune system. In: Curtis ASG (ed) Cell-cell recognition. Symp Soc Exp Biol Vol 32. Cambridge Univ Press, Cambridge London New York Melbourne, pp 411–428

Keegstra K, Talmadge KW, Bauer WD, Albersheim P (1973) The structure of plant cell walls. III. A model of the walls of suspension-cultured sycamore cells based on the interconnections of the macromolecular components. Plant Physiol 51:188–197

Lamport DTA (1970) Cell wall metabolism. Annu Rev Plant Physiol 21:235–270

Linskens HF (1968) Host–pathogen interactions as a special case of interrelations between organisms. Neth J Plant Pathol 74:(Suppl 1), 1–8

Loewenstein WR (1968) Communication through cell junctions. Implications in growth control and differentiation. Dev Biol Suppl 2:151–183

Malcom WM (1966) Biological interactions. Bot Rev 32:243–254

Marchant R, Robards AW (1968) Membrane systems associated with the plasmalemma of plant cells. Ann Bot 32:457–470

Mattsson O, Knox RB, Heslop-Harrison J, Heslop-Harrison Y (1974) Protein pellicle of stigma papillae as a probable recognition site in incompatibility reactions. Nature 247:298–300

Mendelson C, Dufau ML, Catt KJ (1975) Gonadotrophin binding and the stimulation of cyclic adenosine 3′:5′-monophosphate and testosterone production in isolated Leydig cell. J Biol Chem 250:8813–8823

Mirelman ME, Galun E, Sharon N, Lotan R (1975) Inhibition of fungal growth by wheat germ agglutinin. Nature 256:414–416

Morgan TH (1910) Cross- and self-fertilization in *Ciona intestinalis*. Roux Arch Entwicklungsmech 30:206–235

Morgan TH (1942) The genetic and the physiological problems of self-sterility in *Ciona*. J Exper Zool 40:199–228

Morré DJ (1975) Membrane biogenesis. Annu Rev Plant Physiol 26:441–481

Nicolson GL (1974) The interaction of lectins with animal cell surfaces. Int Rev Cytol 39:89–190

Overall RL, Wolfe J, Gunning BES (1983) Intercellular communication in *Azolla* roots: I. Ultrastructure of plasmodesmata. Protoplasma 111:134–150

Person C, Samborski DJ, Rohringer R (1962) The gene-for-gene concept. Nature 204:561–562

Preston RD (1974) The physical biology of plant cell walls. Chapman and Hall, London

Preston RD (1979) Polysaccharide conformation and cell wall function. Annu Rev Plant Physiol 30:55–78

Roberts K, Hills GJ (1976) The crystalline glycoprotein cell wall of the green alga *Chlorogonium elongatum*: a structural analysis. J Cell Sci 21:59–70

Roth S (1973) A molecular model for cell interaction. Quart Rev Biol 48:54–63

Singer SJ, Nicolson GL (1972) The fluid mosaic model of the structure of cell membranes. Science 175:720–731

Subtelny S, Wessells NK (eds) (1980) The cell surface: mediator of developmental processes. 38th Symp soc dev biol. Academic Press, London New York

Wiese L, Wiese W (1975) On sexual agglutination and mating type substances in isogamous dioecious Chlamydomonads. IV. Unilateral inactivation of the sex contact capacity in compatible and incompatible taxa by α-mannosidase and snake venom protease. Dev Biol 43:264–276

Wilson HV (1907) On some phenomena of coalescence and regeneration in sponges. J Exp Zool 5:245–258

3 Evolutionary Aspects of the Eukaryotic Cell and Its Organelles

J.M. WHATLEY and F.R. WHATLEY

3.1 Introduction

Fundamental differences exist between prokaryotic and eukaryotic cells (DOUGHERTY 1957). Of these the defining feature is usually taken to be the presence in eukaryotic cells of the nucleus with its envelope-pore membrane complex, and its absence from prokaryotes. Other major distinctions include the presence in eukaryotes of 80S ribosomes (as opposed to 70S ribosomes in prokaryotes) and of a number of discrete membrane bound organelles like mitochondria and microbodies which specialize in different metabolic activities. CAVALIER-SMITH (1981) lists 22 characters which he considers to be universally present in eukaryotes but absent from prokaryotes. These distinguishing characteristics of the eukaryotic cell fall into three broad groups; (1) those associated with the siting, form, production and expression of nucleic acids (2) the endomembrane system which serves as a means both of communication and of organelle segregation and (3) the systems of actin microfilaments and tubulin microtubules, which in the modern eukaryotic cell provide the basis of plasticity and mobility for the cell as a whole, and for its constituents.

The purpose of this paper is to discuss the possible modes of origin of eukaryotic organelles, their separate functions and the ways in which these functions are integrated into the overall metabolism of the plant cell.

Hypotheses on the origin of eukaryotic organelles fall into two classes; those which suggest (1) that all organelles evolved within the cell itself from components already present (autogenous origin) and those which suggest (2) that some organelles evolved from organisms, once free-living, which were taken up as symbionts by host cells (endosymbiotic origin). However, before it is possible to discuss the evolutionary origin of organelles it is first necessary to consider the origin of the eukaryotic cell itself.

3.1.1 The Origin of the Eukaryotic Cell

Until recently it has been assumed that the first eukaryotic cell evolved directly from a prokaryote. CAVALIER-SMITH (1981), for example, considers that the first eukaryote may have been similar to a non-flagellated (walled) fungus which had evolved from a prokaryotic ancestor, perhaps a purple non-sulphur bacterium. Alternatively, it has been suggested that the prokaryotic ancestor may have resembled either a blue-green alga (CAVALIER-SMITH 1975) or the archaebacterium *Thermoplasma acidophilum* (SEARCY et al. 1978) and that the resulting protoeukaryote may have been either an alga, perhaps similar to a red alga (UZZELL and SPOLSKY 1974) or an amoeba, similar to the simple (wall-less) giant amoeba, *Pelomyxa palustris* (JOHN and WHATLEY 1975, 1977).

By contrast, WOESE (1981) and STACKEBRANDT and WOESE (1981), largely on the basis of detailed analysis of ribosomal RNA's and supported by DNA-DNA hybridization studies, have proposed a division of living organisms originating from a common ancestor (the progenote) into three primary kingdoms, the Archaebacteria, the Eubacteria (both of which today show prokaryotic features as listed below) and a third leading to the eukaryotes. An early divergence of the three groups is proposed, but several different molecular and structural traits are today shared by different pairings of these three primary groups of organisms.

It is not known when, in relation to the earth's geologic past, the first eukaryotic cell evolved. RAFF and MAHLER (1975) and CAVALIER-SMITH (1981) strongly believe that there is good evidence to suggest that the first eukaryote was aerobic. Equally the authors believe that the available evidence points towards its principal energy metabolism being anaerobic (glycolysis), though this may have been combined with a few essential reactions involving oxygen and confined to certain truncated membrane-bound electron transport sequences. There is possibly confusion in applying the term aerobic to organisms that have a requirement for certain synthetic reactions involving O_2 e.g. demethylation of sterols, but which have a fermentative energy metabolism.

3.1.2 The Acquisition of Eukaryotic Features

The sequence of acquisition of the various distinguishing features of the eukaryotic cell and the modes of origin of its organelles are also matters of dispute. Although the presence of a nucleus is usually considered the defining characteristic of a eukaryote (DOUGHERTY 1957), it is likely that an important initial stimulus for the evolution of a eukaryotic cell was the establishment of an endomembrane system capable of exchanging large molecules and aggregates with the external environment by endo- and exo-cytosis (STANIER 1970). Such a system implies at least a localized capacity for change of shape, and hence the early appearance of actin filaments in the cytoplasm. (Actin-like filaments have been identified in the wall-less archaebacterium, *Thermoplasma*.) The characteristic eukaryotic segregation of different metabolic activities into many different types of specialized compartments or organelles (by whatever means these organelles evolved) is directly dependent on the presence of such an endomembrane system. The ramification of this system throughout the cell cytoplasm brings the external environment (contained within its cisternae) into close proximity to all parts of the cell cytoplasm and provides the basis for the characteristically large size of eukaryotic cells.

CAVALIER-SMITH (1981) suggests that the evolution of actin microfilaments, tubulin microtubules and a simple endomembrane system preceded the (autogenous) origin of mitochondria and the nucleus. He places the (autogenous) origin of flagella before the acquisition of a capacity for phagocytosis. MARGULIS

(1981), on the other hand, suggests that the evolution of flagella (from symbiotic spirochaetes) may have taken place either before the acquisition of mitochondria (derived from symbiotic bacteria taken up by phagocytosis) or afterwards (if the free-living bacterial ancestors were themselves predatory, like *Bdellovibrio* today) (STOLP 1979). In spite of some earlier suggestions to the contrary, it is now generally accepted that chloroplasts (derived from photosynthetic endosymbionts) became established after mitochondria (WHATLEY et al. 1979).

3.1.3 The Origin of Eukaryotic Organelles

In considering the origins of eukaryotic organelles it is convenient to divide them into three groups (Figs. 1; 2a). The first group of organelles includes the nuclear envelope, endoplasmic reticulum, the Golgi apparatus or dictyosomes, microbodies and ejectile organelles such as trichocysts, together with a wide range of vacuoles and vesicles with a variety of different contents, e.g. lysosomes and vesicles in which wall scales are assembled. These endomembrane organelles appear to have several features in common. All can be produced de novo; none contains its own DNA; the available information suggests that most (probably all) are derived directly or indirectly from the nuclear envelope or endoplasmic reticulum, though the derived organelle may show little resemblance in structure and content to the "parent" endomembrane; all the derived organelles, with the possible exception of microbodies, seem to release their contents to the exterior of the cell (after fusion with the plasma membrane) or to its topological equivalent, e.g. the main plant vacuole. No individual organelle derived from the endomembrane system is a permanent component of the cell; all are involved in continuing cycles of renewal and degeneration. However, some form of endomembrane system is a universal feature of all eukaryotic cells. We ourselves believe that all endomembrane organelles had an autogenous origin.

The second group of organelles comprises those which we believe to have evolved from endosymbionts. They include chloroplasts (SCHIMPER 1883), mitochondria (ALTMAN 1890, MERESCHOWSKY 1905), hydrogenosomes (MÜLLER 1975), kappa and other particles in *Paramecium* (SONNEBORN 1959) and omikron in *Euplotes* (HECKMAN 1980). These organelles also have several features in common. None of them arise de novo; all of them contain their own prokaryotic-type DNA; all arise from pre-existing organelles of similar structure and content; reproduction is by fission, and once an organelle has been acquired it normally remains as a permanent component of the cell. Of this group of organelles only mitochondria and chloroplasts are found in plant cells. It should, however, be noted that none of the organelles in this group (not even mitochondria) are universal constituents of eukaryotic cells.

The third group of organelles represents a problem category. It includes the nucleus and microtubules, together with microtubule-associated structures such as flagella and the mitotic spindle. These organelles do not fall neatly into either of the two preceding groups. A variety of very different modes of evolutionary origin (both autogenous and endosymbiotic) have been pro-

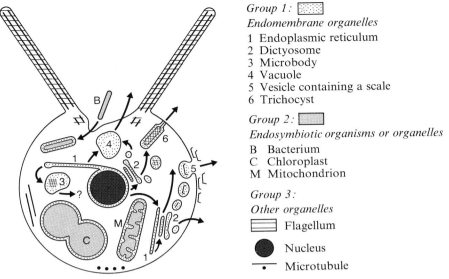

Group 1:
Endomembrane organelles
1 Endoplasmic reticulum
2 Dictyosome
3 Microbody
4 Vacuole
5 Vesicle containing a scale
6 Trichocyst

Group 2:
Endosymbiotic organisms or organelles
B Bacterium
C Chloroplast
M Mitochondrion

Group 3:
Other organelles
Flagellum

Nucleus

Microtubule

Fig. 1. A eukaryotic alga and its organelles

posed for them, but none seems to us to provide a completely satisfactory explanation.

We shall consider each of these three groups of organelles in turn and discuss their possible evolutionary origins, as well as some later modifications which appear to have taken place and selected aspects of their metabolic interactions within the modern eukaryotic plant cell.

3.2 Group 1: Endomembrane Organelles

3.2.1 The Endomembrane System

It is generally believed that the endomembrane system of the cell forms a functional continuum which links the nuclear envelope and the plasmamembrane. This continuum seldom forms a single structural entity but rather it is intermittent and dynamic, its diverse structural components, the contents of their cisternae and the character of their membranes undergoing continual (but highly controlled) and often very rapid modification (MORRÉ and MOLLENHAUER 1974). The main elements of this intermittent continuum are the nuclear envelope, endoplasmic reticulum, dictyosomes with their associated vesicles and the plasmamembrane (Figs. 2 a, c). Additional vesicles (or organelles) of varied content are released from the nuclear envelope and the endoplasmic reticulum (Fig. 1). The structure, function and evolutionary significance of the specialized chloro-

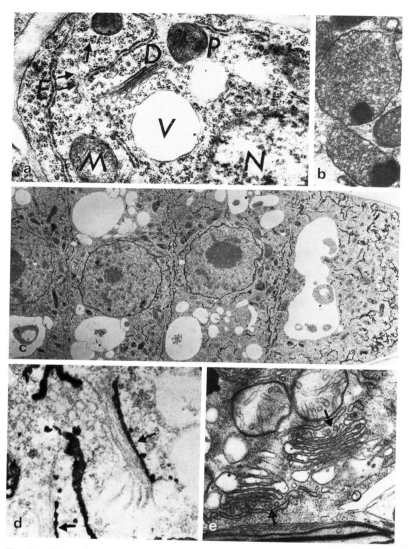

Fig. 2. a Part of a meristematic cell of *Phaseolus vulgaris* showing endoplasmic reticulum (*E*), polyribosomes (*arrow*), a dictyosome (*D*), a mitochondrion (*M*), proplastids (*P*), vacuoles (*V*), microtubules (*double arrow*) and part of the nucleus (*N*). ×27,500. **b** Two microbodies in the fern, *Ophioglossum reticulatum* ×22,000. **c** Cells of the root tip of *Zea mays* which were fixed in 1% glutaraldehyde and 0.5% paraformaldehyde and impregnated with zinc iodide in 2% aqueous osmium tetroxide; the nuclear envelope and endoplasmic reticulum give a similar staining response ×3,200. **d** The same tissue as in **c** at higher magnification shows that cisternae at the forming face of a dictyosome (*arrows*) give a similar staining response to cisternae of endoplasmic reticulum; those at the maturing face do not. (Block for **c** and **d** provided by Dr. C.R. Hawes) ×32,000. **e** Dictyosomes in the brown alga, *Fucus serratus* are associated with strands of endoplasmic reticulum: vesicles derived from the endoplasmic reticulum (*arrows*) appear to fuse to form new dictyosome cisternae ×25,000

Table 1. The possible developmental pathways of some plant endomembrane organelles

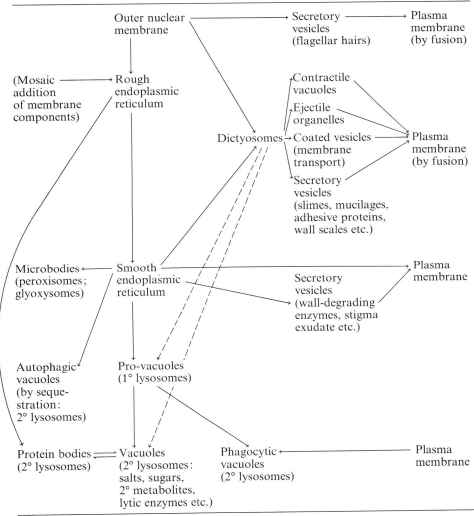

plast endoplasmic reticulum of some chromophyte algae, together with its associated vesicles, have recently been reviewed by Gibbs (1981 b).

Some of the varied types of vesicles or organelles which are believed to be derived from different elements of the endomembrane system are shown in Table 1. These organelles frequently contain proteinaceous material and/or polysaccharides and some have inorganic contents; some provide compartments in which lytic enzymes are kept segregated for a time from other cellular components. Though several different types of endomembrane vesicle are known to fuse with the plasmamembrane and release their varied contents to the exterior of the cell, others, in plants, seem to contribute to the formation of large central vacuoles, which are the topological equivalent of the outside of the cell; the fate of yet other vesicles is either uncertain or completely unknown. Experiments with secretory cells in animals indicate that after vesicles have fused with the plasmamembrane, membrane material may be retrieved by becoming incorporated into Golgi cisternae, condensing vacuoles and lysosomes (Herzog and Miller 1979). Little is known about the retrieval of membranes (or their constituents) in plant cells, though, again, there must be an active cyclical system.

Though it is likely that all the organelles which we have included in this group (Table 1) are derived in one way or another from the basic endomembrane system, this does not account for the evolution of the system itself. Probably the endomembrane network arose initially as an extension of the capacity of the plasmamembrane to form invaginations. However, a fundamental feature of the modern endomembrane network in eukaryotic cells is its association with a high degree of directional movement. The formation of invaginations may well have been quite an early evolutionary development: membrane-bound vesicles and invaginations derived from the plasmamembrane are present in some prokaryotes e. g. mesosomes, photosynthetic vesicles and lamellae in some bacteria. However, in prokaryotes there appears to be no mechanism for controlling movement of vesicles like that associated with actin microfilaments in eukaryotic cells. Nevertheless, actin-like material has been identified in the wall-less archaebacterium, *Thermoplasma acidophillum* in which it may provide a cytoskeleton (Searcy et al. 1978, 1981). Taylor (1978) and Cavalier-Smith (1981) have recently suggested that the first eukaryotes may have resembled some modern non-flagellated eukaryotic fungi (Zygomycetes, Hemiascomycetes, Euascomycetes and Basidiomycetes). These can show several possibly primitive features, e. g. a simple endomembrane system with unstacked Golgi cisternae giving rise to secretory vesicles that release their contents to the exterior.

3.2.2 Dictyosomes

The simplest dictyosomes, or Golgi bodies, are found in some fungi, in which they are represented by single cisternae. More complex dictyosomes are made up of a polarized stack of cisternae. The structure and function of dictyosomes has recently been reviewed by Mollenhauer and Morré (1980). In many animals, algae and fungi the forming face of a dictyosome is often closely associated with a portion of endoplasmic reticulum or nuclear envelope, and it has been suggested that vesicles budding off from the endoplasmic reticulum or, in some algae, from the nuclear envelope may become lined up and fuse to provide the successive cisternae of the forming face (Fig. 2 e). In higher plants such a close association is uncommon though the membranes of cisternae at the forming face (Fig. 2 d) are morphologically and cytochemically similar to endoplasmic reticulum (Mollenhauer and Morré 1980). Both the contents of the cisternae and the cisternal membranes themselves change in structural composition as they develop and, towards the distal face of the dictyosomes, the membranes more closely resemble the plasmamembrane. The area around a dictyosome usually lacks ribosomes and other organelles but includes filamentous

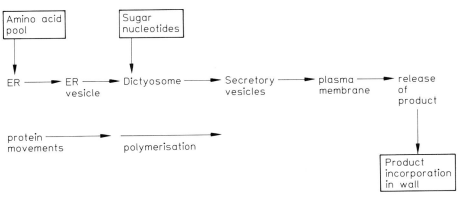

Fig. 3. Endoplasmic reticulum, dictyosomes and the production of cell-wall materials. (After ROBINSON 1977)

material which (like exclusion zones at the cell surface) may contain actin. It was, for example, found that cytochalasin B, which disrupts such filaments, prevented the transfer of vesicles from the Golgi apparatus to the surface of the cell. Microtubules, too, are often found close to dictyosomes. In some animals drugs which inhibit microtubules have been found to affect the Golgi apparatus, but no similar effect of such treatment has so far been found in plant cells (MOLLENHAUER and MORRÉ 1980).

The biochemical activities of the Golgi apparatus concerned with the secretion of plant cell-wall material have been examined principally in pea epicotyls, maize root tips and plant cell protoplasts. Dictyosomes are responsible for the synthesis of pectic substances and mucopolysaccharide with which the microfibrillar cellulose phase (formed at the plasmalemma) is impregnated, using sugar nucleotides formed in the cytoplasm as monosaccharide donors. There is no evidence of the synthesis of the polysaccharides in the rough ER and their subsequent transfer to the dictyosome – though the possibility remains that the enzymes responsible are transferred along this route. Figure 3 illustrates the inter-relations between the endoplasmic reticulum and dictyosomes in the production of cell-wall components. The cell plate at division is initially composed entirely of pectic substances supplied via the dictyosomes, and the cellulose component is only added to subsequent cell-wall layers. The mucopolysaccharides are of particular importance in root tips; they penetrate the microfibrillar phase to become deposited on the outside of the root caps as slimes. As well as their role in synthesizing and secreting these carbohydrates, dictyosomes in several algae may also form cellulosic scales and move them outside the cell by fusion of their secretory vesicles with the plasmamembrane. Estimates of the turnover time for dictyosomes in secretory cells range from 10 to 40 min (MOLLENHAUER and MORRÉ 1980).

The dictyosomes can also have a role in protein transport and secretion (Fig. 3); they may acquire their synthetic enzymes from the polyribosomes that are associated with the forming face, or by transfer from the rough endoplasmic reticulum, and they may well transfer to the plasmamembrane the synthetic enzymes required for cellulose synthesis. In a few cases dictyosomes are concerned with a massive secretion of proteins, as in the transfer of the adhesive protein of zoospores of the alga *Enteromorpha* from Golgi vesicles to the outside of the cell (Evans and Christie 1970). However, the secretion of proteolytic enzymes to the outer surface of *Dionaea* flytraps proceeds either by fusion with the plasmalemma of vesicles budded off directly from the endoplasmic reticulum or by temporary penetration of the plasmalemma by branches of the endoplasmic reticulum (Robins and Juniper 1980). Although the Golgi and endoplasmic reticulum vesicles may be morphologically indistinguishable, their separate origins have been clearly established by radioactive labelling experiments.

The specialization of some dictyosomes in polysaccharide metabolism is analogous to the specialization of some microbodies in photorespiration or fatty acid metabolism; both represent particular manifestations of restricted parts of the endomembrane continuum.

3.2.3 Microbodies

In land plant tissues the most prominent microbodies are in the leaf (as peroxisomes), in which they specialize in photorespiration, or in fatty seeds (as glyoxysomes), in which they specialize in the conversion of fats into succinate. Peroxisomes are vesicles 0.2 to 1.5 μm in diameter bounded by a single membrane and characterized by the possession of catalase; they are widely distributed in eukaryotic cells (Fig. 2b). Using flavoproteins, peroxisomes can oxidize a restricted number of substrates, leading to the production of H_2O_2, and the accumulated H_2O_2 is removed by catalase via peroxidatic or catalatic activities (i.e. secondary oxidations in which H_2O_2 substitutes for O_2 are possible). The study of castor bean endosperm (Beevers 1979) shows that glyoxysomes are produced in large numbers during germination and that they undoubtedly arise from the smooth endoplasmic reticulum by a process of vesiculation (Fig. 1). The phospholipid composition of the glyoxysomes closely resembles that of the endoplasmic reticulum, and pulse chase experiments with labelled choline at the time when the glyoxysomes were being actively formed show that the phospholipids were labelled (synthesized) first in the endoplasmic reticulum and subsequently in the microbodies. Equally, the enzymes characteristic of glyoxysome metabolism (β-oxidation, glyoxylate cycle) were detectable first in association with the endoplasmic reticulum and were later concentrated in the glyoxysomes; this is consistent with the synthesis of the enzymes on ribosomes attached to the endoplasmic reticulum and their subsequent transfer to the glyoxysomes. More recent investigations have indicated that at least some of the matrix proteins of glyoxysomes are made on ribosomes not attached to membranes. These proteins subsequently enter the glyoxysomes from the cytosol (Lord and Roberts 1983). It is characteristic of organelles derived in this way from endoplasmic reticulum that they may be formed temporarily in large numbers and later lost when the metabolic requirement falls.

DE DUVE (1969) proposed that the peroxisomes are very primitive in origin. He pointed to the problem engendered by the production of oxygen by the early photosynthetic prokaryotes that used water as electron donor. Leakage of electrons would have occurred from reduced flavoproteins and thiols towards the newly available acceptor O_2, and H_2O_2 would have been produced. The evolutionary pressure to acquire a system to overcome the accumulation of H_2O_2 must have been intense and the acquisition of an enzyme like catalase to protect against H_2O_2 could have been the solution. DE DUVE also draws attention to the extra energetic advantages that might have followed if O_2 could now be conveniently used to make pyruvate available for further energy-conserving fermentative steps at the level of substrate phosphorylation (e.g. the phosphoroclastic reaction) instead of being required to act merely as the terminal acceptor to balance the essential redox reactions of fermentation. In addition, otherwise non-fermentable substrates like glycerol would also have become available if their excess electrons were first removed by transfer to oxygen. The number of energy sources available would thereby be increased. The concept, then, is of early peroxisomes as having an important primitive oxidative metabolism involving perhaps carbohydrates, fats, amino acids and purines, the whole dependent on catalase to avert the unpleasant consequences of H_2O_2 accumulation. In this view the energy yield in early oxidative metabolism is restricted, because the peroxisomes lacked many of the steps necessary to take such carbon compounds all the way to CO_2, although it is an improvement on simple fermentations. Following the acquisition of mitochondria, with a full complement of tricarboxylic acid cycle enzymes, the relative importance of peroxisomes to energy metabolism would have decreased. However, the peroxisomes have maintained appropriate isoenzymes that enable them to contribute significantly in special situations to carbon metabolism, notably in the conversions of fats into carbohydrates during germination, and in the recovery in photorespiration of carbon otherwise lost to the photosynthetic cycle after the formation of glycolate when O_2 substitutes for CO_2 in the ribulose-bis-phosphate carboxylase reaction. More recently the importance of photorespiration as a way to dissipate excess energy (POWLES and OSMOND 1978), as well as to participate in an active ammonia recycling (LEA and MIFLIN 1980), have emphasized the regulatory importance of leaf peroxisomes. TOLBERT (1981) draws attention to the high degree of specialization shown by microbodies in different tissues and reminds us that they are absent from, or uncommon in, meristematic tissue but are formed in response to the need to regulate the energy balances in mature tissues. He suggests that peroxisomes are associated with higher forms of development rather than being a primitive respiratory organelle. However, microbodies are present in most algae (DODGE 1973; cf. also SILBERBERG 1975; STABENAU 1984). If, as DE DUVE suggests, they did arise primitively in response to toxic oxygen, they have obviously become much more specialized since.

3.2.4 Vacuoles and Lysosomes

Some unicellular algae contain phagocytic vacuoles. Many freshwater and a few marine algae contain contractile vacuoles which are concerned with osmoregulation, and perhaps also with excretion (Dodge 1973). Phagocytic and excretory activities may well have been the initial functions of vacuoles in simple eukaryotic cells.

In a mature plant cell much of the interior is commonly occupied by a large central vacuole. One function of this (and of other types of vesicles or vacuoles) is to contain lytic activity, the release and expression of which can be associated with cell death or with the turnover of cell constituents. The mode of origin of vacuoles is not entirely clear; several different mechanisms may well be involved.

Vacuoles have been described (e.g. Buvat 1971) as developing from enlarging vesicles derived from the endoplasmic reticulum or from dictyosomes. Matile (1974) suggests that the smallest vacuoles (provacuoles) are derived directly from the endoplasmic reticulum and later fuse to form larger vacuoles. He characterizes the provacuoles by their content of hydrolytic enzymes which, together with material added from Golgi vesicles, account for the presence of these enzymes in the various types of vacuole; together all these vacuoles constitute the lysosomal compartment of the plant cell. In addition, lytic enzymes are contained in vesicles derived from the endoplasmic reticulum and possibly from the Golgi apparatus. These vesicles may secrete material across the plasmamembrane to the outside of the cell.

Recently Marty and his associates (1980) have reviewed an alternative hypothesis for the origin of vacuoles in meristematic cells. They suggest that tubular pro-vacuoles (primary lysosomes) develop from specialized smooth endoplasmic reticulum at the maturing face of a dictyosome (Table 1). These pro-vacuoles are then believed to encage areas of cytoplasm, later sequestering them by lateral fusion and subsequently digesting their contents. Marty has speculated that the contents and fused inner membranes break down, and that the resistant outer membranes become the tonoplasts of the developing vacuoles (secondary lysosomes). In older cells, vacuoles can increase in size by fusion with each other or by direct fusion with pro-vacuoles (Fig. 2a, c).

In addition to having lytic activity, vacuoles can accumulate a range of different solutes, e.g. salts, sugars, secondary metabolites and amino acids (Marty et al. 1980), as well as insoluble proteins and salts of organic acids. The specialized protein bodies of seeds may develop in several different ways, directly from vacuoles, from rough endoplasmic reticulum, and by expansion of cytoplasmic vesicles (Lott 1980).

3.2.5 Ejectile Organelles

The ejectile organelles comprise a varied and distinctive but poorly understood group of single-membrane-bound structures which lack DNA and which apparently are derived from the endomembrane system (Table 2). They are found in both plants (unicellular algae) and animals (ciliates and flagellates) and their varied nomenclature, structure and distribution were most recently reviewed by Hausmann (1978). In animal cells these ejectile organelles appear to be used to capture prey: their function in plant cells is uncertain.

Table 2. Ejectile organelles

Algal family	Type of ejectile organelle	Probable site of origin	Probable content
Chrysophyceae	Discobolocyst	Dictyosome	Mucopolysaccharide
Cryptophyceae	Ejectosome (Taeniobolocyst)	Dictyosome	Proteinaceous
Dinophyceae	Nematocyst (Cnidocyst)	?	Proteinaceous
	Trichocyst	Dictyosome	Proteinaceous
Euglenophyceae	Unique "mucocyst"	?	Mucopolysaccharide?
Prasinophyceae	Ejectosome	?	Proteinaceous
Raphidophyceae (Chloromonadophyceae)	Alcontobolocyst (cf. trichocyst)	?	Proteinaceous
	Mucocyst	Dictyosome	Mucopolysaccharide

There are two classes of ejectile organelles, those which have muciferous and those which have proteinaceous contents. Though they seem to be more highly organized structures, the ejectile organelles called discobolocysts and mucocysts, should perhaps be equated with the muciferous bodies and other types of secretory vesicles, which release mucilage to the exterior of the cell. In each of the different types of organelle which have proteinaceous contents, the structure of the proteinaceous core is distinctive. Trichocysts (Fig. 4a), for example, have two components, a para-crystalline main shaft and a narrow neck of twisted fibres. Nematocysts are more complex, having a ribbed capsule containing a coiled tube with a stylet at the tip. Ejectosomes (Fig. 4b) contain a coiled tubular ribbon with a sharply angled tip (DODGE 1973, HAUSMANN 1978).

In animal cells the earliest fibro-granular stages of development, the beginnings of the proteinaceous core, can first be identified in vesicles adjacent to the endoplasmic reticulum; in algal cells the early stages of differentiation have been found in vesicles adjacent to dictyosomes (HAUSMANN 1978). Progressive assembly of the paracrystalline core takes place as the vesicles move towards the cell periphery, a pattern of development similar to that followed during the assembly of algal scales. When a trichocyst, for example, reaches the edge of a dinoflagellate, it takes up a precise position between two thecal plates, at right angles to the plasmamembrane and aligned so that the main shaft lies towards the interior and the neck towards the exterior of the cell. During differentiation of the trichocysts of *Paramecium* and the mucocysts of *Tetrahymena,* five or six rows of small particles appear on their surrounding membrane towards the anterior end. Precise positioning of these ejectile organelles seems to be achieved when these anterior particles become associated with particles which appear on the plasmamembrane in the form of a fusion rosette (SATIR 1974).

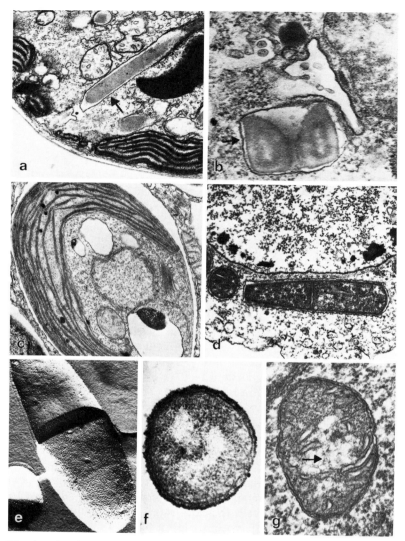

Fig. 4. a A trichocyst (*arrow*) in the dinoflagellate, *Glenodinium* sp. (Block provided by Prof. J.D. Dodge) ×18,000. **b** An ejectosome (*arrow*) in a cryptomonad. (Block provided by Prof. J.D. Dodge) ×50,000. **c** An endosymbiotic eukaryotic green alga, *Chlorella*, lies within a vacuole in the cytoplasm of its animal host, *Hydra*. (Micrograph provided by Dr. C.R. Hawes) ×13,000. **d** Bacterial endosymbionts are tightly enclosed in vacuoles provided by their host, the amoeba, *Pelomyxa palustris* ×10,000. **e** A freeze-fracture image of the bacterium, *Paracoccus denitrificans*. (Material prepared by Dr. S. Knutton) ×25,000. **f** Thin section of the bacterium, *Paracoccus denitrificans* ×50,000. **g** A mitochondrion from *Zea mays,* showing fibrils of DNA (*arrow*) ×44,000

It is curious that some ejectile organelles which appear to be very similar in basic structure are found in phylogenetically quite unrelated organisms. Trichocysts are found in some dinoflagellates and in some ciliates; nematocysts in some dinoflagellates and coelenterates; ejectosomes in some (eukaryotic) cryptomonads, in the prasinophyte alga *Pyramimonas* and (remarkably) as R-bodies in kappa particles (symbiotic bacteria) in *Paramoecium* (DODGE 1973, PREER et al. 1974, PREER 1975). This last most unlikely apparent pairing of similar "organelles" in a eukaryotic alga and in a prokaryotic bacterium has been previously pointed out by PREER et al. (1974). The R-body of *Paramecium* is believed to be associated with a defective phage. A viral origin for the "contents" of ejectosomes and other ejectile organelles should thus be considered (PREER 1975, PREER et al. 1974, TAYLOR 1979): such an origin would at least account for their unusual phylogenetic distribution. Some viruses and some endomembrane organelles follow parallel courses in their progressive assembly within vesicles derived from the endomembrane system. One should therefore perhaps also consider a viral or plasmid origin in connection with other endomembrane organelles, e.g. microbodies. If the contents of any endomembrane organelles did indeed have a viral origin, then this would again represent a form of endosymbiosis, but one which was different from that proposed in the endosymbiotic theory. A plasmid origin has of course been proposed for mitochondria and chloroplasts (RAFF and MAHLER 1975), and viruses have been mentioned in connection with the evolutionary origin of microtubules (TAYLOR 1979).

3.3 Group 2: Mitochondria and Chloroplasts

The organelles in the second group are those which the authors believe to have evolved from organisms that were once free-living but were later taken up as endosymbionts by host cells. In plants, the organelles which fall into this category are mitochondria and chloroplasts. As a result of information which has become available in recent years, the concept of an endosymbiotic origin for chloroplasts has become increasingly accepted (DODGE 1979, 1980, GIBBS 1978, 1981a, WHATLEY et al. 1979, WHATLEY and WHATLEY 1981). However, there is less general acceptance of the idea that mitochondria may have evolved from endosymbionts (CAVALIER-SMITH 1981, MAHLER 1980). Because both mitochondria and chloroplasts have their own prokaryotic-type DNA, never appear to arise de novo and replicate by fission to produce offspring which resemble the parent organelles, then it seems likely that both must have had a similar, i.e. an endosymbiotic, ancestry.

3.3.1 The Endosymbiotic Hypothesis

The endosymbiotic hypothesis proposes that certain forms of free-living organisms were taken up endocytotically by host cells (Figs. 4 c, d; 5). If the alien

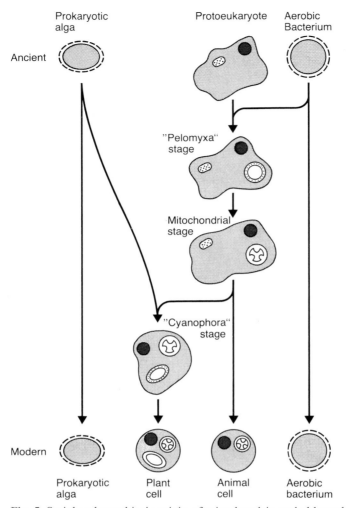

Fig. 5. Serial endosymbiotic origin of mitochondria and chloroplasts

organisms avoided digestion by the host and became capable of maintaining themselves permanently within the host cytoplasm, they could obtain from the host essential metabolic substrates and in return release to the cytoplasm excess material of value to the host. In time the autonomy of the alien organisms was progressively lost and both partners became adapted to an endosymbiotic existence of which the host cell took overall control. Proliferation of the symbiont became synchronized with the cell division of the host: many of its redundant biosynthetic capabilities were lost or taken over by the host, its internal structure was modified and efficient systems of exchange developed between the two partners.

Support for this hypothesis has come from several sources (see for example, Margulis 1981 and Whatley et al. 1979). Investigations of modern symbiotic relationships have provided information about exchange of metabolites (SMITH 1979) and about the mechanisms for maintaining a balanced population (MCAULEY 1981, MUSCATINE and POOL 1979).

The photosynthetic cyanelles of some anomalous algae, like *Cyanophora paradoxa* and *Glaucocystis nostochinearum,* seem to reflect intermediate stages in evolution between endosymbionts and chloroplasts (KIES 1980, TRENCH 1981, WHATLEY and WHATLEY 1981). Free-living organisms have been found which possess many of the characteristics to be expected of ancestral symbionts (JOHN and WHATLEY 1975, 1977, DAYHOFF and SCHWARTZ 1981, FOX et al. 1980, WOESE 1981, LEWIN and WITHERS 1975, NEWCOMBE and PUGH 1975). We can therefore point to modern examples of organisms with many of the features "required" to illustrate the theory of the endosymbiotic origin and the sequential integration of first mitochondria and, later, chloroplasts into host eukaryotic cells (Fig. 5). The only indication of the time scale required for such modifications comes from the work of JEON and JEON (1976). They found that it took less than 10 years for amoebae which accidentally became infected with bacteria to become inviable unless the bacteria were present, thus establishing a new obligatory symbiosis.

As DODSON (1979) in his recent review concludes, it is reasonably probable that mitochondria evolved from endosymbiotic aerobic bacteria and there exists as much proof as may reasonably be expected (for such an ancient event) that chloroplasts evolved from endosymbiotic algae.

3.3.2 Mitochondria – Their Possible Ancestry

Support for a bacterial origin for mitochondria comes from numerous comparative studies on ribosome size (though the RNA/protein ratio in mitochondrial ribosomes is different from that in cytoplasmic ribosomes, thus contributing to an erroneously low estimate of their size), on the pattern of protein synthesis inhibition, on the properties of RNA and DNA and of enzyme proteins, particularly their amino acid sequences (molecular evolution) and on metabolic and structural characteristics. The small size of the mitochondrial chromosome compared with that in free-living bacteria has been a major factor considered as evidence against an endosymbiotic origin. The transfer of genetic information from endosymbiont to host nuclear genome has been questioned in the past, but the work of SCHELL et al. (1979) on plasmid transfer from the soil bacterium, *Agrobacterium* to a plant host nucleus (crown gall disease) provides a possible mechanism. The recent observations that certain codons like UGA and AUA are read differently in the ribosomes of the cytoplasm (prokaryotic *and* eukaryotic) from in those of mitochondria (summarized in SANGER 1981), indicating a change in the mitochondrial genetic code, have been interpreted by some as refuting the endosymbiotic hypothesis. However HECKMAN et al. (1980) interprets this divergence as being possibly due to a necessity for the mitochondrion to "defend" itself against the host genome and being of later selective origin rather than a primitive form of the code.

Although it has generally been assumed that only one uptake of bacteria led to the establishment of mitochondria in animals, plants and fungi, Dayhoff and Schwartz (1981) have pointed to large discontinuities in the c-type cytochrome sequences and suggest that two separate endosymbioses may have preceded this common step and given rise to the mitochondria in the protist *Tetrahymena* on the one hand and to those in the *Crithidia* and *Euglena* group on the other. They also argue that a separate endosymbiosis in the flagellate group is consistent with the lack of mitochondria in the Trichomonadida, an anaerobic flagellate group, but it is worth pointing out that it is in just this group that hydrogenosomes occur. These organelles, perhaps derived by a separate endosymbiosis from anaerobic clostridia, may play the role of anaerobic mitochondria and make ATP available to the host by an extended substrate phosphorylation. Taylor (1978) in particular has pointed to variations in the internal morphology of mitochondria from different phylogenetic groups. Though aware that account must be taken of alterations produced by changes in physiological state, Taylor nevertheless distinguishes between groups with tubular cristae (e.g. cryptomonads, red and green algae, land plants, higher animals and some fungi) and groups with flattened cristae (e.g. ciliates, most chromophyte algae and other fungi). These groups differ from those of Dayhoff.

Since questions of evolutionary origins cannot be answered by repeating the original experiment, one is tempted to look for surviving species that might more or less closely resemble the ancestral participants in the original endosymbiotic act. Using a method rather like numerical taxonomy, John and Whatley (1975, 1977) compared the energetics of free-living bacteria, including *Paracoccus denitrificans,* with mitochondria (Figs. 4 e, f, g). Criteria which were taken into account were electron transport intermediates, particularly cytochromes and quinones, inhibitor sensitivities, including antimycin and rotenone, electron transport control, phosphatidyl choline and other membrane lipid constituents and many features of the phosphorylation and ATPase reactions. They concluded that, of the bacteria they considered, and on the basis of the evidence then available, *Paracoccus* showed the greatest number of similarities to mitochondria.

Although they chose *Paracoccus* as their "working model" for evolutionary speculation, John and Whatley emphasized that in the future this bacterium would be regarded only as one representative of a small group of aerobic bacteria all with obvious affinities with mitochondria. The respiratory chain in light-grown *Rhodopseudomonas spheroides* is closely similar to the mitochondrial one (Dutton and Wilson 1974), and more recently Baltscheffsky and Baltscheffsky (1981) have drawn attention to the many mitochondrial features of *Rhodospirillum rubrum* (but which so far has not been shown to have cytochrome aa_3) and of *Rhizobium trifolii* (especially in its electron transport chain). Of special interest in this comparison are the results of Fox et al. (1980), based on cataloguing of oligonucleotides derived from 16S-RNA, which show a close evolutionary relationship between the purple non-sulphur photosynthetic bacteria *Rhodospirillum,* and *Rhodopseudomonas* spp., and the non-photosynthetic aerobes *Paracoccus* (a denitrifying bacterium) and *Rhizobium* (a specialist in symbiotic nitrogen fixation). A close relationship between the two genera of photosynthetic bacteria and *Paracoccus* had earlier been indicated on the basis of cytochrome sequencing (DAYHOFF and SCHWARTZ 1981).

When they enter the legume nodule the *Rhizobium* bacteroids lose their cell walls (BERINGER et al. 1979). This loss of the wall is a most unusual feature for a prokaryotic symbiont, though it is very common for eukaryotes. However, such a tendency may well have played a part in the evolutionary transformation of a walled bacterium into a wall-less mitochondrion. Leghaemoglobin is formed in the nodule cells of legumes only after infection. The *Rhizobium* bacteroids synthesize the haem group but the globin is a product of the host cell. As the BALTSCHEFFSKYS (1981) remind us, a similar situation occurs in mammals, where haem biosynthesis takes place in the mitochondria but the synthesis of the globin polypeptides occurs on the cytoplasmic ribosomes.

3.3.3 Modification Within the Protoeukaryotic Host

From such comparisons we may therefore guess something of the nature of the ancestral bacterium that perhaps became a mitochondrion. What of the host cell? One living candidate considered to resemble the postulated proto-eukaryote is the unique amoeba *Pelomyxa palustris*. It is nucleated (the "definitive" eukaryotic criterion) but, among other apparently primitive features, it lacks mitochondria. However, it always contains populations of three types of bacteria, which it maintains from one generation to the next; it has been suggested that the symbiotic association may represent a modern model showing some structural and metabolic similarity to the intermediate "required" of the endosymbiotic theory, though the bacteria themselves are not necessarily related to possible mitochondrial precursors (WHATLEY et al. 1979).

Pelomyxa may have an essentially anaerobic metabolism, to which has been added the aerobic capability of the symbiotic bacteria. *Pelomyxa* tolerates only low oxygen tensions and one might speculate that a primary use of the bacterium to the host could be its ability to scavenge toxic oxygen from the cytoplasm. A recent report (CHAPMAN-ANDRESEN and HAMBURGER 1981) concerning the respiration of *Pelomyxa palustris*, however, concludes that the smallest symbiotic bacteria act as the respiratory "organelles", and that the characteristic large bacterial symbionts (Fig. 4 d) are concerned with attacking the glycogen storage granules; it further suggests that the host contributes little to the overall energy metabolism by fermentation. It is, of course, characteristic of modern aerobic eukaryotes that the rate of glycolysis (fermentation) in the cytoplasm is regulated to coincide with the metabolic demands both of the mitochondria and of the rest of the cell, and glycolysis itself contributes only a small proportion of the total ATP requirement. If the conclusion of CHAPMAN-ANDRESEN and HAMBURGER is correct, it implies that the metabolism of the host and its symbiont(s) are more closely integrated than the earlier suggestion of JOHN and WHATLEY (1975) implies.

JOHN and WHATLEY (1975) discussed the biochemical steps that would be required if a *Paracoccus*-like ancestral bacterium were to become a mitochondrion. These included the retention of the ATPase and the constitutive components of the respiratory chain (those directed towards oxygen), the loss of adaptive components (directed towards dissimilatory nitrate reduction and the use of many different carbon sources) and the modification of some of the symport mechanisms of the bacterial plasmamembrane (now the inner mitochondrial membrane). The only new component necessary for an evolutionary transition would have been the acquisition of the adenine nucleotide carrier by the mitochondrial membrane – only when the carrier was acquired would ATP synthesized by the endosymbiont be available to the host. They thought the initial symbiotic advantage for the intermediate stage (which they called the *Pelomyxa* stage: cf. Fig. 5) might have been the use by the bacterium of the lactic acid produced by fermentation of glucose in the host cytoplasm; the increase in size of the ancestral host might well have been limited by its inability to eliminate the fermentative end-product.

What might be the precursor of *Pelomyxa?* Attention has been directed towards the archaebacterium *Thermoplasma acidophilum,* which it has been proposed may resemble the evolutionary ancestor that gave rise to the cytoplasm and nucleus of the eukaryotic cell (SEARCY et al. 1981). Amongst prokaryotes, *Thermoplasma* may have the largest number of eukaryotic features e.g. in the absence of a cell wall, the presence of actin-like filaments with which it may regulate its shape, and the presence of a basic histone-like protein in association with its (circular) DNA – thought to be a feature that protects the DNA from thermal denaturation at 60°. *Thermoplasma* has an essentially anaerobic fermentative metabolism, to which has been added a short electron transport chain (membrane-bound quinone and cytochrome b) to allow it to operate a sulphate-dependent ATP-ase necessary to permit its survival in a medium containing 10^{-1} to 10^{-2}M sulphuric acid (SEARCY and WHATLEY 1983). This type of membrane-bound oxidative chain is characteristic of non-mitochondrial membrane systems in modern eukaryotes. However, the recent proposals of Fox et al. (1980) divide living systems into three kingdoms with a common ancestor, and suggest that the Archaebacteria, Eubacteria and Eukaryotes have equivalent status and should not be thought of as derived one from another.

3.3.4 Chloroplasts

In 1883 SCHIMPER suggested that chloroplasts as semi-autonomous organelles might have evolved from endosymbiotic blue-green algae. More recent investigations have shown that blue-green algae and the chloroplasts of red algae both have chlorophyll a as their primary photosynthetic pigment and phycobilins (phycocyanin and phycoerythrin) as their accessory pigments. Furthermore, in both, the phycobilins are concentrated in phycobilisomes, often seen in electron micrographs as "knobs" protruding from single thylakoids. It is therefore not unreasonable to believe that the chloroplasts of red algae may have evolved from blue-green algal symbionts (Figs. 6 a, b). However, the photosynthetic apparatus in the chloroplasts of other algae and of land plants differs considerably from that in the blue-green algae (Figs. 6; 7). Chloroplasts of green algae,

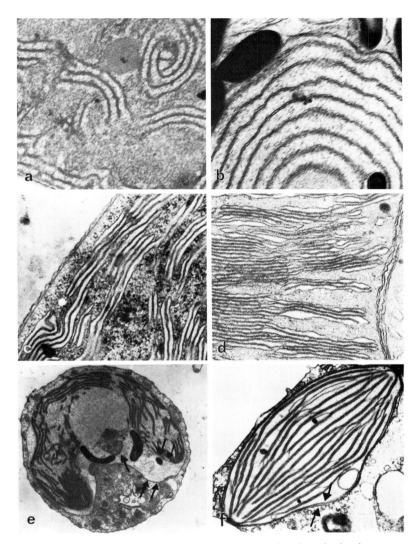

Fig. 6. a The filamentous blue-green alga, *Anabaena cylindrica,* showing single photosynthetic thylakoids. (Block provided by Dr. N.G. CARR) × 32,000. **b** Part of a red alga, *Porphyridium* sp., showing a piece of the chloroplast with single thylakoids × 35,000. **c** Part of a prokaryotic prochloron showing its paired and irregularly stacked thylakoids. (Block provided by Dr. R.A. LEWIN) × 40,000. **d** Part of a chloroplast of the primitive land plant, *Selaginella apus,* showing its irregularly stacked thylakoids × 50,000. **e** A cryptomonad chloroplast, showing the two pairs of surrounding membranes (*double arrows*) and the nucleomorph (*arrow*) in the periplastidial space between them. (Block provided by Prof. J.D. DODGE) × 7,800. **f** A chloroplast from a brown alga, *Fucus serratus,* showing the two pairs of surrounding membranes (*arrows*) with no significant periplatidial space. The thylakoids are regularly associated in bands of three × 16,000

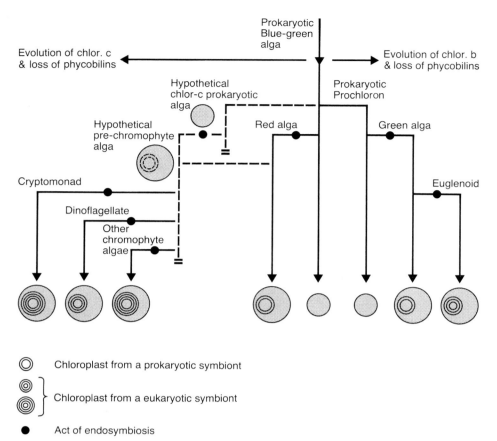

Fig. 7. The polyphyletic origin of chloroplasts

euglenoids and land plants have chlorophyll a and accessory chlorophyll b; their thylakoids form tightly bonded stacks. Chloroplasts of the remaining algae – the chromophyte algae of Christensen (1964) – contain chlorophyll a and accessory chlorophyll c; their thylakoids are more loosely associated in broad, usually triple bands (e.g. Whatley and Whatley 1981).

The absence of possible precursors of these other types of chloroplast was for many years a major point of criticism of the endosymbiotic theory. However, in 1975 the independent discovery by Lewin and Withers and by Newcomb and Pugh of prokaryotic algae (prochlorons), which contained chlorophyll a, accessory chlorophyll b and a thylakoid system which involved stacking, at last provided a modern species perhaps resembling a symbiotic ancestor of the chloroplasts of green algae, euglenoids and land plants (Figs. 6 c, d). It is now only for the chromophyte algae (Figs. 6 e, f) that a possible prokaryotic ancestor is not known. However, the lack of a possible prokaryotic ancestor is not the only problem presented by the chloroplasts of chromophyte algae.

The chloroplasts of red and green algae and land plants are surrounded by two membranes; those of the chromophyte algae (and of the green euglenoids) are surrounded by either three or four membranes. This is difficult to explain by means of the conventional endosymbiotic theory, which envisages prokaryotes as the ancestors of both mitochondria and chloroplasts.

When symbionts are taken up by host cells they are commonly sequestered inside vacuoles, i.e. the symbionts lie within the host endomembrane system and are thus topologically outside the cell (Figs. 1, 4c, d). The vacuolar membrane (part of the host cell) is usually considered to be homologous with the outer membrane of the organelle's double envelope; the plasmamembrane of the symbiotic prokaryote is considered to be homologous with the inner membrane of the organelle. These homologies are acceptable for mitochondria and for chloroplasts of red and green algae and land plants, but are inappropriate for those chloroplasts which are surrounded by more than two membranes. One group of chromophyte algae, the cryptomonads, has suggested at least a partial solution to both the pigment and the membrane problems (DODGE 1979, GIBBS 1981 a, b, WHATLEY and WHATLEY 1981, WHATLEY et al. 1979).

The chloroplasts of cryptomonads are surrounded by four membranes, an inner pair immediately adjacent to the thylakoid system (i.e. the double envelope of the chloroplast proper) and an outer pair, usually designated chloroplast endoplasmic reticulum (Fig. 6e). The outer membrane of the pair making up the chloroplast endoplasmic reticulum has ribosomes attached to it and is continuous with the outer nuclear envelope. The inner membrane of this pair lacks ribosomes and forms a completely separate and closed compartment. The space between the inner and outer pairs of membrane is called the perichloroplastic or periplastidial space. This space contains starch, 80S ribosomes (as opposed to the 70S ribosomes of the chloroplast proper) and a structure called a nucleomorph (GREENWOOD 1974). This nucleomorph is surrounded by two membranes in which there are pores or slits. It contains osmiophilic granular and fibrillar material for which GILLOT and GIBBS (1980) reported a staining response similar to that of the nucleolus (RNA). MORRALL and GREENWOOD (1982) have recently described division (binary fission) of the nucleomorph. Densely staining rod-shaped particles, which in electronmicrographs resemble the condensed chromatin of some nuclei, are closely associated with the nucleomorph envelope; these particles assume an ordered alignment, double in number and separate towards the "poles" of a fibrous band which the authors suggest might be equated with a primitive spindle, perhaps based on a system of actin microfilaments. No microtubules were visible either in the nucleomorph or in the adjacent periplastidial space. Several authors have suggested that the complex cryptomonad chloroplast may have been derived from a eukaryotic (rather than a prokaryotic) endosymbiont, of which the nucleomorph is all that remains of the nucleus and the perichloroplastic space with its 80S ribosomes represents all that remains of the eukaryotic cytoplasm (GREENWOOD 1974, DODGE 1979, GIBBS 1981a, b, WHATLEY and WHATLEY 1981, WHATLEY et al. 1979). The chloroplast sensu stricto, enclosed only by the inner pair of membranes would in that case have been the result of an earlier act of endosymbiosis carried out by this eukaryote and involving a photosynthetic prokaryote.

In other chromophyte algae (Fig. 6 f) the perichloroplastic space is either absent (dinoflagellates and also the green euglenoids) or is much reduced (the remaining chromophyte algae), but the chloroplasts of all are now believed to have evolved in one way or another from eukaryotic symbionts (Dodge 1979, Gibbs 1978, 1981 a, b, Whatley and Whatley 1981, Whatley et al. 1979).

Cryptomonad chloroplasts are also distinctive in their photosynthetic pigments. Chloroplasts of all chromophyte algae including the cryptomonads contain chlorophyll a and, as accessory pigments, the chlorophylls c. Cryptomonad chloroplasts also contain phycobilins and in this respect therefore resemble the chloroplasts of red algae. The cryptomonad phycobilins are, however, different from those of red algae and are located within the thylakoid sacs rather than in protruding phycobilisomes. It is possible that the eukaryotic symbiont which gave rise to the cryptomonad chloroplast (the pre-chromophyte alga of Fig. 5) may have been a modified red alga in which chlorophyll c evolved. Alternatively, chlorophyll c may have evolved in a free-living prokaryotic alga which later became the endosymbiotic source of the prechromophyte algal plastid. Neither of these hypothetical ancestors is known today. However, the concept of the evolution of chloroplasts from both prokaryotic and eukaryotic symbionts appears to have gained wide acceptance and certainly accounts for the polyphyletic derivation of these organelles.

An endosymbiotic origin for the chloroplast is supported by "evolutionary trees" (dendrograms) based on sequencing of nucleotide cataloguing of chloroplast 5S ribosomal RNA's, 16S ribosomal RNA's and transfer RNA's and on the amino acid sequences of the chloroplast proteins plastocyanin, ferredoxin and cytochrome c_6 isolated from a variety of algae and land plants (Doolittle and Bonen 1981, Phillips and Carr 1981, Schwarz and Dayhoff 1981).

3.4 Other Organelles

3.4.1 The Nucleus

Evolution of the modern nucleus has involved much more than just the segregation of chromosomal material within a membrane-delimited compartment. Instead of a single circular prokaryotic type chromosome which on replication segregates as a result of simple membrane growth, the DNA in the eukaryotic nucleus is arranged in linear chromosomes which have come to segregate by the sophisticated mechanism of mitosis. The process of transcription of RNA in the nucleus has been separated from the process of translation on the rough endoplasmic reticulum of the cytoplasm. Ribosomal RNA is now processed and ribosomal sub-units are assembled in the specialized nucleolus. A mechanism has evolved for the transport of rRNA from the nucleoplasm to the cytoplasm across the nuclear envelope; associated with this has been the evolution of the complex system of pores which distinguish the nuclear membrane from other parts of the endomembrane network (Figs. 8 a, b).

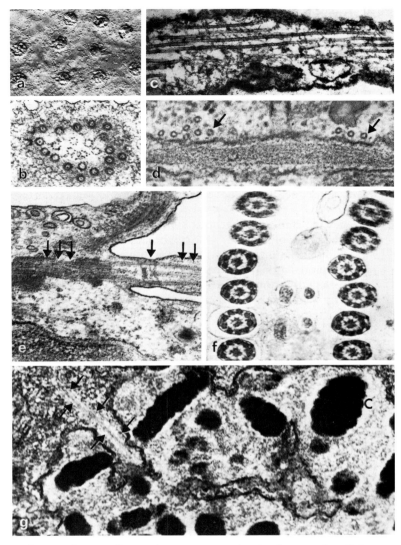

Fig. 8. a Freeze fracture image of a nuclear pore complex in the amoeba, *Pelomyxa palustris*. (Material prepared by Dr. S. KNUTTON) ×30,000. **b** Thin section showing a nuclear pore complex in *Pelomyxa* ×25,000. **c** Cytoplasmic microtubules in *Pelomyxa* seen in longitudinal section ×40,000. **d** Cytoplasmic microtubules in *Phaseolus vulgaris* seen in cross-section (*arrows*) ×75,000. **e** Longitudinal section of part of a cryptomonad showing microtubules in a flagellum (*double arrows*), its basal body (*triple arrows*) and the transition zone (*single arrow*). (Block provided by Prof. J.D. DODGE) ×30,000. **f** Transverse section of flagella (cilia) of a didemnid, (the animal host of prochlorons) showing the 9+2 arrangement of the microtubules ×40,000. **g** The dividing nucleus of a dinoflagellate, *Glenodinium* sp., showing microtubules (*arrows*) lying *outside* the nucleus in a cytoplasmic invagination. The chromosomes (*C*) remain permanently condensed. (Block provided by Prof. J.D. DODGE) ×20,000

3.4.2 Hypotheses on the Origin of the Nucleus

Until WOESE and his associates (FOX et al. 1980, WOESE 1981) suggested that there were three primary kingdoms of organisms, it was generally assumed that the eukaryotic cell evolved directly from the prokaryotic cell and ideas about the origin of the eukaryotic nucleus were based on this assumption. As for mitochondria and chloroplasts, both autogenous and endosymbiotic hypotheses have been put forward (Fig. 9), and six of these are described briefly below:

1. The nucleus had an autogenous origin. That part of a prokaryotic cell which contained DNA was encircled by an extended invagination of the plasmamembrane. The cell thus became separated into two compartments, the DNA-containing nucleus and the DNA-free cytoplasm; the two compartments subsequently underwent divergent evolution (UZZELL and SPOLSKY 1974, 1981).
2. A somewhat similar autogenous hypothesis suggests that the nucleus evolved when the DNA of a prokaryotic cell became enclosed by ribosome-associated intracytoplasmic membranes, either photosynthetic (CAVALIER-SMITH 1975) or non-photosynthetic (CAVALIER-SMITH 1981); the membranes involved arose originally from plasmamembrane invaginations.
3. An early endosymbiotic hypothesis suggested that chromosomes evolved from bacteria taken up by a host cell and sequestered within the nucleus (ALTMANN 1890). As TAYLOR (1980) points out, this idea originated from erroneous light microscope observations. However, the similarity in appearance as seen in the electron microscope between the DNA of some prokaryotes and the permanently condensed histone-free chromosomes of dinoflagellates (Fig. 8 G) has helped to perpetuate this hypothesis, though the occasional discussion of this idea tends to be verbal rather than written.
4. The nucleus evolved from a prokaryote taken up as a symbiont by another organism that used genetic RNA, DNA being provided by the symbiont (JEON and DANIELLI 1971).
5. Because evolution of the nucleus was believed to be associated with the evolution of mitosis with its accompanying microtubular spindle, it was suggested that the eukaryotic cell might have arisen following the uptake of a prokaryotic spirochaete (which contained microtubules) by a *Thermoplasma*-like prokaryote with a capacity for phagocytosis. The relatively large spirochaete genome would then have been the source of the nucleus as well as the flagella and other microtubule-associated structures (KUNICKI-GOLDFINGER 1980).
6. The final hypothesis takes quite a different approach. Some bacteria, e.g. *Clostridium,* sporulate under conditions of environmental stress. The process is distinctive in that asymmetrical cell division is followed by the engulfment of the smaller daughter cell by the larger. The small engulfed cell is separated from its encircling sister cell by two plasmamembranes, between which a wall normally develops. It has been proposed by GOULD and DRING (1979) and discussed in more detail by DAWES (1981) that the eukaryotic cell evolved from a sporulating prokaryote in which the larger cell provided the future eukaryotic cytoplasm and the smaller cell provided the nucleus. The two

Fig. 9. The origin of the nucleus

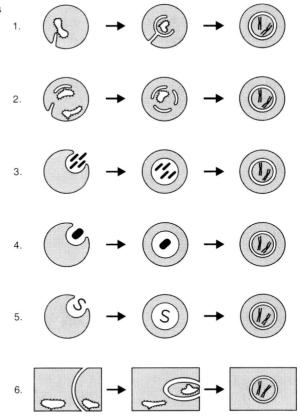

plasmamembranes (between which no wall developed) evolved into the nuclear envelope. The capacity for self-engulfment would later be reflected in the development of an endomembrane system and an ability to carry out phagocytosis.

In the first two (autogenous) hypotheses the nucleus and the cell cytoplasm are seen as two segments of the same original prokaryotic cell which have become separated and have subsequently undergone differential evolution. The other four (endosymbiotic) hypotheses consider the nucleus as an alien organism either of the same parentage as (No. 6) or of a species unrelated to the organism that provided the cell cytoplasm. The two autogenous hypotheses are perhaps the most generally accepted and they, together with the final (partially autogenous) hypothesis based on sporulation, seem to us not unreasonable. On the basis of present evidence, the remaining hypotheses seem to us less probable.

3.4.3 Microtubules

Evolution of the modern mitotic system of nuclear division depended on the participation of microtubules. For the origin of microtubules (Fig. 8 c, d), opposing hypotheses have been put forward. The first hypothesis suggests a symbiotic origin, in which prokaryotic spirochaetes already containing an ordered system of 9 + 2 interconnected tubules became attached to the host plasmamembrane (ectosymbiosis) by means of specialized attachment sites (Sagan 1967, Margulis (Sagan) 1981). Following this hypothetical ancient symbiosis, the main body of the spirochaete evolved into the eukaryotic flagellum with its 9 + 2 arrangement of microtubules and its attachment site evolved into the basal body and flagellar root system (Fig. 8 e, f). The microtubules of the root system later gave rise to microtubular systems elsewhere in the host cell, including those associated with the mitotic spindle. The basal bodies which act as organizing centres for the flagellar microtubules became intermittently dissociated from the flagellar apparatus and entered the main body of cell cytoplasm as centrioles and other more diffuse forms of organizing centres for the formation of cytoplasmic microtubules.

Evidence in support of this hypothesis is provided by the presence in some spirochaetes of tubulin-like material and of tubular structures similar in diameter to microtubules. In addition, numerous spirochaetes associated in pairs and beating in synchrony form an ectosymbiotic association with the protozoan, *Mixotricha paradoxa* i. e. an association similar to that envisaged by the hypothesis (Margulis 1981). Opponents of the idea of a symbiotic origin for flagella have suggested that the apparent presence of tubulin in some spirochaetes results from the transfer of genetic information from eukaryotic cells with which the spirochaetes are normally in close contact. However, Margulis (1981) has recently reported the presence of tubulin-like protein in spirochaetes which live in an essentially eukaryote-free habitat.

It should be noted that if flagella evolved from symbiotic spirochaetes, then the act of symbiosis itself and the subsequent fate of the symbionts are conspicuously different from those envisaged for other (symbiotic) organelles. The spirochaetes were not incorporated into vacuoles within the host cell but formed only a superficial attachment. The resulting organelle, the flagellum, apparently lacks DNA, unlike mitochondria and chloroplasts; it does not undergo a prokaryotic form of division but instead is reassembled de novo from the basal body. The main structural components of the flagellum, the microtubules, are not confined to the "parent" organelle, but are also used for quite different functions elsewhere in the cell. These cytoplasmic tubules, too, arise de novo by progressive assembly from often diffuse organizing centres to which they show no structural similarity.

A further difficulty for the symbiotic hypothesis is the seeming absence in a number of apparently primitive eukaryotes (e.g. red algae, some fungi and some protozoa) of flagella or any flagellar remnant (basal bodies or centrioles) and the presence of cytoplasmic microtubules or of a mitotic spindle. The amoeba, *Pelomyxa palustris,* for example has many apparently primitive features and in this respect is quite unlike any other species of amoeba; for example, it lacks mitochondria, a Golgi apparatus and conventional endoplasmic reticulum. Its method of division remains uncertain and evidence for the presence of flagella or flagellar remnants is inconclusive at this time. However, microtubules are undoubtedly present in the cytoplasm (Fig. 8 c). Absence of other microtubular derivatives might indicate an autogenous rather than an endosymbiotic origin for microtubules.

An autogenous origin for microtubules and flagella has been put forward in several alternative hypotheses. It should, however, be noted that the eukaryotic flagellum (the undulipodium of MARGULIS) is quite different in chemistry, structure and mode of action from the similarly named bacterial flagellum and no direct evolutionary relationship between the two is believed to exist. However, it has been suggested that the self-assembling microtubular proteins may have had a viral origin (TAYLOR 1980).

The first autogenous hypothesis proposes that microtubules evolved inside the membrane-enclosed nucleus and were first involved in the formation of the mitotic spindle (PICKETT-HEAPS 1974). PICKETT-HEAPS believes that in any account of evolutionary origins, the nuclear envelope and the mitotic spindle are inseparable and that centrioles (and hence eukaryotic flagella) evolved from the simpler organizing centres of this earlier nuclear microtubular system. Second autogenous hypothesis suggests that microtubules first evolved in a walled proto-fungus as a mechanism related to the transport of secretory vesicles through the cytoplasm and only later became associated with the nucleus (CAVALIER-SMITH 1981).

During mitosis two categories of microtubules can usually be distinguished, those which mediate the movement of chromosomes which were perhaps initially attached to the nuclear envelope, and those which help to form the spindle proper. Microtubules may initially have been present in the cytoplasm (i.e. outside the permanently closed nuclear envelope as seen in dinoflagellates today – Fig. 8 g), and have participated only in the movement of chromosomes which were either attached to the nuclear envelope (KUBAI 1975) or (even earlier) attached to the prokaryotic plasmamembrane (CAVALIER-SMITH 1981). Inclusion of microtubules in the spindle proper may well have been a much later evolutionary event: the initial spindle may even have been an actin-microfilament-based structure not unlike the fibrous band of the cryptomonad nucleomorph (MORRALL and GREENWOOD 1982). A comparatively late inclusion of microtubules in the eukaryotic spindle may be indicated by the diversity of spindle structures found among modern algae and protozoa (DODGE 1980).

In the unusual prokaryote, *Mycoplasma gallisepticum,* the DNA appears to be attached to a disc which is itself only remotely attached to the plasmamembrane. This disc replicates and, prior to cell division, the two daughter discs come to lie at opposite poles of the mother cell (MANILOFF and MOROWITZ 1972). Preliminary observations by GHOSH et al. (1978) suggest that cell division is blocked by cytochalasin B acting on an actin-like protein. Morphologically, this actin-based system of division resembles the mitotic system but lacks microtubules (HEATH 1980). HEATH suggests (1) that the eukaryotic nucleus and its envelope might have evolved following invagination of a disc-bearing portion of plasmamembrane of a prokaryote like this mycoplasma; (2) that as cell size increased, microtubules evolved and became included in the actin spindle, thus providing greater strength; (3) that the nuclear-membrane-associated discs or plaques which were already attachment sites for the genome also became nucleating sites for the initially continuous microtubules of the spindle; and (4) that microtubules with their organizing centres subsequently evolved cytoplasmic functions.

Hypotheses which suggest an autogenous origin for microtubules in eukaryotes have the advantage of beginning with the simple (i.e. individual microtubules) and proceeding to the more complex (i.e. the flagellar apparatus and the fully fledged mitotic spindle). As Pickett-Heaps points out, the almost universally conservative structure of flagella and centrioles suggests for them an early monophyletic origin. The authors believe that if an autogenous hypothesis of microtubule origin is to be consistent, one must envisage that flagella originated in a primitive eukaryotic cell in which cytoplasmic and chromosome microtubules were already present, but in which evolution of the microtubule-associated spindle was still at an early stage and had not yet undergone much divergence of form (Fig. 10). Though the idea of an autogenous origin for microtubules is attractive, much of what little "evidence" there is in its favour is unsatisfactory. The strongest "evidence" is the de novo origin of microtubular structures. If tubulin self-assembles and is an ancient protein, then its manifestations as microtubules and associated structures would be consistent with, but certainly would not prove, an autogenous origin. Other "evidence" quoted in favour requires assumptions about the primitive or reduced status of certain eukaryotes (e.g. *Pelomyxa*) or is based on the absence (rather than the presence) of features, such as centrioles, that would be expected in a primitive organism if the endosymbiotic hypothesis is correct. Further information on the origin of microtubules may be provided by investigations now in progress on sequencing of the proteins in microtubules of spirochaetes and of eukaryotic species.

3.4.4 Speculations and Sequences of Origin

Speculations about the evolutionary origins of cellular organelles fall into two classes, those which propose an autogenous origin for all organelles and those which propose an endosymbiotic origin for some. The difficulties in trying to reach a satisfactory solution to the problem of the origin of organelles is well illustrated by the very different proposed or implied pathways of sequential origin for microtubule-associated processes (including phagocytosis) and for some organelles (Fig. 10). Margulis (1981), for example, believes (Fig. 10 a) that flagella evolved (from symbiotic spirochaetes) after the acquisition of mitochondria and so the proposed bacterial ancestors of mitochondria must have been taken up by some means other than conventional phagocytosis. The method suggested by Margulis is by predatory entry somewhat similar to that employed by modern prokaryotic bdellovibrios when they invade other prokaryotes (Stolp 1979). This would imply that uptake of the precursors of mitochondria and chloroplasts took place by different symbiotic mechanisms and Margulis (1981) has agreed that an origin for mitochondria after flagella (Fig. 10 b), involving phagocytic uptake, is an acceptable alternative.

Though the parasitic bdellovibrios penetrate the walls of their prokaryotic hosts and enter the space between the walls and the plasmamembrane, they do not subsequently become incorporated in vacuoles within the host cytoplasm.

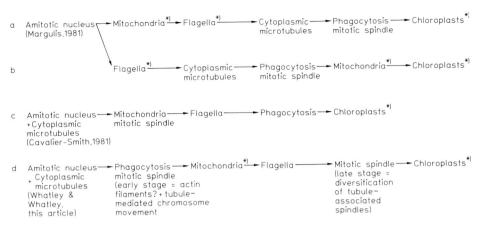

*)Organelles believed by the quoted authors to be of symbiotic origin

Fig. 10. The possible origin of some eukaryotic organelles

Indeed STOLP (1979) states that true endoparasitic (endosymbiotic) associations are restricted to eukaryotic host cells. However, a recent paper by WUJEK (1979) reports the occasional presence (in less than 5% of the population) in the blue-green alga, *Pleurocapsa minor* of what appear to be truly endosymbiotic bacteria, though the author, surprisingly, states his belief that in the (prokaryotic) alga, *Pleurocapsa,* the endosymbiotic bacteria are not contained in vacuoles, as in other (eukaryotic) algae, though they are surrounded by an electron-transparent zone. The mode of uptake of the bacteria is not known. If further more rigorous investigation (including serial sectioning) confirms this unique observation of an intracytoplasmic symbiont within a prokaryotic host, then this would suggest that the evolution of a capacity for true endosymbiosis may not necessarily have depended on the prior evolution of a phagocytic mechanism based on a system of microfilaments and microtubules. (The engulfment of one daughter cell by another during sporulation in clostridia points in a similar direction). A detailed study of the bacteria-*Pleurocapsa* association would be fascinating from the evolutionary point of view.

CAVALIER-SMITH (1981) now believes that a capacity for phagocytosis was a late evolutionary development (Fig. 10 c), which was preceded by the autogenous origin of mitochondria and was followed by the endosymbiotic origin of chloroplasts. CAVALIER-SMITH, TAYLOR (1978) and DODGE (1979) place the origin of mitosis before that of flagella. We prefer to believe (Fig. 10 d) that flagella must have evolved while development of the mitotic spindle was still incomplete.

To a large extent the discrepancies which exist between the various proposed orders of sequential evolution of different organelles result from our lack of information about ancient species, and our inability to identify evolutionary relics with any certainty. Among the so-called primitive species which exist today, we are unable to distinguish between the truly primitive and the reduced; we cannot tell if, say, the absence of flagella is a primary primitive feature or a result of secondary loss. We tend to equate the simple with the primitive but this is particularly dangerous with respect to "simple" parasitic species which are surely anomalous and therefore likely to be unreliable guides to the basic patterns of evolution in the eukaryotic cell.

3.5 Metabolic Interactions

3.5.1 Fermentations and Oxygen

The earliest fermentations may well have been internally balancing redox reactions involving pairs of amino acids as donors and receptors (Clarke and Elsden 1980). We know such fermentations in present-day clostridia as Strickland reactions, and it has been suggested (Miller and Orgel 1974) that the appropriate substrates may well have been available in the "primaeval soup" as substances accumulated abiotically. These fermenters would, of course, be under severe danger of using up all the energetic capital available and an early "escape" must have been the evolution of a light-capturing mechanism (PS I), to allow a bacterial type of photosynthesis to replace the dependence on fermentation. The resulting products of CO_2 fixation not only made available substrates for growth but also offered a source of carbohydrates for exploitation in new fermentations of the type most familiar in modern organisms. The evolution of an auxiliary photosystem (PS II) by blue-green prokaryotes, enabling them to use water to reduce CO_2, made their photosynthesis independent of the limited availability of reducing substrates, at the same time releasing O_2 into the environment. An early response to accumulation of oxygen (initially produced abiotically in small amounts?) was the formation in turn of the Cu/Zn superoxide dismutase, the Fe/Mn superoxide dismutase and the haem-protein catalase, the dismutases serving to overcome the toxic effects of the free radical O_2^- (produced by leakage of electrons from reduced flavoproteins) and the catalase serving to remove the less toxic H_2O_2 as it accumulated. De Duve (1969) has supposed that microbodies arose as catalase-containing primitive respiratory particles with limited electron transport, whose activities served to reinforce fermentation by increasing the number of phosphorylating steps available. In modern eukaryotic cells catalase is indeed confined to microbodies. The importance of microbodies is now in their ability to help balance physiological functions. The archaebacterium *Thermoplasma* has been put forward as a model of a prokaryotic precursor of the primitive protoeukaryote (Searcy et al. 1978, 1981). Among its features is a basically fermentative energy metabolism; but the organism requires some oxygen and the plasmamembrane contains a short flavoprotein and cytochrome-containing electron transport chain, whose function is to pump protons and so supply energy to drive the SO_4^{2-}-exporting ATPase on which its survival depends (it grows only below pH 2 in a medium acidified with sulphuric acid!). One can imagine a primitive protoeukaryote which has, then, an essentially fermentative metabolism but has a secondary requirement for oxygen and is therefore "aerobic" or "microaerobic" (SEARCY and WHATLEY 1982).

3.5.2 Aerobic Organisms, Endosymbiosis and the Integration of Metabolism

In parallel with the evolution of these truncated electron transport chains in *Thermoplasma,* or in microbodies in early eukaryotes, was the development of the obligate aerobic eubacteria e.g. *Paracoccus,* whose constitutive electron transport closely resembles that of the mitochondria. Many other eubacterial facultative aerobes are known, amongst which is the photosynthetic *Rhodopseudomonas.* This again has a mitochondrial type of electron transport chain, but under anaerobic conditions can switch to photosynthesis; it employs a number of common intermediate steps in both its respiratory and photosynthetic chains. Perhaps *Paracoccus* had a photosynthetic ancestor. At this stage in evolution most of the basic biochemical reactions found in the energy metabolism of later eukaryotic systems were already present. According to the endosymbiotic view the uptake of prokaryotes gave rise to eukaryotic organelles, leading sequentially to mitochondria and to green or red algal chloroplasts, and the evolving eukaryote acquired new combinations of metabolic pathways. This clearly gave rise to the need to integrate the old and the new pathways, the transfer of genetic information from the new organelles to the host nucleus allowed control to be exerted by the host on the organelle's reproduction while the acquisition of specific exchange systems allowed only selected components of the metabolic pools maintained in each subcellular compartment to be exchanged in a controlled way between metabolic (organelle) compartments (Fig. 11). An obvious example of integration is the way in which the adenylate transfer system (which is absent from free-living bacteria) was acquired by the mitochondrion, so that the ATP synthesized internally by oxidative phosphorylation is made available on demand to the cytoplasm for use outside the mitochondrial compartment. The rate of electron transport in the mitochondria is controlled by the internal availability of ADP (this is called respiratory control and is also shown by *Paracoccus denitrificans* particles) so that the mitochondrial activity is adjusted to the general energetic demands of the cell. Mitochondria can metabolize only a limited number of compounds and they have therefore a more restricted metabolism than the bacterial precursors from which they may have been derived. They are permeable to pyruvate, some dicarboxylic and tricarboxylic acids and some amino acids, but are impermeable to sugars.

The complete oxidation of sugars requires the participation of the glycolytic enzymes of the cytosol, which in producing pyruvate ($+ NADH_2$) yield a little ATP by substrate phosphorylation. The modern glycolytic sequence presumably represents the maintenance in the modern eukaryote of the fermentative reactions on which the energy metabolism of the protoeukaryote formerly depended. The rate of glycolysis is mainly controlled by the allosteric control of the key enzyme fructose-bis-phosphatase, and the ratio of (ADP + Pi): ATP is critical

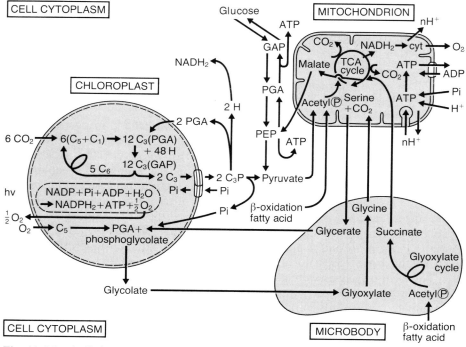

Fig. 11. Metabolic inter-relationships of some cellular compartments

for this control (Atkinson 1966). The cystol is also the site of many other metabolic sequences e.g. fatty acid synthesis from acetate, sugar synthesis from malate and protein synthesis from amino acids.

3.5.3 Protein Synthesis and Nuclear Control

Protein synthesis in the cytoplasm takes place on 80S ribosomes attached to the endoplasmic reticulum, and is under the control of a series of RNA molecules (messenger, ribosomal and transfer RNA's) synthesized by transcription of DNA in the nucleolus and released into the cytoplasm through the nuclear envelope. Chloroplasts and mitochondria each have their own DNA's, which code for a restricted number of polypeptides, especially structural components, and confer a degree of autonomy on these organelles. Control by the nucleus over these organelles is, however, accomplished by confining the synthesis of key electron transport and other components to 80S ribosomes. Thus cytochrome c is synthesized in the cytoplasm as a polypeptide having an additional terminal amino acid sequence that is used to recognize the membrane site where it will be taken into the mitochondrion. The recognition sequence is deleted during the passage of the cytochrome c through the mitochondrial membrane, releasing the cytochrome c inside for incorporation into the structural matrix (inner mitochondrial membrane) that has been made under the instruction of the mitochondrial DNA.

The principal soluble protein in chloroplasts is ribulose-bis-phosphate carboxylase/oxygenase. This is composed of large and small subunits; of these the large subunits are made in the chloroplast on 70S ribosomes under the direction of the chloroplast genome, but the small subunits are made in the cytoplasm on 80S ribosomes under nuclear direction and are subsequently transferred to the inside of the chloroplasts (ELLIS 1981).

3.5.4 Chloroplast Metabolism

Within the chloroplasts the photoreactions (PS I and II) are associated with the thylakoid compartment (Fig. 11); using light energy, the thylakoid enzymes catalyze the formation of NADPH and ATP at the (external) thylakoid surface in contact with the stroma. The stroma contains all the enzymes necessary for CO_2 fixation by the reductive pentose pathway, using the NADPH and ATP to drive the cycle. The CO_2 fixation cycle is autocatalytic (LILLEY and WALKER 1979) and intermediates can either be withdrawn to accumulate temporarily as starch or allowed to pass to the cytoplasm for further metabolism. The loss of intermediates is restricted principally to triose P, which is exchanged for Pi from the cytoplasm at a specific membrane site (HEBER 1974). The Pi for this exchange is derived from the further metabolism of the triose P, which can be used for, say, sucrose synthesis, or for the production of pyruvate; the latter can be oxidized in the mitochondria to CO_2 and H_2O in oxidative phosphorylation. The rate of oxidation of the pyruvate by the mitochondria determines the availability of Pi and hence controls the amount of triose P leaving the chloroplast. A supplementary exchange is that of triose P for phosphoglyceric acid (PGA). If the triose P is oxidized in the cytoplasm only to PGA, which is then taken back into the chloroplast, the net effect is to export reducing equivalents directly to the cytoplasm as NADH; this may be important in synthetic reactions.

3.5.5 Photorespiration: Three Cooperating Organelles

If instead of CO_2 the ribulose-bis-phosphate carboxylase oxygenase uses O_2 as its second substrate, one of the products is P glycollate. This is lost from the chloroplast as glycolic acid (a phosphatase removes the P group) and the glycolate is further metabolized by microbodies (peroxisomes) operating in conjunction with the mitochondria (Fig. 11); this is termed photorespiration (TOLBERT 1981). In the glycolate pathway in the peroxisomes a flavoprotein enzyme oxidizes it to glyoxylate (and the H_2O_2 produced is destroyed by catalase), which is then transaminated to glycine. The glycine is transferred to the mitochondria, where it is oxidized to CO_2 and serine, and the toxic ammonia released is converted to gluatamate. The CO_2 is then available as a photosynthetic substrate in the chloroplasts, and the serine (three quarters of the total carbon photorespired) moves back to the peroxisome, where it is transaminated in the glycerate pathway to hydroxypyruvate and reduced to glycerate. The glycerate can now move into the chloroplast stroma, where it re-enters the reductive pentose cycle after phosphorylation to give PGA. There is an active but cryptic ammonia exchange occurring during this sequence of reactions. Participation of three organelles (microbodies, mitochondria and chloroplasts) is involved in photorespiration, and each one carries out partial reactions of the overall sequence in compartments separated from the general cytoplasm.

Some of the organelles in the eukaryotes are "permanent", like mitochondria and chloroplasts. They have their own genomes and are partly autonomous; they appear to have arisen endosymbiotically. Their metabolic coordination appears to depend largely on the acquisition of specific carrier systems that allow each compartment to carry out special functions separately but to communicate only through a restricted number of compounds. A certain degree of automatic control not requiring the input of additional energy has been attained. Other organelles like microbodies are "inducible" and are produced temporarily from the endomembrane system in response to particular needs of the cell, but are dispensed with when not further required. The turnover of these temporary organelles can be quite rapid.

3.6 Concluding Remarks

In this article we have distinguished within the eukaryotic plant cell three groups of organelles: (1) those which are of autogenous origin and are part of the endomembrane system (2) those which we believe to have had an endosymbiotic origin (mitochondria and chloroplasts) (3) a group including the nucleus and microtubules, whose origin we consider uncertain.

The endomembrane system is clearly an early feature of the eukaryotic cell. It presumably arose by invaginations of the plasmamembrane like those to be found in some prokaryotes as, for example, in mesosomes and chromatophores. Its evolution has permitted the extension of available membrane surfaces for metabolic purposes; it has further allowed the development of larger cell size by making use of the channels of its cisternae, which penetrate all parts of the cytoplasm and facilitate communication with the external environment. Subsequently its ability to separate off parts of itself as specialized temporary organelles has allowed particular metabolic activities to be sequestered, and induced when required, as in peroxisomes, glyoxysomes, lysosomes and dictyosomes. Materials are carried in small endomembrane vesicles through the cytoplasm and across the plasmamembrane and the tonoplast. The phenomenon of exocytosis has led to the possibility of using metabolic products (waste products?) as components of the secondary cell wall. Such movements probably require the cooperation of microfilaments and microtubules and the further development of this combination led to phagocytosis and the possibility of endosymbiosis. However, the compartmentation achieved by the endomembrane system alone is already significant.

The evolution of a capacity for phagocytosis made possible the uptake and retention of symbionts within vacuoles in the host cytoplasm: many modern prokaryotic and eukaryotic symbionts are known. The endosymbiotic hypothesis proposes that the uptake of aerobic prokaryotes led to the formation of mitochondria and the subsequent uptake of photosynthetic prokaryotes and eukaryotes led to the formation of a variety of chloroplasts. We ourselves believe that proplastids and other non-photosynthetic plastids (including etioplasts) represent a later development from photosynthetically active chloroplasts (WHATLEY 1982). Following a long period of "domestication", it appears that the photosynthetic apparatus of plastids, particularly in angiosperms, may have become an inducible feature, rather than a permanent component, although the plastid itself is a permanent organelle. Although it is generally agreed that chloroplasts had an endosymbiotic origin, there continues to be some discussion about mitochondria, for which various autogenous origins are still considered.

The acquisition of each of these organelles in turn provided the host at a stroke (on the endosymbiotic view) with a subcellular compartment containing a completely new capability, respiration or photosynthesis. The integration of these new capabilities with the metabolism of the host led to the production of organelles from the free-living precursors. The integration involved subsequent transfer of genetic information from the symbiont to the nucleus of the host (cf. plasmid transfer) but the circular DNA remaining in the organelles continues to be responsible for directing the synthesis of structural proteins and other components which cannot pass through the membranes surrounding the organelles (e.g. the large but not the small subunit of ribulose-bis-phosphate carboxylase). Mitochondria and chloroplasts (unlike endomembrane organelles) are not induced de novo but are, of course, permanent inhabitants of their respective "host" cells.

The evidence for an endosymbiotic origin is based on items of information from a wide variety of sources, none of which is necessarily convincing on its own. The evidence includes information on (1) structure, as seen in the electron microscope (2) the differential location of particular constituents (e.g. cytochromes) in the electron transport chains of different organelles (3) amino acid sequencing of small molecular weight proteins (e.g. cytochromes and ferredoxins) from prokaryotic and eukaryotic organisms and from organelles (4) the base sequencing of nucleic acids, particularly from ribosomal RNA (5) the occurrence of modern symbionts which serve as models of evolutionary intermediates between free-living organisms and endosymbiotically derived organelles (6) the identification of living organisms which have many of the features expected in the endosymbiotic ancestors of mitochondria and the chloroplasts of red and green algae.

The evolutionary origin of the nucleus and microtubule-associated organelles remains debatable. Of the six proposals that we have summarized, the three which are most attractive to us imply, in one way or another, an autogenous origin. However, the third of these, the suggestion that the nucleus might have arisen by a modification of what happens during bacterial sporulation, also implies a form of endosymbiosis.

The nucleus provides a separate compartment in which nucleic acids (DNA and RNA) are manufactured. The RNA's are usually exported through the nuclear envelope and assembled in the cytoplasm into ribosomes or are used as transfer RNA. The host DNA is retained inside the nucleus and is further protected by combination with histones. In the eukaryotic cell the increase in genome size and the assembly of DNA into chromosomes has been accompanied by, and has required, the evolution of a new efficient system for chromosome segregation involving the participation of microtubules in the mitotic spindle.

Hypotheses concerned with the evolution of microtubules present two completely opposed views. The first hypothesis proposes (1) that microtubules evolved in prokaryotic spirochaetes, where they formed precisely ordered $9+2$ assemblages (2) that the spirochaetes formed a superficial symbiotic attachment to the plasmamembrane of eukaryotic cells and (3) that the symbiotic spirochaetes later evolved into flagella. Subsequently the flagellar microtubules entered the host cytoplasm, where they assumed the other eukaryotic functions with which they are now associated. The second hypothesis suggests that microtubules first appeared in eukaryotes as individual, autogenous structures either within the nucleus from which they later entered the cytoplasm or within the cell cytoplasm where their possible initial function was either as a cytoskeletal element, or to direct the movement of vesicles, or to assist in a simple extranuclear form of chromosome segregation. Subsequently the cytoplasmic microtubules became assembled in more complex arrays to form flagella and the mitotic spindle.

There is insufficient evidence to allow a firm conclusion about the evolutionary origin of microtubules. The problem is accentuated by their diverse functions in the apparently primitive as well as in the complex modern eukaryotic cell. However, the structural similarity of all flagella must indicate an early monophyletic origin for this organelle. The essentially conservative basic structure of the eukaryotic nucleus and its envelope and pore complex also implies an early monophyletic origin, and an early association between chromosomes, microtubules and the nuclear envelope is probable. However, the varied sites and arrangements of the microtubules found in mitotic spindles of different protists and unicellular algae today seem to suggest that full development of these diverse and complex spindles only took place after the evolution of flagella.

In the eukaryotic cell, all the different compartments (however they arose) contribute to the total metabolism of the organism and clearly interact with each other. One feature of the control mechanisms involved is the restriction of the components which are allowed to enter or leave each compartment. Each organelle is responsible for one main function. Mitochondria interact with the cytoplasm by exchanging ATP for ADP and ortho-phosphate, and can take in only a small number of compounds as carbon sources. Mature chloroplasts exchange triose phosphates and phosphoglyceric acid for inorganic phosphate, but are not permeable to most of the intermediates of the reductive pentose cycle. Glycolic acid produced in chloroplasts is transferred to the leaf peroxisomes for the initial steps in photorespiration and the peroxisomes in turn export glycine to the mitochondria. The endomembrane system, microfilaments and microtubules all participate in the directed movement of cellular materials. By being responsible for the manufacture of DNA and RNA, the nucleus maintains overall control of the many and diverse activities of the eukaryotic cell.

The eukaryotic cell contains a variety of distinctive organelles. Some have certainly arisen autogenously; others may have evolved from endosymbionts. There is no reason to suppose that all organelles evolved in the same way. It is now generally accepted that chloroplasts evolved from endosymbiotic algae. We ourselves believe that mitochondria also evolved from endosymbionts. However, the "domestication" of mitochondria within the eukaryotic cell took place a very long time ago and this organelle has been subjected to greater modification than the more recently evolved chloroplasts (PARTHIER 1980). The prolonged period of divergent specialization (autogenous origin) or domestication (endosymbiotic origin) associated with the evolution and subsequent major modification of the nucleus and microtubule-associated structures make it much more difficult to determine how these organelles originated. In the absence of surviving protoeukaryotes or truly primitive eukaryotes, it is impossible to establish how or in what sequence most eukaryotic organelles evolved.

Acknowledgements. We should like to thank Dr. CARR, Professor DODGE, Dr. HAWES, Dr. LEWIN and Dr. KNUTTON for providing us with material for electron micrographs.

References

Altmann R (1890) Die Elementarorganismen und ihre Beziehung zu den Zellen. Veit, Leipzig

Atkinson DE (1966) Regulation of enzyme activity. Annu Rev Biochem 35:85–124

Baltscheffsky H, Baltscheffsky M (1981) Mitochondrial ancestor models: *Paracoccus, Rhizobium* and *Rhodospirillum*. In: Lee CP, Shatz G, Dallner G (eds) Mitochondria and microsomes. Addison-Wesley, London pp 519–540

Beevers H (1979) Microbodies in higher plants. Annu Rev Plant Physiol 30:159–193

Beringer JE, Brewin N, Johnston AWB, Schulman HM, Hopwood DA (1979) The *Rhizobium* – legume symbiosis. Proc R Soc Lond B 204:219–233

Buvat R (1971) Origin and continuity of cell vacuoles. In: Reinert J, Ursprung H (eds) Origin and continuity of cell organelles. Springer, Berlin Heidelberg New York, pp 127–157

Cavalier-Smith T (1975) The origin of nuclei and of eukaryotic cells. Nature 256:463–467

Cavalier-Smith T (1981) The origin and early evolution of the eukaryotic cell. Symp Soc Gen Microbiol 32:33–84

Chapman-Andresen C, Hamburger K (1981) Respiratory studies on the giant amoeba *Pelomyxa palustris*. J Protozool 28:433–440

Christensen T (1964) The gross classification of algae. In: Jackson DF (ed), Algae and Man. Plenum Press, New York London pp 59–64

Clarke PH, Elsden SR (1980) The earliest catabolic pathways. J Molec Evol 15:333–338

Dawes IW (1981) Sporulation in evolution. Symp Soc Gen Microbiol 32:85–130

Dayhoff MO, Schwartz RM (1981) Evidence on the origin of eukaryotic mitochondria from protein and nucleic acid sequences. Ann NY Acad Sci 361:92–103

Dodge JD (1973) The fine structure of algal cells. Academic Press, London New York

Dodge JD (1979) The phytoflagellates: fine structure and phylogeny. In: Levan-Dowsky M, Hutner SH (eds) Biochemistry and physiology of protozoa Vol 1. Academic Press, London New York, pp 7–57

Dodge JD (1980) Morphology and phylogeny of flagellated protists. In: Schwemmler W, Schenk HEA (eds) Endocytobiology. de Gruyter, Berlin, pp 33–50

Dodson EO (1979) Crossing the procaryote–eucaryote border: endosymbiosis or continuous development? Can J Microbiol 25:652–674

Doolittle WF, Bonen L (1981) Molecular sequence data indicating an endosymbiotic origin for plastids. Ann NY Acad Sci 361:248–256

Dougherty EC (1957) Neologisms needed for structures of primitive organisms. J Protozool 4 suppl:14

Dutton PL, Wilson DF (1974) Redox potentiometry in mitochondrial and photosynthetic bioenergetics. Biochim Biophys Acta 346:165–212

Duve de C (1969) Evolution of the peroxisome. Ann NY Acad Sci 168:369–381

Ellis RJ (1981) Chloroplast proteins: synthesis, transport and assembly. Annu Rev Plant Physiol 32:111–137

Evans LV, Christie AO (1970) Studies on the ship-fouling alga *Enteromorpha*. I Aspects of the fine structure and biochemistry of swimming and newly settled zoospores. Ann Bot Lond 34:451–466

Fox GE, Stackebrandt E, Hespell RB, Gibson J, Manioloff J, Dyer TA, Wolfe RS, Balch WE, Tanner RS, Magrum LJ, Zablen LB, Blakemore R, Gupta R, Bonen L, Lewis BJ, Stahl DA, Luehrsen KR, Chen KN, Woese CR (1980) The phylogeny of prokaryotes. Science 209:457–463

Ghosh A, Maniloff J, Gerling DA (1978) Inhibition of mycoplasma cell division by cytochalasin B. Cell 13:57–64

Gibbs SP (1970) The comparative ultrastructure of the algal chloroplast. Ann NY Acad Sci 175:454–473

Gibbs SP (1978) The chloroplasts of *Euglena* may have evolved from symbiotic green algae. Can J Bot 56:2883–2889

Gibbs SP (1981a) The chloroplasts of some algal groups may have evolved from eukaryotic algae. Ann NY Acad Sci 361:193–208

Gibbs SP (1981b) The chloroplast endoplasmic reticulum: structure, function and evolutionary significance. Int Rev Cytol 72:49–99

Gillot MA, Gibbs SP (1980) The cryptomonad nucleomorph: its ultrastructure and evolutionary significance. J Phycol 16:558–568

Gould GW, Dring GJ (1979) On a possible relationship between bacterial endospore formation and the origin of eukaryotic cells. J Theoret Biol 81:47–53

Greenwood AD (1974) The Cryptophyta in relation to phylogeny and photosynthesis. Abstr 8th Int Congr EM Canberra 2:566–567

Hausmann K (1978) Extrusive organelles in protists. Int Rev Cytol 52:197–276

Heath IB (1980) Variant mitosis in lower eukaryotes: indication of the evolution of mitosis? Int Rev Cytol 64:1–80

Heber U (1974) Metabolite exchange between chloroplasts and cytoplasm. Annu Rev Plant Physiol 25:393–421

Heckman K (1980) Omikron, an essential endosymbiont of *Euplotes aediculatus*. In: Schwemmler W, Schenk HEA (eds) Endocytobiology. de Gruyter, Berlin, pp 393–400

Heckman JE, Sarnoff J, Alzner-DeWeerd B, Yin S, BajBhandary UL (1980) Novel features in the genetic code and codon-reading patterns in *Neurospora crassa* mitochondria based on sequences of six mitochondrial tRNA's. Proc Natl Acad Sci USA 77:3159–3163

Herzog V, Miller F (1979) Membrane retrieval in secretory cells. Symp Soc Exp Biol 33:101–116

Jeon KW, Danielli JF (1971) Micrurgical studies with large free-living amoebas. Int Rev Cytol 30:49–89

Jeon KW, Jeon MS (1976) Endosymbiosis in amoebae: Recently established endosymbionts have become required cytoplasmic components. J Cell Physiol 89:337–344

John P, Whatley FR (1975) *Paracoccus denitrificans* and the evolutionary origin of the mitochondrion. Nature 254:495–498

John P, Whatley FR (1977) *Paracoccus denitrificans* as a mitochondrion. Adv Bot Res 4:51–115

Kies L (1980) Morphology and systematic position of some endocyanomes. In: Schwemmler W, Schenk HEA (eds) Endocytobiology. de Gruyter, Berlin, pp 7–19

Kubai DF (1975) The evolution of the mitotic spindle. Int Rev Cytol 43:167–227

Kunicki-Goldfinger WJH (1980) Evolution and endosymbiosis. In: Schwemmler W, Schenk HEA (eds) Endocytobiology. de Gruyter, Berlin, pp 969–984

Lea PJ, Miflin BJ (1980) Transport and metabolism of asparagine and other nitrogen compounds within the plant. In: Stumpf PK, Conn EE (eds) The Biochemistry of Plants, Vol V. Academic Press, London, New York, pp 569–607

Lewin RA, Withers NW (1975) Extraordinary pigment composition of a prokaryotic alga. Nature 256:735–737

Lilley R McC, Walker DA (1979) Studies with the reconstituted chloroplast system. In: Gibbs M, Latzko E (eds) Encyclopedia of plant physiology vol 6. Springer, Berlin Heidelberg New York, pp 41–53

Lord JM, Roberts LM (1983) Formation of glyoxysomes. Int Rev Cytol Suppl 15:115–156

Lott JNA (1980) Protein bodies. In: Stumpf PK, Conn EE (eds). The biochemistry of plants, vol 1. Academic Press, London New York, pp 589–623

McAuley PJ (1981) What determines the population size of the intra-cellular algae in the digestive cells of green hydra? Experientia 37:346–347

Mahler HR (1980) Non-symbiotic hypotheses of mitochondrial origin and their relevance to cell research. In: Schwemmler W, Schenk HEA (eds) Endocytobiology. de Gruyter, Berlin, pp 869–892

Maniloff J, Morowitz HJ (1972) Cell biology of the mycoplasmas. Bacteriol Rev 36:263–290

Margulis L (1981) Symbiosis in cell evolution. Freeman, San Francisco

Marty F, Branton D, Leigh RA (1980) Plant vacuoles. In: Stumpf PK, Conn EE (eds) The biochemistry of plants, Vol 1. Academic Press, London New York, pp 625–658

Matile PH (1974) Lysosomes. In: Robards AW (ed) Dynamic aspects of plant ultrastructure. McGraw-Hill, New York

Mereschowsky C (1905) Über Natur und Ursprung der Chromatophoren im Pflanzenreiche. Biol Zentrbl 25:593–604

Miller SL, Orgel LE (1974) The origins of life on earth. Prentice-Hall, New York

Mollenhauer HH, Morré DJ (1980) The Golgi apparatus. In: Stumpf PK, Conn EE (eds) The biochemistry of plants, Vol I. Academic Press, London New York, pp 437–488

Morrall S, Greenwood AD (1982) Ultrastructure of nucleomorph division in species of cryptophyceae and its evolutionary implications. J Cell Sci 54:311–328

Morré DJ, Mollenhauer HH (1974) The endomembrane concept: a functional integration of endoplasmic reticulum and Golgi apparatus. In: Robards AW (ed) Dynamic aspects of plant ultrastructure. McGraw-Hill, New York, pp 84–137

Müller M (1975) Biochemistry of protozoan microbodies. Annu Rev Microbiol 29:467–483

Muscatine L, Pool RR (1979) Regulation of numbers of intracellular algae. Proc R Soc Lond B 204:131–139

Newcomb EH, Pugh TD (1975) Blue-green algae associated with ascidians of the Great Barrier Reef. Nature 253:533–534

Parthier B (1980) Evolutionary aspects of gene expression organization in macro-compartments. In: Nover L, Lynen F, Mothes K (eds) Cell compartmentation and metabolic channeling. Fischer, Jena, Elsevier/North Holland Biomedical Press, Amsterdam New York, pp 107–121

Phillips DO, Carr NG (1981) Molecular approaches to the endosymbiotic hypothesis. Ann NY Acad Sci 361:298–311

Pickett-Heaps JD (1974) Evolution of mitosis and the eukaryotic condition. Biosystems 6:37–48

Powles SB, Osmond CB (1978) Inhibition of the capacity and efficiency of photosynthesis in bean leaflets illuminated in a CO_2-free atmosphere at low oxygen: a possible role for photorespiration. Aust J Plant Physiol 5:619–629

Preer Jr JR (1975) The hereditary symbionts of *Paramecium aurelia*. Symp Soc Exp Biol 29:125–145

Preer Jr JR, Preer LB, Jurand A (1974) Kappa and other endosymbionts in *Paramoecium aurelia*. Bact Rev 38:113–163

Raff RA, Mahler HR (1975) The symbiont that never was: an inquiry into the evolutionary origin of the mitochondria. Symp Soc Exp Biol 29:41–92

Robins RJ, Juniper BE (1980) The secretory cycle of *Dionaea muscipula*. III. The mechanism of release of digestive secretion. New Phytol 86:313–327

Robinson DG (1977) Plant cell-wall synthesis. Adv Bot Res 5:89–151

Sagan L (1967) On the origin of mitosing cells. J Theoret Biol 14:225–274

Sanger F (1981) Determination of nucleotide sequences in DNA. Bioscience 1:3–18

Satir B (1974) Membrane events during the secretory process. Symp Soc Exp Biol 28:399–418

Schell J, Van Montagu M, De Beuckeleer M, De Block M, Depicker A, De Wilde M, Engler G, Genetello C, Hernalsteens JP, Holsters M, Seurinck J, Silva B, Van Vliet F, Villarroel R (1979) Interactions and DNA plasmid transfer between *Agrobacterium tumefaciens,* the Ti plasmid and the plant host. Proc R Soc Lond B 204:251–266

Schimper AFW (1883) Über die Entwickelung der Chlorophyllkörner und Farbkörper. Bot Z 41:105–114

Schwartz RM, Dayhoff MO (1981) Chloroplast origins: inferences from protein and nucleic acid sequences. Ann NY Acad Sci 361:260–269

Searcy DG, Whatley FR (1982) *Thermoplasma acidophilum* cell membrane: cytochrome b and sulfate-stimulated ATP-ase. Zbl Bakt Hyg, I Abt Orig C3:245–257

Searcy DG, Stein DB, Green GR (1978) Phytogenetic affinities between eukaryotic cells and a thermophilic mycoplasma. Biosystems 10:19–28

Searcy DG, Stein DB, Searcy KB (1981) A mycoplasma-like archaebacterium possibly related to the nucleus and cytoplasm of eukaryotic cells. Ann NY Acad Sci 361:312–323

Silverberg BA (1975) An ultrastructural and cytochemical characterization of microbodies in green algae. Protoplasma 83:269–295

Smith DC (1979) From extracellular to intracellular: the establishment of a symbiosis. Proc R Soc Lond B 204:115–130

Sonneborn TM (1959) Kappa and related particles in *Paramecium.* Adv Virus Res 6:229–338

Stabenau H (1984) Microbodies of different algae. In: Compartments in algal cells and their interaction Symp. Göttingen 1983. Springer, Berlin Heidelberg New York (in press)

Stackebrandt E, Woese CR (1981) The evolution of prokaryotes. Symp Soc Gen Microbiol 32:1–31

Stanier RY (1970) Some aspects of the biology of cells and their possible evolutionary significance. Symp Soc Gen Microbiol 20:1–38

Stolp H (1979) Interactions between *Bdellovibrio* and its host cell. Proc R Soc Lond B 204:211–217

Taylor FJR (1978) Problems in the development of an explicit hypothetical phylogeny of the lower eukaryotes. Biosystems 10:67–89

Taylor FJR (1979) Symbioticisms revisited: a discussion of the evolutionary impact of intracellular symbioses. Proc R Soc Lond B 204:267–286

Taylor FJR (1980) The stimulation of cell research by endosymbiotic hypotheses for the origin of eukaryotes, In: Schwemmler W, Schenk HEA (eds) Endocytobiology. de Gruyter, Berlin, pp 917–942

Tolbert NE (1981) Peroxisomes and glyoxysomes. Annu Rev Biochem 50:133–157

Trench RK (1981) Chloroplasts: presumptive and de facto organelles. Ann NY Acad Sci 361:341–355

Uzzell T, Spolsky C (1974) Mitochondria and plastids as endosymbionts: a revival of special creation? Am Sci 62:334–343

Uzzell T, Spolsky C (1981) Two data sets: alternative explanations and interpretations. Ann NY Acad Sci 361:481–499

Whatley JM (1982) Plastids past, present and future. Int Rev Cytol Suppl 14:329–373

Whatley JM, Whatley FR (1981) Chloroplast evolution. New Phytol 87:233–247

Whatley JM, John P, Whatley FR (1979) From extracellular to intracellular: the establishment of mitochondria and chloroplasts. Proc R Soc Lond B 204:165–187

Woese CR (1981) Archaebacteria. Sci Am 244(6):94–106

Wujek DE (1979) Intracellular bacteria in the blue-green alga *Pleurocapsa minor.* Trans Am Micros Soc 98:143–145

4 Autotrophic Eukaryotic Freshwater Symbionts

W. REISSER and W. WIESSNER

4.1 Introduction

In the freshwater habitat various autotrophic algae and heterotrophic organisms exist in more or less tight partnership. Here we report on so-called endosymbiotic systems in which the heterotrophic partner encloses the autotrophic one. We define the term symbiosis as a permanent co-habitation of dissimilar organisms with physical contact and mutual benefit. Information on such associations has usually been mainly descriptive, based on field observations and light microscopic studies, with few physiological data (BUCHNER 1930, 1953, DROOP 1963, GOETSCH and SCHEURING 1926, KARAKASHIAN 1970, TAYLOR 1973, YONGE 1944). However, during the last 10 to 20 years interest in this field has grown tremendously. This was initiated by the suitability of symbiotic systems for studies on general problems of cell–cell recognition, autoimmunity and the evolution of eukaryotic autotrophic cells. Since recent investigations centre on very few systems, some generalized aspects of this report are derived from a rather small experimental base.

4.2 The Taxonomic Position of Symbiotic Partners

The heterotrophic partners of freshwater endosymbiotic systems belong to various taxonomic groups: Protozoa, Porifera, Cnidaria, Turbellaria, Nemathelminthes, and Mollusca (BUCHNER 1930, 1953, DROOP 1963, GOETSCH and SCHEURING 1926, YONGE 1944) (Table 1).

The autotrophic partners in general are chlorophycean algae, mostly of coccoid organization, for which the term zoochlorella was coined (BRANDT 1882). However, since this nomenclature was used lateron for all types of endosymbiotic green algae, it is taxonomically ambiguous and should be abandoned. Most endosymbiotic algae of freshwater systems belong to the genus *Chlorella*. They do not differ principally by morphology or cytology from free-living chlorellae, but are distinguished from them mainly by physiological characteristics, such as excretion of high amounts of carbohydrates at strain-specific rates (MEWS and SMITH 1982, MUSCATINE et al. 1967, REISSER 1981a). The various chlorellae from different endosymbiotic associations are not identical. Chlorellae isolated from *Spongilla fluviatilis* (strain 211–40c of the Sammlung von Algenkulturen Göttingen, SCHLÖSSER 1982) belong to the *Chlorella sorokiniana* Sh. et Kr. group (KESSLER 1978), whereas algae from *Paramecium bursaria* are re-

Table 1. Features of endosymbiotic associations between eukaryotic algae and heterotrophic partners in the freshwater habitat

Host	Symbiotic alga		Location of algae		Carbohydrates excreted by isolated algae identified as		Comments
	Chlorella sp.	Other	Intracellular	Intercellular	Maltose	Glucose	
Protozoa [a]							
Mayorella viridis	(+)		+				
Climacostomum virens	+		+			+	No vitamin requirement of alga, excretes also traces of xylose
Coleps hirtus	(+)		+				
Euplotes daidaleos	+		+				No vitamin requirement of alga
Euplotes patella	(+)		+				Green *E.p.* probably identical with *E. daidaleos*
Frontonia leucas	(+)		+				
Heleopera sphagni	(+)		+				
Ophrydium versatile	(+)		+				
Paramecium bursaria	+		+	+			Alga needs vitamins B_1, B_{12}, excretes traces of glucose
Psilotricha viridis		(+)	+				Alga reported with stigma (*Chlamydomonas* sp.?)
Stentor polymorphus	+		+	+			Alga needs vitamins B_1, B_{12}
Stentor roeseli	(+)		+				Aposymbiotic occurrence also reported
Vorticella sp.	(+)		+				
Vorticella sp.		(+)	+				Symbiotic alga: *Oocystis* sp.?
Porifera							
Ephydatia fluviatilis	+		+			+	Alga in amoebocytes
Spongilla lacustris	(+)		+				
Spongilla fluviatilis	+		+			+	Alga in amoebocytes, no vitamin requirement
Cnidaria							
Chlorohydra hadleyi	+		+				Aposymbiotic occurrence also reported

[a] Selection of species; for further species, mostly described only as green see: BUCHNER (1930, 1953), DANGEARD (1900), GOETSCH and SCHEURING (1926), KAHL (1930)
( ) = experimental confirmation needed

Table 1 (continued)

Host	Symbiotic alga		Location of algae		Carbohydrates excreted by isolated algae identified as		Comments
	Chlorella sp.	Other	Intracellular	Intercellular	Maltose	Glucose	
Hydra viridis	+		+		+		Alga in digestive cells, needs vitamins
Mollusca							
Anodonta sp.	+		+	+			
Linnaea sp.	(+)			+			
Unio pictorum	(+)			+			
Turbellaria							
Castrada viridis	(+)						
Dalyellia viridis	(+)		+	+			
Phaenocora typhlops	+						
Typhoplana sp.							
Nemathelminthes							
Cephalodella sp.	(+)			+			

lated to *Chlorella vulgaris* Beijerink and *Chlorella sorokiniana* Sh. et Kr. (REISSER 1975).

Occasionally other chlorophycean algae have also been reported to occur in freshwater symbiotic systems, e.g. *Scenedesmus* sp., *Ankistrodesmus* sp., *Pleurococcus* sp., *Oocystis* sp. and *Chlamydomonas* sp. (GEITLER 1947, KAHL 1930). But these data are based mostly on field observations and need careful corroboration under laboratory conditions in order to exclude the possibility that the observed algae have only been taken up as food and not into a mutual relationship with their host.

4.3 The Location of Autotrophic Partners

Autotrophic partners of symbiotic freshwater systems are located within their heterotrophic host either intra- or intercellularly ("endosymbionts"). An intracellular location of algal cells occurs mainly in protozoa and in digestive cells of porifera and cnidaria. There the endosymbionts are situated in the host cytoplasm within vacuoles. At least in more thoroughly studied systems such

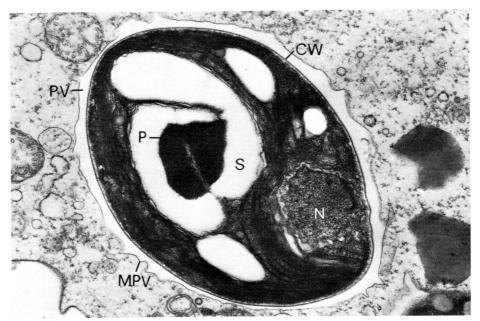

Fig. 1. Symbiotic *Chlorella* enclosed by *Paramecium bursaria* in a perialgal vacuole. *C* chloroplast; *CW* cell wall; *MPV* membrane of perialgal vacuole; *PV* perialgal vacuole; *N* nucleus; *P* pyrenoid; *S* starch. Thin section; ×23,000

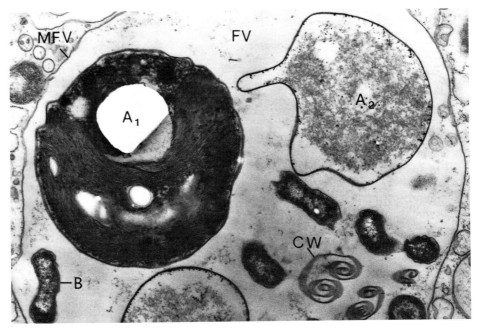

Fig. 2. Infection of alga-free *Paramecium bursaria* with a free-living *Chlorella* sp. Algae are sequestered in food vacuoles and show various states of digestion (A_1, A_2); *B* bacterium; *CW* cell wall; *FV* food vacuole; *MFV* membrane of food vacuole. Thin section; ×17,500

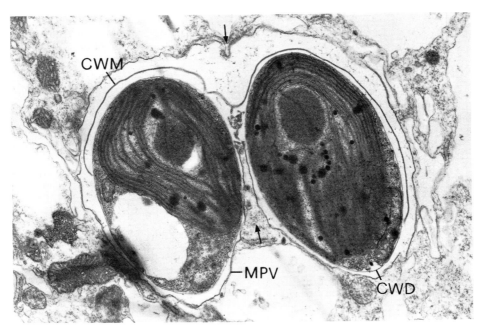

Fig. 3. Formation of a new perialgal vacuole membrane around daughter cells after division of the mother cell (*Chlorella* sp.) within a perialgal vacuole of *Stentor polymorphus*. *Arrows* indicate places where the new membrane has been formed. *CWD* cell wall of daughter cell; *CWM* cell wall of mother cell; *MPV* membrane of perialgal vacuole. Thin section; × 24,000

as *Climacostornum virens, Paramecium bursaria, Stentor polymorphus, Vorticella* sp., *Spongilla fluviatilis,* and *Hydra viridis* with *Chlorella* sp., two types of vacuoles are observed (GRAHAM and GRAHAM 1978, 1980, KARAKASHIAN et al. 1968, OSCHMAN 1967, REISSER 1981a, REISSER et al. 1983). One type, the so-called perialgal vacuole, harbours always only one algal cell, the wall of which is closely attached to the vacuolar membrane of the host (Figs. 1, 4). The other type, the food vacuole, is of varying size and contains not only bacteria but also algae in various stages of digestion and other material (Fig. 2). It is a special feature of a perialgal vacuole that it divides simultaneously with the enclosed alga (Figs. 3, 4) and probably protects it from host lytic enzyme action (HOHMAN et al. 1982, KARAKASHIAN and RUDZINSKA 1981, MEIER et al. 1980a, 1980b; O'BRIEN 1982, REISSER et al. 1980). Labelling with electron-dense markers indicates that in both *Paramecium bursaria* and *Hydra viridis* lysosomes cannot fuse with perialgal vacuoles (HOHMAN et al. 1982, KARAKASHIAN and RUDZINSKA 1981, O'BRIEN 1982). In *Paramecium bursaria* perialgal vacuoles do not undergo the cyclosis typical for food vacuoles. Besides being formed concomitant to algal division, perialgal vacuoles are also formed via phagocytosis during the uptake of suitable algae by both green and aposymbiotic hosts. Both intra- and intercellular location of endosymbiotic algae is reported for

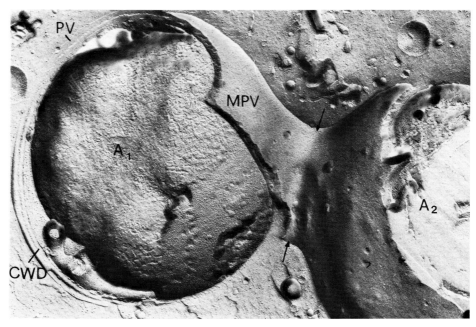

Fig. 4. Division of a perialgal vacuole in *Paramecium bursaria*. After the division of the endosymbiotic *Chlorella* into two daughter cells (A_1, A_2) the perialgal vacuole (*PV*) divides at once (*arrows*). The perialgal vacuole membrane (*MPV*) is closely attached to the cell walls of daughter cells (*CWD*). Freeze fracture; × 18,600. (Photograph kindly provided by R. Meier, University of Göttingen)

turbellaria and mollusca, but these data need electron microscopic confirmation (Eaton and Young 1975, Goetsch and Scheuring 1926, Haffner 1925).

Most endosymbiotic systems are hereditary, e.g. in unicellular hosts algae are distributed to daughter cells during host cell division. In *Spongilla lacustris* algae are propagated in gemmules (Gilbert and Simpson 1976), in *Hydra viridis* in buds, eggs or by adhesion to eggs (Muscatine and McAuley 1982, Thorington et al. 1979). In turbellaria, algae probably adhere to the egg case and so reinfect the young host (Eaton and Young 1975).

4.4 Physiological Features of Symbiotic Partners and Their Association

The specific physiology of endosymbiotic partnerships has been studied so far only for a few associations. Nevertheless, the data which are obtained from different systems have several common characteristics.

The gas exchange between symbiotic partners is characterized by the interaction between autotrophic and heterotrophic metabolism. Oxygen produced by

Table 2. Dependence of respiratory oxygen consumption and photosynthetic oxygen production[a] in alga-free and alga containing *Paramecium bursaria* and in the isolated symbiotic *Chlorella* sp. on different light fluence rates and the addition of glucose to the incubation medium

Light fluence rate	Glucose (0.024 M)	Alga-free *P.b.*	Alga containing *P.b.*	Alga containing *P.b.*	Isolated symbiotic *C.* sp.
W m^{-2}		μl O$_2$ min^{-1} 10^6 cells^{-1}		μl O$_2$ min^{-1} mg Chlorophyll	
0	–	-10.52 ± 0.48	-8.24 ± 0.38	-14.30 ± 0.66	-3.20 ± 0.17
0	+	-27.30 ± 1.35	-9.48 ± 0.40	-16.50 ± 0.70	-3.40 ± 0.18
2.94	–	/	-5.36 ± 0.28	-9.30 ± 0.49	$+4.60 \pm 0.23$
9.84	–	/	-2.68 ± 0.14	-4.60 ± 0.24	$+5.70 \pm 0.27$
19.05	–	/	$+3.27 \pm 0.18$	$+5.60 \pm 0.30$	$+38.30 \pm 1.73$

[a] Calculated for true photosynthesis; / = experiment not done
P.b. = *Paramecium bursaria*; *C.* sp. = *Chlorella* sp.
Photosynthetic compensation point of the isolated symbiotic *Chlorella* sp.: 1.18–2.35 Wm^{-2}
Photosynthetic compensation point of the alga-containing *Paramecium bursaria:* 10.40–15.50 Wm^{-2}

the photosynthesizing autotrophic partner is consumed by the heterotrophic host's respiration, thereby releasing carbon dioxide which is in turn photosynthetically assimilated by the alga. Sufficient light intensities can thus satisfy the host's oxygen requirements and in some systems even allow a net oxygen production for the whole association (Table 2) (*Paramecium bursaria:* PADO 1967, REISSER 1976b, *Spongilla lacustris:* SZUCH et al. 1978, *Hydra viridis:* CANTOR and RAHAT 1982, PHIPPS and PARDY 1982, *Dalyellia viridis:* HEITKAMP 1979, *Phaenocora typhlops:* YOUNG and EATON 1975).

In general, endosymbiotic algae sustain host growth. As could be shown with *Paramecium bursaria* (KARAKASHIAN 1963, WEIS 1974), *Spongilla lacustris* (FROST and WILLIAMSON 1980), *Hydra viridis* (MUSCATINE and LENHOFF 1963), and *Phaenocora typhlops* (YOUNG and EATON 1975) under conditions of food limitation the alga-containing host is more viable than an aposymbiotic one. The carbon metabolism of the alga probably plays a key role in the symbiotic partnership. Symbiotic chlorellae excrete high amounts of maltose, glucose, xylose, and other sugars not only when located within their partner but also when outside the symbiotic system. With the exception of one *Chlorella* strain isolated from *Hydra viridis* (JOLLEY and SMITH 1978), symbiotic chlorellae probably do not need any special host factor triggering carbohydrate excretion. Up to 86% of the totally fixed carbon can be excreted by the algae isolated from *Paramecium bursaria* (MUSCATINE et al. 1967). Within the symbiotic unit the excretion amounts to 57% (REISSER 1976b). Corresponding amounts are 30–40% for *Hydra viridis* (EISENSTADT 1971, MEWS 1980) and 9–17% for *Ephydatia fluviatilis* (WILKINSON 1980).

Probably because of their internal sugar supply, alga-containing *Paramecium bursaria* take up less glucose from the external medium than aposymbiotic ones, as is indicated by the rate of respiratory oxygen consumption (Reisser 1980b) (Table 2). In *Hydra viridis*- and *Paramecium bursaria*-chlorellae, maltose is not solely derived from starch degradation. In the light it is a more direct product of photosynthesis (Cernichiari et al. 1969, Ziesenisz et al. 1981), and there is good evidence in *Paramecium bursaria*-algae for maltose synthesis from glucose precursors by a maltose-phosphorylase (Ziesenisz personal communication).

It is noteworthy that symbiotic chlorellae isolated from *Paramecium bursaria* lack carbon anhydrase activity and possess a ribulose-1,5-bisphosphate carboxylase which has a higher affinity for carbon dioxide than that from corresponding aposymbiotic chlorellae (Reisser and Benseler 1981).

Symbiotic chlorellae probably excrete sugars by an ATP-requiring process, which depends on the external proton activity and proceeds best at acid pH (4.0–5.0) but nearly stops at pH 7.0 (Cernichiari et al. 1969, Muscatine et al. 1967, Ziesenisz et al. 1981).

As symbiotic chlorellae in situ always divide only into two or four daughter cells, their low division rate and concomitant high photosynthetic capacity seem to be prerequisites for sugar excretion at high rates. In this sense, the carbohydrate release of endosymbionts is a good example of the evolutionary advantage of special adaptation to an ecological niche.

Little information exists on nitrogen-, sulphur-, and phosphorus metabolism of freshwater endosymbiotic systems. In *Paramecium bursaria* the algae are supplied by their partner with ammonia and amino acids. When alga-free *Paramecium bursaria* are infected with suitable endosymbiotic chlorellae, which are pale-green due to N-deficient culture, regreening due to chlorophyll synthesis occurs as a result of nitrogen supply from the host (Reisser 1976a).

Green *Paramecium bursaria* take up ammonium ions from the external medium, whereas alga-free *Paramecium bursaria* and other *Paramecium* species excrete ammonium ions (Albers et al. 1982). In *Hydra viridis* a transfer of amino acids and precursors of nucleic acids as well as of sulphur compounds from the host to the algae is reported (Cook 1980, 1981).

4.5 Self-Regulation and Partner-Coordination in Endosymbiotic Systems

The evolutionary success of endosymbiotic systems depends in part on the maintenance of partner stability even under conditions which are disadvantageous for one or both partners, i.e. a common reaction of the unit under stress conditions is necessary. This requires a high degree of coordination. Little is known of the actual mechanisms of such coordination and regulation. Special regulatory substances transmitting information from one partner to the other have not been found. Two examples will illustrate these problems: regulation of endosymbiotic algae number, and the phenomenon of photoaccumulation.

In general, algae are lost from the symbiotic unit when photosynthesis is switched off by continuous darkness, blocking with special inhibitors, photodestruction, etc. (Frost and Williamson 1980, Pardy 1976, Pringsheim 1928,

REISSER 1976 a), and growth of the heterotrophic partner is sustained by feeding. Algae in different stages of digestion can be observed in most endosymbiotic systems. Perhaps a certain percentage of the algae is always digested, and there exists a host-specific upper limit of algal population which induces digestion of algae surpassing a certain numerical threshold. This so-called farming mechanism either requires a selective accessability of lytic enzymes to the endosymbionts or a limited digestive capacity of the host. Probably some kind of evolution of endosymbiotic systems exists. More primitive systems regulate their algal population via digestion by farming (digestive regulation). They are still able to destroy their symbiotic partnership. Advanced systems protect their symbiotic algal population against digestion and achieve regulation of algal number by an expulsion of excess algae or by ecological mechanisms (ecological regulation). The latter maintains an algal population size according to the amount of available nutrients (carbon dioxide, mineral salts, amino acids, etc.) and other external conditions (light, etc.) of the special ecological niche.

In ciliate-algae systems examples of both types of regulation exist; for example in green *Stentor polymorphus* probably farming works (REISSER 1981 a). The percentage of digested algae is much higher than in *Paramecium bursaria*, where symbiotic chlorellae are protected in perialgal vacuoles against lytic host enzymes even under conditions which are extremely unfavourable for host growth (constant darkness, inorganic medium) (MEIER et al. 1980 a, REISSER 1980 a, REISSER et al. 1980). Accordingly, data on the growth behaviour of *Paramecium bursaria* hint at ecological regulation. Growth rates of the partners are well-coordinated. A constant algal population size exists for each given light intensity. Even under conditions favouring optimal photosynthesis and maximum growth of algal cells, a rupture of the host is avoided. When light intensity is increased, the algal number also increases. This results in higher amounts of oxygen and maltose produced by the algae which trigger an increase of the ciliate's division rate until a new constant ratio between the partners is established. When light intensity decreases, the algal population size diminishes and consequently the host's division rate drops (KARAKASHIAN 1963, PADO 1965, REISSER 1980 a, 1981 a, 1981 b, WEIS 1969). In *Hydra viridis* algae are also protected against host digestion and there are conclusive hints as to an ecological regulation type of the algal population size here (MCAULEY and SMITH 1982, MUSCATINE and NECKELMAN 1981).

A second example of coordinated behaviour is the phenomenon of photoaccumulation. Green *Paramecium bursaria* show a step-down photophobic response (see HÄDER DP, Photomovement. In: Physiology of movements. HAUPT W and FEINLEIB ME, eds; Chapter 3.2.1, p. 269) and can gather in a light spot, whereas the isolated partners never do. The intensity of photoaccumulation depends on both the light fluence rate and the size of the algal population. The light stimulus is received by the algal chlorophyll and the action spectrum of the reaction is that of photosynthesis. A so far unknown sensory transduction mechanism transmits signals of the algal photosynthetic apparatus to host cilia. This coordination within the symbiotic association transforms the partners to a unit with a new characteristic (CRONKITE and VAN DEN BRINK 1981, IWATSUKI and NAITHOH 1981, NIESS et al. 1981, 1982, PADO 1972, SAJI and OOSAWA 1974).

Sensory transduction does not depend on any special symbiotic feature of the algae such as sugar excretion: infection studies with aposymbiotic *Paramecium bursaria* and different strains of symbiotic and free-living chlorellae show that photoaccumulation also occurs in artificial systems, i.e. in paramecia where a minimum number of free-living chlorellae are present. Obviously photoaccumulation depends mainly on the presence of any photosynthesizing partner (NIESS et al. 1982). It is also observed in other ciliate-algae systems (REISSER unpublished results) and in green nemathelminthes (GELEI 1927).

4.6 Specifity of Cell Recognition and Symbiotic Partnership

Since stability of endosymbiotic systems requires coordinated reactions, effectivity of interactions between the partners is facilitated by a certain partner specificity. Accordingly endosymbiotic systems are generally characterized by a specificity of their partners which can be demonstrated by reinfection experiments (BOMFORD 1965, HÄMMERLING 1946, HIRSHON 1969, KARAKASHIAN and KARAKASHIAN 1965, MUSCATINE et al. 1975, PARDY and MUSCATINE 1973, PRINGSHEIM 1928). In general, the infection of hosts from experimentally aposymbiotic culture with different strains or species of algae usually results in a new, permanently stable system only when the original symbiotic strain of algae is offered. Thus a symbiotic association between the genus *Paramecium* and the genus *Chlorella,* which is stable even under stress conditions, is not achieved in general, but only when *Paramecium bursaria* is reinfected with its own original strain of symbiotic *Chlorella* sp. This strain obviously differs significantly from other symbiotic and free-living chlorellae. There can even be differences in infectivity between symbiotic chlorellae isolated from different *Paramecium* strains (BOMFORD 1965, SIEGEL 1960, WEIS 1978, 1979).

The success of stable symbiosis formation depends on at least two factors. First, after uptake by the host, algae must escape digestion and must be protected against lytic enzymes (infection). Second, algae must be able to meet the special physiological requirements of their symbiotic milieu so that they can colonize their partner (population). We shall consider these two processes separately.

Infection experiments with both the green *Hydra* and the green *Paramecium* system show that infection requires a cell–cell recognition mechanism which is probably located on the cell surfaces. Algae which are suitable for symbiosis formation are not taken up into perialgal vacuoles but into food vacuoles after their surface has been changed by application of specific antibodies, lectins, proteolytic or carbohydrate degrading enzymes (MEINTS and PARDY 1980, POOL 1979, REISSER et al. 1982). Thus recognition requires specific surface structures of so far unknown composition, and probably also specific surface charges triggering the formation of perialgal instead of food vacuoles (MCNEIL et al.

1981). Special physiological features of the infecting algae, e.g. maltose excretion, seem to play no or only a minor role for positive infection. Both aposymbiotic *Hydra viridis* and *Paramecium bursaria* can be infected not only with their original symbiotic strain of *Chlorella* sp., but also with some free-living *Chlorella* species which do not excrete sugars and which nevertheless can survive within the host for some time (KARAKASHIAN and KARAKASHIAN 1965, JOLLEY and SMITH 1980, MCAULEY and SMITH 1982, NIESS et al. 1982).

The population of the host with endosymbionts, however, relies mainly on the specific physiological features of the latter. More algal strains can infect aposymbiotic partners and escape immediate digestion than really form permanently stable associations. In *Paramecium bursaria* all infective strains and species of *Chlorella* except the original symbiotic one are lost in continuous darkness or under conditions of limited food supply (KARAKASHIAN and KARAKASHIAN 1965, NIESS et al. 1982). Corresponding observations have been made with aposymbiotic *Hydra viridis* (JOLLEY and SMITH 1980).

Specificity in freshwater endosymbiotic systems with hosts other than protozoa or cnidaria has not been studied very thoroughly. Exact quantitative data are lacking (GOETSCH and SCHEURING 1926). Cell recognition in hereditary endosymbiotic systems is not only important during infection or feeding, but perhaps also when algae divide in the established symbiotic association. Division of alga and perialgal vacuole takes place simultanously and a new membrane is formed immediately around each young daughter cell (Figs. 3, 4). The enclosure of only one alga per perialgal vacuole and the close attachment of the perialgal vacuole membrane to the algal cell surface probably not only facilitates the exchange of nutrients between the partners, but is also both the result and prerequisite of recognition processes leading to the formation and assembly of special perialgal membrane material.

Specificity is also a quality of the host and not only of the algal partner. In the genus *Paramecium,* only *Paramecium bursaria* forms a symbiotic unit with *Chlorella* sp., whereas other *Paramecium* species digest every kind of algae offered. Infection studies with green and aposymbiotic *Paramecium bursaria* give some hint of some host susceptibility to different strains of algae which can be lost, but exact information on the nature of any "susceptibility-factor" is not available (KARAKASHIAN 1975). In transplantation experiments with green *Stentor polymorphus* and alga-free *Stentor coeruleus,* the symbiotic chlorellae only survive when *Stentor polymorphus* plasma is predominating (TARTAR 1953). Similar results as to some host plasma-bound factors have been obtained by grafting experiments with *Hydra viridis* and *Chlorohydra hadleyi* (PARK et al. 1967). The probable role of host autoimmunity mechanisms has until now scarcely been taken into account. Whether infection or reinfection is a common event in the natural habitat is not quite clear. There are some reports on ciliates and turbellarians which have been collected as both green and alga-free, but data as to the stability of those green systems excluding any temporal greening by food organisms are lacking (BUCHNER 1930, 1953, KAHL 1930). Thus it cannot be excluded that infection experiments in hereditary associations mean an artificial impact on the systems.

4.7 Ecology of Endosymbiotic Systems

There exist very few exact data on population size, periods of occurrence, requirements of water quality, temperature, etc. in the natural habitat of freshwater endosymbiotic systems. It is therefore necessary to rely mainly on laboratory data and to extrapolate them to the natural milieu. Nevertheless, it is quite reasonable that such data must be taken to indicate the framework of maximum reaction potentials, although the situation is obviously complicated by the fact that natural occurrence depends on many, often diverging, factors. The endogenous oxygen supply in host–algae systems possibly means some advantage in habitats in which oxygen is scarce. Actually alga-containing protozoans are reported to occur mainly in oxygen deficient milieus (KAHL 1930), but one should bear in mind that endogenous oxygen supply by endosymbiotic algae in sufficient light intensities becomes an additional oxygen demand by respiration of algae in the dark. Green *Paramecium bursaria* are reported to occur perferably in clearer, sometimes acid waters (SONNEBORN 1970). Perhaps they are less dependent on external supply with bacteria because of their internal production of carbohydrates by endosymbiotic algae. Endosymbiotic associations of amoebae and green algae have been observed mainly in the upper layers of *Sphagnum* societies where both light and oxygen are abundant (SCHÖN-BORN 1965). In general, the carbon dioxide concentration in the natural milieu is not a limiting factor for algal photosynthesis, and the endosymbiotic environment offers a constant high supply of carbon dioxide to the enclosed algae, favouring a high production of carbohydrates (PHIPPS and PARDY 1982, REISSER 1980b). A tendency seems to exist for some ciliate-algae systems to occur preferably during the colder months of the year, e.g. in autumn and spring (WESEN-BERG-LUND 1909, GELEI 1927) although the time of irradiation with light intensities sufficient for net production of oxygen is then shorter than in summer. In contrast, we have observed green *Paramecium bursaria* throughout the whole year in a β-mesosaprobic pond with the highest population size in June, and other observations suggest only a minor role of temperature in determining population size (LANDIS 1982).

As to nitrogen-, phosphorus-, and sulphur supply, it might be an ecological advantage to the heterotrophic host that the endosymbiotic algae can make nitrate, phosphate, and sulphate available to its metabolism, e.g. by excretion of amino acids, although digestion of bacteria by hosts will have the same result. Another role of endosymbionts could be a sort of sewage-disposal function by metabolizing nitrogen waste products of their hosts. However, other species of the same host genera generally exist without algae and usually aposymbiotic hosts grow under optimal laboratory conditions as well as symbiotic ones (KARAKASHIAN 1963, MEWS and SMITH 1982, MUSCATINE and LENHOFF 1963).

Thus arguments for a physiological necessity of symbiosis formation are rather weak. In most endosymbiotic systems studied so far partners can be separated from each other and cultured independently under laboratory conditions. An independent existence of partners from stable hereditary symbiotic associations in the natural habitat has as yet not been proven beyond doubt. As far as the algae are concerned, they probably can not compete with free-living species because of their high energy loss due to sugar excretion and their low division rate. The situation is somewhat different in systems which are not hereditary, e.g. where eggs must be re-infected by symbiotic algae. These algae must be able to exist for some time in an aposymbiotic milieu unless they adhere to egg cases. Thus from the evolutionary point of view, symbiosis formation could be advantageous for both partners not primarily for physiological but for ecological reasons, since it means a buffer system against unfavourable conditions in their natural habitat.

References

Albers D, Reisser W, Wiessner W (1982) Studies on the nitrogen supply of endosymbiotic chlorellae in green *Paramecium bursaria*. Plant Sci Lett 25:85–90

Bomford R (1965) Infection of alga-free *Paramecium bursaria* with strains of *Chlorella, Scenedesmus* and a yeast. J Protozool 12:221–224

Brandt K (1882) Über das Zusammenleben von Algen und Tieren. Biol Centrbl 1:524–527

Buchner P (1930) Tier und Pflanze in Symbiose. Borntraeger, Berlin 2nd edn

Buchner P (1953) Endosymbiose der Tiere mit pflanzlichen Mikroorganismen. Birkhäuser, Basel Stuttgart

Cantor MH, Rakat M (1982) Regulation of respiration and photosynthesis in *Hydra viridis* and in its separate cosymbionts: Effect of nutrients. Physiol Zool 55:281–288

Cernichiari E, Muscatine L, Smith DC (1969) Maltose excretion by the symbiotic algae of *Hydra viridis*. Proc R Soc Lond Sec B 173:557–576

Cook CB (1980) Sulfur metabolism in the green *Hydra* symbiosis. The incorporation of sulfate-sulfur by symbiotic and aposymbiotic *Hydra viridis*. In: Schwemmler W, Schenk HEA (eds) Endocytobiology 1, Endosymbiosis and Cell Biology. de Gruyter, Berlin, pp 249–257

Cook CB (1981) Adaptions to endosymbiosis in green *Hydra*. Ann NY Acad Sci 361:273–283

Cronkite D, Brink v d S (1981) The role of oxygen and light in guiding "photoaccumulation" in the *Paramecium bursaria-Chlorella* symbiosis. J Exptl Zool 217:171–177

Dangeard PA (1900) Les zoochlorelles du *Paramecium bursaria*. Botaniste 7:161–191

Droop MR (1963) Algae and invertebrates in symbiosis. In: Nutman PS, Mosse B (eds) The 13. Symposium of the society for general microbiology. Cambridge Univ Press, Cambridge, pp 171–199

Eaton JW, Young JO (1975) Studies on the symbiosis of *Phaenocora typhlops* (Vejdovsky) (Turbellaria; Neorhabdocoela) and *Chlorella vulgaris* var. vulgaris, Fott & Nováková (Chlorococcales). Arch Hydrobiol 75:50–75

Eisenstadt E (1971) Transfer of photosynthetic products from symbiotic algae to animal tissue in *Chlorohydra viridissima*. In: Lenhoff HM, Davis LV, Muscatine L (eds) Experimental coelenterate biology. Univ of Hawaii Press, Honolulu Hawaii USA pp 202–208

Frost TM, Williamson CE (1980) In situ determination of the effect of symbiotic algae on the growth of the freshwater sponge *Spongilla lacustris*. Ecology 61:1361–1370

Geitler L (1947) Über die systematische Zugehörigkeit der Zoochlorellen. Sitzungsber oester Akad Wiss 156:357–362

Gelei JV (1927) Angaben zu der Symbiosefrage von *Chlorella*. Biol Zentr 47:449–461

Gilbert JJ, Simpson TL (1976) Gemmule polymorphism in the freshwater sponge *Spongilla lacustris*. Arch Hydrobiol 78:268–277

Goetsch W, Scheuring L (1926) Parasitismus und Symbiose der Algengattung *Chlorella*. Z Morph Ökol Tiere 7:220–253

Graham LE, Graham JU (1978) Ultrastructure of endosymbiotic *Chlorella* in a *Vorticella*. J Protozool 25:207–210

Graham LE, Graham JU (1980) Endosymbiotic *Chlorella* (Chlorophyta) in a species of *Vorticella* (Ciliophora). Trans Am Micros Soc 99:160–166

Hämmerling J (1946) Über die Symbiose von *Stentor polymorphus*. Biol Zentralbl 65:52–61

Haffner KV (1925) Untersuchungen über die Symbiose von *Dalyellia viridis* und *Chlorohydra viridissima* mit Chlorellen. Z Wiss Zool 126:1–69

Heitkamp U (1979) Der Einfluß endosymbiotischer Zoochlorellen auf die Respiration von *Dalyella viridis* (G. Shaw 1791) (Turbellaria, Neorhabdocoela). Arch Hydrobiol 86:499–514

Hirshon JB (1969) The response of *Paramecium bursaria* to potential endocellular symbionts. Biol Bull 136:33–42

Hohmann TC, McNeil PL, Muscatine L (1982) Phagosome–lysosome fusion inhibited by algal symbionts of green *Hydra viridis*. J Cell Biol 94:56–63

Iwatsuki K, Naitoh Y (1981) The role of symbiotic *Chlorella* in photoresponses of *Paramecium bursaria*. Proc Jpn Acad Sci 57:318–323

Jolley E, Smith DC (1978) The green *Hydra* symbiosis. I. Isolation, culture and characteristics of the *Chlorella* symbiont of "European" *Hydra viridis*. New Phytol 81:637–645

Jolley E, Smith DC (1980) The green *Hydra* symbiosis. II. The biology of the establishment of the association. Proc R Soc Lond B 207:311–333

Kahl A (1930) Urtiere oder Protozoa I. Wimpertiere oder Ciliata (Infusoria). In: Dahl F (ed) Die Tierwelt Deutschlands und der angrenzenden Meeresteile und nach ihrer Lebensweise, Vol 18, Fischer, Jena

Karakashian MW (1975) Symbiosis in *Paramecium bursaria*. In: Jennings DH, Lee DL (eds) Symbiosis. Symp Soc Exp Biol 29:145–173

Karakashian SJ (1963) Growth of *Paramecium bursaria* as influenced by the presence of algal symbionts. Physiol Zool 36:52–68

Karakashian SJ (1970) Invertebrate Symbioses with *Chlorella*. In: Chambers KL (ed) Oregon State Univ Biological Colloquium 29 Biochemical Coevolution. Oregon State Univ Press, Corvallis Oregon USA, pp 33–52

Karakashian SJ, Karakashian MW (1965) Evolution and symbiosis in the genus *Chlorella* and related algae. Evolution 19:368–377

Karakashian SJ, Rudzinska MA (1981) Inhibition of lysosomal fusion with symbiont-containing vacuoles in *Paramecium bursaria*. Exept Cell Res 131:387–393

Karakashian SJ, Karakashian MW, Rudzinska MA (1968) Electron microscopic observations on the symbiosis of *Paramecium bursaria* and its intracellular algae. J Protozool 15:113–128

Kessler E (1978) Physiological and biochemical contributions to the taxomomy of the genus *Chlorella*. XII. Starch hydrolysis and a key for the identification of 13 species. Arch Microbiol 119:13–16

Landis WG (1982) The spatial and temporal distribution of *Paramecium bursaria* in the littoral zone. J Protozool 29:159–161

McAuley PJ, Smith DC (1982) The green *Hydra* symbiosis. V. Stages in the intracellular recognition of algal symbionts by digestive cells. Proc R Soc Lond B 216:7–23

McNeil Pl, Hohmann TC, Muscatine L (1981) Mechanisms of nutritive endocytosis. II. The effect of charged agents on phagocytotic recognition by digestive cells. J Cell Sci 52:243–269

Meier R, Reisser W, Wiessner W (1980a) Zytologische Analyse der Endosymbioseeinheit von *Paramecium bursaria* Ehrbg. und *Chlorella* spec. II. Die Regulation der endosymbiontischen Algenzahl in Abhängigkeit vom Ernährungszustand der Symbiosepartner. Arch Protistenkd 123:333–341

Meier R, Reisser W, Wiessner W, Lefort-Tran M (1980b) Freeze-fracture evidence of differences between membranes of perialgal and digestive vacuoles in *Paramecium bursaria*. Z Naturforsch 35c:1107–1110

Meints RH, Pardy RL (1980) Quantitative demonstration of cell surface involvement in a plant–animal symbiosis: lectin inhibition of reassociation. J Cell Sci 43:239–251

Mews LK (1980) The green *Hydra* symbiosis. III. The biotrophic transport of carbohydrate from alga to animal. Proc R Soc Lond B 209:377–401

Mews LK, Smith DC (1982) The green *Hydra* symbiosis. VI. What is the role of maltose transfer from alga to animal? Proc R Soc Lond B 216:397–413

Muscatine L, Lenhoff HM (1963) Symbiosis: On the role of algae symbiotic with *Hydra*. Science 142:956–958

Muscatine L, McAuley PJ (1982) Transmission of symbiotic algae to eggs of green *Hydra*. Cytobios 33:111–124

Muscatine L, Neckelmann N (1981) Regulation of numbers of algae in the *Hydra-Chlorella* symbiosis. Ber Dtsch Bot Ges 94:571–582

Muscatine L, Karakashian SJ, Karakashian MW (1967) Soluble extracellular products of algae symbiotic with a ciliate, a sponge and a mutant *Hydra*. Comp Biochem Physiol 20:1–12

Muscatine L, Cook CB, Pardy RL, Pool RR (1975) Uptake, recognition and maintenance of symbiotic *Chlorella* by *Hydra viridis*. In: Jennings DH, Lee DL (eds) Symbiosis. Symp Soc Exp Biol 29:175–203

Niess D, Reisser W, Wiessner W (1981) The role of endosymbiotic algae in photoaccumulation of green *Paramecium bursaria*. Planta 152:268–271

Niess D, Reisser W, Wiessner W (1982) Photobehaviour of *Paramecium bursaria* infected with different symbiotic and aposymbiotic species of *Chlorella*. Planta 156:475–480

O'Brien T (1982) Inhibition of vacuolar membrane fusion by intracellular symbiotic algae in *Hydra viridis*. J Exptl Zool 223:211–218

Oschman JL (1967) Structure and reproduction of the algal symbionts of *Hydra viridis*. J Phycol 3:221–228

Pado R (1965) Mutual relation of protozoans and symbiotic algae in *Paramaecium bursaria*. I. The influence of light on the growth of symbionts. Folia Biol Cracow 13:(2) 173–182

Pado R (1967) Mutual relation of protozoans and symbiotic algae in *Paramaecium bursaria*. II. Photosynthesis. Act Soc Bot Pol 36:97–108

Pado R (1972) Spectral activity of light and phototaxis in *Paramecium bursaria*. Acta Protozool 11:387–393

Pardy RL (1976) The production of aposymbiotic *Hydra* by the photodestruction of green *Hydra* zoochlorellae. Biol Bull 151:225–235

Pardy RL, Muscatine L (1973) Recognition of symbiotic algae by *Hydra viridis*. A quantitative study of the uptake of living algae by aposymbiotic *H. viridis*. Biol Bull 145:565–579

Park HD, Greenblatt CL, Mattern CFT, Merril CR (1967) Some relationships between *Chlorohydra*, its symbionts and some other chlorophyllous forms. J Exp Zool 164:141–162

Phipps DW, Pardy RL (1982) Host enhancement of symbiont photosynthesis in the *Hydra*-algae symbiosis. Biol Bull 162:83–94

Pool RR (1979) The role of algal antigenic determinants in the recognition of potential algal symbionts by cells of *Chlorohydra*. J Cell Sci 35:367–379

Pringsheim EG (1928) Physiologische Untersuchungen an *Paramaecium bursaria*. Arch Protistenkd 64:289–418

Reisser W (1975) Zur Taxonomie einer auxotrophen *Chlorella* aus *Paramecium bursaria*. Arch Microbiol 104:293–295

Reisser W (1976a) Die stoffwechselphysiologischen Beziehungen zwischen *Paramecium bursaria* Ehrbg. und *Chlorella* spec. in der *Paramecium bursaria*-Symbiose. I. Der Stickstoff- und der Kohlenstoffstoffwechsel. Arch Microbiol 107:357–360

Reisser W (1976b) Die stoffwechselphysiologischen Beziehungen zwischen *Paramecium bursaria* Ehrbg. und *Chlorella* spec. in der *Paramecium bursaria*-Symbiose. II. Symbiose-spezifische Merkmale der Stoffwechselphysiologie und der Cytologie des Symbioseverbandes und ihre Regulation. Arch Microbiol 111:161–170

Reisser W (1980a) The regulation of the algal population size in *Paramecium bursaria*. In: Schwemmler W, Schenk HEA (eds) Endocytobiology 1, Endosymbiosis and Cell Biology. de Gruyter Berlin, pp 97–104

Reisser W (1980b) The metabolic interactions between *Paramecium bursaria* Ehrbg. and *Chlorella* sp. in the *Paramecium bursaria*-symbiosis. III. The influence of different CO_2-concentrations and of glucose on the photosynthetic and respiratory capacity of the symbiotic unit. Arch Microbiol 125:291–293

Reisser W (1981a) The endosymbiotic unit of *Stentor polymorphus* and *Chlorella* sp. Morphological and physiological studies. Protoplasma 105:273–284

Reisser W (1981b) Host-symbiont interaction in *Paramecium bursaria*: Physiological and morphological features and their evolutionary significance. Ber Dtsch Bot Ges 94:557–563

Reisser W, Benseler W (1981) Comparative studies on photosynthetic enzymes of the symbiotic *Chlorella* from *Paramecium bursaria* and other symbiotic and non-symbiotic *Chlorella* strains. Arch Microbiol 129:178–180

Reisser W, Meier R, Wiessner W (1980) Zytologische Analyse der Endosymbioseeinheit von *Paramecium bursaria* Ehrbg. und *Chlorella* spec. I. Der Einfluß des Ernährungszustandes algenfreier *Paramecium bursaria* auf den Verlauf der Infektion mit der aus grünen Paramecien isolierten *Chlorella* sp. Arch Protistenkd 123:326–332

Reisser W, Radunz A, Wiessner W (1982) The participation of algal surface structures in the cell recognition process during infection of aposymbiotic *Paramecium bursaria* with symbiotic chlorellae. Cytobios 33:39–50

Reisser W, Fischer-Defoy D, Staudinger J, Schilling N, Hausmann K (1983) The endosymbiotic unit of *Climacostumum virens* and *Chlorella* sp. I. Morphological and physiological studies on the algal partner and its localization in the host cell. Protoplasma (in press)

Saji M, Oosawa F (1974) Mechanism of photoaccumulation in *Paramecium bursaria*. J Protozool 21:556–561

Schlösser UG (1982) Sammlung von Algenkulturen: List of strains. Ber Dtsch Bot Ges 95:181–276

Schönborn W (1965) Untersuchungen über die Zoochlorellen-Symbiose der Hochmoor-Testaceen. Limnologica 3:173–176

Siegel RW (1960) Hereditary endosymbiosis in *Paramecium bursaria*. Exptl Cell Res 19:239–252

Sonneborn TM (1970) Methods in *Paramecium* research. In: Prescott DM (ed) Methods in cell physiology vol 4. Academic Press, London New York, pp 241–339

Szuch EJ, Studier EH, Sullivan RB (1978) The relationship of light duration to oxygen consumption in the green freshwater sponge *Spongilla lacustris*. Comp Biochem Physiol 60 A:221–223

Tartar V (1953) Chimeras and nuclear transplantations in ciliates, *Stentor coeruleus S. Polymorphus*. J Exp Zool 124:63–103

Taylor DL (1973) Algal symbionts of invertebrates. Annu Rev Microbiol 27:171–187

Thorington G, Berger B, Margulis L (1979) Transmission of symbionts through the sexual cycle of *Hydra viridis*. I. Observation on living organisms. Trans Am Micros Soc 98:401–413

Weis DS (1969) Regulation of host and symbiont population size in *Paramecium bursaria*. Experientia 15:664–666

Weis DS (1974) Sparing effect of light on bacterial consumption of *Paramecium bursaria*. Trans Am Micros Soc 93:135–140

Weis DS (1978) Correlation of infectivity and concanavalin A agglutinability of algae exsymbiotic from *Paramecium bursaria*. J Protozool 25:366–370

Weis DS (1979) Correlation of sugar release and concanavalin A agglutinability with infectivity of symbiotic algae from *Paramecium bursaria* for aposymbiotic *Paramecium bursaria*. J Protozool 26:117–119

Wesenberg-Lund (1909) Beiträge zur Kenntnis des Lebenszyklus der Zoochlorellen. Int Rev Hydrobiol 2:153–162

Wilkinson CR (1980) Nutrient translocation from green algal symbionts to the freshwater sponge *Ephydatia fluviatilis*. Hydrobiologia 75:241–250

Yonge CM (1944) Experimental analysis of the association between invertebrates and unicellular algae. Biol Rev 19:68–80

Young JO, Eaton JW (1975) Studies on the symbiosis of *Phaenocora typhlops* (Vejdovsky) (Turbellaria; Neorhabdocoela) and *Chlorella vulgaris* var. vulgaris, Fott & Nováková (Chlorococcales). II. An experimental investigation into the survival value of the relationship to host and symbiont. Arch Hydrobiol 75:225–239

Ziesenisz E, Reisser W, Wiessner W (1981) Evidence of de novo synthesis of maltose excreted by the endosymbiotic *Chlorella* from *Paramecium bursaria*. Planta 153:481–485

5 Autotrophic Eukaryotic Marine Symbionts

D. L. Taylor

5.1 Introduction

"Zusammenleben ungleichnamiger Organismen" (DeBary 1879). Freely translated as: the association of dissimilarly named organisms (Starr 1975), this definition has proved to be one of the most durable and least well-understood in biology. Here, I will use symbiosis in its larger sense, as the superclass of all other organismal associations, since this will permit a broader appreciation of the range in form that algal-invertebrate symbiosis may take, as well as the breadth and depth of its physiological and ecological consequences.

Although the range of genera and species is not large, several distinctly different marine micro-algae are commonly found in endosymbiotic relationships with a broad range of invertebrate hosts. Such associations occur within the photic zone at all latitudes of the world's oceans. They are most commonly encountered in the shallow-water environments of the tropics and subtropics, where certain associations, notably those involving dinoflagellates and hermatypic (reef-building) corals, have achieved the status of major biogeochemical events in the ancient and modern history of the earth.

Interest in these relationships lies in two areas of general concern to biologists. These are related, but widely separated in scale. They deal with the fundamental cellular interrelationships occurring in metazoa, and the fundamental organismal interrelationships occurring in ecosystems. If one regards organisms as cellular ecosystems (F.J.R. Taylor 1983), then these extremes of scale are, in real practical terms, part of the same organizational/informational continuum. Historically, the study of endosymbiosis has proved to be a suitable source of insights into the functional workings of both ends of this continuum.

In the past decade, the general subject of autotrophic marine symbionts has been reviewed and summarized several times (Taylor 1973a, b, Trench 1979, Cook 1983). More recently, these reviews have dealt with specific, more specialized subareas, such as primary production, nutritional physiology, calcification, etc. (Taylor 1983a, b). These are referred to liberally in the present paper, since it is the purpose here to provide an overview of subjects arising from contemporary research on the biology and ecology of marine algal endosymbiosis.

5.2 Symbiont Genera and Species

There are some very general distinctions that exist between the types of algal symbiont occurring in freshwater environments and those occurring in marine environments. These have been known for some time (BRANDT 1882), and are highlighted by the common contemporary usage of the terms zoochlorellae (freshwater/green) and zooxanthellae (marine/brown). Subsequent reviewers (e.g. DROOP 1963, TAYLOR 1973a, b), further specified these distinctions by suggesting that freshwater symbionts were predominantly algal species belonging to the Chlorophyta (sensu CHRISTENSEN 1965), while marine symbionts were predominantly algal species belonging to the Chromophyta (sensu CHRISTENSEN 1965). As we shall see, contemporary studies of marine endosymbiosis have revealed a broader range of algal genera involved in these associations, several of which occur within the Chlorophyta. Thus, while the distinction appears to hold for freshwater endosymbiotic relationships, it is less clear among marine examples. Nevertheless, the overwhelming majority of marine algal symbionts are chromophytes, and the distinction remains a useful parallel that reflects the dominance of the Chromophyta in the free-living microalgal populations of the oceans.

5.2.1 Range

In the past decade, several new algal classes, genera and species have been added to the list of microalgae that exist primarily as endosymbionts in the tissues of marine invertebrates (Table 1). These additions are significant in that they have been made in the context of more critical and formalized methods for establishing the existence of an endosymbiotic relationship. They are therefore a genuine (as opposed to speculative) addition to our knowledge of these organisms.

Various species of Dinophyceae are by far the most common endosymbionts in marine habitats. Among these, *Symbiodinium (Gymnodinium) microadriaticum* is the most frequently encountered. Dinoflagellates are found to inhabit the cells and tissues of a broad variety of protozoa, sponges, coelenterates, turbellarians and molluscs. As with the free-living species, many symbiotic species are extremely cosmopolitan in their distributions. The Chrysophyceae currently appear to be the second most important source of symbionts. Among these, diatoms are most common. They are presently known in associations involving foraminifera (LEE 1983, LEE et al. 1980, RÖTTGER et al. 1980), and have been reported as symbionts in the acoel turbellarian *Convoluta convoluta* (AX and APELT 1965). They are more restricted with respect to both geographic range and host type than the Dinophyceae, and together with the Dinophyceae constitute the two major classes of marine endosymbiotic algae.

Other endosymbiotic taxa are rare, and generally very restricted with respect to host type. These include *Platymonas convolutae* (Prasinophyceae) inhabiting

Table 1. Marine algal symbionts

Symbiont	Host(s)	Author(s)
1. Dinophyceae		
a) *Symbiodinium (Gymnodinium) microadriaticum*	Various benthic coelenterates	FREUDENTHAL (1962), TAYLOR (1969b, 1971), KEVIN et al. (1969)
b) *Amphidinium klebsii*	*Amphiscolops langerhansi*	TAYLOR (1971)
c) *Amphidinium (Endodinium) chattonii*	*Velella velella*	TAYLOR (1969b)
d) *Amphidinium* sp.	*Collozoum inerme*	TAYLOR (1974, 1983)
e) *Zooxanthella nutricula*	*Collozoum inerme*	HOLLAND and CARRÉ (1974)
f) Various unidentified dinoflagellates possibly *Symbiodinium* spp.	Various foraminifera	LEE (1983)
2. Bacillariophyceae		
a) *Licmophora* sp.	*Convoluta convoluta*	AX and APELT (1965)
b) *Fragilaria shiloi*	*Amphistegina lessonii*	LEE et al. (1982)
c) *Nitzschia laevis*	*Amphistegina lessonii*	LEE et al. (1982)
d) *Nitzschia panduriformis*	*Heterostegina depressa*	LEE et al. (1982)
e) *Nitzschia valdestriata*	*Heterostegina depressa*	LEE et al. (1982)
3. Chlorophyceae		
a) *Chlamydomonas hedleyi*	*Archais angulatus*	LEE (1983)
b) *Chlamydomonas provasolii*	*Cyclorbiculina compressa*	LEE (1983)
4. Prasinophyceae		
a) *Platymonas convolutae*	*Convoluta roscoffensis*	PARKE and MANTON (1967)
	Convoluta psammophila	SARFATTI and BEDINI (1965)
b) *Pedinomonas symbiotica*	*Thallassolampe margarodis*	CACHON and CARAM (1979)
c) *Pedinomonas noctilucae*	*Noctiluca scintillans*	SWEENEY (1971, 1976)
5. Rhodophyceae and *Porphyridium* sp.	*Peneroplis planatus* *Spirulina arietina*	LEE (1980, 1983)

the acoel turbellarians *Convoluta roscoffensis* and *C. psammophila,* (PARKE and MANTON 1967, SARFATTI and BEDINI 1965), *Chlamydomonas hedleyi* (Chlorophyceae) inhabiting the foraminiferan *Archais angulatus* (LEE et al. 1974) and *Porphyridium* sp. (Rhodophyceae) reported in symbiosis with the foraminifera (*Peneroplis planatus* and *Spirulina arietina* (LEE 1980, 1983).

5.2.2 Systematics

There is a clear need for workers in the field to be consistent with the systematic criteria used for other, non-symbiotic algae. In addition, there is a need for agreement with respect to the usefulness of the symbiotic state as a criterion of systematic importance. Many symbiotic genera and species appear closely related to, or fall within, recognized free-living assemblages. Specific examples include the dinoflagellates *Symbiodinium (Gymnodinium) microadriaticum* and *Amphidinium klebsii,* the prasinophyte *Platymonas convolutae,* and the chlorophyte *Chlamydomonas hedleyi* as well as several diatoms recently reported as symbionts of foraminifera (LEE 1980, 1983). It is reasonable to ask whether existence in a symbiotic association is in itself a sufficient reason to place these organisms in separate taxa. This type of placement has been suggested several times for symbiotic dinoflagellates, most recently in a form that resurrects and elevates "zooxanthella" to its former taxonomic status without full consideration of morphology and life history (LOEBLICH and SHERLEY 1979). It has not been an important issue with other taxa to date.

These questions have been discussed elsewhere (TAYLOR 1983a). In those discussions, I have emphasized that the total morphology and life history of the organism, not its habitat, is of primary systematic importance. This approach follows the suggestion of BALL (1968), and serves to bring the systematics of symbiotic species into alignment with that of the free-living microalgae, thereby emphasizing similarities and relationships that are both revealing and useful in the study of the functional aspects of their physiology and ecology. Previously, I have placed the most common endosymbiotic dinoflagellate in the genus *Gymnodinium,* basing this largely upon the morphology of the motile stage and other structural considerations. F.J.R. TAYLOR (1982) correctly questions this placement with arguments based on distinctive features of the life-history, that set this organism apart from the free-living genus *Gymnodinium.* Cell division during the encysted amastigote stage rather than the motile mastigote stage is an important distinction that argues for a separate systematic position. For this reason, I have returned to the use of the genus *Symbiodinium (Gymnodinium) microadriaticum* in this publication. It is clear, however, that a more general revision of all known endosymbiotic dinoflagellates is in order before a final (correct) placement is determined.

Apart from the general, philosophical issues of systematic placement, there are also important questions with regard to speciation that need careful consideration (TAYLOR 1983a). Dinoflagellate symbionts provide the best example of the problems encountered, specifically as they relate to the ubiquitous symbiont, *S. microadriaticum.* The Dinophyceae are recognized as an extremely ancient assemblage, with origins in the mid- to late-Cambrian. Symbiosis with invertebrates is believed to have emerged prominently in the form of dinoflagellate/coral associations during the Triassic. It is in the modern examples of these symbioses that *S. microadriaticum* figures most prominently. Like many free-living dinoflagellates, this organism is extremely cosmopolitan in its distribution. This characteristic may be explained in part through present knowledge of continental drift, notably by the analysis of the positions of continental masses after the break-up of the supercontinents Laurasia and Gondwana. In this geological context, it may be argued that coral host species present suitable conditions for the genetic isolation of their symbionts, and that in time the organism recognized generally as *S. microadriaticum* might evolve towards subraces, varieties and even new species with specific adaptations to specific hosts. Conversely, it can also be argued that the inherent uniformity and stability of the symbiotic state might work against potential speciation, and that the present morphological unity of the *S. microadriaticum* assemblage is an expression of that fact. It has been noted elsewhere that the truth lies somewhere between the extremes of a single pandemic species and multiple species specific to each host type (TAYLOR 1973b, 1974, 1983a). Evidence from electrophoretic studies of *S. microadriaticum* isozymes suggests that there exist host-specific races or varieties (SCHOENBERG and TRENCH, 1980a, b, c). This view is further enhanced by data on diurnal periodicities of motility (FITT et al. 1981) and fatty acid and sterol composition (BISHOP and KENRICK, 1981, KOKKE et al. 1981). While these observations do not bear directly on the question of whether different isolates constitute different species, they do present favorable arguments for regarding symbionts with morphologies similar to *S. microadriaticum* as an assemblage of races and/or varieties, but it is not sufficient for the separation of individual species, as has been proposed (LOEBLICH and SHERLEY 1979).

It has been suggested (TAYLOR 1981a, 1983a) that the juxtaposition of host and symbiont genomes might significantly alter or supress gene expression in the partners. There is even the possibility of direct genetic exchange or the pooling of genetic resources in the intact symbiotic unit (F.J.R. TAYLOR 1983). Because of this, it is best to look upon the symbiotic association as a functional unit, rather than to consider the two partners as independent entities because of their morphological differences. Traits such as symbiont isozyme patterns may be rather a function of gene expression within the total genetic pool of the functional unit than a stable characteristic of the symbiont per se. At present, we do not know the answer to the question contained in these speculations. However, there is evidence that similar events occur in other cellular systems (KOLLAR and FISHER 1980). Genetic mapping, DNA hybridization studies and an analysis of DNA homologies would provide useful criteria for evaluating possible genetic separation among strains of *S. microadriaticum*, and could serve to resolve many of the species-related questions that currently beset symbiotic research.

5.3 Cellular Relationships of Symbionts and Hosts

Under optimum conditions, endosymbiotic algae form autotrophic tissues with the cells of their invertebrate hosts. Success depends upon several factors that arise from the pre-disposition of specific algae and invertebrates and/or mechanisms that have evolved as the symbiotic unit stabilized itself in time. Three factors considered here, (1) recognition and selection, (2) placement and (3) regulation, have a significant impact upon the physiological and ecological success of a given association, and play essential roles in the stability and development of symbiotic associations.

5.3.1 Recognition and Selection

Symbiotic continuity may be ensured through the coordination of symbiont cell division with host reproduction, or by the selective reinfection of each new host generation. The distinction between these two processes is often blurred. Frequently the former condition will be presented with potential symbionts occurring in the external environment, and processes of recognition and selection will ultimately come into play. These processes may include cell-surface phenomena (structures, antigens, attractant chemicals), external chemical cues, or physical signals (e.g. swimming behaviour). Our knowledge of how they may operate in a symbiosis is reasonably good in the case of freshwater endosymbionts, and is comparatively poor for marine examples. Because it is reasonable to generalize from the data on freshwater endosymbionts, the reader should consult REISSER (Chap. 6, this Vol.) for a discussion of current data.

The continuing investigation of the species question in *Symbiodinium microadriaticum* (KINZIE and CHEE 1982, FITT et al. 1981, SCHOENBERG and TRENCH 1980a, b, c) has produced valuable information on specific cell-surface structures and evidence for host-specific selectivity of *S. microadriaticum* strains. It is apparent from this work that the mechanisms of recognition and selection operating here are similar to those occurring with *Chlorella* symbiotic with green *Hydra* and *Paramecium* (POOL 1979, REISSER et al. 1982).

Other marine symbioses where recognition and selection have been examined include the acoel turbellarians *Convoluta roscoffensis* and *Amphiscolops langerhansi* (PROVASOLI et al. 1968, TAYLOR 1971). For each new generation, both hosts rely on re-establishing the symbiosis from wild populations of potential symbionts occurring in the environment. This is accomplished first by a chemical attractant associated with the egg cases (KEEBLE 1910), and second by competitive selection in the host (TAYLOR 1981 b). Cell-surface phenomena may also play a role, but they have not been investigated to date. Both hosts may be presented with a range of alternative algal symbionts. If competitive infections are avoided, these will become established in stable symbioses with varying degrees of success (PROVASOLI et al. 1968, TAYLOR 1971). If natural symbionts (*Platymonas convoluta/C. roscoffensis; Amphidinium klebsii/A. langerhansi*) are presented, they will establish themselves in the host, and out-compete the resident symbiont to its total exclusion. Among the range of experimental symbionts there is a clear hierarchy or "pecking order" that emerges. PROVASOLI et al. (1968) have suggested that ascendancy, or the failure to achieve it, might be interpreted in terms of nutrient-dependent differences in the growth rates of the algae. This has been tested recently (TAYLOR 1981 b), and there are indeed significant differences in nutrient utilization and uptake potential which would serve to explain the competitive phenomena observed in these animals.

Foraminiferan hosts appear to be less selective and as a consequence are found to establish symbioses with a variety of different algal symbionts (LEE 1983). Some analysis of symbiont preference and selective mechanisms has been attempted, but it is not known whether selection is nutritionally based or dependent upon cell-recognition phenomena.

5.3.2 Placement

Symbiont placement in host tissues is either intra- or inter-cellular, the former being most common among acellular hosts, and the latter the rule for most multicellular hosts (TAYLOR 1973a, b). When symbionts are situated within host cells, there exist specialized vacuoles or carrier organelles that isolate the alga and protect it from digestion. These are poorly described in marine symbioses (e. g. TAYLOR 1969a), but appear similar to those described in associations involving *Hydra* and *Paramecium* (see REISSER 1983). Among marine symbioses, they are best known from studies of benthic foraminifera (LEE 1980, 1983).

Adaptations of marine endosymbionts to the inter-cellular conditions vary with the association that is formed, and the algal species involved. Three examples exist which demonstrate the degree of algal modification that can take place. Ultrastructural studies of associations involving *S. microadriaticum* (TAYLOR 1973a, b, TRENCH 1979), *Amphidinium klebsii* (TAYLOR 1971) and *Platymonas convolutae* (PARKE and MANTON 1967) reveal a range of modifications affecting the cell wall – host cell boundary. *S. microadriaticum* develops a thickened amphiesma that may be interpreted as defensive against host digestion. Host cells are closely adpressed and entwined around the symbiont in complex processes. In contrast, *P. convolutae* discards its cell wall and resides as a naked protoplast closely interdigitated among host cells in the epidermis. *A. klebsii* exists in its host apparently without significant modification, but in close contact with adjacent epidermal cells of the host. Each of these conditions represents the best balance between resistance to digestion and host–symbiont cellular intimacy that can be achieved in the symbiosis that is formed. In symbiotic associations involving *S. microadriaticum*, there is evidence that symbiont cell numbers are regulated by host predation and digestion (TAYLOR 1969 unpublished). Symbiont digestion is apparently not a factor in the symbioses involving *A. klebsii* and *P. convolutae*, and the adaptations that appear to exist in the alga are a reflection of this fact.

5.3.3 Regulation

Success of a symbiosis, whether defined in evolutionary or physiological terms, demands a substantial degree of inter-cellular co-ordination and co-regulation. This is essential if a stable, long-term association is to develop. The most obvious requirement for regulation concerns the host's ability to control its algal endosymbionts with respect to their growth and material production. Control of cell growth is achieved either through mechanisms which (1) harvest excess cell production (Cook 1983, Taylor 1983a), (2) limit cell proliferation by control or withdrawal of essential nutrients, or (3) chemically inhibit cell division. Material production, notably carbon fixation, rates of translocation and quality of translocate may also be controlled and regulated through interactions with the host. Mechanisms affecting this aspect are obscure and remain to be investigated.

The processes governing cell proliferation have been examined in a number of distinctly different associations. The data suggest that for all of the cases studied, excess cell production is either harvested as a source of carbon for the host (Cook 1983, Lee 1980, 1983, Muscatine 1980a, b, Trench 1979, Taylor 1983a) or simply expelled into the environment (Taylor 1969a, Provasoli et al. 1968).

Specific regulatory substances which might affect both cell proliferation and material production have not been identified to date. However, it is reasonable to presume that such coordinated multicellular systems would utilize some chemical means for regulation.

5.4 Primary Production

In purely ecological terms, the principal distinction between endosymbiosis occurring in freshwater habitats and that occuring in marine habitats lies in the magnitude of the contribution which marine autotrophic symbioses make to the primary production of the world's oceans. This is largely due to the spectacular success of dinoflagellate–cnidarian symbiosis in tropical and subtropical marine environments, where they exist as the foundation of the coral reef ecosystem. Smith (1978) has estimated the total area of coral reef communities at 15% of the shallow sea floor (0–30 m). Given the abundance of autotrophic symbiosis in these communities, and given dinoflagellate symbiont densities of 10^6 cm^{-2} of reef surface area (Drew 1972, Kawaguti and Nakayama 1973, Lasker 1981, Kevin and Hudson 1979), it is apparent that a significant fraction of the world's oceanic productivity will arise from this source (Muscatine 1980a, Muscatine et al. 1981). A major goal of research on marine endosymbiosis is to provide a quantitative description of primary processes, and integrate this into current knowledge of production in free-living, open-ocean communities. Current research in the field views these associations in the context of the autotrophic functional unit, thereby emphasizing a holistic rather than reductionist perspective in experimental studies.

5.4.1 Photosynthesis

Photosynthesis by algal symbionts is the principal host resource for reduced carbon in the overwhelming majority of associations studied (Taylor 1983a). Even among facultative associations (Taylor 1973a, b), where the host will feed independently, current data indicate that most of the carbon used in growth is derived from photosynthesis (MUSCATINE and PORTER 1977, PORTER personal communication). YONGE's (1958) characterization of the symbionts of corals as "imprisoned phytoplankton" is thus a valuable perspective that relates not only to the growth and nutrition of the functional unit, but also emphasizes the relationship between the primary production of symbiotic micro-algae and that of free-living micro-algae.

The photosynthetic pathways of algal endosymbionts have not received the treatment they deserve. Only the dinoflagellate, *S. microadriaticum,* has received concentrated study (MUSCATINE 1980a, TAYLOR 1981b). The most detailed information available relates to the qualitative aspects of photosynthetic products in vitro and in vivo and their rates of excretion and translocation to the host. Carbon fixation is generally believed to follow the C_3-pathway, but there is increasing evidence that like other dinoflagellates *S. microadriaticum* might exhibit a mixed C_3-C_4 metabolism (TRENCH 1979, TAYLOR 1981b). The most detailed current analysis of photosynthetic products is that of SCHMITZ and KREMER (1977). Their data reveal a substantial and complex range of organic compounds, but fail to distinguish between intra-cellular and extra-cellular algal products, or the products of host metabolism of algal carbon (COOK 1983). Nevertheless, known algal products are readily translocated to the host, where they are metabolized to yield energy and build tissue. Viewed ecologically, the host functions as the primary consumer in a one-step food chain. Alternatively, the complete functional unit may be viewed as an analogue of multicellularity in higher plants, where a significant portion of photosynthetic production goes into the structure of the organism (MEYERS 1980). In a very real, functional sense, these "imprisoned phytoplankton" have successfully made the transition from a unicellular existence, where photosynthetic production emphasizes cell machinery, to a multicellular existence emphasizing structure and the division of labour. The mechanism of this transition has important parallels in current theories of punctuated evolution (STANLEY 1979, F.J.R. TAYLOR 1983), and deserves greater attention.

The actual determination of photosynthetic rates of primary production in situ is fraught with difficulties. The problems have been discussed at length by MUSCATINE (1980a, b, MUSCATINE et al. 1981), and further reviewed elsewhere (MCCLOSKEY et al. 1978, LEWIS 1977, RAVEN 1981). In general, variability in experimental technique is so great that the data available in the literature can barely be reconciled (e.g. DAVIES 1980, FALKOWSKI and DUBINSKY 1980, TAYLOR 1973a, b, PORTER personal communication). Matters are further complicated by the fact that only net photosynthesis of the intact association is measured, and the essential element, net photosynthesis of the symbiont, is ignored or not approached. The result is a serious underestimate of primary production that is further complicated by a lack of a systematic experimental approach. Recently, these questions have been addressed, and a rigorous approach detailed (MUSCATINE et al. 1981). Results are still forthcoming, but one should anticipate a deeper understanding of both the photosynthetic process in situ, and the end product expressed as primary production.

Analysis of photosynthesis as a function of irradiance (P vs. I) in a range of marine endosymbiotic associations reveals a variety of sun and shade adaptations, and little if any photosynthetic inhibition under natural conditions (BARNES and TAYLOR 1973, WETHEY and PORTER 1976, CHALKER and TAYLOR 1978, FALKOWSKI and DUBINSKI 1980). These adaptations are related to changes in chlorophyll a content (FALKOWSKI and DUBINSKY 1980), not in the total number of algal symbionts, that is, related to a change in size not number of photosynthetic units (chl a/P-700). Numerous field studies demonstrate the general adaptability of symbioses to reduced illumination and correlate well with the mechanism described here. Adaptations to higher irradiance rates are less well understood, and may involve protective interactions with the host's tissues that have yet to be examined.

5.4.2 Endosymbiotic Exchange of Carbon and Nitrogen

The success of symbiotic associations involving algae and invertebrates is generally believed to be based upon the ability to acquire, share and conserve nutritional resources between the two partners. In facultative associations, where the host feeds (TAYLOR 1973a, b), the potential sources of carbon and nitrogen and their routes to the cellular milieu are varied. Carbon may be acquired in the symbiosis in three principal ways: (1) symbiont photosynthesis and translocation (TRENCH 1979, McCLOSKY et al. 1978, MUSCATINE 1980a), (2) assimilation of dissolved organic carbon (DOC) by both the host and symbiont (STEPHENS 1962) and (3) ingestion and assimilation of particulate organic carbon (POC) (zooplankton, phytoplankton, detritus) by the host. Most workers now conclude that only the first is of primary significance in the majority of associations (MUSCATINE and PORTER 1977). Similarly, nitrogen may be acquired by (1) assimilation of dissolved nitrogen by the symbiont and (2) ingestion of particulate nitrogen by the host. With the single exception of algal photosynthesis, very little is known about these other processes of nutrient acquisition (COOK 1983, MUSCATINE and PORTER 1977, MUSCATINE 1980b), and even less is known about the mechanisms of internal nutrient sharing and conservation that follow from them.

Present knowledge permits an assessment of the relative importance of the various sources of carbon and nitrogen available to any given symbiosis (TAYLOR 1983a). With respect to carbon, it is clear that symbiont photosynthesis is a major resource, and that the resulting translocation from the alga to the host is both large and significant in the majority of associations studied (TRENCH 1979, MUSCATINE 1973, 1980a, b). Nitrogen acquisition has been studied largely from the standpoint of the uptake of dissolved nitrogen from seawater by the algal symbiont in situ (MUSCATINE 1980a, b, D'ELIA and WEBB 1977, D'ELIA et al. 1983, COOK 1983, TAYLOR 1983b), and through the identification of potential carrier compounds that might be useful as mechanisms for algal-animal translocation (LEWIS and SMITH 1971, SUMMONS and OSMOND 1981). Host-utilization of these materials has been demonstrated in aposymbiotic animals (TAYLOR 1974).

The value of ^{14}C as an isotopic tracer of pathways of photosynthetic carbon has served to emphasize symbiont production as the dominant theme of research on algal-invertebrate associations (McClosky et al. 1978, Muscatine 1980a, Taylor 1983a). Similarly, it has been easier to examine algal pathways for nitrogen acquisition because of available experimental techniques (Taylor 1983b). As a consequence, very little attention has been paid to the question of reverse (host to symbiont) translocation of carbon and nitrogen, despite the fact that it is implicit in, and required by, all current theories of algal-invertebrate nutrient exchange (Smith 1979, Cook 1983). Recently, this question has been examined in experiments with the acoel turbellarian *Amphiscolops langerhansi* symbiotic with *Amphidinium klebsii* (Taylor 1983c). Published data indicate that both carbon and nitrogen derived from labelled food ingested by the host are rapidly translocated to the alga, where they are detected in various algal metabolites. Further studies should serve to amplify both the carrier compounds being translocated and the rates of turnover and exchange in this symbiosis. The value of such studies is obvious both as a test of prevailing theory and an essential input into models of the nutritional balance of endosymbiotic associations.

5.5 Ecology

The theme of algal-invertebrate symbiosis as part of a broader ecological continuum was introduced in the opening paragraphs of this review. The view of the host as a habitat for the symbiont, and the concept of a cellular ecosystem has been raised several times in the past decade (Taylor 1971, 1974, 1981a, Smith 1979, F.J.R. Taylor 1983). From the perspective of cell biology, it is a useful concept that permits fresh insights into both cellular evolution and cellular function. Here, the concern is more with the other end of the continuum – the role of these functional units in the environments where they occur.

Recent reviews of the literature (Trench 1979, Muscatine 1980a, b, Taylor 1983a) strongly suggest a major ecological role for algal-invertebrate symbiosis in marine environments. This is in sharp contrast to the impression gained from studies of freshwater associations, and it is typified most clearly by tropical coral reef communities. These communities are major biogeochemical phenomena with very clear foundations in the synergism gained from the juxtaposition of a primary producer and a heterotroph in cellular symbiosis. Algal-invertebrate symbiosis is found throughout the benthic invertebrate species comprising the core community of the coral reef. It functions primarily to conserve and regulate the flux of nutrients and carbon, confining them to the biomass and preventing loss to the overlying waters. As a result, biologically rich communities come to reside in apparently nutrient-poor waters. The paradox is understood if we consider the role of endosymbiosis in altering the conventional nutrient patterns of aquatic communities.

In specific associations within these same communities, symbiosis also serves to establish both the structure and dynamic architecture of the habitat itself (Taylor 1983a). A consequence of the symbiosis between *Symbiodinium micro-*

adriaticum and the hermatypic corals is a greatly enhanced rate of $CaCO_3$ deposition that is directly dependent upon and proportional to the photosynthetic rate of the symbiont. The very form of the coral reef itself and the dynamic response of its structure to physical effects is a consequence of this association. Light, mediated by symbiont photosynthesis, provides both energy and form to the total coral reef ecosystem.

5.6 Conclusions

While distinctions may be made between freshwater and marine autotrophic eukaryotic symbionts on the basis of their impact on specific ecosystems, or the relative importance of different algal classes in fresh vs. marine environments, they remain one and the same – a part of an extraordinarily important cellular phenomenon that has periodically gained significance in the realm of biological thought. A comparative reading of this review, and that of REISSER (Chap. 6, this Vol.) will confirm this.

One hundred years ago, the study of algal-invertebrate symbiosis flourished, and played a significant role in the development of biological thought, particularly as it pertained to cell theory and general concepts of multicellularity. Since that time, it has always held a special fascination for biologists, students and researchers alike. In the past decade, symbiosis research has been in the ascendant again; a growing number of workers recognize the importance of algal-invertebrate endosymbiosis as experimental systems for examining cellular and organismal problems. Symbiosis is not a specialized event. Rather, it is seen to occur throughout living systems as a way of life and a basic fact of life. It is commonplace to speculate on the evolutionary consequences of these liasons. Hopefully, that speculation will lead to a deeper understanding of cellular origins and function, and the informational (genetic) system that guides living organisms along time's arrow.

5.7 Addendum – Prochloron[1]

In recent years there have been a series of studies of symbiotic associations between didemnid ascidians (protochordates) and a kind of unicellular green prokaryotic alga now assigned to a new genus, *Prochloron* (see review by LEWIN, 1981). Unlike cyanophytes, *Prochloron* lacks phycobilins as accessory photosynthetic pigments: instead, it synthesizes chlorophyll *b* in addition to the chlorophyll *a* found in all green plants. Cells of this algae are almost exclusively found associated with colonies of didemnids, where they generally occur embedded in the cellulosic exoskeleton, in mucilaginous material in the cloaca, or

───────────────

[1] On request of the editors kindly supplied by Dr. R.A. LEWIN, La Jolla, California

loosely attached to strands of mucilage in atria around the pharynx. Experimental evidence indicates that an appreciable proportion of the carbon fixed by photosynthesis in these algal cells is subsequently liberated, and can then be taken up by cells of the animal host. Since *Prochloron* has not yet been grown in sustained laboratory culture, so far most of the information on this subject has had to be obtained from studies carried out *in hospite,* in tropical reef areas where symbiotic didemnids may be relatively abundant. However, it should soon be possible to grow this algae in vitro, apart from its host, and we should then be in a better position to evaluate the respective interactions of these remarkable symbionts.

References

Ax P, Apelt G (1965) Die "Zooxanthellen" von *Convoluta convoluta* (Turbellaria acoela) entstehen aus Diatomeen. Naturwissenschaften 15:444–446

Ball G (1968) Organisms living on and in protozoa. In: Chen TT (ed) Research in protozoology vol 3. Pergamon, Oxford, pp 566–718

Barnes DJ, Taylor DL (1973) In situ studies of calcification and photosynthetic carbon fixation in the coral *Montastrea annularis*. Helgol Wiss Meeresunters 24:284–291

Bishop CA, Kenrick JR (1981) Fatty acid composition of symbiotic zooxanthellae in relation to their hosts. Lipids 15:799–804

Brandt K (1882) Ueber das Zusammenleben von Thieren und Algen. Mitt Zool Stat Neapel 4:191–302

Cachon M, Caram B (1979) A symbiotic green alga, *Pedinomonas symbiotica* sp. nov. (Prasinophyceae), in the radiolarian *Thalassolampe margarodes*. Phycologia 18:177–184

Chalker BE, Taylor DL (1978) Rhythmic variations in calcification and photosynthesis associated with the coral *Acropora* cervicornis (Lamarck). Proc R Soc Lond B 201:179–189

Christensen T (1965) Systematisk Botanik, nr 2 Alger. Botanik 2:1–80

Cook CB (1983) Metabolic interchange in algae–invertebrate symbiosis. Int Rev Cytol Suppl 4:177–210

Davies P (1980) Respiration in some Atlantic reef corals in relation to vertical distribution and growth form. Biol Bull 158:187–194

De Bary A (1879) Die Erscheinung der Symbiose. Trübner, Strassburg

D'Elia CF, Webb KL (1977) The dissolved nitrogen flux of reef corals. Proc 3rd Int Coral Reef Symp 1:325–330

D'Elia CF, Domotor SL, Webb KL (1983) Nutrient uptake kinetics of freshly isolated zooxanthellae. (In press)

Drew EA (1972) The biology and physiology of alga–invertebrate symbioses. II. The density of symbiotic algal cells in a number of hermatypic hard corals and alcyonarians from various depths. J Exp Mar Biol Ecol 9:71–75

Droop MR (1963) Algae and invertebrates in symbiosis. Symp Soc Gen Microbiol 13:171–199

Falkowski P, Dubinsky Z (1980) Light-shade adaptation of *Stylophora pistillata,* a hermatypic coral from the Gulf of Eilat. Nature 289:172–174

Fitt WK, Chang SS, Trench RK (1981) Motility patterns of the different strains of the symbiotic dinoflagellate *Symbiodinium* (= *Gymnodinium*) *microadriaticum* (Freudenthal) in culture. Bull Mar Sci 31:436–443

Freudenthal HD (1962) *Symbiodinium* gen. nov. and *Symbiodinium microadriaticum* sp. nov., a zooxanthella: Taxonomy, life cycle and morphology. J Protozool 9:45–52

Holland A, Carré D (1974) Les xanthelles des radiolaires Sphaerocollides, des acanthaires et de *Vellela vellela:* infrastructure, cytochimie, taxonomie. Protistologica 10:573–601

Kawaguti S, Nakayama T (1973) Population densities of zooxanthellae in reef corals. Biol J Okayama Univ 16:67–71

Keeble FW (1910) Plant animals. Cambridge Univ Press, Cambridge, pp 1–160

Kevin KM, Hudson RCL (1979) The role of zooxanthellae in the hermatypic coral *Plesiastrea urvillei* (Milne-Edwards and Haime) from cold waters. J Exp Mar Biol Ecol 36:157–170

Kevin MJ, Hall WT, McLaughlin JJA, Zahl PA (1969) *Symbiodinium microadriaticum* Freudenthal, a revised taxonomic description, ultrastructure. J Phycol 5:341–350

Kinzie, RA III, Chee GS (1982) Strain-specific differences in surface antigens of symbiotic algae. Appl Environ Microbiol 44:1238–1240

Kokke WCMC, Fenical W, Bohlin L, Djerassi C (1981) Sterol synthesis by cultured zooxanthellae: implications concerning sterol metabolism in the host–symbiont association in Caribbean gorgonians. Comp Biochem Physiol B 68:281–287

Kollar EJ, Fisher C (1980) Tooth induction in chick epithelium: expression of quiescent genes for enamel synthesis. Science 207:993–995

Lasker HR (1981) Phenotypic variation in the coral *Montastrea cavernosa* and its effects on colony energetics. Biol Bull 160:292–302

Lee JJ (1980) Nutrition and physiology of the Foraminifera. In: Levandowsky M, Hutner SH (eds) Biochemistry and physiology of protozoa vol 3. Academic Press, London New York, pp 43–66

Lee JJ (1983) Perspective on algal endosymbionts in larger Foraminifera. Int Rev Cytol Suppl 14:49–77

Lee JJ, Crockett L, Hagen J, Stone R (1974) The taxonomic identity and physiological ecology of *Chlamydomonas hedleyi* sp. nov., algal flagellate symbiont from the foraminifer *Archaias angulatus*. Br Phycol J 9:407–422

Lee JJ, McEnery M, Lee MJ, Reidy JJ, Garrison JR (1980) Algal symbionts in larger Foraminifera. In: Schwemmler W, Schenk H (eds) Endocytobiology. de Gruyter, Berlin, pp 113–124

Lee MJ, Ellis R, Lee JJ (1982) A comparative study of photoadaptation in four diatoms isolated as endosymbionts from larger foraminifera. Mar Biol 68:193–197

Lewin RA (1981) The Prochlorophytes. In: Starr MP (ed) The prokaryotes, edited by M.P. Starr et al. Springer, Berlin Heidelberg New York, pp 257–266

Lewis DH, Smith DC (1971) The autotrophic nutrition of symbiotic marine coelenterates with special reference to hermatypic corals. I. Movement of photosynthetic products between the symbionts. Proc Roy Soc Lond, B 178:111–129

Lewis JB (1977) Experimental tests of suspension feeding in Atlantic reef corals. Mar Biol 36:147–150

Loeblich AR III, Sherley JL (1979) Observations on the theca of the motile phase of free-living and symbiotic *Zooxanthella microadriatica* (Freudenthal) comb. nov. J Mar Biol Assoc UK 59:195–206

Meyers J (1980) On the algae: Thoughts about physiology and measurements of efficiency. In: Falkowski P (ed) Primary productivity in the sea. Plenum, New York London, pp 1–16

McCloskey LR, Wethey DS, Porter JW (1978) The measurements and interpretation of photosynthesis and respiration in reef corals. Monogr Oceanogr Methodol 5:379–396

Muscatine L (1973) Nutrition of corals in: Jones OA, Endean R (eds) Biology and geology of coral reefs vol 2. Academic Press, London New York, pp 77–115

Muscatine L (1980a) Productivity of zooxanthellae in: Falkowski P (ed) Primary productivity in the sea. Plenum, New York London, pp 381–402

Muscatine L (1980b) Uptake, retention and release of dissolved inorganic nutrients by marine algal-invertebrate associations. In: Cook CB, Pappas PW, Rudolph ED (eds) Cellular interactions in symbiosis and parasitism. Ohio State Univ Press, Columbus, pp 229–244

Muscatine L, Porter JW (1977) Reef Corals: Mutualistic symbioses adapted to nutrient-poor environments. Bioscience 27:454–460

Muscatine L, McCloskey LR, Marian RE (1981) Estimating the daily contribution of carbon from zooxanthellae to coral animals respiration. Limnol Oceanogr 26:601–611

Parke M, Manton I (1967) The specific identity of the algal symbiont in *Convoluta roscoffensis*. J Mar Biol Assoc UK 47:445–464

Pool RR (1979) The role of algal antigenic determinants in the recognition of potential algal symbionts by cells of *Chlorohydra*. J Cell Sci 35:367–379

Provasoli L, Yamasu T, Manton I (1968) Experiments on the resynthesis of symbiosis in *Convoluta roscoffensis* with different flagellate cultures. J Mar Biol Assoc UK, 48:465–479

Raven JA (1981) Nutritional strategies of submerged benthic plants: The acquisition of C, N and P by rhizophytes and haptophytes. New Phytol, 88:1–30

Reisser W (1983) Autotrophic eukaryotic freshwater symbionts. In: Linskens HF (ed) Encyclopedia of Plant Physiology. Springer, Berlin Heidelberg New York, vol 17, pp 59–74 (this vol.)

Reisser W, Radunz A, Wiessner W (1982) The participation of algal surface structures in the cell recognition process during infection of aposymbiotic *Paramecium bursaria* with symbiotic chlorellae. Cytobios 33:39–50

Röttger R, Irwan A, Schmaljohann R, Franzisket L (1980) Growth of symbiont-bearing foraminifera *Amphistegina lessonii* d'Orbigny and *Heterostegina depressa* d'Orbigny (Protozoa). In: Schwemmler W, Schenk HEA (eds) Endocytobiology, vol I. de Gruyter, Berlin-New York, pp 125–132

Sarfatti G, Bedini C (1965) The symbiont of the flatworm *Convoluta psammophila* Berkl Observed at the electron microscope. Caryologia 18:207–223

Schmitz K, Kremer BP (1977) Carbon fixation and analysis of assimilates in a coral dinoflagellate symbiosis. Mar Biol 42:305–313

Schoenberg DA, Trench RK (1980a) Genetic variation in *Symbiodinium* (= *Gymnodinium*) *microadriaticum* Freundenthal, and specificity in its symbiosis with marine invertebrates. I. Isoenzyme and soluble protein patterns of axenic cultures of *Symbiodinium microadriaticum*. Proc R Soc Lond B 207:405–427

Schoenberg DA, Trench RK (1980b) Genetic variation in *Symbiodinium* (= *Gymnodinium*) *microadriaticum* Freudenthal, and specificity in its symbiosis with marine invertebrates. II. Morphological variation *Symbiodinium microadriaticum*. Proc R Soc Lond B 207:429–444

Schoenberg DA, Trench RK (1980c) Genetic variation in *Symbiodinium* (= *Gymnodinium*) *microadriaticum* Freudenthal, and specificity in its symbiosis with marine invertebrates. III. Specificity and infectivity of *Symbiodinium microadriaticum*. Proc R Soc Lond B 207:445–460

Smith DC (1979) From extracellular to intracellular: the establishment of a symbiosis. Proc Soc Lond B 204:115–130

Smith SV (1978) Coral reef area and the contributions of reefs to processes and resources of the world's oceans. Nature 273:225–226

Stanley SM (1979) Macroevolution: Pattern and process. Freeman, San Francisco

Starr MP (1975) A generalized scheme for classifying organismic associations. In: Jennings DH, Lee DL (eds) Symbiosis. Symp Soc Exp Biol 29:1–20

Stephens GC (1962) Uptake of organic material by aquatic invertebrates. I. Uptake of glucose by the solitary coral *Fungia scutaria*. Biol Bull 123:64–659

Summons RE, Osmond CG (1981) Nitrogen assimilation in the symbiotic marine alga *Gymnodinium microadriaticum*: Direct analysis of ^{15}N incorporation by GC-MS methods. Phytochemistry 20:575–578

Sweeney BM (1971) Laboratory studies of green *Noctiluca* from New Guinea. J Phycol 7:53–58

Sweeney BM (1976) *Pedinomonas noctilucae* (Prasinophyceae), the flagellate symbiotic in *Noctiluca*. J Phycol 12:460–464

Taylor DL (1969a) On the regulation and maintenance of algal numbers in zooxanthellae–coelenterate symbiosis. J Mar Biol Assoc UK 49:1057–1065

Taylor DL (1969b) Identity of zooxanthellae isolated from Pacific Tridacnidae. J Phycol 5:336–340

Taylor DL (1971) On the symbiosis between *Amphidinium klebsii* and *Amphiscolops langerhansi*. J Mar Biol Assoc UK 51:301–313

Taylor DL (1973a) Cellular interactions of algal-invertebrate symbiosis. Adv Mar Biol 11:1–56

Taylor DL (1973b) Algal symbionts of invertebrates. Annu Rev Microbiol 27:171–187

Taylor DL (1974) Symbiotic marine algae: taxonomy and biological fitness. In: Vernberg WB and Vernberg FJ (eds) Symbiosis in the sea. Univ S. Carolina Press, Columbia pp 245–262

Taylor DL (1981a) The evolutionary impact of intracellular symbiosis. Ber Dtsch Bot Ges 94:583–590

Taylor DL (1981b) Nutrient competition as a basis for symbiont selection in algal-invertebrate symbiosis. In: Schwemmler WS, Schenk H (eds) Endocytobiology. de Gruyter, Berlin pp 279–291

Taylor DL (1983a) Coral/algal symbiosis. In: Goff L (ed) Handbook in phycology vol 3. Cambridge Univ Press, Cambridge (in press)

Taylor DL (1983b) Symbioses. In: Carpenter EJ, Capone DG (eds) Nitrogen in the marine environment. Academic Press, London New York (in press)

Taylor DL (1983c) Translocation of carbon and nitrogen from the turbellarian host *Amphiscolops langerhansi* (Turbellaria: Acoela) to its algal endosymbiont *Amphidinium klebsii* (Dinophyceae). Proc Linn Soc Lond (in press)

Taylor FJR (1982) Symbiosis in marine microplankton. Ann Inst Oceanogr 58:61–90

Taylor FJR (1983) Some eco-evolutionary aspects of interacellular symbiosis. In: Joen KW (ed) Intracellular symbiosis. Int Rev Cytol Suppl 14:1–28

Trench RK (1979) The cell biology of plant–animal symbiosis. Annu Rev Plant Physiol 30:485–531

Wethey DS, Porter JW (1976) Sun and shade differences in productivity of reef corals. Nature 262:281–282

Yonge CM (1958) Ecology and physiology of reef-building corals. In: Buzzati-Traverso AA (ed) Perspectives in marine biology. Univ California Press, Berkeley, pp 117–135

6 Endosymbiotic Cyanobacteria and Cyanellae

W. Reisser

6.1 Introduction

Cyanobacteria are outstanding among prokaryotes because of their ability to exploit water-bound protons for photosynthesis and to fix molecular nitrogen. Hence they are successful in nearly all types of biotopes and are also frequently observed in close relationship with other organisms.

In this report, a survey is given of endosymbiotic associations of cyanobacteria living in a unicellular host, or inter- or intracellularly within the tissue of a multicellular host. This frame excludes lichens, as well as more ecologically defined relationships of cyanobacteria, e.g. in attaching to the outer surface of a host. For surveys of this field see PASCHER (1914), GEITLER (1959), and WHITTON (1973).

According to PASCHER (1914, 1929a), symbiotic relationships of cyanobacteria and hosts are called syncyanoses, and the individual association cyanom. PASCHER originally (1929a) proposed the term cyanella for all kinds of symbiotic cyanobacteria, but later on this was restricted to unicellular ones (GEITLER 1959). For reasons of continuity in this report a unicellular symbiotic cyanobacterium living intracellularly will be called a cyanella whereas a filamentous one, and one living intercellularly, will be referred to as symbiotic cyanobacterium.

It must be stressed that the term cyanella has no taxonomic value, and does not imply any decision as to the cyanobacterian or organellar nature of its holder. The term zoocyanella (MCLAUGHLIN and ZAHL 1966) is as confusing as zoochlorella and should be avoided.

6.2 Taxonomy and Localization

Syncyanoses will be classified for taxonomic characterization according to host organization, i.e., whether multicellular or unicellular.

6.2.1 Multicellular Hosts

In multicellular partners, symbiotic cyanobacteria can be localized both intra- and intercellularly. The host range comprises chlorophycean algae, bryophytes, pteridophytes and spermatophytes, as well as sponges, echiuroids and tunicates. Symbiotic cyanobacteria in bryophytes and cormophytes show the typical cyanobacterian cell-wall structure (Lang 1968), are generally filamentous, and can usually be isolated from their host and grown independently. Most of them presumably belong to the genera *Nostoc* and *Anabaena,* although exact taxonomic classification is often hampered by changes in cell anatomy after isolation from hosts (Sect. 4.1). In the marine milieu there are relatively few described syncyanoses, and the multicellular hosts belong to very different taxa.

6.2.1.1 Bryophytes

A great variety of bryophyte taxa is reported to harbour symbiotic cyanobacteria, which usually live intercellularly and belong to the genus *Nostoc* (Duckett et al. 1977, Granhall and Hofsten 1976, Peirce 1906, Rodgers and Stewart 1977, Table 1). *Anthoceros* sp. and *Blasia pusilla* have been investigated most intensively. Here mostly *Nostoc sphaericum* and *Nostoc calcicola* are located in special cavities in the ventral side of the gametophyte, into which they presum-

Table 1. Symbiotic cyanobacteria and cyanellae in multicellular and unicellular hosts

Host	Symbiotic cyanobacterium[a]	Commentary
A. Multicellular hosts		
Porifera		
Class Calcarea, Class Demospongiae	*Aphanocapsa* (I, E), *Phormidium* (E)	Different species see (1), *A.* (I) in vacuoles
Echiurida		
Bonellia fuliginosa, Ikedosoma gogoshinense	? (I)	In special cells of epidermal tissue
Tunicata		
Didemnum, Didemnum, Lissoclinum, Trididemnum, Diplosoma	*Prochloron didemni* (E), *Prochloron* (E)	In cloacal cavity and in grooves around cloacal aperture
Chlorophyceae		
Oedogonium, Codium bursa	? (Filamentous) (I), *Phormidium* (I)	In oogonia
Bryophyta		
Anthoceros, Blasia pusilla	*Nostoc sphaericum* (E), *Nostoc calcicola* (E)	In ventral cavities of gametophyte
Aneura, Cavicularia, Diplolaena, Drepanocladus, Pellia, Riccia, Riccardia	*Nostoc* (E)	
Sphagnum	*Nostoc* (E), *Hapalosiphon* (E)	In hyaline cells
Pteridophyta		
Azolla	*Anabaena azollae* (E)	Probably in all species, in leaf cavities

Table 1 (continued)

Host	Symbiotic cyanobacterium[a]	Commentary
Gymnospermae		
Cycas, Dioon, Encephalartos, Macrozamia, Zamia	*Nostoc punctiforme* (E), *Nostoc cycadeae* (E), *Anabaena* (E)	In coralloid roots, in *Macrozamia communis* also (I)
Angiospermae		
Gunnera	*Nostoc punctiforme* (I)	In stem nodules, vacuoles
Lemna trisulca	*Nostoc* (I)	
Trifolium alexandrinum	*Nostoc punctiforme* (I)	In root nodules, vacuoles
B. Unicellular hosts		
Phycomycetes		
Geosiphon pyriforme	*Nostoc sphaericum* (I)	In vacuole
Diatoms		
Denticulata vanheurcki, Epithemia sorex, E. turgida, E. zebra, Rhopalodia gibba, R. gibberula	Cyanellae	Freshwater
Rhizoselenia, Hemiaulus membranaceus	*Richelia intracellularis* (I)	Marine
Roperia	? (*Nostoc*?) (I)	Marine
Rhizopods		
Paulinella chromatophora	Cyanellae	
Apoplastidal protists		
Cyanophora paradoxa	Cyanellae (*Cyanocyta korschikoffiana*)	
Glaucocystis nostochinearum	Cyanellae (*Skujapelta nuda*)	
Chalarodora azurea, Chroomonas gemma, Cryptella cyanophora, Cyanomastix morgani, Cyanoptyche gloeocystis, Glaucocystis cingulata, G. duplex, G. geitleri, G. incrassata, G. oocystiforma, Glaucosphaera vacuolata, Gloeochaete wittrockiana, Peliaina cyanea	Cyanellae	Proposed to group together as Glaucophyta
Others		
Cyanidium caldarium	Cyanella? plastid?	Grows in 50–60 °C, 1 n H_2SO_4

[a] Location, I = Intracellular, E = Extracellular; (1) = WILKINSON (1980)

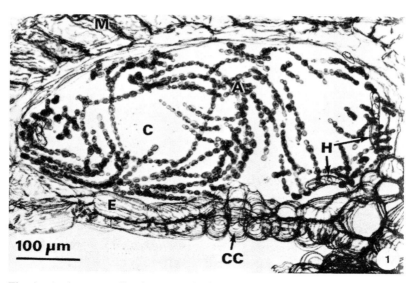

Fig. 1. *Anabaena azollae* in young leaf cavity of *Azolla. A Anabaena azollae* filaments; *C Azolla* leaf cavity; *CC* cavity closure area; *E* epidermis; *H* hair; *M* mesophyll (Calvert and Peters 1981)

ably can penetrate through "stomata" or special pores. The cavities are filled with slime which is probably produced by the *Nostoc* spp., and are penetrated by special multicellular filaments of host origin. In *Blasia pusilla* cavities, filaments show a characteristic transfer cell organization with typical wall ingrowths. Similar filaments are also described in *Anthoceros* spp. cavities, but their transfer cell status is not equally obvious. In *Anthoceros* spp., cyanobacteria also live intracellularly in the same filaments.

In *Sphagnum riparium, Nostoc* sp. has been described as living in hyaline cells.

6.2.1.2 Pteridophytes

Among pteridophytes, by far the best-studied cyanom is that of *Azolla* spp. with *Anabaena azollae* (Calvert and Peters 1981, Duckett et al. 1975, Holst and Yopp 1979a, 1979b, Kawamatu 1965a, 1965b, Moore 1969, Peters and Mayne 1974a, 1974b, Peters et al. 1978). *Anabaena azollae* has been found in all species of *Azolla* (Table 1) in cavities of the fern's dorsal leaf lobe (Fig. 1). The cavities are filled with slime and penetrated by multicellular hair-like filaments showing typical transfer cell features. In mature leaves, the cavities are sealed. They are probably infected by penetration of *Anabaena azollae* through the apex of young leaves.

6.2.1.3 Gymnosperms

Different taxa of cycads are reported to contain cyanobacteria in special mucus-rich so-called algal zones of coralloid roots (Grilli Caiola 1972a, 1972b,

1975a, 1975b, 1980, Nathanielsz and Staff 1975, Neumann 1977, Obukowicz et al. 1981, Schaede 1944, Wittmann et al. 1965). *Nostoc* and *Anabaena* species (Table 1) infect coralloid roots via muculaginous spaces in the tips, and are usually located intercellularly. In *Macrozamia communis,* both intercellular and intracellular localization occurs in an extra coralloid-root area. In *Cycas revoluta,* cells of the algal zone form special protrusions, and sometimes *Anabaena* sp. is observed within host cells.

6.2.1.4 Angiosperms

Among angiosperms, the best-known cyanom is formed by *Gunnera* sp. and *Nostoc punctiforme,* which is located intracellularly in special stem nodules (Miehe 1924; Neumann et al. 1970, Schaede 1951, Silvester and McNamara 1976, Table 1). These nodules develop from secretory glands producing muculaginous substances, by the aid of which *Nostoc punctiforme* invades the glands. It penetrates intercellularly through gland tissue, and at the base of the gland gets into thin-walled meristematic cells by a special mechanism. This results in the formation of a host-plasmalemma-derived membrane around the intracellular cyanobacteria.

Other less intensively studied angiosperm-cyanobacterian cyanoms are reported from *Trifolium alexandrinum* and *Nostoc punctiforme,* which is located in root nodules (Bhaskaran and Venkatamaran, 1958), and from *Lemna trisulca* harbouring intracellularly growing cyanobacteria of different taxa (Cohn 1872).

6.2.1.5 Sponges

Symbiotic cyanobacteria have been reported in a great variety of marine sponges (Sara 1971, Wilkinson 1980). Unicellular cyanobacteria probably belong to *Aphanocapsa* sp. and are situated both intercellularly and intracellularly in host vacuoles. Filamentous cyanobacteria are usually located intercellularly in sponge tissues exposed to light. Most of them have been tentatively classified as *Phormidium* sp.

6.2.1.6 Tunicates

In tropical ascidians, *Prochloron* sp. is located usually intercellularly in the cloacal cavity, and in radial grooves around the cloacal aperture (Chapman and Trench 1982, Cox and Dwarte 1981, Lewin 1975, 1977, 1981, Lewin and Cheng 1975, Lewin and Withers 1975, Stackebrandt et al. 1982, Sybesma et al. 1981, Whatley 1977). *Prochloron* (see also this volume, chapter 5, p. 86) is a unicellular photosynthetic prokaryote containing chlorophyll a and b and lacking phycobilisomes. It has been studied intensively in relation to its taxonomic position as a possible transition state between cyanobacteria and the plastids of chlorophytes, or alternatively as an autonomous line of evolution which developed parallel to cyanobacteria from some ancient cyanobacterium. Unlike *Prochloron* itself, the symbiotic relationship with tunicates has been studied in far less detail. *Prochloron didemni* has been described as occurring in

Didemnum sp. *Prochloron* has also been described in other tunicates another species perhaps being involved (Table 1). Frequently ascidians can be devoid of symbionts.

6.2.1.7 Others

There exist some sporadic reports of cyanobacteria–multicellular host associations which should be mentioned here for reasons of completeness.

In marine echiuroid worms such as *Bonellia fuliginosa* and *Ikedosoma gogoshinense,* cyanobacteria live intracellularly in special cells of the epidermal tissue (Kawaguti 1971). In *Oedogonium* oogonia (Reinsch 1879) and in *Codium bursa* (Jacob 1961), different species of filamentous cyanobacteria have been observed.

6.2.2 Unicellular Hosts

Associations of cyanobacteria endosymbiotic in unicellular hosts are observed predominantly in freshwater. Hosts are either diatoms with photosynthesizing chromatophores, or heterotrophic organisms such as phycomycetes, amoebae and algal-like apoplastidal cells of uncertain taxonomic position.

6.2.2.1 *Geosiphon pyriforme*

The phycomycete *Geosiphon pyriforme* and *Nostoc sphaericum* form the only known endosymbiotic association of fungus and cyanobacterium. The system was described first by v. Wettstein (1915) and later studied by light (Knapp 1933, Mollenhauer 1970) and electron microscopy (Schnepf 1964). *Nostoc sphaericum* is enclosed in a host plasmalemma-derived membrane, and shows a complete, unreduced, cyanobacterian wall structure. Heterocysts are formed frequently, but physiological studies are lacking.

6.2.2.2 Amoebae and Apoplastidal Algae

The endocyanoms of *Paulinella chromatophora* (Thecamoebae), *Cyanophora paradoxa, Glaucocystis nostochinearum, Glaucosphaera vacuolata,* and *Gloeochaete wittrockiana* (apoplastidal alga-like hosts) have been well studied by both light and electron microscopy. For other associations which have been studied less thoroughly, or have been observed only once during field studies, see Table 1 (Bourdu and Lefort 1967, Geitler 1924, 1927, 1959, Giddings et al. 1983, Hall and Claus 1963, 1967, Kies 1974, 1976, 1979, 1980, Kies and Kremer 1979, Korschikoff 1930, Lefort 1965, Lefort and Pouphile 1967, McCracken et al. 1980, Mignot et al. 1969, Pascher 1929a, 1929b, Pickett-Heaps 1972, Richardson and Brown 1970, Robinson and Preston 1971, Schenk 1977, Schmidt et al. 1979, Schnepf 1965, Schnepf et al. 1966, Trench 1982a, 1982b, Trench et al. 1978). In the above-mentioned systems, cyanellae are surrounded by a membrane, and have a cyanobacterian-type cell wall structure but with a more or less reduced number of different layers. In *Glaucosphaera vacuolata*-cyanellae, an electron-dense wall is lacking, and in other cyanellae the cell walls can be lyzed by lysozyme. Cyanellae can be of various shape,

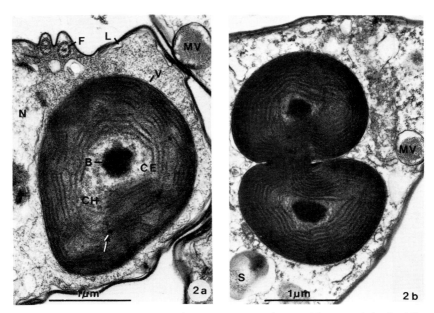

Fig. 2. a Cyanella in *Cyanophora paradoxa*. *CE* centroplasm with central body (*B*): *CH* chromatoplasm with thylakoids and osmiophilic droplets (*arrow*); *F* flagellar apparatus; *L* lacuna; *MV* mastigoneme vesicle; *N* nucleus; *V* host vacuolar membrane. **b** Division of cyanella in *Cyanophora paradoxa*; *S* starch, for other details see **a**. (**a, b** kindly provided by D.G. ROBINSON, University of Göttingen)

from spherical to rod-shaped and they show a number of typical cyanobacterian features (Figs. 2a, b, 3–6). They divide by binary fission (Fig. 2b). Cells are separated in an inner colourless zone (centroplasm) and an outer zone (chromatoplasm) with concentrically arranged thylakoids carrying phycobilisomes. In the centroplasm, polyphosphate granules and conspicuous electron-dense bodies of different size and form are frequently observed. They are probably free of DNA, and are possibly analogous to the carboxysomes which have been described in aposymbiotic cyanobacteria. Between thylakoids, lipid droplets can generally be observed, and in *Glaucosphaera vacuolata* they can contain carotenoids. Polyglucan and cyanophycin granules are usually absent.

The taxonomic position of hosts other than *Paulinella chromatophora* is uncertain, mainly because of lack of plastids and sexual processes. They probably do not belong to any of the known algal taxa and seem to hold a very isolated position, with no recent parallels in heterotrophic or photosynthesizing algae. *Cyanophora paradoxa, Glaucocystis nostochinearum, Glaucosphaera vacuolata,* and *Gloeochaete wittrockiana* have in common a conspicuous lacunae system, and all but *Glaucosphaera vacuolata* possess a flagellar apparatus (Fig. 2a). Other ultrastructural features hint at some relationship between *Gloeochaete wittrockiana, Glaucocystis nostochinearum, Cyanophora paradoxa, Glaucosphaera vacuolata*. Similar uncertainties exist as to the taxonomic position of cyanellae which are variously considered to be either symbiotic cyanobacteria or de facto

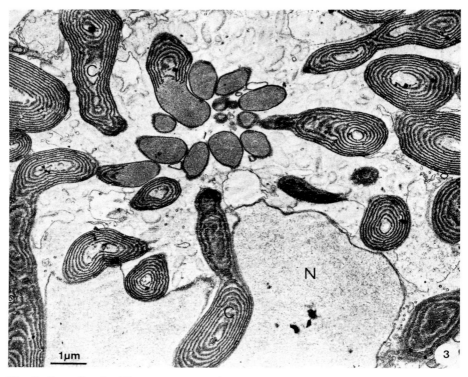

Fig. 3. Cyanellae in *Glaucocystis nostochinearum*. *C* cyanella; *N* nucleus. Permanganate fixation does not allow visualizing the phycobilisomes but concentric thylakoids are preserved. The distal ends of cyanellae are devoid of photosynthetic membranes and converge in a point of the host cell. (Lefort 1965)

plastids. The organellar properties are stressed by Korschikoff (1930) and Skuja (1954), who grouped cyanellae systems together in the new algal taxon of *Glaucophyta*. Hall and Claus accentuate the symbiotic character of cyanellae by endowing them with a proper designation, e.g. cyanellae in *Cyanophora paradoxa* as *Cyanocyta korschikoffiana* (1963) and in *Glaucocystis nostochinearum* as *Skujapelta nuda* (1967). Since these problems cannot be decided solely on ultrastructural features they will be discussed later (Sect. 6.5).

6.2.2.3 Diatoms

Among unicellular hosts of cyanobacteria, diatoms hold an outstanding position because they form associations with two different photosynthetic units in one cell. In the freshwater habitat, the best-studied endocyanoms are formed by *Rhopalodia gibba* and *Rhopalodia gibberula* (Drum and Pankratz 1965, Floener and Bothe 1980). In these associations, one to five cyanellae are enclosed individually in membranes and have a prominent five-layered cell wall, comparable with the organization of aposymbiotic unicellular cyanobacterian cell walls. Thylakoids are not centrically arranged as in *Cyanophora paradoxa*

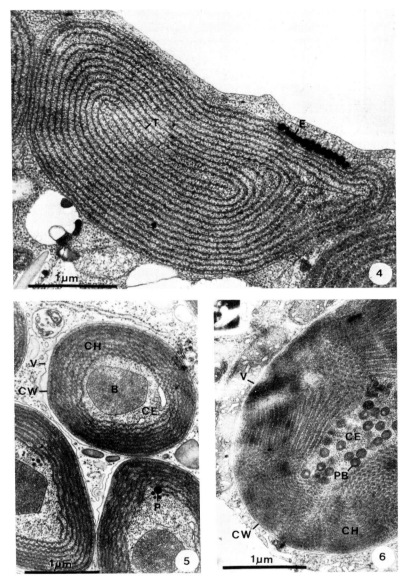

Fig. 4. Cyanella in *Glaucosphaera vacuolata*. Thylakoids (*T*) in the cyanella are concentrically arranged, a prominent cell wall is lacking. *E* eye-spot-like structure

Fig. 5. Cyanelles in *Gloeochaete wittrockiana*. *CW* cell wall of cyanella; *P* polyphosphate granule, for other details see Fig. 2a

Fig. 6. Cyanella in *Paulinella chromatophora*. *PB* polyhedral body (=carboxysome?), for other details see Figs. 2a and 5. (Figs. 4–6 kindly provided by L. KIES, University of Hamburg)

cyanellae, but are disposed transversely, and phycobilisomes are lacking. Other freshwater endocyanoms with cyanellae are formed by *Epithemia sorex, E. turgida, E. zebra* and *Denticulata vanheurcki* (GEITLER 1977). In the marine milieu, *Rhizoselenia styliformis* and *Hemiaulus membranaceus* are reported to contain a filamentous heterocyst-forming cyanobacterium which has been identified as *Richelia intracellularis* (KIMOR et al. 1978). In *Roperia* spp., a filamentous cyanobacterium has also been observed. It seems to be *Nostoc*-like, but detailed ultrastructural studies are lacking.

6.2.2.4 *Cyanidium caldarium*

Cyanidium caldarium has been studied thoroughly during the last decade because of its unusual habitat (Sect. 6.4.2) and its taxonomic position. It is a eukaryotic unicellular alga with chlorophyll a and phycobilisomes the chloroplast being surrounded by one membrane. The thylakoidal organization resembles that of cyanobacteria. Hence discussion centres on the question whether *Cyanidium caldarium* is an endocyanom with a still unknown host and cyanellae, or a primitive rhodophycean-type cell with true organelles. The discussion will not be reported here in detail, since it is often obviously confused by the different growth conditions applied. For detailed information see SECKBACH et al. (1981), who designate *Cyanidium caldarium* as a primitive rhodophyte, and KREMER and FEIGE (1979), who show by $^{14}CO_2$-studies that labelling of fixation products in *Cyanidium caldarium* is different from that observed in rhodophytes.

6.3 Physiology

In comparison to the ultrastructure, the physiology of syncyanoses has not been adequately studied. This failure mainly originates from the difficulties of obtaining enough material for mass cultures, and in separating and culturing the partners independently under defined conditions. Thus it is often necessary to rely on indirect conclusions and on comparison with aposymbiotic cyanobacteria under different environmental conditions. Although the situation is far from being analyzed in every detail the data indicate that syncyanoses can be divided according to physiological features into two groups. One group comprises photosynthesizing hosts, e.g. diatoms, bryophytes, cormophytes and symbiotic cyanobacteria with no or minor photosynthetic activity but prominent fixation of molecular N_2, and the other group comprises apoplastidal hosts, e.g. sponges, tunicates, amoebae, algae (Sect. 6.2.1.7) with symbiotic cyanobacteria which supply the hosts predominantly with photosynthetic fixation products but show no or only minor N_2-fixation.

6.3.1 Nitrogen Metabolism

A conspicuous feature of filamentous symbiotic cyanobacteria is the usually higher frequency of heterocysts they produce when living in syncyanoses than after isolation from them, in comparison with aposymbiotic cyanobacteria of the same genus. So in *Nostoc* spp. living in *Anthoceros punctatus* or *Blasia pusilla,* the heterocyst frequency is about 43% and 30% respectively, and after isolation from the system, only about 3–6% (RODGERS and STEWART 1977). Phycobilisomes are lacking in situ (DUCKETT et al. 1977). In *Azolla* the corresponding rates are 20–30% and 6% (HILL 1975). In the *Azolla* spp., syncyanose heterocyst frequency could be increased by supplying fructose and rhamnose (TEL-OR et al. 1983). In cycads, heterocyst frequencies are reported up to 90% (GRILLI CAIOLA 1972a), and in *Gunnera* up to 80% (SILVESTER and MCNAMARA 1976).

These observations hint at molecular nitrogen fixation in symbiotic cyanobacteria. N_2-fixation has been experimentally confirmed in the syncyanoses of *Anthoceros* and *Blasia pusilla* (BOND and SCOTT 1955, RODGERS and STEWART 1977, STEWART and RODGERS 1977, WATANABE and KIYOHARA 1963), *Cavicularia densa* (WATANABE and KIYOHARA 1963), *Sphagnum* (BASILIER 1980, GRANHALL and HOFSTEN 1976), *Azolla* (HOLST and YOPP 1979a, PETERS and MAYNE 1974b, PETERS et al. 1977), *Gunnera* spp. (SILVESTER and SMITH 1969), cycads (GRILLI CAIOLA 1980, WATANABE and KIYOHARA 1963), and *Rhizoselenia styliformis* (WEARE et al. 1974). For cyanellae it has been confirmed in sponges (WILKINSON and FAY 1979) and *Rhopalodia gibba* (FLOENER and BOTHE 1980).

Comparably few data are available for transfer of fixed nitrogen to hosts. The *Azolla–Anabaena azollae* cyanom grows without ammonium or nitrate when molecular nitrogen is supplied (HOLST and YOPP 1979a); without cyanobacteria *Azolla* needs a complex nitrogen-containing substrate (PETERS and MAYNE 1974a). Nitrogen is probably transferred from cyanobacteria to *Azolla* as ammonium and glutamic acid (NEWTON and SELKE 1981, RAI et al. 1981, RAY et al. 1978). In the *Anthoceros* and *Blasia* syncyanoses, about 98% of fixed nitrogen is reported to be transferred to hosts as ammonium (STEWART and RODGERS 1977). Among syncyanoses of apoplastidal algae and cyanellae, *Cyanophora paradoxa* is the best-studied system. It is unable to fix molecular nitrogen (FLOENER et al. 1982), and cyanophycin granules are absent (KIES 1980). Nitrogen supply probably works via a NADH-dependent nitrate reductase, which has been detected exclusively in host cytoplasm (BÖTTCHER et al. 1982, FLOENER et al. 1982). Other aspects of nitrogen metabolism in *Cyanophora paradoxa* are controversial. FLOENER et al. (1982) report on a ferredoxin-dependent nitrite-reductase activity in cyanellae, and propose for *Cyanophora paradoxa* an assimilatory pathway of nitrate reduction, as is usual in eukaryotic photosynthesizing organisms, the cyanellae supplying the host with ammonium. BÖTTCHER et al. (1982) report on nitrite-reductase activity also in host cytoplasm, and doubt the proposed key-role of cyanellae.

6.3.2 Carbon Metabolism

In syncyanoses formed by *Gunnera* spp. and cycads, cyanobacteria synthesize photosynthetic pigments, but significant CO_2-fixation or O_2-production has not been reported. In *Blasia pusilla* no significant CO_2-fixation or O_2-production could be measured, and phycobilisomes are absent in situ (STEWART and

Rodgers 1977). In the *Azolla–Anabaena azollae* cyanom, *Anabaena* contains phycobilisomes and a low rate of CO_2-fixation can be detected, but it is probably insignificant for meeting the energy supply necessary for N_2-fixation (Peters 1975, Ray et al. 1979).

High heterocyst frequencies (Sect. 6.3.1) and N_2-fixation rates lead to the conclusion that cyanobacteria must be supplied by their partner with photosynthates. In the *Azolla* cyanom, possibly sucrose and fructose are transferred to cyanobacteria (Ray et al. 1979, Tel-Or et al. 1983), in *Anthoceros* and *Blasia* cyanoms, sucrose has also been suggested as an energy supply unit for cyanobacterian N_2-fixation (Stewart and Rodgers 1977). Comparable data for diatom cyanoms are lacking, but it is noteworthy that in *Rhopalodia* cyanellae are devoid of phycobilisomes (Floener and Bothe 1980).

In syncyanoses with apoplastidal hosts, there are conclusive hints of a carbohydrate supply by cyanobacterian partners. In *Cyanophora paradoxa, Glaucocystis nostochinearum, Glaucosphaera vacuolata* and *Gloeochaete wittrockiana*, cyanellae show ribulose-1,5-bisphosphate-carboxylase activity and fix CO_2 via the reductive pentose-phosphate cycle. They have similar patterns of CO_2-fixation products. Most intensely labelled are glucose, fructose, and maltose. The data hint at the excretion of glucose, which can be converted by hosts to maltose and polyglucans. Accordingly, starch granules are reported to occur in host cytoplasm (Codd and Stewart 1977, Kremer et al. 1979, Schnepf and Brown 1971) (Fig. 2b). Excretion of glucose provides an interesting parallel with symbiotic cyanobacteria in lichens, and with symbiotic chloroplasts in *Elysia viridis* which also excrete this sugar into their hosts (Smith et al. 1969).

The physiology of *Cyanophora paradoxa* has been studied very thoroughly. It has an obligate photoautotrophic type of carbon-metabolism, i.e. it needs besides mineral salts only vitamins and light energy for growth. It is unable to degrade exogenic substrates via respiration; the photosynthetic compensation point is about 400 lx, and the saturation point between 7,500 and 10,000 lx. The isolated cyanellae excrete about 20% of the photosynthetically fixed carbon as glucose, sucrose, and maltose (Böttcher et al. 1982, Bothe and Floener 1978, Floener and Bothe 1982, Kremer et al. 1979, Trench et al. 1978).

In sponges 5–12% of carbon fixed photosynthetically by cyanobacteria is reported to be transferred to the host as glycerol (Wilkinson 1980). Corresponding rates for *Prochloron* are 7–26% as glycerol (Fisher and Trench 1980), but probably this is only of minor significance for annual host biomass production (Johns et al. 1981).

6.4 Symbiosis-Specific Features

Partners in symbiotic systems are generally of a different genome type, and share a common microhabitat by either intracellular association or at least close physical contact. Hence evolutionary success requires a certain degree of coordination, resulting in the development of common symbiosis-specific features on different levels of symbiotic organization.

6.4.1 Morphology

In syncyanoses of multicellular hosts with filamentous cyanobacteria the morphological integration of partners is generally of low degree. After isolation from hosts the cyanobacteria usually show some changes in the arrangement of thylakoids, cell shape, length of filaments and heterocyst frequency, but these changes are modificatory, resulting from the different environmental conditions in situ and in vitro.

As to hosts, in cycads the formation of coralloid roots is not induced by cyanobacteria (GRILLI CAIOLA 1980, SCHAEDE 1944) and is not a symbiosis-specific feature as the nodules of legumes (WERNER et al. 1980) are.

The conspicuous formation of mucus in *Gunnera* and cycad systems probably does not depend on the presence of cyanobacteria (NEUMANN 1977, SILVESTER and McNAMARA 1976), whereas in *Blasia*- and *Anthoceros* cyanoms there are some hints of slime formation by *Nostoc* spp. (DUCKETT et al. 1977). The physiological function of mucus is not clear; perhaps it serves as a sort of substrate for cyanobacteria (OBUKOWICZ et al. 1981), or has some attractant function (SCHAEDE 1951).

The formation of transfer cells in leaf cavities of *Azolla* sp. (Fig. 1) (PETERS et al. 1978) is also observed in symbiont-free ferns, whereas in some cycads it seems to be a symbiosis-specific feature which is induced by cyanobacteria (OBUKOWICZ et al. 1981). A corresponding specific induction is discussed for "protrusions" in *Blasia* and *Anthoceros* cyanoms (DUCKETT et al. 1977).

In unicellular hosts cyanellae are enclosed by a membrane. Such an enclosure seems to be a general symbiosis-specific feature which can also be observed in other endosymbiotic systems such as of algae and invertebrates (REISSER and WIESSNER, Chap. 4 this Vol). As in those systems, membranes are closely attached to the partner surface in syncyanoses, probably thus facilitating the exchange of metabolites by maximum surface contact. Accordingly, after division of cyanellae they are at once divided from each other by the formation of enclosing membranes (MIGNOT et al. 1969).

Whereas in *Gunnera* spp. the process of enclosing of *Nostoc* spp. in a host plasmalemma-derived membrane could be studied by electron microscopy (NEUMANN et al. 1970), the origin of the enclosing membrane around cyanellae in such systems as *Cyanophora paradoxa* or diatom cyanoms is a matter of discussion. In analogy to other symbiotic systems (REISSER and WIESSNER, Chap. 4, this Vol.) it can be speculated that it is derived from some phagocytotic processes which host cells were able to perform before the evolution of the present symbiotic system. Unfortunately ultrastructural as well as biochemical data for a comparison of host plasmalemma and membranes around cyanellae are lacking.

Another symbiosis-specific morphological feature of cyanellae is the reduction of cell-wall layers as compared to free-living unicellular cyanobacteria. This reduction means a degree of morphological adaptation which is generally not observed in other symbiotic systems (REISSER and WIESSNER, Chap. 4, this Vol.). An interesting yet unrelated parallel can be observed in *Convoluta roscoffensis*. This marine flatworm harbours chlorophycean algae (*Tetraselmis convolu-*

tae) which form protoplasts when living within their partner. But this system is not hereditary, since eggs of *Convoluta roscoffensis* need to be infected by temporarily free-living algal partners (OSCHMAN 1966).

6.4.2 Physiology, Behavioural and Ecological Features

There seems to be a basic difference in metabolic integration in syncyanoses of autotrophic and heterotrophic hosts. Cyanobacteria symbiotic in autotrophic hosts mainly serve as N_2-fixing units metabolically, and are supplied with carbohydrates, whereas in heterotrophic hosts cyanellae mainly function as photosynthetic units. Interestingly, photosynthesizing cyanobacteria seem to be far more integrated into host metabolism and morphology than N_2-fixing ones, which usually can be separated from hosts and grown independently, thus showing no symbiosis-specific physiological features. As yet, independent culture of cyanellae has not been achieved.

Integration of cyanellae is demonstrated in *Paulinella chromatophora*, which has never been observed to form food vacuoles (KIES 1974) and in *Cyanophora paradoxa*. When cyanellae in *Cyanophora paradoxa* are poisoned by prokaryote-specific antibiotics, the host metabolism is not able to compensate for the loss by uptake of external organic substrates (TRENCH 1982b). There are conclusive indications of a cooperation of partners in biosynthesis of chlorophyll (TRENCH 1982b) and differentiation of thylakoids (MARTEN et al. 1982).

Some of the most interesting consequences of coordinated host-symbiont interaction are the common behavioural features of the whole symbiotic system, as is observed in photobehaviour in ciliate-algae systems (REISSER and WIESSNER Chap. 4, this Vol.). Unfortunately, corresponding experiments with syncyanoses such as *Cyanophora paradoxa* are lacking. PASCHER (1929a) notes observations on "phototaxis" of motile cells of *Cyanophora paradoxa, Glaucocystis nostochinearum, Gloeochaete wittrockiana, Cyanoptyche gloeocystis,* and *Peliaina cyanea,* and KAWAGUTI (1971) reports a "light-dependent behaviour" in associations of cyanobacteria with echiuroid worms, but experimental data are not available.

There are also very few systematic studies of syncyanoses in respect to the ecological consequences of symbiosis formation. The *Azolla-Anabaena azollae* cyanom, as well as bryophyte-cyanobacterian associations, are observed frequently in nitrogen-depleted habitats (ASHTON and WALMSLEY 1976), and the *Gunnera–Nostoc punctiforme* system is reported to "occupy pioneer situations" (SILVESTER and SMITH 1969). Endocyanoms such as *Cyanophora paradoxa* are reported to occur preferably in mesotrophic or eutrophic ponds and lakes among plankton and "Aufwuchs" organisms (KIES 1980).

The ecology of *Cyanidium caldarium* has been well studied. It has been isolated from hot springs (50–60 °C), and can grow in $1n\ H_2SO_4$ (SECKBACH et al. 1981).

6.4.3 Regulation and Host–Symbiont Specificity

The stability of symbiotic systems requires mechanisms of regulation mastering such seemingly trivial problems as the coordination of partner growth rates.

In the *Azolla–Anabaena azollae* cyanom, outgrowth is avoided by an intimate developmental parallelism of fern and cyanobacteria (HILL 1975, 1977). In *Cyanophora paradoxa* the number of cyanellae is not constant, indicating that the division rates of host and partners are not completely coordinated. Under certain culture conditions the host cell shows a higher division rate than the cyanellae, but an ultimate outgrowth is not possible (TRENCH 1982b). A similar situation seems to prevail in *Glaucocystis nostochinearum* (COLWELL and WICK-STROM 1976). Only cursory observations are available for other syncyanoses. In *Peliaina cyanea* (PASCHER 1929a), *Gloeochaete wittrockiana* and *Glaucocystis nostochinearum* (KIES 1976, 1980), cyanellae sometimes can be digested, but the observations are not substantiated by physiological data as measurements of digestive enzyme activity. In *Gloeochaete wittrockiana*, cyanellae "can be lost" during host cell division (KIES 1976). PASCHER (1929a) observed the extrusion of intact cyanellae from *Cyanoptyche gloeocystis*.

Regulation in symbiotic systems is facilitated by a certain degree of partner specificity, which is characteristic at least for most endosymbiotic systems of eukaryotes (REISSER and WIESSNER, Chap. 4, this Vol.). Partner specificity in endocyanoms such as *Cyanophora paradoxa* or *Rhopalodia* spp. has not been tested because isolation of cyanellae and subsequent infection of hosts with other symbionts was not possible. In some sponges, two different cyanobacterian partners can be present in the same host (WILKINSON 1980). In bryophyte and cormophyte syncyanoses reports on host–symbiont specificity are contradictory, and often lack tests for the efficiency of new artificially induced associations. Symbiont-free *Blasia* and *Anthoceros* could be infected with *Nostoc* isolated from *Cycas* and *Gunnera* (DUCKETT et al. 1977, ENDERLIN and MEEKS 1983). In another experiment, *Blasia* did not form a new association with free-living *Nostoc* and *Anabaena* or *Nostoc* isolated from *Collema* (RODGERS and STEWART 1977). Aposymbiotic *Gunnera* could be infected with *Nostoc* isolated from *Anthoceros, Cycas* and *Peltigera* but usually not with *Anabaena azollae* and free-living *Nostoc* and *Anabaena* (BONNET and SILVESTER 1981, NEUMANN et al. 1970). Different cycads probably also have different *Nostoc* (GRILLI CAIOLA 1980). *Anabaena* isolated from different *Azolla* show a close relationship of surface structures, as could be demonstrated by specific antibody techniques (GATES et al. 1980, TEL-OR et al. 1983). These surface structures possibly participate in cellular recognition processes, thus playing a role for host–symbiont specificity similar to that been proposed for lichens (BUBRIK and GALUN 1980) and demonstrated for algae-invertebrate systems (POOL 1979, REISSER et al. 1982).

6.5 Conclusions

A great deal of scientific attention has been given during the last few years to the study of syncyanoses for two reasons: the idea of exploiting the ability of cyanobacteria to fix molecular nitrogen by constructing artificial systems of cyanobacteria and plant cells, and the revival of the endosymbiosis hypothesis speculating on the development of chloroplasts from symbiotic cyano-bacterian-like ancestors in apoplastidal hosts.

In artificial syncyanoses, the theoretical advantages – e.g. exploitation of nitrogen-deficient soils, reducing the requirements for nitrogen fertilizers – are tempting. The *Azolla–Anabaena azollae* cyanom is reported to bind about 100 kg N ha^{-1}·a^{-1} (TEL-OR et al. 1983), the *Gunnera* sp. system about 72 kg N ha^{-1}·a^{-1} (SILVESTER and MCNAMARA 1976). Corresponding data for plant–rhizobia associations for *Glycine max* are about 100 kg N ha^{-1}·a^{-1} and for *Medicago sativa* about 300 kg N ha^{-1}·a^{-1} (BURNS and HARDY 1975). In attempts to synthesize artificial syncyanoses, usually protoplasts of higher plant cells such as *Nicotiana tabacum, Daucus carota, Allium cepa,* and *Zea mays* are infected with free-living cyanobacteria such as *Anabaena variabilis* (MEEKS et al. 1978), *Gloeocapsa* (BURGOON and BOTTINO 1976) or cyanellae isolated from *Glaucocystis nostochinearum* (BRADLEY 1979, 1980). Different methods of infection are applied using special "fusion media" or micromanipulation techniques. Results are still rather unsatisfactory. Usually only a small percentage of protoplasts takes up the offered cells, which are subsequently either expelled or disintegrated. In a few cases uptake seemed to be stable, but experiments were not continued, so that enough material for physiological measurements could not be obtained.

Syncyanoses of apoplastidal hosts and cyanellae have been studied for a long period mainly in relation to taxonomy. After the endosymbiosis hypothesis (FAMINTZIN 1907, MERESCHKOWSKY 1905, SCHIMPER 1883) had been revived and substantiated (for references see MARGULIS 1981), they became extremely attractive in the search for the missing link between free-living cyanobacteria and plastids. Since a tremendous amount of literature has been accumulated during the last few years (Sect. 6, 2.2.2), this problem will not be discussed here in detail, but some comments on general trends are given.

A great deal of effort in applying both electron microscopic and biochemical techniques centres on the question whether cyanellae are symbiotic cyanobacteria, or presumptive or de facto plastids. Indeed, it is tempting to construct a line of succession according to reduction states of cyanellae from *Paulinella chromatophora* (prominent cell wall) to *Glaucocystis nostochinearum* (no wall) and *Cyanidium caldarium* (plastid in rhodophyte?). But it is obvious that this can only be a hypothetical model, because of the different taxonomic position of hosts and the lack of any recent photosynthesizing algae related to them. Nevertheless, endocyanoses with cyanellae could prove to be extremely helpful in studying the evolution of plastid–cytoplasm–nucleus relationships. Failure to grow partners independently of each other can probably be overcome by improved experimental techniques, but it reflects both adaptation and specialization in the system. The genome size of cyanellae in *Cyanophora paradoxa* is about 5–10% of that in aposymbiotic cyanobacteria, and thus comparable to that of the chloroplasts of chlorophytes (HERDMANN and STANIER 1977, MUCKE et al. 1980). On the other hand, $^{14}CO_2$-assimilatory patterns of cyanellae in *Cyanophora paradoxa, Gloeochaete wittrockiana, Glaucocystis nostochinearum* and *Glaucosphaera vacuolata* show a far more significant similarity to the corresponding patterns of aposymbiotic cyanobacteria than to rhodophycean plastids (KREMER et al. 1979).

The photosynthetic fixation pattern of *Prochloron* shows striking similarities to that of symbiotic plastids, but none to that of cyanobacteria or plastids of chlorophytes such as higher plants (KREMER et al. 1982).

Possibly plastids have evolved in different ways (WHATLEY et al. 1979): to rhodophycean plastids via cyanella-like ancestors, and to plastids of chlorophytes via endosymbiotic *Prochloron*-like forms.

References

Ashton PJ, Walmsley RD (1976) Die Symbiose zwischen dem Wasserfarn *Azolla* und der Blaualge *Anabaena*. Endeavour 35:39–44

Basilier K (1980) Fixation and uptake of nitrogen in *Sphagnum* blue-green algal associations. Oikos 34:239–242

Bhaskaran S, Venkataramaran GS (1958) Occurrence of a blue-green alga in nodules of *Trifolium alexandrinum*. Nature 181:277–278

Böttcher U, Brandt P, Müller B, Tischner R (1982) Physiologische Charakterisierung der Endocyanelle *Cyanocyta korschikoffiana* Hall & Claus. I. Photosynthetische und N-assimilatorische Eigenschaften in der symbiontischen Assoziation *Cyanophora-Cyanocyta*. Z Pflanzenphysiol 106:167–172

Bond G, Scott GD (1955) An examination of some symbiotic systems for fixation of nitrogen. Ann Bot 19:67–77

Bonnett HT, Silvester WB (1981) Specificity in the *Gunnera-Nostoc* endosymbiosis. New Phytol 89:121–128

Bothe H, Floener L (1978) Physiological characterization of *Cyanophora paradoxa,* a flagellate containing cyanelles in endosymbiosis. Z Naturforsch 33c:981–987

Bourdu R, Lefort M (1967) Structure fine, observée en cryodécapage des lamelles photosynthétiques des cyanophycées endosymbiotiques: *Glaucocystis nostochinearum* Itzigs, et *Cyanophora paradoxa* Korschikoff. CR Acad Sci Paris 265:37–40

Bradley PM (1979) Micromanipulation of cyanelles and a cyanobacterium into higher plant cells. Physiol Plant 46:293–298

Bradley PM (1980) Co-culture of carrot cells and a green alga on medium deficient in nitrogen. Z Pflanzenphysiol 100:65–67

Bubrick P, Galun M (1980) Symbiosis in lichens: Differences in cell wall properties of freshly isolated and cultured phycobionts. FEMS Microbiol Lett 7:311–313

Burgoon AC, Bottino PJ (1976) Uptake of the nitrogen-fixing blue-green algae *Gloeocapsa* into protoplasts of tobacco and maize. J Hered 67:223–226

Burns RC, Hardy RW (1975) Nitrogen fixation in bacteria and higher plants. Springer, Berlin Heidelberg New York

Calvert HE, Peters GA (1981) The *Azolla–Anabaena azollae* relationship. IX. Morphological analysis of leaf cavity hair populations. New Phytol 89:327–335

Chapman DJ, Trench RK (1982) Prochlorophyceae: Introduction and bibliography. In: Rosowski JR, Parker BC (eds) Selected papers in phycology II. Phycological Society of America, pp 656–658

Codd GA, Stewart WDP (1977) Quaternary structure of the D-ribulose-1,5-diphosphate carboxylase from the cyanelles of *Cyanophora paradoxa*. FEMS Lett 1:35–38

Cohn F (1872) Ueber parasitische Algen. Beitr Biol Pflanz 1:87–108

Collwell GL, Wickstrom CE (1976) Cell, cyanelle, and chlorophyll a relationships in *Glaucocystis nostochinearum* Itz. J Phycol 12:11

Cox D, Dwarte DM (1981) Freeze-etch ultrastructure of a *Prochloron* species – the symbiont of *Didemnum molle*. New Phytol 88:427–438

Drum RW, Pankratz S (1965) Fine structure of an unusual cytoplasmic inclusion in the diatom genus, *Rhopalodia*. Protoplasma 60:141–149

Duckett JG, Toth R, Soni SL (1975) An ultrastructural study of the *Azolla, Anabaena azollae* relationship. New Phytol 75:111–118

Duckett JG, Prasad AKSK, Davis DA, Walker S (1977) A cytological analysis of the *Nostoc*-bryophyte relationship. New Phytol 79:349–362

Enderlin CS, Meeks JC (1983) Pure culture and reconstitution of the *Anthoceros-Nostoc* symbiotic association. Planta 158:157–165

Famintzin A (1970) Die Symbiose als Mittel der Synthese von Organismen. Biol Zentralbl 27:353–364

Fisher CR, Trench RK (1980) In vitro carbon fixation by *Prochloron* sp. isolated from *Diplosoma virens*. Biol Bull 159:636–648

Floener L, Bothe H (1980) Nitrogen fixation in *Rhopalodia gibba,* a diatom containing blue-greenish inclusions symbiotically. In: Schwemmler W, Schenk HEA (eds) Endocytobiology 1. de Gruyter, Berlin, pp 541–552

Floener L, Bothe H (1982) Metabolic activities in *Cyanophora paradoxa* and its cyanelles. II. Photosynthesis and respiration. Planta 156:78–83

Floener L, Danneberg G, Bothe H (1982) Metabolic activities in *Cyanophora paradoxa* and its cyanelles. I. The enzymes of assimilatory nitrate reduction. Planta 156:70–77

Gates JE, Fisher RW, Goggin TW, Azrolan NI (1980) Antigenic differences between *Anabaena azollae* fresh from the *Azolla* fern leaf cavity and free-living cyanobacteria. Arch Microbiol 128:126–129

Geitler L (1924) Der Zellbau von *Glaucocystis nostochinearum* und *Gloeochaete wittrockiana* und die Chromatophoren-Symbiosetheorie von Mereschkowsky. Arch Protistenkd 47:1–24

Geitler L (1927) Bemerkungen zu *Paulinella chromatophora*. Zool Anz 73:333–335

Geitler L (1959) Syncyanosen. In: Ruhland E (ed) Handbuch der Pflanzenphysiologie. Springer, Berlin Heidelberg New York 11, pp 530–545

Geitler L (1977) Zur Entwicklungsgeschichte der Epithemiaceen *Epithemia, Rhopalodia* und *Denticula* (Diatomophyceae) und ihre vermutlich symbiontischen Sphäroidkörper. Plant Syst Evol 128:259–275

Giddings TH, Wasmann C, Staehelin LA (1983) Structure of the thylakoids and envelope membranes of the cyanelles of *Cyanophora paradoxa*. Plant Physiol 71:409–419

Granhall U, Hofsten AV (1976) Nitrogenase activity in relation to intracellular organisms in *Sphagnum* mosses. Physiol Plant 36:88–94

Grilli Caiola M (1972a) Cell morphology of the blue-green algae under culture conditions from *Cycas revoluta* isolated. I. Light microscope observations. Caryologia 25:137–145

Grilli Caiola M (1972b) Cell morphology of the blue-green algae under culture conditions from *Cycas revoluta* isolated. II. An electron microscope study. Caryologia 25:147–161

Grilli Caiola M (1975a) A light and electron microscopic study of blue-green algae growing in the coralloid roots of *Encephalartos altensteinii* and in culture. Phycologia 14:25–33

Grilli Caiola M (1975b) Structural and ultrastructural aspects of blue-green algae growing in the coralloid-roots of *Dioon edule* and in culture. Phykos 14:29–34

Grilli Caiola M (1980) On the phycobionts of the cycad coralloid roots. New Phytol 85:537–544

Hall WT, Claus G (1963) Ultrastructural studies on the blue-green algal symbiont in *Cyanophora paradoxa* Korschikoff. J Cell Biol 19:551–564

Hall WT, Claus G (1967) Ultrastructural studies on the cyanelles of *Glaucocystis nostochinearum* Itzigsohn. J Phycol 3:37–51

Herdman M, Stanier RY (1977) The cyanelle: chloroplast or endosymbiotic prokaryote? FEMS 1:7–12

Hill DJ (1975) The pattern of development of *Anabaena* in the *Azolla–Anabaena* symbiosis. Planta 122:179–184

Hill DJ (1977) The role of *Anabaena* in the *Azolla–Anabaena* symbiosis. New Phytol 78:611–616

Holst RW, Yopp JH (1979a) Effect of various nitrogen sources on growth and the nitrate-nitrite reductase system of the *Azolla mexicana–Anabaena azollae* symbiosis. Aquat Bot 7:359–367

Holst RW, Yopp JH (1979b) Studies of the *Azolla–Anabaena* symbiosis using *Azolla mexicana*. I. Growth in nature and laboratory. Am Fern J 69:17–25

Jacob F (1961) Zur Biologie von *Codium bursa* (L.) Agardh und seiner endophytischen Cyanophyceen. Arch Protistenkd 105:345–406

Johns RB, Nichols PD, Gillan FT, Perry GJ, Volkman JK (1981) Lipid composition of a symbiotic prochlorophyte in relation to its host. Comp Biochem Physiol 69:843–849

Kawaguti S (1971) Blue-green algae in echiuroid worms. In: Cheng TC (ed) Aspects of the biology of symbiosis. Baltimore Univ Park Press, pp 265–273

Kawamatu S (1965a) Electron microscope observations on blue-green algae in the leaf of *Azolla imbricata* Nakai. Cytologia 30:75–79

Kawamatu S (1965b) Electron microscope observations on the leaf of *Azolla imbricata* Nakai. Cytologia 30:80–87

Kies L (1974) Elektronenmikroskopische Untersuchungen an *Paulinella chromatophora* Lauterborn, einer Thekamöbe mit blau-grünen Endosymbionten (Cyanellen). Protoplasma 80:69–89

Kies L (1976) Untersuchungen zur Feinstruktur und taxonomischen Einordnung von *Gloeochaete wittrockiana*, einer apoplastidalen capsalen Alge mit blaugrünen Endosymbionten (Cyanellen). Protoplasma 87:419–446

Kies L (1979) Zur systematischen Einordnung von *Cyanophora paradoxa*, *Gloeochaete wittrockiana* und *Glaucocystis nostochinearum*. Ber Dtsch Bot Ges 92:445–454

Kies L (1980) Morphology and systematic position of some endocyanoms. In: Schwemmler W, Schenk HEA (eds) Endocytobiology 1. de Gruyter, Berlin, pp 7–19

Kies L, Kremer BP (1979) Function of cyanelles in the thecamoeba *Paulinella chromatophora*. Naturwissenschaften 66:578

Kimor B, Reid FMH, Jordan JB (1978) An unusual occurrence of *Hemiaulus membranaceus* Cleve (Bacillariophyceae) with *Richelia intracellularis* Schmidt (Cyanophyceae) off the coast of Southern California in October 1976. Phycologia 17:162–166

Knapp E (1933) Über *Geosiphon pyriforme* Fr. Wettst., eine intrazellulare Pilz-Algen-Symbiose. Ber Dtsch Bot Ges 51:210–215

Korschikoff AA (1930) *Glaucosphaera vacuolata*, a new member of the glaucophyceae. Arch Protistenkd 70:217–222

Kremer BP, Feige GB (1979) Accumulation of photoassimilatory products by phycobili-protein-containing algae with special reference to *Cyanidium caldarium*. Z Naturforsch 34c:1209–1214

Kremer BP, Kies L, Rostami-Rabet A (1979) Photosynthetic performance of cyanelles in the endocyanomes *Cyanophora*, *Glaucosphaera*, *Gloeochaete*, and *Glaucocystis*. Z Pflanzenphysiol 92:303–317

Kremer BP, Pardy R, Lewin RA (1982) Carbon fixation and photosynthates of *Prochloron*, a green alga symbiotic with an ascidian, *Lissoclinum patella*. Phycologia 21:258–263

Lang NJ (1968) The fine structure of blue-green algae. Annu Rev Microbiol 22:15–46

Lefort M (1965) Sur le chromatoplasma d'une cyanophycée endosymbiotique: *Glaucocystis nostochinearum* Itzigs. CR Acad Sci Paris 261:233–236

Lefort M, Pouphile M (1967) Données cytochimiques sur l'organisation structurale du chromatoplasma de *Glaucocystis nostochinearum*. CR Soc Biol 5:992–998

Lewin RA (1975) A marine *Synechocystis* (Cyanophyta, Chroococcales) epizoic on ascidians. Phycologia 14:153–160

Lewin RA (1977) *Prochloron*, type genus of the Prochlorophyta. Phycologia 16:217

Lewin RA (1981) The Prochlorophytes. In: Starr MP, Stolp H, Trüper HG, Balows A, Schlegel HG (eds) The Prokaryotes. Springer, Berlin Heidelberg New York pp 257–266

Lewin RA, Cheng L (1975) Associations of microscopic algae with didemnid ascidians. Phycologia 14:149–152

Lewin RA, Withers NW (1975) Extraordinary pigment composition of a prokaryotic alga. Nature 256:735–737

Margulis L (1981) Symbiosis in cell evolution. Freeman, San Francisco

Marten, S, Brandt P, Wiessner W (1982) On the developmental dependence between *Cyanophora paradoxa* and *Cyanocyta korschikoffiana* in symbiosis. Planta 155:190–192

McCracken DA, Nadakavukaren MJ, Cain JR (1980) A biochemical and ultrastructural evaluation of the taxonomic position of *Glaucosphaera vacuolata* Korsh. New Phytol 86:39–44

McLaughlin JJA, Zahl PA (1966) Endozoic algae. In: Henry SM (ed) Symbiosis I. Academic Press, New York pp 257–297

Meeks, JC, Malmberg RL, Wolk CP (1978) Uptake of auxotrophic cells of a heterocyst-forming cyanobacterium by tobacco protoplasts, and the fate of their associations. Planta 139:55–60

Mereschkowsky C (1905) Über Natur und Ursprung der Chromatophoren im Pflanzenreiche. Biol. Zentralbl 25:593–604

Miehe H (1924) Entwicklungsgeschichtliche Untersuchungen der Algensymbiose bei *Gunnera macrophylla* Bl. Flora 117:1–15

Mignot JP, Joyaon L, Pringsheim EG (1969) Quelques particularités structurales de *Cyanophora paradoxa* Korsch., protozoaire flagellé. J Protozool 16:138–145

Mollenhauer D (1970) Botanische Notizen Nr. 1: Beobachtungen an der Flechte *Geosiphon pyriforme*. Nat Mus 100:213–223

Moore AW (1969) *Azolla:* Biology and agronomic significance. Bot Rev 35:17–34

Mucke H, Löffelhardt W, Bohnert HJ (1980) Partial characterization of the genome of the "endosymbiotic" cyanelles from *Cyanophora paradoxa*. FEBS Lett 111:347–352

Nathanielsz CP, Staff IA (1975) On the occurrence of intracellular blue-green algae in cortical cells of the apogeotropic roots of *Macrozamia communis* L. Johnson. Ann Bot 39:363–368

Neumann D (1977) Ultrastrukturelle Untersuchungen zur Symbiose von Cyanophyceen mit Cycadeen (*Cycas circinnalis* L., *Zamia furfuracea* L.). Biochem Physiol Pflanz 171:313–322

Neumann D, Ackermann M, Jacob F (1970) Zur Feinstruktur der endophytischen Cyanophyceen von *Gunnera chilensis* Lam. Biochem Physiol Pflanz 161:483–498

Newton JW, Selke ES (1981) Assimilation of ammonia by the *Azolla–Anabaena* symbiosis. J Plant Nutr Soil Sci 3:803–811

Obukowicz M, Schaller M, Kennedy GS (1981) Ultrastructure and phenolic histochemistry of the *Cycas revoluta–Anabaena* symbiosis. New Phytol 87:751–759

Oschman JL (1966) Development of the symbiosis of *Convoluta roscoffensis* Graff and *Platymonas* sp. J Phycol 2:105–111

Pascher A (1914) Über Symbiosen von Spaltpilzen und Flagellaten mit Blaualgen. Ber Dtsch Bot Ges 32:339–352

Pascher A (1929a) Studien über Symbiosen. I. Über einige Endosymbiosen von Blaualgen in Einzellern. Jahrb Wiss Bot 71:386–462

Pascher A (1929b) Über die Natur der blaugrünen Chromatophoren des Rhizopoden *Paulinella chromatophora*. Zool Anz 81:189–194

Peirce GJ (1906) *Anthoceros* and its *Nostoc* colonies. Bot Gaz 24:55–59

Peters GA (1975) The *Azolla–Anabaena azollae* relationship. III. Studies on metabolic capabilities and a further characterization of the symbiont. Arch Microbiol 103:113–122

Peters GA, Mayne BC (1974a) The *Azolla–Anabaena azollae* relationship. I. Initial characterization of the association. Plant Physiol 53:813–819

Peters GA, Mayne BC (1974b) The *Azolla–Anabaena azollae* relationship. II. Localization of nitrogenase activity as assayed by acetylene reduction. Plant Physiol 53:820–824

Peters GA, Toia RE, Lough SM (1977) The *Azolla–Anabaena azollae* relationship. V. $^{15}N_2$ fixation, acetylene reduction, and H_2 production. Plant Physiol 59:1021–1025

Peters GA, Toia RE, Raveed D, Levine NJ (1978) The *Azolla–Anabaena azollae* relationship. VI. Morphological aspects of the association. New Phytol 80:583–593

Pickett-Heaps J (1972) Cell division in *Cyanophora paradoxa*. New Phytol 71:561–567

Pool RR (1979) The role of algal antigenic determinants in the recognition of potential algal symbionts by cells of *Chlorohydra*. J Cell Sci 35:367–379

Rai AN, Rowell P, Stewart WDP (1981) Glutamate synthase activity in symbiotic cyanobacteria. J Gen Microbiol 126:515–518

Ray TB, Toia RE, Mayne BC (1978) *Azolla–Anabaena* relationship. VII. Distribution of ammonia-assimilating enzymes, protein, and chlorophyll between host and symbiont. Plant Physiol 62:463–467

Ray TB, Mayne BC, Toia RE, Peters GA (1979) *Azolla–Anabaena* relationship. VIII. Photosynthetic characterization of the association and individual partners. Plant Physiol 64:791–795

Reinsch PF (1879) Beobachtungen über entophyte und entozoische Pflanzenparasiten. Bot Ztg 37:33–43

Reisser W, Radunz A, Wiessner W (1982) Participation of algal surface structures in the cell recognition process during infection of aposymbiotic *Paramecium bursaria* with symbiotic chlorellae. Cytobios 33:39–50

Richardson FL, Brown TE (1970) *Glaucosphaera vacuolata*, its ultrastructure and physiology. J Phycol 6:165–171

Robinson DG, Preston RD (1971) Studies on the fine structure of *Glaucocystis nostochinearum* Itzigs. J Exptl Bot 22:635–643

Rodgers GA, Stewart WDP (1977) The cyanophyte-hepatic symbiosis. I. Morphology and physiology. New Phytol 78:441–458

Sara M (1971) Ultrastructural aspects of the symbiosis between two species of the genus *Aphanocapsa* (Cyanophyceae) and *Ircinia variabilis* (Demospongiae). Mar Biol 11:214–221

Schaede R (1944) Über die Korallenwurzeln der Cycadeen und ihre Symbiose. Planta 34:98–124

Schaede R (1951) Über die Blaualgensymbiose von *Gunnera*. Planta 39:154–170

Schenk HEA (1977) Inwieweit können biochemische Untersuchungen der Endocyanosen zur Klärung der Plastiden-Entstehung beitragen? Arch Protistenkd 119:274–300

Schimper AFW (1883) Über die Entwicklung der Chlorophyllkörner und Farbkörper. Bot Ztg 41:105–112

Schmidt B, Kies L, Weber A (1979) Die Pigmente von *Cyanophora paradoxa*, *Gloeochaete wittrockiana* und *Glaucocystis nostochinearum*. Arch Protistenkd 122:164–170

Schnepf E (1964) Zur Feinstruktur von *Geosiphon pyriforme*. Ein Versuch zur Deutung cytoplasmatischer Membranen und Kompartimente. Arch Mikrobiol 49:112–131

Schnepf E (1965) Structur der Zellwände und Cellulosefibrillen bei *Glaucocystis*. Planta 67:213–224

Schnepf E, Brown RM (1971) On relationships between endosymbiosis and the origin of plastids and mitochondria. In: Reinert J, Ursprung H (eds) Origin and continuity of cell organells. Springer, Berlin Heidelberg New Yprk, pp 299–322

Schnepf E, Koch W, Deichgräber G (1966) Zur Cytologie und taxonomischen Einordnung von *Glaucocystis*. Arch Mikrobiol 55:149–174

Seckbach J, Hammermann IS, Hanania J (1981) Ultrastructural studies of *Cyanidium caldarium*: contribution to phylogenesis. Ann NY Acad Sci 361:409–425

Silvester WB, McNamara PJ (1976) The infection process and ultrastructure of the *Gunnera–Nostoc* symbiosis. New Phytol 77:135–141

Silvester WB, Smith DR (1969) Nitrogen fixation by *Gunnera–Nostoc* symbiosis. Nature 224:1231

Skuja H (1954) Glaucophyta. In: Melchior H, Werdermann E (eds) A. Engler's Syllabus der Pflanzenfamilien I. Borntraeger, Berlin, pp 56–57

Smith D, Muscatine L, Lewis D (1969) Carbohydrate movement from autotrophs to heterotrophs in parasitic and mutualistic symbiosis. Biol Rev 44:17–90

Stackebrandt E, Seewaldt E, Fowler VJ, Schleifer KH (1982) The relatedness of *Prochloron* sp. isolated from didemnid ascidian hosts. Arch Microbiol 132:216–217

Stewart WDP, Rodgers GA (1977) The cyanophyte–hepatic symbiosis. II. Nitrogen fixation and the interchange of nitrogen and carbon. New Phytol 78:459–471

Sybesma J, van Duyl FC, Bak RPM (1981) The ecology of the tropical compound ascidian *Trididemnum solidum*. III. Symbiotic association with unicellular algae. Mar Ecol Prog Ser 6:53–59

Tel-Or E, Sandovsky T, Kobiler D, Arad C, Weinberg R (1983) The unique symbiotic properties of *Anabaena* in the water fern *Azolla*. In: Papageorgiou GC, Packer L (eds) Photosynthetic procaryotes. Elsevier/North-Holland, Amsterdam Oxford New York, pp 303–314

Trench RK (1982a) Cyanelles. In: Schiff JA (ed) On the origins of chloroplasts. Elsevier/North-Holland, Amsterdam Oxford New York, pp 56–76

Trench RK (1982b) Physiology, biochemistry, and ultrastructure of cyanellae. In: Round FE, Chapman DJ (eds) Progress in phycological research 1. Elsevier/North-Holland, Amsterdam Oxford New York, pp 257–288

Trench RK, Pool RR, Logan M, Engelland A (1978) Aspects of the relation between *Cyanophora paradoxa* (Korschikoff) and its endosymbiotic cyanelles *Cyanocyta korschikoffiana* (Hall & Claus). I. Growth, ultrastructure, photosynthesis and the obligate nature of the association. Proc R Soc Lond B 202:423–443

Watanabe A, Kiyohara T (1963) Symbiotic blue-green algae of lichens, liverworts and cycads. In: Studies on microalgae and photosynthetic bacteria. Jap Soc Plant Pysiologists, Univ of Tokyo Press, pp 189–196

Weare NM, Azam F, Mague TH, Holm-Hansen O (1974) Microautoradiographic studies on the marine phycobionts *Rhizoselenia* and *Richelia*. J Phycol 10:369–371

Werner D, Wilcockson J, Stripf R, Mörschel E (1980) Differentiation of *Rhizobium japonicum* to bacteroids in the symbiosis of soybean nodules and in vitro. In: Schwemmler W, Schenk HEA (eds) Endocytobiology 1. de Gruyter, Berlin, pp 473–490

Wettstein F v (1915) *Geosiphon* Fr. Wettst., eine neue interessante Siphonee. Öster Bot Z 65:145–156

Whatley JM (1977) The fine structure of *Prochloron*. New Phytol 79:309–313

Whatley JM, John P, Whatley FR (1979) From extracellular to intracellular: the establishment of mitochondria and chloroplasts. Proc R Soc Lond B 204:165–187

Whitton BA (1973) Interactions with other organisms. In: Carr NG, Whitton BA (eds) The biology of blue-green algae. Blackwell, Oxford, pp 415–433

Wilkinson CR (1980) Cyanobacteria symbiotic in marine sponges. In: Schwemmler W, Schenk HEA (eds) Endocytobiology 1. de Gruyter, Berlin, pp 553–563

Wilkinson CR, Fay P (1979) Nitrogen fixation in coral reef sponges with symbiotic cyanobacteria. Nature 279:527–529

Wittmann W, Bergersen FJ, Kennedy GS (1965) The coralloid roots of *Macrozamia communis* L. Johnson. Aust J Biol Sci 18:1129–1134

7 Epiphytism at the Cellular Level
with Special Reference to Algal Epiphytes

S. C. DUCKER and R. B. KNOX

7.1 Introduction

There are two kinds of habitat where the limiting factor for plant development is the availability of a place in sufficient light, (1) the forests of the tropics, and (2) the littoral zone of lakes and in the intertidal zone of the sea. It is therefore in these places that we find the greatest development of one class of plants which struggles for "standing room only" in the photic zone, namely the epiphytes.

The term epiphyte was first applied to grass growing on trees by J.D. HOOKER in his *Himalayan Journals* (1854), although LINNAEUS had earlier named an epiphytic orchid *Epidendron* (1753). Until the middle of the 19th century, marine and other epiphytes were called parasites; thus HARVEY (1855) remarks on the large number of "parasitical" algae on the Australian seagrass *Amphibolis antarctica*. GOEBEL (1889), writing the first review on epiphytism, agrees with the concept of Hooker, i.e. that of a plant growing on another plant species, and gives a large number of examples, dealing mainly with terrestrial tropical plants. However, OLTMANNS (1923) distorted the concept of epiphytism by assuming that most epiphytes are plants which grow on a substratum like stone, wood or the chitin-cover of Bryozoa and other animals. In the literature, epiphytes are usually defined as plants which, although growing on another plant, do not derive nutrition from their hosts (LINSKENS 1976). They have been referred to as space parasites by some authors (JAAG 1945). We believe that the term light parasite is more applicable, but ask the question whether it is true that these plants are only finding a convenient substrate or are specific and selective in their host, with which they often appear to have a physiological relationship.

Epiphytes are found among ferns, mosses, liverworts and algae and in many families of the higher plants, but there are no records of epiphytic conifers. They occur in two distinct environments:

1. Terrestrial plants are the hosts, and the epiphytes are bryophytes, algae, lichens, pteridophytes, or in the flowering plant families Orchidaceae, Araceae, Bromeliaceae, Piperaceae and Asclepiadaceae.

2. Freshwater and marine environments. The hosts are freshwater plants, mangroves, seagrasses, and macroalgae which are colonized by micro- and macroalgae.

A detailed terminology for the different relationships of hosts and epiphytes embracing both lower and higher plants has been proposed by LINSKENS (1963b). He differentiated between the spatial and functional relationships of the partners, and defined epiphytes and basiphytes (the hosts). Holo-epiphytes are epiphytes that are attached to the outer layers of the basiphyte, whereas amphi-epiphytes are anchored deeply in the tissues of the basiphyte. He showed that the grades of relationship vary from association to association.

The most important part of his work is the definition of the functional relationships between the epiphytes and basiphytes. He discussed the "Symbiose" or "Zusammenleben" of De Bary (1879), and said that there are two kinds of such a Zusammenleben, the indifferent kind, where neither partner benefits from the association, and the symbiotic kind, where the relationship is of mutual benefit to both partners. He discussed allelo-parasitism, a term coined by Schaede and Meyer (1962), where the relationship of the two partners is symbiotic, but is formed by different types of organisms (Ducker and Knox 1978). During the last two decades a large amount of literature has accumulated dealing with the relationships of epiphytic marine algae, and describing examples from all groups of algae. Apart from ecological observations, different investigatory techniques have been employed to elucidate these relationships. Some of the techniques used and ideas put forward may be applicable to other groups in the plant kingdom. Accordingly this work is considered in greater detail in this chapter, as it throws considerable light on the diverse intercellular interactions between epiphyte and basiphyte. No attempt, however, is made to cover the entire spectrum of algal epiphytism and only a few selected examples are considered.

7.2 Dispersal and Contact of Algal Epiphytes

In the marine environment the dispersal of epiphytic algae can occur in three different ways, through vegetative propagules, thallus fragmentation, and spores. Consequently, there are three modes of attachment and adhesion processes which initiate the epiphytic interaction.

7.2.1 Vegetative Propagules

Excellent examples of vegetative propagules which make a quick start possible on a new substratum, both epiphytic and epilithic, are found in the genus *Sphacelaria* (Phaeophyta). These multicellular propagules are formed during periods of unusually high and low temperatures, and provide for asexual reproduction independent of sexual reproduction. They exist in bi- and triradiate forms which attach to larger brown algae and corallines, where they quickly give rise to a new epiphyte (Van den Hoek and Flinterman 1968). The Australian brown alga *Zonaria diesingiana* J. Ag. forms minute multicellular plantlets (Fig. 1) which, distributed by the currents, settle and can lodge among other algae to give rise to new epiphytic *Zonaria* plants (Allender and Kraft 1983).

7.2.2 Thallus Fragmentation as Dispersal Mechanisms

Thallus fragmentation has been reported for the epiphytic *Bornetia secundiflora* in the English Channel, where sexual reproduction of this species does not

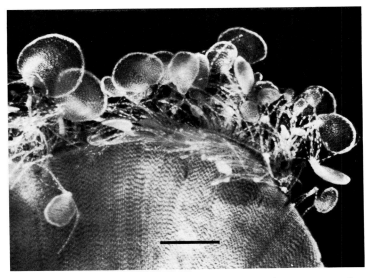

Fig. 1. *Zonaria diesingiana* forming multicellular platelets which, acting as vegetative propagules, initiate new epiphytic plants. Bar 1 mm. (Photo by C.O'Brien)

occur (DIXON 1965). FUNK (1927) discussed the *Bryopsis–Halymenia* relationship, where fragments of *Bryopsis* attach to the basiphyte. We have confirmed the observations (FUNK 1955) that fragments of such genera as *Cladophora, Bryopsis, Polysiphonia* and *Ceramium* become dislodged and by twisting around the branches of other algae start a new epiphytic relationship.

There are also epiphytic forms with special thallus modifications adapted for attachment, e.g. the barbed branches of *Asparagopsis armata* Harv., the hooked branches of *Hypnea episcopalis* Hookf. and Harv., *Mychodea hamata* Harv., *Champia zostericola* (Harv.) REEDMAN and WOMERSLEY, and of *Plocamium leptophyllum* KUETZING. It is evident that these red algal epiphytes can regenerate from attached thallus fragments. Once these fragments are anchored to a basiphyte, haustorial pads are formed which attach to the host. Attachment of *Champia zostericola* is "at first by means of a small discoid holdfast with one to several axes, later attaching by small adventitious multicellular pads to the seagrass" (REEDMAN and WOMERSLEY 1976, p. 87). The brown alga *Lobospira bicuspidata* J.E. Aresch., a frequent epiphyte in Australian waters, also attaches by hooked branches to the basiphyte.

7.2.3 Spores

Characteristically, all macroscopic marine algae have at certain points in their life history a spore stage sensu lato, making the spore a dispersal agent. The unflagellated spores of the Rhodopyta can be tetra- or carposporic, or may form within poly- and monosporangia. NGAN and PRICE (1979) examined a large number of red algal spores. They showed that the spores are of many sizes, but are all surrounded by a conspicuous hyaline mucilaginous sheath. Apparently there is a relationship between spore size and substrate, epiphytic species having generally larger spores than epilithic species. For instance, *Dicranema revolutum* has very large carpospores ($20-30 \times 55$ μm, Kraft 1977) and is restricted as an epiphyte to the stems of *Amphibolis antarctica*. The largest tetrasporangia formed by articulated coralline algae have been reported by Ducker et al. (1976) in the tiny epiphyte *Jania pusilla* ($100-130 \times 200-350$ μm). Epiphytic algae such as *Leveillea jungermanniales* and *Champia forsteri* have larger spores than related epilithic species. There is also a larger volume of mucilage surrounding the spores of epiphytic species. Boney (1981) showed not only that the space/volume relationship but also the mucilage sheath is consistently larger in epiphytic Rhodophyta, a feature that is especially impressive in *Polysiphonia lanosa*.

The mucilaginous sheath at the surface of algal spores is strategically sited to play a role in adhesion of the spores to the basiphyte. However, Boney (1975) has shown that its primary role may be spore buoyancy rather than adhesion, since the production of adhesive mucilage is a special event following spore release in certain red algae. This conclusion has been reached from a number of ultrastructural studies in which the cytology of the Golgi apparatus and its activity in the spore cytoplasm has been taken as an index of mucilage secretion. Specific Golgi activity has been associated with spore release in *Ceramium* (Chamberlain and Evans 1973) and *Jania rubens* (Moorjani and Jones 1972) and with synthesis of adhesive mucilage in the green alga *Enteromorpha* (Callow and Evans 1974).

The mucilage in the carposporangium of the red alga *Polysiphonia lanosa* stains with alcian blue, an indicator of acidic mucopolysaccharides (Pearse 1972). The presence of mucilage is otherwise difficult to demonstrate, although the use of Indian ink-seawater suspensions has been effectively employed by Boney (1975) to detail the mucilage sheath of *Polysiphonia* spores. More recently, the mucilage layers in the mature tetrasporangium of *Palmaria palmata*, although not an epiphyte, have been shown to be structurally and chemically dimorphic with an outer thick layer that is periodic acid-Schiff (PAS)-positive, and an inner layer around the spores that is PAS-negative. The layer staining with PAS is likely to be composed of neutral polysaccharides, with vicinal glycol groups (Pearse 1972). The inner layer of polysaccharides lacks these groups, as do many acidic polysaccharides, especially pectins, which are stained by ruthenium red and alcian blue. Moss (1981a) has observed that the zygotes of the two lithophytes *Fucus* and *Ascophyllum* attach first by the exudation of mucilage, and then by the rapid growth of a primary rhizoid which forms an adhesive "foot" on the substrate. There is a need for more precise histochemical characterization of epiphytic algal spore mucilages, and for observation of the events that occur at release, and at germination. Certainly, "successful adhesion is a crucial life-history event" (Boney 1981).

Members of the Ectocarpales are among the most successful ubiquitous epiphytes in all seas. Their dispersal is by flagellated zoids. Clayton (1974) has grown a number of the southern Australian epiphytic Ectocarpales in culture, observing the settlement of the zoids and their primary divisions into prostrate and upright filaments. She does not record any special adhesive mechanisms of the zooids.

In the sea the dispersal of algal spores, whether flagellated or unflagellated, is certainly completely fortuitous. It initiates one of the most intriguing sequences of events in the life history of an epiphyte. Is settlement specific or are survival and development restricted to certain hosts? At what stage of the spore settling and development is a selection for the substrate or basiphyte possible? These questions were discussed by MOORJANI and JONES (1972) and JONES and MOORJANI (1973) when they compared two articulated corallines, *Corallina officinalis*, a lithophyte, with *Jania rubens*, an epiphyte on brown and larger red algae, both from the same environment. They showed that the spores of *Jania* were attached securely after 4.5 h to glass slides, but that the spores of *Corallina* needed 48 h for attachment to glass. The spores of *Jania* produced more mucilage than those of *Corallina*. Negatively staining particles adhered to the surface of all settled spores. Positive staining with methylene blue and alcian blue indicated that the adhesive was probably a mucopolysaccharide. In the *Jania* spores sodium alizarin sulphonate produced no colour, indicating that calcification did not occur at the newly settled stage; these spores could not be detached with dilute hydrochloric acid. However, acid treatment caused a detachment of the spores of *Corallina*, calcification having been indicated by a red colouration with sodium alizarin sulphonate. Consistent with earlier observations, the larger spores of *Jania* were 24-celled in 1–2 days after attachment; the smaller *Corallina* spores also produced 24–36-celled sporelings in the same time which were, however, heavily calcified at this stage. Development after attachment was faster in *Jania*, reaching the typical basal disc in 8 days; but the basal crust of *Corallina* needed 12 weeks for development. There seems to be a specific mechanism by which spores of epiphytic algae germinate more rapidly than those of epilithic algae. WOELKERLING et al. (1983) found that the spores of epiphytic Corallinaceae, such as taxa of *Fosliella, Jania* and *Metagoniolithon*, settle and germinate rapidly (within 60 min) on glass slides when grown in a defined culture medium.

CHAMBERLAIN (1978), in discussing the relationships amongst epiphytic, crustose Corallinaceae, stressed the adaptations of these forms to the transitory nature of their substrate, which necessitates rapid settlement, germination and reproduction. She further stressed the very restricted host range of some of these epiphytes (e.g. *Fosliella lejolisii* on *Zostera marina*), although other crustose epiphytes are found on a range of algae.

7.3 Specificity of Epiphytic Relationship

7.3.1 Specificity of Brown Algal Epiphytes and Basiphytes

Most of the ubiquitous members of the Ectocarpales, e.g. *Feldmannia, Ectocarpus* or *Giffordia* seem to have no specialized requirements of hosts or substrates. They are, however, well adapted to a quick response, once settled on another

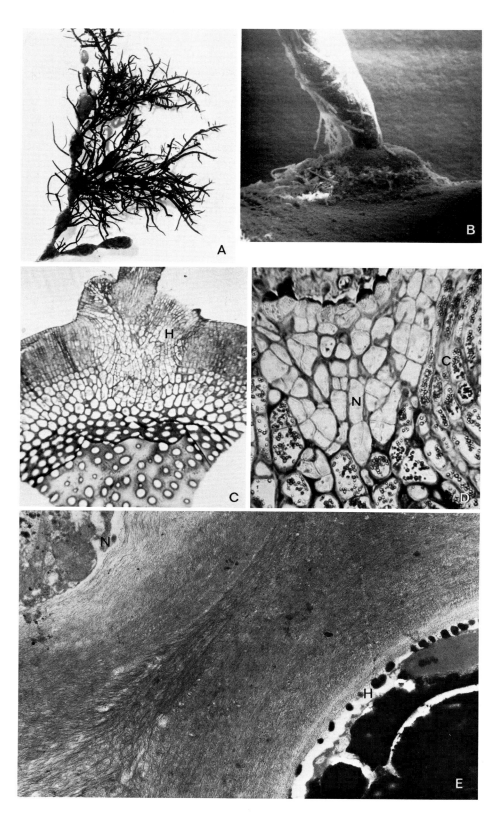

alga. SCAGEL (1966) noted that the whole matter of epiphytism, endophytism and parasitism in the brown algae is in need of a critical study.

A new and positive contribution has been made by HALLAM et al. (1980). *Notheia anomala* seems to be almost entirely host-specific on the Australian fucoid *Hormosira banksii*. This epiphyte grows only on *Hormosira* (apart from a doubtful record on *Xiphophora*). *N. anomala* could not be grown in culture, even though the medium employed could be used successfully for most other brown algae. If, however, fresh homogenized extract of the basiphyte was added to the culture solution, plants of *Notheia* became quite solid and more paren-chymatous, and grew up to 5 mm in diameter (PARISH 1976). *Hormosira* is dioecious, and *Notheia* infects both the antheridial and oogonial conceptacles (Fig. 2). The usual infection site is close to the ostiole of a conceptacle, and the basal part of the epiphyte sinks into the thallus due to the bending away of the cortical cells and the development of a wedge-shaped growth by the epiphyte. Both the epi- and basiphyte show normal chloroplasts. While there are numerous plasmodesmata between the cortical cells of *Hormosira,* these were not observed between the cells of the host and the obligate epiphyte (Fig. 2); although ultrastructural study has shown that there is a very close contact between the adjacent cell walls of the basiphyte and the epiphyte. The interface area of *Notheia* and *Hormosira* is a wall-to-wall boundary, with a loose and fibrous lamella between the walls of the two species (Fig. 2).

RHYS WILLIAMS (1965) reported two specific brown algal epiphytes from Scotland. *Litosiphon laminariae* grows almost exclusively on *Alaria esculenta,* where it is claimed to form galls initiated in the cryptostomata of the basiphyte. She was able to grow the epiphyte in "Erdschreiber" medium alone, but the addition of *Alaria* stimulated the growth of the epiphyte in culture for several generations. Another brown alga, *Spongonema tomentosum,* occurs on two fu-coids, *Fucus vesiculosus* and *Himanthalia elongata,* where it is attached by a single layered disc, with some filaments penetrating into the tissue of the basiphyte.

7.3.2 Specificity of Red Algal Epiphytes on Brown Algal Basiphytes

Epiphytism is particularly well-developed in the Rhodophyta, some species or even genera depending entirely on this mode of existence. Many red algal epi-phytes are restricted to a few or even to one brown algal basiphyte, and are then regarded as obligate epiphytes (DIXON 1973). DIXON suggested that LINS-KENS' hypothesis (1963a) of the surface tension of the host being the deciding

---

←

Fig. 2 A–E. The specific epiphyte *Notheia anomala* growing on *Hormosira banksii.* (HAL-LAM et al. 1980). **A** Appearance of epiphyte on basiphyte ($\times 0.8$). **B** Scanning electron micrograph of young *Notheia* plant on the thallus of the basiphyte ($\times 90$). **C** Transverse section through infection site, showing the displaced cortical cells of *Hormosira* (*H*) ($\times 100$). **D** Detail of infection wedge, showing the lateral compression of *Hormosira* corti-cal cells (*C*) and the thinner-walled *Notheia* cells (*N*) penetrating into the tissue ($\times 380$). **E** Transmission electron micrograph of the interface between cell walls of *Notheia* (*N*) and *Hormosira* (*H*) ($\times 9000$)

factor for epiphyte settlement is not proven on a chemical or physical basis, and needs investigating further. It is, however, evident that certain conditions must be fulfilled for a successful settlement of an epiphyte on a brown alga.

Members of the order Ceramiales frequently colonize the Australian fucoid *Cystophora* and other brown algae. Three species of the genus *Heterothamnion* are described from southern Australia, each of which occurs on a single host plant: *H. episiliquosum* on *Cystophora siliquosa* and *H. muelleri* and *H. sessile* on *C. platylobium*. Other members of the Heterothamniae, like *Perithamnion* and *Tetrathamnion,* are also epiphytes on brown algae, and Wollaston (1967) reported that these genera are attached by rhizoidal structures which penetrate deeply into the host tissue. Parsons (1975) recorded *Haplodasya urceolata* as epiphytic on six Australian brown algae. Gordon (1972) reported a number of Ceramiaceae restricted to brown algal basiphytes among these *Interthamnion attenuatum*. This species is normally an obligate epiphyte on *Zonaria spiralis,* but Huisman (1980) grew it successfully to maturity in PES (Provasoli's enriched seawater) without the addition of an extract of the host tissue.

Many Corallinaceae are epiphytes on larger brown algae. The host specificity of *Jania rubens* in British waters was discussed by Duerden and Jones (1974, 1981) who argued that there is a metabolic dependency between *J. rubens* and the host *Cladostephus spongiosus,* and that extra cellular metabolites produced by the basiphyte may influence the initial stages of colonisation by the epiphyte. Ducker et al. (1976) reported that the endemic *Jania pusilla* is restricted to *Cystophora siliquosa* in southern Australia. Chamberlain (1978) discussed the taxonomic relationships amongst epiphytic crustose Corallinaceae, many of which colonize brown algae.

Ende and Linskens (1962) showed that certain epiphytes have preferential settling areas on the thallus of *Himanthalia elongata*. Of special interest are the two epiphytes *Ceramium rubrum* and *Ectocarpus draparnaldioides,* which settle both on the annual and sexually mature thallus parts of the brown alga. They do not overlap on the thallus; *Ceramium* occupies the upper parts and does not penetrate into the thallus of the basiphyte (>70 μm). However, *Ectocarpus* situated on the lower parts of the fertile fronds forms a wedge-shaped foot (up to 1000 μm deep) into the thallus of *Himanthalia*. A third alga, *Rhodymenia,* is restricted to the sterile holdfast region of *Himanthalia*. Ende and Linskens (1974) observed also that these specific epiphytes are replaced by unspecific epiphytes when the basiphytes are old and moribund. Ende and Linskens argued that *Ceramium* is a true epiphyte, while the foot of *Ectocarpus* has a parasitic tendency. No explanation for these epiphytic reactions is suggested.

The relationship of *Polysiphonia lanosa* to the basiphyte *Ascophyllum nodosum* is discussed in Section 3.1.2.

7.3.3 Specificity of Red Algae as Epiphytes and Basiphytes

Evans et al. (1978) in their paper on "parasitic" red algae, suggested that in this case it is particularly important to establish the boundaries between epiphytism and parasitism, and say that the term epiphyte should be restricted to algae which are totally independent of their hosts for their nutrient requirements. Many examples of these red/red algal associations can be quoted, but in no instance do we know of any detailed cellular investigations. It is a field of investigation wide open for research, particularly in the Australian region, where the greatest number of endemic Rhodophyta are found (WOMERSLEY 1981).

A few examples may be given. WOMERSLEY (1965) reported that the mono-typic genus *Sonderella linearis* is restricted to *Ballia callitricha*, and is usually attached to the older axes of the basiphyte by down-growing rhizoidal filaments of elongated cells. Later these rhizoids grow along and extend all around the *Ballia* axis. In another instance a very small species (up to 2 mm high) of *Polysiphonia, P. haplodasyae*, is reported to be confined to the basiphyte *Haplodasya urceolata*, and is attached by a basal, cellular disc on the corticated axis of the host. The elongate cells of the epiphyte intermingle with the cells of the host. This epiphyte lacks rhizoids, but other epiphytic species of *Polysiphonia* which are not restricted to one host show a variety of prostrate filaments with scattered rhizoids which can be single-celled or consist of multicellular pads (WOMERSLEY 1979). WOMERSLEY (1978) also reports on the variable rhizoids of the genus *Ceramium*, which can be both multicellular and branched, or uniseriate in epiphytic and epilithic species. JONES and DUERDEN (1972) found that there are variations in the morphology of the primary rhizoid in *Ceramium rubrum* and used these differences for the elucidation of some of the fundamental principles involved in the host–epiphyte relationship. Here two modes of attachment are recorded differing in their structure and chemistry. One is typified by the formation of a short, much-divided, holdfast attaching by means of a secreted adhesive (lithophyte), while the other forms an elongate primary rhizoid producing no cementing agent but relying on the tendrillar activity of both itself and the secondary rhizoids to achieve attachment (epiphytic colonisation).

7.3.4 Seagrasses and Algal Epiphytes

There is a large literature on the epiphytes of seagrasses (see HARLIN 1975, 1980 for review), and, as DUCKER et al. (1977) pointed out earlier, it is a most striking feature of seagrasses that their epiphytes are diverse and abundant. The seagrass epiphytes fall into two distinct groups according to their host specificity.

7.3.4.1 Non-Specific Epiphytes

These grow also on other substrates (for example macroscopic algae, stones, wood and shells), and the seagrass essentially only provides a surface for additional settling room in an illuminated area. These epiphytes show generally no special adaptations of their holdfast area and are ubiquitous algae of the littoral zone. For example, several algae belonging to this group occur on *Amphi-*

bolis antarctica, Bryopsis plumosa, Ulva lactuca, Chaetomorpha darwinii, Dictyota dichotoma, Colpomenia sinuosa and *Laurencia filiformis* (Ducker et al. 1977).

7.3.4.2 Specific Epiphytes

The most interesting group of algae are those which are completely restricted to their seagrass host, for example, on the genus *Amphibolis, Dudresnaya australis, Dicranema cincinnalis, D. revolutum* and *Laurencia cruciata.*

Of particular interest are two coralline algae, the articulated *Metagoniolithon stelliferum* and the crustose *Fosliella cymodoceae* (Jones and Woelkerling 1984), which colonize both the stems and the leaves of *Amphibolis* (Ducker 1979). The two corallines are the faithful companions of *Amphibolis* throughout its range (Ducker and Knox 1978). The interesting fact is that the *Metagoniolithon* spores do not germinate directly on the surface of *Amphibolis,* but germinate within the ostiole of conceptacles of *Fosliella.* Both species exhibit, therefore, a dual specificity for *Amphibolis.* However, Woelkerling et al. 1983 have successfully germinated both coralline partners on glass slides in a specially designed medium, although the sexual cycle of neither has been completed.

The primary colonizer, *Fosliella,* succeeds in eroding the cuticle of the leaves and stem of the seagrass and forms a pad on which the *Metagoniolithon* spores can settle and germinate (Ducker and Knox 1978). It is through the *Fosliella* cells that *Metagoniolithon* makes direct contact with the epidermal cells of the seagrass (Fig. 3). The sites of contact in the epidermal cell walls of the basiphyte are stained purple with toluidine blue, in contrast to the green staining of the cell walls. Purple-stained strands radiate across the wall to the convoluted plasma membranes. These are reminiscent of the wall ingrowths of transfer cells known to occur at sites of cell–cell transfer of soluble substances (Gunning and Steer 1975). The pathways leading from *Metagoniolithon* through *Fosliella* to the epidermal cells of the seagrass are stained brilliantly (Fig. 3). We believe that these represent the sites of connections between the red algae and the seagrass epidermal cells which provide a physical means of contact and may be a pathway for informational exchange between basiphyte, and the two epiphytes. Channels through the leaf-surface cuticle of seagrasses have been previously observed in electron micrographs of epidermal cells of *Cymodocea* and *Thalassia* (Doohan and Newcomb 1976) where electron-opaque, osmiophilic material is secreted on the surface. Algal antibiotics are known to be released by seagrasses (Burkholder 1973) and may regulate the frequency and composition of epiphytes.

Harlin (1980) has given an interesting account of the relationship of *Smithora naiadum,* a red alga which is found exclusively on the seagrasses *Phyllospadix scouteri* and *Zostera marina* in the northern parts of the Pacific Ocean. *Smithora* is said to have an obligate association with seagrasses. Harlin (1980) observed the production of amoeboid cells from spores produced from the winter stage of *Smithora.* The amoeboid cells attached to glass coverslips in culture. On the seagrass, however, the amoeboid cells form a basal disc which is divided into upper and lower cell layers. The lower layer differentiates into pseudorhizoids orientated towards the basiphyte. The epidermal cells of the seagrass

Fig. 3 A, B. Stem of the seagrass *Amphibolis antarctica,* with two specific epiphytes *Metagoniolithon stelliferum* and *Fosliella cymodoceae.* **A** *Fosliella* is the primary colonizer and in this transverse section is seen eroding the cuticle of the host and forming direct connections with the cell wall; (*c*=cuticle, *e*=epidermal layer). **B** Basal crust of *M. stelliferum* (*M*) traversing the intermediate epiphyte, *F. cymodoceae* (*H*) and forming connections with the cell wall of the basiphyte. (DUCKER and KNOX 1978)

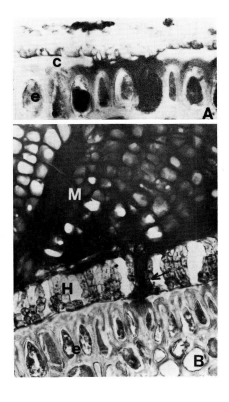

show wall ingrowths, which again resemble those of transfer cells (GUNNING and PATE 1969, PATE and GUNNING 1969). HARLIN (1971 b, 1973 a) did show, however, that *Smithora* can be grown under field conditions on an artificial substrate, although it does not grow to maturity.

7.4 Ion Exchange and Nutrient Transfer Between Basiphyte and Epiphyte

7.4.1 Experimental Translocation of Radioactive Compounds

7.4.1.1 Algae as Epiphytes and Basiphytes

LINSKENS (1963 b) was the first to investigate experimentally whether epiphytes derive substances from their hosts. Using mainly *Codium dichotomum* as a basiphyte and diverse epiphytic algae, he submerged these in solutions containing ^{32}P phosphate. Both the basiphyte and the epiphytes were submerged in the different experiments. LINSKENS (1963 b) demonstrated that transport is in an acropetal direction (base → apex), even in meristem-free thallus segments. But it is generally assumed that epiphytic plants do not have a "taking and giving" relationship when compared with relationships among parasitic plants. Linskens showed, however, that the epiphytes take up substances from the basiphytes. As these are autotrophic plants, they should be regarded as semiparasitic. In the reciprocal experiment, transport from epiphyte to basiphyte – also occurred although at a lesser rate but differed quantitatively according to the type of epiphyte. He demonstrated that the amount of translocation varied in different species of algae. Linskens (1968) also suggested that there need not be direct contact by rhizoids between the host and epiphyte.

Harlin (1973b) confirmed Linskens' (1968) prediction by observing the translocation of ^{32}P as phosphate and $^{14}CO_2$ in experiments with diverse red algae and their epiphytes. She found that the transfer of these isotopes does not depend on the penetrating rhizoids of the epiphyte; it is the mere proximity of the two partners which makes the exchange possible. The epiphytes are seemingly capable of utilizing useful exudates produced by the basiphytes before dilution by the bathing seawater.

Interesting observations have been made recently by Court (1980) on the relationship of *Janczewskia gardneri* and *Laurencia spectabilis* which has hitherto been quoted as the classical example of algal parasitism. Ultrastructural studies established that the "parasite" had functioning typical red algal chloroplasts, that both host and "parasite" incorporated $^{14}CO_2$ into similar compounds such as sugars and amino acids, and that there was no exchange of products between the two partners of the association. He concluded that the mature *Janczewskia gardneri* is not a parasite but an obligate epiphyte.

7.4.1.2 The Special Case of *Ascophyllum nodosum* and *Polysiphonia lanosa*

This association has been investigated by different workers using diverse techniques. Citharel (1972a, b) discussed the relationship of the brown alga *Ascophyllum* with its epiphyte *Polysiphonia lanosa* which is anchored by a primary rhizoid in the cortical cells of the host. Using ^{14}C-glutamic acid, he established that *P. lanosa* used the organic substances of the host and should be regarded as a true epiphyte sensu Fritsch (1948). He found that certain cells of *Ascophyllum* are broken down when they come in contact with the rhizoids of *P. lanosa*, so he regarded this species as a true parasite.

The distribution of photosynthate in *Ascophyllum* and its relationship to the epiphytic *P. lanosa* have also been investigated by Harlin and Craigie (1975, 1981). They found that the basiphyte is not a major source of organic carbon for this epiphyte. The maximum amount of radioactive carbon compounds lost from the host during the experiments, is less than 0.3% of the total ^{14}C fixed by the alga, and only 5% could have moved through the frond. The remaining fraction of the ^{14}C lost from the thallus was released into the bathing seawater. They found that *P. lanosa* is perfectly capable of fixing its own carbon. They made an important point, however, namely, that certain growth factors are needed for spore germination or cell division in the early stages of the life history of the epiphyte. They also suggested that certain cortical surface characteristics of the host are necessary for the settlement of the *Polysiphonia* spores.

The association of these two marine algae was also investigated using radio-isotopes by PENOT (1974), PENOT and PENOT (1981), TOOTLE and HARLIN (1972) and TURNER and EVANS (1977). All of these authors found some transfer of material from basiphyte to epiphyte, but all were doubtful about the active translocation of ^{14}C-labelled products of photoassimilation from *Ascophyllum* to *Polysiphonia*. It is known that the aerial parts of *P. lanosa* are able to utilize exogenously supplied ^{14}C-glucose (TURNER and EVANS 1977).

RAWLENCE and TAYLOR (1970) investigated the complex rhizoid formation of *P. lanosa*, which forms both a primary sporeling rhizoid and later secondary rhizoids. Some rhizoids showed a protoplasmic protrusion which extended through the rhizoid wall near the apex, but there was no evidence that this formed any haustorial connection with the *A. nodosum* cells. Fragments of *P. lanosa*, when cultured in close proximity to *A. nodosum* tissue, did not penetrate the basiphyte tissue.

In an ultrastructural study of the same relationship, RAWLENCE (1972) found that the rhizoids of *P. lanosa* digest their way into the host tissue. The endoplasmic reticulum and vacuole membranes of the host cells broke down on first contact with the epiphyte, even before the rhizoids actually touched the cell. When the digestion was complete, *A. nodosum* material was incorporated into the wall of the *Polysiphonia* rhizoid. A complex rhizoidal system later incorporated more host cells which were digested in contact with the protoplast of the intrusive rhizoid. It is believed that the rhizoids produce digestive enzymes and that the system is analogous to that of parasitic fungi in angiosperm tissue.

The anti-fouling mechanism in *Ascophyllum nodosum* is discussed in Section 7.5.

7.4.1.3 Algae Epiphytic on Seagrasses

HARLIN (1971 a) recorded that *Smithora naiadum* epiphytic on two seagrasses shows an active movement of ^{32}P and products of ^{14}CO$_2$ fixation. This exchange is two-directional. Part of the transfer occurs through the seagrass–algal interface, and part by leakage from the basiphyte into the bathing seawater and subsequent uptake by the epiphyte. One of the translocated materials is sucrose.

McRoy and GOERING (1974) also found that there is a direct transfer of carbon and nitrogen from *Zostera* to the epiphytes. Using ^{14}CO$_2$ and ^{15}NO$_3^-$, they established that the substances taken up by the roots of the *Zostera* plants could finally be detected in the epiphytes on the leaves. This work suggests that a transfer of nitrogen occurs from the mud substrate of *Zostera* to the epiphytes.

7.4.1.4 Effect of Epiphytes on Eelgrass Photosynthesis

The effect of diverse epiphytic diatoms on the photosynthesis of *Zostera marina* has been measured at varying light intensities, using a ^{14}C technique (SAND-JENSEN 1977). The crust of epiphytes may be several layers thick, and it reduces the rate of photosynthesis of the leaves by acting as a barrier to carbon dioxide uptake and by reducing the light intensity in the leaf. The epiphytes reduced the *Zostera* photosynthesis by up to 31% at optimum light conditions and ambient HCO_3^- concentrations. It was suggested that one of the effects of pollution producing high nutrient levels in the waters surrounding the seagrass will be a thick crust of epiphytes, which will contribute to the consequent thinning and final disappearance of the seagrass from potential habitats. It appears that aquatic macrophytes can reduce the effect of epiphytic cover by either forming new photosynthetic tissues, or by checking algal growth by excreting antibiotics.

7.4.2 Implications of Seasonality of Epiphytes

In southern Australia we have noted over a number of years that there is a seasonal change of epiphytic algae on the seagrass *Amphibolis* (DUCKER et al. 1977); the non-host-specific algae cover old leaves particularly in the late summer (February–March) period. These leaves will then be shed, the young leaves appearing vigorously in March and April. At the same time the new growth is covered by the sporelings of the specific epiphytes. We have observed furthermore that a higher percentage of spores of epiphytic Corallinaceae will germinate on glass slides in the autumn months; this has been confirmed by W. WOELKERLING (personal communication). BRAUNER (1975), working with the epiphytes on *Zostera marina* at Beaufort, North Carolina, found a strong seasonality of the different seaweeds due to the wide temperature and salinity ranges of the environment. Only 10% of the 80 epiphytes, species of *Fosliella, Heteroderma, Dermatolithon* and *Myrionema, Enteromorpha,* are present all the year round, and certain groups flourish at different times of the year.

7.5 Defence of Basiphytes Against Epiphytes

The colonization by epiphytes of seagrasses and algae occurs at two different stages: either when the basiphyte is young and actively growing, or when the basiphyte is old and moribund or the tissue is injured.

We have stated earlier that colonization of young plants or young parts of plants is seemingly carried out only by specific epiphytes. In contrast, nonspecific epiphytes are able to invade, part or all of old basiphytes. These observations have not been experimentally confirmed. The only evidence comes from work by RUSSELL (1982), who has shown that spores settle and develop more easily on parts of the thalli which have scars or other uneven sites in some Laminariaceae and Fucaceae.

One may speculate that certain metabolites favour the settling of spores of the specific algae, while the spores of the non-specific forms are unable to penetrate a defensive metabolic barrier. FOGG (1981) has shown that glycollate is liberated from the cells of aquatic plants. The extent to which this occurs varies greatly according to the physiological condition of the plant or tissue, but is conceivable that a certain threshold concentration of this, or a derivative, in the microsphere of the basiphyte, favours the settling of specific epiphytes while inhibiting the settling of non-specific epiphytes. In support of this conclusion, ENDE and LINSKENS (1962) record that young sporelings of *Himanthalia* are free of epiphytes, while detached and old walls are heavily colonized. Thus the exchange of substances, either promotory or inhibitory between epiphyte and basiphyte, may take place within the microsphere of the basiphyte, without any host penetration (see LINSKENS 1968, and Fig. 4, box 5).

7.5.1 Self-Cleaning Process of Actively Growing Basiphytes

Moss (1981 b) has investigated by SEM and TEM the control of epiphytes on the brown alga *Halydrys siliquosa*. The rich flora of bacteria, microalgae and juvenile macroalgae scattered over the entire thallus of the basiphyte is held in check by a mechanism for shedding these epiphytes by loss of a surface layer or "skin". Ultrastructural studies showed that the skin consists of the outermost layers of the cell walls of the meristoderm of the brown algal basiphyte. This skin accumulated in laboratory cultures, but is continually sloughed off in the sea, thus exposing a new clean thallus surface. Moss (1982) points out that this process of self-cleaning is only possible as long as the meristoderm of the brown alga remains active, and a continued production of vesicular material between microfibrils pushes older layers of the meristoderm cell wall to the outside. The vesicular material disintegrates in the seawater, thus liberating the outermost layers of cell wall, together with their attached epiphytes. Similarly, in the fucoids *Himanthalia elongata* and *Ascophyllum nodosum,* the continuous shedding of the outermost layers of the meristoderm cell walls acts as a self-cleaning process in the actively growing parts of the algae. Moss (1982) does not agree with FILION-MYKLEBUST and NORTON (1981), who claim that whole cells are shed as a cleaning process by members of the Fucales.

A self-cleaning process seems to be active also in the green alga *Enteromorpha intestinalis* and may account for its ecological success. This alga can shed outer layers of its cell wall together with adhering epiphytes, because the inner layers of the cell wall are continuously replaced (MCARTHUR and MOSS 1977). SIEBURTH and TOOTLE (1981) show that the heavily fouled epidermis of *Chondrus crispus* is exposed to the underlying surface by the sloughing off of the outer layer carrying the epiphytes. Earlier, SIEBURTH and THOMAS (1973) showed that *Zostera marina* is lightly colonized by diatoms when the leaves are young, but carries a heavy crust of epiphytic diatoms in early autumn, which is then lost in late autumn.

7.5.2 Production of Antibiotics by Basiphytes

SIEBURTH and TOOTLE (1981) discussed microbial fouling of brown and red algae. They claim it shows a seasonal variation, correlated with water temperatures and/or algal antibiotics. The production of antibiotics by marine algae (*Laminaria* and *Himanthalia*) might be responsible for their successful protection against epiphytes according to AL-OGILY and KNIGHT-JONES (1977). MOSS (1982) suggests that the phenols produced by several members of the Fucales reported by CRAIGIE and MCLACHLAN (1964) might prevent the settling of epiphytes. The phenols might act as antibiotics.

7.6 Conclusions

Plants have a complex relationship with each other in any environment, each plant exploiting every opportunity to propagate its own kind. These relationships include epiphytic interactions which are sometimes looked upon as neutral relationships (LINSKENS 1968). But as LINSKENS also points out "living cells are open systems which maintain their balance by continuous exchange of constituents". The epiphytic systems which we have reviewed fall into two distinct classes, those which display no specificity and are opportunistic epiphytes, (in the sense of HOOKER 1854 and FRITSCH 1948), and those which potentially display a functional interaction of the epiphyte and basiphyte.

The host-specific epiphytes, in their mutual interactions with the basiphyte, appear to operate at different cellular levels (Fig. 4). Attachment to the host usually involves a specific secreted adhesive rather than relying entirely on physical processes. This is followed by germination and growth which might lead to host penetration of varying degrees and finally to exchange of substances. Alternatively, the exchange can occur by functional interaction in the microsphere without penetration of the basiphyte. The possibility that these mutual interactions lead to transport of substances from the basiphyte to epiphyte, and vice versa, is suggested by the few physiological studies so far carried out. The crossings of the barriers 1 to 5 shown in Fig. 4 appear to be crucial to the success of the mutual interactions in these specific epiphytes. All healthy vigorous plants produce defence or promoting substances which may inhibit or favour the interaction at any of the contact points in the epiphyte–basiphyte relation. Coordination of these interlocking processes is necessary in both partners.

The opportunistic epiphytes settle and attach themselves usually by physical means (Fig. 4, box 1 a). Attachment is often facilitated by the maturity or ageing processes of the basiphyte, and by the absence of defence substances in the basiphyte. These neutral epiphytes appear to have an uncoordinated relationship with the basiphyte. When compared with the specific epiphytes, it is apparent that both attach to the host, but that the interaction is usually terminated with the settling or even germination of the epiphyte (Fig. 4, box 2 a).

How does a basiphyte prevent growth of an undesirable epiphyte? This could be readily achieved by defence substances (e.g. antibiotics), preventing in the first instance settlement or attachment (barrier 1) or later, germination (barrier 2). There is at present no information concerning the first possibility,

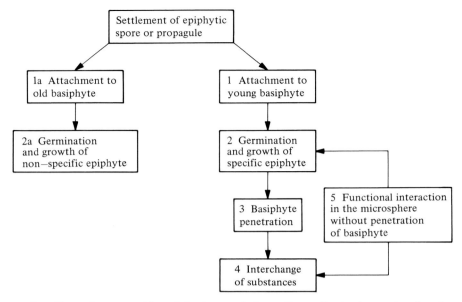

Fig. 4. Specific and non-specific epiphytic associations. Generalized scheme showing the events of the cellular interactions between *specific algal epiphytes* and their *basiphytes* based on the limited evidence reviewed in the text. Barriers where the interaction may be terminated are indicated by numbers *1* to *5* (see text). At present, knowledge is fragmentary, and this scheme may need revision as new information becomes available

but only circumstantial evidence arising from the absence of settled spores or other propagules and the monitoring of the production of antibiotics. The second possibility is supported by evidence from different brown and green algae, where germinated spores and small epiphytes are sloughed off through a self-cleaning process of the outer cellular layer.

An interesting observation, relevant to basiphyte–epiphyte interactions, is that all opportunistic or non-specific epiphytes are only able to colonize mature basiphytes or the older parts of basiphytes. These epiphytes do not colonize young, active basiphytes. This suggests that young, developing organisms have active defence mechanisms operating to prevent settling of epiphytes (ENDE and LINSKENS 1974). It is a striking observation that the specific epiphytes are the only ones able to colonize young shoots. This indicates that the specific epiphytes are able to interlock with the cell recognition mechanisms of the basiphyte. Perhaps this is the expression of a gene-for-gene relationship of the kind considered by LINSKENS (1963b, 1968). The possibility exists that the defence mechanism involves a chemical recognition process of the key-in-lock type, analogous with mammalian blood group substances or the antibodies of the immune system (see review by CLARKE and KNOX 1978). It is evident that the specific epiphytes have been able to key-in to the recognition process, while the non-specific epiphytes have not been able to form such an interaction.

Acknowledgements. Partial support by the Australian Research Grants Committee is acknowledged. We thank Drs. Margaret Clayton, Gerry Kraft, Betty Moss and Bill Woelkerling for helpful discussion and unpublished data. We thank Mr. Chris O'Brien for the use of his unpublished photograph and Drs. Neill Hallam and Margaret Clayton for kind permission to reproduce Fig. 2.

References

Allender BM, Kraft GT (1983) The marine algae of Lord Howe Island (New South Wales). I. The Dictyotales and Cutleriales (Phaeophyta). Brunonia 6:73–130

Al-Ogily SM, Knight-Jones EW (1977) Antifouling role of antibiotics produced by marine algae and protozoans. Nature 265:728–782

Bary HA de (1879) Die Erscheinung der Symbiose. Trübner, Strassburg

Boney AD (1975) Mucilage sheaths of spores of red algae J Mar Biol Ass UK 55:511–518

Boney AD (1981) Mucilage: the ubiquitous algal attribute. Br Phycol J 16:115–132

Brauner JF (1975) Seasonality of epiphytic algae on *Zostera marina* at Beaufort, North Carolina. Nova Hedwigia 26:123–133

Burkholder PR (1973) The ecology of marine antibiotics and coral reefs. In: Jones OA, Endean R (eds) Biology and Geology of Coral Reefs Vol II. Academic Press, London New York, pp 117–182

Callow ME, Evans LV (1974) Studies on the ship-fouling alga *Enteromorpha*. III. Cytochemistry and autoradiography of adhesive production. Protoplasma 80:15–27

Chamberlain AHL, Evans LV (1973) Aspects of spore production in the red alga *Ceramium*. Protoplasma 76:139–159

Chamberlain YM (1978) Taxonomic relationships amongst epiphytic, crustose Corallinaceae. In: Irvine DEG, Price JH (eds) Modern approaches to the taxonomy of red and brown algae. Academic Press, London New York, pp 223–246

Citharel J (1972a) *Polysiphonia lanosa* (L.) Tandy, est-il un simple epiphyte? Comp Rend D 274:1904–1906

Citharel J (1972b) Contribution à l'étude du métabolism azoté des algues marines. Utilisation metabolique d'acide glutamique – ^{14}C par *Ascophyllum nodosum* (Linné) Le Jolis et *Polysiphonia lanosa* (Linné) Tandy. Bot Mar 15:157–161

Clarke AE, Knox RB (1978) Cell recognition in flowering plants. Quart Rev Biol 53:3–28

Clayton MN (1974) Studies on the development, life history and taxonomy of the Ectocarpales (Phaeophyta) in southern Australia. Aust J Bot 22:743–813

Court GJ (1980) Photosynthesis and translocation studies of *Laurencia spectabilis* and its symbiont *Janczewskia gardneri* (Rhodophyceae). J Phycol 16:270–279

Craigie J, McLachlan J (1964) Excretion of coloured ultraviolet-absorbing substances by marine algae. Can J Bot 42:23–33

Dixon PS (1965) Perennation, vegetative propagation and algal life histories, with special reference to *Asparagopsis* and other Rhodophyta. Bot Gothob 3:67–74

Dixon PS (1973) Biology of the Rhodophyta. Oliver Boyd, Edinburgh

Doohan ME, Newcomb EH (1976) Leaf ultrastructure and δ ^{13}C values of three seagrasses from the Great Barrier Reef. Aust J Plant Physiol 3:9–23

Ducker SC (1979) Australian articulated coralline algae: the genus *Metagoniolithon* Weber-van Bosse (Corallinaceae, Rhodophyta). Aust J Bot 27:67–101

Ducker SC, Knox RB (1978) Alloparasitism between a seagrass and algae. Naturwissenschaften 65:391–392

Ducker SC, Le Blanc JD, Johansen HW (1976) An epiphytic species of *Jania* (Corallinaceae:Rhodophyta) endemic to southern Australia. Contrib Herb Aust 17:1–8

Ducker SC, Foord NJ, Knox RB (1977) Biology of Australian seagrasses: the genus *Amphibolis* C. Agardh (Cymodoceaceae). Aust J Bot 25:67–95

Duerden RC, Jones WE (1981) The host specificity of *Jania rubens* (L.) Lamour. in British waters. In: Fogg GE, Jones WE (eds) Proc 8th Int Seaweed symp. Mar Sci Lab Menai Bridge, pp 313–319

Duerden RC, Jones WE (1974) The host specificity of *Jania rubens* (L.) Lamour in British waters. 8th Int Seaweed Symp Bangor pA 36

Ende G van den, Linskens HF (1962) Beobachtungen über den Epiphytenbewuchs von *Himanthalia elongata* (L.) S.F. Gray. Biol Zbl 81:173–181

Ende G van den, Linskens HF (1974) Cutinolytic enzymes in relation to pathogenesis. Ann Rev Phytopathol 12:247–258

Evans LV, Callow JA, Callow ME (1978) Parasitic red algae: an appraisal. In: Irvine DEG, Price JH (eds) Modern approaches to taxonomy of red and brown algae. Systematics Association Vol 10. Academic Press, London New York, pp 87–109

Filion-Myklebust C, Norton TA (1981) Epidermis shedding in the brown seaweed *Ascophyllum nodosum* (L.) Le Jolis and its ecological significance. Mar Biol Lett 2:45–51

Fogg GE (1981) The ecology of an extracellular metabolite of seaweeds. In: Fogg GE, Jones WE (eds) Proc 8th Int Seaweed Symp. Mar Sci Lab Menai Bridge, pp 46–53

Fritsch FE (1948) The structure and reproduction of the algae. Cambridge Univ Press, Cambridge London New York Melbourne

Funk G (1927) Die Algenvegetation des Golfes von Neapel. Pubbl Staz Zool Napoli 7:(Suppl)

Funk G (1955) Beiträge zur Kenntnis der Meeresalgen von Neapel. Pubbl Staz Zool Napoli 25:(Suppl)

Goebel K (1889) Epiphyten. Pflanzenbiol Schilderung Elwert, Marburg, 1:147–329

Gordon EM (1972) Comparative morphology and taxonomy of the Wrangelieae, Sphondylothamnieae, and Spermothamnieae (Ceramiaceae, Rhodophyta). Aust J Bot (Suppl) 4:1–180

Gunning BES, Pate JS (1969) "Transfer cells" – plant cells with wall ingrowths, specialized in relation to short-distance transport of solutes – their occurrence, structure and development. Protoplasma 68:107–133

Gunning BES, Steer M (1975) Ultrastructure and the biology of plant cells. Arnold, London

Hallam ND, Clayton MN, Parish D (1980) Studies on the association between *Notheia anomala* and *Hormosira banksii* (Phaeophyta). Aust J Bot 28:239–248

Harlin MM (1971a) Translocation between marine hosts and their epiphytic algae. Plant Physiol 47 (suppl):41

Harlin MM (1971b) An obligate marine algal epiphyte: can it grow on a synthetic host? J Phycol 7:4s

Harlin MM (1973a) "Obligate" algal epiphyte: *Smithora naiadum* grows on a synthetic substrate. J Phycol 9:230–232

Harlin MM (1973b) Transfer of products between epiphytic marine algae and host plants. J Phycol 9:243–248

Harlin MM (1975) Epiphyte–host relations in seagrass communities. Aquat Bot 1:125–131

Harlin MM (1980) Seagrass epiphytes. In: Phillips RC, McRoy CP (eds) Handbook of seagrass biology. Garland, New York, pp 117–131

Harlin MM, Craigie JS (1975) The distribution of photosynthate in *Ascophyllum nodosum*. J Phycol 11:109–113

Harlin MM, Craigie JS (1981) The export of organic carbon from *Ascophyllum nodosum* as it relates to epiphytic *Polysiphonia lanosa*. In: Fogg GE, Jones WE (eds) Proc 8th Int Seaweed Symp. Mar Sci Lab Menai Bridge, p 193

Harvey WH (1855) Some account of the marine botany of the colony of Western Australia. Trans R Ir Acad 22:525–566

Hoek C van den, Flinterman A (1968) The life-history of *Sphacelaria furcigera* Kütz. (Phaeophyceae). Blumea 16:193–242

Hooker JD (1854) Himalayan Journals. Murray, London

Huisman JH (1980) Taxonomic and culture studies on the Ceramiaceae (Rhodophyta) of Portsea Jetty, Port Phillip Bay. BSc (Hons) Thesis, Univ Melbourne

Jaag O (1945) Epiphytismus, Parasitismus und Symbiose bei Pflanzen. Pathol Microbiol 8:463–485

Jones PL, Woelkerling WJ (1984) An analysis of trichocyte and spore germination attributes as taxonomic characters in the *Pneophyllum-Fosliella* complex (Corallinaceae, Rhodophyta). Phycologia 23 (2) in press

Jones WE, Duerden RC (1972) Variation in the morphology of the primary rhizoid in *Ceramium rubrum* (Huds.) C. Ag. Br Phycol J 7:281

Jones WE, Moorjani SA (1973) The attachment and early development of the tetraspores of some coralline red algae. Spl Publ Mar Biol Ass India 1973:293–204

Kraft GT (1977) Studies of marine algae in the lesser-known families of the Gigartinales (Rhodophyta). II. The Dicranemaceae. Aust J Bot 25:219–267

Linnaeus C (1753) Species plantarum ... Facsimile Edition 1959. Ray Soc, London

Linskens HF (1963a) Oberflächenspannung an marinen Algen. Proc K Med Akad Wet C 66:205–217

Linskens H (1963b) Beitrag zur Frage der Beziehungen zwischen Epiphyt und Basiphyt bei marinen Algen. Pubbl Staz Zool Napoli 33:274–293

Linskens H (1968) Host–pathogen interaction as a special case of interrelations between organisms. Neth J Plant Pathol 74 (suppl) 1:1–8

Linskens HF (1976) Specific interactions in higher plants. In: Wood RKS, Graniti A (eds) Specificity in plant diseases. Plenum, New York London, pp 311–325

McArthur DM, Moss BL (1977) The ultrastructure of cell walls in *Enteromorpha intestinalis* (L.) Link Br Phycol J 12:359–368

McRoy CP, Goering JJ (1974) Nutrient transfer between the seagrass *Zostera marina* and its epiphytes. Nature 248:173–174

Moorjani S, Jones WE (1972) Spore attachment and development in some coralline algae. Br Phycol J 7:282 (abstr)

Moss BL (1981a) Attaching mechanisms of zygotes and embryos of the Fucales. In: Fogg GE, Jones WE (eds) Proc 8th Int Seaweed Symp. Mar Sci Lab Menai Bridge, pp 117–124

Moss BL (1981b) The control of epiphytes by *Halidrys siliquosa* (L.) Lyngb. 13 Int Bot Congr (abstr) p 170

Moss BL (1982) The control of epiphytes by *Halidrys siliquosa* (L.) Lyngb. (Phaeophyta, Cystoseiraceae) (Note) Phycologia 21:185–188

Ngan Y, Price IR (1979) Systematic significance of spore size in the Florideophyceae (Rhodophyta). Br Phycol J 14:285–303

Oltmanns F (1922–1923) Morphologie und Biologie der Algen. Vol 1–3, Fischer, Jena

Parish D (1976) Aspects of the *Notheia anomala, Hormosira banksii* association. Unpub Hons Report, Bot Dept, Monash University, Clayton Victoria

Parsons MJ (1975) Morphology and taxonomy of the Dasyaceae and the Lophothalieae (Rhodomelaceae) of the Rhodophyta. Aust J Bot 23:549–713

Pate JS, Gunning BES (1969) Vascular transfer cells in angiosperm leaves: a taxonomic and morphological survey. Protoplasma 68:135–156

Pearse AG (1972) Histochemistry, theoretical and applied, 2nd edn. Churchill, London

Penot M (1974) Ionic exchange between tissues of *Ascophyllum nodosum* (L.) Le Jolis and *Polysiphonia lanosa* (L.) Tandy. Z Pflanzenphysiol 73:125–131

Penot M, Penot M (1981) Ion transport and exchange between *Ascophyllum nodosum* (L.) Le Jolis and some epiphytes. In: Fogg GE, Jones WE (eds) Proc 8th Int Seaweed Symp. Mar Sci Lab Menai Bridge, pp 217–223

Rawlence DJ (1972) An ultrastructural study of the relationship between rhizoids of *P. lanosa* (L.) Tandy (Rhodophyceae) and tissues of *Ascophyllum nodosum* (L.) Le Jolis (Phaeophyceae). Phycologica 11:279–290

Rawlence DJ, Taylor ARA (1970) The rhizoids of *Polysiphonia lanosa*. Can J Bot 48:607–611

Reedman DJ, Womersley HBS (1976) Southern Australian species of *Champia* and *Chylocladia* (Rhodymeniales: Rhodophyta). Trans R Soc S Aust 100:75–104

Rhys Williams P (1965) Observations on two epiphytic members of the Phaeophyceae. Br Phycol Bull 2:390–391

Russell G (1982) Epiphytism: some spatial and developmental relationships with hosts. Scientific Programme and Abstracts, First Internat Phycol Congr St John's, p 42

Sand-Jensen K (1977) Effect of epiphytes on eelgrass photosynthesis. Aqua Bot 3:55–63

Schaede R, Meyer FH (1962) Die pflanzlichen Symbiosen, 3rd edn. Fischer, Stuttgart

Sieburth JM, Thomas CD (1973) Fouling on eelgrass (*Zostera marina* L.). J Phycol 9:46–50

Sieburth JM, Tootle JL (1981) Seasonality of microbial fouling on *Ascophyllum nodosum* (L.) Lejol., *Fucus vesiculosus* L., *Polysiphonia lanosa* (L.) Tandy and *Chondrus crispus* Stackh. J Phycol 17:57–64

Tootle JL, Harlin MM (1972) Investigation of biochemical interaction between *Ascophyllum nodosum* and its obligate epiphyte, *Polysiphonia lanosa*. J Phycol 8:11s

Turner CHC, Evans LV (1977) Physiological studies on the relationship between *Ascophyllum nodosum* and *Polysiphonia lanosa*. New Phytol 79:363–371

Woelkerling WJ, Spencer KG, West JA (1983) Studies on selected Corallinaceae (Rhodophyta) and other algae in a defined marine culture medium. J Exp Mar Biol Ecol 67:61–77

Wollaston EM (1967) Morphology and taxonomy of southern Australian genera of Crouanieae Schmitz (Ceramiaceae, Rhodophyta). Aust J Bot 16:217–417

Womersley HBS (1965) The morphology and relationships of *Sonderella* (Rhodophyta, Rhodomelaceae). Aust J Bot 13:435–450

Womersley HBS (1978) Southern Australian species of *Ceramium* Roth (Rhodophyta). Aust J Mar Freshwat Res 29:205–257

Womersley HBS (1979) Southern Australian species of *Polysiphonia* Greville. (Rhodophyta). Aust J Bot 27:459–528

Womersley HBS (1981) Biogeography of Australian marine macroalgae. In: Clayton MN, King RJ (eds) Marine botany: an Australian perspective. Longman, Cheshire Melbourne, pp 292–307

8 Genetics of Recognition Systems in Host-Parasite Interactions

P. R. DAY

8.1 Introduction

The formal genetics that govern the recognition between different races of parasites and different varieties of hosts began with R.H. BIFFEN's discovery that resistance to yellow rust in wheat is simply inherited (BIFFEN 1905, 1912). Parasites have been explored somewhat less than hosts, first because breeding crop plants for disease resistance has generated much of our information on hosts and, secondly, because many parasites are largely unknown genetically, especially those in which technical difficulties such as obligate parasitism or complexity of life cycles interfere with genetical analysis. In writing this chapter I have deliberately narrowed the choice of examples to demonstrate principles and opportunities. The next major advances in understanding host–parasite interactions will come from molecular genetics. I hope this chapter will indicate the range of opportunities.

I am principally concerned here with plant hosts and fungal parasites with some mention of bacteria. The parasites described are also pathogens in that they produce disease and damage or even kill their hosts. I use the terms virulent, or avirulent, to describe the ability, or lack of ability, of a given race or strain of a parasite to cause disease on a plant known to carry genes for resistance to that parasite species. Thus a race of the rust *Puccinia striiformis* is virulent on one wheat cultivar, avirulent on another and non-pathogenic on tomato. The tomato is not usually regarded as a host for *P. striiformis,* and although it may have genes for resistance to this rust the adjective "avirulent" is inappropriate for the rust until these have been defined. The question of non-host resistance will be treated later.

Interaction is normally observed when host and parasite are brought into contact by inoculation. However, these interactions may be examined at several different levels which range from molecules, through organelles, cells, organs, whole organisms to populations that may be as large as those on a continent. Most of our detailed biochemical knowledge has emerged from dissecting the events at or below the level of the whole organism. Genetic studies of fungal parasites and their hosts are usually conducted in greenhouses or growth chambers where space limits the population sizes. However, the measurement of virulence gene frequencies in pathogen populations may involve sampling over very large areas.

8.2 Gene-for-Gene Hypothesis

As genetic studies of virulence in parasites began to parallel studies in host plants, it was inevitable that the two genetic systems should be compared. HAROLD FLOR, working in Fargo, North Dakota, in the 1940's was the first to do this. Using pot-grown cultures of flax and its rust (*Melampsora lini*), he confirmed that avirulence in the rust and resistance in the host were generally determined by single dominant genes. From these studies FLOR concluded that a resistance gene was only effective when it encountered a rust culture with a corresponding avirulence allele. Since resistance was epistatic to susceptibility no matter what alleles for virulence the rust culture carried at other loci, it was avirulent on every flax variety with that particular resistance gene. FLOR's discovery, called the gene-for-gene hypothesis (FLOR 1942), has since been shown to apply to a wide range of host–parasite interactions (SIDHU 1975), not all of which show the clear-cut distinctions between phenotypic classes observed by FLOR. Indeed ELLINGBOE (1981) has argued convincingly that in principle even resistance that is governed by many genes of individually small effect will, on careful analysis, be shown to be matched by a comparable control of "virulence" in the pathogen. The biological context of the gene-for-gene concept assumes that a basic compatibility between host and parasite has evolved upon which the interactions controlled by genes for avirulence in the parasite and resistance in the host are superimposed (DAY 1974, ELLINGBOE 1976). The interaction disrupts this compatibility, causing resistance or incompatibility. HEATH (1981 b) has stressed the importance of distinguishing between the adaptations on the part of host and parasite that are responsible for their basic compatibility and the mechanisms that govern cultivar resistance and race specificity. As she points out, aspects of the former could easily confuse analysis of the latter. The gene-for-gene hypothesis suggests that the allele for avirulence in the parasite and the corresponding allele for resistance in the host play a key role in recognition, since they determine that the interaction will be incompatible. The interaction can only be restored to compatibility if the avirulence allele in the parasite is changed to virulence, or if the allele for resistance in the host is changed to susceptibility. Clearly recognition will involve the interaction of the products of a resistance gene and an avirulence gene to determine incompatibility.

To describe host susceptibility as "compatibility" may sound like a contradiction in terms since the end result is often massive destruction and death of host tissue. It is rather a term that describes the early stages of parasite growth when its hyphae freely invade the host tissue with little or no response from the host and little or no macroscopic damage to its tissues. Resistance or "incompatibility" is accompanied by rapid changes in host metabolism that often result in localized host tissue necrosis and the arrest of invading hyphae. The precise point of arrest will vary with the system studied. Resistance may prevent host penetration, the invading pathogen producing little more than a short germ tube, or it may allow appreciable pathogen development and sporulation, merely reducing the extent of invasion compared with a compatible interaction.

FLOR's hypothesis had several consequences. In race taxonomy it eventually became clear that the plant pathologist and plant breeder needed to know the frequencies of those virulence genes in the pathogen population that threatened the resistance genes deployed in their crops. Long, complicated race names became unnecessary since the complete spectrum of virulence that could be detected in each pathogen sample was usually irrelevant. Races are now most often referred to by formulae that reflect their virulence phenotypes (DAY 1976a). The methods of data collection and analysis used to study variation in populations of plant pathogens need to be chosen and applied with care to avoid drawing incorrect conclusions about the best strategies for deploying resistance (WOLFE and KNOTT 1982). With the advent of growing mixtures of cultivars carrying different resistance genes to exploit the damping effect on the build-up of epidemics (WOLFE and BARRETT 1980) virulence gene frequencies have become very important. The composition of the mixtures must be changed at intervals determined by survey to prevent the selection of races that are virulent on more than one of the components.

For the last 40 years the gene-for-gene hypothesis has heightened expectation that the investigation of such apparently simple control systems should soon reveal the mechanisms responsible for host–parasite recognition and specificity. In fact progress in this direction has been disappointing, principally because the mechanism of specificity determination is a good deal more complex than the genetic controls imply.

8.3 Genetic Fine Structure

The most complete example of genetic fine structure analysis comes from work in Adelaide by LAWRENCE et al. (1981a). These workers recognized the importance and potential of the flax–flax rust system and, beginning in the 1960's, undertook a number of experiments designed to continue and extend the exploration of the genetic controls begun earlier by FLOR. The rather detailed account which follows highlights some technical problems and how they were overcome.

8.3.1 Rust Resistance in Flax

Some 29 different dominant genes are known in flax which confer resistance to rust. These are in four clusters of closely linked or allelic genes that include 13 in group L, seven in M, three in N and five in P. The remaining locus K is represented by only one gene. A study of four of the alleles in group M (M, $M4$, $M3$ and $M1$) revealed rare recombinants among nearly 37,000 segregants from 51 families (MAYO and SHEPHERD 1980). These were selected by crossing flax lines homozygous for the genes to be compared. The $F1$'s were then test-crossed to a universal susceptible flax variety and the test cross-progenies were screened by sequentially inoculating them with two rust races, each virulent to one but avirulent to the other *M* parent. Two classes of rare recombinants were recovered, either susceptible to both races, or resistant to both races. The latter class were test-crossed to a line homozygous for a third *M* gene to demonstrate not only the reappearance of the two original genes due to rare recombination, but also further recombination involving the added locus. The availability of races that discriminate the various classes of segregants limits the extent to which such analyses are possible. The phenotypes of the pairs of genes obtained in coupling (*MM*3, *MM*4 and *M*4*M*3) were identical to the respective genotypes in repulsion. The authors therefore concluded that these four genes represent separate, closely linked loci that function independently of each other.

An earlier report of similar tests of genes of group *L* had revealed only doubly recessive recombinants susceptible to both tester races (SHEPHERD and MAYO 1972). In one example from the cross *L*2 × *L*10, six such recombinants were selfed and one of them generated the parental gene *L*10. Evidently *L*2 and *L*10 in coupling have a null phenotype similar to the doubly recessive form. Hence it is likely that *L*2 and *L*10 are different alleles of the same functional gene and that these two specificities cannot be accommodated functionally in a single gene product.

8.3.2 The Rust *Melampsora lini*

8.3.2.1 Sexuality

Genetic studies with the rust are somewhat more complicated than with the host. Although some rusts, including *M. lini,* can be cultured on defined media, these techniques have not yet been extended to genetic studies. A further complication of some rusts is that they produce different stages of their life cycles on different host plants. Flax rust has the advantage of producing all stages on flax.

Rust cultures are normally maintained by transferring dikaryotic uredospores to leaves of fresh host plants. Long-term storage is by uredospores held in liquid nitrogen. Dikaryotic teliospores are produced on the dead stems of the host. When the teliospores germinate, their paired nuclei are fused. The diploid nucleus undergoes meiosis and haploid basidiospores are formed. These are air-dispersed and infect leaves, producing flask-shaped pycnia. The pycnia produce a sugary fluid (nectar) containing uninucleate pycniospores interspersed with occasional hair-like cells that grow out through the mouth of the pycnium. Fertilization occurs when the nectar of one pycnium encounters the hairs of another of different mating type. The product is a dikaryon which forms a pustule of aeciospores close to the fertilized pycnium. Until recently rusts were considered to show a simple two-allele (+vs-)heterothallism. On this basis both selfing and crossing should give 50% compatibility. However, LAWRENCE (1980) found that on selfing (carried out by cross-fertilizing sibling pycnia), one flax rust culture showed only 25% compatibility while another showed 35%. In a cross between these cultures 100% of fertilizations were successful. He suggested that heterothallism in *M. lini* is determined by a two-locus system with multiple alleles where selfing would be expected to result in 25% compatibility. He proposed that if one or both loci are made up of two subunits, as observed in a number of hymenomycetes (see FINCHAM et al. 1979), then recombination between them would generate new specificities. This recombination, according to its frequency, could account for up to 50% compatibility on selfing, as observed earlier by FLOR (1942, 1946). The highest frequency of inter-subunit recombination observed in *Schizophyllum commune* was close to 23% for the *A* locus (RAPER et al. 1960). A frequency of 20% at both of the two loci postulated in *M. lini* would lead to about 41% compatibility. If 50% compatibility is common on selfing then there may be other explanations. For example, in some populations a simple one-locus, two-allele system may operate, while in others a two-locus or tetrapolar system may be found, as suggested by LAWRENCE's data.

8.3.2.2 Genetics of Virulence

The aeciospores are products of sexual fusion. Because they are dikaryotic, they behave in a way similar to diploids. Genes controlling avirulence are expressed in both haploid and dikaryon, but since pycniospores cannot be used to propagate the haploid rust clonally, evaluation of virulence is normally only carried out with dikaryotic cultures.

The genetic studies of the rust in Adelaide confirmed an important modification of the gene-for-gene relationship. This was that in the rust two genes interact to determine virulence on the *M*1 resistance gene and that two genes also determine virulence on each of the resistance genes *L*1, *L*7 and *L*10 (LAWRENCE et al. 1981a). These findings explained anomalies that had been detected earlier by FLOR. Thus virulence on the cultivar Williston Brown (*M*1) occurs in the presence of a dominant allele for aviulence *A-M*1 when a dominant allele *I-M*1 at a second locus is present. When *I-M*1 is absent (*i-M*1 *i-M*1) *A-M*1 produces an avirulent reaction on Williston Brown. The alleles *I-L*1, *I-L*7 and *I-L*10 have similar effects, and the data do not exclude the possibility that these three alleles are identical and closely linked with *I-M*1. Although *I* was called an inhibitor gene, there is no evidence to indicate its biochemical mode of action. The authors draw attention to several examples of similar interactions in other rusts and wisely suggest caution in inferring rust genotypes from their phenotypes.

In a second paper LAWRENCE et al. (1981b) described a fine structure analysis of four tightly linked rust genes *A-P, A-P*1, *A-P*2 and *A-P*3 controlling virulence on the four cultivars Koto (*P*), Akmolinsk (*P*1), Abyssinian (*P*2) and Leona (*P*3). A heterozygous culture of the rust was obtained with the genotype *A-P a-P*1 *a-P*2 *a-P*3/*a-P A-P*1 *A-P*2 *A-P*3. This was crossed with a homozygous recessive culture virulent on all four cultivars.

The crosses were made by pooling nectar from 25 pycnia produced by basidiospores from the homozygous virulent parent and transferring it to 120–140 well-separated pycnia produced by basidiospores from the heterozygous parent.

Of some 5073 fertilized pycnia, 3160 produced aeciospores. That only 62.3% of these crosses were successful was attributed to suboptimal growth conditions. The progenies, each consisting of an aecial pustule, were pooled in groups of from 60 to 110 which were then inoculated to a set of three host tester lines with the genotypes $PP1$, $PP2$ and $PP3$ obtained by crossing the cultivars carrying these genes. A total of 39 bulk collections were tested in this way. Because of the tight linkage, parental progeny carried either A-P or the combination A-$P1$ A-$P2$ A-$P3$ and hence were avirulent on all three testers. Half of the recombinants were expected to be virulent on one or more of the testers. Three presumptive recombinants were recovered after eliminating more than 200 other cultures that were presumed to have arisen either as contaminants or by nuclear exchange.

In another experiment, plants of the three host tester lines were directly inoculated with basidiospores from the heterozygous parent. Expectations were similar to those from the first experiment using aeciospores, namely that parental gametes would be avirulent but that half of the recombinant gametes would be virulent on one of the host testers. In theory some might be virulent on more than one of the testers, but this could not be tested at the haploid level. The first two attempts to carry out this experiment were unsuccessful apparently because the genetic backgrounds of the tester stocks did not favour recovery of pycnia from what were in fact very dilute inocula, since the frequency of virulent recombinants was expected to be very low. However, by two generations of backcrossing to the extremely and universally susceptible variety Hoshangabad, testers were obtained on which seven pycnia developed. These were used to fertilize pycnia of the homozygous virulent strain to produce aeciospores for further tests.

The analyses of the total of 10 recombinants recovered from the two sets of experiments indicates the order A-P A-$P2$ A-$P1$ A-$P3$ and suggests that the alleles belong to the same cistron. However, the possibility that they belong to separate but contiguous cistrons cannot be excluded.

Unfortunately such analyses are laborious and time-consuming. Although they show that host and parasite genes have structures like those of better-known organisms such as maize and *Neurospora* there seems little prospect at present that the flax–flax rust system can be exploited to better understand the molecular biology of the genetic controls. This point will be treated again later.

8.4 Mutation Studies

8.4.1 Host Plants

The importance of the study of induced mutants for the development of microbial genetics hardly needs emphasis. Mutational blocks in metabolic pathways revealed the steps of biosynthesis and much about their controls. In host plants mutation has been used principally as a practical tool to supplement the resistance available from related and wild material. An International Atomic Energy

Authority symposium (Anonymous 1977) held several years ago illustrates the extent, but rather limited, success of this approach. Relatively few crops make use of resistance genes recovered by induced mutation. In any case it seems that they are just as likely as naturally occurring resistance genes to select new virulence genes in the pathogens (see Day 1974). In view of how undeveloped our knowledge of gene action is in plants it is not surprising that the study of induced resistant mutants has not led very far.

However, in comparing host–parasite interaction with other systems that govern specificity, it seems at first sight that induced mutation in the host may create new specificities in the form of disease resistance. Mutations affecting self-incompatibility in angiosperms or mating-type specificity in basidiomycetes never confer new functional specificities. Unfortunately few studies have rigorously eliminated pollen or seed contamination as a possible source of resistant "mutants", and so far as I know, apart from mutations of the *ml-o* locus in barley (Simons 1979), none have demonstrated that the new specificity occurs at a locus which is known to carry other alleles that govern resistance. Mutations of the loci governing rust resistance in flax would be especially interesting, given the extent of current information in this system discussed in Section 8.3.

8.4.2 Mutation in Pathogens

Mutational studies in pathogens were at first directed to confirm that virulence could indeed arise in this way. Several studies with economic pathogens have shown this to be so (see Day 1974). However, the possibilities inherent in this approach have hardly begun to be explored. In Section 8.5 some of these are discussed in relation to bacterial pathogens.

In theory the gene-for-gene hypothesis could be tested by establishing that all or most mutations from avirulence to virulence on a host with a single gene for resistance occur at the same locus in the pathogen (Day 1974). That such a test has still to be carried out probably reflects the technical difficulties in the genetic analysis of most plant pathogens.

8.5 The Recognition Mechanism

The large class of host–parasite associations that conform to the gene-for-gene relationship, as mentioned earlier, all demonstrate a highly evolved relationship between the two components which is disrupted by the confrontation between a gene for resistance in the host and a corresponding gene for avirulence in the pathogen. The nature of the confrontation and the features of both genes that determine their "correspondence" are important aspects of specificity. When their mode of action is understood we should be able to explain in biochemical terms the differences between resistant and susceptible hosts and avirulent and virulent pathogens. At the time of writing it is not possible to do this for any host–parasite interaction of this kind. Consequently all we can do is to consider some hypotheses and discuss how they might be tested.

In addition to the genetic information discussed in Section 8.3 any hypothesis must accommodate the following features of most systems:

1. resistance is not preformed but appears shortly after the incompatible reaction begins;
2. the materials produced when a resistant reaction is invoked inhibit virulent as well as avirulent parasites;
3. the reaction is generally localized at the site of interaction and in neighbouring cells. Although systems have been described where preinoculation can initiate changes in leaves and organs some distance from the site (KuĆ 1981) these are uncommon;
4. it should be testable in terms of identifying the products of the genes involved and demonstrating their complementarity or specificity in vivo and in vitro.
5. different resistance genes may or may not have different effects.

A widely accepted model is based on the concept that resistance is induced, that the pathogen molecules, or elicitors, responsible for induction are products of avirulence genes, and that the genes for resistance are responsible for specific receptors such that only the combination of a specific receptor and elicitor will initiate a chain of reactions producing an incompatible phenotype.

8.5.1 Models Involving Phytoalexin Synthesis

Much attention has been paid to phytoalexins, or substances formed by host tissue "... in response to injury, physiological stimuli, infectious agents or their products ... (that) accumulate ...to levels which inhibit the growth of microorganisms" (KuĆ 1972). However, that they play a crucial role in determining specificity has been disputed.

For example, one has only to examine the evolution of models that assign a primary role to phytoalexins to appreciate the problems. At first it was suggested that only avirulent pathogens are inhibited by phytoalexins. However, in most examples both are inhibited equally. The alfalfa pathogen, *Stemphylium botryosum,* was reported to be able to degrade its host's phytoalexin (HIGGINS and MILLAR 1969). In theory this property could account for race specificity. A current model suggests that virulent isolates do not elicit phytoalexin formation but that avirulent isolates do. For example, recent work by MAYAMA et al. (1982) established a correlation between the amount of the oat phytoalexins, avenalumins I and II, and the degree of development of crown rust (*Puccinia coronata avenae*) in inoculated leaves. Inoculations of 21 different oat cultivars, each carrying a different single gene for crown rust resistance with two rust races, showed that only trace amounts of the avenalumins were detectable in compatible interactions. Concentrations known to be inhibitory were formed in reactions that were either moderately or highly resistant and were proportional to the degree of incompatibility observed.

In other systems not only avirulent races but other microorganisms and a bewildering array of chemicals, elicit phytoalexins. If a gene for avirulence is the determinant of specificity, one would expect that the role of its producer as an elicitor would be equally specific. Accordingly there has been an intensive search for compounds produced by avirulent pathogens, not present in virulent strains, that elicit phytoalexins in resistant tissues. Elicitor activity may be assayed by showing that resistant plant tissue produces a phytoalexin when treated with the elicitor. However, the amounts of phytoalexin formed may be insufficient to account for the in vivo incompatible response or, alternatively, may be active in vivo because they are highly localized but not detectable in an assay.

Although still subject to rigorous confirmation, it appears likely that the soybean pathogen, *Phytophthora megasperma,* produces extracellular glycoproteins that behave as specific elicitors for resistant soybeans (WADE and ALBERS-HEIM 1979, KEEN and LEGRAND 1980).

An interesting variant of the elicitor hypothesis suggests that when both avirulent and virulent races elicit phytoalexins, differential production of these defense compounds occurs because the virulent pathogen produces a suppressor which overrides the non-specific elicitor. An example of this comes from *Phytophthora infestans* on potato (GARAS et al. 1979, DOKE et al. 1980). DOKE and TOMIYAMA (1980) found that the incompatible response of potato tuber protoplasts to cell-wall components of *P. infestans* is to some extent specifically suppressed by water-soluble glucans extracted from compatible pathogen races. In two other host–parasite systems pathogen-produced suppressors have also been found to override the expression of resistance. HEATH (1981a) showed that extracts of the bean rust fungus, or of rust-infected bean tissue, promoted haustoria formation in older, more resistant, bean leaves and also increased the frequency of haustoria formation by cowpea rust which is normally a non-pathogen. OKU et al. (1980) also reported that *Mycosphaerella pinodes* produces a low molecular weight peptide that suppresses pisatin formation in pea leaves. BUSHNELL and ROWELL (1981) have put forward a model for host–parasite specificity that stresses the role of suppressors.

8.5.2 Models and Their Implications

ELLINGBOE (1982) has discussed the genetical implications of four models to explain the mechanism of host–parasite specificity. The first and simplest proposes that the product of an avirulence gene interacts with the product of a resistance gene to form a dimer molecule responsible for incompatibility. The monomers are inactive and the formation of the dimer involves specific recognition – a lock and key interaction is implied – between the gene products. How the dimer governs incompatibility is not pursued.

The second model supposes that the avirulence gene determines the specificity of a glycoprotein which interacts specifically with a host glycoprotein or carbohydrate. The specificity might, for example, involve the insertion of a short side chain on an already complex glycoprotein molecule whose synthesis is directed by other genes. Again how the product of pathogen glycoprotein and host receptor governs incompatibility is not stated. ELLINGBOE dismisses this model on the grounds that since a number of genes govern the synthesis of the pathogen molecule, there is an implicit departure from the expectation of a one-gene-to-one-gene relationship demonstrated by the analysis of gene-for-gene host–parasite interactions.

In the two other models the specific interaction of gene products causes synthesis of either one phytoalexin or several phytoalexins. Again, since several genes are concerned with the synthesis of one, or several, such compounds in the host these two models are dismissed on similar grounds. ELLINGBOE rightly points out that defects in carbohydrate synthesis are not necessarily lethal in fungi and hence failure to produce a specific avirulence product could be expected to have more than one genetic basis in the second model. What is questionable, however, is whether there have been sufficient genetic analyses in plant pathogens of virulent strains of independent origin to be certain that this never happens. This is precisely what a comparison of a number of independently induced virulent mutants advocated in Section 4.2 might establish. In dismissing the last two models ELLINGBOE is on even less firm ground, for it is clear that defects in the synthesis of phytoalexins would be almost certain to reduce host fitness and hence be eliminated by natural selection. They would make the host susceptible to any pathogen carrying avirulence genes that invoke synthesis of those phytoalexins. Such forms might well be found after mutagen treatment in the protected environment of a greenhouse or experimental plot. However, mutation experiments invariably seek resistant mutants by treating susceptible material and not susceptible mutants of resistant material.

8.5.3 Some Simpler Bacterial Systems

Evidently the host–parasite systems that have been explored to date are too complex to have allowed a complete explanation of the mechanism in terms of the gene products that interact. This is certainly the case for flax–flax rust, where it has been shown that production of the phytoalexins coniferyl alcohol and coniferyl aldehyde occurred in incompatible reactions (KEEN and LITTLE-FIELD 1979). It is now becoming clearer that this problem of understanding the mechanism of specificity will be more easily solved by investigating bacterial plant pathogen systems since the tools of molecular genetics can so readily be applied. The information gained could then be used to explore the more complex fungal pathogens. Some candidate systems include bacterial blight of cotton *Xanthomonas campestris* pv *malvacearum* (BRINKERHOFF 1970), and a bacterial disease of pepper (*Capsicum annuum*) *Xanthomonas campestris* pv *vesicatoria* (DAHLBECK and STALL 1979). In both examples virulent mutants were recovered following inoculation of leaves of resistant host cultivars with avirulent strains of the two bacteria. Although in neither case has a formal genetic analysis like that in flax–flax rust been carried out, this will surely not be long in coming.

Recent studies with another plant pathogenic bacterium, *Pseudomonas syringae* pv *savastanoi* by COMAI and KOSUGE (1980, 1982), although not addressing the problem of host specificity, demonstrate the possibilities for *Xanthomonas*.

P.s. pv *savastanoi* is a pathogen of olive and oleander producing tumour-like outgrowths called knots on infected twigs and leaves. The production of host tumours depends on the synthesis by the bacterium of the plant growth hormone indoleacetic acid (IAA). This is formed from L-tryptophan (trp) via the intermediate indoleacetamide. Two enzymes are responsible; trp-monooxygenase, which carries out the first step, and indoleacetamide hydrolase, which converts the intermediate to IAA. The enzyme trp-monooxygenase confers resistance to the toxic tryptophan analogue 5-methyl tryptophan (5-mt) converting it to the non-toxic 5-methyl indoleacetamide. Resistance to 5-mt is thus a convenient way of detecting trp-monooxygenase activity. The structural gene for this enzyme (iaaM) was shown to be located on a plasmid of 52 kb (pIAA1) and when bacterial cells were cured of the plasmid by treatment with acridine orange they were both sensitive to 5-mt and non-pathogenic. Cured cells, co-transformed with pIAA1 and a wide host range plasmid RSF1010 carrying drug resistance markers, were able to synthesize IAA and were pathogenic. They could not be selected by screening for resistance to 5-mt because this phenotype is produced by other mechanisms and arises more frequently than by transformation. Since transformation is normally limited by the competence of the recipient cells to take up DNA selection for streptomycin resistance carried by RSF1010, usually recovered cotransformants carrying pIAA1 as well.

A constructed plasmid (pLUC1) made up of RSF1010 carrying a 2.75 kb EcoR1 fragment of pIAA1 which contains iaaM also confers pathogenicity on cured strains even though they only form indoleacetamide (COMAI and KOSUGE 1982). The EcoR1 site of the vector plasmid RSF1010 occurs in a transcriptional unit between the genes for sulfonamide and streptomycin resistance, and expression of iaaM was shown to be dependent on the vector promoter. The position of iaaM was mapped by selecting Tn1 insertions in the plasmid pLUC1. The transposon Tn1 is 5.3 kb in size and is carried by the plasmid RP4 and confers ampicillin resistance on cells that harbour plasmids containing it. RP4 was introduced into *E. coli* cells containing pLUC1 by conjugation with donor cells and cells containing both plasmids were selected, grown for 24 h and the plasmid DNA extracted. The DNA was treated with the restriction endonuclease Xho1 and transformed in *E. coli* cells which were selected for ampicillin resistance. Since RP4 has two Xho1 sites and pLUC1 has none, most of the resistant cells had pLUC1 with a Tn1 insertion. By selecting Tn1 insertions that blocked expression of either sulfonamide or streptomycin resistance and relating this to iaaM activity and the position of the insertions on restriction maps, the authors were able to locate iaaM on the plasmid with some accuracy and determine its orientation.

The prospect of such analyses in bacterial plant pathogens was remote even a few years ago. In both pathovars of *Xanthomonas campestris* it should be possible to characterize restriction fragments of either chromosomal or plasmid DNA that carry an avirulence gene by transforming virulent cells and testing them for avirulence. If transposon insertions inactivate the gene and restore virulence, then it can be precisely mapped, it could be excised and expressed in minicells to allow a product to be harvested, and it could even be sequenced for comparison with candidate gene products or elicitors. Once identified these can be used to locate and characterize the host receptor site coded by the resistance gene. Identification of the product of this gene will then assist in establishing the identity of the gene and eventually provide a means of selecting it either from libraries of the host plant genomic DNA or from cDNA's prepared from mRNA formed when resistant tissue is challenged by an avirulent race.

8.5.4 Application of Molecular Genetics to Fungal Systems

In the corn smut pathogen, *Ustilago maydis,* DAY et al. (1971) selected, independently of the host, mutants of the *b* locus that governs smut pathogenicity. In this smut, following sexual fusion, the developmental sequence controlled by *b* leads to the production of a pathogenic dikaryon provided the two nuclei carry different alleles. Diploid cells homozygous for *b* are non-pathogenic and, on suitable media, form yeast-like colonies. As a result of mutation at *b* the cells form an aerial mycelium and can be readily detected at low frequencies (10^{-5}–10^{-6}) against a yeast-like background. As the authors predicted, the mutant diploids were pathogenic when tested on corn seedlings. Like mutants affecting mating type specificity in other organisms these mutants simulate, but not completely so, the phenotype resulting when different alleles react to produce compatibility. Clearly the *U. maydis* system lends itself to the kinds of techniques discussed earlier that would lead to isolation and identification of the *b* gene products and knowledge of how they control pathogenicity. This is all the more feasible now that transformation is possible in *U. maydis* (R. HOLLIDAY personal communication).

Although the action of *b* in *U. maydis* involves a different form of specificity than that governing cultivar-race specificity, the latter is well known in other smuts.

8.6 Non-Host Resistance

The importance of distinguishing the mechanisms that have evolved to achieve compatibility from those that govern cultivar-race specificity has already been briefly discussed. The absence of such mechanisms is responsible for what has been termed non-host resistance. HEATH (1981b) has referred to the lack of information on whether host and non-host resistance occur through different mechanisms. I have already referred to tomato as a non-host of *Puccinia striiformis*. I am tempted to follow HEATH's usage and call it an "indisputable non-host". However, I am reluctant to do so in the absence of experimental tests of this hypothesis. For example, if tomato was a non-host of this rust because of the toxic effects of an alkaloid found in the leaves the situation would not hold for a non-alkaloid-forming mutant of the host or for an alkaloid-resistant form of the pathogen. Although deliberately oversimplified, such an argument introduces an element of specificity into the concept of non-host resistance.

Clearly where many such effects are superimposed to form multiple barriers to host–parasite interaction, it is unlikely that they would ever be breached. However, even such different hosts as barley and melon can support haustorial development of the mildew non-pathogens *Sphacelotheca fuliginea* and *Erysiphe graminis* f.sp. *hordei* respectively in experiments carried out by OUCHI et al. (1979). For example, barley leaves were preinoculated with conidia of a compatible race of *E.g. hordei* and the superficial mycelium removed by a cotton swab 48 h later. When conidia of *S. fuliginea* were applied the non-pathogen was readily able to colonize host cells that harboured compatible haustoria but could not colonize cells four or more cells away. Similar results were reported from inoculations with an incompatible race of *E.g. hordei* after inoculation of the barley with compatible *E.g. hordei*. Evidently induction of accessibility, the term applied by OUCHI et al., overcomes barriers that operate against both pathogen (cultivar-race specific resistance) and non-pathogen (non-host, non-specific resistance).

In 1976 I was optimistic that non-host resistance could have great potential value once methods for genetic transformation in plants are available (DAY 1976b). This could well be the case, but I would also predict that much of what we now think of as non-host resistance will prove race-specific once exposed in the field.

References

Anonymous (1970) Induced mutations against plant diseases. AEA, Vienna
Biffen RH (1905) Mendel's laws of inheritance and wheat breeding. J Agric Sci 1:4–48
Biffen RH (1912) Studies in inheritance in disease resistance II. J Agric Sci 4:421–429
Brinkerhoff LA (1970) Variation in *Xanthomonas malvacearum* and its relation to control. Ann Rev Phytopathol 8:85–110
Bushnell WR, Rowell JB (1981) Suppressors of defense rections: a model for roles in specificity. Phytopathology 71:1012–1014
Comai L, Kosuge T (1980) Involvement of plasmid deoxyribonucleic acid in indoleacetic acid synthesis in *Pseudomonas savastanoi*. J Bacteriol 143:950–957
Comai L, Kosuge T (1982) Cloning and characterization of iaaM, a virulence determinant of *Pseudomonas savastanoi*. J Bacteriol 149:40–46
Dahlbeck D, Stall RE (1979) Mutations for change of race in cultures of *Xanthomonas vesicatoria*. Phytopathology 69:634–636
Day PR (1974) Genetics of host–parasite interaction. Freeman, San Francisco
Day PR (1976a) The taxonomy of physiologic races. In: Subramaniam CV (ed) Taxonomy of fungi. Univ Madras, pp 164–169
Day PR (1976b) Gene functions in host–parasite systems. In: Wood RKS, Graniti A (eds) Specificity of plant diseases. Plenum, New York London, pp 65–73
Day PR, Anagnostakis SL, Puhalla JE (1971) Pathogenicity resulting from mutation at the *b* locus of *Ustilago maydis*. Proc Natl Acad Sci USA 68:533–535
Doke N, Garas NA, Kuć J (1980) Effect on host hypersensitivity of suppressors released during the germination of *Phytophthora infestans* cystospores. Phytopathology 70:35–39
Doke N, Tomiyama T (1980) Suppression of the hypersensitive response of potato tuber protoplasts to hyphal wall components by water soluble glucans isolated from *Phytophthora infestans*. Physiol Plant Pathol 16:177–186
Ellingboe AH (1976) Genetics of host–parasite interactions. In: Heitefuss R, Williams

PH (eds) Physiological Plant Pathology. Encyclopedia of Plant Physiology New Ser Vol 4, Springer, Berlin Heidelberg New York, pp 762–778

Ellingboe AH (1981) Changing concepts in host–pathogen genetics. Ann Rev Phytopathol 19:125–143

Ellingboe AH (1982) Genetical aspects of active defense. In: Wood RKS (ed) Active defense mechanisms in plants. Plenum, New York London, pp 179–192

Fincham JRS, Day PR, Radford A (1979) Fungal genetics, 4th edn. Blackwell, Oxford

Flor HH (1942) Inheritance of pathogenicity in *Melampsora lini*. Phytopathology 32:653–669

Flor HH (1946) Genetics of pathogenicity in *Melampsora lini*. J Agric Res 73:335–357

Garas NA, Doke N, Kuć J (1979) Suppression of the hypersensitive reaction in potato tubers by mycelial components from *Phytophthora infestans*. Physiol Plant Pathol 15:117–126

Heath MC (1981a) Nonhost resistance. In: Staples RC, Toenniessen GH (eds) Plant disease control: resistance and susceptibility. Wiley and Sons, New York, pp 201–217

Heath MC (1981b) A generalized concept of host–parasite specificity. Phytopathology 71:1121–1123

Higgins VJ, Millar RL (1969) Degradation of alfalfa phytoalexin by *Stemphylium botryosum*. Phytopathology 59:1500–1506

Keen NT, Legrand M (1980) Surface glycoproteins: evidence that they may function as the race-specific phytoalexin inhibitors of *Phytophthora megasperma* f. sp. *glycinea*. Physiol Plant Pathol 17:175–192

Keen NT, Littlefield LJ (1979) The possible associations of phytoalexin with resistance gene expression in flax to *Melampsora lini*. Physiol Plant Pathol 14:265–281

Kuć J (1972) Phytoalexins. Ann Rev Phytopathol 10:207–232

Kuć J (1981) Multiple mechanisms, reaction rates, and induced resistance in plants. In: Staples RC, Toennissen GH (eds) Plant disease control: resistance and susceptibility. Wiley and Sons, New York, pp 259–272

Lawrence GJ (1980) Multiple mating-type specificites in the flax rust *Melampsora lini*. Science 209:501–503

Lawrence GJ, Mayo GME, Shepherd KW (1981a) Interactions between genes controlling pathogenicity in the flax rust fungus. Phytopathology 71:12–19

Lawrence GJ, Shepherd KW, Mayo GME (1981b) Fine structure of genes controlling pathogenicity in flax rust, *Melampsora lini*. Heredity 46:297–313

Mayama S, Matsuura Y, Iida H, Tani T (1982) The role of avenalumin in the resistance of oat to crown rust, *Puccinia coronata* f. sp. *avenae*. Physiol Plant Pathol 20:189–199

Mayo GME, Shepherd KW (1980) Studies of genes controlling specific host–parasite interactions in flax and its rust. Heredity 44:211–227

Oku H, Shiraishi T, Ouchi S, Ishiura M, Matsueda R (1980) A new determinant of pathogenicity in plant disease. Naturwissenschaften 67:310–311

Ouchi S, Hibino C, Oku H, Fujiwara M, Nakabayashi H (1979) The induction of resistance or susceptibility. In: Daly JM, Uritani I (eds) Recognition and specificity in plant host–parasite interactions. Jpn Sci Soc Press, Tokyo, pp 49–65

Raper JR, Baxter MG, Ellingboe AH (1960) The genetic structure of the incompatibility factors of *Schizophyllum commune*: the A factor. Proc Natl Acad Sci USA 46:833–842

Shepherd KW, Mayo GME (1972) Genes conferring specific plant disease resistance. Science 175:375–380

Sidhu GS (1975) Gene-for-gene relationships in plant parasitic systems. Sci Prog, Oxford 62:467–485

Simons MD (1979) Modification of host–parasite interactions through artificial mutagenesis. Ann Rev Phytopathol 17:75–96

Wade M, Albersheim P (1979) Race-specific molecules that protect soybeans from *Phytophthora megasperma* var. *sojae*. Proc Natl Acad Sci USA 76:4433–4437

Wolfe MS, Barrett J (1980) Can we lead the pathogen astray? Plant Disease 64:148–155

Wolfe MS, Knott DR (1982) Populations of plant pathogens: some constraints on analysis of variation in pathogenicity. Plant Pathology 31:79–90

9 The Mycorrhizal Associations

J. L. HARLEY

9.1 Introduction

Many books and reviews have been written during recent years which describe the structure and physiology of mycorrhizal symbioses (some examples are: HARLEY 1969, MARKS and KOZLOWSKI 1973, SMITH 1974, SANDERS et al. 1975, HARLEY 1978a, b, MIKOLA 1980, SMITH 1980, HARLEY and SMITH 1983). This chapter will therefore, after a general introduction, restrict itself to a few special aspects of the interaction between mycorrhizal fungi and their hosts which are either under active investigation or about which much new information is needed before advances can be made.

Mycorrhizal symbiosis is so common in the plant kingdom that one can say with confidence that mycorrhizas are more common as nutrient-absorbing organs than uninfected roots. They are present in most species of all the groups of Pteridophyta (perhaps less common in Filicales than in others), in all groups of Gymnospermae and in almost all families of Angiospermae (the species of certain families such as Cruciferae, Caryophyllaceae and Cyperaceae are commonly non-mycorrhizal, as are many species in very wet places).

There are several very common types of mycorrhizas upon which much research has been done, and a few others less frequent and less studied. Mycorrhizal fungi belong to all groups, Phycomycetes, Ascomycetes and Basidiomycetes. Although the perfect stages of all are not known, each can still confidently be ascribed to one or other of these large groups.

The mycorrhizal Phycomycetes belong to the family Endogonaceae, of which the species of a number of genera form a kind of mycorrhiza called vesicular-arbuscular mycorrhiza, with members of the majority of species in the phyla listed above. This kind of mycorrhiza is the commonest and the most widespread ecologically. None of the fungi has yet been brought into pure axenic culture, although their spores may germinate in artificial media and form a mycelium which will infect susceptible roots if they are placed in the culture and grow between root and root. These endogonaceous fungi were, till about 1974, all referred to the genus *Endogone,* but in that year GERDEMANN and TRAPPE (1974, 1975) reconsidered the family Endogonaceae. They concluded that the vesicular-arbuscular mycorrhizal species belonged to several genera, of which perhaps the most important was *Glomus,* and that the genus *Endogone* itself did not form vesicular-arbuscular mycorrhiza. Since then there has been a considerable volume of research on the ecology and taxonomy of the group the reader is also referred to Vol. 12C of this Encyclopedia, pp. 391–421. Ecophysiology of mycorrhizal symbioses, by M. MOSER and K. HASELWANDTER). A key to the Endogonaceae was published by HALL and FISH (1978). Some species of the

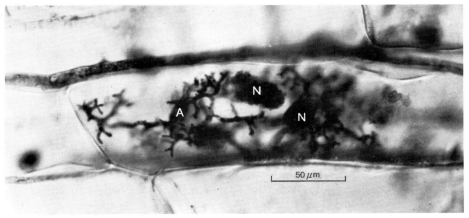

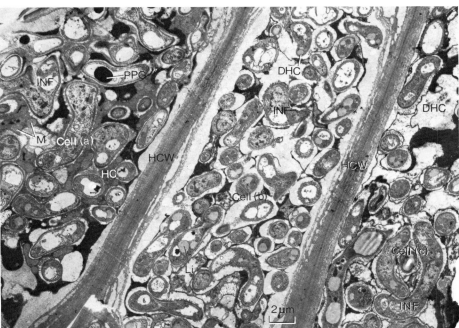

Fig. 1. **A** Section of root of onion showing an arbuscule within a cortical cell. Note the enlarged nucleus (*N*) and the dichotomous arbuscule. Bar 50 µm (Photo B. MOSSE). **B** T.E.M of a root of *Rhododendron ponticum* showing L. S. of three adjacent cortical cells. Cell (*a*) contains living hyphae within the living host cell. In cell (*b*) healthy living hyphae are associated with degenerating host cytoplasm. In cell (*c*) both host cytoplasm and fungal cytoplasm are degenerating. *DHC* degenerating host cytoplasm; *HC* host cytoplasm; *HCW* host cell wall; *INF* intracellular fungus; *PPG* polyphosphate granule. Bar 2 µm. (Photo J. DUDDRIDGE)

genus *Endogone* (sensu stricto) which do not produce vesicular-arbuscular mycorrhiza, but form a kind of ectomycorrhiza with trees, have recently been
cultured (WARCUP 1975), so there is perhaps hope of success in the future
with species of *Glomus* and the related genera of vesicular-arbuscular mycorrhiza. In the soil these latter fungi colonize the roots of plants, grow along
their surfaces and form appressoria from each of which a hypha penetrates
a cell. Within the tissues the hyphae may form coils inside the cells. They
pass from cell to cell, often growing in the middle lamella region of the walls,
and form complex, frequently dichotomous haustoria called arbuscules within
the cortical cells, most often in the inner cortex (Fig. 1a). The mycelium on
and around the surface of the root, as well as in the tissues and the cells,
forms swellings called vesicles, which contain several nuclei and many oil drops.
Sporocarps, chlamydospores and azygospores may also be produced on the
extraradical mycelium. This kind of mycorrhiza formed on the primary roots,
and on young achlorophyllous tissue of rootless plants, consists therefore of
three phases: extraradical mycelium in the soil extending several centimetres
at least from the surface of the root, an intercellular hyphal system in the
cortex and an intracellular hyphal system of which the dichotomous arbuscules
form a very large surface of contact with the host.

The other kinds of mycorrhiza (with the very minor exception of those
formed by *Endogone* spp.) involve septate mycelial fungi belonging to the Ascomycetes and Basidiomycetes. The hosts of each of these kinds of mycorrhiza
are taxonomically more restricted, but the ectomycorrhizas are the most widespread and occur on the roots of some gymnosperms and angiosperms. These
characteristically form complex organs of fungal and host tissue, so that a
thick pseudoparenchymatous tissue of fungus, the sheath or mantle, encloses
the host rootlet overarching its apex. From it hyphae pass into the soil. Connections with the soil may be prolific, consisting of many branching hyphae, or
of hyphal strands or rhizomorphs. In other cases connections are less extensive
and consist of short hyphae, or setae, or sometimes the sheath may be almost
smooth. From the inner side of the fungal sheath hyphal systems penetrate
within the cell walls of the cortex in the middle lamella region, to form what
is called the Hartig net (Fig. 2A). This is not a network of simple hyphae
but a complicated cellular branch system developed from the hyphae penetrating
between the cells. It was described by MANGIN (1910) as consisting of "pal-
metti", by STRULLU (1976) of "lames fongiques" and by NYLUND and UNISTAM
(1982) of "labyrinthine" hyphal systems. It represents a considerable develop-

---

Fig. 2. A S. E. M. showing a surface view of the cut end of a transverse section of an
ectomycorrhiza of *Pinus sylvestris*. *HN* Hartig net; *CC* cortical cell; *S* fungal sheath.
Bar 10 μm. (Photo J. DUDDRIDGE). **B** T. E. M. showing details of the Hartig net region
of mycorrhiza of *P. sylvestris* showing hyphae (*HN*) and adjacent cortical cells of the
host (*CC*). The cortical cell wall has lost its integrity and is indistinguishable from the
matrix in which the Hartig net hypha is embedded and forms the "involving layer"
(*IL*). Host cytoplasm (*HC*); Golgi body (*G*); rough endoplasmic reticulum (*RER*); mitochondrion (*M*); host plasmalemma (*HPL*); fungal cytoplasm (*FC*); fungal plasmalemma
(*FPL*); fungal nucleus (*N*); tannin vacuole (*TV*). Bar 1 μm. (Photo J. DUDDRIDGE)

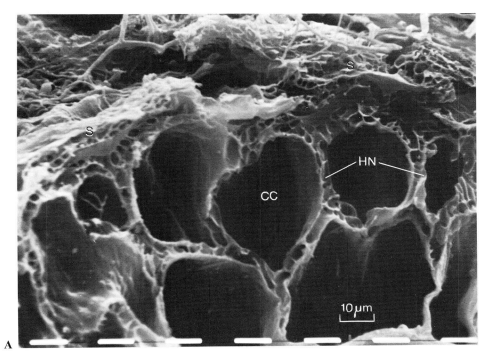

A

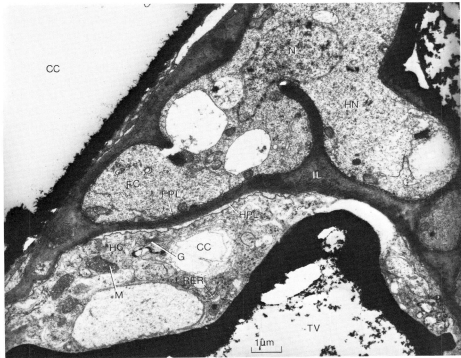

B

ment of the fungus, giving a large area of contact between the cells of the
two symbionts without penetration by the hyphae into the living active cells
of the host (Fig. 2 B). This kind of mycorrhiza is found in arborescent gymnosperms belonging to the *Pinaceae* (only extremely rarely in other Gymnosperm
families) and in arborescent and shrubby Angiosperms of many diverse families
but especially in Fagales, Juglandales, Myrtaceae, Dipterocarpaceae and some
woody legumes, especially of the Caesalpinoidae. It is rare in herbaceous angiosperms but has been reported in *Polygonum viviparum* and in the genus
Kobresia amongst others. The fungi form epigeous or hypogeous fruiting bodies
and sometimes sclerotia in considerable quantities.

Three further types of mycorrhiza are found in the Ericales in which the
septate fungi belong to the Ascomycetes and Basidiomycetes. Many of the Ericaceae and Epacridaceae, especially those with fine or "hair" roots, are associated
with fungi, usually Ascomycetes, which form a considerable weft of mycelium
around the root and on its surface but not a tight fungal tissue or sheath.
This type of mycorrhiza is called ericoid. The hyphae penetrate between and
into the cells of the root of the host to form extensive coils within them (Fig. 1 B).
Other ericaceous species (Arbutoideae) with coarser root systems, such as those
of *Arctostaphylos* and *Arbutus,* form a different kind of mycorrhiza usually
with basidiomycetous fungi, reminiscent of ectomycorrhiza. Indeed the same
fungal species may form ectomycorrhiza with conifers or broad-leaf hosts and
"arbutoid" mycorrhiza with species of *Arctostaphylos* and *Arbutus* (MOLINA
and TRAPPE 1982). In these there is a firm fungal sheath, a Hartig net and
also extensive penetration into the living host cells to form hyphal coils. The
arbutoid mycorrhizas therefore have common features with both ericoid mycorrhizas and with ectomycorrhizas. LARGENT et al. (1980) have reported that some
Ericales form ectomycorrhizas in addition to ericoid and arbutoid mycorrhizas.
It is of great interest that a kind of mycorrhiza somewhat similar to the arbutoid
has been described in species of *Pinus* and other ectomycorrhizal hosts especially
in the seedling stages by LAIHO and MIKOLA (1964), MIKOLA (1965), LAIHO
(1965), WILCOX (1968a, b, 1971), WILCOX and GANMORE NEUMANN (1974) and
WILCOX et al. (1975). These ectendomycorrhizas are formed by imperfect fungi
probably of ascomycetous affinity. In them the sheath may not be as well
developed as in ectomycorrhizas or arbutoid mycorrhizas but a similar Hartig
net and penetration into living cells occurs. The same is true of the rather
similar mycorrhizas of the green Pyrolaceae, in which there is also intercellular
and intracellular penetration. The achlorophyllous species of *Monotropa* and
related genera have a different kind of mycorrhiza again. The fungal sheath
is well-developed and a Hartig net penetrates the walls of the epidermis, but
in addition hyphae penetrate into the cells of the host by simple hyphal pegs.
These haustoria, which appear simple under the light microscope, undergo complex development and finally degeneration. Around them the wall of the host
cell is stimulated to form complex carbohydrate ingrowths and so seems to
resemble a transfer cell (DUDDRIDGE and READ 1982). The fungus isolated from
Monotropa hypopitys by BJÖRKMAN (1960) was found to be capable of forming
mycorrhiza with ectomycorrhizal trees. There are in consequence in this symbiosis three partners – a green photosynthesizing tree host, its ectomycorrhizal

fungus, and also the achlorophyllous *Monotropa* (KAMIENSKI 1981, BJÖRKMAN 1960, FURMAN 1966). Other species of *Monotropa* are described as associating with parasites of woody plants like *Armillaria mellea* (CAMPBELL 1971) and therefore have similarities with the saprophytic or partially saprophytic Orchidaceae (see HARLEY 1973).

The mycorrhizal associations of Orchidaceae involve fungi, mostly basidiomycetes, which are either active in hydrolysis of cellulosic or lignified materials or are parasites of living hosts, as well as being symbiotic with orchid plants. The hyphae penetrate the cells of the absorbing organs whether root or rhizome, forming coils within the living cells. Initial infection or orchid seedlings encourages them to grow, carbon compounds as well as other nutrients being derived from the fungus. Infection of the protocorms and roots of orchid plants is essential in the case of seedlings in the phase before photosynthesizing organs are formed, which may be very prolonged, and for achlorophyllous orchids throughout their lives.

9.2 Initiation of Infection

In all kinds of mycorrhiza, except those of achlorophyllous plants such as Monotropaceae and Orchidaceae, the fungal symbiont derives carbon compounds from the photosynthesizing host plant (MELIN and NILSSON 1957, LEWIS and HARLEY 1965a, b, c, HO and TRAPPE 1973, BEVEGE et al. 1975, SNELLGROVE et al. 1982). Amongst the fungi that have been grown and tested in culture few have more than a very limited ability to use carbon polymers such as cellulose. For this reason alone they are ecologically restricted to habitats where a supply of simple soluble carbohydrate is available. This is, of course, also assumed of those which have not so far been successfully grown in culture. In the soil the most important habitats where simple compounds are available are on the surfaces of roots and in the rhizosphere region of green plants. Indeed the initial phase of infection by mycorrhizal fungi is the development of mycelium on the root surface, where not only a supply of carbohydrate is released but also other potential nutrients such as organic and amino acids and a variety of vitamins, and growth substances which many mycorrhizal fungi need to have exogeneously supplied (see ROVIRA 1965, see SCOTT RUSSELL 1977). The mycorrhizal fungi are therefore part of the rhizosphere or root surface population of microorganisms for this part of their existence, but the extent to which they are specifically limited to the environs of potential mycorrhizal hosts in this phase has not been sufficiently investigated. As we shall see, the specificity of vesicular-arbuscular mycorrhizas to potential hosts is very low and most tested species are extremely catholic in their choice of mycorrhizal partners, so they may be able to spread in any root region. A very few species of ectomycorrhizal fungi have been tested in respect of their possible existence in the rhizosphere of non-hosts, by THEODORU and BOWEN (1971). Some showed an ability to spread in the rhizosphere of plants which do not form extomycorrhiza at all. It seems possible that some of the fungi of ericoid mycorrhiza may have the like property. There is much doubt whether in the early stages of association there is any specific influence generated by a potential host plant to attract compatible fungi. In the case of vesicular-arbuscular mycorrhiza it has been observed that hyphae from germinating propagules do not grow towards a host root but at random till within 1–2 mm of it. However, in some cases at that kind of distance, the hyphal tip branches repeatedly close to the root, so that a cluster of appressoria form on the root surface (Mosse and Hepper 1975, Powell 1976). This reaction requires study to determine what generates it. The fungi of ectomycorrhiza have been found to be stimulated in their growth by products of roots, but not necessarily by those of potential hosts alone. This subject was reviewed by Melin (1959) in volume XI of the First Series of this Encyclopedia and few advances in the subject have been made since. However, recently Fries (1981) and Birraux and Fries (1981) have investigated the reaction of the spores of species of ectomycorrhizal fungi to the proximity of roots. Not only did they confirm the conclusions of others that the basidiospores germinate more readily in the environs of roots, but that some, e. g. *Thelephora terrestris* were stimulated greatly to germinate only by roots of potential hosts and not by others.

Since the mycorrhizal fungi are at best slow-growing in culture, their ability to spread against the competition of other micro-organisms in the rhizosphere poses problems. As has been said, growth factors, such as the M-factor of Melin, are not sufficiently specific to them to explain their apparent ability to compete with other, more rapidly growing, denizens of that habitat. A few ectomycorrhizal fungi have been shown to produce antibiotics in culture (see Marx 1972, 1973, 1975, and see Harley and Smith 1983), but this is neither a common feature to most of them nor to mycorrhizal fungi in general.

It is not uncommon to describe the fungi of mycorrhiza as ecologically obligate symbionts and yet to imply that the host plant is in some way different, i.e. less obligately symbiotic. This is not really so. It is unusual to find any mycorrhiza-forming host plant always uninfected in a natural habitat. Of course almost all of them can be grown uninfected in soil-culture given adequate nutrients, water and light; indeed in rich surroundings the extent of development of mycorrhiza may be reduced even when an inoculum is present. However, in the natural competition of plant associations they are mycorrhizally infected and derive nutrients from the soil mainly through their mycorrhizal fungus. In the same sense the mycorrhizal fungi appear to be ecologically obligate symbionts. The vesicular-arbuscular fungi are indeed only known as symbionts, but many species from other kinds of mycorrhiza have been grown in culture, although some need special nutrition, including the vitamins and unknown root-derived growth factors. The extent of their ability to persist free in the population of micro-organisms of the root surface is unknown, but since their specificity to particular hosts is low (see later), it is not as important an ecological consideration as was once thought.

The references given in the opening lines of this article will fully cover the points made, they will also describe the nutritive relationships between the symbionts. Here let it be shortly stated that the fungal hyphae in the soil absorb nutrients by active metabolic means and translocate them throughout their my-

Table 1. Rates of inflow (mol cm^{-1} root length s$^{-1} \times 10^{-14}$) of phosphate into mycorrhizal and non-mycorrhizal roots of onion. The hyphal inflow, in the same units, is calculated from the percentage of the root length which is infected. The flux in the hyphae is calculated assuming a total cross-sectional area of 4.7×10^{-6} cm^2 of hyphae cm^{-1} root length based upon estimates of the number and diameter of hyphae entering the roots. (Data of SANDERS and TINKER 1973)

	Total inflow		Infected length (%)	Hyphal inflow	Hyphal flux
	Mycorrhizal	Non-mycorrhizal			
Experiment A	13.0	4.2	50	17.6	3.8
Experiment B	11.5	3.2	45	18.5	3.8

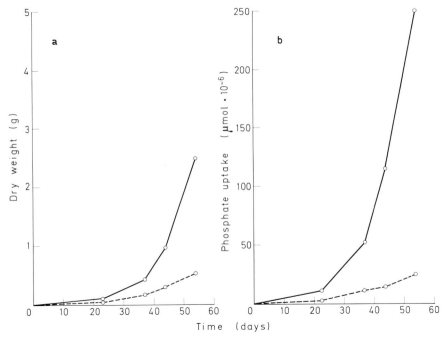

Fig. 3. a The dry weight of mycorrhizal (*full lines*) and non-mycorrhizal (*dashed lines*) onion seedlings during a growth period of 54 days. **b** The amounts of phosphate in μmol per pot of seedlings absorbed by the same populations of mycorrhizal (*full lines*) and non-mycorrhizal (*dashed lines*) onion. (After SANDERS and TINKER 1973)

celium. This process is especially important for the absorption of nutrients such as phosphate, ammonium, potassium, etc. which are relatively immobile and have low movement coefficients in the soil (see NYE and TINKER 1977, SANDERS and TINKER 1973). Such ions, being rapidly absorbed by the roots of a plant, become deficient in their immediate environs where a zone of deficiency forms, so that the rate of uptake by the absorbing surface tends to depend more and more upon the rate of diffusion of these ions and less upon the properties of the absorbing surface. The hyphae serve to cross this deficiency zone, and the rates of translocation of nutrient ions through them to any sink such as the root cells exceeds that of diffusive movement in the soil, and has been shown to be sufficient to explain the greater rates of uptake of nutrients like phosphate into the mycorrhizal host (see Table 1 and Fig. 3a and b). The fungus, for its part, receives carbon compounds from the host. This has been demonstrated by experiments in which $^{14}CO_2$ was photosynthesized by the host and products have been traced to the fungal structures (see HARLEY 1969, Ho and TRAPPE 1973, STRIBLEY and READ 1974, BEVEGE et al. 1975, SNELLGROVE et al. (1982). There is a net movement of carbon compounds from the photosynthesizing host to the fungus, and this is widely believed to result from the maintenance of a sink in the fungal mycelium and tissues. The sink may be the result of rapid accumulation of storage products in the fungus (LEWIS and HARLEY 1965a, b, c) or of rapid growth and consumption. Nevertheless, movement of particular carbon compounds in the reverse direction into the host is to be expected and has been shown to occur in the transfer of amino acids (LEWIS 1976); but the net movement is to the fungus.

9.3 Structure and Development of Mycorrhizal Roots

The demonstration of a bidirectional transfer of nutrients between mycorrhizal symbionts immediately raises questions about the structure of the interface between them. It is often tacitly assumed that this interface is constant in structure during the life of a mycorrhizal organ, but this is not true, as HARLEY and SMITH (1983) have emphasized. All mycorrhizal organs go through a sequence of development, maturation and senescence in which the relationship between the symbionts changes. Moreover, a whole root system is mycorrhizally infected and different parts of it, different orders of root in the hierarchy of branches, may have different relationships with the mycorrhizal fungus. The organs called mycorrhizas are essentially those organs in which there is a relationship between metabolically active symbionts, and they often seem to be recognizable as separate joint organs, or mature states of the joint absorbing system. This is especially true of ectomycorrhizas, and it is therefore easiest to introduce the subject by reference to them.

The root systems of ectomycorrhizal plants consist of main roots or their branches, which continue growing indefinitely, having an apex capable of continuous cell division and growth. These are the permanent axes of the system and serve to extend its zone of exploitation. They bear branches, some of which repeat the continued activity of their parent axis and some are restricted in extent of growth. The ultimate rootlets of the system are very restricted in length of life (a few months to perhaps a year) and in linear extent. They become fully and permanently enclosed in the fungal sheath and form a Hartig net. They are called mycorrhizas – joint fungus–root absorbing organs – the feeding roots of ectomycorrhizal trees. Nevertheless all the roots of the system are infected by the fungus in some manner or at some period of their lives. This has been further illustrated by the work of WILCOX (1968a, b), PICHÉ et al. (1981) and CHILVERS and GUST (1982). The surfaces and tissues of the long roots are colonized in various ways by the mycorrhizal fungus. A weft of hyphae or hyphal strands or even a compact sheath may colonize the surface, and a Hartig net may be formed in the cortex. ROBERTSON (1954) showed that the long roots of *Pinus sylvestris* developed a Hartig net within the cortex but no sheath, and this has been found to be true of many other conifers. CLOWES (1951) described the fungal sheath on the long roots of *Fagus* but no Hartig net within them, and since then a superficial development of a fungal sheath without penetration into the cortex has been observed in the species of other genera such as *Nothofagus, Eucalyptus* and in some conifers. The branches of the long roots either become infected as they meet the fungus in the Hartig net or sheath as they grow out into the soil, or sometimes they remain uninfected. Uninfected root branches, according to the size of their meristems, grow for greater or lesser time before their rate of growth decreases (WILCOX 1968a, b, WARREN WILSON 1951). Their apical structure changes as their growth rate decreases and they become infected, forming sheath and Hartig net. The changes in the apex, described in detail first by WARREN WILSON (1951), result in a reduction of the size of the group of initials and development of mature tissue of cortex and stele near to the apex and the initiation of a histological appearance like that of the host tissue of ectomycorrhiza. It is important to note that these changes may occur in soil lacking mycorrhizal fungi and are not generated by fungal influence, as WARREN WILSON showed, but when mycorrhizal fungi are present they are followed by the establishment of mycorrhizal sheath and Hartig net on the apical region. The shortest lateral roots become totally converted to mycorrhizal organs, the longer ones develop full mycorrhizal structure at the apex, and may remain uninfected in their proximal regions. The extreme development of this differentiation in the root system, this heterorhizy, is seen in some species of *Pinus* where the ultimate rootlets have extremely small meristems and rootcaps, and simple, even monarch, steles. They often grow little after penetrating through the cortex of their mother root and may soon abort if they remain uninfected. They grow and branch dichotomously after infection, but their dichotomy is not essentially a property conferred by fungal infection, for uninfected short roots may dichotomize in culture (BARNES and NAYLOR 1959a, b, FAYE et al. 1982, see HARLEY 1969) in the absence of fungi. In some ectomycorrhizal species of tree almost all the roots therefore are associated with the mycorrhizal fungus in some way. The higher orders of branch become fully enclosed in the fungal sheath and yet continue to grow and remain fully enclosed, for the growth of the sheath keeps pace with that of the host. At times of rapid root growth especially, the host axis of the largest of these fully developed mycorrhizas may break through the sheath and grow free and uninfected for a period, but those of the highest orders remain fully enclosed permanently. If the structural relationship between fungus and host be examined, it shows, as ATKINSON (1975) illustrated, a sequential change from the apex to proximal region of an axis in the middle or lower orders in the hierarchy (Figs. 4 and 5). At the apex the

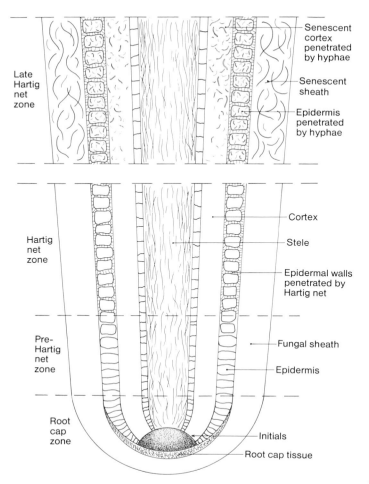

Late
Hartig
net
zone

Hartig
net
zone

Pre-
Hartig
net
zone

Root
cap
zone

Senescent
cortex
penetrated
by hyphae

Senescent
sheath

Epidermis
penetrated
by hyphae

Cortex

Stele

Epidermal walls
penetrated by
Hartig net

Fungal sheath

Epidermis

Initials

Root cap tissue

Fig. 4. Diagram to illustrate the zones of contact of fungus and host in ectomycorrhiza as described by ATKINSON (1975). In the root-cap zone the fungus is in contact with the senescent cells of the root cap. In the pre-Hartig net zone the fungal sheath is in contact with the living epidermis. In the Hartig net zone the fungus also penetrates between the epidermal and sometimes the cortical cells as living hyphal systems. In the late Hartig net zone the epidermal and cortical cells become penetrated by fungal hyphae and the sheath begins to senesce. Secondary thickening begins to form secondary stelar and cortical tissues

fungal sheath overlies the root cap which consists of relatively few degenerating cells and a layer of living products of the cap meristem. Here the fungus is in contact with non-living degenerating cells. Behind this root-cap zone the fungal sheath overlies the living epidermal cells of the root – a zone called by ATKINSON the pre-Hartig-net zone. Proximal to this again, beginning in a region where the cells of the axis are reaching their mature size, is the Hartig-net zone where living active fungal cells of sheath and Hartig net, and living

Fig. 5. Part of the root system of *Fagus sylvatica* from the humus layer of forest soil. The main axis, which is secondarily thickened, bears branches which are completely enclosed in fungal sheath. These further bear four lower orders of branch, all enclosed. The main branches and some of the first order of their laterals may display all the zones described in Fig. 4. The later orders do not usually display the late Hartig net zone, and are relatively short-lived. Those branches that do possess the late Hartig net zone have grown and may continue to grow for two or more seasons

cortical cells are closely and extensively in contact. Behind this again secondary thickening of the stele and of the cortex results in incipient senescence of the primary cortex. Here the living fungal hyphae penetrate the senescent primary cortical cells. This, ATKINSON called the late Hartig-net zone. The extent of the latter zone seems to depend on the position in the hierarchy of the root examined. In those of the highest orders there is a tendency to a more general senescence. There seems to occur, as STRULLU (1976) pointed out, a senescence often beginning with the outer layers of the fungal sheath and proceeding centri-petally. In any event there is a change of relationship between the symbionts so that finally, in the late Hartig-net zone, the hyphae proliferate in senescent dead host cells, although there is prolonged phase of close inter-relationship between living active associates in the Hartig net zone. The lengths and temporal extents of the zones of differing relationship change with time of year and with position in the hierarchy of root branches and are different in different species of host.

We can compare this sequence with that in vesicular-arbuscular mycorrhiza which has been particularly observed in the early phases of the growth of the hosts. After spreading in the root region, the apices of the hyphae either give rise to appressoria on the root surface or branch repeatedly near the surface to form a pre-infection fan of subsidiary hyphae from which many appressoria are formed, and penetration occurs (Mosse and Hepper 1975, Powell 1977). This process of infection occurs on the roots of primary structure. Mathematical modelling of the infection process by Smith and Walker (1981) has shown that penetration occurs in *Trifolium* with a ten times greater probability in the apical region than elsewhere in the root system of 35-day-old plants. Within the root the hyphae spread longitudinally, mostly intercellularly, to form an "infection unit" arising from a single entry. Following this, penetration of the cells results in the formation of arbuscules very rapidly within them. In arbuscule-containing cells the association is between metabolically active fungal hyphae and metabolically active host cells at first. The life of the arbuscule is very short and in various combinations of different hosts and fungi it has been estimated that the period of their activity is between about 2 and 15 days. The collapse and degeneration of the arbuscule which follows is a phase of association of a degenerating hypha within an active host cell. But the cell remains alive and may sometimes be susceptible to re-infection by the fungus. Although not fully studied as a sequence the secondary thickening of the roots results in the axes losing their infected cortex.

The progress of the relationship of host and fungus in ericoid mycorrhizas has some similarities with that of vesicular-arbuscular mycorrhizas but there are important differences. The hair roots of many Ericaceae are much more transitory than most of those with vesicular-arbuscular infection. In them the fungus penetrates, without forming appressoria, near their apices, entering directly into the cells or into the cortex via the interstices between the cortical cells. The cortex is usually of a single layer of cells which become heavily colonized by hyphal coils. After a period when the active hypha and host cell are in contact, the latter senesces and the fungus persists for a time in the dead cell (Fig. 1 B). Not all the adjacent cells in the region of the cortex are necessarily senescent together, but the whole hair root is short-lived and there is no re-infection of host cells because they lose their contents before the hyphae senesce (Bonfante Fasolo and Gianinazzi-Pearson 1979, Duddridge 1980, Bonfante Fasolo et al. 1981a). In this respect there is some resemblance to the senescent phase of ectomycorrhizas. In *Rhododendron* the root system is more robust. In this there is, according to the description of Peterson et al. (1980), a further resemblance to ectomycorrhiza because some of the main axes of the system are more persistent and undergo secondary thickening, so producing senescent primary cortical cells in which the fungus may persist for a time.

In the Orchidaceae, as is well known, the fungal hyphae or fungal coils degenerate in physiologically active cells just as the arbuscules do in vesicular-arbuscular mycorrhiza. BURGEFF (1932, 1936) described several patterns of "digestion", it being thought that an active enzymic activity on the part of the host engendered the fungal collapse. This was interpreted by various observers either as a defensive activity on the part of the host or as a nutritional activity that released essential nutrients from the fungal body. However, the cells of the orchid remain alive and may be re-infected by hyphae once more and again (BURGEFF 1932, STRULLU 1976).

It should be noted that the repeated colonization of the cells of the orchid by the fungal hyphae suggests that some signal of attraction is produced by the cells. This has been interpreted as additional evidence that the orchids are "parasitic" on their mycorrhizal fungi from which they obtain carbon compounds and other nutrients. This kind of repetitive penetration and degeneration of hyphae in the cells has also been described as occurring in some vesicular-arbuscular infections. However, in that case the fungus has been shown to receive carbon compounds from the host and, what is more, cannot live in any known conditions in the absence of its host, whereas the potentially free-living fungi of orchids release carbon to their host. There is therefore no virtue in attaching names like parasite to the host where degeneration of some of the hyphal structures takes place in the tissues. What is needed is further investigation of the physiology of degeneration and of the reactions of the participants over the whole period of associated growth.

There are certain common features in the onset and development of infection in all types of mycorrhiza. Penetration of the tissues and of the living and active cells usually takes place in a region behind the growing point of the root. This region was described in ectomycorrhiza by MARKS and FOSTER (1973) as the mycorrhizal infection zone. The actual meristem is not invaded in any kind of mycorrhiza in normal conditions, but there is doubt whether this is due to its possessing an "immunity" or to the fact that in roots it is constantly renewed and transitory in position, always moving forward in the substrate. Two points, however, speak for a degree of immunity. In orchid embryos and protocorms the behaviour is similar, the mature cells at the suspensor end of the embryo and in the cortical region become infected by the fungi but the meristem does not. In ectomycorrhiza the fungal sheath may overarch and enclose the apex, but the meristem does not become invaded by the Hartig net. In any event, there is in all mycorrhizas an apical zone of uninvaded tissue which includes the apical meristem and the region where cells are dividing and expanding to their full size. Behind this, where the full size of cells has been or is being achieved, invasion into the middle lamella region and sometimes through the cell walls occurs.

There clearly exists in all mycorrhizas a period or zone where physiologically active cells of both participants are intimate contact. As HARLEY and SMITH (1983) emphasize, it is in that phase and in that phase only when we would expect bi-directional movement of substances to occur between the symbionts. They present to one another then, and only then, the maximum area of contact between living cells. Before that, the contact is poorer and less extensive; after that either the fungus degenerates or the host cells senesce, or both. It is therefore in that phase that interest centres on the detailed nature of the contact between the symbionts such as can be revealed by fine structural study.

9.4 Fine Structural Relationship Between the Symbionts

The contact between the physiologically active symbionts at the cellular level shows great similarity in certain particulars in different kinds of mycorrhiza, in pathogenic biotrophic infections of green plants by fungi and in bacterial and actinomycete symbioses. Considerable study in recent years has been made of the contact region in vesicular-arbuscular infections (Scannerini and his colleagues 1967–1979, Bonfante-Fasolo 1978, Bonfante-Fasolo and Scannerini 1977, Bonfante-Fasolo et al. 1981 a, b, Cox and Sanders 1974, Cox et al. 1975, Dexheimer et al. 1979, Gianinazzi-Pearson et al. 1981, Strullu 1976, 1978, Holly and Peterson 1979, Kinden and Brown 1975 a, b, c–1976, and many others). As the penetrating fungal hypha enters the living host cell, the wall of the cell and its plasmalemma appears to invaginate, so that the hypha is surrounded by a wall derived from the host. The penetrating hyphae do not break through the host plasmalemma, which remains intact, surrounding the developing host wall that encloses the intracellular hyphae. Near the point of entry the hypha is surrounded by a layer of wall which seems to be similar to the normal wall of the host cell, but as the hypha or arbuscule grows this becomes thinner and between the host plasmalemma and the hyphal wall a fairly wide matrix, the interfacial matrix, develops in addition or instead. This matrix surrounds the whole hypha, whether coil or arbuscule, even in its finest branches. It contains numerous vesicles apparently derived from the plasmalemma of the host, and scattered polysaccharide fibres. The infection of Ericaceae has many similarities. Again the host wall and plasmalemma seem to invaginate as the mycorrhizal hypha penetrates. The cytoplasm of the host is separated by its plasmalemma and a fibrilla interfacial matrix from the wall of the hypha (Nieuwdorp 1969, Bonfante-Fasolo and Gianinazzi-Pearson 1979, Duddridge 1980, Peterson et al. 1980). An interfacial matrix of similar structure and origin is present in *Arctostaphylos uva-ursi* as described by Scannerini and Bonfante-Fasolo (1983), Duddridge (1980), and in orchids as shown by Strullu (1976). In all of these endomycorrhizas, as Strullu emphasized, there are cytoplasmic changes in the host, particularly the increase in volume of the nucleus and often an increase of cytoplasm and the organelles.

Although the fungal hyphae in ectomycorrhizas do not penetrate into the cells of their host, the relationship between the symbionts in the Hartig net region is not entirely different. The fungus penetrates into the cell walls of the host through the middle lamella region, and although some have described unmodified host walls and hyphae in juxtaposition (Foster and Marks 1966), others have observed modifications of the wall of the host and also that of the hyphae (Fig. 2B). Strullu (1974) and Strullu and Gerault (1977) describe a modified "contact" zone between ascomycetous hyphae and host in *Betula* and *Pseudotsuga,* and SCANNERINI (1968) an "involving layer", probably polyuronides, as a modification of the wall of *Pinus strubus* in contact with *Tuber.* ATKINSON (1975) and DUDDRIDGE (1980) both observed that modified or unmodified interfaces might exist between the symbionts of ectomycorrhizas in the Hartig net region.

In all these cases, as HARLEY and SMITH (1983) emphasized, the area of contact involves the living plasmalemmae of both participants separated by an apoplastic space which consists of the hyphal wall which is sometimes modified, and a greater or lesser modification of the wall of the host. As the association ages, it is as if the fungus becomes progressively less able to maintain or generate the modification of the wall of the host in vesicular-arbuscular and orchid mycorrhizas. The host plasmalemma continues to release polysaccharide-containing vesicles and to form fibrous polymers, but recovers its ability to organize them into a coherent wall structure. The arbuscules and the hyphal coils in vesicular-arbuscular mycorrhiza, and the fungal coils in orchid mycorrhiza become progressively encapsulated in a host wall system to form the "digestion clumps" or "sporangioles" of the older workers, in which the degenerating fungus is sequestered from the living host cell. In ericoid mycorrhizas this does not happen for the fungus and host cell either senesce together, or the fungus continues for a time to inhabit and absorb the contents of the defunct host cell (Fig. 1 B). In ectomycorrhiza the fungus penetrates the disrupted and senescing cortex and also appears to absorb the contents of the cells.

HARLEY and SMITH (1983) have pointed out the probability that the problems concerning the penetration of the walls by fungi which do not seem to produce pectinases or cellulases may be explicable by these observations on the initiation and development of the contact zones, interfacial matrix, etc. The picture provided is one where the fungus appears to inhibit or modify the enzymic activities of the host in respect of cell-wall production or maturation. The modification of the other activities of the host cells appears to be more extensive in endomycorrhizas than in ectomycorrhizas, for in them considerable changes in nuclear volume are involved, as well as in the enzymes associated with the host plasmalemma especially in the arbuscules of vesicular-arbuscular mycorrhiza. The membrane-bound ATPases of the plasmalemma become concentrated around the arbuscules as they mature (DEXHEIMER et al. 1979, GIANINAZZI et al. 1979, GIANINAZZI-PEARSON et al. 1981, MARX et al. 1982). These are believed to have an important function in the movement of materials between the symbionts. In addition, as these authors have emphasized, there is a considerable change in the enzymes involved in phosphate metabolism in infected cells. The mechanism of these membrane and enzyme changes and their full import cannot receive further useful comment pending a fuller description of their nature, which is being actively sought in several laboratories.

By contrast no such membrane changes have been described in ectomycorrhizas in which similar exchanges of both carbohydrate and inorganic nutrients take place between the partners. In these, theorizing has involved both the possibilities that the natural leakiness of roots and the exfoliation of cells by them may account for movement of substances to the fungus, or that substances or "factors", especially of a hormonal nature, may be produced by one or both partners and influence the membranes of the other. The natural leakiness of roots of seedlings has been shown to be very extensive in many species of plant (see for instance SCOTT-RUSSELL 1977) so that there is some doubt whether an additional mechanism encouraging loss of carbon compounds is necessary to mycorrhizal roots. Certainly the "loss" is not of a very different order of magnitude from that calculated as the gain of carbon in mycorrhizal plants (see HARLEY and SMITH 1983). On the other hand, as WEDDING and HARLEY (1976) indicated, the common fungal carbohydrate, mannitol, if it were to pass in adequate small quantities from fungus to host in ectomycorrhizas, would so affect the activity of the carbohydrate enzymes as to increase the concentration of hexoses within the host cells and hence might increase the leakage of carbohydrates, without a necessity for a change in structure or permeability of the membrane of the host. They suggested that common fungal products of like activity should be sought and investigated, before recourse were made to the assumption that special leakage of hormones or factors was involved.

There have been various views put forward concerning the movement of phosphate to the host that involve the metabolism of polyphosphates which have now been reported to be present in all kinds of mycorrhiza (STRULLU and GOURRET 1973, ASHFORD et al. 1975, LING LEE et al. 1975, COX et al. 1975, CHILVERS and HARLEY 1980, STRULLU et al. 1981, etc.). WOOLHOUSE (1975) suggested that polyphosphate breakdown, coupled with polysaccharide synthesis in the hyphae, might constitute both a system of phosphate release from, and a mechanism of maintaining a carbohydrate sink in, the fungal hyphae. HARLEY and SMITH (1983) in commenting on and elaborating such a mechanism, pointed out that such a coupling would presumably require a quantitative relationship between the reciprocal movements of phosphate and hexose between the partners, as well as a continued uptake of phosphate and generation of polyphosphate in the fungal mycelium. CAPACCIO and CALLOW (1982) have identified polyphosphate glucokinase (as well as polyphosphate kinase and endo- and exo-polyphosphatases) in the tissues of onion root mycorrhiza with *Glomus mosseae*. That enzyme, the product of which is glucose-6-P, may well be involved as a stage in the maintenance of a sink for glucose in the fungal tissue, but as they aver, present information is not sufficient to clarify the mechanism of uptake, translocation and transfer of phosphate or of transfer of carbohydrate.

The present knowledge of the kinds of substances transferred between partners has been fully discussed by HARLEY and SMITH (1983), but as they have pointed out, more investigation of the interface between the symbionts, especially in terms of both its structure and biochemistry and of potentially mobile metabolites by use of labelling techniques, is required for further advances to be made.

Emphasis is now being placed, in endomycorrhizas as it has always been in ectomycorrhizas, more upon the possibility of movement of substances between living cells than on the processes which used to be called digestion. Even in the achlorophyllous host, digestion of the fungus has been more and more discounted as a source of supply of nutrients. Not only has it been suggested by MOLLISON (1943) and HADLEY and WILLIAMSON (1971) that the growth of infected orchid protocorms precedes digestion of the fungus in the cells, but it has been calculated by COX and TINKER (1976) that digestion in vesicular-arbuscular mycorrhiza could only account for a tiny part (1%) of the absorption of phosphate by the host if the arbuscules contained a similar amount of phosphate per unit volume as the hyphae. In vesicular-arbuscular mycorrhiza there is no question of a net transfer of carbon compounds from the fungus to the host by digestion, but this is often surmised to occur in Orchidaceae and other partial or total saprophytes. However, as STRULLU (1976) and others have shown, the walls of the hyphae which might be supposed to be the most important source of carbon compounds, are not disintegrated but encapsulated in a carbonaceous layer by the host. Hence it seems likely that transfer of carbon compounds to the host takes place between adjacent metabolically active hyphae and cells.

The behaviour of the fungus in the holosaprophyte *Monotropa hypopitys* has recently been described by DUDDRIDGE and READ 1982). Their account greatly amplifies that of LUTZ and SJOLUND (1973) with *M. uniflora.* The well-known hyphal pegs which penetrate the cortical cells of the host are at first contained by the invaginated cell wall and plasmalemma of the host. As the shoots of *Monotropa* begin to show aboveground, the pegs produce carbohydrate ingrowths from the invaginated cell walls of the host into the cortical cells. The whole peg and ingrowths are totally surrounded by the plasmalemma of the host. DUDDRIDGE and READ describe the modified structure of the cortical cell of the host as strikingly similar to the transfer cells of GUNNING and PATE (1969). The ingrowths of the wall of the cells of *Monotropa* increase to fill the greater part of the volume of the cell, glycogen disappears from the fungal sheath and Hartig net, presumably indicating a movement of carbohydrate to the site of infection. Eventually at the flowering stage, the tips of the hyphal pegs burst and lose their contents into the host cell. This stage is followed by the degeneration of the Hartig net and fungal sheath. In this symbiosis the degeneration of the fungus is very clearly the terminal stage only of a process in which materials pass between metabolically active partners. It is also clear that there must be one or more factors or conditions which stimulate the fungus to penetrate the cortical cell, develop its peg and then to release carbon compounds, derived from glycogen in the sheath, into the host, so that growth and development of the flowering scape as well as the formation of ingrowths of the transfer cells can occur.

The similarity of the mature cortical cells containing fungal pegs and ingrowths to transfer cells is of interest because the interfacial matrix of the ingrowths is, like those of arbuscules, of great surface area and covered by plasmalemma of the host. It differs in being generated as a carbohydrate matrix of complex surface produced by the host between plasmalemma and a simple fungal surface, compared with a matrix covering the complexly branched hyphae of the arbuscule. These observations recall the description of mycorrhiza in Bryophyta, especially in *Calypogeia trichomanis,* by NEMEC (1899 and 1904) and in other bryophytes by STAHL (1949). In these the hyphae penetrated the rhizoids and formed a kind of pseudoparenchyma, reminiscent of the fungal sheath, at its basal cell. From this pseudoparenchyma, simple (zapfenähnliche), hyphal pegs passed into the overlying cell of the thallus. It is of great interest also that HADLEY et al. (1971) and HADLEY (1975) described hyphal protruberances on the surfaces of the hyphae of *Rhizoctonia* within the host cells of orchid mycorrhiza which were calculated to increase the contact area between the symbionts by about 15%. A more recent development is that of ASHFORD and ALLWAY (1982), who discovered a form of mycorrhiza in *Pisonia grandis* in which an extensive fungal sheath was formed but the Hartig net was not well developed. The cells of the host abutting on the fungal sheath formed wall ingrowths which were often branched. The structure of these cells was reminiscent of those of transfer cells, and it should be noted that they do not involve the invagination and elaboration of a host wall and plasmalemma round a penetrating fungal hypha, as do the other structures resembling transfer cells already described.

In all these interfacial structures between the symbionts the important feature seems to be, however, that the plasmalemma of the host is greatly increased in area. This gives added point to the need, expressed by HARLEY and SMITH (1983), for much further work on the microchemial identification of enzymes in the interfacial region, especially on the membranes of the symbionts. The work of MARX et al. (1982), GIANINAZZI-PEARSON and GIANINAZZI (1976, 1978), GIANINAZZI et al. (1979) on localization of enzymes, especially ATPases and phosphatases on the cell membrane of host and endophyte, has been especially valuable in emphasizing that enzymatic changes are associated with the arbuscule and presumably with phosphate transfer.

9.5 Specificity in Mycorrhizal Symbiosis

Although the mycorrhizal condition is usual in land plants, there are no known distinguishing features held in common by all mycorrhizal fungi which are peculiarly theirs. None of the mycorrhizal fungi which are symbiotic with photosynthesizing green plants seems to possess the potential for hydrolysis of carbon polymers such as cellulose rapidly enough to grow at normal rate on the products. There are, however, some which can hydrolyze cellulose to a minor degree, but those genotypes of ectomycorrhizal fungi which have been found to be very active in cellulose hydrolysis have also been found not to be mycorrhiza formers (LUNDEBERG 1970, LINDEBERG and LINDEBERG 1977).

The absence of clear positive characters delineating mycorrhiza fungi as a group is paralleled by a similar lack of diagnostic characters distinguishing those forming each particular kind of mycorrhiza. The fungi of ectomycorrhiza have received detailed study in culture. They belong to families in the Phycomycetes, Ascomycetes and Basidiomycetes, but they possess no known defined physiological or biochemical character which groups them together. Amongst them some form ectomycorrhiza with one host and another kind of mycorrhiza

Table 2. Some examples of mycorrhizal fungi which form either two different kinds of mycorrhiza with different hosts or a mycorrhizal association with one and a pathogenic association with another

Fungus	Mycorrhizal associations		Pathogenic association	Authority
	1	2		
28 spp. of Basidomycete including *Boletus edulis* *Pisolithus tinctorius, Rhizopogon* spp. etc.	Ectomycorrhiza of conifers	Arbutoid mycorrhiza of *Arbutus* and *Arctostaphylos*	–	MOLINA and TRAPPE (1982a)
Boletus sp.	Ectomycorrhiza of pine	Monotropoid mycorrhiza *Monotropa hypopitys*	–	BJÖRKMAN 1960
"E-type" fungi	Ectendomycorrhiza of *Pinus* seedlings	Ectomycorrhiza of *Picea* seedlings	–	MIKOLA 1965
Rhizoctonia solani ≡ *Thanatephorus cucumeris*	Orchid mycorrhiza *O. purpurella* and others	–	Destructive parasite of many herbs	DOWNIE 1957
Rhizoctonia Goodyerae-repentis ≡ *Ceratobasidium cornigerum*	Orchid mycorrhiza *Goodyera*	–	Destructive parasite of herbs	PÉREMBOLON and HADLEY 1965)
Armillaria mellea	Orchid mycorrhiza of diverse orchids (achlorophyllous)	Monotropid mycorrhiza *Monotropa uniflora*	Destructive parasite of woody and herbaceous plants	KUSANO 1911 CAMPBELL 1970, 1971

with another (Table 2). The striking feature is that with one host the fungus may penetrate into the living cells, whereas with another it does not. In both cases there is a period of prolonged association of metabolically active cells of the symbionts within the mycorrhizal organ. In the contrary case, a single species of host plant may form more than one kind of mycorrhiza, each with a different fungus. A single host may form vesicular-arbuscular mycorrhiza or ectomycorrhiza, another ectendomycorrhiza or ectomycorrhiza, and another arbutoid or ectomycorrhiza (Table 3). This kind of variability of behaviour of both hosts and fungi increases the difficulty of penetrating the problems of specificity, recognition, compatibility and effectiveness.

It has been clear for a long time that specificity at the taxonomic level, that is species-to-species restriction of fungus and host, does not commonly exist in mycorrhizal symbiosis and there is great doubt that it exists at all.

Table 3. A few examples of host genera in which some species have been reported to form more than one kind of mycorrhiza alternatively or simultaneously

A. Ectomycorrhizas and vesicular-arbuscular mycorrhizas

Acacia	*Crataegus*	*Helianthemum*	*Myrica*	*Pyrus*
Acer	*Cupressus*	*Juniperus*	*Populus*	*Salix*
Alnus	*Eucalyptus*	*Leptospermum*	*Prunus*	*Shepherdia*
Casuarina	*Fraxinus*	*Malus*		

B. Ectomycorrhizas and arbutoid mycorrhizas or ectendomycorrhizas

Arbutus	*Arctostaphylos*	*Pinus*

C. Ectomycorrhizas and ericoid mycorrhizas

Kalmia	*Ledum*	*Vaccinium*

The data compiled by TRAPPE (1962, 1964, 1969, 1971) for ectomycorrhiza, by GERDEMANN and TRAPPE (1974) for vesicular-arbuscular mycorrhiza and the general discussions of specificity in Orchidaceae, Ericaceae and other groups by HARLEY (1969) and HARLEY and SMITH (1983) clearly show this. MOLINA and TRAPPE (1982) have very recently examined specificity in arbutoid mycorrhizas. Out of 28 isolates of mycorrhizal fungi obtained from mycorrhizas of many species of host, all but three formed arbutoid mycorrhiza with *Arbutus menziesii* and *Arctostaphylos uva-ursi,* but ectomycorrhizas with conifers. That is, the fungi were not restricted to the formation of one kind of mycorrhiza only, nor were they specific as to hosts in forming either kind of mycorrhiza. Table 4 gives examples of the hosts which may be associated with a single fungal species. Since they were compiled there has been much increased information further emphasizing the same point.

As there great physiological and biochemical variation in those mycorrhizal fungi which have been grown in culture, there might be specificity at a finer level, genotype of fungus with genotype of host, but there is no evidence for it, as HARLEY and SMITH (1983) emphasized. MOLINA and TRAPPE (1982b) have classified patterns of specificity in conifer mycorrhiza based upon their own experiments with seven species of host and 29 species of fungus. They recognized three broad categories of specificity. Fungi shown in culture to have a broad host range and whose sporocarps are found in nature in association with many diverse hosts; fungi of intermediate host range in culture whose sporocarps tend to be limited in nature to the stands of a few hosts; fungi forming mycorrhizas in culture with few species of host and whose sporocarps are specific to a small range of particular hosts or to species of one genus of host in natural ecosystems. Examples of the first group include very many of the well-known mycorrhizal species such as *Amanita muscaria, Boletus edulis, Lactarius deliciosus, Laccaria laccata, Paxillus involutus* and *Pisolitus tinctorius,* etc. At the other extreme in the third group, *Alpova diplophloeus* seems to form ectomycorrhiza with species of the genus *Alnus* only, but examples of specificity as restricted as this are at present few. Certain genera of fungi are restricted or mainly restricted to few genera of host. In a study of *Rhizopogon,* a genus restricted

Table 4. Illustrating the general lack of close specificity in mycorrhizal associations. None of the lists is in any sense exhaustive

Fungus	Recorded hosts	Proved hosts by inoculation
A. Examples of wide host range ectomycorrhizal fungi derived from TRAPPE 1962		
Amanita citrina	11	4
Amanita muscaria	24	8
Boletus edulis	24	2
Lactarius deliciosus	22	5
Lactarius helvus	5	3
Scleroderma aurantium	20	5
Suillus granulatus	31	4
Suillus variegatus	7	4

B. Examples of fungi, the same strain of which forms ectomycorrhiza readily with *Pseudotsuga menziesii*, *Picea sitchensis*, *Tsuga heterophylla*, *Larix occidentalis*, *Pinus contorta*, *P. monticola* and *P. ponderosa*. (MOLINA and TRAPPE 1982b)

Amanita muscaria	*Lactarius deliciosus*	*Pisolithus tinctorius*
Astraeus pteridis	*Melanogaster intermedium*	*Scleroderma hypogeaeum*
Cenococcum geophilum	*Paxillus involutus*	*Tricholoma flavivirens*
Laccaria laccata		

C. Examples of some of the hosts of *Glomus fasciculatus* and *G. mosseae* which have been categorically demonstrated in experiments

1 *G. fasciculatus*

Agropyron junciforme	*Geum* sp.	*Plantago lanceolata*
Allium cepa	*Gossypium* sp.	*Potentilla* sp.
Ammophila arenaria	*Hypochaeris radicata*	*Rubus spectabilis*
Bouteloua gracilis	*Liquidambar styraciflua*	*Rubus ursinus*
Citrus spp.	*Liriodendron tulipifera*	*Sitanion hystrix*
Clintonia uniflora	*Lolium perenne*	*Stachys mexicana*
Coprosmia robusta	*Lotus uliginosus*	*Taxus brevifolia*
Crataegus douglasii	*Lycopersicon* spp.	*Thuja plicata*
Deschampsia dianthioides	*Maianthemum dilatata*	*Trifolium repens*
Epilobium watsonii	*Malus* sp.	*Vigna unguiculata*
Fragaria vesca	*Medicago sativa*	*Zea mais*
Fragaria chiloensis	*Ornithogalum umbellatum*	

2 *G. mosseae*

Allium cepa	*Lolium perenne*	*Sorghum bicolor*
Centrosema pubescens	*Lycopersicon* sp.	*Stylosanthes gayanensis*
Citrus spp.	*Medicago sativa*	*Trifolium parviflorum*
Festuca occidentalis	*Nardus stricta*	*Trifolium repens*
Fragaria vesca	*Nicotiana* sp.	*Trifolium pratense*
Glycine max	*Nyssa sylvatica*	*Trifolium subterraneum*
Liquidambar styraciflua	*Pittosporum suspensum*	*Triticum aestivum*
Liriodendron tulipifera	*Podocarpus macrophylla*	*Zea mais*

mainly to conifers, MOLINA and TRAPPE (1982b) observed general similarities in host preference in species included within a section of the genus, but often overlapping preferences in closely related groups of species. These authors suggested however that mycorrhizal specificity and compatibility, as well as sporocarp specificity, might possibly be found to be important characters in delimiting subgenera of *Rhizopogon,* but there was no close specificity of species or subspecies to particular hosts.

HARLEY and SMITH (1983) in a detailed consideration of the implications of the relative unspecificity of mycorrhizal fungi to particular hosts, compared and contrasted them with other endophytes, mutualistic and antagonistic. They emphasized that a lack of selective pressure towards resistance by the host or towards incompatibility by the fungus is to be expected. Any genetic change in these directions will, as VANDERPLANK (1978) indicated, tend to break the advantageous symbiosis and be selected against. This is in direct contrast with antagonistic symbiosis or pathogenicity by biotrophic fungi, where selection of resistant host genotypes and then fungal pathotypes to overcome them will occur, thus giving rise to close, gene-for-gene, compatibility or resistance. The lack of specificity that results from this absence of selection for resistance in the mycorrhizal plants has very important implications with respect to the process of recognition of one another by the symbionts and to their ecological behaviour alone and in association.

9.6 Effectiveness of Mycorrhizal Infection

The specificity so far discussed has concerned the ability of a host and a fungus to form together a joint mycorrhizal organ of one of the usual kinds. In the experimental work on this subject no consideration has generally been given to the estimation of the effectiveness[1] of the symbiosis in physiological terms. In this respect the work on mycorrhiza usually differs from that on pathogenic symbiosis, where the death of plants or their organs or the development of recognizable disease symptoms in the host may provide a physiological measure of specificity and effectiveness, or that on *Rhizobium,* where both the formation of the nodular structure and the efficiency of nitrogen fixation on some such basis as per unit weight of nodule can be more readily assessed.

The reason for the difference is that physiological effectiveness of mycorrhizal symbiosis is difficult to estimate, and varies much with the conditions under which the symbionts are kept. In any event its normal assessment has usually been by measurement of the size, weight or length of mycorrhizal plants compared with an uninfected control during growth over a period of weeks at least, in controlled conditions. The work on specificity of orchid mycorrhiza is an exception, for there the ability of a fungus to stimulate the growth and development of germinating seedlings has usually been the estimate of the compatibility of the symbionts rather than the detailed structure of the association. The fungi isolated from the roots of adult plants have usually been tested for

---

1 *Effectiveness* of a symbiotic union is used in this essay to denote the extent to which the union increases the growth rate, nutrient uptake, reproduction, etc. of the symbionts. The term effectiveness has sometimes been used to denote only the extent of development of mycorrhizas on the root system of a plant or a population of plants, when inoculated with a fungus under standard test conditions

so-called symbiotic activity with the seeds of various orchids including those of the host of origin. As with other mycorrhizas, no clear specificity but a degree of preference has been found (see HARLEY 1969, HADLEY 1970, WARCUP 1971, 1981). It has also been observed that ecological preference of many of the fungi gives the impression of specificity to one or few hosts at the ecological level similar to the "sporocarp specificity" of some ectomycorrhizal fungi. In a recent paper, ALEXANDER and HADLEY (1983) have examined both the structural and physiological relationships between numerous isolated of *Rhizoctonia goodyerae-repentis* and *Goodyera repens* in the germinating phase. *R. goodyerae-repentis* (*Ceratobasidium cornigerum* (Bourd.) Rogers, as perfect stage, see WARCUP and TALBOT 1966) is a very variable species. All the isolates used formed compatible intracellular infections with the germinating seeds under the cultural conditions employed, but they varied greatly in their stimulation of growth. Some isolates failed to stimulate infected seedlings to grow more than the uninfected control, but at the same time did not show any pathogenic tendency. However, there was a great variation in the rate of growth of protocorms within every culture. Some of the protocorms in a single culture, although not parasitized, grew as slowly as the control, while others grew far faster. The fungal species used may well be heterokaryotic and this might explain not only this variation, but also the changes in compatibility and ability to stimulate growth, which occur during the preservation of the fungus in culture for 1 or 2 years. It is of interest that the common pathogen of crop plants *Rhizoctonia solani* (*Thanatephorus cucumeris* (Frank) Donk.) which is also a mycorrhizal fungus of orchids (e. g. *Dactylorchis*), is heterokaryotic, and single basidiospore isolates differ considerably in pathogenicity (FLENTJE et al. 1970). WILLIAMSON and HADLEY (1970) found that several strains of this fungus, all pathogenic in different degrees on crucifers, were compatible with and stimulated growth of seedlings of *Dactylorchis purpurella*. Their variation in pathogenicity was not correlated with their variability in stimulating growth of *D. purpurella* seedlings.

9.7 Compatibility and Effectiveness

The relative unspecificity of fungi and hosts in all kinds of mycorrhiza comes as a surprise to those who might be led to expect that organisms which form permanent mutualistic associations would be more likely to be highly specific than pathogenic biotrophs whose association is temporary, ending perhaps with the premature death of the host. Indeed it might further be expected that ectomycorrhizal fungi, symbiotic with long-lived perennials, might be more specific than those of annual herbs, but that is not so either. The explanation lies, as has been mentioned, in the selective advantage of the associated state and a consequent selection against variation towards incompatibility. In addition to the nutritional advantages mentioned, there is for the fungus a further selective advantage arising out of non-specificity. There will always be, in any soil or in any ecosystem, a suitable habitat, i.e.: suitable host roots, for a non-specific fungal symbiont. The problem of dispersal and preservation will be simplified for those that are unspecific. The Endogonaceae of vesicular-arbuscular mycorrhiza seem to be the least specific, for they are most widely compatible with land plants of all phyla. They are known as fossils in the Rhynie plants of the Devonian and in later groups in the Carboniferous, Mezozoic and Quaternary eras. Despite wide temporal, spatial and taxonomic distribution of most of the fungi, some, e.g. *Glomus gerdemannii,* are reported to be very restricted in their host range, but these merit detailed taxonomic and physiological investigation. The various species of Endogonaceae exhibit degrees of effectiveness with different hosts in different conditions. This has been fully realized since the work of DAFT and NICOLSON (1966) who showed that different isolates of vesicular-arbuscular mycorrhizal fungi (probably different species) had different effects on the growth in dry weight of a single genotype of host in uniform conditions. GERDEMANN (1975) defined mycorrhizal dependency as "the degree to which a plant is dependent on the mycorrhizal condition to produce maximum growth or yield at a given level of fertility" of soil. MENGE et al. (1978) used as numerical measure of mycorrhizal dependency, the dry weight of a mycorrhizal plant as a percentage of the dry weight of a non-mycorrhizal plant at a given level of soil fertility. They demonstrated a great variation of dependency in *Citrus* seedlings using *Glomus fasciculatus* in artificial soil cultures of different fertility. As they pointed out, variation of dependency may arise from many causes, including the structure of the mycorrhizal roots themselves. Similar results have been obtained with other kinds of mycorrhiza. Not only are different hosts differently dependent on ectomycorrhizal infection, but different fungal species or isolates have different effects on the growth of a single host (see HARLEY and SMITH 1983). So far, inadequate investigation has been made of the causes of this variation in terms of the structure, physiology or biochemistry of the mycorrhizas. STONE, as long ago as 1950, showed that in ectomycorrhizas of pine the extent of the extramatrical mycelium was important in nutrient absorption and indeed, with hindsight, this is clearly to be expected, but the explanations of mycorrhizal dependency of the host or mycorrhizal effectiveness might be found to arise from many causes. These two concepts are essentially similar. "Mycorrhizal dependency" is essentially a crop- or host-orientated concept where interest centres on the growth or development of the host rather than of the fungus. "Mycorrhizal effectiveness" considers the effectiveness of the mycorrhizas as organs active in some physiological processes e.g. nutrient absorption, but it is often measured in similar terms, of dry weight increment per unit time compared to an uninfected control, as mycorrhizal dependency. It might, however, be measured in terms of nutrient uptake per unit of carbohydrate consumed, or nutrient transfer to the host per unit of carbohydrate consumed by the fungus. The factors affecting these measures of mycorrhizal efficiency are reminiscent of those affecting the nodules of legumes, where effectiveness may depend on the structure and rates of growth and senescence of the nodules, as well as upon their biochemical properties of nitrogen fixation per unit mass of active bacterial tissue.

Real advances in these matters have yet to be made. They will depend not only on the detailed descriptions of the mycorrhizal systems in morphological and histological terms, but also on measurement of rates of their physiological processes such as absorption and movement of ions, respiration and carbohydrate consumption and the effects of external factors upon them. To make any practical use of efficient host–fungus combinations, selection of desired genotypes will be essential. At the moment almost nothing is known of the mating systems and the sources of genetic variation, sexual, heterokaryotic and parasexual, of mycorrhizal fungi except by analogy with species belonging to the same major taxon, e.g. Agaricales, Boletales, Tuberales, Pezizales, Mucorales etc., and this is very unsatisfactory indeed. Clearly the whole subject of reproductive biology of the mycorrhizal fungi needs much investigation so that selected and experimentally produced genotypes may be compared upon clones of hosts in controlled conditions.

9.8 Recognition

Very little experimental work has been performed on the subject of the mutual recognition of mycorrhizal fungi and their hosts. As we have seen, their specificity is not close, a single fungal isolate may form vesicular-arbuscular mycorrhiza with a wide range of species of host of all the phyla of land plants in laboratory experiments. In ectomycorrhizal fungi there are grades of specificity but it is possible to compare the ecological distribution of sporophores and of mycorrhizal organs of a particular structure with their expected distribution, on the basis of laboratory synthesis of mycorrhiza. Specificity seems to be closer in the competition of natural vegetation than in pure culture. The extent to which this impression is real is questionable. For instance, the unreliability of the absence of sporophores as indicators of the absence of Basidiomycetes and Ascomycetes is well known. LAMB and RICHARDS (1970) studied mycorrhiza of *Pinus radiata* and *P. elliottii* var. *elliottii* in Australian plantations. They isolated 186 cultures of Basidiomycetes, representing 40 species, of which only four could be identified. From subsequent tests of some of them on their hosts, they concluded that there may be, even in a restricted area, many more fungi forming ectotrophic mycorrhizas with pines than is commonly believed or suggested by the presence of sporophores, and that the species which form large fruiting bodies and are regularly cited as major mycorrhiza-formers may, in fact, have a minor role in mycorrhiza formation in pine plantations. Similarly, much more investigation of the variation with habitat of mycorrhizas formed by known pairs of host and fungus seems to be needed before full reliance can be put on their identification in ecological conditions. At all events, any mycorrhizal host is usually compatible with a wide range of fungi and each mycorrhizal fungus with a wide range of hosts. Moreover at a single time one individual plant may associate with several species of fungi (Table 5). The questions therefore arise as to whether hosts recognize compatible fungi or whether they recognize incompatible ones, whether mycorrhizal fungi recognize potential hosts or non-hosts. Alternatively it is possible that the relatively few plants,

Table 5. Mycorrhizal infection in species of *Festuca*. Species of mycorrhizal fungus identified on each species and their mean number per community and per plant. (Data of MOLINA et al. 1978)

Host *Festuca* species	Fungus[b]											Number of fungal species	
	GLFA	GLTS	GLMA	GLMG	GLMO	GLMI	GICA	ACLA	ACSC	SCRU	AC	Per community	Per plant
Viridula (12)[a]	+	+	+	–	–	+	+	+	+	–	–	3.6	2.7
Idahoensis (6)	+	+	+	–	+	+	+	+	–	–	–	2.7	2.4
Scabrella (5)	+	+	+	–	+	+	+	+	–	–	–	3.2	2.3
Arazonica (2)	+	–	+	+	–	+	+	–	+	–	–	5.0	3.5
Thurberi (1)	+	–	+	–	–	–	+	+	–	–	+	4.0	2.7
Ovina (1)	+	+	–	–	–	+	–	–	–	–	–	4.0[c]	2.3

[a] Numbers in parenthesis indicate numbers of communities sampled

[b] GLFA, *Glomus fasiculatus*; GLTE, *G. tenuis*; GLMA, *G. macrocarpus*; GLMG, *G. macrocarpus var. macrocarpus*; GLMO, *G. mosseae*; GLMI, *G. m. var. geosporus*; GLMO, *G. mosseae*; GLMI, *G. microcarpus*; GICA, *Gigaspora calospora*; ACLA, *Acaulospora laevis*; ACSC, *A. scrobiculata*; SCRU, *Sclerocystis rubiformis*; AC, *Acaulospora* sp.

[c] Unidentified species probably *A. trappei* present

such as species of families like Cruciferae or Caryophyllaceae, which rarely
or never form any sort of mycorrhiza, possess some property or produce some
substance that inhibits or antagonizes the fungi. This latter explanation receives
some support from the fact that some plants that are usually non-mycorrhizal
can be made to accept mycorrhizal fungi if grown with strongly mycorrhizal
hosts in the same culture. *Lupinus* is remarkable amongst the legumes in being
a genus whose species are non-mycorrhizal or little infected by vesicular-arbuscular endophytes. TRINICK (1977) has observed increased infection in species
of *Lupinus* grown with *Trifolium pratense*. Similar effects of mycorrhizal plants
on members of the Chenopodiaceae were observed by HIRRELL et al. (1978).
Ferns are often little infected by mycorrhizal fungi in the sporophyte stage,
but COOPER (1976) and IQBAL et al. (1981) have shown that in natural communities they may accept vesicular-arbuscular endophytes or ectomycorrhizal fungi
according to the infection of the plants associated with them. A more cogent
piece of evidence in this problem is found in the many reports that non-mycorrhizal plants such as those of *Lupinus* or the Cruciferae diminish the mycorrhizal
development of plants grown with them (MORLEY and MOSSE 1976, HAYMAN
et al. 1975). As has been stated, some fungi and hosts show a much closer
specificity than the general. For instance, *Alpova diplophloeus* seems to be restricted to species of *Alnus*, and *Suillus grevillei* has a very small range of
hosts. Species of *Alnus* also form mycorrhizas with few genera of ectomycorrhizal fungi though some, e.g. *Alnus alnus*, form vesicular-arbuscular mycorrhiza.
This might suggest that very selective species like some of those of *Alnus* produce
an inhibitor which is overcome by or inactive upon very few potential mycorrhiza fungi.

If we accept this argument and its conclusions we are driven to accept
the view that a very large number of fungi are compatible with a very large
number of host plants because there is some very common property that is
held by these fungi, which allows them to associate with most land plants
provided the latter do not possess some property antagonistic to them. We
must assume that the majority of land plants are not antagonistic to vesicular-
arbuscular fungi and amongst these there are some also that do not antagonize
ectomycorrhizal fungi in addition. A large number of species of arborescent
Angiospermae and all the Pinaceae do not antagonize ectomycorrhizal fungi,
some of them also allow infection from vesicular-arbuscular fungi. The Arbutoideae accept ectomycorrhizal fungi which in them actually penetrate into the
cells non-pathogenically, although they do not do so in ectomycorrhizal hosts.
This catalogue can be extended to all types of mycorrhiza, and suggests that
mycorrhizal fungi have extensive compatibility with potential hosts which is
perhaps only limited by the inhibitory properties in the host, which itself can
be universal or selective in response to the fungi. MOLINA (1981) describes a
strong host reaction by roots of *Alnus* to the wide-host-range mycorrhizal fungus
Paxillus involutus. The epidermal and cortical cells in contact with the fungus
deposited polyphenolic compounds in their walls. This seemed to prevent exploitation of the cortex to form a Hartig net. Similar reactions to incompatible
mycorrhizal fungi have been observed by DUDDRIDGE (personal communication)
in *Betula*, and by MALAJCZUK et al. (1982, and see MOLINA 1981) in *Eucalyptus*.

In considering these problems, HARLEY and SMITH (1983) point out that the roots of every plant are colonized by a root-surface population of micro-organisms which includes the mycorrhizal fungi. In the first stage of mycorrhizal infection these fungi colonize the surface in competition with other members of the population. The members of the root-surface populations rely for growth particularly on the presence of essential nutrients including carbohydrates, organic and amino acids and a multiplicity of other compounds including vitamins and growth factors in the root exudates and sloughed remains of cells. There is as yet no general explanation of the ability of mycorrhizal fungi which, even when they can be grown in culture, are slow-growing and poor competitors, to compete successfully with the many, often rapidly growing other denizens of the root surface. Constituents of the root surface such as the so-called M-factor of MELIN (see 1959) are not sufficiently specific to provide an adequate explanation. The matter, however, needs further scrutiny in view of NYLUND'S (1981) claim of a degree of specificity towards *Piloderma croceum* of the exudate of *Picea abies,* and of the observation by BIRRAUX and FRIES (1981) that the spores of *Thelephora terrestris* are stimulated to germinate only by the proximity of the roots of potential hosts. However, although growth on the root surface is not a unique property of mycorrhizal fungi, non-pathogenic penetration into the cortex of the root is. The mycorrhizal fungi must all possess the ability to penetrate in the middle lamella region of the cortical tissues and they must have some common property that allows them to do this, and that property differentiates them from the other denizens of the root surface which do not penetrate.

HARLEY and SMITH (1983) have considered this subject in detail and they point out that the available evidence suggests that penetration occurs in a region of the root where the cells of the epidermis and cortex are reaching or have reached their full size and are in the process of laying down their cell walls. The electron microscopic pictures of many mycorrhizas show evidence of an "interfacial matrix", a "contact zone" or an "involving layer" derived from the wall of the host (and perhaps that of the fungus also) separating the hypha from the plasmalemma of the host, both in intercellular and intracellular infections. This indicates an action of the fungus upon the process of condensation of polymers to form the cell walls of the host. Such an action could arise, as VANDERPLANK (1978) proposed in the case of pathogens, by the proteins (enzymes) of the fungal walls complexing with the condensing enzymes forming the cell walls of the host and inhibiting them. It could also arise if the fungus produced on or from its surface an inhibitor of the host enzymes. The appearance of the interfacial matrix in endomycorrhizas gives the impression, as DEX-HEIMER et al. (1979) suggest, of the host plasmalemma continuing to secrete vesicles and to form carbohydrate fibres, but being unable to organize them as a coherent wall. The ability to form a wall is recovered when the fungus senesces, degenerates and becomes encapsulated within the cell. It is a corollary of the views of HARLEY and SMITH that variations of the constitutions of the enzymes which bring about wall-building of the host, or of the proteins of the fungal surface, will provide a primary explanation of the limitation of ecto-mycorrhizal fungi to particular groups of host, ericoid mycorrhiza fungi to others, and vesicular-arbuscular mycorrhiza fungi, to a wide range of hosts.

This hypothesis clearly needs detailed examination, especially into the nature of the enzymes or surface proteins or glycoproteins of the walls or developing walls of both host and fungus. It is implicit in it that wall-synthesizing enzymes of the host are affected in their activity by substances on the fungal wall. In addition, a special examination is needed of the exact location where infection occurs and the condition there of the host cells and their walls. The growth pouch technique of FORTIN et al. (1980), which allows the process of infection to be followed under the light microscope and subsequent sampling and fixation of particular stages, should be of great value in this problem. The hypothesis, however, has the great merit of explaining how it is that fungi which do not secrete enzymes hydrolyzing carbon polymers such as pectins or cellulose penetrate the cortex and into the cells.

9.9 Ecological Aspects of Mycorrhizal Infection

Theorizing about the ecological implications of mycorrhizal infection usually involves the erroneous assumption that plants may very frequently lack mycorrhiza in natural conditions. This arises from the older literature, in which much research was aimed at problems arising from the need to establish exotic trees in habitats far outside their natural ranges. Although the intensity of infection may vary, plants lacking mycorrhiza are not commonly found. Geographically, vesicular-arbuscular infection is worldwide, so that no such problem arose in cultivating plants requiring that kind of infection in new habitats. The problems arose in ectomycorrhizal trees in particular, when they were planted outside their usual ecological range. Ectomycorrhiza is geographically mainly restricted in distribution to the northern and southern temperate and subarctic regions, to montane regions in the tropics and to some seasonal rainfall areas, and it is possessed with minor exceptions only by arborescent or shrubby species. The arbutoid mycorrhiza has a more restricted but somewhat similar range, but the ericoid mycorrhiza is particularly an infection of shrubby plants on the poorest soils in rigorous climates with short growing seasons. HARLEY and SMITH (1983) have discussed this subject following READ (1983), who especially emphasized the importance of ericoid mycorrhiza in the most rigorous habitats. The researches of READ and his colleagues have emphasized the importance of the ericoid mycorrhizal infection in the uptake of nitrogenous as well as phosphatic compounds from the soil and also the resistance to or avoidance of heavy metal toxicity (see READ 1983, for references, and BRADLEY et al. 1981, 1982). It seems likely that the species in Epacridaceae which inhabit ecological situations similar to those of many species of Ericaceae are similar in their mycorrhizal properties.

The intervention of mycorrhizal infection in the processes of nutrient absorption from the soil, which has been so widely and convincingly demonstrated in laboratory and field experiments, must also be fundamental to ecological behaviour. Stated simply, the hyphal connections of the mycorrhiza with the soil enable an extensive and changing exploitation of the soil by the symbiotic system. The hyphae cross the region of nutrient deficiency of relatively immobile nutrients near the root surface and ramify in regions hitherto unexploited. Deficiency zones will form round the hyphae but branching and exploitation of new zones of hyphal growth is less costly per unit-absorbing area than the growth and branching of roots. The carbon compounds necessary for growth and branching as well as for the energy required for active nutrient uptake are derived by the fungus from the photosynthetic process of the host. The fungus is therefore in a competitively strong position as compared with other micro-organisms that have to compete for carbon in the soil. This has an important effect on soil processes, as GADGIL and GADGIL (1975) have shown. The preferential removal of nutrients such as nitrogenous and phosphatic substances from the humus layer may decrease the rate of humus breakdown, for that is usually limited by the availability of inorganic nutrients to the micro-organisms.

HARLEY and SMITH have also pointed out that the low specificity of the symbionts appears to introduce a new dimension into the consideration of competition between plants. Mycorrhizal infection undoubtedly aids the exploitation of soil-derived nutrients, so that a mycorrhizal plant would be expected to compete with advantage against a non-mycorrhizal one. However, in a natural plant community or ecosystem, most of the plants will be mycorrhizally infected in one way or another. All those of whatever species which could form, for instance, vesicular-arbuscular mycorrhiza, might be linked together by one or several mycelia of the species of mycorrhizal Endogonaceae present. Similarly the ectomycorrhizal plants or the ericoid mycorrhizal plants in that habitat would be linked together by their own peculiar coterie of fungi (Fig. 6). Each group could compete with plants infected in a different way. Competition between the hosts infected in a single way would seem to reside more in their ability to act as a sink for the soil-derived nutrients absorbed by the fungus and to release less carbohydrate to the fungus than their competitors, than in direct competition. These aspects of ecology have received very little consideration so far. FITTER (1977) showed that mycorrhizal infection by a common endophyte resulted in improved competition of *Holcus lanatus* against *Lolium perenne*. This resulted at least in part from the reduction in the length of root of *Lolium perenne* caused by infection and an increase in its extent of infection in the presence of *H. lanatus*.

These considerations emphasize the great importance of further examination of specificity and effectiveness in order to elucidate not only their physiological and biochemical bases but also their ecological impact. Effectiveness of infection for the host as defined in terms of MENGE's "mycorrhizal dependency" of the host (MENGE et al. 1978) is based on the increment in weight which is dominantly increment of carbon compounds. It will therefore depend greatly not only upon photosynthetic rate, but also inversely upon the proportion of photosynthetic products which pass to the fungus. It may be assumed that mycorrhizal infection, provided it only improves the supply of soil-borne nutrients to the host, will effectively increase relative growth rate only when photosynthesis is limited by nutrient supply from the soil. If relative growth rate were limited by light intensity, or CO_2 supply for instance, mycorrhizal infection

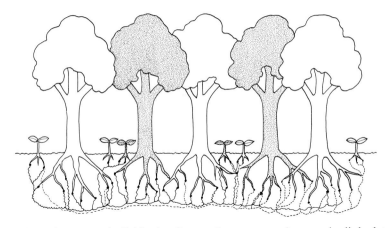

Fig. 6. Diagram to show that many individuals of more than one species may be linked by one or more than one species of fungal mycelium. The carbon compounds released by the dominant may be such as to reduce the demand on the plants under their shade, and competition between species depends upon this relationship with the fungi (see text)

would not increase it but, by the removal of the photosynthate by the fungus, might decrease it. There are many examples where experimental mycorrhizal plants have been observed to be lower in dry weight than their uninfected controls when grown in soils sufficient in nutrients. STRIBLEY et al. (1980) pointed out that plants infected with vesicular-arbuscular mycorrhizas consistently have higher internal phosphate concentrations. These could be explained by the drain on photosynthesis by the infecting fungus.

The quantities of carbon derived by mycorrhizal fungi from the photosynthetic process of their hosts has been estimated in several approximate ways. HARLEY (1973) used the estimates, derived from various authors, of fruiting body production, weight of fungal sheath and its respiration, as well as estimates of photosynthesis, respiration and growth of *Pinus cembra* by TRANQUILLINI (1964) to obtain some idea of the cost in carbon of ectomycorrhizal fungi to their hosts. A conservative amount of the order of one-tenth of that going into wood production in ectomycorrhizal trees was suggested. More recently the part played by mycorrhizal infection in the circulation of carbon in ecosystems has been more intensively studied. This is especially important because only a proportion (about 30–60%) of the CO_2 released by "soil respiration" can be explained by breakdown of detritus (VOGT et al. 1980, and see HARLEY 1973). FOGEL and HUNT (1979) and VOGT et al. (1982) have calculated the proportion of the annual turnover of carbon which is due to mycorrhizal hyphae, sheath, sclerotia and fruit bodies in forests dominated by ectomycorrhizal trees. Values between 14% and 50% are obtained. Approximate or rough as these estimates may be, this valuable and painstaking work indicates that the intervention of mycorrhizal fungi in the carbon budget of ecosystems is important enough to deserve much greater study than it has so far received. In a similar way the cycling of other nutrients through mycorrhizal fungi (see Fogel 1980) needs attention.

Acknowledgements. I am extremely grateful for the assistance given by Dr. S.E. Smith, with whom all the ideas in this article have been discussed. She is not responsible for any errors of fact or of opinions which may have been expressed. I also wish to thank Dr. Jane Duddridge and Dr. Barbara Mosse for allowing me to use the photographs in Figures 1 and 2, and Mrs. E.L. Harley for help in the preparation of the text and bibliography.

References

Alexander C, Hadley G (1983) Variation in symbiotic activity of *Rhizoctonia* isolates from *Goodyera repens* mycorrhizas. Trans Brit Mycol Soc 80:99–106
Ashford AE, Allway WC (1982) A sheathing mycorrhiza on *Pisonia grandis* R.Br. (Nyctaginaceae) with development of transfer cells rather than a Hartig net. New Phytol 90:511–519
Ashford AE, Ling Lee M, Chilvers GA (1975) Polyphosphate in eucalypt mycorrhizas: a cytochemical demonstration. New Phytol 74:447–457
Atkinson MA (1975) The fine structure of mycorrhizas. D Phil Thesis, Univ Oxford
Barnes RL, Naylor AW (1959a) Effect of various nitrogen sources on the growth of isolated roots of *Pinus serotina*. Physiol Plant 12:82–89
Barnes RL, Naylor AW (1959b) In vitro culture of pine roots and the use of *Pinus serotina* in metabolic studies. For Sci 5:158–163
Bevege DI, Bowen GD, Skinner MF (1975) Comparative carbohydrate physiology of ecto- and endo-mycorrhizas. In: Sanders FE, Mosse B, Tinker PB (eds) *Endomycorrhizas*. Academic Press, London New York, pp 152–174
Birraux D, Fries N (1981) Germination of *Thelephora terrestris* basidiospores. Can J Bot 59:2062–2064
Björkman E (1960) *Monotropa hypopitys* L. an epiparasite on tree roots. Physiol Plant 13:308–327
Bonfante-Fasolo P (1978) Some ultrastructural features of the vesicular-arbuscular mycorrhiza in the grape vine. Vitis 17:386–305
Bonfante-Fasolo P, Gianinazzi-Pearson V (1979) Ultrastructural aspects of endomycorrhiza in the Ericaceae. I. Naturally infected hair roots of *Calluna vulgaris* L. Hull. New Phytol 83:739–744
Bonfante-Fasolo P, Scannerini S (1977) A cytological study of the vesicular-arbuscular mycorrhiza in *Ornithogalum umbellatum* L. Allionia 22:5–21
Bonfante-Fasolo P, Berta G, Gianinazzi-Pearson V (1981a) Ultrastructural aspects of endomycorrhizas in the Ericaceae. II. Host endophyte relationships in *Vaccinium myrtillus*. New Phytol 89:219–224
Bonfante-Fasolo P, Dexheimer J, Gianinazzi S, Gianinazzi-Pearson V, Scannerini S (1981b) Cytochemical modifications in the host–fungus interface during intracellular interactions in vesicular-arbuscular mycorrhizae. Plant Sci Lett 22:13–21
Bradley R, Burt AJ, Read DJ (1981) Mycorrhizal infection and resistance to heavy metal toxicity in *Calluna vulgaris*. Nature 292:335–337
Bradley R, Burt AJ, Read DJ (1982) The biology of mycorrhiza in the Ericaceae. VIII. The role of mycorrhizal infection in heavy metal resistance. New Phytol 91:197–201
Burgeff H (1932) Saprophytismus und Symbiose. Fischer, Jena
Burgeff H (1936) Samenkeimung der Orchideen. Fischer, Jena
Campbell EO (1970) Morphology of the fungal associations of *Corallorhiza* in Michigan. Mich Bot 9:108–113
Campbell E (1971) Notes on the fungal associations of two *Monotropa* sp. in Michigan. Mich Bot 10:63–67

Capaccio LCM, Callow JA (1982) The enzymes of polyphosphate metabolism in vesicular-arbuscular mycorrhizas. New Phytol 91:81–91

Chilvers GA, Gust LW (1982) The development of mycorrhizal populations on pot-grown seedlings of *Eucalyptus St. Johnii*. R.T. Bak. New Phytol 90:667–690

Chilvers GA, Harley JL (1980) Visualization of phosphate accumulation in beech mycorrhizas. New Phytol 84:319–326

Clowes FAL (1951) The structure of mycorrhizal roots of *Fagus sylvatica*. New Phytol 50:1–16

Cooper KM (1976) A field survey of mycorrhizas in New Zealand ferns. NZ J Bot 14:168–181

Cox G, Sanders FE (1974) Ultrastructure of the host–fungus interface in a vesicular arbuscular mycorrhiza. New Phytol 73:901–912

Cox G, Tinker PB (1976) Translocation and transfer of nutrients in vesicular-arbuscular mycorrhizas. I. The arbuscule and phosphorus transfer: a quantitative ultrastructural study. New Phytol 77:371–378

Cox G, Sanders FE, Tinker PB, Wild JA (1975) Ultrastructural evidence relating to host endophyte transfer in a vesicular-arbuscular mycorrhiza. In: Sanders FE, Mosse B, Tinker PB (eds) Endomycorrhizae. Academic Press, London New York, pp 297–312

Daft MJ, Nicolson TH (1966) Effect of *Endogene* mycorrhiza on plant growth. New Phytol 65:343–350

Dexheimer J, Gianinazzi S, Gianinazzi-Perason V (1979) Ultrastructural cytochemistry of the host-fungus interfaces in the endomycorrhizal association *Glomus mosseae/Allium cepa*. Z Pflanzenphysiol 92:191–206

Downie DG (1957) *Corticium solani* as an orchid endophyte. Nature 179:160

Duddridge JA (1980) A comparative ultrastructural analysis of a range of mycorrhizal associations. Ph D Thesis, Univ Sheffield

Duddridge JA, Read DJ (1982) An ultrastructural analysis of the development of mycorrhizas in *Monotropa hypopitys* L. New Phytol 92:203–214

Faye M, Rancillac M, David A (1982) Determination of the mycorrhizogenic root formation in *Pinus pinaster* Sol. New Phytol 67:557–565

Fitter AH (1977) Influence of mycorrhizal infection on competition for phosphorus and potassium by two grasses. New Phytol 79:119–125

Flentje NT, Stretton HM, McKenzie AR (1970) Mechanism of variation in *Rhizoctonia solani*. In: Parmeter J (ed) *Rhizoctonia solani*, biology and pathology. Univ California Press, Berkeley pp 52–65

Fogel R (1980) Mycorrhizae and nutrient cycling in natural forest ecosystems. New Phytol 86:199–212

Fogel R, Hunt G (1979) Fungal and arboreal biomass in a western Oregon Douglas fir ecosystem: distribution patterns and turnover. Can J For Res 9:245–256

Fortin JA, Piché Y, Lalonde M (1980) Technique for observation of early morphological changes during ectomycorrhiza formation. Can J Bot 58:361–365

Foster RC, Marks GC (1966) The fine structure of the mycorrhizas of *Pinus radiata* D. Don. Austr J Biol Sci 18:1027–1038

Fries N (1981) Effects of plant roots and growing mycelia on basidiospore germination in mycorrhiza-forming fungi. In: Laursen GA, Amirati YF (eds) Arctic & alpine mycology. Proc 1st Int Symp (FISAM), Barrow, Alaska Aug 1980, pp 1–10

Furman TE (1966) Symbiotic relationship of *Monotropa*. Am J Bot 53:627

Gadgil RL, Gadgil PD (1975) Suppression of litter decomposition by mycorrhizal roots of *Pinus radiata*. NZ J For Sci 5:35

Gerdemann JW (1975) Vesicular-arbuscular mycorrhizae. In: Torrey JG, Clarkson DT (eds) The development and function of roots. Academic Press, London New York, pp 575–591

Gerdemann JW, Trappe JM (1974) The Endogonaceae in the Pacific Northwest. Mycol Mem 5:76

Gerdemann JW, Trappe JM (1975) Taxonomy of Endogonaceae. In: Sanders FE, Mosse B, Tinker PB (eds) Endomycorrhizas. Academic Press, London New York, pp 35–51

Gianinazzi-Pearson V, Gianinazzi S (1976) Enzymatic studies on the metabolism of vesicular-arbuscular mycorrhiza. I. Effect of mycorrhiza formation and phosphorus nutrition on soluble phosphatase activities in onion roots. Physiol Vég 14:833–841

Gianinazzi-Pearson V, Gianinazzi S (1978) Enzymatic studies on the metabolism of vesicular-arbuscular mycorrhiza. II. Soluble alkaline phosphatase specific to mycorrhizal infection in onion roots. Physiol Plant Pathol 12:45–53

Gianinazzi S, Gianinazzi-Pearson V, Dexheimer J (1979) Enzymatic studies on the metabolism of vesicular-arbuscular mycorrhiza. III. Ultrastructural localisation of acid and alkaline phosphatase in onion roots infected with Glomus mosseae (Nicol & Gerd.) New Phytol 82:127–132

Gianinazzi-Pearson V, Morandi D, Dexheimer J, Gianinazzi S (1981) Ultrastructural and ultracytochemical features of a Glomus tenuis mycorrhiza. New Phytol 88:633–639

Gunning BES, Pate JS (1969) "Transfer cells". Plant cells with wall ingrowths, specialized in relation to short distance transport of solutes – their occurrence, structure and development. Protoplasma 68:107–133

Hadley G (1970) Non-specificity of symbiotic infection in orchid mycorrhiza. New Phytol 69:1015–1023

Hadley G (1975) Fine structure of orchid mycorrhiza. In: Sanders FE, Mosse B, Tinker PB (eds) Endomycorrhizas. Academic Press, London New York, pp 335–351

Hadley G, Williamson B (1971) Analysis of pest infection growth stimulus in orchid mycorrhiza. New Phytol 70:445–455

Hadley G, Johnson RPC, John DA (1971) Fine structure of the host fungus interface in orchid mycorrhiza. Planta 100:191–199

Hall IR, Fish BJ (1978) A key to the Endogonaceae. Trans Br Mycol Soc 73:26k–270

Harley JL (1969) The biology of Mycorrhiza. McGraw-Hill New York

Harley JL (1973) Symbiosis in the ecosystem. J Nat Sci Counc Sri Lanka 1:31–48

Harley JL (1978a) Ectomycorrhizas as nutrient-absorbing organs. Proc R Soc Lond B 203:1–21

Harley JL (1978b) Nutrient absorption by ectomycorrhizas. Physiol Vég 16:533–545

Harley JL, Smith SE (1983) Mycorrhizal symbiosis. Academic Press, London New York

Hayman DJ, Johnson AM, Ruddlesdin I (1975) The influence of phosphate and crop species on Endogone spores and vesicular-arbuscular mycorrhiza under field conditions. Plant Soil 43:489–495

Hirrell MC, Mehravaran H, Gerdemann JW (1978) Vesicular-arbuscular mycorrhiza in Chenopodiaceae and Cruciferae – do they occur? Can J Bot 56:2813–2817

Ho I, Trappe JM (1973) Translocation of ^{14}C from Festuca plants to their endomycorrhizal fungi. Nat New Biol 244:30–31

Holley JD, Peterson RL (1979) Development of a vesicular-arbuscular mycorrhiza in bean roots. Can J Bot 57:1960–1978

Iqbal SH, Yousaf M, Jounus M (1981) A field survey of mycorrhizal association in ferns in Pakistan. New Phytol 87:69–79

Kamienski F (1881) Die Vegetationsorgane der Monotropa hypopitys L. Bot Z 29:458

Kinden DA, Brown MF (1975a) Electron microscopy of vesicular-arbuscular mycorrhizae of yellow poplar. I. Characterisation of endophytic structures by scanning electron steroscopy. Can J Microbiol 21:989–993

Kinden DA, Brown MF (1975b Electron microscopy of vesicular-arbuscular mycorrhizas of yellow poplar. II. Intracellular hyphae and vesicles. Can J Microbiol 21:1768–1780

Kinden DA, Brown MF (1975c) Electron microscopy of vesicular-arbuscular mycorrhizae of yellow poplar. III. Host–endophyte interactions during arbuscular development. Can J Microbiol 21:1930–1939

Kinden DA, Brown MF (1976) Electron microscopy of vesicular-arbuscular mycorrhizae of yellow poplar. IV. Host–endophyte interactions during arbuscular deterioration. Can J Microbiol 22:64–75

Kusano S (1911) Gastrodia elata and its symbiotic association with Armillaria mellea. J Agric Tokyo 4:1–66

Laiho O (1965) Further studies on the ectendotrophic mycorrhiza. Acta For Fenn 79:1–35

Laiho O, Mikola P (1964) Studies on the effects of some eradicants on mycorrhizal development in forest nurseries. Acta For Fenn 77:1–34

Lamb RJ, Richards BN (1970) Some mycorrhizal fungi of *Pinus radiata* and *P. elliottii* var. *elliottii* in Australia. Trans Br Mycol Soc 54:371–378

Largent DL, Sugihara N, Wishner C (1980) Occurrence of mycorrhizae on ericaceous and pyrolaceous plants in northern California. Can J Bot 58:2274–2279

Lewis DH (1976) Interchange of metabolites in biotrophic symbioses between angiosperms and fungi. In: Sunderland N (ed) Perspectives in experimental biology, Vol 2. Pergamon, London New York, pp 207–219

Lewis DH, Harley JL (1965a) Carbohydrate physiology of mycorrhizal roots of beech. I. Identity of endogenous sugars and utilization of exogenous sugars. New Phytol 64:224–237

Lewis DH, Harley JL (1965b) Carbohydrate physiology of mycorrhizal roots of beech. II. Utilization of exogenous sugars by uninfected and mycorrhizal roots. New Phytol 64:238–256

Lewis DH, Harley JL (1965c) Carbohydrate physiology of mycorrhizal roots of beech. III. Movement of sugars between host and fungus. New Phytol 64:256–269

Lindeberg G, Lindeberg M (1977) Pectinolytic ability of some mycorrhizal and saprophytic hymenomycetes. Arch Microbiol 115:9–12

Ling Lee M, Chilvers GA, Ashford AE (1975) Polyphosphate granules in three different kinds of tree mycorrhiza. New Phytol 75:447

Lundeberg G (1970) Utilization of various nitrogen sources, in particular bound soil nitrogen, by mycorrhizal fungi. Stud For Suec 79:1–95

Lutz RW, Sjolund RD (1973) *Monotropa uniflora*: Ultrastructural details of its mycorrhizal habit. Am J Bot 60:339–345

Malajczuk N, Molina R, Trappe JM (1982) Ectomycorrhiza formation in *Eucalyptus*. I. Pure culture synthesis, host specificity and mycorrhizal compatability with *Pinus radiata*. New Phytol 91:467–482

Mangin L (1910) Introduction à l'étude des mycorrhizes des arbres forestières. Nouv Arch Mus Hist Nat Paris Ser 5, 2:245–260

Marks GC, Foster RC (1973) Structure, morphogenesis and ultrastructure of ectomycorrhizae. In: Marks GC, Kozlowski TT (eds) Ectomycorrhizae. Academic Press, London New York, pp 1–41

Marks GC, Kozlowski TT (1973) Ectomycorrhizae. Academic Press, London New York

Marx C, Dexheimer J, Gianinazzi-Pearson V, Gianinazzi S (1982) Enzymatic studies on the metabolism of vesicular-arbuscular mycorrhiza. IV. Ultracytoenzymological evidence (ATPase) for active transfer processes in the host arbuscule interface. New Phytol 90:37–43

Marx DH (1972) Ectomycorrhizae as biological deterrents to pathogenic root infections. Annu Rev Phytopathol 10:429–454

Marx DH (1973) Mycorrhizae and feeder root diseases. In: Marks GC, Kozlowski TT (eds) Ectomycorrhizae. Academic Press, London New York, pp 351–382

Marx DH (1975) Role of ectomycorrhizae in the protection of pine from root infection by *Phytophthora cinnamomi*. In: Bruehl GW (ed) Biology and control of soil-borne plant pathogens. Am Phytopathol Soc, pp:112–115

Melin E (1959) Mycorrhiza. In: Ruhland W (ed) Encyclopedia of plant physiology. Springer, Berlin Heidelberg New York, pp 605–638

Melin E, Nilsson H (1957) Transport of C^{14}-labelled photosynthate to the fungal associate of pine mycorrhiza. Svensk Bot Tidsk 51:166–186

Menge JA, Johnson ELV, Platt RG (1978) Mycorrhizal dependency of several citrus cultivars under three nutrient regimes. New Phytol 81:553–559

Mikola P (1965) Studies on ectendotrophic mycorrhiza of pine. Acta For Fenn 79:1–56

Mikola P (1980) Tropical mycorrhiza research. Clarendon, Oxford

Molina R (1981) Ectomycorrhizal specificity in the genus *Alnus*. Can J Bot 59:325–334

Molina R, Trappe JM (1982a) Lack of mycorrhizal specificity by the ericaceous hosts *Arbutus menziesii* and *Arctostaphylus uva-ursi*. New Phytol 90:495–509

Molina R, Trappe JM (1982b) Patterns of ectomycorrhizal host specificity and potential among Pacific northwest conifers and fungi. For Sci 28:423–457

Molina RJ, Trappe JM (in preparation) Cultural descriptions, ectomycorrhiza formation and specificity in the genus *Rhizopogon*

Molina RJ, Trappe JM, Strickler GS (1978) Mycorrhizal fungi associated with *Festuca* in western United States and Canada. Can Bot 56:1691–1695

Mollison JE (1943) *Goodyera repens* and its endophyte. Trans Bot Soc Edin 33:391–403

Morley CD, Mosse B (1976) Abnormal vesicular-arbuscular mycorrhizal infections in white clover induced by lupin. Trans Br Mycol Soc 67:510–513

Mosse B, Hepper CM (1975) Vesicular-arbuscular mycorrhizal infections in root organ cultures. Physiol Plant Pathol 5:215–223

Nemec B (1899) Die Mykorrhiza der Lebermoose. Ber Dtsch Bot Ges 17:311–317

Nemec B (1904) Über die Mykorrhiza bei *Calypogaea trichomanis*. Beih Bot Zentralbl 16:253–268

Nieuwdorp PJ (1969) Some investigations on the mycorrhiza of *Calluna, Erica* and *Vaccinium*. Acta Bot Neerl 18:180–196

Nye PH, Tinker PB (1977) Solute movement in the soil–root system. Blackwell, Oxford

Nylund JE (1981) The formation of ectomycorrhiza in conifers. Structural and physiological studies with special reference to the mycobiont *Piloderma croceum* Erikss. & Agorts. J. Ph D Thesis, Univ of Uppsala

Perémbolon M, Hadley G (1965) Production of pectic enzymes by pathogenic and symbiotic *Rhizoctonia* strains. New Phytol 64:144–151

Peterson TA, Mueller WC, Englander L (1980) Anatomy and ultrastructure of *Rhododendron* root-fungus association. Can J Bot 58:244–263

Piché Y, Fortin JA, Lafontaine JG (1981) Cytoplasmic phenols and polysaccharides in ectomycorrhizal and non-mycorrhizal short roots of pine. New Phytol 88:596–703

Powell CL (1976) Development of mycorrhizal infections from *Endogone* spores and infected root segments. Trans Br Mycol Soc 66:439–445

Powell CL (1977) Mycorrhizas in hill country soils. N Z J Agric Res 20:53–57

Read DJ (1983) The biology of mycorrhiza in the Ericales. Can J Bot 61:985–1004

Robertson NF (1954) Studies on the mycorrhiza of *Pinus sylvestris*. New Phytol 53:253–283

Rovira AD (1965) Plant root exudates and their influence upon soil micro-organisms. In: Baker KF, Snyder WC (eds) Ecology of soil-borne plant pathogens. Univ of California Press, Berkeley, pp 710–786

Sanders FE, Tinker PB (1973) Phosphate flow into mycorrhizal roots. Pestic Sci 4:385–395

Sanders FE, Mosse B, Tinker PB (1975) Endomycorrhizas. Academic Press, London New York

Scannerini S (1968) Sull' ultrastrattura delle ectomicorrize. II. Ultrastruttura di una micorriza di ascomycete *Tuber albidum* × *Pinus strobus*. Alliona 14:77–95

Scannerini S (1972) Ultrastruttura delle endormicorrize di *Ornithogalum umbellatum* L. all' inizio dell' attivata vegetativa. Allionia 18:129–150

Scannerini S, Bellando M (1967) Some ultrastructural features of endotrophic mycorrhiza in *Ornithogalum umbellatum*. Giorn Bot Ital 101:313–324

Scannerini S, Bonfante-Fasolo P (1983) Comparative ultrastructural analysis of mycorrhizal associations. Can J Bot 61:917–943

Scannerini S, Bonfante-Fasolo P (1975) Dati preliminari sull' ultrastruttura di vesicole intracellulari nell'endomicorriza di *Ornithogalum umbellatum* L. Atti Accad Sci Torino 109:519–621

Scannerini S, Bonfante-Fasolo P (1976) Ultrastructural features of a vesicular-arbuscular mycorrhiza. Proc VI Eur Congr Electron Microsc 2:492–494

Scannerini S, Bonfante-Fasolo P (1979) Ultrastructural cytochemical demonstration of polysaccharides and proteins within the host-arbuscle interfacial matrix in endomycorrhiza. New Phytol 83:87–94

Scannerini S, Bonfante PF, Fontana A (1975) An ultrastructural model for the host–symbiont interaction in the endotrophic micorrhizae of *Ornithogalum umbellatum* L.

In: Sanders FE, Mosse B, Tinker PB (eds) *Endomycorrhizae*. Academic Press, London New York, pp 313–324

Scott Russell R (1977) Plant root systems. McGraw Hill, New York

Smith SE (1974) Mycorrhizal fungi. Crit Rev Microbiol 3:275–313

Smith SE (1980) Mycorrhizas of autotrophic higher plants. Biol Rev 55:475–510

Smith SE, Walker NA (1981) A quantitative study of mycorrhizal infection in *Trifolium*: separate determination of the rates of infection and of mycelial growth. New Phytol 89:225–240

Snellgrove RC, Splittstoesser WE, Stribley DP, Tinker PB (1982) The carbon distribution and the demand of the fungal symbiont in leek plants with vesicular-arbuscular mycorrhizas. New Phytol 92:75–81

Stahl M (1949) Die Mykorrhiza der Lebermoose mit besonderer Berücksichtigung der thallösen Formen. Planta 37:103–148

Stone EL (1950) Some effects of mycorrhizae on the phosphorus nutrition of Monterey pine seedlings. Proc Soil Sci Soc Am 14:340–345

Stribley DP, Read DJ (1974) The biology of mycorrhiza in Ericaceae. III. Movement of Carbon-14 from host to fungus. New Phytol 73:731–741

Stribley DP, Tinker PB, Rayner J (1980) Relation of internal phosphorus concentration and plant weight in plants infected by vesicular-arbuscular mycorrhizas. New Phytol 86:261–266

Strullu DG (1974) Étude ultrastructurale du réseau de Hartig d'une ectomycorrhize à ascomycètes de *Pseudotsuga menziesii*. C R Acad Sci Paris Ser D 278:2138–2142

Strullu DG (1976) Contribution à l'étude ultrastructurale des ectomycorrhizes a basidiomycètes de *Pseudotsuga menziesii* (Merb.). Bull Soc Bot Fr 123:5–16

Strullu DG (1978) Histologie et cytologie des endomycorrhizes. Physiol Vég 16:657–669

Strullu DG, Gerault A (1977) Étude des ectomycorrhizes à basidiomycètes et à ascomycètes de *Betula pubescens* Ehrh. en microscopie électronique. C R Acad Sci Paris Ser D 284:2243–2244

Strullu DG, Gourret JP (1973) Étude des mycorrhizes ectotrophe de *Pinus brutia* Ten. en microscopie électronique à balayage et à transmission. CR Acad Sci Paris, ser D 227:1757–1760

Strullu DG, Gourret JP, Garrec JP, Fourey A (1981) Ultrastructure and electron probe analysis of the metachromatic vacuolar granules occurring in *Taxus* mycorrhizas. New Phytol 87:537–545

Theodoru C, Bowen GD (1971) Effects of non-host plants in growth of mycorrhizal fungi of radiata pine. Aust Forest 35:17–22

Tranquillini W (1964) Photosynthesis and dry matter production of trees at high altitudes. In: Zimmermann MH (ed) Formation of wood in forest trees. Academic Press, London New York, pp 505–518

Trappe JM (1962) Fungus associates of ectotrophic mycorrhizae. Bot Rev 28:538–606

Trappe JM (1964) Mycorrhizal hosts and distribution of *Cenococcum graniforme*. Lloydia 27:100–106

Trappe JM (1971) Mycorrhiza-forming ascomycetes. In: Hacskaylo E (ed) *Mycorrhizae*. US Govt Printing Office, Washington

Trinick MJ (1977) Vesicular-arbuscular infection and soil phosphorus utilization in *Lupinus* spp. New Phytol 78:297–304

Vanderplank JE (1978) Genetic and molecular basis of plant pathogenesis. Springer, Berlin Heidelberg New York

Vogt KA, Edmunds RL, Antos GC, Vogt DJ (1980) Relationship between CO_2 evolution, ATP concentration, and decomposition in four forest ecosystems in Western Washington. Oikos 35:72–79

Vogt KA, Grier CC, Meier CE, Edmunds RL (1982) Mycorrhizal role in net primary production and nutrient cycling in *Abies amabilis* (Dougl.) Forbes ecosystems in Western Washington. Ecology 63:370–380

Vozzo JH, Hacskaylo E (1971) Inoculation of *Pinus caribaea* with ectomycorrhizal fungi in Puerto Rico. For Sci 17:239–245

Warcup JH (1971) Specificity of mycorrhizal associations in some Australian orchids. New Phytol 70:41–46

Warcup JH (1975) A culturable *Endogone* associated with eucalypts. In: Sanders FE, Mosse B, Tinker PB (eds) Endomycorrhizae. Academic Press, London New York, pp 53–63

Warcup JH (1981)The mycorrhizal relationships of Australian orchids. New Phytol 87:371–387

Warcup JH, Talbot PHB (1966) Perfect states of some rhizoctonias. Trans Br Mycol Soc 49:427–435

Warren Wilson J (1951) Micro-organisms in the rhizosphere of beech. Ph D Thesis, Univ Oxford

Wedding RT, Harley JL (1976) Fungal polyol metabolites in the control of carbohydrate metabolism of mycorrhizal roots of beech. New Phytol 77:675–688

Wilcox HE (1968a) Morphological studies of the roots of red pine *Pinus resinosa*. I. Growth characteristics and branching patterns. Am J B 55:247–254

Wilcox HE (1968b) Morphological studies of the roots of red pine *Pinus resinosa*. II. Fungal colonization of roots and development of mycorrhizae. Am J Bot 55:686–700

Wilcox HE (1971) Morphology of ectendomycorrhizae in *Pinus resinosa*. In: Hacskaylo E (ed) Mycorrhizae. US Gov Print Office, Washington DC pp 53–68

Wilcox HE, Ganmore–Neumann R (1974) Ectendomycorrhizae in *Pinus resinosa* seedlings. I. Characteristics of mycorrhiza produced by a black imperfect fungus. Can J Bot 52:2145–2155

Wilcox HE, Ganmore–Neumann R, Wang CJK (1975) Ectendomycorrhizas in *Pinus resinosa* seedlings. II. Characteristics of two fungiproducing ectendomycorrhizae in *Pinus resinosa*. Can J Bot 52:2279–2282

Williamson B, Hadley G (1970) Penetration and infection of orchid protocorms by *Thanatephorus cucumeris* and other *Rhizoctonia* isolates. Phytopathology 60:1092

Woolhouse HW (1975) Transport problems in endomycorrhizas. In: Sanders FE, Mosse B, Tinker PB (eds) Endomycorrhizas. Academic Press, London New York, pp 209–239

10 Plant–Bacterial Interactions

L. SEQUEIRA

10.1 Introduction

The present emphasis on increased food and fibre production has underscored the need for new methods to control plant diseases and to fertilize crops in ways that are less energy-intensive than those in current practice. Although many approaches to these problems are possible, one area that offers much promise for the future is the study of plant–bacterial interactions. There are several arguments in support of this statement. First, bacteria cause numerous, devastating diseases of plants in the field and of plant products in storage, for which no adequate controls have been devised. Chemical treatments are either too expensive or are ineffective, and breeding for disease resistance is often constrained by the highly variable nature of plant pathogenic bacteria. Thus, alternative means for control must be devised. Second, bacteria take part in associative and symbiotic relationships with plants that are extremely important because they allow better plant growth and endow certain plants with the capacity to fix atmospheric nitrogen. Yet it is clear that the host range of these bacteria and their efficiency under a wide range of environmental conditions are limited, but could be improved substantially if we knew more about the details of their interactions with plants. Thus, this review will attempt to: (a) assess the present status of the field of bacterial plant relationships, with particular emphasis on the nature of the host and parasite/symbiont surface components that come in contact at the initial stages of the interaction, and (b) to point out how this knowledge may lead to new methods for disease control and for increased productivity.

Microbes preceded plants during evolution and, as higher plants made their appearance on land, interactions between the two groups of organisms became established along widely divergent or, in some cases, closely parallel lines. Since bacteria surrounded both aerial and subterranean plant parts, parasitic or symbiotic relationships probably existed very early in evolution, although the geologic record is not explicit on this point. From our present view, the very surprising fact is that, in spite of such long co-evolution, so very few bacteria have developed the capacity to invade plant tissues. Very few species of bacteria, representing only five major genera, parasitize plants. In addition, it is surprising that most Gram-positive bacteria, which constitute a large portion of the soil microflora surrounding the roots, contain no pathogens of significance with the exception of certain members of the Corynebacteriae. Even fewer species, belonging to only one major genus, have developed a symbiotic relationship that allows them to multiply within plant cells. This is a highly complex relationship, for the bacteria must draw on the host's nutrients but not sufficiently to damage the cells or to cause an overt host response that might destroy the invader.

Because of the large number of openings that are present on the plant's surface (stomata, hydathodes, lenticels, etc.) and the numerous wounds that are caused by insects, the use of agricultural implements, wind action, the abrasive qualities of soil particles, etc., it is surprising indeed that the bacteria that gain entrance to the plant are, for the most part, unable to multiply. Since dead plant tissues are a good substrate for most soil bacteria, it is evident that the living plant has the capacity to ward off or inhibit the development of bacteria that gain access to the intercellular spaces, the interior of wounded cells, or the lumina of xylem vessels. A priori, it can be conjectured that such mechanisms for resistance must be of two types, one of general action against the vast majority of the bacteria that enter the plant, and one specific against particular strains of a pathogen or symbiont. Alternatively, only specific mechanisms of resistance may be active in those instances when compatible pathogens or symbionts avoid and/or inhibit the general mechanisms of resistance.

The relationships of bacteria with higher plants may be classified in three general groups: (a) epiphytic, (b) parasitic, and (c) symbiotic. The epiphytic relationships are characteristic of the rhizosphere and phyllosphere and involve a wide variety of bacteria that are highly adapted to the environment on the surface of the plant. Although these relationships are important in terms of such phenomena as nutrient uptake, frost damage, biological control of plant pathogens, etc. and are fully deserving of consideration in this review, space constraints are such that the discussion that follows must be limited to the *parasitic* and *symbiotic* relationships.

The relationships between plants and parasitic bacteria are not unlike those of animals and their potential pathogens. For many years, animal microbiologists have considered two general groups of pathogenic bacteria: extracellular and intracellular. The extracellular parasites owe their virulence to their ability to resist or inhibit phagocytosis. The intracellular parasites, on the other hand, are highly adapted to ingestion by phagocytes and even use phagocytes as a means for spread within the host. There is an evident analogy between these two types of animal pathogens and those that attack plants. The vast majority of phytopathogenic bacteria are extracellular parasites and apparently owe their virulence to an ability to prevent attachment to the host cell walls. The bacteria that penetrate plant cells, such as rhizobia, or those that owe their virulence to their ability to transfer part of their genome to the plant cell, such as *Agrobacterium tumefaciens* (SCHELL 1982), benefit from the reaction of the host and thus are similar to the intracellular parasites of mammalian cells. Although the latter relationship is ultimately symbiotic in nature, the distinction between symbiosis and parasitism is never entirely clear. For example, certain strains of *Rhizobium japonicum* produce toxins that affect plant growth, whereas some strains of *Agrobacterium rhizogenes* cause no overt damage to the host. These general concepts are useful to our understanding of the mechanisms that plants employ to prevent colonization and of those that pathogens employ to inhibit or bypass the plant's defense mechanisms.

A further generalization may be made regarding the nature of the relationships between plants and bacteria. In most parasitic systems, there has been intense selective pressure for compatibility in the part of the parasite and for incompatibility in the part of the host. The marked specificity of the bacteria that cause foliage diseases of plants indicates a relatively narrow adaptation to nutritional and toxic components of the host environment. The response of the host to the presence of a potential pathogen reaches maximum expression in the hypersensitive reaction (HR). The successful bacterial parasite must avoid or prevent the HR. Contact between the host and parasite apparently is essential for induction of the HR. By analogy with other biological systems, the specificity of the HR probably is the result of a recognition phenomenon in which surface components of both host and parasite play an important role. Thus, this review will deal very largely with the nature of these surface components that determine, to a large extent, the ultimate fate of the interaction between bacteria and their plant hosts.

10.2 Incompatible Interactions

It has been known for many years that plant tissues undergo rapid necrosis when inoculated with high concentrations of incompatible, plant–pathogenic bacteria (DYE 1958). The significance of this phenomenon, however, did not become apparent until KLEMENT (1963) demonstrated that: (a) the collapse of the host cells occurred only in incompatible interactions, and (b) the rapidity of the host response was typical of hypersensitive-type reactions. The development of relatively simple techniques to introduce bacteria into plant leaves, specially of tobacco, opened up innumerable avenues for the study of the physiology and biochemistry of the hypersensitive response. The facts that tobacco cells are extremely sensitive to the presence of incompatible bacteria and react in a rapid, predictable fashion, have made this host a popular choice for experimental work involving the HR.

10.2.1 Characteristics of the Hypersensitive Reaction

When tobacco leaves are infiltrated with at least 5×10^6 cells ml^{-1} of an incompatible strain of a phytopathogenic pseudomonad, collapse of the infiltrated area follows within 6 h at 28 °C. The same reaction is obtained with avirulent variants of compatible strains. Wild-type, compatible strains, on the other hand, may not cause overt symptoms for 48 h and, by the time necrosis appears, the bacterium usually has multiplied beyond the infiltrated area. Saprophytic bacteria, in general, cause no visible symptoms when they are applied to tobacco leaves in the same manner.

Tissues undergoing the HR lose water rapidly, become bleached, and are paper-thin within 24–48 h after infiltration. The collapsed area becomes sharply delimited from the surrounding healthy tissues and the bacteria are effectively contained within the area where they were placed initially.

It is significant that the HR becomes irreversible after an induction period of approximately 3 h in most incompatible host–parasite interactions (SEQUEIRA 1976, MEADOWS and STALL 1981). It is evident, therefore, that what transpires between host and pathogen during this initial period is of great significance in terms of the ultimate effects on the host cell. Yet, most of the research emphasis has been placed on the structural, physiological details of the latent period, the time necessary for completion of the changes that lead to cell collapse. Predictably, marked changes occur in the host plasmalemma, which invaginates and releases numerous vesicles and electron-opaque granular material in the space between the plasmalemma and the cell wall (SEQUEIRA et al. 1977). In addition, there is an increase in the amount of electrolytes that leak out of the cells (GOODMAN 1968) and, as the HR proceeds to completion, other organelles become deranged. Mitochondria lack well-defined cristae and the internal membranes of chloroplasts lose their integrity (GOODMAN and PLURAD 1971).

In addition to the changes in conductivity that precede tissue collapse during the latent period, other important physiological changes occur. Respiratory metabolism increases, maximal rate occurring just before tissue collapse (NEMETH et al. 1969). The activities of polyphenol oxidase, peroxidase, and other oxidative enzymes, which could contribute to the increase in oxygen consumption, do not increase significantly, however.

10.2.2 Factors that Affect Expression of the Hypersensitive Response

The bacteria-induced HR is affected by numerous environmental factors, such as light, temperature, humidity, as well as by the age and concentration of the inoculum, and the age of the inoculated leaf (SEQUEIRA 1979). Of particular interest is the effect of temperature. The HR appears at progressively shorter intervals when tobacco plants are exposed to a temperature range from 16° to 28 °C (LOZANO and SEQUEIRA 1970). As the temperature increases to 37 °C, however, the HR is suppressed, even by exposures to this temperature as short as 4 h immediately after infiltration (SÜLE and KLEMENT 1971). Exposures to high temperatures inhibit the HR only during the induction period. Since the metabolic activity of the bacterium is required during this period, DURBIN and KLEMENT (1977) concluded that high temperature represses the HR by interfering with some temperature-sensitive factor of the bacterium rather than of the plant. The most important factor appears to be the inability of most plant-pathogenic bacteria to multiply at 37 °C, although a requirement for cell division cannot be clearly separated from that of active metabolic activity.

Light appears to affect the HR differently in different host–parasite combinations. In the tobacco–*P. solanacearum* interaction, long exposures to darkness suppressed the HR and shifted a normally incompatible relationship to a compatible one (LOZANO and SEQUEIRA 1970). The contrary appears to be the case in soybean leaves incubated in the dark for 24 h after infiltration with *Erwinia amylovora;* symptoms of the HR were more severe in plants subjected to that treatment (GIDDIX et al. 1981). Others report that the HR is not affected by illumination during the induction and lag periods (KLEMENT and GOODMAN 1968, LYON and WOOD 1976, SASSER et al. 1974, SMITH and KENNEDY 1970). These ambiguous effects suggest that certain products of the photosynthesizing host are inhibitory to bacteria only in some specific instances. In the vast majority of the interactions that have been studied, these products appear to have little or no effect on bacterial growth within the host tissues. This is surprising, in view of the fact that the HR is an energy-demanding process (KEEN et al. 1981) which should be intensified under the higher levels of ATP generated by photophosphorylation.

The evidence that protein synthesis is required for expression of the HR is mostly indirect, however. It is based on inhibition of the HR in leaf tissues pretreated with substances that are known to inhibit protein synthesis in other plant systems. The dangers of this approach are evident; it is often difficult to separate the effects of the inhibitor on the plant from those on the bacteria. In addition, there are non-specific effects of the inhibitor on plant tissues. For example, the HR was inhibited in soybean leaves infiltrated with blasticidin S at the time of, or up to 9 h after inoculation with incompatible strains of *Pseudomonas syringae* pv. *glycinea* (KEEN et al. 1981). Since the incompatible strain multiplied in blasticidin S-treated tissues to the same extent as the compatible strain in untreated tissues, the authors concluded that inhibition of the HR resulted from an effect of the antibiotic on the host. However, considerable multiplication of S aprophytes also occurred in blasticidin S-treated tissues, indicating a high level of toxicity of the antibiotic to plant tissues. Growth of saprophytes would be expected only in dead or dying tissues. Thus, it could not be concluded with certainty that the HR was the mechanism normally preventing multiplication of incompatible bacterial pathogens.

The bacteria-induced HR is also affected by the concentration of divalent cations in the medium used for suspending the inoculum. When inoculum of the pepper strain of *Xanthomonas campestris* pv. *vesicatoria* was amended with Ca^{2+} solutions, or Ca^{2+} was used as a pre-inoculation treatment, electrolyte loss in resistant pepper cultivars was reduced (COOK and STALL 1971). These results are in general agreement with the concept that deterioration of the plasmalemma is the first observable effect of the HR. It is surprising, therefore, that Ca^{2+} enhances electrolyte loss from pepper leaves inoculated with the tomato strain of *X. campestris* pv. *vesicatoria* (COOK 1973). It is difficult to interpret these results, but it must be assumed that the two distinct pathotypes affect the host and induce the HR in quite different ways.

It is interesting that the HR induced by bacteria is inhibited by non-permeating (mannitol) or slow-permeating (sucrose) plasmolyzing agents (GULYAS et al. 1979). Since the development of the HR was inhibited gradually as a function of the concentration of these agents, the HR does not appear to be dependent on cell turgor. This is in contrast with the viral HR, in which discontinuous effects of the plasmolyzing agents, perhaps brought about by breaking of plasmodesmata, indicate a strong dependence on cell turgor.

10.2.3 Attachment of Bacteria to Plant Cell Walls:
A Requirement for Induction of the HR?

The HR induced by bacteria is akin to the self-incompatibility responses in higher plants in which pollen of the inappropriate mating group is prevented from germinating on the stigma, or the development of the germ tube is arrested in the style (Heslop-Harrison 1978). In both systems, there is now considerable evidence that the recognition phenomenon that allows the host cell to distinguish between "self" and "non-self" is dependent on complementary interactions of surface molecules of each partner. For this interaction to take place, there must be close contact between the two surfaces. In the case of phytopathogenic bacteria, incompatibility is established once they come to rest on the surface of mesophyll cells of the host. If this close contact is prevented by maintaining the intercellular spaces continuously water-soaked, or by suspending the bacteria in agar, the HR is inhibited (Cook and Stall 1977, Stall and Cook 1979).

In certain hosts, such as tobacco, the initial "docking" of the bacterial cell appears to be followed by firm attachment to the plant cell wall. In resistant bean plants, on the other hand, there is no evidence that firm attachment is causally related to induction of the HR (Daub and Hagedorn 1980). Although irreversible binding to the host may be unrelated to HR induction, it is an interesting phenomenon that can be used to advantage in attempts to elucidate the nature of the surface components that interact. These studies indicate clearly that attachment is a relatively complex phenomenon. In *P. solanacearum* the initial evidence for attachment was obtained from studies at the electron microscope level in which the incompatible (B1) variant was shown to be enveloped readily by a host surface pellicle. The parental, compatible strain (K60) remained free and multiplied in the intercellular spaces (Sequeira et al. 1977). The processes of attachment and envelopment are relatively rapid; they appear to be complete by the time the first symptoms of the HR appear, usually within 6 h after infiltration. The initial response of the host cell to the presence of the bacterium is an invagination of the plasmalemma in the area immediately below the point of attachment of the bacterium on the cell wall. There is an accumulation of membrane-bound vesicles in the space between the plasmalemma and the cell wall; thereafter, granular and fibrillar material accumulate in this space as well as around the bacterium, which by now is surrounded by a pellicle of apparent cuticular origin. Similar responses have been reported in tobacco inoculated with *P. syringae* pv. *pisi* (Goodman et al. 1976) and in resistant cotton leaves inoculated with *Xanthomonas campestris* pv. *malvacearum* (Cason et al. 1978). That attachment and immobilization of bacteria are not limited to HR-inducing bacteria is clear from evidence that saprophytic bacteria are attached to tobacco (Sequeira et al. 1977) and bean leaf cell walls (Sing and Schroth 1977).

An evaluation of the information on attachment of bacteria to plant cell walls led Hildebrand et al. (1980) to conclude that these phenomena are artefacts resulting from infiltration and fixation procedures. They suggested that in bean leaves the water used as the suspending medium for the inoculum dissolves materials from the cell surface and that, as the water evaporates or is absorbed by the cell, the bacteria are physically trapped by this film. Compatible, pathogenic bacteria, on the other hand, may avoid entrapment by maintaining a liquid phase as they multiply in the intercellular spaces. Although there is no question that bacteria become entrapped in the interstices between adjoining mesophyll cells, there are many reasons to question this physical interpretation of the process of attachment. Many of these arguments were presented by WHATLEY and SEQUEIRA (1981) and there is little point in describing them in detail here. It should be sufficient to indicate that: (a) the response of bean leaves appears to be substantially different from that of tobacco leaves to the same organisms, (b) there is incontrovertible evidence for an active host response to attachment, and (c) polystyrene balls or asbestos fibres that settle on the cell wall surface do not elicit attachment or envelopment in tobacco. Perhaps the most convincing argument in favour of attachment as an active process is the demonstration that it occurs in tobacco cells grown in callus tissue (HUANG and VAN DYKE 1978) and cell suspension culture (DUVICK and SEQUEIRA 1981).

The development of quantitative methods to measure attachment by *P. solanacearum* B1 has allowed an examination of the kinetics of the phenomenon (DUVICK and SEQUEIRA 1981 and DUVICK unpublished results). To assay for attachment, ^{14}C-labelled bacteria are incubated with suspension culture cells (SCC) or isolated tobacco cell walls, and then filtered through a Miracloth or nylon disc. The SCC or cell walls are retained by the disc, but not the bacteria that are free. Thus, counts retained on the discs provide a measure of the bacteria that attach to the cells or cell walls. With this assay, it was shown that B1 cells attach fairly rapidly, reaching saturation at about 90 min on SCC and 30 min on cell walls. Under the same conditions, virulent (K60) bacterial cells showed no significant attachment. The results indicate that attachment occurs in two steps: "docking" and irreversible binding. These then are followed by envelopment under certain conditions. It may be useful to consider each of these phenomena in some detail.

The initial "docking" of bacteria on SCC apparently is an ionic interaction, for it is inhibited by high concentrations of phosphate buffer during the first 15 min and the bacteria can be dislodged by washing with this buffer. Thereafter, however, attachment becomes increasingly tight and by 30–90 min it cannot be reversed significantly by repeated washing with 1.0 M phosphate buffer or with detergents such as Triton X-100. Irreversible binding is strongly temperature-dependent, with a maximum at 28 °C and a Q_{10} of approximately 1.8, and occurs at pH values between 4.0 and 8.0.

Attachment of B1 cells to SCC or cell walls apparently is not dependent on the presence of a metabolically active bacterium, although a temperature-sensitive component is involved. Pretreating the bacteria at 45 °C for 25 min, for instance, totally removed their ability to attach. The kinetics of bacterial cell death, however, differed substantially from those of loss of ability to attach. Bacteria killed by UV light or by rifamycin retained their ability to attach. DUVICK concluded that attachment is dependent on a bacterial heat-labile component, but he showed that this component is pronase and trypsin-resistant. It is not known how virulent and avirulent cells of *P. solanacearum* differ in relation to this component.

It is interesting that envelopment appears to occur only in the intact plant. It is perhaps significant that SCC do not respond hypersensitively to the B1 strain, indicating that they are metabolically different from intact cells. A recent report (ATKINSON and HUANG 1982) indicates that tobacco SCC do exhibit the HR when inoculated with *P. syringae* pv. *pisi;* it would be useful to know if envelopment can be detected in this host-parasite system.

An additional, significant difference between the intact plant and SCC is the fact that heat-killed bacteria appear to attach in the case of the former but not in the latter system. In fact, heat-killed bacteria are enveloped at the tobacco cell wall in much the same way that live cells are. The response of the host is even more marked in terms of the appositions that develop below the point of attachment on the cell wall, presumably because the HR is not induced and the host cell does not die. The reasons for this marked difference between SCC and leaf cells are not entirely clear, but some rather obvious factors include the media in which the cells have to function and the different origin of the host cells.

It should be indicated that there is evidence that saprophytic bacteria may bind even more effectively to tobacco cell walls than incompatible forms of phytopathogenic bacteria. When cells of *P. fluorescens,* a saprophyte, *P. syringae* pv. *pisi,* a pea pathogen, and pv. *tabaci,* a tobacco pathogen, were infiltrated into tobacco leaves and then recovered by centrifugation, the percentage recoveries of the latter two pathogens were substantially higher than those of the saprophyte (ATKINSON et al. 1981). The results suggest that bacteria that produce extracellular slime may be difficult to remove from plant tissues by low speed centrifugation but do not, to my mind, lead to the conclusion that efficient absorption of bacterial cells is not necessary for induction of the HR. It is known that a single bacterial cell is capable of inducing the collapse of a single host cell (TURNER and NOVACKY 1974) and, at the level of inoculum (10^7–10^8 cfu ml^{-1}) used by ATKINSON et al. (1981), the recoveries of 8.7% to 10% of the bacteria after 2 h suggest that enough bacteria may have remained attached to the tissue to cause confluent necrosis.

10.2.4 Cell-Wall Components Involved in Attachment

In the intact tobacco leaf, attachment of avirulent bacteria does not necessarily lead to the HR. Heat-killed bacteria, for instance, are attached but do not induce the HR. Until very recently, therefore, it was thought that the surface component that provided HR-inducing strains of *P. solanacearum* with the ability to attach to tobacco cells had to be a heat-stable component of the bacterial cell wall. The most obvious components were the extracellular polysaccharide (EPS) and the lipopolysaccharide (LPS) component of the outer membrane. As with many other phytopathogenic bacteria, the EPS of *P. solanacearum* is present in the form of a highly soluble slime, rather than as an insoluble capsule. EPS is produced profusely by virulent strains of *P. solanacearum* but not at all by avirulent ones; the slime-less strains invariably are found to be avirulent. Since the shift from virulence (K60) to avirulence (B1) endows the bacterium with the capacity to induce the HR, it was initially thought that EPS played a major role in preventing recognition by blocking receptor sites on the cell wall. However, virulent, race 2 strains of *P. solanacearum* (from banana) produce copious amounts of EPS, yet are incompatible in tobacco leaves and induce a rapid HR (LOZANO and SEQUEIRA 1970). There are no differences in composition of EPS from races 1 and 2 (WHATLEY unpublished data). In addition, the HR induced by B1 strain cannot be prevented by pretreating tobacco leaves with EPS, or by inoculating leaves with B1 cells suspended in a solution containing EPS. These and other criteria indicated that EPS probably does not play an important function in preventing attachment of bacteria to plant cell walls. Other properties of EPS, such as its ability to cause water soaking (EL BANOBY and RUDOLPH 1979) or to cause vascular dysfunction (AYERS et al. 1979) are probably more important for pathogenicity. The initial hypothesis that plant pathogenic bacteria, like many extracellular, mammalian pathogens, utilize EPS to prevent recognition, thus bypassing the defense reaction of the host, does not appear to be supported by the experimental evidence.

The arguments presented above led to a consideration of LPS as a component of the mechanism for recognition that leads to the HR in tobacco leaves. The location of LPS, extending outward from the outer membrane of Gram-negative bacteria, and the highly variable nature of the O-specific antigen at the projecting end of the hetero-polysaccharide chain, indicate that LPS is highly suited as an agent for specificity.

Support for a role of LPS in the induction of the HR by *P. solanacearum* was obtained from a comparison of the composition of the LPS from the K60 and B1 strains of *P. solanacearum*. The parental, K60 strain appears to have complete ("smooth") LPS; it consists of a lipid A region, a core region (containing ketodeoxyoctonoic acid, heptose, glucose, and rhamnose) and an O-polysaccharide chain (containing rhamnose, xylose, and N-acetyl glucosamine). The B1 strain, on the other hand, appears to lack the O-polysaccharide chain and thus has incomplete ("rough") LPS (BAKER and SEQUEIRA 1981). When a series of virulent and avirulent strains of *P. solanacearum* were examined, there was a strong correlation between LPS structure and ability to induce the HR (WHATLEY et al. 1980). In general, strains with "smooth" LPS did not induce the HR; strains with "rough" LPS did. These differences in LPS composition could be demonstrated most readily by gel electrophoresis and LPS in the presence of sodium dodecyl sulfate. The LPS from smooth strains migrated slowly, whereas that from the rough strains migrated rapidly, reflecting differences in both charge and molecular weight. More recent work with a large number of avirulent mutants of K60, selected for resistance to LPS-dependent bacteriophages (HENDRICK and SEQUEIRA 1981), indicates that their LPS is similar in carbohydrate composition and electrophoretic migration patterns to those of HR-inducing strains, yet, contrary to our expectations, none of them was able to induce the HR (HENDRICK unpublished data). None of these strains, however, attach to SCC to the same extent that the B1 strain does (DUVICK unpublished data). These data do not support the original conclusion that mere loss of the O-polysaccharide chain in LPS uncovers cryptic polysaccharide sequences that are an important prerequisite for firm binding to cell walls and thus for HR induction. Evidently, very subtle changes, in addition to loss of the O-polysaccharide and EPS, must occur in the shift from K60 to the B1 variant. As indicated before, there is evidence that a temperature-sensitive component of the bacterium is involved in tight binding to host cell walls. Mutations that alter LPS composition often are accompanied by marked changes in the arrangement and composition of proteins in the outer membrane (NIKADO and NAKAE 1979). Of particular interest, therefore, are the pili, proteinaceous extensions from the outer membrane that are involved in attachment of bacteria to mammalian cells (BEACHEY 1981).

LPS is a negatively charged molecule, in part due to the high content of ketodeoxyoctonoic acid (KDO) and phosphate ions, and would be expected to interact ionically with positively charged groups at the plant cell wall. These groups are provided, for the most part, by hydroxyproline-rich glycoproteins (HPRG's) that, until recently, were thought to function primarily as structural components (LAMPORT 1980). Proteins that agglutinate bacteria have been extracted from many different plants (BOHLOOL and SCHMIDT 1974, DAZZO et al. 1978, SEQUEIRA and GRAHAM 1977, FETT and SEQUEIRA 1980, ROMEIRO et al. 1981), but only those that are located on the cell wall are of particular significance in the early interactions between bacteria and their hosts. The cell wall HPRG's are of particular interest in this regard. First, their highly basic nature (generally associated with a high content of lysine) suggests that they can bind bacterial LPS radily and, indeed, it has been demonstrated that they do so in vitro (SEQUEIRA and GRAHAM 1977, DUVICK et al. 1979). Second, levels of HPRG's have been reported to increase in plant tissues in response to wounding (CHRISPEELS 1976, STUART and VARNER 1980) or infection by pathogens (ESQUERRÉ-TUGAYÉ and LAMPORT 1979).

Initially it was thought that the glycoprotein from potato tubers that agglutinated B1 but not K60 cells of *P. solanacearum* was potato lectin because agglutination could be reversed by chitotriose, a hapten for that lectin. A detailed examination of this interaction, however, indicates that the component that strongly agglutinates bacterial cells in potato extracts is a glycoprotein that is separable from potato lectin (LEACH 1981). Like potato lectin, the agglutinin is rich in hydroxyproline, but its physical and chemical properties are very different from those of the lectin. In particular, the bacterial agglutinin has very weak haemagglutinating activity, whereas potato lectin is a very strong haemagglutinin.

The agglutinin from potato is located only at the cell wall, as demonstrated by immunofluorescence procedures (LEACH 1981). There, it can interact with bacteria that invade the intercellular spaces. The interaction appears to be a non-specific, charge–charge interaction, rather than a specific, lectin–carbohydrate interaction as proposed earlier. This does not imply that the agglutinin does not play an important role in the initial interaction of host and parasite. The agglutinin may be extremely important in the initial "docking" of the bacterium on the cell wall, providing loose binding until other reactions, presumably mediated by enzymatic breakdown of the host cell wall, can provide irreversible binding.

What is the nature of the "transducer" that relays the information from the site of attachment of the bacterium on the cell wall to the target organelle responsible for the physiological changes that lead to the HR? We can only speculate as to the sequence of events that initiates the host response. Some researchers consider that bacterial toxins are involved, but the evidence is not convincing (CROSTHWAITE and PATIL 1978). Alternatively, one can envision that tight binding to the cell wall is due to enzymes produced by the bacterium when in contact with the cell wall and capable of degrading specific cell-wall polymers. It is well known that specific fragments of cell-wall polysaccharides have phytoalexin elicitor activity, a phenomenon closely linked to HR induction (HAHN et al. 1981). Perhaps similar fragments of the cell wall are involved in HR induction by plant pathogenic bacteria.

10.3 Compatible Interactions

Compatibility with the potential host is an essential feature of intracellular parasites (or symbionts) or of those that must have close contact with the host in order to transfer part of their genome to the host cell. These types of parasite, therefore, benefit from recognition, for it is to their advantage to be attached rapidly to the host cell wall. There they can initiate a morphological response that will in turn be important for their multiplication. In the case of plant parasites (or symbionts), one can predict that they must avoid or inhibit responses from the plant, such as the HR, which would hinder their multiplication.

Compatible interactions between plants and bacteria have been studied in detail only in two systems: the crown gall disease caused by *Agrobacterium tumefaciens* and the root nodules caused by different species of *Rhizobium* in leguminous hosts. Since the two genera are closely related, it is not surprising perhaps that the details of their interactions with the host offer some interesting parallels.

10.3.1 The *Agrobacterium tumefaciens*–Plant Interaction

A great deal of attention has been devoted in recent years to the details of the transformation phenomenon whereby *A. tumefaciens* transfers its oncogenic properties, governed by genes in a portion of the bacterial Ti plasmid, to the host cell (NESTER and KOSUGE 1981). From our perspective, the important phenomena concern the interactions that take place before genetic information is transferred. For transfer to take place, prior attachment to cell walls of injured tissues must take place, and there is only a relatively short "window" of susceptibility following injury. LIPPINCOTT and LIPPINCOTT (1969) demonstrated that in pinto bean leaves, attachment of the bacterium to a specific site is essential as a first step in tumour induction. This was based entirely on indirect evidence from studies in which tumour induction was found to be inhibited by heat-killed virulent *A. tumefaciens* cells or living avirulent cells. The binding process appeared to be complete within 15 min after virulent cells were added, for avirulent cells were not inhibitory if added beyond this period.

Demonstration that binding of *A. tumefaciens* to plant tissue culture cells occurred was obtained by means of live cell counts (MATTHYSSE et al. 1978) or by measuring the retention of ^{32}P-labelled bacteria (OHYAMA et al. 1979). Attachment appeared to be maximal within 2 h of incubation. On carrot tissue culture cells, attachment is specific for strains that contain the Ti plasmid (MATTHYSSE et al. 1978) and does not require the living carrot cell.

Initial studies carried out by WHATLEY et al. (1976) indicated that LPS preparations from virulent and from binding, avirulent strains of *A. tumefaciens* were highly inhibitory in the tumour-inhibition assay on pinto bean leaves. As might be expected, inhibition occurred only when LPS was added before or simultaneously with the inoculum, but not when added 15 min later. The LPS from non-binding strains of *A. radiobacter* did not have inhibitory properties. Only the core-O-polysaccharide component appeared to be involved, for the lipid A portion, even when solubilized by complexing with bovine serum albumin, did not have an inhibitory effect on tumour formation (WHATLEY and SEQUEIRA 1981). When the Ti plasmid was introduced into non-binding, avirulent strains, these strains became oncogenic and their LPS became inhibitory in the tumour-induction assay (WHATLEY et al. 1978). However, when strains containing the Ti plasmid were cured of the plasmid, they retained the site-binding ability, indicating that the information necessary for binding must be contained in the chromosome.

The question of a possible receptor for *A. tumefaciens* on host cell walls has been given some attention, but the data do not allow any firm conclusions. LIPPINCOTT and LIPPINCOTT (1978) suggested, on the basis of inhibition of tumour induction in pinto bean leaves, that the receptor was present in dicots but not in monocots, which correlates with the broad host range of the pathogen. Assays with isolated cell wall components indicated that polygalacturonic acid was highly inhibitory, and pectin only mildly inhibitory to tumour induction (LIPPINCOTT and LIPPINCOTT 1980). The authors associated this difference in inhibition with differences in degree of methylation. Non-binding cell walls, such as those from monocots, embryonic tissue, or crown gall tumours, became inhibitory when treated with pectin methylesterase. Thus, it was concluded that polygalacturonic acid in the primary cell wall provided the binding site for the bacterium.

More recent work allows other possible interpretations of these results. The LPS of *A. tumefaciens* is a negatively charged molecule and it may interact with basic components (such as HPRG's) on the cell wall. Acidic polymers, such as polygalacturonic acid, would be expected to interfere with binding at concentrations that saturate the binding sites. In fact, based on direct infectivity assays on potato tuber discs, PUEPPKE and BENNY (1981) concluded that the capacity to inhibit tumorigenesis was not limited to polygalacturonic acid but was associated with acidic polysaccharides in general. Several uronic acid-containing carbohydrates were shown to inhibit tumorigenesis very markedly.

That the receptor for *A. tumefaciens* may, in fact, be a protein is suggested by recent work reported by GURLITZ and MATTHYSSE (1982). Receptors for the bacterium on the surface of carrot tissue culture cells were removed by digestion with trypsin and chymotrypsin, or with detergents. Carrot embryo cells, which lack binding sites for the bacterium, also lacked at least two surface proteins. It would be interesting to determine if these proteins are related to the HPRG isolated from carrot discs by STUART and VARNER (1980). We know that this HPRG is a strong bacterial agglutinin (DUVICK unpublished data). If an HPRG is the receptor for bacterial attachment, however, it must be very heat resistant, for exposure of carrot cells to 65 °C for 20 min did not alter their ability to bind virulent cells of *A. tumefaciens* (MATTHYSSE et al. 1982).

Wounding of the host tissue is a prerequisite for tumor induction and it is interesting that treatments that degrade the cell wall (such as rubbing with carborundum) may be effective because they expose the plasmalemma, as they appear to do for infection by mechanically transmitted viruses. Thus, it is significant that cells of *A. tumefaciens* bind effectively to carrot protoplasts and that this binding is highly specific (MATTHYSSE et al. 1982).

There is some information as to the specific carbohydrate components of the LPS of *A. tumefaciens* that may be important in the attachment of the bacterium to host cells. The quenching of the fluorescence emission of labelled LPS was reversed by preincubating with N-acetyl galactosamine and α-D(+)-galactose. Although the method is not highly reliable, BANERJEE et al. (1981) concluded that these sugars could be part of the O-antigen portion of *A. tumefaciens* LPS and that they might play a role in attachment of the bacterium to the host cell wall. More direct methods for chemical analysis of LPS are available to determine whether these carbohydrates are indeed part of the O-antigen portion of *A. tumefaciens* LPS and whether they play a role in attachment of the bacterium to the host cell wall.

As in the case of *P. solanacearum* described previously, it seems likely that the initial "docking" of *A. tumefaciens* on plant cell walls is a rather non-specific phenomenon mediated by bacterial LPS and proteins on the cell wall. In both systems, binding appears to become irreversible within 15 min after initial contact and reaches saturation within 30–90 min (DUVICK and SEQUEIRA 1981, MATTHYSSE et al. 1982). Tight binding in the case of *A. tumefaciens* is correlated with the synthesis of cellulose fibrils by the bacterium (MATTHYSSE et al. 1981). By scanning electron microscopy, it was shown that after the bacteria come in contact with the carrot wall surface cellulose fibrils surround the cells and anchor them to the plant cell wall. The fibrils then entrap additional bacteria, resulting in the formation of large clusters within a few hours of incubation. That cellulose fibril formation may be essential for firm binding to host cells was apparent from experiments with cellulose-deficient mutants of *A. tumefaciens* (MATTHYSSE and LAMB 1982). These mutants could be dislodged readily from the plant surface by washing and showed a much reduced level of tumor formation.

It is surprising that there has been so very little emphasis on the possible role of pili (fimbriae) in the attachment of bacteria to plant cells, and specifically in attachment of *A. tumefaciens,* since certain types of pili are specifically involved in attachment of bacteria to mammalian cells (Beachey 1981). In fact, "F" type pili are involved in bacterial conjugation and it is tempting to suggest that transfer of the Ti plasmid of *A. tumefaciens* to host cells may be mediated by similar, conjugative pili. Recent work by Stemmer and Sequeira (1981) confirms that *A. tumefaciens* produces pili in profusion, but it is not known if their synthesis is coded by a portion of the Ti DNA. The fact that there is a strain-dependent, temperature-sensitive phase in the early stages of tumorigenesis (Rogler 1981) is consistent with the hypothesis that pili play an important role in bacterial attachment and/or gene transfer in the crown gall disease.

10.3.2 The *Rhizobium*–Legume Interaction

The highly specific interactions of certain strains of *Rhizobium* with the roots of legumes, leading to nodulation and nitrogen fixation, have been the subject of extensive investigations for decades. In particular, an extensive literature has accumulated on the details of possible recognition phenomena, presumably mediated by surface components of both host and symbiont. The subject of recognition has been reviewed recently and comprehensively by Graham (1981) and Bauer (1981). It would be pointless, therefore, to review this information again. In this section, my objective will be to illustrate differences and/or similarities between recognition phenomena in the *Rhizobium*–legume system and those in the host–parasite systems described in previous sections. Nature exhibits a great deal of economy in the mechanisms that it employs to transfer information; it is not surprising, therefore, that the informational potential contained in the surface polysaccharides of both symbiotic and pathogenic bacteria appears to be utilized in similar ways, although the response of the host may differ substantially in each interaction.

10.3.2.1 Attachment of Rhizobia to Plant Cell Walls

As with *Agrobacterium tumefaciens,* rhizobia must be in close contact with the host cell wall in order to exert the specific morphological responses that are required for multiplication of the bacteria. These responses involve a characteristic shepherd's crook deformation of the root hair cell, invagination of the plasmalemma at the point of contact with the bacterium, and formation of the infection thread. Before these processes can be initiated, however, a highly specific recognition phenomenon must take place after initial contact, for only certain hosts will respond to bacterial attachment. Some workers suggest that attachment itself is a specific phenomenon (Stacey et al. 1980), but others point to the fact that some species of *Rhizobium* bind to non-hosts quite readily (Chen and Phillips 1976). The controversy appears to result from the lack of adequate methods to measure attachment quantitatively or to determine whether irreversible attachment has taken place. It seems likely that with *Rhizobium,* as with other bacteria, the initial binding to the plant host surface is rather loose, and probably the result of non-specific, ionic interactions. Tight, irreversible binding, however, may occur only in certain compatible interactions that lead to nodulation.

Again, as in the case of *A. tumefaciens,* tight binding of *Rhizobium* to the surface of root hairs appears to be accompanied by the formation of cellulose fibrils (NAPOLI et al. 1975). Similar fibrils are produced by the bacteria in culture and, in the plant, these structures apparently help anchor the bacterium to the surface of the host cell. Since rhizobia appear to attach in an end-on, polar fashion (STACEY et al. 1980) it must be assumed that the cellulose fibrils are also produced polarly, but this is not at all clear.

It should be pointed out that many rhizobia are not restricted in their host range and can nodulate many species of legumes. These so-called promiscuous rhizobia apparently attach and cause the same type of specific responses in hosts from different cross-inoculation groups (SHANTHARAM and WONG 1982). Whatever mechanisms are proposed to explain specific attachment of rhizobia, they must account for the existence of these promiscuous strains.

It is evident that specificity is determined at many points in the complex processes from attachment to nodulation, and that attempts to explain specificity on the basis of a single factor are not likely to succeed. A second point that may be made is that binding of rhizobia to a compatible host usually occurs on root hairs as well as on main roots, although different species differ in this regard. Yet, in the case of *R. japonicum,* only certain specific root cells from the developing root epidermis, destined to become root hairs by further morphological alterations, are infectible (BHUVANESWARI et al. 1980, 1981). These facts indicate that specificity is not likely to be determined at the early steps of attachment.

10.3.2.2 The Role of Extracellular Polysaccharides in Specificity of Rhizobia

Perhaps no other aspect of rhizobial–plant interactions has been the subject of more controversy than the nature of the bacterial cell-wall component involved in host recognition. The accepted dogma is that bacterial EPS, generally present in the form of an organized capsule, interacts specifically with a host protein (lectin). Most strains of *Rhizobium* produce capsulate cells, but both capsulate and non-capsulate cells are present at all stages of the growth cycle. Ultrastructural studies with *R. japonicum,* however, indicate that only those cells that have an amorphous extracellular capsule bind to soybean lectin (SBL) and that they constitute a very small (1%) proportion of the total population in culture (BAL et al. 1978). The proportion of such cells may be much higher in the rhizosphere but, nevertheless, there is no incontrovertible evidence that only capsulate cells attach to the root hairs of the appropriate leguminous host.

The clearest case for the role of EPS involves the *R. japonicum*–soybean interactions. As we shall see later, there is now some evidence that SBL, or a protein very much like SBL, is present on the surface of soybean roots (GADE et al. 1981). Thus, the evidence that ferritin-labelled SBL binds specifically to the capsular material of *R. japonicum* (CALVERT et al. 1978) attains particular significance. The capsular material consists of a pentasaccharide repeating unit in which the backbone is made up of one residue of mannose, two residues of glucose, and one residue of galacturonic acid, and a side chain that consists of one residue of either galactose or 4-O-methylgalactose (MORT and BAUER 1980, 1982). Since galactose provides the potential hapten for SBL, it is significant that the ratio of galactose to 4-O-methylgalactose changes with age of the culture and that this change is associated with changes in the ability of the cells to bind SBL (MORT and BAUER 1980).

The role of EPS of *R. japonicum* in the entire process of nodulation may be quite different from that originally envisioned, however. Pretreatment of soybean roots with EPS from *R. japonicum* enhances the infectibility of the epidermal cells that will become root hairs (BAUER et al. 1979). It seems unlikely that this effect is mediated by the same lectin that binds the bacterial cells, for one would expect that pretreatment with excess EPS would inhibit such binding.

In spite of the evidence that EPS plays an important role at some stage in nodulation of *R. japonicum,* recent evidence suggests that EPS does not have to be present in the form of a capsule, contrary to the results by BAL et al. (1978) that were discussed previously. LAW et al. (1982) determined that many non-capsulate mutants of *R. japonicum* were still capable of nodulating soybean plants, but that their nodulating ability was correlated with the amount of soluble EPS that was released into the medium. This is a rather obvious but important point that may not have been considered by others who have classified mutants as EPS⁻ merely because the cells were not capsulate. This may be the case in the work of SANDERS et al. (1981), who concluded that EPS production by *R. leguminosarum* and *R. phaseoli* is not necessary for nodulation.

One of the problems in the interpretation of much of the literature on the role of EPS is related to the fact that the assays are based entirely on nodulation, rather than on attachment. Thus, it is not possible to determine whether non-capsulate mutants, EPS⁺ or EPS⁻, differ in their ability to attach to the host. Quantitative assays, such as those used to measure attachment in the case of *A. tumefaciens* and *P. solanacearum,* should be applied to the *Rhizobium*–legume interactions. A second problem is related to difficulties in obtaining reasonably pure preparations of capsular EPS, which is often contaminated with LPS and other polysaccharides that are extruded from the cells. Several adequate methods for purification of the EPS from several rhizobia have been published recently (MORT and BAUER 1980, ROBERTSON et al. 1981 etc.) and these should be used more extensively in studies involving attachment and/or nodulation.

10.3.2.3 The Role of Lipopolysaccharides in Specificity of Rhizobia

The possibility that LPS has a role in specificity of rhizobia is made more attractive by the well-known fact that in *Salmonella* the arrangement of the sugars in the O-antigen portion of the molecule is associated with differences in virulence to animals (WILKINSON 1977). Since the O-antigen is highly variable in most bacteria, and is highly exposed in bacteria that are non-capsulate, LPS appears to be a suitable structure for specific interactions with host components.

This hypothesis received initial support from a study of the interaction of LPS from four different rhizobia with lectins from the four corresponding hosts. Only the LPS from the appropriate bacterium was bound by each lectin (WOL-PERT and ALBERSHEIM 1976). Similarly, the binding of host lectins to *R. japonicum* and *R. leguminosarum* could be inhibited by LPS extracted from these bacteria (KATO et al. 1979). With *R. japonicum,* differences in composition of the O-polysaccharide appeared to be related to the ability of different strains to nodulate soybean (MAIER and BRILL 1978).

The confusion that arises from the use of impure preparations of bacterial polysaccharides is exemplified by the very extensive work dealing with the specific interactions of *R. trifolii* with white clover roots. The putative receptor for binding of these bacteria to white clover is a protein, trifoliin, which has been purified and described by DAZZO et al. (1978). The initial results indicated that the antigenic determinant on the bacterium that bound to the surface of the clover root was an acidic polysaccharide that was free of KDO and of endotoxin activity (DAZZO and HUBBELL 1975). Thus, the authors assumed that their preparation contained no LPS. Further examination, however, revealed that the preparation responsible for binding *R. trifolii* to clover root hairs was serologically identical to the O-antigen of the LPS of *R. trifolii* (DAZZO and BRILL 1979).

It is perhaps significant that the relative amounts of several glycosyl components from *R. trifolii* LPS change with age of the culture, similar to results previously reported for *R. japonicum* EPS by MORT and BAUER (1980). The influence of growth phase of the bacterium may be one of the reasons why, as HRABAK et al. (1981) suggest, there are conflicting reports from different laboratories as to the lectin-binding properties of different bacterial cell-wall preparations. Although the data tend to support the notion that LPS is the important recognition molecule in the *R. trifolii*–clover interaction, the possibility that other polysaccharides, perhaps unrelated to LPS or capsular EPS, play this role should be considered. *Rhizobium leguminosarum,* for instance, produces a polysaccharide, different from either EPS or LPS, which appears to play an important role in the binding of the bacterium by pea lectin (PLANQUÉ and KIJNE 1977).

As it should be apparent to the reader, the evidence for the involvement of LPS in rhizobial–legume interactions is not particularly strong and this impression is reinforced by studies of the LPS from *R. leguminosarum, R. phaseoli,* and *R. trifolii* that indicate no correlation between chemical composition of the LPS and nodulating group (CARLSON et al. 1978). The authors concluded that the data did not lend support to the hypothesis that LPS is involved in recognition. Since the arrangement of sugar residues in LPS can vary in many different ways, compositional analyses may not be very useful for comparative studies, however.

A possible explanation for the conflicting results obtained in different laboratories is that specificity in different strains of *Rhizobium* is controlled by extrachromosomal genes located on plasmids. Since different laboratories use strains that may carry different plasmids, the results of assays for attachment, nodulation, etc. predictably will be different also. This is emphasized in a recent report by RUSSA et al. 1982, who determined that the LPS composition in nodulating and non-nodulating strains of *R. trifolii* differed in the content of fucose, glucose, and quinovosamine, and that these differences were correlated with the presence of the plasmid pUCS202. Quinovosamine, for instance, was absent in LPS of a non-nodulating strain, but reappeared simultaneously with the ability to nodulate after introduction of pUCS202 in the same strain. These effects were reversed when the plasmid was removed by treatment with acridine orange. The results, although limited to very few strains, support the hypothesis that LPS is involved in specific recognition of rhizobia by their hosts.

10.3.2.4 The Lectin Receptor Hypothesis

Whether LPS, EPS, or other polysaccharide constituents of the bacterial cell wall ultimately will be shown to be essential in recognition of rhizobia, the receptor at the host cell wall must be able to interact specifically with polysaccharides. Lectins are proteins that bind to specific sugars and thus are particularly well-suited for the role of receptors. As indicated before, however, plant cell walls also contain proteins that interact non-specifically with bacteria. Thus, it is important to determine not only that lectins are located on the cell wall, but that their interaction with bacteria is hapten-reversible.

Support for the lectin hypothesis was given initially by BOHLOOL and SCHMIDT (1974), who reported that fluorescent SBL (see p. 201) bound specifically to 22 of 25 nodulating strains of *R. japonicum*. The binding sites on the bacterium were found to be polar (BOHLOOL and SCHMIDT 1976), a fact that agreed with the finding that *R. japonicum* attached end-on to the surface of soybean roots (STACEY et al. 1980). However, rhizobia bind to solid matrices in a consistently polar manner, thus there may be nothing specific about their polar orientation on roots (MARSHALL et al. 1975). The role of SBL as the receptor, however, obtained considerable support from extensive studies at Bauer's laboratory, in which the early observations of BOHLOOL and SCHMIDT were confirmed and hapten reversibility of attachment was demonstrated (BHUVANESWARI et al. 1977). In addition, many of the nodulating strains that did not bind SBL were shown to develop receptors for SBL when grown in soybean root exudates (BHUVANESWARI and BAUER 1978).

For a long time, a major criticism of SBL as a receptor was the fact that it could not be detected on the surface of soybean roots, particularly at the stage when they could still be nodulated (PUEPPKE et al. 1978). There is recent evidence, however, that roots of the soybean cultivar Chippewa do contain a lectin with properties that are very similar to those of SBL (GADE et al. 1981). Whether this lectin is present on the surface of root hairs, however, is still an open question.

A second problem that has not been resolved satisfactorily is the lack of SBL in the seeds and roots of soybean lines that are nodulated by *R. japonicum* (SU et al. 1980). It has been suggested that the lectin might be more tightly bound in these lines and thus more difficult to extract than in ordinary soybean cultivars (WHATLEY and SEQUEIRA 1981). Several alternative explanations are possible, but they all hinge on the accuracy and sensitivity of the assays for SBL.

The receptor for *R. trifolii* in clover roots is also thought to be a protein with lectin-like properties. The protein, named trifoliin, was initially isolated from clover seed by DAZZO et al. (1978) and was found to bind both to the bacteria and to clover roots; thus it was thought to form a bridge between the bacterium and the host cell wall (DAZZO 1980). Immunofluorescence was used to demonstrate that trifoliin is present on the root surfaces of white clover (DAZZO et al. 1978) and that it is released in the growth medium of intact white clover roots (DAZZO and HRABAK 1981). Fluorescence of bacterial cells bound to trifoliin was decreased if 2-deoxy-D-glucose but not other sugars was added, indicating that this sugar was a possible hapten. This sugar also eluted the lectin from clover seedling roots. The relationship to the hapten presumably present on the LPS of *R. trifolii*, however, is not clear. The most recent evidence indicates that 2-deoxy-D-glucose is not present in the LPS of the bacterium. Rather, quinovosamine (2-amino-2,6-dideoxyglucose) appears to be the potential hapten (HRABAK et al. 1981, RUSSA et al. 1982) but this has not been confirmed.

One of the problems in much of the work that implicates lectins as recognition molecules on the surface of legume roots is that there has been no serious attempt to establish the relationship between agglutination, attachment, and nodulation. It should be possible to show that only those bacteria that bind to the lectin in vitro are those that attach to the host cell wall and eventually induce nodulation. In the few cases where this has been attempted, the correlations are poor. *Rhizobium leguminosarum*, for instance, does not nodulate clover and is not agglutinated by clover lectin (DAZZO 1980) but will bind to clover roots (CHEN and PHILLIPS 1976). Similarly, the lectins from lentil and pea agglutinate some *R. leguminosarum* strains that infect these plants but will also agglutinate strains of *R. japonicum* that do not infect either plant (KATO et al. 1981, WONG 1980). There are several other examples that indicate that binding of lectins from different legumes to rhizobia is not correlated with the ability of these bacteria to nodulate these plants (LAW and STRIJDOM 1977, PUEPPKE et al. 1980).

Perhaps the strongest evidence for the involvement of lectins in rhizobial infections is based on the fact that the presumed haptens of the lectin interfere with attachment of the bacteria to the root surface of the appropriate host (DAZZO et al. 1978, STACEY et al. 1980). Even this work, however, loses some impact because we do not know precisely what the natural haptens are and because different laboratories have difficulty demonstrating hapten reversibility of attachment (BAUER 1981).

10.4 Concluding Remarks – The Potential for Manipulating Plant–Bacterial Interactions

At the outset, it was stated that one of the objectives of this article was to show that the details of plant–bacterial interactions can provide insight into new and improved methods for disease control and to improve plant productivity. It is perhaps self-evident that without this basic knowledge no substantial progress can be made towards manipulating plant–bacterial interactions for the benefit of man. It could be argued, however, that we do not need to know all the details before the information can be used to resolve specific problems. Plant breeders, for instance, have been extremely successful in manipulating genes for resistance to bacterial plant diseases or for improved nitrogen fixation in plants without any detailed knowledge as to what specific processes are controlled by these genes. Even at the risk of incurring the wrath of many of my colleagues who feel that the matters discussed in this paper are far too esoteric and inconsistent to lead to any possible practical applications, may I suggest that we follow the example of plant breeders and apply what we know.

The evidence summarized in this paper indicates that bacteria adhere to the host surface by means of specific surface components and that tight binding must occur before either compatible or incompatible responses are elicited. Would it be possible to manipulate these responses by blocking or enhancing the process of attachment? There are, of course, no data from the work with plants that would permit an answer to this question. Work in other fields, however, suggests that the answer would be yes.

First, there is evidence that bacterial infection of animals may be prevented by blocking the adherence of bacterial cells to epithelial surfaces (Beachey 1981). The binding of certain strains of *E. coli* to the urinary tract in mice, for instance, could be prevented by adding alpha-methylmannoside, a hapten for the lectin on the bacterial pili (Aronson et al. 1979). Several other results support the concept that analogues of the receptor molecule interfere with microbial adherence and limit further colonization of the tissue (see review by Beachey 1981). To follow this argument, therefore, it should be possible to develop strains of rhizobia that attach to the host roots even in the presence of specific hapten analogues which may eliminate attachment by other, less efficient strains.

Alternatively, it should be possible to select host cultivars that are nodulated by strains of *Rhizobium* that combine greater efficiency for nitrogen fixation and the ability to bind irreversibly to the surface of the root hair.

In terms of disease control, the ultimate goal would be the use of compounds of bacterial origin that elicit the plants' own defense system. Evidently, induction of the HR by bacterial cell-wall components would not be desirable. However, the finding that bacterial LPS induces systemic, resistant responses in the plant (Graham et al. 1977) does provide a useful lead as to how the plant's response may be modulated.

It is probably wise not to speculate further as to the different directions that research on manipulation of plant–bacterial interactions may take in the next decade. It is obvious, however, that the continued identification of the molecules involved in recognition between the bacterium and the plant host will enable us in the near future to modify this initial interaction.

References

Aronson M, Medalia O, Schori L, Mirelman D, Sharon N, Ofek I (1979) Prevention of colonization of the urinary tract of mice with *Escherichia coli* by blocking adherence with methyl-D-mannopyranoside. J Infect Dis 139:329–332

Atkinson MM, Huang JS (1982) Hypersensitive-like reaction in tobacco suspension cultures to *Pseudomonas pisi*. (Abstr) Phytopathology 72:354

Atkinson MM, Huang J-S, Van Dyke CG (1981) Adsorption of pseudomonads to tobacco cell walls and its significance to bacterium–host interactions. Physiol Plant Pathol 18:1–5

Ayers AR, Ayers SB, Goodman RN (1979) Extracellular polysaccharide of *Erwinia amylovora*: a correlation with virulence. Appl Env Microbiol 38:659–666

Baker CJ, Sequeira L (1981) Further characterization of the lipopolysaccharide (LPS) produced by *Pseudomonas solanacearum*. (Abstr) Phytopathology 71:201

Bal AK, Shantharam S, Ratnam S (1978) Ultrastructure of *Rhizobium japonicum* in relation to its attachment to root hairs. J Bact 133:1393–1400

Banerjee D, Basu M, Choudhury I, Chatterjee GC (1981) Cell surface carbohydrates of *Agrobacterium tumefaciens* involved in adherence during crown gall tumor initiation. Biochem Biophys Res Commun 100:1384–1388

Bauer WD (1981) Infection of legumes by rhizobia. Annu Rev Plant Physiol 32:407–449

Bauer WD, Bhuvaneswari TV, Mort AJ, Turgeon BG (1979) The initiation of infections in soybean by *Rhizobium*. 3. *R. japonicum* polysaccharide pretreatment induces root hair infectibility. Plant Physiol 63 (Suppl):135

Beachey EH (1981) Bacterial adherence: adhesin–receptor interactions mediating the attachment of bacteria to mucosal surfaces. J Infect Dis 143:325–345

Bhuvaneswari TV, Bauer WD (1978) Role of lectins in plant–microorganism interactions 3. Influence of rhizosphere-rhizoplane culture conditions on soybean lectin-binding properties of rhizobia. Plant Physiol 62:71–74

Bhuvaneswari TV, Pueppke SG, Bauer WD (1977) Role of lectins in plant–microorganism interactions. 1. Binding of soybean lectin to rhizobia. Plant Physiol 60:486–491

Bhuvaneswari TV, Turgeon BG, Bauer WD (1980) Early events in the infection of soybean (*Glycine max* Merr.) by *Rhizobium japonicum*. 1. Localization of infectible root cells. Plant Physiol 66:1027–1031

Bhuvaneswari TV, Bhagwat AA, Bauer WD (1981) Transient susceptibility of root cells in four common legumes to nodulation by rhizobia. Plant Physiol 68:1144–1149

Bohlool BB, Schmidt EL (1974) Lectins: a possible basis for specificity in the *Rhizobium*–legume root nodule symbiosis. Science 185:269–271

Bohlool BB, Schmidt EL (1976) Immunofluorescent polar tips of *Rhizobium japonicum*: possible site of attachment or lectin binding. J Bacteriol 125:1188–1194

Calvert HE, Lalonde M, Bhuvaneswari TV, Bauer WD (1978) Role of lectins in plant–microorganism interactions. 4. Ultrastructural localizations of soybean lectin binding sites on *Rhizobium japonicum*. Can J Microbiol 24:785–793

Carlson RW, Sanders RE, Napoli C, Albersheim P (1978) Host–pathogen interactions 13. Purification and partial characterization of *Rhizobium* lipopolysaccharides. Plant Physiol 62:912–917

Cason ET, Richardson PE, Essenberg MK, Brinkerhoff LA, Johnson WM, Venere RJ (1978) Ultrastructural cell wall alterations in immune cotton leaves inoculated with *Xanthomonas malvacearum*. Phytopathology 68:1015–1021

Chen AP, Philips DA (1976) Attachment of *Rhizobium* to legume roots as the basis for specific interactions. Physiol Plant 38:83–88

Chrispeels M (1976) Biosynthesis, intercellular transport and secretion of extracellular macro-molecules. Annu Rev Plant Physiol 27:19–38

Cook AA (1973) Characterization of hypersensitivity in *Capsicum annuum* induced by the tomato strain of *Xanthomonas vesicatoria*. Phytopathology 63:915–918

Cook AA, Stall RE (1971) Calcium suppression of electrolyte loss from pepper leaves inoculated with *Xanthomonas vesicatoria*. Phytopathology 61:484–487

Cook AA, Stall RE (1977) Influence of watersoaking on development of the hypersensitive response in pepper leaves caused by *Xanthomonas vesicatoria*. Phytopathology 67:1101–1103

Crosthwaite LM, Patil SS (1978) Isolation of an endotoxin from *Pseudomonas phaseolicola* which induces cell collapse in bell pepper. Phytopathol Z 91:80–90

Daub M, Hagedorn D (1980) Growth kinetics and interactions of *Pseudomonas syringae* with susceptible and resistant bean cultivars. Phytopathology 70:429–436

Dazzo FB (1980) Adsorption of microorganisms to roots and other plant surfaces. In: Bitton G, Marshall KC (eds) Adsorption of microorganisms to surfaces. Wiley and Sons, New York, pp 253–316

Dazzo FB, Brill WJ (1979) Bacterial polysaccharide which binds *Rhizobium trifolii* to clover root hairs. J Bacteriol 137:1362–1373

Dazzo FB, Hrabak EM (1981) Presence of trifoliin A, a *Rhizobium*-binding lectin, in clover root exudate. J Supramol Struct Cell Biochem 16:133–138

Dazzo FB, Hubbell DH (1975) Cross-reactive antigens and lectins as determinants of symbiotic specificity in the *Rhizobium*–clover association. Appl Microbiol 30:1017–1033

Dazzo FB, Yanke WE, Brill WJ (1978) Trifoliin – a *Rhizobium* recognition protein from white clover. Biochim Biophys Acta 539:276–286

Durbin RD, Klement Z (1977) High-temperature repression of plant hypersensitivity to bacteria: a proposed explanation. In: Kiraly E (ed) Current topics in plant pathology. Akademiai Kiado, Budapest pp 239–242

Duvick J, Sequeira L (1981) Binding of *Pseudomonas solanacearum* to tobacco and potato suspension culture cells (Abstr). Phytopathology 71:872

Duvick J, Sequeira L, Graham TL (1979) Binding of *Pseudomonas solanacearum* surface polysaccharides to plant lectins in vitro. (Abstr) Plant Physiol 63:134

Dye WW (1958) Host specificity in *Xanthomonas*. Nature 182:1813–1814

El-Banoby FE, Rudolph K (1979) A polysaccharide from liquid cultures of *Pseudomonas phaseolicola* which specifically induces water soaking in bean leaves (*Phaseolus vulgaris* L.) Phytopathol Z 95:38–50

Esquerré-Tugayé M-T, Lamport DTA (1979) Cell surfaces of plant–microorganism interactions. I. A structural investigation of cell wall hydroxyproline-rich glycoproteins which accumulate in fungus-infected plants. Plant Physiol 64:314–319

Fett WF, Sequeira L (1980) A new bacterial agglutinin from soybean. II. Evidence against a role in determining pathogen specificity. Plant Physiol 66:853–858

Gade W, Jack MA, Dahl JB, Schmidt EL, Wold F (1981) The isolation and characterization of a root lectin from soybean [*Glycine max* (L.) cultivar Chippewa]. J Biol Chem 256:12905–12910

Giddix LR, Lukezic FL, Pell EJ (1981) Effect of light on bacteria-induced hypersensitivity in soybean. Phytopathology 71:111–115

Goodman RN (1968) The hypersensitive reaction in tobacco: a reflection of changes in host permeability. Phytopathology 58:872–873

Goodman RN, Plurad SR (1971) Ultrastructural changes in tobacco undergoing the hypersensitive reaction caused by plant pathogenic bacteria. Physiol Plant Pathol 1:11–16

Goodman RN, Huang PY, White JA (1976) Ultrastructural evidence for immobilization of an incompatible bacterium, *Pseudomonas pisi*, in tobacco leaf tissue. Phytopathology 66:754–764

Graham TL (1981) Recognition in the *Rhizobium*–legume symbiosis. Int Rev Cytol (Suppl)13:127–148

Graham TL, Sequeira L, Huang TR (1977) Bacterial lipopolysaccharides as inducers of disease resistance in tobacco. Appl Environ Microbiol 34:424–432

Gulyas A, Barnes B, Klement Z, Farkas GL (1979) Effect of plasmolytica on the hypersensitive reaction induced by bacteria in tobacco: a comparison with the virus-induced hypersensitive reaction. Phytopathology 69:121–124

Gurlitz RHG, Matthysee AG (1982) Receptors for *Agrobacterium tumefaciens* on the surface of carrot tissue culture cells. Proc Ann Meeting Am Soc Microbiol p 22

Hahn MG, Darvill AG, Albersheim P (1981) Host–pathogen interactions XIX. The endogenous elicitor, a fragment of a plant cell wall polysaccharide that elicits phytoalexin accumulation in soybeans. Plant Physiol 68:1161–1169

Hendrick CA, Sequeira L (1981) Lipopolysaccharide-defective mutants of *Pseudomonas solanacearum*. (Abstr) Phytopathology 71:880

Heslop-Harrison J (1978) Cellular recognition systems in plants. Arnold, London

Hildebrand DC, Alosi MC, Schroth MN (1980) Physical entrapment of Pseudomonads in bean leaves by films formed in air–water interfaces. Phytopathology 70:98–109

Hrabak E, Urbano MR, Dazzo FB (1981) Growth-phase-dependent immunodeterminants of *Rhizobium trifolii* lipopolysaccharide which bind trifoliin A, a white clover lectin. J Bacteriol 148:697–711

Huang JS, Van Dyke CG (1978) Interaction of tobacco callus tissue with *Pseudomonas tabaci, P. pisi,* and *P. fluorescens.* Physiol Plant Path 13:65–72

Kato G, Maruyama Y, Nakamura M (1979) Role of lectins and lipopolysaccharides in the recognition process of specific legume–*Rhizobium* symbiosis. Agric Biol Chem 43:1085–1092

Kato G, Maruyama Y, Nakamura M (1981) Involvement of lectins in *Rhizobium*–pea recognition. Plant Cell Physiol 22:759–771

Keen NT, Ersek T, Long M, Bruegger B, Holliday M (1981) Inhibition of the hypersensitive reaction of soybean leaves to incompatible *Pseudomonas* spp. by blasticidin S, streptomycin or elevated temperature. Physiol Plant Pathol 18:325–337

Klement Z (1963) Method for the rapid detection of the pathogenicity of phytopathogenic Pseudomonads. Nature 199:299–300

Klement Z, Goodman RN (1968) The hypersensitive reaction to infection by bacterial plant pathogens. Ann Rev Phytopathol 5:17–44

Lamport DTA (1980) Structure and function of plant glycoproteins. In: Preiss J (ed) Biochemistry of plants. Vol 3. Academic Press, London New York, pp 501–542

Law IJ, Strijdom BW (1977) Some observations on plant lectins and *Rhizobium* specificity. Soil Biol Biochem 9:79–84

Law IJ, Yamamoto Y, Mort AJ, Bauer WD (1982) Nodulation of soybean by *Rhizobium japonicum* mutants with altered capsule synthesis. Planta 154:100–109

Leach J (1981) Localization, characterization, and quantification of a bacterial agglutinin from potatoes. Ph D Thesis, Univ Wisconsin, Madison

Lippincott BB, Lippincott JA (1969) Bacterial attachment to a specific wound site as an essential stage in tumor initiation by *Agrobacterium tumefaciens.* J Bacterial 97:620–628

Lippincott JA, Lippincott BB (1978) Cell walls of crown-gall tumors and embryonic plant tissues lack *Agrobacterium* adherence sites. Science 199:1075–1077

Lippincott JA, Lippincott BB (1980) Microbial adherence in plants. In: Beachey EH (ed) Bacterial adherence. Chapman and Hall, London, pp 377–397

Lozano JC, Sequeira L (1970) Differentiation of races of *Pseudomonas solanacearum* by a leaf infiltration technique. Phytopathology 60:833–838

Lyon F, Wood RKS (1976) The hypersensitive reaction and other responses of bean leaves to bacteria. Ann Bot 40:489–491

Maier R, Brill WJ (1978) Involvement of *Rhizobium japonicum* O antigen in soybean nodulation. J Bacteriol 133:1295–1299

Marshall K, Cruikshank R, Bushby H (1975) The orientation of certain root-nodule bacteria at interfaces including legume–root hair surfaces. J Gen Microbiol 91:198–200

Matthysse AG, Lamb PW (1982) The role of bacterial cellulose fibrils in the interaction of *Agrobacterium tumefaciens* with plant host cells. Ann Meet Am Soc Microbiol, p 22

Matthysse AG, Wyman PM, Holmes KV (1978) Plasmid-dependent attachment of *Agrobacterium tumefaciens* to plant tissue culture cells. Infect Immun 22:516–522

Matthysse AG, Holmes KV, Gurlitz RHG (1982) Binding of *Agrobacterium tumefaciens* to carrot protoplasts. Physiol Plant Pathol 20:27–33

Meadows ME, Stall RE (1981) Different induction periods for hypersensitivity in pepper to *Xanthomonas vesicatoria* determined with antimicrobial agents. Phytopathology 71:1024–1027

Mort AJ, Bauer WD (1980) Composition of the capsular and extracellular polysaccharides of *Rhizobium japonicum.* Changes with culture age and correlations with binding of soybean seed lectin to the bacteria. Plant Physiol 66:158–163

Mort AJ, Bauer WD (1982) Application of two new methods for cleavage of polysaccharides into specific oligosaccharide fragments. Structure of the capsular and extracellular polysaccharides of *Rhizobium japonicum* that bind soybean lectin. J Biol Chem 257:1870–1875

Napoli C, Dazzo FB, Hubbell D (1975) Production of cellulose microfibrils by *Rhizobium*. Appl Microbiol 30:123–132

Nemeth J, Klement Z, Farkas GL (1969) An enzymological study of the hypersensitive reaction induced by *Pseudomonas syringae* in tobacco leaf tissues. Phytopathol Z 65:267–278

Nester EW, Kosuge T (1981) Plasmids specifying plant hyperplasias. Ann Rev Microbiol 35:531–565

Nikaido H, Nakae T (1979) The outer membrane of Gram-negative bacteria. Advances in Microbial Physiology. Vol 20. Academic Press, London New York

Ohyama K, Pelcher LE, Schaeffer A (1979) In vitro binding of *Agrobacterium tumefaciens* to plant cells from suspension culture. Plant Physiol 63:382–387

Planqué K, Kijne JW (1977) Binding of pea lectins to a glycan type polysaccharide in the cell walls of *Rhizobium leguminosarum*. FEBS Lett 73:64–66

Pueppke SG, Benny UK (1981) Induction of tumors on *Solanum tuberosum* L. by *Agrobacterium*: quantitative analysis, inhibition by carbohydrates, and virulence of selected strains. Physiol Plant Pathol 18:169–179

Pueppke SG, Bauer WD, Keegstra K, Ferguson AL (1978) Role of lectins in plant microorganism interactions. II. Distribution of soybean lectin in tissues of *Glycine max* (L.) Merr. Plant Physiol 61:779–784

Pueppke SG, Freund TG, Schultz BC, Friedman HP (1980) Interaction of lectins from soybeans and peanut with rhizobia that nodulate soybean, peanut, or both plants. Can J Microbiol 26:1489–1497

Robertsen BK, Aman P, Darvill AG, McNeil M, Albersheim P (1981) Host–symbiont interactions. V. The structure of acidic extracellular polysaccharides secreted by *Rhizobium leguminosarum* and *Rhizobium trifolii*. Plant Physiol 67:389–400

Rogler CE (1981) Strain-dependent temperature-sensitive phase in crown gall tumorigenesis. Plant Physiol 68:5–10

Romeiro R, Karr A, Goodman R (1981) Isolation of a factor from apple that agglutinates *Erwinia amylovora*. Plant Physiol 68:772–777

Russa R, Urbanik T, Kowalczuk E, Lorkiewicz Z (1982) Correlation between the occurrence of plasmid pUCS202 and lipopolysaccharide alterations in *Rhizobium*. FEMS Microbiol Lett 13:161–165

Sanders R, Raleigh E, Signer E (1981) Lack of correlation between extracellular polysaccharide and nodulation ability in *Rhizobium*. Nature 292:148–149

Sasser M, Andrews AK, Doganay ZV (1974) Inhibition of photosynthesis diminishes antibacterial action of pepper plants. Phytopathology 64:770–772

Shell J (1982) The Ti-plasmids of *Agrobacterium tumefaciens* In: Parthier B, Boulter D (eds) Encyclopedia of plant physiology, Vol 14B, Berlin Heidelberg New York, Springer pp 455–474

Sequeira L (1976) Induction and suppression of the hypersensitive reaction induced by phytopathogenic bacteria: specific and nonspecific components. In: Wood RKS, Graniti A (eds) Specificity in Plant Diseases Vol 10, NATO ASI Ser. Plenum, New York London, pp 289–306

Sequeira L (1979) Bacterial hypersensitivity. In: Durbin RD (ed) *Nicotiana*: procedures for experimental use. USDA Tech Bull 1586, pp 111–120

Sequeira L, Graham TL (1977) Agglutination of avirulent strains of *Pseudomonas solanacearum* by potato lectin. Physiol Plant Pathol 11:43–54

Sequeira L, Gaard G, de Zoeten GA (1977) Attachment of bacteria to host cell walls: its relation to mechanisms of induced resistance. Physiol Plant Pathol 10:43–50

Shantharam S, Wong PP (1982) Recognition of leguminous hosts by a promiscuous *Rhizobium* strain. Appl Environ Microbiol 43:677–685

Sing VO, Schroth MN (1977) Bacteria–plant cell surface interactions: active immobilization of saprophytic bacteria in plant leaves. Science 197:759–761

Smith MA, Kennedy BW (1970) Effect of light on reactions of soybean to *Pseudomonas glycinea*. Phytopathology 60:723–725

Stacey T, Paau AS, Brill WJ (1980) Host recognition in the *Rhizobium*–soybean symbiosis. Plant Physiol 66:609–614

Stall RE, Cook AA (1979) Evidence that bacterial contact with the plant cell is necessary for the hypersensitive reaction but not the susceptible reaction. Physiol Plant Pathol 14:77–84

Stemmer W, Sequeira L (1981) Pili of plant pathogenic bacteria (Abstr). Phytopathology 71:906

Stuart DA, Varner JE (1980) Purification and characterization of a salt-extractable hydroxyproline-rich glycoprotein from aerated carrot discs. Plant Physiol 66:787–792

Su LC, Pueppke SG, Friedman HP (1980) Lectins and the soybean–*Rhizobium* symbiosis. I. Immunological investigation of soybean lines, the seeds of which have been reported to lack the 120,000 dalton soybean lectin. Biochim Biophys Acta 629:292–304

Süle S, Klement Z (1971) Effect of high temperature and the age of bacteria on the hypersensitive reaction of tobacco. Acta Phytopathol Ac Sc Hungarica 6:119–122

Turner JG, Novacky A (1974) The quantitative relation between plant and bacterial cells involved in the hypersensitive reaction. Phytopathology 64:885–890

Whatley MH, Sequeira L (1981) Bacterial attachment to plant cell walls. In: Loewus F, Ryan C (eds) Recent advances in phytochemistry. Vol 5. Plenum Press, New York London, pp 213–240

Whatley MH, Bodwin JS, Lippincott BV, Lippincott JA (1976) Role for *Agrobacterium* cell envelope lipopolysaccharide in infection site attachment. Infect Immun 13:1080–1083

Whatley MH, Margot JB, Schell J, Lippincott BB, Lippincott JA (1978) Plasmid and chromosomal determination of *Agrobacterium* adherence specificity. J Gen Microbiol 107:395–398

Whatley MH, Hunter N, Cantrell MA, Hendrick CA, Keegstra K, Sequeira L (1980) Lipopolysaccharide composition of the wilt pathogen, *Pseudomonas solanacearum*: correlation with the hypersensitive response in tobacco. Plant Physiol 65:557–559

Wilkinson SG (1977) Composition and structure of bacterial lipopolysaccharides. In: Sutherland I (ed) Surface carbohydrates of the prokaryotic cell I. Academic Press, London New York

Wolpert JS, Albersheim P (1976) Host–symbiont interactions. I. The lectins of legumes interact with the O-antigen containing lipopolysaccharides of their symbiont rhizobia. Biochem Biophys Res Commun 70:729–737

Wong PP (1980) Interactions between rhizobia and lectins of lentil, pea, broad bean, and jackbean. Plant Physiol 65:1049–1052

11 Cellular and Molecular Recognition Between Higher Plants and Fungal Pathogens

J.A. CALLOW

11.1 Introduction

A cell that reacts in a special way in consequence of an association with another cell or its products, does so because it acquires information, information that is conveyed through chemical or physical signals in the processes of recognition. "Recognition" in this sense implies an ability to discriminate between materials in the environment. Molecular recognition, at the cell surface, is considered in terms of interactions between receptors (or cognors, BURKE et al. 1980) binding to complementary ligands or cognons. Cell recognition implies a series of cellular and biochemical events triggered by molecular recognition.

Specificity in disease may be considered to be a consequence of recognition. On the one hand, a potential pathogen may recognize features of a plant which signal the suitability of that plant for parasitism. How, one might ask, do the spores of a pathogen landing on a leaf detect that surface as being suitable for germination and parasitism, as compared with an inanimate plane surface? Are there mechanisms by which propagules may be attracted to the vicinity of a suitable host surface? On the other hand, the potential host plant may be able to detect or recognize a potential fungal pathogen as foreign or "non-self" and use the initial act of recognition to trigger a range of induced resistance mechanisms (ALBERSHEIM and ANDERSON-PROUTY 1975, CALLOW 1977, 1982, 1983, KEEN 1982). How does the plant detect the vast majority of potential pathogens in its environment as "non-self", and how does the successful pathogen escape detection to cause disease?

The interaction of a pathogen with a plant may be considered as a multistep process with recognition mechanisms at various levels, any one, or a combination of which may control specificity (Table 1). The primary aim of this review is to consider current thinking in this field and to discuss recent developments at the molecular and cellular levels by which partial answers to the types of question posed above may be given.

11.2 Recognition of the Host by the Parasite

11.2.1 Spore Germination

A well-adapted spore must be capable of detecting and responding to signals indicating a favourable external environment – i.e. in the context of pathogene-

Table 1. General summary of levels of recognition in host plant–fungal pathogen interactions. (After WYNN and STAPLES 1981)

Recognitional event, tropism or taxis	Response	Stimulus
1. Chemotaxis of zoospores	Directional swimming to host root surface or specific structures e.g. stomata	Chemical aspecific[a] or contact/chemical[b, h]
2. Binding of zoospores	Attachment to root surface	Recognition of carbohydrate ligands of root surface slime[c]
3. Directional emergence of germ tube from spore[d]	Growth towards epidermal cell wall junctions	Contact[d]
4. Adhesion of germ tubes and hyphae	Growth in close contact with plant exterior surface	Contact/chemical may involve "mucilage" of unknown (CHO?) composition
5. Germ tube orientation	Directional growth across surface ridges or lines[e], or towards or along epidermal cell wall junctions[f]	Contact
6. Appressorium formation	Location over stomata[e] or over or near epidermal cell wall junctions[f]	Contact and chemical[g]
7. Adhesion of haustorial[e] mother cell	Close contact with mesophyll cell wall	Contact/chemical[h]
8. Directional emergence of penetration peg	Growth downwards through stomata[e] or cuticle[f]	Contact/chemical[h]
9. Acceptance or rejection of penetration pegs and incipient intracellular infection structures	Triggering or suppression of mechanisms of disease resistance involving de novo RNA and protein synthesis	Chemical, modulated through parasite surface polysaccharides and glycoproteins interacting with host receptors

[a] For zoosporic, phycomycetous root pathogens
[b] For zoosporic pathogens that encyst on the stomatal pore e.g. *Pseudoperonospora humuli* (ROYLE and THOMAS 1973)
[c] *Phytophthora cinnamomi* (HINCH and CLARKE 1980)
[d] *Botrytis squamosa* (CLARK and LORBEER 1976)
[e] Stomatal-penetrating rusts
[f] Direct-penetrating fungi
[g] *Rhizoctonia solani* and *Puccinia graminis tritici* differentiate appressoria in response to specific host compounds (DODMAN and FLENTJE 1970, GRAMBOW and GRAMBOW 1978)
[h] Contact/chemical implies that the precise stimulus is not well understood

sis, the presence of a host. This is achieved in many pathogenic fungi by a delicate interplay between endogenous inhibitors and stimulators of the spore, and stimulatory chemicals on the plant surface acting as host signals (ALLEN 1976, BLAKEMAN 1981, TRIONE 1981). Leaf surface compounds, either leached from the host cells or produced by the surface microflora, may signal host-presence by stimulating spore germination through some relatively non-specific nutritional effect, or by counteracting endogenous spore-germination inhibitors. For example, the endogenous inhibitors of rusts are cis-isomers of cinnamic acid derivatives, and appear to function by preventing digestion of the spore germ-pore plugs, a prerequisite for germ-tube growth. Removal of the inhibitor by leaching in water droplets leads to germ-tube growth but chemical stimulants released by the host such as aldehydes, ketones, alcohols, esters, terpenes and steroidal saponins (ALLEN 1976) may counteract the endogenous inhibitors and thus play a potentially important role in the early stages of recognition and pathogenesis. SWINBURNE (1981) has suggested that spore germination in some necrotrophs may be stimulated by removal of inhibitory iron through chelation by siderophores on the leaf surface. In powdery mildews comparisons between germination on host or non-host surfaces and inert surfaces such as glass appear to be somewhat controversial, but on balance it appears likely that powdery mildew conidia can discriminate between plant and inert surfaces through leaf surface stimulants such as cuticular waxes (WHEELER 1981). In certain cases, some degree of specificity may even be observed. STAUB et al. (1974) showed that only 40% of conidia of barley powdery mildew (*Erysiphe graminis* f. sp. *hordei*) germinated on non-host cucumber leaves, whereas 80% germination was obtained on barley leaf surfaces. On the other hand, several fungi germinate just as well on inanimate surfaces as on host or non-host surfaces (e.g. downy mildew of lettuce, CRUTE and DICKINSON 1976), and it is clear that our current understanding of the signals which allow a pathogen spore to recognize a potential host, and the significance of this in pathogenesis is very limited.

11.2.2 Chemotaxis and Recognition

Chemotactic responses of zoospores of phycomycetous parasites to plant roots are readily demonstrated in vitro (see MITCHELL 1976 for a review). It is consistently observed that the main zone of attraction and binding to the roots is the region of elongation, the implication being that this zone is most active in the initiation of chemical gradients through the secretion of attractive substances. However, zoospore binding to root surfaces involves at least two components which are not always distinguished. Chemotaxis sensu stricto implies a directional swimming response by the zoospore (FRAENKEL and GUNN 1961, and is then followed by binding, or attachment to the root surface and encystment. The actual attachment process is likely to be controlled by quite different mechanisms from those initiating chemotaxis (see Sect. 11.2.3). Thus, for example, while ZENTMEYER (1961) claimed species-specific chemo-attraction of zoospores of *Phytophthora* spp. to roots of avocado and citrus seedlings, in fact species-specific binding was being demonstrated which does not necessarily imply a specific chemo-attractive response.

The chemical nature of attractive substances is generally studied in root models based on glass capillaries containing test compounds in free solution or agar. Five species of *Phytophthora* all responded positively and non-specifically to a wide range of compounds commonly contained in plant root exudates, including vitamins, phenolics, nucleotides, purine and pyrimidine bases, growth regulators, sugars, organic acids and amino acids (KHEW and ZENTMEYER 1973). Amino acids were the most effective, inducing a distinct directional movement to the mouth of the capillary followed by attachment, whereas other compounds merely caused spore accumulation by trapping without inducing positive directional movement. Arginine, aspartate, glutamate and methionine were the most effective, the minimal effective concentration being in the range. $1.4-32 \times 10^{-5}$ M. Various metabolic inhibitors and surfactants at concentrations which did not affect motility were ineffective in preventing chemotaxis. In an attempt to examine the precise nature of the spore chemoreceptors, "desensitization" experiments were carried out. Spores pretreated with amino acids were somewhat less sensitive, but nevertheless still responded to gradients of amino acids, suggesting that the chemoreceptors may be relatively non-specific.

While chemotaxis can be observed in vitro, its precise role in the soil is debatable. However, ALLEN and NEWHOOK (1973) found that zoospores of *Phytophthora cinnamomi* were attracted to capillaries containing ethanol. Ethanol is known to be secreted from plant roots under anaerobic conditions such as those encountered in waterlogged soils, conditions which favour the initiation of "damping-off" disease characteristic of this group of pathogens. It was suggested that rather than functioning through precise chemoreceptors, ethanol might cause physical changes in electrostatic charge distribution on the flagella, leading to flagellar reorientation and altered direction of movement.

Zoospores of some downy mildews such as *Pseudoperonospora humuli* show a positive taxis towards open stomata of hop leaves. The response appears to involve both the physical stimulus of the open stoma and chemical stimuli dependent on active photosynthesis (ROYLE and THOMAS 1973). However, the actual release of *P. humuli* zoospores from sporangia on the leaf surface is largely environmentally determined and occurs equally well on glass slides as on host leaves (COHEN 1981).

11.2.3 Zoospore Binding to Root Surfaces

Chemotaxis of zoospores to plant roots is only a prelude to their adhesion, encystment and the subsequent initiation of the processes of infection. Surprisingly, the biochemistry of zoospore binding and recognition at the plant root surface has been little studied. The binding of zoospores of *Phytophthora cinnamomi* to maize roots (HINCH and CLARKE 1980) and *Pythium aphanidermatum* to cress roots (LONGMAN and CALLOW unpublished) is inhibited if the root surface slime polysaccharides are oxidized with periodate. Mild trypsinization of *Pythium* spores, at concentrations which had no effect on motility, also inhibited binding. HINCH and CLARKE (1980) further showed that pretreatment of maize roots with the L-fucose-specific lectin from *Ulex europaeus,* or selective hydrolysis of terminal L-fucosyl residues from the root surface slime, also inhibited binding. Lectins with specificities for other sugar residues in slime polysaccharides, notably galactose, glucose and mannose, were relatively less effective in reducing spore binding. The specific ability of zoospores of *P. cinnamomi* to recognize terminal fucosyl residues was further demonstrated by binding to group 0 erythrocytes (terminal fucose) and fucosyl-Sepharose beads (but not to beads of Sephadex or Sepharose containing terminal glucose and galactose residues respectively; CLARKE personal communication). The evidence then, from the small amount of work done, is consistent with a model which suggests that the zoospore surface bears cognors with a specificity for complementary, saccharide-containing cognons of the root surface slime, and that in *P. cinnamomi* at least, fucosyl residues are important. However, root slimes of other species, such as wheat, do not contain terminal fucose (WRIGHT and NORTHCOTE 1974), and clearly it would of some interest to examine the sugar specificity of zoospore binding in these species.

11.2.4 Recognition and Tropisms

Following spore germination the subsequent growth of germ tubes and differentiation of infection structures is profoundly influenced by tropistic responses to a variety of chemical (chemotropic) and contact (thigmotropic) stimuli (Table 1 and WYNN and STAPLES 1981). Chemotropic stimuli are fairly easily defined, WYNN and STAPLES (1981) defined thigmotropism as a response to a surface, the mechanism of which is not understood, but involving large insoluble host polymers such as those in the cuticle. It was suggested that contact responses should also be restricted to those that occur on an inert surface. The rather negative definition reflects the paucity of advanced biophysical and biochemical work on this aspect of fungal physiology. Clearly the distinction between the two types of stimuli may be far from absolute.

Germ tube growth in close contact with the host cuticular surface, directional growth towards structural features of the host such as stomata and the differentiation of appressoria appear to be wholly, or at least mainly, determined by contact stimuli. Most studies have been carried out with rust fungi. DICKINSON (1949, 1970, 1971, 1972) showed that germ tubes of several fungi would either grow adpressed to the surface or in the air, depending on the nature of the artificial membranes used. Artificial hydrophobic surfaces were particularly effective in inducing close contact and WYNN (1976, and in WYNN and STAPLES 1981) showed that removal of hydrophobic cuticular waxes from leaves abolished close contact and oriented thigmotropic growth. Rust germ tubes grow at right angles to structural features of the host surface such as epidermal cell-wall ridges and in response to repetitive structures of artificial membranes (DICKINSON 1969). Repetitive contact stimuli appear to be translated into particular patterns of orientation of hyphal wall microfibrils, thus determining the direction of growth (DICKINSON 1977), although it is not known how the mechanical stimuli are translated into the appropriate biochemical responses to effect these patterns. LEWIS and DAY (1972) suggested that the periodicity of the crystal lattice lines of epicuticular waxes provided the contact stimuli for directional growth towards stomata.

Many experiments using synthetic leaf replicas, other inert surfaces, or even isolated cuticles, have shown that appressorial development is also determined primarily by a contact stimulus. In the case of most rusts that penetrate from appressoria positioned over stomata, the parasite appears to recognize the specific surface topography provided by the stomatal pore lips (WYNN 1976) and this is translated into a trigger for differentiation involving nuclear DNA synthesis and nuclear division (STAPLES et al. 1975, WYNN and STAPLES 1981). However, studies with synthetic membranes (DICKINSON 1970) have revealed that differentiation can be induced by irregularities as small as 120×1.2 nm, and the precise surface feature recognized is still far from clear. *Puccinia graminis* f. sp. tritici is unusual in that appressorial differentiation appears to involve volatile and non-volatile fractions of the host which are thought to be part of the chemical environment of stomata (GRAMBOW and GRAMBOW 1978). Phenolics which are active in inducing differentiation may be released from host cuticle and cell walls as a consequence of the interaction of the fungus and volatile components of the host yet to be identified.

Stomatal penetration by rusts is followed by differentiation of infection hyphae and haustorial mother cells. At least some components of non-host immunity or resistance appear to involve a failure of the invading heterologous rust to form haustorial mother cells in close contact with mesophyll (HEATH 1974, 1977), implying a recognition and attachment process of the host wall by a homologous rust. WYNN and STAPLES (1981) suggest that this is purely a contact response, and this view is supported by experiments which show that induction of haustoria on artificial membranes in vitro occurs in response to specific structural features, although these are different from those involved in appressorium formation (DICKINSON 1971, 1972). However, a chemical basis for mesophyll wall recognition cannot be ruled out. MENDGEN (1978) labelled a particulate fraction from germ-tube walls of *Uromyces phaseoli* with a fluorescent probe, and treated frozen-sections of various host and non-host plants with the labelled wall preparation. No differences were detected in the binding of the preparation to resistant and susceptible cultivars of the host *Phaseolus vulgaris*. However, tissues of non-host plants in which fewer haustoria formed bound much smaller amounts, and no binding was detected to tissues of those non-hosts in which no haustorial development occurs. This demonstration, although it was carried out at a low level of resolution (cell walls could not be distinguished from cytoplasm) and with a crude preparation from germ tubes rather than infection hyphae, suggests that specific host tissue may be recognized by the fungus through chemical associations involving fungal wall polysaccharides and glycoproteins.

11.3 Recognition of the Pathogen by the Host: Resistance Triggering

Plants, like other living organisms, have evolved immune or resistance mechanisms of various types by which they can counter the advance of foreign organisms. The result is that disease, tends to be specific, a given pathogen usually only infecting a distinct range of host plants. Resistance may be divided into "general" or "non-host" resistance, where a plant is clearly outside the range of hosts a given parasite may infect (resistance at the species level) and "specific" resistance, the result of genetic modification of a potential host, either naturally or through plant breeding, rendering it resistant to pathogens that would otherwise infect it (DAY 1974). For example, potato and wheat are "hosts" and "non-hosts" respectively of the potato late blight pathogen, *Phytophthora infestans,* but within the host species *Solanum tuberosum* there occur cultivars which have specific resistance to races of this pathogen.

Every plant possesses a diverse array of resistance mechanisms operating at different levels, and successful infection is dependent on the ability of the pathogen to avoid or overcome each in turn (see the "switching-point" model of HEATH 1974). Functionally the mechanisms may be either constitutive (pre-formed) or inducible (active or facultative). Induced defences are considered to involve synthesis of phytoalexins, hypersensitive cell death or the formation of structural barriers. In some cases at least it seems that these mechanisms are accompanied by changes in gene expression involving de novo RNA and protein synthesis (DIXON and LAMB 1979, KEEN and BRUEGGER 1977, LEGRAND 1983, VAN LOON and CALLOW 1983, YOSHIKAWA et al. 1977, YOSHIKAWA 1983), although the evidence for a causal relationship in fungal disease is equivocal (VAN LOON and CALLOW 1983). It is presumed that these inducible defences are triggered or mediated by an initial recognition event between the plant and the pathogen involving the detection of biochemical features of the pathogen surface by surface receptors of the host, enabling it to distinguish "self" from "non-self". This presumption is based on four principal lines of evidence.

a) The rapidity and localization of physiological responses in incompatible cells. Responses may be detected within a few hours of initial host–parasite contact, and are often limited to the cell being invaded (e. g. KITIZAWA et al. 1973, MACLEAN and TOMMERUP 1979).

b) The high degree of specificity in host-parasite systems and its genetic control.

c) Logical considerations of the evolution of resistance mechanisms.

d) Increasing indications that the activity of certain macromolecular components of pathogens can induce biochemical events associated with resistance.

11.3.1 The Genetic Argument

Most systems of cultivar- and race-specific resistance are controlled by one, or at most, a few genes of readily detectable effect (oligogenic or major gene systems), and where they have been clearly analyzed, they conform to a so-called "gene-for-gene" relationship (DAY 1974, ELLINGBOE 1976, 1981). In such systems, genes for resistance and susceptibility in the host are associated with complementary genes for avirulence and virulence in the pathogen, as gene pairs. Genes for resistance, which are usually dominant, are only expressed when matched with an appropriate, complementary gene for avirulence (also usually dominant) in the pathogen (Fig. 1). Furthermore, in those host–patho-

Fig. 1. The quadratic check in gene-for-gene interactions. *R* and *r* represent host genes for resistance (dominant) and susceptibility (recessive) at one gene locus. *A* and *a* represent parasite genes for avirulence (dominant) and virulence (recessive). Compatible and incompatible responses are denoted by + and − respectively

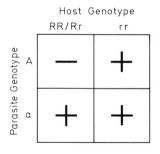

Fig. 2. Simple lock–key interpretation of the molecular implications of the gene-for-gene relationship. Products of host resistance/susceptibility alleles are shown as surface-localized cognors with defined binding-site configurations. Products of parasite avirulence/virulence alleles are shown as molecules of substantially common structure but specific terminal cognons. Only in the case R/A can correct binding and recognition take place, resulting in resistance. (After CALLOW 1982)

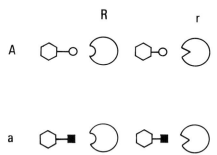

gen combinations where a number of different gene loci for resistance exist and are matched by complementary loci for virulence in the parasite, a gene pair specifying incompatibility is generally epistatic to gene pairs specifying compatibility. It has been suggested (CALLOW 1977, ELLINGBOE 1976, 1981, 1982, PERSON and MAYO 1974) that the major implication of this is that in systems of race- and cultivar-specific resistance there is specific recognition between host and parasite in the incompatible combination, and that this combination represents the active response. The combination between R and A, or rather between their gene products, is seen as the definitive reaction, constituting in effect a "stop-signal", since it triggers incompatibility (Fig. 2). Compatibility is seen as the failure to invoke defence. The high degree of specificity exhibited by gene-for-gene systems suggests highly selective host receptors capable of detecting specific features of parasite races.

11.3.2 Evolution of Resistance and Its Implications for Recognition

It is unlikely that each plant has a specific and different gene to control the recognition of each of the myriad of potential pathogens in a manner analogous to the gene-for-gene relationship. It seems more probable that each plant species has evolved the ability to detect or recognize the majority of pathogens through common surface components, the resulting molecular interaction serving to trigger resistance mechanisms. Specificity at the species level may be explained by the ability of successful or compatible pathogens to suppress or divert this

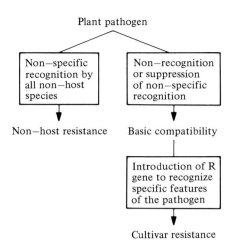

Plant pathogen

Non—specific recognition by all non—host species

Non—recognition or suppression of non—specific recognition

Non—host resistance   Basic compatibility

Introduction of R gene to recognize specific features of the pathogen

Cultivar resistance

Fig. 3. Summary of the possible origins of non-host and cultivar resistance. (After HEATH 1982)

general or non-specific recognition so that the fungus is recognized as "self", resulting in disease (Fig. 3; HEATH 1981a, 1982, OUCHI and OKU 1981). In this case the reactions conferring specificity occur in the compatible or suscepti-ble reaction. Cultivar and race-specific resistance is then seen as being superim-posed on systems of basic compatibility (ELLINGBOE 1976, PERSON and MAYO 1974). Once a system of basic compatibility has been established the plant host must experience strong selection pressure to evolve effective resistance. By ran-dom mutation, or through the incorporation of single genes for resistance from wild relatives in the process of plant breeding, a new variety is created with a restored ability to recognize the pathogen, resulting in resistance through the same, or a different set of resistance mechanisms from those involved in non-specific resistance. The host gene controlling this recognition is a gene for resistance, and the fungal gene controlling the synthesis of the recognized product of the fungus has, in effect, become the gene for avirulence. The ele-ments of a gene-for-gene relationship are thus established (ALBERSHEIM and ANDERSON-PROUTY 1975, CALLOW 1977, HEATH 1981a, 1982).

11.3.3 Pathogen Cognons as Modulators of Host Defence

Implicit in the discussions of host–pathogen specificity above is the concept of molecular recognition, that through receptors (cognors) on the host surface (although other locations are possible, see Sect. 3.4), plants may be able to detect complementary, surface-localized or secreted cognons of potential patho-gens, and use these as signals or triggers for the induction of resistance mecha-nisms. To account for the observed specificity and degree of variability in host–parasite systems, the fungal cognons must have a high information content. Whilst one tends to think of proteins in this context, it has been pointed out (CLAMP 1974, CLARKE and KNOX 1978) that polysaccharide structures have a much greater potential informational content. ALBERSHEIM and ANDERSON-

PROUTY (1975), CALLOW (1977), CLARKE and KNOX (1978), SEQUEIRA (1978) have reviewed the role of polysaccharide structures in many systems of biological recognition and there is now a growing body of evidence consistent with the general idea that plants can recognize pathogens as "non-self" through the regulatory activity of polysaccharides and glycoproteins. A range of substances has been isolated from several fungal pathogens, which when applied to plants either elicit defence reactions or modulate them in some way. The term "cognon" will be used here to describe collectively these chemically and functionally diverse molecules recognized by plants.

11.3.3.1 Polysaccharide Elicitors

Most plants respond to invasion by pathogenic microorganisms by accumulating antimicrobial substances known as phytoalexins (the voluminous literature is best consulted through BAILEY and MANSFIELD 1981). The term "elicitor" was introduced initially (KEEN 1975) to define substances that can cause phytoalexin accumulation in plants, but the term has since often been applied to molecules with activity in the induction of any facet of resistance, including tissue browning, callose accumulation and electrolyte leakage. Many elicitor preparations have been described, but discussion here will be largely confined to those in which at least a partial characterization of the active molecules has been carried out. Elicitors are isolated from mycelial or cell-wall extracts of the fungus or its cell-free culture filtrates and are generally either certain wall polysaccharides or glycoproteins (see WEST 1981 for a comprehensive list).

The cell walls of the majority of fungal pathogens contain the structural polysaccharides, chitin (poly-N-acetylglucosamine), amorphous β-linked glucans (predominantly β-$(1 \rightarrow 3)$-linked with β-$(1 \rightarrow 6)$-branches), α-$(1 \rightarrow 3)$- and α-$(1 \rightarrow 4)$-linked glucans and certain heteropolymers such as galactomannans (ARONSON 1981). In addition, particular groups may contain cellulose or chitosan (a β-linked, non-acetylated polymer of glucosamine). Of these, to date only β-glucans and chitosan have been shown to have elicitor activity, and by far the greater part of our information has been obtained for the β-glucans of *Phytophthora megasperma* f. sp. glycinea (syn. *P. megasperma* f. sp. sojae, soybean root and stem rot) by ALBERSHEIM and associates (ALBERSHEIM and VALENT 1978, AYERS et al. 1976a, AYERS et al. 1976b, AYERS et al. 1976c, EBEL et al. 1976). Several fractions which elicited the soybean phytoalexin, glyceollin, were obtained from hot water extracts of mycelia, the most active being principally a polydisperse, branched $\beta(1 \rightarrow 3)$ glucan (MW average = 100,000) containing small amounts of other neutral sugars. Activity in all fractions appeared to be due to glucan, and similar eliciting fractions were detected in cell-free culture filtrates, probably released from the fungal wall by autolysis. Elicitor activity was heat and trypsin-stable, but sensitive to periodate oxidation and digestion with soybean exo-β-glucanase. Activity was dependent on the branched structure rather than the $\beta(1 \rightarrow 3)$ linkage *per se* since sparsely branched $\beta(1 \rightarrow 3)$ glucans such as laminaran were much less effective. Partial acid and glucanase hydrolysis released active oligosaccharides and the smallest active fragment isolated so far contains nine glucosyl residues (ALBERSHEIM and VALENT 1978). Glyceollin elicitation was accompanied by an expansion in host secondary metabolism, as shown by increased activity of phenylalanine ammonia lyase. The glucan elicitors are highly active, less than 0.2 μg (10^{-11} mol) per hypocotyl being sufficient to elicit enough glyceollin to inhibit the growth of *P. megasperma* f. sp. glycinea, in vitro.

Glucan elicitors have been isolated from filtrates and mycelial extracts of virulent and avirulent races of *Colletotrichum lindemuthianum* and non-pathogenic *Colletotrichum* species (ANDERSON 1978a, b), and *Fusarium oxysporum* (ANDERSON 1980a). The *Fusarium* glucans were very similar to those of *P. megasperma* f. sp. *glycinea,* being β-(1 → 3) and (1 → 6) linked but *Colletotrichum* glucan elicitors were β-(1 → 3) and (1 → 4) linked and were not degraded by endoglucanase. Thus plants appear to be able to recognize more than one type of glucan structure. In fact glucan elicitors appear to be aspecific since they are active on both compatible and incompatible host cultivars. Furthermore, glucans are generally active when tested on non-host plants. Green bean (*Phaseolus vulgaris*), for example will respond not only to the β-(1 → 3), (1 → 4)-linked glucans of its pathogen *C. lindemuthianum,* but also to the β-(1 → 3), (1 → 6)-linked glucans of the non-pathogens *P. megasperma* f. sp. glycinea and *Fusarium oxysporum* (ANDERSON 1980a, CLINE et al. 1978). Likewise, active, non-specific glucans can be extracted from baker's yeast (HAHN and ALBERSHEIM 1978). It has been suggested therefore, that the widely occurring β-glucans in fungal walls serve as recognition cues in general or non-host resistance at the species level (ALBERSHEIM and VALENT 1978).

The ability of plants to respond to glucan elicitors is not totally universal. *Fusarium* and *Colletotrichum* glucans appear to be ineffective in tomato stems and fruit (ANDERSON 1980a) and glucan elicitors of *P. megasperma* f. sp. glycinea are only weak elicitors in potato (LISKER and KUĆ 1977). Although it was reported that the phytoalexin elicitors of autoclaved sonicates of *Phytophthora infestans* were glucans (LISKER and KUĆ 1977) other chemically diverse elicitors have been isolated from this fungus (see Sect. 11.3.3.3) and KURANTZ and ZACHARIUS (1981) showed that neither glucan nor a lipid fraction act as elicitors separately while a combination of the two gave substantial activity. A wider role for β-glucans in pathogenesis, other than as phytoalexin elicitors, is discussed by WOODWARD et al. (1980).

HADWIGER and BECKMAN (1980) have recently implicated chitosan as an active elicitor of phytoalexin accumulation in pea. Chitosan is a minor component of *Fusarium* walls (1.5%) and induces protection against *F. solani* f.sp. pisi if applied with, or prior to, inoculation with the fungus. Histochemical and immunocytochemical localization (HADWIGER et al. 1981) suggest that chitosan fragments, probably released from the fungus by pea enzymes, penetrate into and are recognized by the pea cells. In addition to their eliciting activity, chitosan and its hydrolysis products also inhibit germination and growth of the fungal pathogen.

11.3.3.2 Protein and Glycoprotein Elicitors

Some eight elicitor preparations containing polypeptide are listed by WEST (1981). Monilicolin A, a small (MW 8000) polypeptide isolated from mycelia of *Monilinia fructicola* (CRUIKSHANK and PERRIN 1968), elicits the phytoalexin phaseollin in the host, french bean, but is ineffective against peas and broad beans. Most other polypeptide-containing elicitors are glycoproteins, and some are clearly derived from surface glycoproteins of the cell wall. About 10% of the cell wall of *Pyricularia oryzae* (rice blast pathogen) can be solubilized to yield a proteoheteroglycan containing 90% carbohydrate and 10% protein (NAKAJIMA et al. 1977a, NAKAJIMA et al. 1977b). The major neutral sugars are D-mannose, D-glucose and D-galactose (6:2:1) arranged in the form of a main chain of α-(1 → 6)-D-mannopyranosyl residues with side chains of one to four α-D-mannopyranosyl units joined by α-(1 → 2) linkages and in some cases terminated by D-glucopyranosyl or D-galactofuranosyl residues. The polysaccharide is attached to the protein through O-mannosyl serine or O-mannosyl threonine. Recent studies in the author's lab (BIRD and CALLOW unpublished data) show that fractions containing this surface proteoheteroglycan are active but non-specific elicitors of phytoalexins in rice and other plants, and may thus serve as a non-specific recognition cue in general defence, as suggested for the glucan elicitors. In fact, *P. oryzae* cell walls also contain β-(1 → 3), (1 → 6)-linked glucans (NAKAJIMA et al. 1972, 1980) that also have elicitor activity (BIRD and CALLOW, unpublished data), and it remains far from clear what the precise significance of multiple eliciting components is in vivo (see also Sect. 11.3.4).

A polydisperse, phosphorylated proteogalactoglucomannan (av. MW 70,000) of similar general structure to that of *P. oryzae* has also been isolated from mycelia, walls and culture filtrates of *Cladosporium fulvum* (tomato leaf mould, DEWIT and ROSEBOOM 1980, DEWIT and KODDE 1981, DOW and CALLOW 1979a, LAZAROVITS et al. 1979). These separate studies highlight some of the problems in this area of research. The precise chemical composition (proportions of neutral sugars, protein etc.) varies with the culture medium used, fungal growth phase and source of material (cell wall, whole mycelium or culture filtrates). For example, the cell-wall glycoprotein isolated by DEWIT and KODDE (1981) has a mannose/galactose ratio of 1.2:1, with a trace of glucose, whereas one of the several glycoprotein fractions isolated from culture filtrates by DOW and CALLOW (1979a), with analogous biological activity, has a mannose/galactose/glucose ratio of 21:3:1. Furthermore the isolated materials may be polydisperse, representing a "family" of related molecules with different degrees of glycosylation and phosphorylation (DOW and CALLOW 1979a). However, it seems likely that all studies on *C. fulvum* involve the same biologically active principle, and activities at concentrations in the range 10–100 µg cm^{-3} have been obtained for phytoalexin accumulation, electrolyte leakage, necrosis and callose formation (CALLOW and DOW 1980, DEWIT and ROSEBOOM 1980, DOW and CALLOW 1979b, LAZAROVITS and HIGGINS 1979, LAZAROVITS et al. 1979). None of these activities is race- or cultivar-specific and the glycoprotein will also elicit phytoalexins in non-host plants (DEWIT and ROSEBOOM 1980). Whether these activities are primary or secondary consequences of the action of the glycoproteins is unknown, and an alternative view is that the glycoproteins may serve as toxins to promote pathogenesis. Activity in semi-pure preparations is sensitive to periodate, proteases and alkaline degradation of the protein-carbohydrate linkages, suggesting that both protein and saccharide moieties are essential for full activity. Activity is specifically degraded by α-mannosidase but other carbohydrases are either relatively or totally ineffective suggesting an important role for mannosyl residues. Although binding of labelled glycoprotein to tomato leaf cells has been studied (Dow and CALLOW 1979b), the precise nature of the recognized ligands could not be determined.

Glycoprotein elicitors have been isolated from *Phytophthora megasperma* f. sp. glycinea and *Colletotrichum lindemuthianum* and offer some experimental support for the specific-elicitor hypothesis of recognition and specificity (see Sect. 11.3.5). KEEN and LEGRAND (1980) semi-purified glycoproteins from alkaline extracts of cell-walls of nine races of *P. megasperma* f. sp. glycinea. The glycoproteins elicited glyceollin accumulation in two near-isogenic soybean cultivars with the same relative specificity as the fungal races from which they were obtained. However, the ratio of effectiveness on incompatible/compatible cultivars was only 2:3. The fractions contained glucose and mannose, were bound by the lectin Con A, and were completely inactivated by periodate. Pronase digestion, on the other hand, had little effect on activity, suggesting that the glycan moiety bears the active residues.

ANDERSON (1980b), in a preliminary study, isolated glycoproteins from three races of *C. lindemuthianum* and tested them on one cultivar of *Phaseolus vulgaris*. Preparations from the avirulent race were 100-fold and 10-fold more active in eliciting glyceollin than those from virulent and intermediate races respectively. There was also a correlation between activity and complexity, the most active preparation containing fewer glycoproteins. The active principles were not purified, however, and clearly the specificity observed could be due to a number of factors. The preparations may contain specific glycoprotein elicitors. Alternatively, non-specific elicitors may be present, but masked or suppressed by other factors in the preparations from virulent races.

Certain extracellular fungal enzymes can act as elicitors. SWINBURNE (1975) isolated a protease-rich elicitor preparation from sap expressed from apple tissue rotted by *Nectria galligena*. The endopeptidase appeared to be of fungal origin and co-purified with elicitor. Proteases from other biological sources also elicited the apple phytoalexin benzoic acid. STEKOLL and WEST (1978) isolated both high and low molecular weight elicitors of the castor bean phytoalexin casbene from culture filtrates of *Rhizopus stolonifer*. The partially purified high molecular weight component was heat-, pronase- and periodate-sensitive, implicating a glycoprotein. Further purification (LEE and WEST 1981a, b) showed that the elicitor is a polygalacturonase of MW 32,000. The glycosylated portion contains predominantly mannose, and enzyme activity appears to be crucial to elicitor activity. Polygalacturonases are widespread and important determinants of pathogenicity in many fungi, and thus the possibility is raised that elicitors based on polygalacturonase may be more widespread. WEST (1981) suggests a number of ways in which catalytic activity may confer elicitor activity. Although non-specific mechanisms based on gross cell damage induced by the enzyme cannot be excluded, a more subtle mechanism based on the release of certain polysaccharides from the host cell wall acting as secondary messengers or elicitors was proposed (see Sect. 11.3.4).

11.3.3.3 Other Elicitors

Application of lipid-containing extracts from *P. infestans* to potato tissue confers resistance to the pathogen and elicits potato phytoalexins (ERSEK 1977). BOS-TOCK et al. (1981) have identified arachidonic and eicosapentaenoic acids as active fractions. A role for carbohydrate residues in the *P. infestans* elicitor(s) cannot be excluded, however. GARAS and KUĆ (1981) were able to precipitate an elicitor with potato lectin (specific for N-acetyl-glucosamine residues); the synergism between lipid and glucan elicitors has already been referred to (Sect. 11.3.1.2).

11.3.3.4 Specificity Factors

Two instances have been reported of preparations which appear to contain race- and cultivar-specific determinants of resistance without being phytoalexin elicitors. Partially purified extracellular glycoproteins from incompatible races of *Phytophthora megasperma* f. sp. glycinea, but not those from compatible races, induced protection against living, compatible races of the fungus when injected into soybean hypocotyls (WADE and ALBERSHEIM 1979). The preparations did not have phytoalexin-inducing activity and their mode of action remains un-known. Tentative results appear to suggest that the glycosylated portions of the active fractions are more effective than the intact glycoproteins (ALBERSHEIM et al. 1981). The sugar compositions of the glycoproteins varied quantitatively, as did the glycosyl moieties of mannan-containing extracellular invertases puri-fied from the same fungal races (ZIEGLER and ALBERSHEIM 1977). However, it is not clear what exactly the relationship is between the invertase glycoproteins and the protecting glycoproteins.

Specific elicitors of necrosis without phytoalexin accumulation have been detected in the intercellular fluids of compatible combinations of tomato culti-vars and races of the tomato leaf mould pathogen, *Cladosporium fulvum* (DEWIT and SPIKMAN 1982). When injected into resistant, but not susceptible cultivars, the preparations specifically induced necrosis and chlorosis, tissue responses characteristic of incompatible reactions in this system. This demonstration may imply that the pathogen surface carries or secretes specific resistance elicitors which are presumably not detected in a compatible gene-for-gene host. If so, then they appear to be only expressed in plants since they could not be detected in in vitro cultures of the pathogen. An alternative hypothesis based on the presence of specific suppressors is less likely, although not totally ruled out by experiments in which intercellular fluids were injected into cultivars contain-ing more than one gene for resistance. The specificity of these materials was tested in an impressive number of race–cultivar combinations, and studies on the composition of the intercellular fluids and active principles contained therein are awaited with interest.

An alternative and more specific role for fungal glucans has been recently suggested (reviewed by CURRIER 1981). A crude mixture of high molecular weight components from hyphal walls of *Phytophthora infestans* will elicit hyper-

sensitive necrosis and phytoalexin accumulation, both in resistant and susceptible cultivars (Varns et al. 1971). The resistant responses induced by either an
incompatible race of the pathogen or hyphal wall extracts can be prevented
or "suppressed" by water-soluble β-glucans isolated from compatible races.
Similar fractions from incompatible races are less effective (Garas et al. 1979).
The hyphal-wall elicitors are bound by crude membrane fractions of potato
(Doke et al. 1979) or by protoplasts in which they induce cytoplasmic aggregation and lysis, responses taken to be characteristic of the hypersensitive reaction
(Doke and Tomiyama 1980a). Protoplast responses were also specifically suppressed by glucans from compatible races (Doke and Tomiyama 1980b). The
suppressive glucans have been partially characterized as β-(1 → 3)- and (1 → 6)-
linked, containing 17–23 glucose residues and with both neutral and phosphorylated forms (Doke et al. 1979). It has been suggested that non-specific elicitors
in the hyphal wall extract (see Sect. 3.3.3 above) may bind to receptors (Sect.
11.3.4) on the host surface and that the glucan suppressor may block the binding
sites in a competitive and specific manner. The chemical basis of specificity
is unknown. Indeed, Keen (1982) has questioned whether such low molecular
weight homopolymers with a restricted range of linkages would exhibit sufficient
structural variability to account for the large range of specificities encountered
in this host–parasite interaction.

11.3.4 Recognition of Pathogen Cognons and Their Modes of Action

The interaction of fungal cognons with the plant cell is poorly understood
and critical evaluation of the physiological roles of these substances is urgently
required. Unfortunately, precise information at the biochemical and cell levels
is available only for glucan and to a lesser extent, chitosan elicitors. The rest
remains highly speculative, and the dangers of over-generalization are obvious.
One of the major obstacles to progress in this area of recognition research
is the lack of good, defined cell systems compared with, say, studies on recognition of regulators by animal cells. Too many studies in biochemical plant pathology appear to utilize inadequately characterized, heterogeneous preparations
and depend on bioassays which are often highly variable and which may be
of unknown relevance to the actual resistance mechanisms used by the plant.
Even with standardization (Hahn et al. 1981), comparison of different elicitors
is still fraught with problems. Although the concentration of an elicitor applied
to the plant tissue may be known, this has little relevance to the physiological
concentration of effective cognons at receptor sites.

The initial host receptors or cognors for pathogen cognons are unknown.
The diversity of cognons suggests qualitatively different receptors, and three
basic types of interaction have been suggested:

a) binding by complementary surface receptors, possibly involving lectins
or lectin-like proteins, with signal transmission to the rest of the cell via secondary messengers (Albersheim and Anderson-Prouty 1975, Callow 1977),

b) complexing with plant DNA (Hadwiger and Loschke 1981),

c) enzymic release of exogenous elicitors (West 1981).

The plasma membrane offers a logical site for the receipt and translation of chemical stimuli originating from pathogens, and although host receptors for pathogen cognons have yet to be isolated there are a number of lines of circumstantial evidence supporting their existence. A variety of plant protoplasts will bind β-glucosyl Yariv antigens through the so-called β-lectins (LARKIN 1977). Cell wall β-glucans of *P. infestans* altered the membrane potential of potato cells (KOTA and STELZIG 1977) and agglutinated potato protoplasts (PETERS et al. 1978), implying the presence of surface glucan receptors. Preliminary direct evidence for surface-localized β-glucan receptors has recently been obtained (YOSHIKAWA 1983). A homogeneous ^{14}C-labelled β-$(1 \rightarrow 3)$-glucan elicitor bound to a membrane fraction containing "predominantly" plasma membranes. Binding was stated to be specific, being inhibited by unlabelled glucan only, and a single class of binding sites appeared to be present.

Glucan elicitors appear to induce changes in gene expression and metabolic regulation comparable to those involved in resistance to pathogens, including de novo RNA and protein synthesis (DIXON and LAMB 1979, HAHLBROCK et al. 1981). By analogy with regulatory mechanisms proposed for signal recognition and transmission in animal cell systems therefore, an orderly sequence of events following elicitor binding has been suggested, involving discrete receptors, membrane transduction, and signal transmission via secondary messengers (WOODWARD et al. 1980). However, β-glucans are known to bind and solubilize membrane sterols (PILLAI and WEETE 1975), and the possibility cannot be excluded that the primary action of glucan elicitors is to induce changes in membrane integrity, and that the changes in gene expression and metabolic regulation leading to phytoalexin accumulation are secondary consequences.

Shortly after infection of potato tissue with *P. infestans* the host plasma membrane appears to be tightly bound to the fungal infection structure, both in compatible and incompatible interactions (NOZUE et al. 1979). Isolated host membrane preparations, which were reported to contain plasma membranes (although unequivocal evidence that this is so was not provided), bound to fungal infection structures in the presence of potato lectin (FURUICHI et al. 1980). Pretreatment with chitobiose, a disaccharide hapten for potato lectin, inhibited binding and also inhibited hypersensitive cell death of tuber cells inoculated with an incompatible race of the pathogen (NOZUE et al. 1980). No other sugars appeared to be effective although MARCAN et al. (1979) reported that laminaribiose and methyl-β-D-glucopyranoside caused significant inhibition of cell death and browning of tuber tissue treated with mycelial sonicates of the same pathogen. The cause of this discrepancy is not clear but may lie in the use of sonicates by the latter authors rather than living fungus. GARAS and KUĆ (1981) further demonstrated that potato lectin binds and precipitates elicitors of phytoalexin accumulation extracted from the pathogen, although the lectin–elicitor complexes retained activity. These results raise the possibility that potato lectin may play some role in the potato blight system, in the recognition of pathogen surface components including the non-specific elicitor. Such an interaction might then be subject to specific suppressors (see Sect. 11.3.3.4). However, before such a role for potato lectin can be considered to be established, further information is required on the chemical characteristics of the elicitor (see Sect. 3.3.3), its localization in relation to the fungal surface, details of the lectin–elicitor binding characteristics, and evidence on the location of the interacting potato lectin molecules. The isolation of pure plasma membranes is a notoriously difficult problem, and more convincing evidence needs to be provided to support claims (FURUICHI et al. 1980) that plasma membranes are a major component of their membrane fractions.

A second model for elicitor action has been proposed for chitosan elicitors (reviewed by HADWIGER and LOSCHKE 1981). Chitosan elicitors released from the pathogen were detected inside host cells by immuno- and cytochemical techniques (HADWIGER et al. 1981), and appeared to accumulate in the nucleus. Chitosan appears to bind to DNA, altering its physical properties, as do several other chemical agents which elicit phytoalexins. It was proposed that binding of chitosans to regulatory regions of heterochromatin results in changes in gene expression leading to synthesis of phytoalexins and/or host enzymes capable of releasing inhibitory levels of chitosan from fungal walls. Further experimental and cytogenetic proof is required before this hypothetical scheme can be assessed.

A quite different receptor mechanism is envisaged to account for the activity of enzymic elicitors such as the polygalacturonase of *Rhizopus stolonifer* and possibly the many other fungal pathogens that secrete such enzymes (WEST 1981). Although non-specific mechanisms based on gross cell damage cannot be excluded, a more subtle mechanism has been proposed in which the plant receptor is a pectic polysaccharide substrate in the cell wall, processed or degraded by the enzyme to release oligogalacturonides which act as secondary messengers. These "endogenous elicitors" may be recognized in turn by secondary receptors in the plant cell to signal phytoalexin accumulation.

Preliminary evidence for pectic fragments as endogenous elicitors has been obtained in extracts of soybean cell walls (HAHN et al. 1981) and the same oligogalacturonides appear to elicit phytoalexins in castor bean (WEST 1981). An extension of this model, involving the activation of latent plant polygalacturonases, may account for reports of low molecular weight "endogenous" or "constitutive" elicitors of phytoalexin accumulation released from french bean cells following injury or infection (HARGREAVES and BAILEY 1978, HARGREAVES 1979). Further evidence relating to these models is reviewed by ALBERSHEIM et al. (1981) and WEST (1981).

A major argument against a physiological role for elicitors and other modulators is that several are insoluble wall polymers, normally extracted by harsh treatments such as hot water or alkali, and that little cognizance is taken of the availability of these molecules to plant receptors under physiological conditions. For example, whilst imunocytochemistry has shown that proteoheteroglycan elicitors of *Pyricularia oryzae* are localized on the hyphal surface in vitro (NAKAJIMA et al. 1980), the glucan elicitors of the fungus probably occupy inner wall layers. In the case of multiple elicitors in fungal walls, which, if any, are physiologically relevant?

Two ways in which host access to pathogen cognons may be obtained have been suggested. Firstly, many elicitors may be extracted both from fungal walls and culture fluids, suggesting natural release. However, this may be an artefact of culture in vitro, the released elicitors probably representing the products of wall autolysis. The physiological relevance of the cultured state is of course questionable. DEWIT and SPIKMAN (1982) for example, were able to extract specific necrogens only from intercellular fluids of tomato leaves infected by compatible races of *C. fulvum*. These factors appear not to be expressed in vitro. On the other hand, DOKE et al. (1980) noted that specific glucan suppressors were released into germination fluids of *P. infestans* cystospores.

Secondly, plant carbohydrases may mediate release of elicitors. High molecular weight soluble glucomannan elicitors of glyceollin were released from cell walls and living hyphae of *Phytophthora megasperma* f. sp. glycinea after only 2 min incubation with wounded soybean cotyledons and enzyme preparations from them (YOSHIKAWA 1983, YOSHIKAWA et al. 1979, 1981). Two endo-β-$(1 \rightarrow 3)$-glucanases were purified from soybean cotyledons and shown to cause release of the elicitors. Increased activity of plant glycan hydrolases including chitinase and β-$(1 \rightarrow 3)$-glucanases appears to be a consistent response of plant tissues to microbial activity (WOODWARD et al. 1980).

11.3.5 Conclusions: Models for Specific Recognition

The studies reviewed here are mainly consistent with, but do not prove, the conceptual views on the evolution of resistance expressed in Section 3.2. Non-specific elicitors in their various forms could well act as non-specific recognition cues, signalling the presence of a potential pathogen and thus forming the basis of non-host resistance. The structure of the polymers which have elicitor activity is probably fundamental, and it is unlikely therefore that pathogens can escape detection by altering the structures of such components to non-recognized forms. Furthermore, plants appear to have a broad capacity to respond to chemically diverse elicitors – french bean (*Phaseolus vulgaris*), for example, can respond to glucan, glycoprotein and peptide elicitors (Sects. 11.3.3.1 and 11.3.3.2).

The biochemical evidence for suppressors in the induction of basic compatibility at the species level is less good. Circumstantial evidence for their existence is supported by many experiments in which it appears that prior inoculation of a plant with a compatible pathogen permits a susceptible reaction to an otherwise incompatible pathogen (reviewed by OUCHI and OKU 1981). Preparations which inhibit or suppress various resistance responses at the level of species specificity have been reported (SHIRAISHI et al. 1978, OKU et al. 1980, HEATH 1981 b), but until the active components of these preparations are isolated, characterized, and their interaction with other components of the system studied, it is difficult to assess their precise significance as determinants of basic compatibility at the species level. One obvious hypothesis is that they interfere with the correct recognition of non-specific elicitors (HEATH 1981a, BUSHNELL and ROWELL 1981, Fig. 4).

If basic compatibility at the species level truly represents a form of suppressed resistance, as seems logical (and HEATH 1981a suggests that this would not be in conflict with the gene-for-gene hypothesis), then mutation or the incorporation of major genes for resistance into an otherwise compatible host species

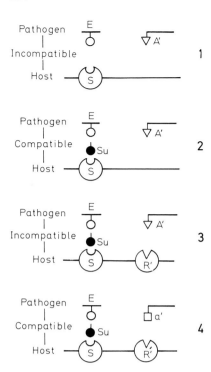

Fig. 4. Hypothetical model depicting the possible relationships between elicitor cognons, non-specific suppressors and host cognons in the co-evolution of race and cultivar-specificity. *State 1.* Incompatibility at the species level (non-host resistance). Non-specific elicitor cognons of the pathogen (E) are recognized by non-specific host cognors or "sensors" (S), triggering induced resistance. *State 2.* Basic compatibility at the species level. Successful pathogen produces suppressor (Su) which competes with elicitor-sensor binding and therefore inhibits resistance triggering. *State 3.* Cultivar resistance. Host acquires a gene for resistance R which codes for primary product R', a surface cognor with the ability to recognize cognons of specific surface molecules of the pathogen (A' the cognon product of pathogen gene A, which in effect therefore becomes a gene for avirulence). The complex R'/A' triggers resistance. *State 4.* Race and cultivar-compatibility. Mutation of A to a in the pathogen creates a new race with an altered cognon (a') no longer recognized by R'. Resistance not triggered

in effect creates a new cultivar with a restored ability to recognize the pathogen as "non-self" once more. The primary products of genes for resistance are thus seen as receptors (cognors) for parasite recognition "cues" (Fig. 4). In theory, the recognition of any parasite surface component would suffice, providing it has the necessary degree of potential structural variation to confer specificity, and there is evidence to show that the glycan moieties of surface glycoproteins in fungi exhibit a great deal of species- and strain-specificity (Ballou 1976, in relation to yeast mannan-glycoproteins, Ziegler and Albersheim 1977 in relation to surface invertase glycoproteins of *P. megasperma* f.sp. *glycinea*). In effect, genes coding for the recognized moiety (or cognon) such as a specific saccharide structure in a glycoprotein, synthesized under the control of a specific glycosyl transferase (Albersheim and Anderson-Prouty 1975) become the genes for avirulence and their products would then acquire an additional function as specific elicitors of phytoalexins (Keen and Legrand 1980), necrosis (Dewit and Spikman 1982) or specific protecting factors (Wade and Albersheim 1979). Since this function of the parasite gene then creates a strong selective disadvantage, the parasite is clearly under pressure to escape detection by the host cultivar containing the new gene for resistance. Mutation to virulence would presumably create sufficient structural differences in the cognon to prevent its recognition by the host cultivar, thus creating the elements of a gene-for-gene system (Fig. 4). If the cognon was part of an important enzyme, or structural polymer, this alteration, while promoting some aspects of parasite success, could also reduce overall fitness if the functional properties of the molecules were also affected. This might explain the observation that unnecessary genes for avirulence tend to be selected against (VANDERPLANK 1968, LEONARD and CZOCHOR 1980).

An alternative scenario, supported mainly by the biochemical studies on the *P. infestans* glucans, is that specific suppressors are determinants of compatibility. This would be at variance with the gene-for-gene hypothesis, since it would require that specificity is the active function of genes for virulence and susceptibility. However, BUSHNELL and ROWELL (1981) have recently presented a model which seeks to reconcile the biochemical and genetic evidence by suggesting that the incorporation of resistance genes into a basically compatible host permits the new host cultivar to recognize the pathogen as "non-self" through recognition of the suppressors determining basic compatibility.

At this stage, all hypotheses and models must be treated with great caution, since the evidence on which they are based is exceedingly fragmentary, occasionally controversial (KEEN 1982), and limited in scope to a handful of host–pathogen systems. Indeed, ELLINGBOE (1982) has recently sounded a strong dissenting note to the currently accepted dogma be rejecting on genetic grounds certain key elements common to most models seeking to explain host–pathogen specificity. As discussed in this chapter, the body of experimental evidence is largely consistent with the view that host-pathogen recognition triggers an expressive phase of resistance involving changes in gene expression leading to resistance mechanisms such as phytoalexin synthesis. This would involve a number of enzymes and genes. Furthermore, most models implicate specific saccharide structures in glycoproteins and polysaccharides as recognition "cues", signalling parasite presence. The synthesis of specific saccharide structures is controlled by glycosyl transferases and again, therefore, is under multigene control. ELLINGBOE (1982) argues that 95% of the genetic variability observed in host–pathogen interactions is of the one-parasite-gene–one-host-gene type, and is therefore inconsistent with any hypothesis relying on the contribution of a number of genes. It is proposed, rather, that the actual event of recognition between the primary products of host and parasite genes itself constitutes the mechanism of incompatibility in a way as yet entirely unknown.

There is then a vast discrepancy between the corpus of experimental evidence, much of it circumstantial and fragmentary, and some aspects of the genetic evidence. At the present time it is not easy to reconcile these fundamentally different views, and top priority in this field must be the determination of the complementary primary products of genes for resistance and avirulence through the procedures of molecular genetics. Not only will this answer many of the controversial problems in biochemical plant pathology, but a fundamental basis will be laid for future efforts to create new resistant cultivars through recombinant DNA technology.

References

Albersheim P, Anderson-Prouty AJ (1975) Carbohydrates, proteins, cell surfaces and the biochemistry of pathogenesis. Annu Rev Plant Physiol 26:31–52

Albersheim P, Valent BS (1978) Host-pathogen interactions in plants. Plants, when exposed to oligosaccharides of fungal origin, defend themselves by accumulating antibiotics. J Cell Biol 78:627–643

Albersheim P, Darvill AG, McNeil M, Valent BS, Hahn MG, Sharp JK, Desjardins AE, Spellman MW, Ross LM, Robertsen BK, Aman P, Franzen LE (1981) Structure and function of complex carbohydrates active in regulating plant–microbe interactions. Pure Appl Chem 53:53:79–88

Allen PJ (1976) Control of spore germination and infection structure formation in the fungi. In: Heitefuss R, Williams PH (eds) Encyclopedia of plant physiology, New Ser. Physiological plant pathology. Springer, Berlin Heidelberg New York, pp 51–85

Allen RN, Newhook FJ (1973) Chemotaxis of zoospores of *Phytophthora cinnamomi* in capillaries of soil pore dimensions. Trans Br Mycol Soc 61:287–302

Anderson AJ (1978a) Initiation of resistant responses in bean by mycelial wall fractions from three races of the bean pathogen, *Colletotrichum lindemuthianum*. Can J Bot 56:2247–2251

Anderson AJ (1978b) The isolation from three species of *Colletotrichum* of glucan containing polysaccharides that elicit a defense response in bean (*Phaseolus vulgaris*). Phytopathology 68:189–194

Anderson AJ (1980a) Studies on the structure and activity of fungal glucans. Can J Bot 58:2343–2348

Anderson AJ (1980b) Differences in the biochemical compositions and elicitor activity of extracellular components produced by three races of the fungal plant pathogen, *Colletotrichum lindemuthianum*. Can J Microbiol 26:1473–1479

Aronson JM (1981) Cell wall chemistry, ultrastructure and metabolism. In: Cole GT, Kendrick B (eds) Biology of conidial fungi Vol 2. Academic Press, London, pp 459–507

Ayers AR, Ebel J, Finelli F, Berger N, Albersheim P (1976a) Host-pathogen interactions. IX. Quantitative assays of elicitor activity and characterization of the elicitors present in the extracellular medium of cultures of *Phytophthora megasperma* var sojae. Plant Physiol 57:751–759

Ayers AR, Ebel J, Valent B, Albersheim P (1976b) Host–pathogen interactions. X. Fractionation and biological activity of an elicitor isolated from the mycelial walls of *Phytophthora megasperma* var. sojae. Plant Physiol 57:760–765

Ayers AR, Valent B, Ebel J, Albersheim P (1976c) Host–pathogen interactions. XI. Composition and structure of wall-released elicitor fractions. Plant Physiol 57:766–774

Bailey JA, Mansfield JW (eds) (1981) Phytoalexins. Blackie, Glasgow

Ballou CE (1976) Structure and biosynthesis of the mannan component of the yeast cell envelope. Adv Microb Physiol 14:93–158

Blakeman JP (ed) (1981) Microbial ecology of the phylloplane. Academic Press, London New York

Bostock RM, Kuć J, Laine RA (1981) Eicosapentaenoic and arachidonic acids from *Phytophthora infestans* elicit fungitoxic sesquiterpenes in the potato. Science 212:67–69

Burke D, Mendonca-Previato L, Ballou CE (1980) Cell-cell recognition in yeast: purification of *Hansenula wingei* 21-cell sexual agglutination factor and comparison of the factors from three genera. Proc Natl Acad Sci USA 77:318–322

Bushnell WR, Rowell JB (1981) Suppressors of defense reactions: a model for roles in specificity. Phytopathology 71:1012–1014

Callow JA (1977) Recognition, resistance and the role of plant lectins in host-parasite interactions. Adv Bot Res 4:1–49

Callow JA (1982) Molecular aspects of fungal infection. In: Smith H, Grierson D (eds) The molecular biology of plant development. Blackwell, Oxford, pp 467–497

Callow JA (ed) (1983) Biochemical plant pathology. Wiley, New York Chichester Brisbane Toronto

Callow JA, Dow JM (1980) The isolation and properties of tomato mesophyll cells and their use in elicitor studies. In: Ingram DS, Helgeson JP (eds) Tissue culture methods for plant pathologists. Blackwell Oxford, pp 197–202

Clamp JR (1974) Analysis of glycoproteins. Biochem Soc Symp 40:3–16

Clark CA, Lorbeer JW (1976) Comparative histopathology of *Botrytis squamosa* and *B. cinerea* on onion leaves. Phytopathology 66:1279–1289

Clarke AE, Knox RB (1978) Cell recognition in flowering plants. Quart Rev Biol 53:3–28

Cline K, Wade M, Albersheim P (1978) Host-pathogen interactions. XV. Fungal glucans which elicit phytoalexin accumulation in soybean also elicit the accumulation of phyto-alexins in other plants. Plant Physiol 62:918–921

Cohen Y (1981) The processes of infection of downy mildews on leaf surfaces. In: Blakeman JP (ed) Microbial ecology of the phylloplane. Academic Press, London New York, pp 115–133

Cruikshank IAM, Perrin DR (1968) The isolation and partial characterization of monili-colin A, a polypeptide with phaseollin-inducing activity from *Monilinia fructicola*. Life Sci 7:449–458

Crute IR, Dickinson CH (1976) The behaviour of *Bremia lactucae* on cultivars of *Lactuca sativa* and on other composites. Ann Appl Biol 82:433–450

Currier WW (1981) Molecular controls in the resistance of potato late blight. Trends. Biochem Sci 6:191–194

Day P (1974) Genetics of host–parasite interaction. Freeman, San Francisco

Dewit PJGM, Kodde E (1981) Further characterisation and cultivar-specificity of glyco-protein elicitors from culture filtrates and cell walls of *Cladosporium fulvum* (syn. *Fulvia fulva*). Physiol Plant Pathol 18:297–314

Dewit PJGM, Roseboom PHM (1980) Isolation, partial characterization and specificity of glycoprotein elicitors from culture filtrates, mycelium and cell walls of *Cladosporium fulvum* (syn. *Fulvia fulva*). Physiol Plant Pathol 16:391–408

Dewit PJGM, Spikman G (1982) Evidence for the occurrence of race and cultivar–specific elicitors of necrosis in intracellular fluids of compatible interactions of *Cladosporium fulvum* and tomato. Physiol Plant Pathol, 21:1–11

Dickinson S (1949) Studies in the physiology of obligate parasitism. I. The stimuli deter-mining the direction of growth of the germ tubes of rust and mildew spores. Ann Bot NS 13:89–104

Dickinson S (1969) Studies in the physiology of obligate parasitism. VI. Directed growth. Phytopathol Z 66:38–49

Dickinson S (1970) Studies in the physiology of obligate parasitism VII. The effect of a curved thigmotropic stimulus. Phytopathol Z 69:115–124

Dickinson S (1971) Studies in the physiology of obligate parasitism. VIII. An analysis of fungal responses to thigmotropic stimuli. Phytopathol Z 70:62–70

Dickinson S (1972) Studies in the physiology of obligate parasitism. IX. The measurement of thigmotropic stimulus. Phytopathol Z 73:347–358

Dickinson S (1977) Studies in the physiology of obligate parasitism. X. Induction of responses to a thigmotropic stimulus. Phytopathol Z 89:97–115

Dixon RA, Lamb CJ (1979) Stimulation of de novo synthesis of L-phenylalanine am-monia-lyase in relation to phytoalexin accumulation in *Colletotrichum lindemuthianum* elicitor-treated cell suspension cultures of french bean (*Phaseolus vulgaris*) Biochem Biophys Acta 586:453–463

Dodman RL, Flentje NT (1970) The mechanism and physiology of plant penetration by *Rhizoctonia solani*. In: Parmeter JR (ed) *Rhizoctonia solani*: biology and pathology. Univ of California Press, Berkeley, California, pp 149–160

Doke N, Tomiyama K (1980a) Effect of hyphal wall components from *Phytophthora infestans* on protoplasts of potato tuber tissue. Physiol Plant Pathol 16:169–176

Doke N, Tomiyama K (1980b) Suppression of the hypersensitive response of potato tuber protoplasts to hyphal wall components by water-soluble glucans isolated from *Phytophthora infestans*. Physiol Plant Pathol 16:177–186

Doke N, Garas NA, Kuć J (1979) Partial characterization and aspects of the mode of action of a hypersensitivity-inducing factor (HIF) isolated from *Phytophthora infes-tans*. Physiol Plant Pathol 15:127–146

Doke N, Garas NA, Kuć J (1980) Effect on host hypersensitivity of suppressors released during the germination of *Phytophthora infestans* cystospores. Phytopathology 70:35–39

Dow JM, Callow JA (1979a) Partial characterization of glycopeptides from culture filtrates of *Fulvia fulva* (Cooke) Ciferri (syn. *Cladosporium fulvum*) the tomato leaf mould pathogen. J Gen Microbiol 113:57–66

Dow JM, Callow JA (1975b) Leakage of electrolytes from isolated leaf mesophyll cells of tomato induced by glycopeptides from culture filtrates of *Fulvia fulva* (Cooke) Ciferri (syn. *Cladosporium fulvum*). Physiol Plant Pathol 15:27–34

Ebel J, Ayers AR, Albersheim P (1976) Host–pathogen interactions. XI. Response of suspension-cultured soybean cells to the elicitor isolated from *Phytophthora megasperma* var. sojae, a fungal pathogen of soybean. Plant Physiol 57:775–779

Ellingboe AH (1976) Genetics of host–parasite interactions. In: Heitefuss R, Williams PH (eds) Encyclopedia of plant physiology Vol 4. Physiological plant pathology. Springer, Berlin Heidelberg New York, pp 761–778

Ellingboe AH (1981) Changing concepts in host–pathogen genetics. Annu Rev Phytopathol 19:125–134

Ellingboe AH (1982) Genetical aspects of active defense. In: Wood RKS (ed) Active defence mechanisms in plants. Plenum, London New York, pp 179–192

Ersek T (1977) A lipid component from *Phytophthora infestans* inducing resistance and phytoalexin accumulation in potato tubers. Curr Top Pathol:73–76

Fraenkel GS, Gunn DL (1961) The orientation of animals. Dover New York

Furuichi N, Tomiyama K, Doke N (1980) The role of potato lectin in the binding of germ tubes of *Phytophthora infestans* to potato cell membranes. Physiol Plant Pathol 16:249–256

Garas NA, Doke N, Kuć J (1979) Suppression of the hypersensitive reaction in potato tubers by mycelial components from *Phytophthora infestans*. Physiol Plant Pathol 15:117–126

Garas NA, Kuć J (1981) Potato lectin lyses zoospores of *Phytophthora infestans* and precipitates elicitors of terpenoid accumulation produced by the fungus. Physiol Plant Pathol 18:227–237

Grambow HJ, Grambow GE (1978) The involvement of epicuticular and cell-wall phenols of the host plant in the in vitro development of *Puccinia graminis* f. sp. tritici. Z Pflanzenphysiol 90:1–9

Hadwiger LA, Beckman JM (1980) Chitosan as a component of pea–*Fusarium solani* interactions. Plant Physiol 66:205–211

Hadwiger LA, Loschke DC (1981) Molecular communication in host-parasite interactions: Hexosamine polymers (chitosan) as regulator compounds in race-specific and other interactions. Phytopathology 71:756–762

Hadwiger LA, Beckman JM, Adams MJ (1981) Localization of fungal components in the pea–*Fusarium* interaction detected immunochemically with antichitosan and antifungal cell wall antisera. Plant Physiol 67:170–775

Hahlbrock K, Lamb CJ, Purwin C, Ebel J, Fautz E, Schaffer E (1981) Rapid response of suspension-cultured parsley cells to the elicitor from *Phytophthora megasperma* var. sojae. Plant Physiol 67:768–773

Hahn MG, Albersheim P (1978) Host-pathogen interactions. XIV. Isolation and partial characterization of an elicitor from yeast extract. Plant Physiol 62:107–111

Hahn MG, Darvill AG, Albersheim P (1981) Host-pathogen interactions. XIX. The endogenous elicitor, a fragment of plant cell wall polysaccharide that elicits phytoalexin accumulation in soybeans. Plant Physiol 68:1161–1169

Hargreaves JA (1979) Investigations into the mechanisms of mercuric chloride stimulated phytoalexin accumulation in *Phaseolus vulgaris* and *Pisum sativum*. Physiol Plant Pathol 15:279–287

Hargreaves JA, Bailey JA (1978) Phytoalexin production by hypocotyls of *Phaseolus vulgaris* in response to constitutive metabolites released by damaged bean cells. Physiol Plant Pathol 13:89–100

Heath MC (1974) Light and electron microscope studies of the interactions of host and non-host plants with cowpea rust *Uromyces phaseoli* var. *vignae* Physiol Plant Pathol 4:403–414

Heath MC (1977) A comparative study of non-host interactions with rust fungi. Physiol Plant Pathol 10:73–78

Heath MC (1981a) A generalized concept of host–parasite specificity. Phytopathology 71:1121–1123

Heath MC (1981b) The suppression of the development of silicon-containing deposits in french bean leaves during growth of the bean rust fungus. Physiol Plant Pathol 18:149–145

Heath MC (1982) The absence of active defence mechanisms in compatible host–pathogen interactions. In Wood RKS (ed) Active defence mechanisms in plants. Plenum, London New York, pp 143–156

Hinch JM, Clarke AE (1980) Adhesion of fungal zoospores to root surfaces is mediated by carbohydrate determinants of the root slime. Physiol Plant Pathol 16:303–307

Keen NT (1975) Specific elicitors of plant phytoalexin production: determinants of race specificity in pathogens? Science 187:74–75

Keen NT (1982) Mechanisms conferring specific recognition in gene-for-gene plant parasite systems. In: Wood RKS (ed) Active defence mechanisms in plants. Plenum, London New York, pp 67–84

Keen NT, Bruegger B (1977) Phytoalexins and chemicals that elicit their production in plants. In: Hedin P (ed) Host plant resistance to pests. Am Chem Soc Symp Ser 62:1–26

Keen NT, Legrand M (1980) Surface glycoproteins: evidence that they may function as the race-specific phytoalexin elicitors of *Phytophthora megasperma* f. sp. glycinea. Physiol Plant Pathol 17:175–192

Khew KL, Zentmeyer GA (1973) Chemotactic responses of zoospores of five species of *Phytophthora*. Phytopathology 63:1511–1517

Kitizawa K, Inagaki H, Tomiyama K (1973) Cinephotomicrographic observations on the dynamic responses of protoplasm of a potato plant cell to infection by *Phytophthora infestans*. Phytopathol Z 76:80–86

Kota DA, Stelzig DA (1977) Electrophysiology as a means of studying the role of elicitors in plant disease reactions. Proc Am Phytopathol Soc 4:216–217

Kurantz MJ, Zacharius RM (1981) Hypersensitive response in potato tuber: elicitation by combination of non-eliciting components from *Phytophthora infestans*. Physiol Plant Pathol 18:67–77

Larkin PJ (1977) Plant protoplast agglutination and membrane-bound β-lectins. J Cell Sci 26:31–46

Lazarovits G, Higgins VJ (1979) Biological activity and specificity of a toxin produced by *Cladosporium fulvum*. Phytopathology 69:1056–1061

Lazarovits G, Bhullar BS, Sugiyama HJ, Higgins VJ (1979) Purification and partial characterization of a glycoprotein toxin produced by *Cladosporium fulvum*. Phytopathology 69:1062–1068

Lee SC, West CA (1981a) Polygalacturonase from *Rhizopus stolonifer,* an elicitor of casbene synthetase activity in castor bean (*Ricinus communis* L.) seedlings. Plant Physiol 67:633–639

Lee SC, West CA (1981b) Properties of *Rhizopus stolonifer* polygalacturonase, an elicitor of casbene synthetase activity in castor bean (*Ricinus communis* L.) seedlings. Plant Physiol 67:640–645

Legrand M (1983) Phenylpropanoid metabolism and its regulation in disease. In: Callow JA (ed) Biochemical plant pathology. Wiley, Chichester, pp 367–384

Leonard KJ, Czochor RJ (1980) Theory of genetic interactions among populations of plants and their pathogens. Annu Rev Phytopathol 18:237–258

Lewis BG, Day JR (1972) Behaviour of uredospore germ tubes of *Puccinia graminis tritici* in relation to the fine structure of wheat leaf surfaces. Trans Br Mycol Soc 58:139–145

Lisker N, Kuć J (1977) Elicitors of terpenoid accumulation in potato tuber slices. Phytopathology 67:1356–1359

MacLean DJ, Tommerup IC (1979) Histology and physiology of compatibility and incompatibility between lettuce and the downy mildew fungus *Bremia lactucae* Regel. Physiol Plant Pathol 14:291–312

Marcan H, Jarvis MC, Friend J (1979) Effect of methylglucosides and oligosaccharides on cell death and browning of potato tuber discs induced by mycelial components of *Phytophthora infestans*. Physiol Plant Pathol 14:1–5

Mendgen K (1978) Attachment of bean rust cell wall material to host and non-host plant tissue. Arch Microbiol 119:113–117

Mitchell JE (1976) The effect of roots on the activity of soil-borne plant pathogens. In: Heitefuss R, Williams PH (eds) Enclyclopedia of plant physiology, New Ser, Physiological plant pathology. Springer, Berlin Heidelberg New York, pp 94–128

Nakajima T, Tamari K, Matsuda K, Tanaka H, Ogasawara N (1972) Studies on the cell wall of *Pyricularia oryzae*. III. The chemical structure of the *β*-D-glucan. Agric Biol Chem 36:11–17

Nakajima T, Tamari K, Matsuda K (1977a) A cell-wall proteoheteroglycan from *Pyricularia oryzae*: Isolation and partial structure. J Biochem (Tokyo) 82:1647–1655

Nakajima T, Sasaki H, Sato M, Tamari K, Matsuda K (1977b) A cell-wall proteoheteroglycan from *Pyricularia oryzae*: Further studies of the structure. J Biochem (Tokyo) 82:1657–1662

Nakajima T, Tamari K, Matsuda K (1980) Structural studies on the cell wall polysaccharides from *Pyricularia oryzae*: a pathogenic fungus of rice blast disease. In: Sandford PA, Matsuda K (eds) Fungal polysaccharides. Am Chem Soc Symp 126:15–34

Nozue M, Tomiyama K, Doke N (1979) Evidence for adherence of host plasmalemma to infecting hyphae of both compatible and incompatible races of *Phytophthora infestans*. Physiol Plant Pathol 15:111–115

Nozue M, Tomiyama K, Doke N (1980) Effect of N,N'-diacetyl-D-chitobiose, the potato lectin hapten and other sugars on hypersensitive reaction of potato tuber cells infected by incompatible and compatible races of *Phytophthora infestans*. Physiol Plant Pathol 17:221–227

Oku H, Shiraishi T, Ouchi S, Ishiura M, Matsuda R (1980) A new determinant of pathogenicity in plant disease. Naturwissenschaften 67:310

Ouchi S, Oku H (1981) Susceptibility as a process induced by pathogens. In: Staples RC, Toenniessen GH (eds) Plant Disease Control. Wiley, New York, pp 33–44

Person CO, Mayo GME (1974) Genetic limitations on models of specific interactions between a host and its parasite. Can J Bot 52:1339–1347

Peters BM, Cribbs DH, Stelzig DA (1978) Agglutination of plant protoplasts by fungal cell-wall glucans. Science 201:364–365

Pillai CGP, Weete JD (1975) Sterol-binding polysaccharides of *Rhizopus arrhizus, Penicillium roquefortii* and *Saccharomyces carlsbergensis*. Phytochemistry 14:2347–2351

Royle DJ, Thomas GG (1973) Factors affecting zoospores responses towards stomata in hop downy mildew (*Pseudoperonospora humuli*) including some comparisons with grapevine downy mildew (*Plasmopara viticola*). Physiol Plant Pathol 3:405–417

Sequeira L (1978) Lectins and their role in host–pathogen specificity. Annu Rev Phytopathol 16:453–481

Shiraishi T, Oku H, Yamashita M, Ouchi S (1978) Elicitor and suppressor of pisatin induction in spore germination fluid of pea pathogen, *Mycosphaerella pinodes*. Ann Phytopathol Soc Jpn 44:659–665

Staples RC, App AA, Ricci P (1975) DNA synthesis and nuclear division during formation of infection structures by bean rust uredospore germlings. Arch Microbiol 104:123–127

Staub TH, Dahmen H, Schwinn FJ (1974) Light and scanning electron microscopy of cucumber and barley powdery mildew on host and non-host plants. Phytopathology 64:364–372

Stekoll MS, West CA (1978) Purification and properties of an elicitor of castor bean phytoalexin from culture filtrates of the fungus *Rhizopus stolonifer*. Plant Physiol 61:38–45

Swinburne TR (1975) Microbial proteases as elicitors of benzoic acid accumulation in apple. Phytopathol Z 82:152–162

Swinburne TR (1981) Iron and iron chelating agents as factors in germination, infection and aggression of fungal pathogens. In: Blakeman JP (ed) Microbial ecology of the phylloplane. Academic Press, London New York, pp 227–243

Trione EJ (1981) Natural regulators of fungal development. In: Staples RC, Toenniessen GH (eds) Plant disease control, resistance and susceptibility. Wiley, New York Chichester Brisbane Toronto, pp 85–102

Vanderplank JE (1968) Disease resistance in plants. Academic Press, London New York

Van Loon K, Callow JA (1983) Transcription and translation in the diseased plant. In: Callow JA (ed) Biochemical plant pathology. Wiley, Chichester, pp 385–414

Varns JL, Kuć J, Williams EB (1971) Terpenoid accumulation as a biochemical response of the potato tuber to *Phytophthora infestans*. Phytopathology 61:174–177

Wade M, Albersheim P (1979) Race-specific molecules that protect soybeans from *Phytophthora megasperma* f. sp. sojae. Proc Natl Acad Sci USA 76:4433–4437

West CA (1981) Fungal elicitors of the phytoalexin response in higher plants. Naturwissenschaften 68:447–457

Wheeler BEJ (1981) Biology of powdery mildews on leaf surfaces. In: Blakeman JP (ed) Ecology of the phylloplane. Academic Press, London New York, pp 69–84

Woodward JR, Keane PJ, Stone BA (1980) *β*-glucans and *β*-glucan hydrolases in plant pathogenesis with special reference to wilt-inducing toxins from *Phytophthora* species. Am Chem Soc Symp 126:113–141

Wright K, Northcote DH (1974) The relationship of the root cap slimes to pectins. Biochem J 139:525–534

Wynn WK (1976) Appressorium formation over stomates by the bean rust fungus: response to a surface contact stimulus. Phytopathology 66:708–714

Wynn WK, Staples RC (1981) Tropisms of fungi in host recognition. In: Staples RC, Thoenniessen GH (eds) Plant disease control: resistance and susceptibility. Wiley, New York, pp 45–69

Yoshikawa M (1983) Macromolecules, recognition and the triggering of resistance. In: Callow JA (ed) Biochemical plant pathology. Wiley, Chichester, pp 267–295

Yoshikawa M, Masago H, Keen NT (1977) Activated synthesis of poly(A)-containing messenger RNA in soybean hypocotyls infected with *Phytophthora megasperma* var. sojae. Physiol Plant Pathol 10:125–138

Yoshikawa M, Yoshizawa T, Matama M, Masago H (1979) Contact of insoluble cell walls of *Phytophthora megasperma* var. sojae with wounded soybean cotyledons results in rapid liberation of a phytoalexin elicitor. Ann Phytopathol Soc Jpn 45:535

Yoshikawa M, Matama M, Masago H (1981) Release of a soluble phytoalexin elicitor from mycelial walls of *Phytophthora megasperma* var. sojae by soybean tissues. Plant Physiol 67:1032–1035

Zentmeyer GA (1961) Chemotaxis of zoospores for root exudates. Science 133:1591–1596

Ziegler E, Albersheim P (1977) Host–pathogen interactions. XIII. Extracellular invertases secreted by three races of a plant pathogen are glycoproteins which possess different carbohydrate structures. Plant Physiol 59:1104–1110

12 Mating Systems in Unicellular Algae

L. WIESE

12.1 Introduction

Sexual processes in unicellular algae provide outstanding opportunities for the study of various aspects of cell interaction and intercellular communication, such as unilateral or mutual inductions, signalling and signal perception, chemotaxis, membrane adhesion and confluence, cell recognition and cell and nuclear fusion. Because the continuity of their existence has been protected by asexual reproduction, these algae have been able to explore their full evolutionary potential with respect to sexual differentiation and reproduction, leading to a great variety in their sex phenomena and to the evolution of highly efficient interrelations. The different modes of gamete interactions they reveal illustrate component mechanisms in model-like simplicity, and offer the opportunity for experimental analysis of their physiological, genetic, molecular and ultrastructural basis. The variation in sex phenomena arises from the type of fertilization (iso-, aniso- and oogamy; gametangiogamy; potential for parthenogenesis), of sex distribution (monoecy vs. dioecy), of sex determination (genetic vs. modificatory), and of meiosis (gametic, zygotic or intermediate). Additional variation depends on the types of gametes (motile vs. immotile) and on special features of the particular organism. A normal sex act involves an ordered sequence of interrelated steps between and within the partners, the harmonious coordination of which is assured by the principles mentioned at the outset. Any environmental, experimental or genetic disturbance may lead to a developmental block, sterility, sexual isolation, or, occasionally, a shunt into another reproduction mode such as parthenogenesis or automixis, in which the activated cell somehow compensates for the absence of a partner.

This review of mating systems in eukaryotic unicellular algae with emphasis on *Chlamydomonas* will deal only with aspects related to cellular interactions. Some relevant data from multicellular algae are included.

12.2 Cellular Interactions Effecting Gametic
Differentiation. Erogens. Contact-Induced Gametogenesis

Success of mating normally requires the simultaneous presence of both gamete types. Generation of only one kind of sex cell results in a loss of reproduction potential if that gamete type is not capable of parthenogenesis or automixis.

Synchronization of ♂ and ♀ gametogenesis in dioecious and self-fertile mono-ecious species may be effected by (1) a species-typical response to certain environmental conditions (nutrient supply, light and temperature, salinity, pH, tides) (cf. RAYBURN 1974), (2) dependence on an endogenous rhythm or a cell-cycle-stage control of gametic differentiation (KATES and JONES 1964, SCHMEISSER et al. 1973), (3) a membrane-mediated induction of sexual differentiation on random collision of potential sex partners in their vegetative stages, or (4) unilateral or mutual induction of the sexual phase by means of diffusible inducer substances, *erogens*. These pheromones are produced and emitted by individuals of one sex and trigger sexual differentiation in specimens of the same and/or the other sex. Erogens have been detected only after close inspection in the laboratory and in relatively few algae, but they may, in fact, be of wide-spread occurrence. In order to respond, the affected cells may have to be in a special stage of competence (cf. STARR 1971, SANDGREN 1981). Erogens may be of entirely different chemical natures (cf. MIYAKE 1981), macromolecular or not, hydrophobic or hydrophilic (JAENICKE 1982). In some cases in which organisms enter the sexual phase only if both sexes in their vegetative stages are cultivated together, it is not yet decided whether sexualization is contact- or erogen-induced (cf. COLEMAN 1979).

12.2.1 Erogens

Erogens in algae were detected by DIWALD (1938) in *Glenodinium lubliensiforme* (Dinophyta). Two sex-specific, pH- and temperature-sensitive compounds, secreted by vegetative ♂ and ♀ cells, have been isolated and shown to induce the sexual phase in vegetative cells of the opposite sex.

Modern research on erogens started with DARDEN's detection of diffusible substances inducing sexual differentiation in *Volvox* (DARDEN 1966), which led to their isolation, purification and chemical characterization and to information on the mode and site of their synthesis and on their mode of action (JAENICKE 1982). In the various *Volvox* species, erogens may be produced by one or the other, mostly the male, sex and may affect the same sex, the opposite one, or both sexes. In some *Volvox* species no inducing substances seem to exist. The inducers are strictly species-specific, high-molecular substances, most of which have been recognized as glycoproteins (STARR and JAENICKE 1974, KOCHERT and YATES 1974). They are active in very low concentrations of about 10^{-14} or 10^{-16} M. In a remarkable exception, sexual differentiation in the monoecious *V. capensis* is triggered by nM amounts of L-glutamic acid (STARR et al. 1980). Whereas this taxon grows equally well in the light as in the dark with acetate, the formation of sexual colonies occurs only in the light. L-glutamic acid acts upon an early stage competent for sexual induction at the release of young daughter colonies, and is probably produced by a proteolytic enzyme hydrolyzing the parental matrix (JAENICKE and WAFFENSCHMIDT 1979). The induction systems of different taxa vary with respect to the affected, induction-competent stage. Further variation arises with the type of sex distribution in the various taxa. The sex distribution modes in colonial oogamous Volvo-

Table 1. Sex distribution, sexual differentiation, and erogens in various *Volvox* species. (Modified after A.W. Coleman 1979)

Characteristic	*V. carteri* f. nagariensis	*V. carteri* f. weismannia	*V. obversus*	*V. gigas*	*V. dissipatrix* (India)	*V. dissipatrix* (Australia)	*V. rousseletii*	*V. aureus*
Heterothallic, dioecious	×	×	×	×	×		×	
Homothallic								
Dioecious								×
Monoecious						×		
Erogen made by	♂	♂	♂	♂	♂, ♀	☿	♂	♂
Erogens induce	♂, ♀	♂, ♀	♂, ♀	♂, ♀	♂, ♀	☿	♂, ♀	♂, par.

Dioecy: egg and sperm are formed by different individuals. Monoecy: egg and sperm are formed by the same individual. Heterothallism: egg- and sperm-producing individuals are always produced by two different clones, i.e. heterothallism is necessarily connected with dioecy. Homothallism: egg- and sperm-producing (dioecious) individuals or monoecious individuals are formed within one clone. par.: Taxon reproduces parthenogenetically. The erogen-producing cells and the affected sexes are listed

cales require a special nomenclature (cf. Starr 1971, Coleman 1979) (see legend Table 1).

In the homothallic dioecious *V. aureus* strain M5, ♂ colonies produce a ♂-inducing substance (MIS), which causes differentiation of young vegetative colonies into sexual ♂ colonies. The latter lack gonidia, leading to asexual reproduction, but produce androgonidia (male initials), which divide by a series of mitoses into packets of spermatozoa. Young gonidia of vegetative colonies serve as eggs. MIS$_{aureus}$ is sensitive to proteases and to mercaptoethanol and cystein (Darden and Sayers 1971). Even high concentrations of MIS do not induce more than 50% ♂ colonies in a clone culture, thus guaranteeing the persistence of the strain by vegetative reproduction and the availability of young gonidia to serve as eggs, and indicating some differential disposition to respond to the inducer.

In a parthenogenetic strain of *V. aureus* (DS), also homothallic and dioecious, ♂♂ are formed only exceptionally, and sexual reproduction occurs by parthenospores. The strain produces MIS, however, as demonstrable with test cultures of M 5, and has thus obviously lost responsiveness to MIS, the potential for parthenogenesis compensating for the absence of ♂♂. In other *Volvox* species unfertilized eggs are not capable of parthenogenesis.

Sexual isolation between two subspecies of the heterothallic dioecious *Volvox carteri,* f. weismannii and f. nagariensis, involves a taxon-specificity of their inducers and of the responsiveness of the ♀ cells. Activation in each taxon occurs only in response to its own inducer, and only its own inducer is adsorbed by ♂ and ♀ colonies of each taxon, as demonstrated by ^{125}I labelling (Noland et al. 1977). Both inducers, glycoproteins (MW 27,500) with 40% carbohydrate and 60% protein (KOCHERT and YATES 1974, STARR and JAENICKE 1974), are produced in both subspecies by the ♂♂, and trigger sexual differentiation in the ♂ and the ♀ sex. The responsiveness of the young gonidium in f. nagariensis is eliminated by concanavalin A, which is said to alter the membrane organization, so preventing induction (KURN 1981). In taxa that require an inducer for the triggering of ♂ and ♀ differentiation and in which this inducer is only produced by sexual ♂♂, the problem of the appearance of the first sexual male arises. STARR (1971) assumes that such males may arise by a mutation to "spontaneousness". Another interpretation assumes that some epigenetic event may trigger spontaneous sexualization.

The inducer is formed by the sperm packets at their maturation phase as shown by experiments using fluorescent antibodies and by Ouchterlony techniques. The glycoprotein emerges with the release of the sperm packets; however, the polypeptide chain is transcribed many hours earlier, and only the glycosylation step occurs at the time of sperm packet release. The inducer function seems to be associated with carbohydrate moieties of the erogen, since periodate and certain glycosidases inactivate the principle, and tunicamycin blocks its biosynthesis (GILLES et al. 1981).

In the colonial dioecious and anisogamous chrysophycean *Dinobryon cylindricum,* genetically female cells continuously secrete an erogen of unknown nature that induces gametogenesis in vegetative cells of ♂ lines. Each activated ♂ cell, by means of an unequal mitosis, forms only one motile androgamete, which swims towards the immotile gynogamete and fuses to form the zygote (SANDGREN 1981).

In *Eudorina,* a high molecular weight erogen secreted by vegetative ♂ colonies induces sperm packet formation in other ♂ colonies; i.e. it controls the switch from the vegetative to the sexual stage (SZOSTAC et al. 1973).

In the monoecious anisogamous *Chlamydomonas zimbabwiense,* each vegetative cell either divides into sperm-like androgametes or differentiates into one biflagellated naked gynogamete. The latter represents the undivided protoplast hatched from vegetative cells. The hatching, following partial cell wall lysis, occurs only in the presence of sperm, and can be traced to a sperm-secreted, heat-labile, protease-sensitive factor (MW 35,000). The role of this factor as a lysin or an erogen has still to be investigated (HEIMCKE and STARR 1979).

12.2.2 Contact-Induced Gametogenesis

Whereas erogens induce sexual differentiation, i.e. gametogenesis, over a distance, other organisms enter the sexual phase only if potential sex partners, in their vegetative states, collide with each other. It is demonstrably the actual cell–cell contact that activates the potential to undergo gametogenesis. The individuals activating each other need not become the eventual mates. Contact may be achieved by collision between swimming partners, or by growth processes. No case of a membrane-mediated induction has as yet been studied in all detail. Contact may not only trigger the realization of the entire gametogenic programme but may be needed for the passing of certain checkpoints effecting the coordination and synchronization of the partner programmes (Sect. 12.5).

A telemorphic mutual induction of copulation papillae occurs between sexually different cells of *Spirogyra majuscula*. The approximation of the out-growing papilla to form the copulation tube is based upon chemotropism. The mode of action and the nature of the pheromone(s) (erotropin) involved have not yet been investigated (GROTE 1977).

One of the most fascinating cell–cell interactions results in the contact-induced papilla formation and a collusive sex determination between two genetically identical cells at the intraclonal conjugation of *Sirogonium*. Two cells adhere as progametangia, each one forming a conjugation papilla at the contact site. By an unequal mitosis each progametangium divides into one smaller vegetative cell and a gametangium proper, containing the contact site. The contact must be in connection with, or preceded by, an alternative sex determination, one gametangium undergoing ♂, the other one ♀, sex differentiation. Spherical gametes, the contracted single protoplasts of the gametangia, fuse and exhibit physiological anisogamy: the ♂ gamete migrates into the ♀ gametangium. How the two partners consent to alternative pathways of sexual differentiation and when and where this sex determination occurs are unknown (HOSHAW 1965).

12.3 Chemotaxis. Erotactins (Sirenins). Chemotropism. Chemotropins

Contact between motile sex cells or between a motile gamete type and its sessile partner may occur on their random collision or after chemotactically guided approximation. Special pheromones, erotactins or sirenins, emanate from gametes of one type (sometimes from the gametangium as in the *Oedogonium* system, cf. HOFFMAN 1973, RAWITSCHER-KUNKEL and MACHLIS 1962, or in *Chlamydomonas suboogama* (TSCHERMAK-WOESS 1959, 1963), and by directing the locomotion of the opposite gamete type, enhance the chance for a successful encounter of potential mating partners. In unicellular algae, chemotaxis is known from the observation of the actual phenomenon (cf. TSUBO 1961, *Sphaeroplea*, PASCHER 1931), but the exact nature of the sirenins and their mode of action are largely unknown. They may be hydrophilic or hydrophobic. Only in the *Phaeophyceae* has the chemistry of the erotactins been elucidated in detail (cf. MÜLLER 1967, 1976, 1981; see also BEAN 1979), the natural components having been gas chromatographically isolated, purified, characterized, and even synthesized.

Chemotaxis occurs practically exclusively in systems in which the emitting partner is either immotile or significantly slower than its mate. An interesting phase specificity of the chemotactic attraction exists in *Cutleria*, in which the flagellated, originally motile, macrogamete settles down and only then attracts the androgametes.

Whether cases with mutual attraction are realized is at present unknown (BRANDHAM 1967).

12.4 Sex Cell Contact

In anisogamous (= heterogamous) and oogamous species, the proper sex cell contact, occurring after random meeting or after chemotaxis, has been studied in only a few cases, and mostly with multicellular algae (*Oedogonium,* HOFFMAN 1973; brown algae, cf. CALLOW et al. 1981). The analysis of the contact event in lower algae has concentrated on *Volvox,* and especially on isogamous, dioecious chlamydomonads in which biflagellated gametes initiate copulation by adhering with their flagella tips. Gametogenesis occurs under special environmental conditions and can be experimentally triggered by interference with their nitrogen metabolism (SAGER and GRANICK 1954). However, gametogenesis is not, as is often claimed, induced by a general N-depletion. In *C. eugametos* and *C. moewusii,* the cell's potential to become a gamete is realized in the absence and suppressed in the presence of NH_4; NO_3 does not prevent gametogenesis. In the simplest possible cases, in species with hologamy, each vegetative cell differentiates directly into one gamete by attaining the capacity to copulate. In other cases, gametogenesis is or may be, in one sex or in both, connected with gametogenic divisions. In any case, gametic differentiation in dioecious taxa creates a bipolarity between sex cells. This bipolarity is expressed in the appearance of a series of complementary structures, and leads by finely tuned interactions to the formation of the zygote.

In the oogamous *Volvox* (Chlorophyceae), the eggs are located and fertilized inside the ♀ colony. Sperm bundles, produced by multiple mitoses from androgonidia, adhere to ♀ spheroids after random collision, no chemotaxis being involved. The sperm packets are released by the action of a jelly-dissolving enzyme (JAENICKE and WAFFENSCHMIDT 1979). The initial contact ensues between the flagella of the sperm bundles and of *somatic* ♀ cells (*V. carteri* f. weismannia, COGGIN et al. 1979). Only in a second step do the sperm bundles contact the jelly proper, dissociate, and release the jelly-dissolving enzyme, the activity of which can be blocked by trypsin inhibitor. The individual spermatozoa obtain access to the eggs through the locally dissolved jelly, the enzyme forming a "fertilization pore". The striking transitory and auxiliary sperm-binding by the flagella of somatic ♀ cells is sex-, species-, and phase-specific, since no binding occurs to ♂ colonies, ♀ colonies in the asexual stage or ♀ colonies of other species. De-flagellated spheroids do not bind sperm bundles; preparations of isolated ♀ flagella do. Flagella regenerated after de-flagellation of ♀ colonies also possess sperm-binding capacity. The contact capacity of the ♀, and apparently of the ♂ flagella as well, is trypsin-sensitive. ♀ colonies incapacitated by trypsin do not regenerate new sperm-binding sites (COGGIN et al. 1979). The production of the jelly-dissolving enzyme by the sperm bundles is sensitive to inhibitors of protein and RNA synthesis (HUTT and KOCHERT 1971). In *V. rousseletti,* a function in sperm adhesion is ascribed to a peripheral layer of a sulfated polymer of the colony, presumably a mucopolysaccharide, demonstrated electron microscopically after staining with 3,3-diaminobenzidine tetrahydrochloride (BURR and McCRACKEN 1973).

The molecular basis of gamete contact in monoecious algae has not yet been analyzed. Monoecy in unicells is manifested by intraclonal copulation, i.e. by the realization of both sexes on the basis of the same genotype (Starr 1971, Wiese 1981). In the case of morphological gamete dimorphism, as in anisogamy or oogamy, the contact mechanism will be of the same type as that in dioecious forms, based on a bipolar complementarity. Obviously, some developmental switch mechanism channels gametogenesis into the alternative pathways in the same manner as the sex-determining alleles in dioecious organisms. Anisogamy can be expressed to very different degrees, from a mere minor size difference between andro- and gynogametes, to a considerable one culminating in oogamy, with extreme size differences and a highly divergent organization of the androgamete as a spermatozoon and of the gynogamete as an immotile egg.

A very special problem, however, emerges in isogamous haploid monoecists in which each vegetative cell differentiates into one gamete, and genetically identical and morphologically indistinguishable gametes copulate within a clone (cf. Wiese 1981).

12.4.1 The Mating Type Reaction and Its Molecular Basis

The initial sex cell contact in dioecious isogamous chlamydomonads occurs in the mating type reaction (MTR), a species-specific flagella agglutination between colliding gametes of different sex. The agglutination mechanism must be highly efficient, leading to a selective adhesion in a split second without trial-and-error attempts. In denser gamete suspensions, many ($+$) and ($-$) gametes may agglutinate by a common engagement of their flagella (group formation, clumping). The flagella-interconnected cells exhibit a characteristic trembling movement (jittering). Within these clusters two cells of different sex join definitively to form a pair in a species-typical fashion by lateral or head-on fusion. The flagella of both cells now lose their contact capacities, disadhere and disengage from other gametes. In this manner gamete clusters disintegrate into pairs. Each pair subsequently forms one zygote.

Especially when studied comparatively with sexually compatible and incompatible species, this MTR represents an outstanding model for cell recognition and cell adhesion, and for the analysis of the underlying molecular basis and its biosynthesis. The gamete agglutinins responsible, flagella-integrated or flagella-associated mating type substances (MTSs), establish an interacting complementarity; they are synthesized and exposed only during gametogenesis, since vegetative cells of sexually different strains do not agglutinate, and the attainment of gametic adhesiveness is sensitive to actinomycin D, cycloheximide and tunicamycin. The absence of any interaction between gametes of identical sex proves that the phase- and species-specific complementarity must be assymmetrical in type – i.e. the MTS's are sex-specific. Sexual isolation between species is caused by non-matching systems of MTS's, preventing any interaction between gametes of incompatible species. In dioecious species, sex is manifested in the ($+$) and ($-$) differentiation at gametogenesis and is genetically determined by factor-segregation at meiosis (Smith and Regnery 1950, Pascher 1918,

SAGER 1955, GILLHAM 1969). Sex is expressed only under gametogenic conditions. The MTS's must be located on the flagella, especially at the flagella tip; their arrangement on the flagella seems to vary from species to species (TSCHERMAK-WOESS 1963) and may vary within a species during the MTR (contact migration, GOODENOUGH).

The gamete contact in the MTR of isogamous dioecious chlamydomonads must thus be based upon an interaction between sex-specific, flagella-bound components, the MTS's, which establish a species-typical bipolarity. In addition, the mechanism must be dynamic in its ability to permit exchange between partners, and must be programmed to be terminated upon signal (Sect. 12.7).

In monoecious isogamous chlamydomonads, in which morphologically and genetically identical gametes copulate, the nature of the contact mechanisms and the type of sex determination are completely unknown. The problem is accentuated because of the uncertainty as to whether gametogenesis produces like gametes, each capable of copulating with every other, or whether bipolarly different, i.e. $(+)$ and $(-)$ gametes arise by a phenocopy of a genetic sex determination.

The structures postulated to be connected with the contact phenomenon may be defined as follows:
1. The isoagglutinins (IA's). Sex-, species-, and phase-specific systems of membrane vesicles studded with MTS's. Origin and functional role, if any, unknown.
2. The mating type substances (MTS's). Sex-, species-, and phase-specific gamete agglutinins, integrated into or associated with the flagella membrane and representing partners of the contact-establishing complementarity.
3. The ligands. Component structures within the MTS's effecting the contact proper by carrying the actual contact sites. The functioning of individual contact sites may depend on their molecular environment, since they are part of special molecular groups (determinants).

As membrane-bound agglutinins effecting adhesion between gametes of different sex, the MTS's do not fit the concept of gamones (gamete hormones) as developed by HARTMANN (cf. 1956). The term gamones, if still used, should be restricted to principles acting over distances, such as erogens, erotactins, diffusable inducer substances, etc., which taken together are better designated as pheromones. The hypothesis that the complementary contact components interact in a glycosylacceptor–glycosyltransferase relationship (BOSMAN and MCLEAN 1975, MCLEAN and BOSMAN 1975) could not be substantiated (KOEHLE et al. 1980). It rather appears that the increased glycosyltransferase activity, observed at mixing of $(+)$ and $(-)$ gametes, reflects the induced neosynthesis of MTS's which are inactivated in the MTR (see Sect. 12.5).

12.4.1.1 The Isoagglutinins (IA's)

In isogamous chlamydomonads, highly active gametes of either sex shed principles into the medium that, when added to the respectively opposite gamete type, cause a homotypic agglutination (isoagglutination) among gametes of identical sex. Such isoglutinins (IA's) were detected in *Tetraspora* as early as 1931 by GEITLER. In *Chlamydomonas,* the isoagglutination phenomenon was described but falsely interpreted by MOEWUS (1933) (cf. FÖRSTER and WIESE 1954a, GOWANS 1976).

IA's (particle weight 100×10^6 and more) were thought to be batteries of associated glycoproteinaceous MTS's (cf. WIESE 1974) but they actually represent membrane vesicles (MCLEAN et al. 1974), which are studded with MTS-molecules (GOODENOUGH 1977, MUSGRAVE et al. 1981, WILLIAMS 1982). Because of the sex- and species-specific MTS's, integrated into the IA's, the isoagglutinations "copy" the proper gamete contact, and are thus as specific as the gamete-to-gamete agglutination. Functionally, the IA's must be multivalent, capable of interacting with two or more cells. In contrast to the isolated sex contact components in yeast (CRANDALL 1978), the IA's in *Chlamydomonas* do not neutralize each other in vitro, which might reflect a more complex complementarity (WIESE and WIESE 1978) or, more likely, indicate the need of the IA for a live partner (cf. GOODENOUGH 1977).

The availability of the spontaneously shed IA's as carriers of the specific contact capacity offers unique experimental advantages in quantitative and qualitative studies of the contact phenomenon and as raw material for the isolation of the proper MTS's of glycoprotein nature. The existence of these IA's furthermore permits analytical studies under non-physiological conditions with respect to pH, chemicals and temperature, and allows long-lasting enzyme sensitivity experiments. The exact mode of the emergence of the IA's is unexplained. Their production might simply be a physiological surplus, secondarily detached from the flagella membrane or the flagella tips, or they may be synthesized and shed without being properly integrated into the flagellar membrane. The IA's may possibly originate from long amorphous processes of more than a cell length, protruding from the flagella tips as in *C. moewusii* (BROWN et al. 1958) and *C. eugametos* (WIESE unpublished). The IA can be isolated from gamete suspensions by fractional centrifugation, and quantitized by titre determination. From 5 l dense $(-)$ gamete suspension in *C. eugametos,* a 2-ml concentrate can be isolated with a titre of 64×10^9; 0.0001 ng IA in 1 ml still produces a distinct isoagglutination.

Under standardized conditions and with IA preparations of known titre it can be demonstrated in *C. eugametos* and *moewusii* that the responsiveness of the complementary $(+)$ gametes fluctuates by a factor of 1000 and more during the day. A diurnal rhythm in mating competence is known to exist, and this and the light-dependency of the agglutinability and its maintenance require further analysis. This may offer decisive insights into hitherto neglected aspects of the gametic contact phenomenon (FÖRSTER and WIESE 1954, FÖRSTER 1957, 1959, STIFTER 1959, HARTMANN 1962).

12.4.1.2 The Mating Type Substances

The MTS's were first characterized by the resistence of the gametic contact capacity to various enzymes, to concanavalin A, and to periodate (WIESE and coworkers 1969, 1970, 1972, 1975, 1978; MCLEAN and BROWN 1974, GOODEN-OUGH 1977, WILLIAMS 1982). The reagents are applied to live gametes or, a great experimental advantage, to the respective IA's, and their effect is checked with live complementary test gametes. In an informative sex- and species-specific pattern, the gametic agglutinability of various isogamous and dioecious species is sensitive to different proteases, periodate, lectins, and endo- and exoglycosidases. The sensitivity pattern indicates the glycoprotein nature of the MTS's, and reveals common and differential features of their functional structure (cf. WIESE and WIESE 1978).

The sensitivity of the contact capacity to exoglycosidases appears to be of special importance, proving that (1) the contact function depends crucially on carbohydrate moieties of the glycoproteins, and (2) carbohydrate-binding is an essential feature in the contact event (WIESE and WIESE 1978, WILLIAMS 1982, WIESE and GROSSBARD in preparation). While in two syngens (see Sect. 12.8, p. 255) of the clade (see Sect. 12.6, p. 251, footnote) *C. moewusii* both (+) sexes are sensitive to α-exomannosidase, the corresponding (−) sexes are insensitive, but they are differently sensitive to two other exoglycosidases: *eugametos* (−) contact is sensitive to α-exogalactosidase and to α-exo-N-acetylgalactosaminidase; contact in *syngen II* (−) gametes is insensitive to the latter, but sensitive to the first enzyme.

The carbohydrate binding is thus not only crucial for contact but apparently also for its selectivity. Sexual isolation between syngens and between species may be based upon the individuality of their contact mechanisms. It is remarkable and difficult to explain that such crucial terminal sugars as α-linked galactose and N-acetylgalactosamine will not react with specific lectins. There is no blocking of the (−) *gametic* contact capacity by *Bandeiraea,* peanut, or soybean lectin, and no adsorption of the isoagglutinins or of (−) gametes to the immobilized lectins (WILLIAMS 1982). Monosaccharides such as α-methylmannoside do not prevent agglutination from occuring nor, when added to agglutinated gametes, do they reverse the MTR. It appears that these sugars are crucial component structures within determinants specifying a certain complexity that is recognized by the receptor on the complementary gamete agglutinin, but they are not accessible to the lectins.

The complementarity between the interacting agglutinins, however, is not restricted to peripheral determinants in mutual receptor function. The attainment and the maintenance of gametic contact capacity in several related and unrelated *Chlamydomonas* species is sensitive to tunicamycin (TUM) in one but not in the other sex (MATSUDA et al. 1981, WIESE and MAYER 1982). This sensitivity pattern reveals some fundamental bipolarity in the gametes' contact mechanism, common to various species, and permits, on this basis, a certain homologization between sexes of incompatible isogamous taxa (WIESE et al. 1983). Since TUM affects post-translational differentiation by interfering with the N-glycosylation of glycoproteins (cf. STRUCK and LENNARZ 1980), the detected differential sensitivity can be most simply interpreted as a direct effect upon the MTS's (MATSUDA et al. 1981, SNELL 1981, WIESE and MAYER 1982). It is postulated that the contact capacity in the TUM-sensitive sex resides in a ligand of carbohydrate nature, which is N-glycosidically linked to the polypeptide. In the current state of knowledge, it is not clear whether the ligand in the TUM-resistant sex is an 0-linked carbohydrate structure or a sugar-binding sequence of amino acids in a sugar-binding protein. This conclusion is supported by the sensitivity of the established contact capacity to α-D-N-acetylgalactosaminyl oligosaccharidase (Williams 1982).

The MTS's thus differ in principle because they possess decisive functional ligands in a different linkage type. The proper determinants for gamete contact and the specificity it expresses must be superimposed on this fundamental bipolarity, most probably as a patterned arrangement of molecular groupings in a mutual receptor function. The biosynthesis of the MTS's in both sexes must furthermore involve two entirely different pathways of differentiation, the lipid-linked one in the ($+$) sex and the TUM-insensitive one in the ($-$) sex.

The alternative pathways of gametic differentiation as revealed in the unilateral TUM-sensitivity may account for sex-limited mutations affecting one or the other contact capacity. Because of the differences between the lipid-linked ($+$) pathway and the TUM-insensitive ($-$) pathway (Wiese et al. 1983), it may be surmised that the synthesis of the complex glycoproteinaceous MTS's must involve many steps at which mutations might interfere with the attainment of the sex-specific contact capacity. Mutations related to sex-specific alternate pathways will exhibit sex-linked or sex-limited inheritance. The alternative differentiation pattern may represent the marked asymmetry in the genetic specification of the ($+$) and ($-$) agglutination system as postulated by Hwang et al. (1981).

In *C. reinhardii,* the glycoprotein and polypeptide composition of gametic and vegetative flagella has been extensively studied with the aim of identifying and characterizing the gamete agglutinins and the components involved in the signalling process (Bergmann et al. 1975, Goodenough et al. 1980, Monk et al. 1982). Fractions with the gametic contact capacity have been prepared either from isolated gametic flagella by means of octylglucoside or from live cells by EDTA-extraction. Sex differences between the ($+$) and ($-$) agglutinins could not be detected by physicochemical methods. In the more thoroughly studied ($+$) sex, the gametic contact capacity is bound to a t-, protease-, and IO_4-sensitive high molecular weight component, a glycoprotein or a protein-carbohydrate complex. As in *C. eugametos,* the isolated component is physiologically univalent, i.e., it does not cause isoagglutination; however, it can, after adsorption to a slide or EM-grid, readily be demonstrated and quantitatively assayed in the sex-specific adhesion of complementary gametes. Vegetative test cells do not respond, and the agglutinin cannot be extracted from vegetative cells, or from mutants incapable of agglutination. On sucrose density-gradient centrifugation, the component, evidently a polymer, possesses a 12S sedimentation coefficient. The component is not identical with the major flagella membrane glycoprotein, which is considered a possible participant in the signalling event. From gel filtration studies the component's molecular weight is assumed to be between 1 and 4×10^6. It is concluded that the component is an extrinsic glycoprotein, and that only a few copies per flagellum exist (Adair et al. 1982). It should be pointed out that this component not only establishes contact specifically, but also triggers cell-wall lysis and mating-structure activation, events for which the existence of individual and spatially separated activation sites have been postulated and demonstrated.

12.5 Inactivation of Agglutinins During Gamete Contact

Gamete adhesion represents a dynamic interaction and endures only until a gamete has found its mate. Under gametogenic conditions, the gametes' agglutinability is potentially unlimited, and is terminated only upon a "signal" from the completed pairing. Termination of the MTR causes an inactivating alteration of the MTS's of unknown nature. Also, the maintenance of flagella contact, permitting by its dynamic nature partner exchange within a gamete cluster and change of contact sites between two partners, is connected with an inactivation of the contact-establishing MTS's. The long-lasting potential of gametes to agglutinate is based upon a continuous new production and insertion of MTS's, which replace inactivated units and thus compensate for the inactivation.

In systems with just the primary set of MTS's and incapable of fresh synthesis, such as isolated flagella or IA's, the restricted amount of MTS present is soon inactivated in a reaction with live gametes of the opposite sex. In agglutinating gametes, the need for the replenishment of MTS can be convincingly demonstrated with antibiotics that block the synthesis of glycoproteins (Ishiura and Iwasa 1973). Significantly, such antibiotics interfere not only with the attainment but also with the maintenance of the contact capacity. The latter effect emerges most distinctly when the contact capacity is challenged with complementary gametes, isolated flagella or IA's under conditions that prolong the MTR, such as impotent mutants or enzymes that selectively block pairing and thus prevent termination of the MTR.

The inactivation of the gametic contact components during flagella agglutination can be demonstrated by the consumption of the ($+$) and ($-$) IA in *C. eugametos* (Förster and Wiese 1954b). In both possible combinations, the added IA disappears with time from the isoagglutinated cells' supernatant, and the gametes emerge monodispersed. Such gametes reagglutinate upon addition of fresh IA. The inactivation of an IA occurs only in actual contact with the opposite gamete type of the same species (cf. Wiese and Wiese 1978). The duration of an isoagglutination depends on the ratio between the number of responding cells and the amount of IA added. In a given gamete suspension, a small amount of IA causes a short, and a larger amount a longer, isoagglutination, revealing some stoichiometric relation between IA and its receptor structures. By continuous addition of new ($-$) IA, a ($+$) gamete suspension can be kept in constant isoagglutination. The occurring inactivation of the IA is temperature-dependent (Wiese 1974).

In *C. reinhardii,* isolated gametic flagella of one sex lose their adhesiveness in reaction with gametes of the opposite sex, which, after cessation of the flagella-induced isoagglutination, still exhibit an undiminished contact capacity (Snell 1976). With improved techniques (Coulter Counter determination of agglutination, radioactive labelling), Snell and Roseman (1979) proved that flagella inactivation occurs only in direct contact with the opposite gamete type. The establishment of contact is temperature-independent; the time span before a flagella-induced isoagglutination expires reflecting the inactivation of the contact components, is distinctly temperature-dependent. Most importantly, in the interaction of (−) gametes with labelled (+) flagella, radioactive material of unknown nature is released that does not appear when (+) flagella are exposed to (+) gametes. The appearance of this material may be directly connected with the inactivation of the gamete contact of the (+) sex.

The responsiveness of cells in gamete–flagella mixtures can likewise be exhausted when each gamete is exposed to several flagella in the presence of cycloheximide (CH). Without the antibiotic, gametes can be agglutinated repeatedly, always inactivating the added flagella. The antibiotic obviously interferes with the normally occurring replacement of MTS's that were inactivated during the isoagglutination. The neosynthesis required for the maintenance of the adhesiveness emerges also in gamete-gamete interaction using an impotent mt^+ mutant (able to agglutinate, unable to pair). Imp mt^+ gametes combined with wild-type mt^- gametes undergo an agglutination for several hours, whereas in CH-incubated samples the adhesiveness disappears after one hour (SNELL and MOORE 1980). SNELL (1981) also observed that the maintenance of the gametic adhesiveness is sensitive to tunicamycin (TUM) and wondered whether glycosylated molecules are directly involved in flagella agglutination. Adhesion and subsequent dis-adhesion of flagella in the presence of CH are closely correlated to a difference in the ultrastructure of the flagella tip, the reversible flagella tip activation (cf. Sect. 12.6).

In *C. moewusii syngen II,* thermolysin or chymotrypsin prevents pairing but does not impede gametic adhesiveness. In the presence of either enzyme the MTR is thus unlimited (WIESE and WIESE 1978). When fully differentiated (+) and (−) gametes are mixed in the presence of thermolysin and TUM or CH is added immediately, the gametes at first agglutinate intensively. In assays with the two antibiotics, agglutination declines after 45 and 60 min, respectively, but is uninfluenced in controls without TUM or CH. When the redispersed cells in both antibiotic assays are checked with (+) or (−) test gametes, the cells in CH do not respond, whereas in TUM an intensive reagglutination occurs with added (+) but not with (−) gametes. The maintenance of the potential for a long-lasting agglutination obviously requires constant fresh synthesis of MTS. Blocking the biosynthesis unilaterally (TUM) or bilaterally (CH) terminates the MTR precociously, compared with controls (WIESE et al., in preparation).

All these data suggest that the interacting components in the MTR do effect contact and control its specificity, but subsequently become altered in a yet unknown fashion, losing their contact capacity. Gametic agglutinability when challenged is sustained by constant neosynthesis of MTS's. The mode of inactivation of the MTS's during flagella contact is of unknown nature, and it is an open question whether termination of flagella contact at pairing occurs through the same processes.

12.6 Correlations Between Mating Type Reaction and Pairing

The two spatially separated steps in isogamete copulation are intimately related. As already mentioned, the MTR triggers the pairing, and pairing terminates the MTR by eliciting disadhesion of the paired cells' flagella.

The interrelation demands the existence of at least two communications, "signals" in a broad sense: from the agglutinated flagella tips to the fusion sites on the gamete bodies, and from the fusing gametes back to the flagella. Pairing can be experimentally separated from the MTR by interference with its induction, the signal or its transmission, or by enzymatic inactivation of the molecular basis for fusion. The two events may also be separated by mutations that affect the potential for pairing at an undisturbed MTR (FOREST et al. 1978, GOODENOUGH et al. 1976, MATSUDA et al. 1981). Contacts in the MTR and at pairing are based upon different molecular interactions (MESLAND and VAN DEN ENDE 1979, WIESE and WIESE 1978, MATSUDA et al. 1982). The MTR may trigger further component steps, different from species to species, in the course of copulation, such as, for instance, enzymatic softening and shedding of the gametes' envelopes, formation of special mating structures involved in pairing, etc. The responsiveness of the area predestined for gamete fusion may be time-restricted once flagella contact has been made (*Ulva*, LØVLIE and BRYHNI 1976).

It has long been known that the MTR is a prerequisite for pairing. Details of an actual induction have only recently been investigated in *C. eugametos* and, especially, in *C. reinhardii*.

Pairing in the clade *C. moewusii*[1] occurs at the flagella basis of agglutinated gametes by an outgrowth of a papilla on each gamete, which penetrates the cell wall, fuses with its counterpart, and thereby permanently interconnects one (+) and one (−) gamete head on (cf. LEWIN 1954, BROWN et al. 1968, MESLAND 1976, MESLAND and VAN DEN ENDE 1979). A longitudinal association or a spiralization of the partners' flagella brings the two prospective papillar areas into correct juxtaposition. The fusion of the two papillae is based upon a complementarity other than that responsible for the MTR. In *C. moewusii syngen II*, the outgrowth is sensitive to thermolysin or chymotrypsin, the MTR is not (WIESE and WIESE 1978), and the (−) capacity to pair but not the (−) flagella adhesiveness is sensitive to TUM. This TUM-sensitivity of pairing, detected by MATSUDA and his coworkers in *C. reinhardii*, exists also in three taxa of the clade *C. moewusii* and, significantly, as in *C. reinhardii*, it is always the (−) sex in which the attainment of the flagella agglutinability is TUM-insensitive (STEWART and WIESE in preparation). In all these cases, however, it is as yet unresolved whether the block to pairing is connected with the actual contact mechanism or with structures involved in the signalling process. Whereas the induction of the papilla outgrowth is neatly demonstrated and analyzed by TEM and SEM, the type of the signal, its transmission, perception, and mode of action and the molecular basis of papilla fusion are still unknown. OsO_4- or formaldehyde-killed gametes do not trigger papilla activation in the complementary gamete type. Papilla outgrowth can also be induced by the interaction between the IA of one gamete type and live cells of the other sex (MESLAND and VAN DEN ENDE 1978). Isolated gametic flagella and flagella appendages (disc-like structures associated with the flagella membrane, MESLAND 1977) both cause intensive isoagglutination among gametes of the opposite sex. While (−) flagella, separated from their appendages, do not induce papilla formation in

---

1 The taxon *Chlamydomonas moewusii* can be evaluated as a clade, a complex of sibling species which are morphologically indistinguishable but sexually isolated from each other by non-matching sex contact systems

(+) gametes, the appendages do so, indicating a patterned arrangement of special papilla-activation sites (MESLAND and VAN DEN ENDE 1978).

In *C. reinhardii,* pairing occurs by means of special mating structures on both gamete types. A copulation tubule on the (+) gamete fuses with a specialized area, the "choanoid" body, on the (−) gametes (FRIEDMANN et al. 1968, CAVALIER-SMITH 1975, TRIEMER and BROWN 1975, WEISS et al. 1977, GOODENOUGH et al. 1982). As in many chlamydomonads and in other algae, each *reinhardii* gamete is enclosed in a cell-wall. Upon flagella contact this wall is shed after a partial dissolution at its apical pole due to the activation of a highly efficient autolysine (CLAES 1971, 1975, 1977, 1980, SCHLÖSSER 1976, 1981), which can be isolated from (+) and (−) gametes, from the medium of agglutinating gametes, and from ultrasonicated vegetative and sporulative cells. It is the MTR or the isoagglutination of one gamete type by isolated flagella of the other sex that triggers the activation of the autolytic system. Trypsin-inactivated gametes, incapable of agglutinating, do not slip out of their walls. The liberation of the enzyme within 30 s after mixing of the gametes indicates that it is synthesized during gametogenesis and is only released at mixing of gametes, or that it exists in an inactive form requiring activation (cf. CLAES 1977, SCHLÖSSER 1981). The activation of the autolytic system can be triggered by ionophore A 23187 in the presence of calcium and by antisera against gametic and vegetative flagella (GOODENOUGH and JURIVICH 1978). By combining live gametes with glutaraldehyde-fixed gametes of the other sex (which retain their agglutinability) or with isolated flagella of the opposite sex, MATSUDA et al. (1978) demonstrated that flagella contact induces autolysin production in the (+) gamete only. The cell wall of (−) gametes is lysed by enzyme secreted from (+) gametes. RAY et al. (1978) report, however, that glutaraldehyde-fixed gametes and gametic flagella only agglutinate with live gametes of the other sex but do not activate the lytic system. They assume the existence of two spatially separated sexual recognition sites on the flagella, one mediating flagella agglutination, the other, the lytic activation site, responsible for wall lysis.

SOLTER and GIBOR (1977) studied flagella regeneration of deflagellated gametes to clarify the induction of wall lysis by the MTR, and to demonstrate a role of the flagella for signal transmission. When (+) and (−) gametes are deflagellated only two minutes after mixing, wall lysis occurs as completely as in non-deflagellated controls: the autolytic system requires less than 2 min for activation. In a mixture of (+) and (−) gametes deflagellated at the moment of mixing, however, activation of the autolytic system does not occur until after 40 min, when the growing flagella have reached half of their normal length, whereas the capacity to agglutinate reappears after only 10 min, with much shorter flagella. A similar differential delay in the reappearance of the flagella functions occurs after trypsin inactivation of the gametic agglutinability. The capacity to induce pairing emerges considerably later than the reappearance of the capacity to agglutinate.

As in *C. eugametos,* flagella contact also activates the mating structures that initiate lateral fusion of the gamete bodies. The normal triggering of pairing by the MTR can be copied. Two UV-induced *reinhardii* (−) mutants (*imp* 10 and 12) have lost the agglutination capacity but will, in spite of the absent

MTR, pair with (+) gametes if gamete contact is brought about artificially by inducing flagellar contact with antiflagellar-serum. The substitute contact activates the autolytic system and the mating structures, and leads to true pairing (HWANG et al. 1981). This can be shown also for non-agglutinating (+) mutants (GOODENOUGH and JURIVICH 1978, GOODENOUGH et al. 1980). Actual cross-linkage between the flagella of sex-different gametes is required since univalent antibodies, just by interacting with the antigenic site, do not elicit any activation. A series of different data can best be interpreted by the assumption that autolysis of cell wall and activation of mating structures are initiated by different signals. Short-term treatment with chymotrypsin does not interfere with the MTR and wall lysis, but prevents selectively activation of the mating structures, either by affecting the fusion mechanism directly or by eliminating signal perception (MESLAND et al. 1980).

Colchicine (MESLAND et al. 1980) prevents pairing without interfering with the gametes' motility and agglutinability. It selectively prevents tip locking, wall lysis and mating structure activation. The colchicine-caused phenomenon is copied in a conditional mutant (gam-1) (FOREST et al. 1978). "Tipping" is obviously required for the creation and emission of the signal for pairing. It is argued that colchicine may affect a demonstrated exposed glycosylated tubulin, which is supposed to play a yet undetermined role in the mating process.

Tipping is associated with flagella tip activation (FTA) involving characteristic, non-sex-specific changes in the ultrastructure of the flagella tips. Non-activated flagella, i.e. gametic flagella before or after agglutination, possess smooth tapered tips, whereas activated tips are bulbous and characterized by clusters of fibrous material of unknown nature underneath the flagella membrane. FTA also involves a differential arrangement of the fibrils; activated tips have their A microfibrils extended and an altered axonema structure. The elongation of the A tubules causes a distinct lengthening of the flagella tip. The elicitation of FTA by gamete contact can be copied by cross-linking agents such as Con A and antiflagella sera. Since mating structure activation occurs only after FTA, it is concluded that FTA represents a necessary event in the sexual signalling sequence. The signal created must be transmitted via the flagella shaft (MESLAND et al. 1980).

Fusion (pairing) in *C. reinhardii* has been analyzed electron microscopically by GOODENOUGH et al. (1980). The mating structure in differentiated but non-activated (−) gametes consists of a curved, denser layer, the membrane zone, which is coated with fuzzy material, the "fringe". Upon activation with a non-fusing (+) mutant, the (−) mating structure increases in size, forms a pronounced bud stage and finally the "domed" stage, Friedmann's choanoid body. The nonactivated (+) structure is larger, with a more prominent fringe. Activation induces formation of a bud-like structure (containing actin fibrils), which expands to the fertilization tubule, its tip covered with fringe. Initiating the fusion proper, the fringe zones of the (+) and (−) structures adhere on the basis of a (+) and (−) complementarity that is not identical with that of the MTS's, as in *C. moewusii*. This initial contact is succeeded by the fusion proper by means of a yet unknown mechanism causing membrane coalescence between the interconnected naked gametes.

Non-fusing mutants may be impotent because of the absence of a well-developed fringe, detectable by the EM, or, in the presence of a fringe, possibly because of an altered chemical complementarity.

The non-identity of the two complementarities responsible for MTR and pairing emerges from their differential sensitivity to TUM. In addition to the unilateral TUM-sensitivity of the adhesiveness of *reinhardii* (+) gametes, Matsuda et al. (1982) report the existence of a TUM-sensitive compound on the (−) gametes associated with pairing. (−) gametes exposed to TUM during gametogenesis agglutinate but do not pair with untreated (+) gametes. Also, trypsin-inactivated gametes regain flagella adhesiveness after addition of trypsin-inhibitor; in the presence of TUM the (−) but not the (+) cells restore their contact capacity. Such (−) gametes, however, lack the potential to pair. Finally, *gam*-1 *mt⁻* gametes (able to agglutinate, unable to pair), transferred from restrictive (35 °C) to permissive (25 °C) conditions, attain the capacity to pair. This differentiation is blocked by TUM. The TUM-sensitive component could function in the proper fusion mechanism or in the signalling process inducing pairing after MTR.

12.7 Termination of the Mating Type Reaction by Pairing

Gametes of different sex agglutinate only until pairing occurs. However, the gametic potential to sustain flagella agglutination is in principle unlimited under appropriate conditions (light, N-depletion, pH, etc.) (cf. Wiese and Wiese 1978). In this context it should be emphasized that the copulation medium is identical with the induction medium, i.e. gametes are constantly under gametogenic conditions. The unrestricted agglutination potential can be demonstrated in the isoagglutination phenomenon, which "copies" flagella contact. By constant addition of (−) IA in *C. eugametos,* (+) gametes can be kept in permanent agglutination for more than 100 h. During normal agglutination it is the ensuing pairing that terminates flagella adhesion after 5–45 min. The disagglutinated flagella are no longer adhesive to third gametes and, freely motile, move the zygote. In the vis-à-vis pair of *C. moewusii,* an additional regulation exists: the flagella of only one gamete continue to beat; those of his partner trail passively (Lewin 1952, 1954). Regulation of the duration of the MTR by pairing is clearly recognizable from simple observation of the mating process and can be experimentally demonstrated. If in *C. moewusii syngen II* the capacity to pair is eliminated by thermolysin or chymotrypsin, neither of which interferes with the MTR of this taxon, mixed gametes of different sex stay agglutinated indefinitely (Wiese and Wiese 1978). Significantly, mutants that block pairing establish a long-enduring MTR. Matsuda et al. (1978) describe a mutant that blocks pairing, and, indicatively, flagella disadhesion.

The loose and dynamic adhesion between gametes of different sex, based upon non-covalent bonding and probably also on long-distance interactions, must provide for a quickly inducible termination. Upon some yet unknown signal from the completed pairing, the MTS's appear to be chemically altered and rendered inactive. This is in contrast to mating in yeasts (cf. CRANDALL 1978), in which the membrane-integrated complementary agglutinins form a tight complex that can be secondarily dissociated into the individual components in active form.

12.8 Sexual Incompatibility Between Species Resulting from Taxon-Specific or Defective Intercellular Communication and Contact Systems

In unicellular algae, a species is often difficult to define because the rather simple organization lacks identifying characteristics and may, in addition, be subjected to considerable environmental modification. No definition problem normally arises with species that reproduce sexually, the species being conceived as a reproductive unit. Often, however, such reproductive units, sexually isolated from others, have been detected on closer inspection within morphologically well-defined species. To avoid unnecessary nomenclature, T.M. SONNEBORN (1957) coined for these subspecies the term syngen (specified by Roman numbers); a syngen represents the sum of all individuals that could potentially share a common gene pool. Sexual isolation between syngens may be caused by very different types of barriers such as different responsiveness to sex-inducing environmental conditions, taxon-specific inducers of the sexual phase (cf. STARR 1971), taxon-specific erotactins (MÜLLER 1976), and taxon-specific contact systems (TRAINOR 1958, WIESE and WIESE 1977, COLEMAN 1959, 1979). Syngen formation can furthermore be connected with or caused and stabilized by differential karyotypes. Syngens have been detected in unicellular and colonial Chlorophyceae and many other algae.

Note Added in Proof. In *C. m. eugametos* and *C. reinhardti*, the agglutinins proper have been isolated and characterized. The Dutch team of Dr. VAN DEN ENDE detected no major differences between the (+) and (−) isoagglutinin besides of a distinctly different sugar composition. The mentioned unilateral sensitivity of the agglutinins' biosynthesis to TUM, however, indicates the presence of an essential N-linked ligand in the (+) but not in the (−) sex, i.e. the agglutinins proper are different in principle (MATSUDA et al. 1981; WIESE et al. 1983). An extensive analysis of the *eugametos* (−) agglutinin was performed by the Dutch team (cf. VAN DEN ENDE 1981). The glycoproteins of vegetative and gametic cells and of the (−) IA were extracted and compared (MUSGRAVE et al. 1979). Among several gamete-specific glycoproteins, the adhesiveness seems to reside in a large, flagella-bound glycoprotein of $20–40 \times 10^6$ daltons (MUSGRAVE et al. 1981). In contrast to the isoagglutinins, the isolated component is functionally univalent, incapable to cause an isoagglutination. Recombination of the extracted and inert membrane vesicles with the isolated glycoprotein restores multivalent principles (HOMAN et al. 1980). A specific receptor-blocking power of the univalent component could not be stated. In addition of carrying the contact sites, the component also induces the twitching activity and the flagella tip activation (FTA) in (+) gametes (MUSGRAVE et al. 1981). FTA is thought to play a decisive role at the induction of the pairing process. The twitching-inducing activity and the contact capacity can both the blocked by (−)-specific antibodies (LENS et al. 1982). The meaning of the twitching is presently completely unknown. The Dutch team claims that the (−) agglutinin is an extrinsic membrane glycoprotein bound to an intrinsic protein receptor within the membrane. The component is rich in hydroxyproline and in arabinose and galactose.

WILLIAMS (1982) likewise isolated the highmolecular glycoprotein from *eugametos* (−) IA as a univalent principle which reestablished a functional multivalency only after being crosslinked to polylysine. This artificially synthesized principle possesses the identical sex- and species specificity as the *eugametos* (−) gametes. As mentioned above, the isolated component is sensitive to α-D-N-acetylgalactosaminyl oligosaccharidase and to α-exogalactosidase and α-exogalactosaminidase.

References

Adair WS, Monk BC, Cohen R, Hwang C, Goodenough UW (1982) Sexual agglutinins from the *Chlamydomonas* flagellar membrane. Partial purification and characterization. J Biol Chem 257:4593–4603

Bean B (1979) Chemotaxis in unicellular eucaryonts. In: Haupt W, Feinleib ME (eds) Physiology of movements. Springer, Berlin Heidelberg New York, pp. 335–354

Bergman K, Goodenough UW, Goodenough DA, Jawitz J, Martin H (1975) Gametic differentiation in *Chlamydomonas reinhardtii*. II. Flagellar membranes and the agglutination reaction. J Cell Biol 67:606–622

Bosmann HB, McLean RJ (1975) Gametic recognition: lack of enhanced glycosyltransferase ectoenzyme system activity of non sexual cells and sexually incompatible gametes of *Chlamydomonas*. Biochem Biophys Res Comm 63:323–328

Brandham PE (1967) Time-lapse studies of conjugation in *Cosmarium botrytis*. II. Pseudoconjugation and an anisogamous mating behaviour involving chemotaxis. Can J Bot 45:483–493

Brown RM Jr, Johnson C, Bold HC (1968) Electron and phase-contrast microscopy of sexual reproduction in *Chlamydomonas moewusii*. J Phycol 4:100–120

Burr FR, McCracken Md (1973) Existence of a surface layer on the sheath of *Volvox*. J Phycol 9:345–346

Callow JA, Bolwell GP, Evans LV, Callow ME (1981) Isolation and preliminary characterization of receptors involved in egg-sperm recognition in *Fucus serratus*. In: Levering T (ed) Proceedings X intern seaweed symp. de Grueter, Hawthorne

Cavalier-Smith T (1975) Electron and light microscopy of gametogenesis and gamete fusion in *Chlamydomonas reinhardii*. Protoplasma 86:1–18

Claes H (1971) Autolyse der Zellwand bei den Gameten von *Chlamydomonas reinhardii*. Arch Mikrobiol 78:180–188

Claes H (1975) Influence of concanavalin A on autolysis of gametes of *Chlamydomonas reinhardii*. Arch Mikrobiol 103:225–230

Claes H (1977) Non-specific stimulation of the autolytic system in gametes from *Chlamydomonas reinhardii*. Exp Cell Res 108:121–229

Claes H (1980) Calcium-ionophore-induced stimulation of secretory activity in *Chlamydomonas reinhardii*. Arch Mikrobiol 124:84–86

Coggin SJ, Hutt W, Kochert G (1979) Sperm bundle–female somatic cell interaction in the fertilization process of *Volvox carteri* f. weismannia (Chlorophyta). J Phycol 15:247–251

Coleman AW (1959) Sexual isolation in *Pandorina morum*. J Protozool 6:249–264

Coleman AW (1979) Sexuality in colonial green flagellates. In: Lewandovsky M, Hutner SH (eds) Biochemistry and physiology of protozoa (2nd edn), Vol I. Academic Press, London New York, pp 307–340

Crandall M (1978) Mating type interactions in yeasts. In: Curtis ASG (ed) Cell-Cell Recognition. Cambridge University Press, Cambridge London New York Melbourne, pp 105–123

Darden WH (1966) Sexual differentiation in *Volvox aureus*. J Proozool 13:239–255

Darden WD, Sayers ER (1971) The effect of selected chemical and physical agents on the male-inducing substance from *Volvox aureus* M5. Microbios 3:209–214

Diwald K (1938) Die ungeschlechtliche und geschlechtliche Fortpflanzung von *Glenodinium lubiniensiforme* sp. nov. Flora 132:174–192

Ende H van den (1980) Sexual interactions in the green alga *Chlamydomonas eugametos*.

In: O'Day D, Horgen PA (eds) Sexual interactions in eucaryotic microbes. Academic Press, London New York, pp 102–134

Forest CL, Goodenough DA, Goodenough UW (1978) Flagella membrane agglutination and sexual signaling in the conditional gam-1 mutant of *Chlamydomonas reinhardii.* J Cell Biol 79:74–84

Förster H (1957) Das Wirkungsspektrum der Kopulation von *Chlamydomonas eugametos.* Z Naturforsch 12b:765–770

Förster H (1959) Die Wirkungsstärken einiger Wellenlängen zum Auslösen der Kopulation von *Chlamydomonas moewusii.* Z Naturforsch 14b:479–480

Förster H, Wiese L (1954a) Untersuchungen zur Kopulationsfähigkeit von *Chlamydomonas eugametos.* Z Naturforsch Teil B 9:470–471

Förster H, Wiese L (1954b) Gamonwirkungen bei *Chlamydomonas eugametos.* Z Naturforsch 9b:548–550

Friedmann I, Colwin AL, Colwin LH (1968) Fine structural aspects of fertilization in *Chlamydomonas reinhardi.* J Cell Sci 3:115–128

Geitler L (1931) Untersuchungen über das sexuelle Verhalten von *Tetraspora lubrica.* Biol Zentrbl 51:173–187

Gilles R, Bittner C, Jaenicke L (1981) Site and time of formation of the sex-inducing glycoprotein in *Volvox carteri.* FEBS Lett 124:57–61

Gillham NW (1969) Uniparental inheritance in *Chlamydomonas reinhardti.* Am Nat 103:355–388

Goldstein M (1964) Speciation and mating behavior in *Eudorina.* J Protozool 11:317–344

Goodenough UW (1977) Mating interactions in *Chlamydomonas.* In: Reissig JL (ed) Receptors and recognition, Ser B Vol 3. Chapman and Hall, London, pp 323–351

Goodenough UW, Jurivich D (1978) Tipping and mating structure activation induced in *Chlamydomonas* gametes by flagellar membrane antisera. J Cell Biol 79:680–693

Goodenough UW, Weiss RL (1975) Gametic differentiation in *Chlamydomonas reinhardti.* J Cell Biol 67:623–637

Goodenough UW, Hwang C, Martin H (1976) Isolation and genetic analysis of mutant strains of *Chlamydomonas reinhardti,* defective in gametic differentiation. Genetics 82:169–186

Goodenough UW, Adair WS, Caligor E, Forest CL, Hoffman JL, Mesland DAM, Spath S (1980) Membrane–membrane and membrane–ligand interactions in *Chlamydomonas* mating. In: Gilula NB (ed) Membrane–membrane interactions. Raven Press, New York, pp 131–152

Gowans CS (1976) Publications of Franz Moewus on the genetics of algae. In: Lewin RA (ed) The genetics of algae. Blackwell Scientific Publications, Oxford, pp 310–332

Grote M (1977) Untersuchungen zum Kopulationsablauf bei der Grünalge *Spirogyra majuscula.* Protoplasma 91:71–82

Hartmann M (1956) Sex problems in algae, fungi and protozoa. Am Nat 89:321–346

Heimcke JW, Starr RC (1979) The sexual process in several heterogamous *Chlamydomonas* strains in the subgenus *Pleiochloris.* Arch Protistenkd 133:20–42

Hoffman L (1973) Fertilization in *Oedogonium.* I. Plasmogamy. J Phycol 9:62–85

Homann WL (1981) Sexual cell recognition in *Chlamydomonas eugametos.* In: Hansen TC (ed) Lectins, biology, biochemistry, clinical biochemistry. Vol I:51–58

Homann WL, Musgrave A, Molenaar EM, Ende H van den (1980) Isolation of monovalent sexual binding components from *Chlamydomonas eugametos* flagellar membranes. Arch Microbiol 128:120–125

Hoshaw RW (1965) A cultural study of sexuality in *Sirogonium melanosporum.* J Phycol 1:134–138

Hutt W, Kochert G (1971) Effects of some protein and nucleic acid inhibitors on fertilization in *Volvox carteri.* J Phycol 7:316–320

Hwang C, Monk BC, Goodenough UW (1981) Linkage of mutations affecting (−) flagellar membrane agglutinability to the mt⁻ mating-type locus of *Chlamydomonas.* Genetics 99:41–47

Ishiura M, Iwasa K (1973) Gametogenesis in *Chlamydomonas.* II. Effect of cycloheximide on the induction of sexuality. Plant Cell Physiol 14:923–933

Jaenicke L (1982) Volvox biochemistry comes of age. Trends Biochem Sci

Jaenicke L, Waffenschmidt S (1979) Matrix lysis and release of daughter spheroids in *Volvox carteri* – a proteolytic process. FEBS Lett 107:250–253

Kates JR, Jones RF (1964) The control of gametic differentiation in liquid cultures of *Chlamydomonas*. J Cell Comp Physiol 63:157–164

Kochert G (1968) Differentiation of reproductive cells in *Volvox carteri*. J Protozool 15:438–452

Kochert G, Yates I (1974) Purification and partial characterization of a glycoprotein sexual inducer from *Volvox carteri*. Proc Natl Acad Sci USA 71:1211–1214

Köhle D, Lange W, Kauss H (1980) Agglutination and glycosyltransferase activity of isolated gametic flagella from *Chlamydomonas reinhardii*. Arch Microbiol 127:239–243

Kurn N (1981) Altered development of the multicellular alga *Volvox carteri* caused by lectin binding. Cell Biol Int Rep 5:867–875

Lens PF, Briel W van den, Musgrave A, Ende H van den (1980) Sex-specific glycoproteins in *Chlamydomonas* flagella. An immunological study. Arch Microbiol 126:77–81

Lens PF Olofson F, Nederbragt A, Musgrave A, Ende H van den (1982) An antiserum against a glycoprotein functional in flagellar adhesion between *Chlamydomonas eugametos* gametes. Arch Microbiol 131:241–246

Lewin RA (1952) Studies on the flagella of algae. I. General observations on *Chlamydomonas meowusii* Gerloff. Biol Bull (Woods Hole) 102:74–79

Lewin RA (1954) Sex in unicellular algae. In: Wenrich DH (ed) Sex in microorganisms. Amer Assoc Adv Science, Publ Washington DC, pp 100–133

Løvlie A, Bryhni E (1976) Signal for cell fusion. Nature 263:779–781

McLean RJ, Bosmann HB (1975) Cell-cell interactions: Enhancement of glycosyltransferase ectoenzyme systems during *Chlamydomonas* gametic contact. Proc Natl Acad Sci USA 72:310–313

McLean RJ, Brown RM (1974) Cell surface differentiation of *Chlamydomonas* during gametogenesis. I. Mating and concanavalin A agglutinability. Dev Biol 36:279–285

McLean RJ, Laurendi CJ, Brown RM Jr (1974) The relationship of gamone to the mating reaction in *Chlamydomonas moewusii*. Proc Natl Acad Sci USA 71:2610–2613

Martin NC, Goodenough UW (1975) Gametic differentiation in *Chlamydomonas reinhardtii*. I. Production of gametes and their fine structure. J Cell Biol 67:587–605

Matsuda Y, Tamaki S, Tsubo Y (1978) Mating specific induction of cell wall lytic factor by agglutination of gametes in *Chlamydomonas reinhardtii*. Plant Cell Physiol 19:1253–1261

Matsuda Y, Sakamoto K, Mizuochi T, Kobata A, Tamura G, Tsubo T (1981) Mating type-specific inhibition of gametic differentiation of *Chlamydomonas reinhardtii* by tunicamycin. Plant Cell Physiol 22:1607–1611

Matsuda Y, Sakamoto K, Kiuchi N, Mizuochi T, Tsubo Y, Kobata A (1982) Two tunicamycin-sensitive components involved in agglutination and fusion of *Chlamydomonas reinhardtii* gametes. Arch Mikrobiol 131:87–90

Mesland DAM (1976) Mating in *Chlamydomonas eugametos*. A scanning electron microscopical study. Arch Microbiol 109:31–35

Mesland DAM (1977) Flagellar surface morphology of *Chlamydomonas eugametos*. Protoplasma 93:311–323

Mesland DAM, Ende H van den (1979) The role of flagellar adhesion in sexual activation of *Chlamydomonas eugametos*. Protoplasma 98:115–129

Mesland DAM, Ende H van den (1978b) An inhibitor of cell fusion in *Chlamydomonas eugametos*. Arch Mikrobiol 117:131–134

Mesland DAM, Hoffman JL, Caligor E, Goodenough UW (1980) Flagella tip activation stimulated by membrane adhesion in *Chlamydomonas* gametes. J Cell Biol 84:599–617

Miyake A (1981) Physiology and biochemistry of conjugation in Ciliates. In: Levandowsky M, Hutner SH (eds) Biochemistry and physiology of protozoa (2nd edn). Academic Press, New York, pp 126–198

Moewus F (1933) Untersuchungen über die Sexualität und Entwicklung von Chlorophyceen. Arch Protistenkd 80:469–520

Monk BC, Adair WS, Cohen RA, Goodenough UW (1982) *Chlamydomonas* topography: Polypeptide components of the gametic flagellar membrane and cell wall. (in press)

Müller DG (1967) Generationswechsel, Kernphasenwechsel und Sexualität der Braunalge *Ectocarpus siliculosus*. Planta 75:39–54

Müller DG (1976) Sexual isolation between a European and an American population of *Ectocarpus siliculosus* (Phaeophyta). J Phycol 12:252–254

Müller DG (1981) The role of olefine hydrocarbons in sexual reproduction in brown algae. In: Levring T (ed) Proceed intern seaweed symp X. deGrueter, Hawthorne

Musgrave A, Homan WL, Briel ML van den, Lelie N, Schol D, Ero L, Ende H van den (1979) Membrane glycoproteins of *Chlamydomonas eugametos*. Planta 145:417–425

Musgrave A, Eijk Ed van, Welscher R, Brockman R, Lens P, Homan W, Ende H van den (1981) Sexual agglutination factor from *Chlamydomonas eugametos*. Planta 153:362–369

Noland T, Yates J, Kochert G (1977) Binding characteristics of ^{125}J-labeled sexual inducer to *Volvox carteriensis* f. weismannia. J Phycol 13:49

Pascher A (1918) Über die Beziehung der Reduktionsteilung zur Mendelschen Spaltung. Ber Dtsch Bot Ges 36:163–168

Pascher A (1931) Über einen neuen einzelligen und einkernigen Organismus mit Eibefruchtung. Bot Centr Beih A 48:466–480

Rawitscher-Kunkel E, Machlis L (1962) The hormonal integration of sexual reproduction in *Oedogonium*. Am J Bot 49:177–183

Ray DA, Gibor A (1982) Tunicamycin-sensitive glycoproteins involved in the mating of *Chlamydomonas reinhardi*. Exp Cell Res 141:245–252

Ray DA, Solter KM, Gibor A (1978) Flagellar surface differentiation. Evidence for multiple sites involved in mating of *Chlamydomonas reinhardti*. Exp Cell Res 114:185–189

Rayburn WR (1974) Sexual reproduction in *Pandorina unicocca*. J Phycol 10:258–265

Sager R (1955) Inheritance in the green alga *Chlamydomonas reinhardi*. Genetics 40:476–489

Sager R, Granick S (1954) Nutritional control of sexuality in *Chlamydomonas reinhardti*. J Gen Physiol 37:729–742

Sandgren CR (1981) Characteristics of sexual and asexual resting cyst formation in *Dinobryon cylindricum* (Chrysophyta). J Phycol 17:199–210

Schlösser UG (1976) Entwicklungsstadien- und sippenspezifische Zellwand-Autolysine bei der Freisetzung von Fortpflanzungszellen in der Gattung *Chlamydomonas*. Ber Dtsch Bot Ges 89:1–56

Schlösser UG (1981) Algal wall-degrading enzymes–autolysins. In: Tanner W, Loewus FA (eds) Encyclop plant physiol, New Ser Vol 13B. Springer, Berlin Heidelberg New York, pp 333–353

Schmeisser ET, Baumgartel DM, Howell SH (1973) Gametic differentiation in *Chlamydomonas reinhardti:* Cell cycle dependency and rates in attainment of mating competency. Dev Biol Suppl 31:31–37

Snell WJ (1976) Mating in *Chlamydomonas:* A system for the study of specific cell adhesion. J Cell Biol 68:48–69

Snell WJ (1981) Flagella adhesion and deadhesion in *Chlamydomonas* gametes. Effects of tunicamycin and observations on flagellar tip morphology. J Supramol Struct Cell Biochem 16:371–376

Snell WJ, Moore WS (1980) Aggregation-dependent turnover of flagellar adhesion molecules in *Chlamydomonas* gametes. J Cell Biol 84:203–210

Snell WJ, Roseman S (1979) Kinetics of adhesion and deadhesion of *Chlamydomonas* gametes. J Biol Chem 254:10820–10829

Solter KM, Gibor A (1977) Evidence for the role of flagella as sensory transducers in mating of *Chlamydomonas*. Nature 265:444–445

Solter KM, Gibor A (1978) Removal and recovery of mating receptors on flagella of *Chlamydomonas reinhardti*. Exp Cell Res 115:1175–1181

Sonneborn TM (1957) Breeding systems, reproductive methods and species problems in Protozoa. In E Mayr (ed) The species problem. AAAS Publ 50:155–324

Starr RC (1954) Heterothallism in *Cosmarium botrytis* var. subtumidium. Am Bot 41:601–607

Starr RC (1971) Control of differentiation in *Volvox*. In: Runner M (ed) Changing synthesis in development. Symp Soc Dev Biol, Dev Biol Suppl 4:59–100

Starr RC, Jaenicke L (1974) Purification and characterization of the hormone initiating sexual morphogenesis in *Volvox carteri* f. nagariensis. Proc Natl Acad Sci USA 71:1050–1054

Starr RC, O'Neil RM, Miller CF (1980) L-Glutamic acid as a mediator of sexual morphogenesis in *Volvox capensis*. Proc Natl Acad Sci USA 77:1025–1028

Stifter I (1959) Untersuchungen über einige Zusammenhänge zwischen Stoffwechsel und Sexualphysiologie an dem Flagellaten *Chlamydomonas eugametos*. Arch Protistenkd 104:364–388

Struck DK, Lennarz WJ (1980) The function of saccharide-lipids in synthesis of glycoproteins. In: Lennarz WJ (ed) The biochemistry of glycoproteins and proteoglycans. Plenum Press, London New York, pp 35–83

Szostak JW, Sparkuhl J, Goldstein ME (1973) Sexual induction in *Eudorina*. J Phycol 9:215–218

Trainor FR (1958) A comparative study of sexual reproduction in four species of *Chlamydomonas*. Am J Bot 46:65–70

Triemer RE, Brown RM Jr (1975) Fertilization in *Chlamydomonas reinhardti*, with special reference to the structure, development and fate of the choanoid body. Protoplasma 85:99–107

Tschermak-Woess E (1959) Extreme Anisogamie und ein bemerkenswerter Fall der Geschlechtsbestimmung bei einer neuen *Chlamydomonas*-Art. Planta 52:606–622

Tschermak-Woess E (1963) Das eigenartige Kopulationsverhalten von *Chloromonas saprophila*, einer neuen Chlamydomonacee. Oesterr Bot   Z 110:294–307

Tsubo Y (1961) Chemotaxis and sexual behavior in *Chlamydomonas*. J Protozool 8:114–121

Weiss RL, Goodenough DA, Goodenough UW (1977) Membrane differentiations at sites specialized for cell fusion. J Cell Biol 72:144–160

Wiese L (1965) On sexual agglutination and mating type substances (gamones) in isogamous heterothallic Chlamydomonads. I. Evidence of the identity of the gamones with the surface components responsible for sexual flagellar contact. J Phycol 1:46–54

Wiese L (1969) Algae. In: Metz CB, Monroy A (eds) Fertilization, comparative morphology, biochemistry and immunology vol 2. Academic Press, London New York, pp 135–188

Wiese L (1974) Nature of the sexspecific glycoprotein agglutinins in *Chlamydomonas*. Ann NY Acad Sci 234:383–395

Wiese L (1981) On the evolution of anisogamy from isogamous monoecy and on the origin of sex. J Theor Biol 89:573–580

Wiese L, Hayward PC (1972) On sexual agglutination and mating type substances in isogamous dioecious chlamydomonads. III. The sensitivity of sex cell contact to various enzymes. Am J Bot 59:530–536

Wiese L, Mayer RA (1982) Unilateral tunicamycin sensitivity of gametogenesis in dioecious isogamous *Chlamydomonas* species. Gamete Res 5:1–9

Wiese L, Metz CB (1969) On the trypsin sensitivity of fertilization as studied with living gametes in *Chlamydomonas*. Biol Bull 136:483–493

Wiese L, Shoemaker D (1970) On sexual agglutination and mating type substances in isogamous heterothallic chlamydomonads. Biol Bull 138:88–95

Wiese L, Wiese W (1975) On sexual agglutination and mating type substances in isogamous dioecious Chlamydomonads. IV. Unilateral inactivation of the sex contact capacity in compatible and incompatible taxa by α-mannosidase and snake venom protease. Dev Biol 43:264–276

Wiese L, Wiese W (1977) On speciation by evolution of gametic incompatibility. A model case in *Chlamydomonas*. Am Nat 111:732–742

Wiese L, Wiese W (1978) Sex cell contact in *Chlamydomonas*, a model for cell recognition. In: Curtis ASG (ed) Cell interactions, symposia of the society for experimental biology vol 32. Cambridge Univ Press, Cambridge London New York Melbourne, pp 83–104

Wiese L, Wiese W, Edwards DA (1979) Inducible anisogamy and the evolution of oogamy from isogamy. Ann Bot 44:131–139 (London)

Wiese L, Williams L, Baker DL (1983) A general and fundamental biopolarity of the sex cell contact mechanism as revealed by tunicamycin in *Chlamydomonas*. Amer Natural 122:806–816

Williams LA (1982) The functional structure of the (−) mating type substance in *Chlamydomonas*. Diss Florida State Univ

13 Colony Formation in Algae

R. C. STARR

13.1 Introduction

In the algae the term "colony" is used for a wide variety of organisms which have the single common attribute of being two or more cells associated regularly in a fashion such that this association is recognized as the morphological type of a genus and species. The cells in a colony may arise at the same time from the same cell, or they may be the accumulated product of cell divisions or spore production over several generations. The cells in a colony may be held in association by being embedded in gelatinous matrices, by pads or tubes at certain places on the cells, by the remains of parental cell walls, or by specialized areas on the walls.

Those colonies in which the cell number increases through time are said to be non-coenobic. In contrast there are coenobic colonies in which the cell number is fixed at the time the colony is formed, no additional cells being added during the lifetime of that colony. The variety of forms seen in these coenobic colonies is the result of the particular pattern of development each undergoes in its formation from the products of a single cell. It is with these patterns of development in coenobic colonies that this chapter is most concerned; however, the non-coenobic colonies do show an interesting variety of forms (BOURRELLY 1966, 1968, 1970), even though they result in the most part from a passive retention of successive cell generations by some secreted polysaccharide.

13.2 Non-Coenobic Colonies

Among the simplest non-coenobic colonies are those of green algae included in the order Tetrasporales, in which successive generations of cells have reproduced by motile or non-motile spores which are either retained in the gelatinous matrix (palmelloid colonies) or remained attached to one another by gelatinous stalks. The fact that flagellated genera such as *Chlamydomonas* may at times lose their flagella, secrete a gelatinous matrix, and form amorphous palmelloid colonies often makes it a problem to decide whether palmelloid colonies observed in natural collections are the temporary state of a motile alga or a truly palmelloid genus. Some palmelloid colonies are more easily identified by certain distinctive attributes. For example, the genera *Tetraspora* (Fig. 1a, b), *Apiocystis* (Fig. 2), and *Paulschulzia* (Fig. 3) are representatives of those forms

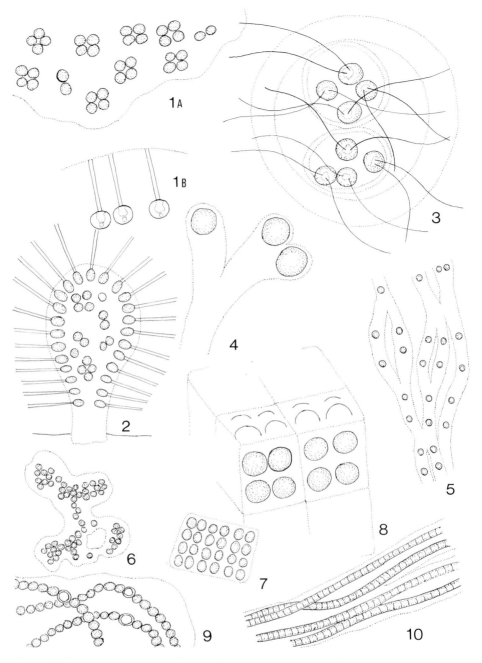

Fig. 1. A *Tetraspora* sp.
B *Tetraspora* sp., detail
showing cells with
pseudocilia

Fig. 2. *Apiocystis* sp.

Fig. 3. *Paulschulzia* sp.

Fig. 4. *Hormotilopsis* sp.

Fig. 5. *Gloeodendron* sp.

Fig. 6. *Microcystis* sp.

Fig. 7. *Merismopedia* sp.

Fig. 8. *Eucapsis* sp.

Fig. 9. *Nostoc* sp.,
trichomes in matrix

Fig. 10. *Schizothrix* sp.,
trichomes in matrix

in which the cells have two or more fine cytoplasmic processes known as pseudo-cilia. In *Hormotilopsis* (Fig. 4) the unipolar secretion of mucilage by the cells results in branching colonies, while in *Gloeodendron* (Fig. 5) the bilateral secretion of mucilage results in an array of gelatinous threads that appear to have separated and fused at different points, but which actually indicate the successive production of spores and the subsequent bipolar secretion of mucilage (THOMPSON 1951, as in *Schizodictyon*). The most recent taxonomic treatments of the order Tetrasporales are by BOURRELLY (1966) and FOTT (1972).

Palmelloid colonies are not limited to the green algae. In the blue-green algae one finds a variety of colonies varying from the amorphous ones of *Micro-cystis* (Fig. 6) to the flat plates of *Merismopedia* (Fig. 7) and the cubical ones of *Eucapsis* (Fig. 8), the form of the genera being a reflection of the changing directions of successive cell divisions. The coalescing sheaths of the trichomes in *Nostoc* (Fig. 9) and *Schizothrix* (Fig. 10) are examples of other blue-greens forming gelatinous colonies (GEITLER 1932). The branching gelatinous filaments of *Mischococcus* (Fig. 11) in the yellow-green algae (Xanthophyceae); the perforated, saccate cylinders of *Phaeosphaera* (Fig. 12) in the golden brown-algae (Chrysophyceae), and the gelatinous colonies of *Gloeodinium* (Fig. 13) in the dinoflagellates (Dinophyceae) are examples of parallel evolution of simple non-coenobic colonies in various classes of the algae.

In some algae, gelatinous strands, secreted at single spots on the cells, are used to attach the cells to physical substrates. As the cells divide, each product continues to secrete a strand, giving the appearance that the gelatinous strands have divided. Continued division of the cells results in the formation of colonies of cells attached to a substrate by multiply branching strands. *Colacium* (Fig. 14) in the Euglenophyceae and *Gomphonema* (Fig. 15) in the Bacillariophyceae are only two examples in two different evolutionary groups. Gelatinous strands that connect the posterior ends of cells are seen in the colonial flagellate *Synura* (Chrysophyceae) (Fig. 16), additional cells being added to the colony as cells undergo division. Fragmentation of larger colonies produces new, smaller ones.

Less common colonial forms occur in other algae. In *Dinobryon* (Fig. 17), the golden-brown flagellate, situated in the bottom of a conical lorica, undergoes cell division, one or both products then migrating from the bottom of the conical lorica to its rim where it comes to rest and secretes a new lorica. Continued multiplications of the cells results in an arborescent colony made up of loricas. In *Polykrikos* (Fig. 18), a colonial member of the dinoflagellates (Dinophyceae), the colony is formed by a series of incomplete cleavages of the cell body during cell division. There is a replication of all cell organelles, including the nucleus and the pair of characteristic flagella, but the individual cells fail to separate.

The organisms named and illustrated do not exhaust the colonial genera that occur among the various divisions and classes of the algae, but they show the great diversity of form resulting from the simple phenomenon of cell populations arising through successive cell division or spore production and failing to be dispersed in the surrounding medium due largely to the restricting forces of extracellular products secreted in the form of gelatinous sheaths or tubes. In most instances, the chemical nature of the gelatinous material is unknown,

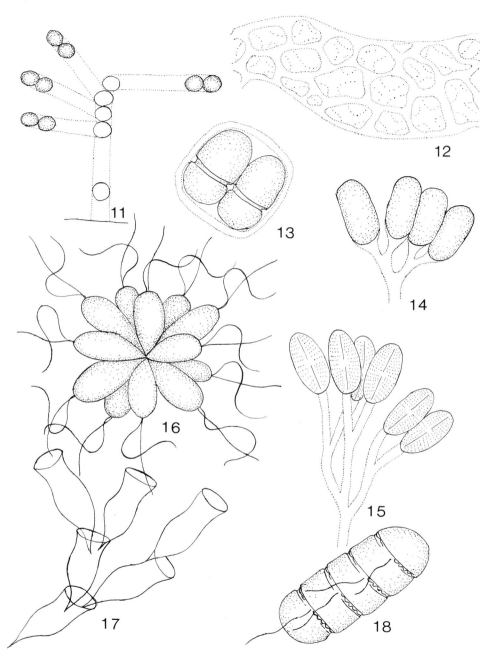

Fig. 11. *Mischococcus* sp.　　　**Fig. 14.** *Colacium* sp.　　　**Fig. 17.** *Dinobyron* sp.

Fig. 12. *Phaeosphaera* sp.　　　**Fig. 15.** *Gomphonema* sp.　　　**Fig. 18.** *Polykrikos* sp.

Fig. 13. *Gloeodinium* sp.　　　**Fig. 16.** *Synura* sp.

but where investigated, the chief component has been found to be a polysaccharide (MACKIE and PRESTON 1974).

13.3 Coenobic Colonies

A coenobium is defined as a colony in which the cell number is fixed at the time of its formation. The contents of a parental cell undergo mitosis and cytoplasmic cleavage to form uninucleate entities, either motile or non-motile, that aggregate in various ways so as to form the type of colony characteristic of the species. Coenobic colonies occur only in the green algae (Chlorophyceae); motile colonies are classified in the order Volvocales, non-motile ones in the order Chlorococcales.

13.3.1 Motile Coenobic Colonies

These colonial green algae are separated into two or more families on the basis of cell arrangement and differences in embryogenesis (BOURRELLY 1966, BOLD and WYNNE 1978, STARR 1980).

13.3.1.1 Family Volvocaceae

The largest family, recognized by all authorities, is commonly known as the *Gonium–Volvox* series. It is characteristically described as forming an evolutionary series, beginning with the 4- to 32-celled species of *Gonium,* in which the flattened coenobia are composed of cells alike in appearance and reproductive potential, and ending with *Volvox* with cell numbers approaching 50,000 of which only a very few have retained the reproductive potential. Whether the various genera are as closely related   to each other as is often implied will remain a matter of conjecture until there is further investigation which will, or will not, educe biochemical evidence of their having arisen from the same line or closely related lines in evolution. On the other hand, from the point of view of development the *Gonium–Volvox* series, be it an artificial or real grouping in the phylogenetic sense, does illustrate increasing complexity in a common pattern of cleavage and cell differentiation. In the more recent literature on *Volvox,* the individual is referred to as a simple multicellular organism in view of the complex sequence of cell divisions and cell movements during embryogenesis which result in a coenobium with differentiated cell types. The same point of view might well be applied to other members of the *Gonium–Volvox* series, even those without different cell types.

The *Gonium-Volvox* series includes the following genera in order of their increasing complexity:

A) Gonium (Fig. 19) (POCOCK 1955, STEIN 1958): a plate of 4 to 32 cells in which the biflagellate cells are pointing in the same direction such that the

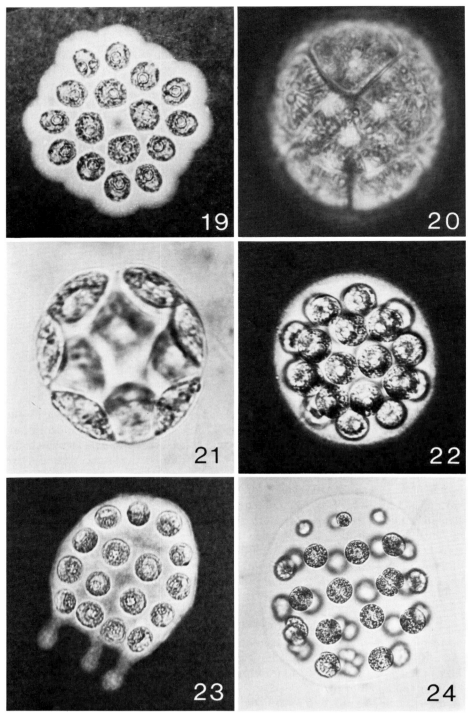

Fig. 19. *Gonium pectorale.* × 590

Fig. 20. *Pandorina morum.* × 600

Fig. 21. *Volvulina steinii.* × 940

Fig. 22. *Eudorina unicocca.* × 280

Fig. 23. *Platydorina caudata.* × 345

Fig. 24. *Pleodorina illinoisensis.* × 230

flat side of the plate is projected forward and rotates as it swims. All cells function in asexual and sexual reproduction. Sexual reproduction is isogamous, the cells of the coenobia acting as gametes.

B) Pandorina (Fig. 20) (COLEMAN 1959): coenobia of 16 or 32 cells in which the cells are arranged to form spheroids; the cells may or may not touch to form a solid or hollow sphere, respectively. All cells function in asexual and sexual reproduction. Sexual reproduction is isogamous, the cells of the coenobia acting as gametes.

C) Volvulina (Fig. 21) (CAREFOOT 1966): coenobia of 16 cells whose hemispherical shape is distinctive. All cells function in asexual and sexual reproduction. Sexual reproduction is isogamous, the cells of the coenobia acting as gametes.

D) Eudorina (Fig. 22) (GOLDSTEIN 1964): coenobia of 16 or 32 cells arranged to form a hollow sphere. All cells function in asexual and sexual reproduction. Sexual reproduction is heterogamous (oogamous); packets of biflagellate sperm are formed.

E) Platydorina (Fig. 23) (HARRIS and STARR 1969): coenobia of 16 or 32 cells arranged to form a hollow sphere which flattens during development. All cells function in asexual and sexual reproduction. Sexual reproduction is heterogamous (oogamous) with special packets of biflagellate sperm.

F) Pleodorina (Fig. 24, 25) (GERISCH 1959, GOLDSTEIN 1964): coenobia of 16 to 128 cells arranged to form a hollow sphere. Different species of the genus show loss of reproductive function to varying degrees in anterior portion of the sphere. Sexual reproduction is heterogamous (oogamous) with special packets of biflagellate sperm.

G) Volvox (Fig. 26, 27) (SMITH 1944): coenobia of 1000 to 50,000 cells arranged to form a hollow sphere. Only a few cells retain the asexual and sexual reproductive function. Sexual reproduction is oogamous with packets of biflagellate sperm.

Certain other genera, *Chlorcorona* (formerly *Corone*), *Lundiella, Stephanoon,* and *Mastigosphaera,* are included in the family Volvocaceae with the *Gonium–Volvox* series, but these genera are known only from the original descriptions, and little is known of their colony formation (STARR 1980).

In all members of the *Gonium–Volvox* series, with the exception of certain species of *Volvox,* embryogenesis to form new colonies occurs only after the asexual reproductive cells have reached maturity, i.e. maximum size under the existing environmental conditions. These enlarged cells will have achieved the necessary mass for little or no growth to occur during the series of cleavages that results in the formation of a new colony. It has been shown that, at the onset of embryogenesis, the asexual reproductive cell in *Volvox carteri* has enough ctDNA in the chloroplast (COLEMAN and MAGUIRE 1982) and enough rRNA (KOCHERT 1975) to distribute to all the cells of the embryo. In contrast, the nuclear DNA is replicated prior to each cleavage during embryogenesis.

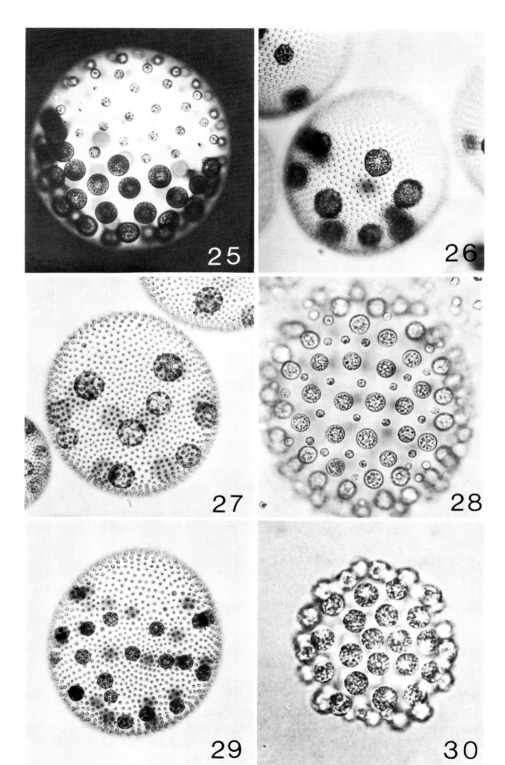

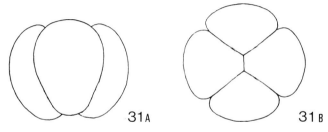

Fig. 31. **A** *Volvox* embryo, four-celled stage from side. **B** *Volvox* embryo, four-celled stage from posterior, showing that only two of the four cells touch

In all members of the series, the first cleavage of the asexual reproductive cells is in a plane parallel to its longitudinal axis, but the second division, although longitudinal, is slightly oblique, such that the division planes of this second division do not intersect at the posterior end. This results in only two of the four cells touching, a pattern characteristic of the four-celled stages of all these colonial genera (Fig. 31 B). Inasmuch as the pattern of cell division is the same for all members of the *Gonium–Volvox* series, the details of embryogenesis are described only for *Volvox,* the most complicated member.

The spheroids of *Volvox* are composed of two types of cells, somatic and reproductive, arranged as a single peripheral layer and embedded in a gelatinous matrix whose composition includes both glycoproteins and polysaccharides (LAMPORT 1974, JAENICKE and WAFFENSCHMIDT 1979, 1981). The asexual reproductive cells, known as gonidia, may begin to divide when they are small and the spheroid, of which they are a part, still young (Fig. 26) (*V. capensis, V. rousseletii, V. aureus, V. globator, V. dissipatrix*); or the gonidium may undergo a long period of cell enlargement prior to the first cleavage (Fig. 27) (*V. carteri, V. obversus, V. gigas, V. tertius, V. spermatosphaera, V. africanus*). The latter pattern of cell enlargement prior to the initiation of cleavage occurs in all other genera of the *Gonium–Volvox* series. Embryogenesis has been studied in a variety of *Volvox* species. The classic studies by JANET (1923) on *V. aureus* and by POCOCK (1933) on *V. capensis* and *V. rousseletii* concerned species in which the gonidia begin to divide when small; successive divisions alternate with periods of growth. More recently *V. carteri* has been the subject of detailed studies.

Fig. 25. *Pleodorina californica.* × 170

Fig. 26. *Volvox aureus.* × 75

Fig. 27. *Volvox carteri,* asexual spheroid with gonidia. × 75

Fig. 28. *Volvox carteri,* male spheroid showing 1:1 ratio of small somatic cells and large sperm-producing cells. × 200

Fig. 29. *Volvox carteri,* female spheroid with eggs. × 85

Fig. 30. *Astrephomene gubernaculifera.* × 320

In this species the gonidia enlarge greatly before the onset of cleavage. Unless otherwise indicated, the following description of embryogenesis is concerned exclusively with *V. carteri* f. nagariensis, and is based on light microscopy (Starr 1969, 1970, Gilles and Jaenicke 1982), and both scanning and transmission electron microscopy (Viamontes and Kirk 1977, Viamontes et al. 1979, Green and Kirk 1981, 1982, Green et al. 1981, Kirk et al. 1982).

In *V. carteri* f. nagariensis, there are three different forms of spheroids, asexual (Fig. 27), male (Fig. 28), and female (Fig. 29). These are distinguished by the type, number, and positioning of the reproductive cells in the mature spheroid, all of which result from specific patterns of cleavage and cell differentiation in embryogenesis. The stages of embryogenesis of asexuals, males, and females have the same morphological appearance (Figs. 32–37) except for the differentiating division when unequal cleavages occur and form the reproductive initials. In individuals growing under optimal conditions, the unequal cleavages occur at the division of the 32-celled stage in the asexual embryo (Figs. 38, 39), at the division of the 64-celled stage in the female embryo, and at the last division in the male embryo.

The first division in embryogenesis follows a slight flattening on the side of the gonidium toward the outside of the parental spheroid, thereby shortening the longitudinal axis of the gonidium and ultimately the developing embryo. The first cleavage is in a plane parallel to this shortened longitudinal axis; and although the division of the two resulting cells is again longitudinal (Fig. 31 A), each longitudinal plane is slightly oblique such that the two division planes in this second division do not intersect at the posterior end (Fig. 31 B). The third series of cleavages is again longitudinal and oblique such that in the eight-celled embryo there is the initiation of the alternating tiers of cells so characteristic of the spherical members of the *Gonium–Volvox* series. Due to the overlapping of the two alternating tiers in the eight-celled stage, the

→

Figs. 32–37. *Volvox carteri* f. nagariensis, early embryogenesis as seen in scanning electron-microscopy. Bar = 10 µm, × 1000. (Photographs courtesy of Dr. David Kirk; from Green and Kirk 1981)

Fig. 32. Two-celled embryo, as seen from anterior end

Fig. 33. Four-celled embryo, as seen from anterior end

Fig. 34. Eight-celled embryo, as seen from anterior end. Four cells are migrating anteriorly to form a hollow sphere, with the phialopore at the anterior end

Fig. 35. 16-celled embryo, as seen from the anterior end. The embryo consists of four tiers of four cells each as indicated by the numbers; however, tier #4 cannot be seen from the anterior end

Fig. 36. 16-celled embryo, as seen from the side. Tiers #1 and #3 have originated by oblique divisions of the anterior cells of the eight-celled embryo (Fig. 34); tiers #2 and #4, by oblique divisions of the posterior cells

Fig. 37. 32-celled stage, as seen from the anterior end. Each cell of the 16-celled embryo has divided to give an anterior (*a*) and a posterior (*p*) derivative

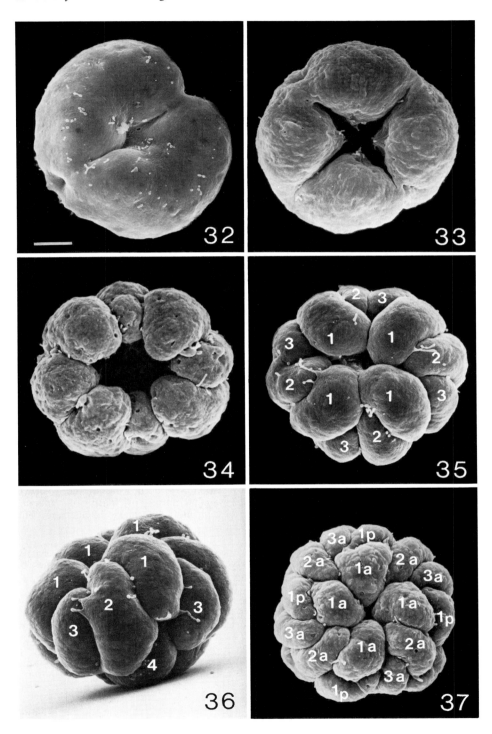

longitudinally oblique division of the cells in this stage result in a 16-celled embryo composed of four tiers of four cells each; however, tiers #1 and #3 were derived from the division of the anterior tier of the eight-celled stage, while tiers #2 and #4 were derived from the division of the posterior tier (Figs. 36, 50). Even at the 16-celled stage the four tiers of cells overlap considerably. Longitudinally, slightly oblique cleavages result in a 32-celled embryo.

The various species of *Volvox* typically show asexual reproductive cells, the gonidia, in predictably small numbers and positions, but in only a few can one see the differentiating divisions that delimit them during embryogenesis. In *Volvox carteri* f. nagariensis the gonidial initials result from unequal cleavages early in embryogenesis. Starr (1969, 1970) noted that at the division of the 32-celled stage, the 16 cells in the anterior part of the embryo that have originated from tiers #1 and #2 of the 16-celled stage divide unequally, each forming a small cell to the anterior and a large cell to the posterior (Figs. 38, 39, 50C). The large cells are the gonidial initials, and the precise pattern by which they are formed is reflected in the arrangement of the gonidia in the mature spheroid. Green and Kirk (1981) studied cell lineages in this species using scanning electron microscopy and reported that the gonidial initials were delimited from derivatives of tiers #1, #2 and #3 of the 16-celled stage, rather than from #1 and #2 alone, as had been reported by Starr (1969, 1970). The importance of understanding cell lineages in formulating biochemical hypotheses to explain this act of differentiation by unequal cell division prompted a re-study of the

Figs. 38–41. *Volvox carteri* f. nagariensis, later stages in embryogeny as observed in scanning electron microscopy. Bar = 10 μm, × 1000. (Photographs courtesy of Dr. David Kirk, Figs. 38, 40–41 from Green and Kirk 1981, Fig. 39 from Green and Kirk 1982)

Fig. 38. 64-celled embryo, as seen from the anterior end. Each cell of the 32-celled embryo (Fig. 37) has divided; cells *1a, 1p, 2a* and *2p* have divided unequally to form large gonidial initials and small somatic initials, while cells *3a, 3p, 4a, and 4p* have divided equally to form only somatic initials. In this photograph the large gonidial initials are indicated by the *numbers* and *letters* corresponding to the cells of the 32-celled embryo from which they have been formed; gonidial initials originating by unequal cleavage of the 2p cells cannot be seen as they lie in the posterior hemisphere. The small somatic initials are not designated by numbers

Fig. 39. 64-celled embryo, as seen from the side. Compare with anterior view seen in Fig. 38. Note large gonidial initials that have been formed by unequal cleavages of the *1a, 1p, 2a,* and *2p* cells. Cells derived from tiers #3 and #4 are all somatic initials

Fig. 40. 128-celled embryo showing continued unequal cleavage of gonidial initials

Fig. 41. Embryo just prior to inversion. Note gonidia in original positions but now separated by large numbers of somatic cells. Phialopore appears as intersecting slits

Fig. 42. *V. carteri* f. nagariensis. Young embryo has been broken open to show large number of cytoplasmic bridges. Bar = 100 μm, × 1600. (SEM photomicrograph courtesy of Dr. David Kirk, from Green and Kirk 1981).

Fig. 43. *V. carteri* f. nagariensis. Cells of embryo with multiple bridges. Bar = 5 μm, × 4000. (SEM photomicrograph courtesy of Dr. David Kirk, Green and Kirk 1981)

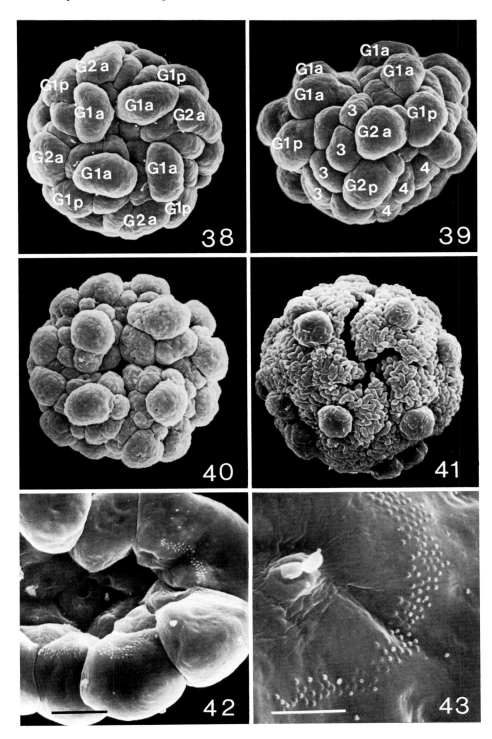

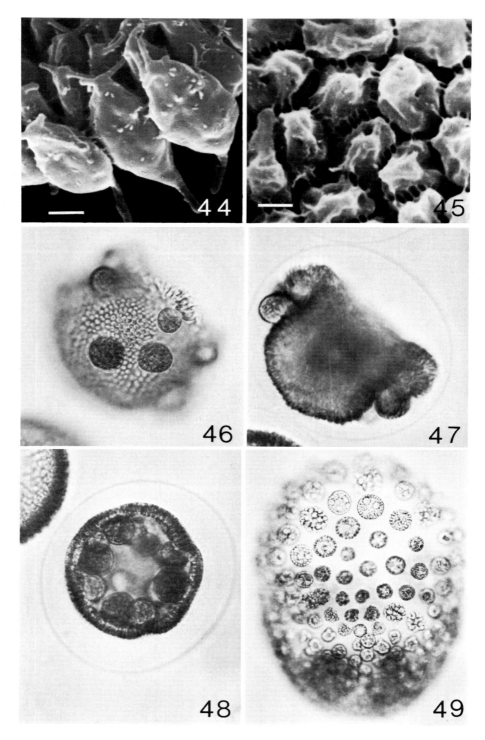

problem by the writer and by GILLES and JAENICKE (1982). Using living material under light microscopy, the successive stages of division in a single embryo could be followed, and again it was concluded that cell lineages showed delimitation of the gonidial initials only from derivatives of tiers #1 and #2 of the 16-celled stage. More recently GREEN and KIRK (1982) have re-evaluated their findings with SEM and now agree with the earlier observations (STARR 1969, 1970) that gonidial initials do indeed arise only from derivatives of tiers #1 and #2.

GILLES and JAENICKE (1982) point out that although the development of 16 gonidia may be expected in the embryogenesis of *V. carteri* f. nagariensis, the full complement is not always realized. The patterns of gonidial arrangement in those spheroids with fewer than 16 gonidia are such as to indicate an intrinsic mechanism which affects the differentiation events rather than a random loss of gonidia. For example, in a spheroid with 13 gonidia, one may predict that 3 of the 4 gonidia derived from the anterior derivatives of tier #2 will be missing. In Fig. 50C GILLES and JAENICKE show the derivation of the 4 gonidia in one quarter of a 64-celled embryo with 16 gonidia. In Fig. 50B one sees that the derivatives from tier #2 of the 16-celled stage have resulted in only a single gonidium rather than two; thus if this had happened in only one of the four quarters, a spheroid with 15 gonidia would have been formed. However, the development in each quarter is independent of that in the other, and so spheroids with 14, 13, or 12 gonidia will result as additional quarters show this loss. In Fig. 50A the derivatives from both tiers #1 and #2 have in each case formed only a single gonidial initial, resulting in a spheroid with only eight gonidia. This extreme reduction is usually seen in populations of *V. carteri* f. nagariensis only under sub-optimal conditions of growth, while in *V. carteri* f. weismannia (KOCHERT 1968) the eight gonidiate condition occurs even in slightly suboptimal conditions, with 12 gonidiate spheroids being the maximum seen in optimal conditions.

◄

Figs. 44, 45. *V. carteri* f. nagariensis. Bar = 1 μm. (SEM photomicrographs courtesy of Dr. DAVID KIRK, from VIAMONTES et al. 1979)

Fig. 44. Side view of cells from an embryo just prior to inversion, showing spindle shapes and cytoplasmic bridges

Fig. 45. Surface view of embryo just prior to inversion. Slight shrinkage of the cells reveals the cytoplasmic bridges

Figs. 46–48. *V. carteri* f. nagariensis. Inversion of the embryo. × 520

Fig. 46. Early inversion, the lips of the phialopore beginning to bend back

Fig. 47. Mid-inversion

Fig. 48. Young embryo just after inversion

Fig. 49. *V. carteri* f. nagariensis. Asexual spheroid carrying somatic regenerator locus. All somatic cells have formed young spheroids. × 40

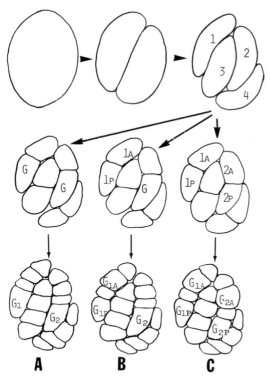

Fig. 50. A–C Diagrams showing the derivation of gonidial initials in one quarter of an embryo of *V. carteri* f. nagariensis. Each quarter divides obliquely (eight celled embryo) followed by another oblique cleavage such that cells #1 and #2 (tiers #1 and #2 in the 16-celled embryo) are derived from different cells of the eight-celled stage. **A** Gonidial initials formed by unequal cleavages of tiers #1 and #2 of the 16-celled stage. **B** Gonidial initials formed by unequal cleavage of tier #2 at division of 16-celled stage and by unequal cleavage of derivatives of tier #1 at division of the 32-celled stage. **C** Gonidial initials formed by unequal cleavage of derivatives of tiers #1 and #2 at division of 32-celled stage. (After GILLES and JAENICKE 1982)

When the pattern seen in Fig. 50C is found, it is said that differentiation occurs at the division of the 32-celled stage, while in Fig. 50A differentiation is said to occur at the division of the 16-celled stage. GILLES and JAENICKE (1982) consider that in the pattern of Fig. 50B differentiation occurs at two different stages, derivatives of tier #2 differentiating at the division of the 16-celled stage, and those derived from tier #1 at the division of the 32-celled stage.

Cleavages of both the gonidial and the somatic initials continue (Fig. 40) until there are, in *V. carteri* f. nagariensis, 2000–4000 cells, at which time cell divisions cease (Fig. 41). Examination of this hollow sphere of cells will show that the chloroplast pole of each cell is to the outside of the sphere, while the nuclear pole (and the site of flagellar extrusion later in development) is towards the inside. Examination with both scanning and transmission electron microscopy shows that the cells are connected by an elaborate cytoplasmic bridge system (Figs. 42–45) (VIAMONTES et al. 1979, GREEN and KIRK 1981, GREEN et al. 1981). The cytoplasmic bridge system is the result of incomplete cytoplasmic cleavage at sites where vesicles fail to fuse. It is present from the first cleavage of the gonidium throughout successive divisions, so that at the end of the divisions the cells are held together by bridges at all sites on the hollow sphere except at its anterior end, where a pair of intersecting slits comprise the phialopore (Fig. 41). This pore has remained more or less obvious throughout development since it was first formed by the upward migration of the top four cells in the formation of the eight-celled stage in early embryogenesis (Fig. 34).

Following cessation of cell division the process of inversion occurs by which the embryo turns inside out (Figs. 46–48). Although *Volvox* has been known for nearly 300 years, it was not until this century that inversion was first described. POCOCK (1933) pointed out that POWERS (1908) was the first to be aware of the process, but he failed to interpret correctly the sequence of events. KUSCHAKEWITSCH (1922, republished 1931) described the process in its entirety, showing how the spherical embryo turned inside out. POCOCK (1933) gave a detailed account of inversion as observed with light microscopy, noting the different shapes of cell in various regions of the inverting embryo. More recently Dr. DAVID KIRK and his associates (VIAMONTES and KIRK 1977, VIAMONTES et al. 1979, GREEN and KIRK 1981, GREEN et al. 1981, KIRK et al. 1982) have studied inversion in *V. carteri* using both TEM and and SEM, and they have proposed that the movements during inversion are driven by the changes in cell shape. The cytoplasmic bridge system which they describe provides the structural integrity needed for the movements of the cell layer. FULTON (1978b) and MARCHANT (1977a) have described colony formation, including inversion, in *Pandorina* and *Eudorina*, respectively; they also demonstrated the presence of basal cell connections during the movements in inversion.

After inversion has been completed, the colonial matrix is secreted. FULTON (1978a) has described this process in *Pandorina*, but as yet this has not been done in *Volvox*. As in *Pandorina*, the cells of *V. carteri* do not remain joined by cytoplasmic bridges after inversion, and so the colonial matrix insures the integrity of the colony. In a "dissociator" mutant of *V. carteri* (HUSKEY et al. 1979, DAUWALDER et al. 1980) no matrix material is secreted and so after inversion has been completed and the cytoplasmic connections have broken, the embryo dissociates into single cells. Even normal spheroids of *V. carteri* can be dissociated into single cells, and the gonidia and somatic cells may be separated for use in experimental procedures (KOCHERT 1975). Gonidia removed even at an early stage of enlargement will continue to enlarge and undergo embryogenesis in a normal manner. This permitted the time-lapse photography of the process of embryogenesis in the film on *Volvox carteri* by STARR and FLATEN[1].

---

1 This color film with English narration is available in two versions (15 min and 25 min) from the Indiana University Audio-Visual Center, Bloomington, Indiana 47401, on a sale or rental basis

Various mutants which affect colony and cell differentiation in *V. carteri* have been described by Starr (1970) and Sessoms and Huskey (1973). More recently, these and other mutant loci have served in a genetic investigation by Huskey and his associates (Huskey et al. 1979). One of the most interesting of these nuclear genetic lesions results in the somatic cells becoming capable of enlarging and functioning reproductively, both asexually and sexually (Fig. 49).

Although most species of *Volvox* do not show in their embryogeny the striking formation of the reproductive cells by the unequal cell divisions so obvious in *V. carteri*, it is important to note that the number and arrangement of the reproductive cells in the spheroids of all species are very predictable and characteristic of the species. Thus it is evident that a differentiating mechanism is operating in these species as in *V. carteri*. One might question whether the unequal cleavage is truly the differentiating step in *V. carteri*, inasmuch as Starr (1970) showed that, in male spheroids carrying the mutant locus which results in somatic cells being capable of enlarging and acting reproductively, even the somatic cells of the male form sperm packets. Thus in this case the unequal cleavages in the male gave rise to the expected two different sizes of cells; yet both were capable of forming sperm. The unequal cleavages in *V. carteri* may be a secondary development whereby a large amount of cellular material is allocated to the reproductive initials, the primary biochemical events of differentiation being common to all species.

Little is known of the biochemical basis of differentiation in *Volvox*, although the induction of sexual morphogenesis by chemical inducers has been studied in a variety of species: *V. aureus* (Darden 1966), *V. carteri* f. weismannia (Kochert 1968, 1975, Kochert and Yates 1974), *V. carteri* f. nagariensis (Starr 1969, 1970, Starr and Jaenicke 1974), *V. rousseletii* (McCracken and Starr 1970), *V. gigas* (Vande Berg and Starr 1971), *V. obversus* (Karn et al. 1974), and *V. capensis* (Miller and Starr 1981). In *V. capensis* the inducer has been identified as L-glutamic acid, effective at very low concentrations (6.8×10^{-8} M), produced during population growth as a by-product of the digestion of the parental matrix as the young individuals escape. In all other species the inducers are species-specific glycoproteins produced by the sperm (Gilles et al. 1981). Kochert (1981) has pointed out that we know little about the mode of action of any *Volvox* inducer nor do we know of any "biochemical parameter to measure which gives us an indication of primary effects".

In *V. carteri* the control of the differentiating divisions has been attributed to unidentified morphogenetic substances (Kochert and Yates 1970), to a change in certain qualities of the cell membrane (Wenzl and Sumper 1981), and to cell size (Pall 1975). It would appear that cell size is important in activating the differentiating division, as can be seen in the varying times of differentiation in embryos with less than 16 gonidia (Gilles and Jaenicke 1982), but the ability of particular cells to respond must lie in some other, as yet unknown, quality. Coleman and Maguire (1982) have observed that in somatic cells of *V. carteri* there is no further replication of the chloroplast DNA, but whether this is the cause or the effect of somatic cell differentiation is unknown. *Volvox* and several species of *Pleodorina* (Figs. 24, 25) are unique in having a developmental pattern such that from 10% (*P. illinoisensis*) to more than 99% (*Volvox*) of the cells produced in the embryogeny of the colony are incapable of significant growth and reproduction, yet they apparently serve no unique, essential role in the survival of the colony. It may well be incorrect to homologize these somatic cells in *Volvox* with the somatic cells in animals which continue to grow and divide during the lifetime of the individual; thus the events of ageing in somatic cells of *Volvox* may not be comparable to ageing in the somatic cells of animals, as is implied in the study by HAGEN and KOCHERT (1980).

Various aspects of the sexual phenomena, including mating-type isolation, in the colonial green flagellates of the *Gonium-Volvox* series are the subject of a masterful review by COLEMAN (1979).

13.3.1.2 Family Astrephomenaceae

In contrast to the members of the *Gonium–Volvox* series, in which a complex inversion process is needed to complete development of the colony, the colonial green spheres of *Astrephomene* (Fig. 30) (POCOCK 1953) develop without inversion. These coenobic colonies are composed of 128 or fewer hemispherical cells, each surrounded by it own matrix; there is no common matrix surrounding the colony as is found in the *Gonium–Volvox* series. Most of the cells are the same size and are capable of reproducing both asexually and sexually, but there may be two to seven small cells in the posterior of the colony which are sterile. Colony formation has been studied with light and electron microscopy (HOOPS and FLOYD 1982).

It is evident as early as the eight-celled stage that the pattern of cleavage in *Astrephomene* (Fig. 51 A–D) is different from that in *Volvox*. Furthermore, the cells in the developing embryo are contiguous near their bases and angle outward rather than inward as in *Volvox*. As the cell number increases, the plate of cells continues to bend back so that at the end of cell division the embryo already has the flagella poles of its cells pointing outward. *Astrephomene* is placed in a monotypic family, the Astrephomenaceae, by some authors.

13.3.1.3 Family Spondylomoraceae

A third coenobic family, the Spondylomoraceae, includes those colonial green flagellates in which the cells in a colony are arranged so as to form grapelike clusters. The most common genus, *Pyrobotrys* (as *Chlamydobotrys* in most of the literature) (Fig. 52) has 8 to 16 biflagellate cells. *Pascherina* with 4 biflagellate cells (KORSCHIKOFF 1928) and *Spondylomorum* with 8 to 16 quadriflagellate cells are rarely seen. The cells in the colonies are arranged in alternating tiers, but a detailed study of colony formation has not been made (SCHULZE 1927, PRINGSHEIM 1960).

13.3.1.4 Family Haematococcaceae

This family is composed of only two genera, the unicellular *Haematococcus* and the colonial *Stephanosphaera*. The beautiful colonies of *Stephanosphaera*

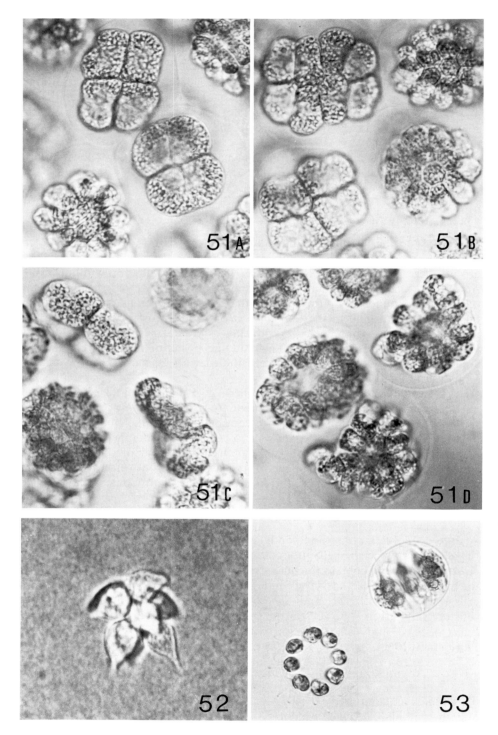

(Fig. 53) with eight *Haematococcus*-like cells arranged in a ring and enclosed by a gelatinous matrix are rarely encountered but can be easily grown in culture. A definitive study of colony formation has not been made, but it is of interest that the first division in colony formation has been reported to be transverse rather than longitudinal as in the *Gonium–Volvox* series (DROOP 1956).

13.3.2 Non-Motile Coenobic Colonies

These colonies are restricted to genera in the order Chlorococcales of the green algae and may be separated into two groups on the basis of whether or not a motile spore stage exists in the ontogeny of the colony. The arrangement of the cells in the colony is not indicative of its formation by motile or non-motile cells, but depends on other factors such as cleavage pattern, the method of cell aggregation, the pattern of cell maturation, and, in some genera, environmental factors, especially nutrition.

13.3.2.1 Reproduction by Non-Motile Spores

In those genera in which motile spores are not a part of colony ontogeny, the cells, either longer than broad or isodiametric, are usually arranged in colonies of four or eight cells; in a few species of *Scenedesmus* and *Coelastrum* there may be 16 cells. In some genera the four- or eight-celled coenobia formed in successive generations may form larger aggregates ("non-coenobic colonies") by being retained in gelatinous matrices (*Crucigenia*, Fig. 55; *Pectodictyon*, Fig. 59). Colonies may have the cells radiating from a common point (*Actinastrum*, Fig. 57), in a line side-by-side (*Scenedesmus*, Fig. 58), in a hollow sphere (*Coelastrum*, Fig. 54), in a flat plate (*Crucigenia*, Fig. 55), in a group of four elongate cells in two tiers (*Tetradesmus*, Fig. 56), and in a cube (*Pectodictyon*, Fig. 59).

The most widely investigated genus of this group is *Scenedesmus*, which occurs in a great variety of forms as a regular component of the phytoplankton in freshwater. Its ease of cultivation made it an early subject for culture studies, but the changes in morphology evoked under different environmental regimes convinced some investigators that the morphology and life histories of algae grown in culture were at best suspect (FRITSCH 1935 pp. 181–182). The branched or unbranched chains of fusiform cells described as *Dactylococcus* by NAEGELI (1849) were considered to be a stage in the development of *Scenedesmus* by GRUNOW (1858), and later workers (CHODAT 1926, GRINTZESCO 1902a, 1902b)

Fig. 51A–D *Astrephomene gubernaculifera*. Various stages in cleavage of cells to form new colonies. Note that in early stages the plate of cells bends back on itself; thus the embryo is formed with the flagellar ends pointing outward. × 900

Fig. 52. *Pyrobotrys* sp. × 825

Fig. 53. *Stephanosphaera* sp. × 250

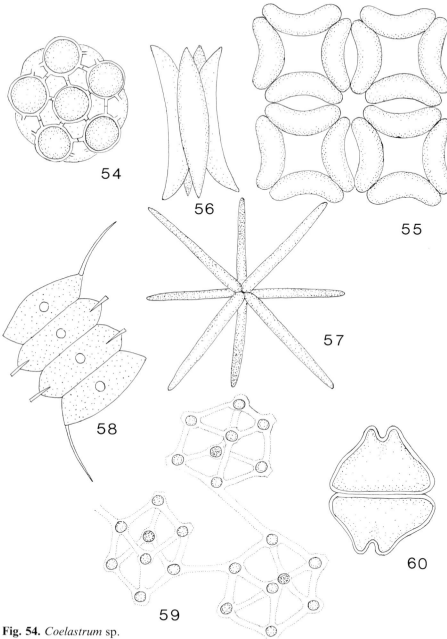

Fig. 54. *Coelastrum* sp.

Fig. 55. *Crucigenia* sp.

Fig. 56. *Tetradesmus* sp.

Fig. 57. *Actinastrum* sp.

Fig. 58. *Scenedesmus* sp.

Fig. 59. *Pectodictyon* sp.

Fig. 60. *Euastropsis* sp.

did, indeed, find such stages in their cultures of *Scenedesmus*. SMITH'S (1914c) study of *Scenedesmus* produced no *Dactylococcus* stages, and his successful cultivation of an isolate of *Dactylococcus* from a natural collection convinced him they were two distinct genera. The more recent investigations of F.R. TRAINOR and his associates on *Scenedesmus* have shown that *Dactylococcus* stages can be evoked in some, but not all, species by growing them in media of varying composition (TRAINOR 1963, 1964), while in others single-celled forms identical with species of *Chodatella* and *Franceia* were formed (TRAINOR 1971). TRAINOR and SHUBERT (1974) developed a series of media containing various salt concentrations and showed that some strains of *Scenedesmus* would form colonies only in a medium of extreme dilution, with nitrogen and phosphorus levels similar to those found in natural waters.

Although the process of colony formation in *Scenedesmus* had been studied earlier (NAEGELI 1849, GRINTZESCO 1902a, b), SMITH (1916a) provided the most detailed account, based on studies of the material in culture. Each cell in a coenobium forms a new coenobium independently of the others, so it is not unusual to find some cells dividing before others in the same coenobium. Following nuclear division, there is a transverse division of the cytoplasm; this is followed by a second nuclear division and a second cytoplasmic division, this time at right angles to the first cleavage plane. The uninucleate protoplasts then elongate until they touch the ends of the parental cell wall. The young colony is liberated through a longitudinal split in the parental cell wall, and the packet of four cells unrolls to form the flat colony typical of the genus. Eight- and 16-celled colonies are formed by having the protoplast undergo further divisions prior to the elongation and liberation of the young colony. PICKETT-HEAPS (1975), in a beautiful SEM and TEM study of *Scenedesmus quadricauda,* has shown that the cells in the colony are held together by adhesive sites on their contiguous walls. The remains of the adhesive sites are probably the bridges described by GRINTZESCO (1902a, b) and TRAINOR (1963) in the *Dactylococcus* stage of the genus. The complex fine structure of the surface and its ornamentations of spines, bristles and warts has been studied by a number of investigators (BISALPUTRA and WEIER 1963, BISALPUTRA et al. 1964, HIGHAM and BISALPUTRA 1970, PICKETT-HEAPS and STAEHELIN 1975, STAEHELIN and PICKETT-HEAPS 1975).

13.3.2.2 Reproduction by Motile Spores

The formation of motile spores in the ontogeny of non-motile coenobic colonies is seen in four genera. The two-celled colonies of *Euastropsis* (Fig. 60), the flat plates of *Pediastrum* (Fig. 61), the colonies with radiating cells in *Sorastrum* (Fig. 62), and the netlike colonies of *Hydrodictyon* (Fig. 65) would at first appear to have been formed by radically different methods, but as MARCHANT and PICKETT-HEAPS (1974) have noted in their studies of the last three genera, the principle involving adhesion of the zoospores at predictable sites on their surface is common to all.

The first step in the formation of a new colony is the production of motile spores by a cell; as in *Scenedesmus* all cells of a colony do not necessarily

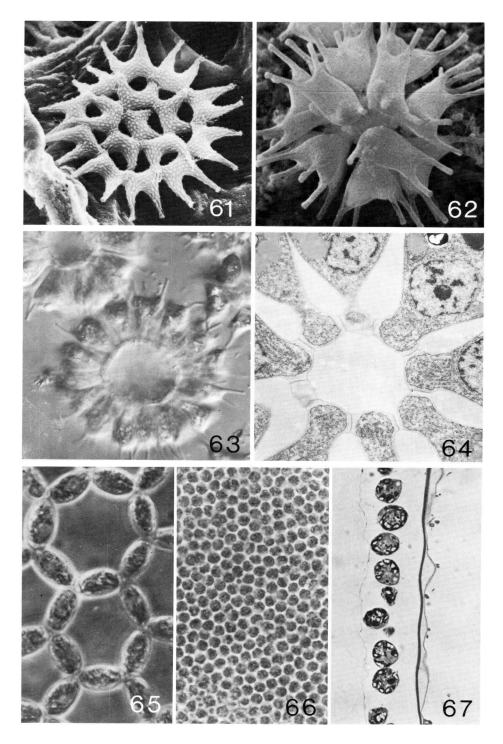

reproduce at the same time. In *Euastropsis* this is accomplished by successive bipartition, a cytoplasmic cleavage occurring after each mitosis (SMITH 1950). In *Sorastrum* there is a series of mitotic divisions prior to the beginning of cleavage, but mitosis continues in the cleaving protoplast (MARCHANT 1974). In *Pediastrum* and *Hydrodictyon* there is a progressive cleavage of a multinucleated protoplast, this condition having arisen by multiple mitoses in the uninucleate cells of *Pediastrum* at the onset of sporogenesis (SMITH 1916b) and during cell growth and enlargement in *Hydrodictyon* (POCOCK 1960).

In *Euastropsis, Pediastrum* and *Sorastrum* the motile spores are released from the parental cell by a rupture in the wall, but they remain enclosed in a thin vesicle formed from the innermost layer of the parental wall. The cells swim within the vesicle and then aggregate to form the colonies typical of the genus. In *Hydrodictyon* the motile spores are formed as a single layer sandwiched between the parental cell wall and the vacuolar membrane which remains intact (Figs. 66, 67). The spores are not released from the parental cell and thus are restricted by the wall and the vacuolar membrane, aggregating as a single layer of cells in the form of the cell from which they were produced, i.e. a closed cylinder. MARCHANT and PICKETT-HEAPS (1974) list the following basic similarities of zoospore aggregation in the genera *Pediastrum, Sorastrum* and *Hydrodictyon;* it is reasonable to expect a similar situation in *Euastropsis* when material is available for study:

A) There are peripheral bands of microtubules in the zoospores which underlie the initial sites of contact during aggregation. In *Pediastrum* and in *Hydrodictyon,* the latter in effect a flat plate, these microtubules are oriented predominantly in the plane of the developing colonies.

────────

Figs. 61–67. Through the courtesy of Dr. HARVEY MARCHANT; *Pediastrum* from MARCHANT 1974; *Sorastrum* and *Hydrodictyon* from PICKETT-HEAPS 1975

Fig. 61. *Pediastrum boryanum.* SEM photomicrograph of colony. × 1870

Fig. 62. *Sorastrum* sp. SEM photomicrograph of colony. × 2130

Fig. 63. *Sorastrum* sp. Optical section of young colony showing cells adhering at bases to form radiating colony. × 2000

Fig. 64. *Sorastrum* sp. TEM photomicrograph of colony to show attachment of cell bases. × 9690

Fig. 65. *Hydrodictyon reticulatum.* Portion of young colony to show cells arranged in form of net. × 1200. (Two films, with English narration, of the vegetative and generative life-cycles of *Hydrodictyon reticulatum* are available on sale or rental basis from the Institute for the Scientific Film (IWF), D-3400 Göttingen; nrs. 1042 and 1043)

Fig. 66. *H. reticulatum.* Portion of cell undergoing asexual reproduction, showing zoospores in layer at cell surface. × 700

Fig. 67. *H. reticulatum.* TEM photomicrograph of cell such as that in Fig. 66, showing a single layer of zoospores sandwiched between the cell wall and the vacuolar membrane. × 1200

B) Zoospores change their shape prior to aggregation; in *Pediastrum* and *Hydrodictyon* they become rhomboidal or nearly rectangular while in *Sorastrum* they are somewhat conical.

C) Zoospores link up at generally predictable sites on their surface.

D) The arrangement of the cells in a developing colony is set before flagella are retracted by the zoospores.

E) In *Pediastrum* and *Sorastrum* the horns develop at certain sites on the cells if such sites are not in contact with other cells.

The association of the microtubules with the contact sites on the aggregating zoospores is not understood, but colchicine-treated zoospores do not develop microtubules, do not change shape, and do not aggregate as expected.

The basic steps in the formation of colonies from motile zoospores as outlined above are based on observations from many investigators before 1974 and reconfirmed since then: *Pediastrum* (Braun 1851, Smith 1916b, Harper 1918, Davis 1964, Gawlik and Millington 1969, Hawkins and Leedale 1971, Honda 1973, Pickett-Heaps 1975, Marchant 1974, 1979, Rogalski et al. 1977, Millington 1981). *Sorastrum* (Davis 1966, Marchant and Pickett-Heaps 1974, Pickett-Heaps 1975). *Hydrodictyon* (Artari 1890, Klebs 1890, 1891, Timberlake 1902, Palik 1928, Pocock 1960, Marchant and Pickett-Heaps 1970, 1971, 1972a, 1972b, 1972c, 1972d, 1974, Hawkins and Leedale 1971, Pickett-Heaps 1975).

13.4 Summary

The great array of colony shapes seen in the algae are the products of relatively few basic patterns of morphogenesis. The formation of non-coenobic colonies, both motile and non-motile, results from the retention of the successive products of spore formation or cell division, usually by some secreted polysaccharide. The variation in colony appearance is due usually either to cell shape or to the directions of the cleavage planes which formed the cells. In motile coenobic colonies, the different groups reflect a few basic patterns of cleavage, with cell number, cell differentiation, and patterns of sexual reproduction serving as the basis for generic distinction. In non-motile coenobic colonies particular sites of cell adhesion in the non-motile spores forming certain genera, and in the motile zoospores aggregating to form other genera, constitute a common basis for colony formation.

Acknowledgements. I wish to thank Drs. David Kirk and Harvey Marchant for furnishing excellent photographs as indicated elsewhere. I would also like to thank Dr. Harold C. Bold for his helpful comments on the manuscript and Mr. John Allensworth for his help with the illustrations.

References

Artari A (1890) Zur Entwicklungsgeschichte des Wassernetzes *Hydrodictyon reticulatum* Roth. Bull Soc Imp Nat Moscou, NS 4:269–287

Bisalputra T, Weier TE (1963) The cell wall of *Scenedesmus quadricauda*. Am J Bot 50:1011–1019

Bisalputra T, Weier TE, Risley E, Engelbrecht A (1964) The pectic layer of the cell wall of *Scenedesmus quadricauda*. Am J Bot 51:548–551

Bold HC, Wynne MJ (1978) Introduction to the algae. Prentice-Hall, New Jersey

Bourrelly P (1966) Les algues d'eau douce. Tome I. Les algues vertes. Boubée, Paris

Bourrelly P (1968) Les algues d'eau douce. Tome II. Les algues jaunes et brunes, Chryso-phycées, Pheophycées, Xanthophycées et Diatomées. Boubée, Paris

Bourrelly P (1970) Les algues d'eau douce. Tome III. Les algues bleues et rouges, les Eugleniens, Peridiniens, et Cryptomonadines. Boubée, Paris

Braun A (1851) Erscheinungen der Verjüngung in der Natur. Freiburg

Carefoot JR (1966) Sexual reproduction and intercrossing in *Volvulina steinii*. J Phycol 2:150–156

Chodat R (1926) *Scenedesmus*. Etude de génétique, de systématique expérimentale et d'hydrobiologie. Rev Hydrobiol 3:71–258

Coleman AW (1959) Sexual isolation in *Pandorina morum*. J Protozool 6:249–264

Coleman AW (1979) Sexuality in the colonial flagellates. In: Hutner SH, Levandowsky M (eds) Physiology and biochemistry of the Protozoa, Vol 1, 2nd edn. Academic Press, London New York, pp 307–340

Coleman AW, Maguire MJ (1982) A microspectrofluorometric analysis of nuclear and chloroplast DNA in *Volvox*. Dev Biol 94:441–450

Darden WH (1966) Sexual differentiation in *Volvox aureus*. J Protozool 13:239–255

Dauwalder M, Whaley WG, Starr RC (1980) Differentiation and secretion in *Volvox*. J Ultrastruct Res 70:318–335

Davis JS (1964) Colony form in *Pediastrum*. Bot Gaz 125:129–131

Davis JS (1966) Akinetes, reproduction and colony form in the green alga *Sorastrum*. Trans Ill State Acad Sci 59:275–280

Droop MR (1956) *Haematococcus pluvialis* and its allies. I. The Sphaerellaceae. Rev Algol NS II:53–71

Fott B (1972) Chlorophyceae (Grünalgen) Ordnung: Tetrasporales. In: Huber-Pestalozzi (ed) Das Phytoplankton des Süsswassers. Binnengewässer Bd XVI(6):1–116 (with 47 plates)

Fritsch FE (1935) The structure and reproduction of the algae. Vol I. Cambridge Univ Press, Cambridge London New York Melbourne

Fulton AB (1978a) Colonial development in *Pandorina morum*. I. Structure and composition of the extracellular matrix. Dev Biol 64:224–235

Fulton AB (1978b) Colonial development in *Pandorina morum*. II. Colony morphogenesis and formation of the extracellular matrix. Dev Biol 64:236–251

Gawlik SR, Millington WF (1969) Pattern formation and the fine structure of the developing cell wall in colonies of *Pediastrum boryanum*. Am J Bot 56:1084–1093

Geitler L (1932) Cyanophyceae, in L. Rabenhorst, Kryptogamen-Flora von Deutschland, Österreich und der Schweiz Vol. 14. Akademische Verlagsges, Leipzig, pp 14:1–1056

Gerisch G (1959) Die Zelldifferenzierung bei *Pleodorina californica* Shaw und die Orga-nisation der Phytomonadinenkolonien. Arch Protistenkd 104:292–358

Gilles R, Jaenicke L (1982) Differentiation in *Volvox carteri*: Study of pattern variation of reproductive cells. Z Naturforsch 37c:1023–1030

Gilles R, Bittner C, Jaenicke L (1981) Site and time of formation of the sex-inducing glycoprotein in *Volvox carteri*. FEBS Lett 124:57–61

Goldstein ME (1964) Speciation and mating behavior in *Eudorina*. J Protozool 11:317–344

Green KJ, Kirk DL (1981) Cleavage patterns, cell lineages and development of a cytoplas-mic bridge system in *Volvox* embryos. J Cell Biol 91:743–755

Green KJ, Kirk DL (1982) A revision of the cell lineages recently reported for *Volvox carteri* embryos. J Cell Biol 94:741–742

Green KJ, Viamontes GI, Kirk DL (1981) Mechanism of formation, ultrastructure and function of the cytoplasmic bridge system during morphogenesis in *Volvox*. J Cell Biol 91:756–769

Grintzesco J (1902a) Recherches expérimentales sur la morphologie et la physiologie de *Scenedesmus acutus* Meyen. Bull Herb Boiss II:217–264

Grintzesco J (1902b) Recherches expérimentales sur la morphologie et la physiologie de *Scenedesmus acutus* Meyen. Bull Herb Boiss II:406–429

Grunow A (1858) Die Desmidiaceen und Pediastreen einiger österreichischen Moore, nebst einigen Bemerkungen über beide Familien im Allgemeinen. Verh Zool Bot Ges Wien 8:489–502

Hagen G, Kochert G (1980) Protein synthesis in a new system for the study of senescence. Exp Cell Res 127:451–457

Harper RA (1918) Organization, reproduction, and inheritance in *Pediastrum*. Proc Am Phil Soc 57:375–438

Harris DO, Starr RC (1969) Life history and physiology of reproduction of *Platydorina caudata*. Arch Protistenkd 111:138–155

Hawkins AF, Leedale GF (1971) Zoospore structure and colony formation in *Pediastrum* spp. and *Hydrodictyon reticulatum* (L.) Lagerheim. Ann Bot 35:201–211

Higham MT, Bisalputra T (1970) A further note on the surface structure of *Scenedesmus* coenobium. Can J Bot 48:1839–1841

Honda H (1973) Pattern formation of the coenobial alga *Pediastrum biwae* Negoro. J Theor Biol 42:461–481

Hoops HJ, Floyd GL (1982) Mitosis, cytokinesis and colony formation in the colonial green alga *Astrephomene gubernaculifera*. Br Phycol J 17:297–310

Huskey RJ, Griffin BE, Cecil PO, Callahan AM (1979) A preliminary genetic investigation of *Volvox carteri*. Genetics 91:229–244

Janet C (1923) Le *Volvox*, troisième mémoire. Macon, Protat Freres, Paris

Jaenicke L, Waffenschmidt S (1979) Matrix-lysis and release of daughter spheroids in *Volvox carteri* – a proteolytic process. FEBS Lett 107:250–253

Jaenicke L, Waffenschmidt S (1981) Liberation of reproductive units in *Volvox* and *Chlamydomonas*: Proteolytic processes. Ber Dtsch Bot Ges 94:375–386

Karn RC, Starr RC, Hudock GA (1974) Sexual and asexual differentiation in *Volvox obversus* (Shaw) Printz, strains Wd3 and Wd7. Arch Protistenkd 116:142–148

Kirk DL, Viamontes GI, Green KJ, Bryant JL Jr (1982) Integrated morphogenetic behavior of cell sheets: *Volvox* as a model. In: Subtelny S, Green PB (eds) Developmental order: Its origin and regulation. 40th Symposium Soc Dev Biol Liss, New York, pp 247–274

Klebs G (1890) Über die Vermehrung von *Hydrodictyon utriculatum*. Ein Betrag zur Physiologie der Fortpflanzung. Flora 73:351–410

Klebs G (1981) Über die Bildung der Fortpflanzungszellen bei *Hydrodictyon utriculatum* Roth. Bot Zeit 49:789–798, 805–817, 821–835, 837–846, 853–862

Kochert G (1968) Differentiation of reproductive cells in *Volvox carteri*. J Protozool 15:438–452

Kochert G (1975) Developmental mechanisms in *Volvox* reproduction. Dev Biol Suppl 8:55–90

Kochert G (1981) Sexual pheromones in *Volvox* develpment. In: O'Day DH, Hargen PA (eds) Sexual interactions in eukaryotic microbes. Academic Press, London New York, pp 73–93

Kochert G, Yates I (1970) A UV-labile morphogenetic substance in *Volvox carteri*. Dev Biol Suppl 23:128–135

Kochert G, Yates I (1974) Purification and partial characterization of a glycoprotein sexual inducer from *Volvox carteri*. Proc Natl Acad Sci USA 71:1211–1214

Korschikoff AA (1928) On two new Spondylomoraceae: *Pascheriella tetras* n. gen. et sp., and *Chlamydobotrys squarrosa* n. sp. Arch Protistenkd 61:223–238

Kuschakewitsch K (1922) Zur Kenntis der Entwicklungsgeschichte von *Volvox*. Bull Acad Sci Oukraine 1:31–36 (reprinted in Arch Protistenkd 73:323–330 1931)

Lamport DTA (1974) The role of hydroxyproline-rich proteins in the extracellular matrix of plants. In: Hay ED, King TJ, Papaconstantinou J (eds) Macromolecules regulating growth and development. Academic Press, London New York, pp 113–130

Mackie W, Preston RD (1974) Cell wall and intercellular region polysaccharides. In:

Stewart WDP (ed) Algal physiology and biochemistry, Botanical Monographs Vol 10. Blackwell, Oxford, pp 40–85

Marchant HJ (1974) Mitosis, cytokinesis, and colony formation in the green alga *Sorastrum*. J Phycol 10:107–120

Marchant HJ (1977a) Colony formation and inversion in the green alga *Eudorina elegans*. Protoplasma 93:325–339

Marchant HJ (1979) Microtubular determination of cell shape during colony formation by the alga *Pediastrum*. Protoplasma 98:1–14

Marchant HJ, Pickett-Heaps J (1970) Ultrastructure and differentiation of *Hydrodictyon reticulatum*. I. Mitosis in the coenobium. Aust J Biol Sci 23:1173–1186

Marchant HJ, Pickett-Heaps J (1971) Ultrastructure and differentiation of *Hydrodictyon reticulatum*. II. Formation of zooids within the coenobium. Aust J Biol Sci 24:471–486

Marchant HJ, Pickett-Heaps J (1972a) Ultrastructure and differentiation of *Hydrodictyon reticulatum*. III. Formation of the vegetative daughter net. Aust J Biol Sci 25:265–278

Marchant HJ, Pickett-Heaps J (1972b) Ultrastructure and differentiation of *Hydrodictyon reticulatum*. IV. Conjugation of gametes and the development of zygospores and azygospores. Aust J Biol Sci 25:279–291

Marchant HJ, Pickett-Heaps J (1972c) Ultrastructure and differentiation of *Hydrodictyon reticulatum*. V. Development of polyhedra. Aust J Biol Sci 25:1187–1197

Marchant HJ, Pickett-Heaps J (1972d) Ultrastructure and differentiation of *Hydrodictyon reticulatum*. VI. Formation of the germ net. Aust J Biol Sci 25:1199–1213

Marchant HJ, Pickett-Heaps J (1974) The effect of colchicine on colony formation in the algae *Hydrodictyon, Pediastrum,* and *Sorastrum*. Planta 116:291–300

McCracken MD, Starr RC (1970) Induction and development of reproductive cells in the K-32 strains of *Volvox rousseletii*. Arch Protistenkd 112:262–282

Miller CE, Starr RC (1981) The control of sexual morphogenesis in *Volvox capensis*. Ber Dtsch Bot Ges 94:357–372

Millington WF (1981) Form and pattern in *Pediastrum*. In: Kiermayer O (ed) Cytomorphogenesis in plants, Cell biol monographs Vol 8. Springer, Berlin Heidelberg New York, pp 99–118

Naegeli C (1849) Gattungen einzelliger Algen, physiologisch und systematisch bearbeitet. Neue Denkschr Allg Schweiz Ges Naturw 10:1–139

Palik P (1928) *Hydrodictyon*-Studien. Mat Nat Anz Ung Akad Wiss 45:20–47

Pall ML (1975) Mutants of *Volvox* showing premature cessation of division: Evidence for a relationship between cell size and reproductive cell differentiation. In: Macmahon D, Fox CF (eds) Developmental biology: Pattern formation, gene regulation. ICN-UCLA Symposia on Molecular and Cellular Biology Vol 2. Benjamin, Menlo Park, CA, pp 148–156

Pickett-Heaps JD (1975) Green algae. Structure, reproduction, and evolution in selected genera. Sinauer, Sunderland

Pickett-Heaps JD, Staehelin LA (1975) The ultrastructure of *Scenedesmus* (Chlorophyceae). II. Cell division and colony formation. J Phycol 11:186–202

Pocock MA (1933) *Volvox* in South Africa. Ann S Afr Mus 16:523–646

Pocock MA (1953) Two multicellular motile green algae, *Volvulina* Playfair and *Astrephomene,* a new genus. Trans Roy Soc S Afr 4:103–127

Pocock MA (1955) Studies in the North American Volvocales. I. The genus *Gonium*. Madrono 13:49–64

Pocock MA (1960) *Hydrodictyon:* a comparative biological study. J S Afr Bot XXVI (pts III & IV) 167–327

Powers JH (1908) Further studies in *Volvox,* with descriptions of three new species. Tran Am Microsc Soc 28:141–175

Pringsheim EG (1960) Zur Systematik und Physiologie der Spondylomoraceen. Österr Bot Z 107:425–438

Rogalski AA, Overton J, Ruddat M (1977) An ultrastructural and cytochemical investigation of the colonial green alga *Pediastrum* during zoospore formation. Protoplasma 91:93–106

Schulze B (1927) Zur Kenntnis einiger Volvocales (*Chlorogonium, Haematococcus, Stephanosphaera,* Spondylomoraceae, and *Chlorobrachis*). Arch Protistenkd 58:508–576

Sessoms AH, Huskey RJ (1973) Genetic control of development in *Volvox:* Isolation and characterization of morphogentic mutants. Proc Natl Acad Sci USA 70:1335–1338

Smith GM (1914) The cell structure and colony formation in *Scenedesmus.* Arch Protistenkd 32:278–297

Smith GM (1916a) A monograph of the algal genus *Scenedesmus* based upon pure culture studies. Trans Wis Acad Sci 18:422–530

Smith GM (1916b) Cytological studies in the Protococcales II. Cell structure and zoospore formation in *Pediastrum boryanum.* (Turp.) Menegh. Ann Bot 30:467–479

Smith GM (1944) A comparative study of the species of *Volvox.* Trans Am Microsc Soc 63:265–310

Smith GM (1950) Freshwater algae of the United States, 2nd edn. McGraw-Hill, New York

Staehelin LA, Pickett-Heaps JD (1975) The ultrastructure of *Scenedesmus* (Chlorophyceae). I. Species with the "reticulate" or "warty" type of ornamental layer. J Phycol 11:163–185

Starr RC (1969) Structure, reproduction and differentiation in *Volvox carteri* f. nagariensis Iyengar, strains HK9 and 10. Arch Protistenkd 111:204–222

Starr RC (1970) Control of differentiation in *Volvox.* Dev Biol Suppl 4:59–100

Starr RC (1980) Colonial chlorophytes. In: Cox E (ed) Phytoflagellates. Elsevier/North Holland, Amsterdam New York, pp 147–163

Starr RC, Jaenicke L (1974) Purification and characterization of the hormone initiating sexual morphogenesis in *Volvox carteri* f. nagariensis Iyengar. Proc Natl Acad Sci USA 71:1050–1054

Stein JR (1958) A morphologic and genetic study of *Gonium pectorale.* Am J Bot 45:664–672

Thompson RH (1951) *Schizodictyon,* a new genus in the Palmellaceae. Am J Bot 38:780–783

Timberlake HG (1902) Development and structure of the swarmspores of *Hydrodictyon.* Trans Wis Acad Sci 13:486–522

Trainor FR (1963) The occurrence of a *Dactylococcus*-like stage in an axenic culture of a *Scenedesmus.* Can J Bot 41:967–968

Trainor FR (1964) The effect of composition of the medium on morphology in *Scenedesmus obliquus.* Can J Bot 42:515–518

Trainor FR (1971) Development of form in *Scenedesmus.* In: Parker BC, Brown RM (eds) Contributions in phycology. Allen, Lawrence, Kansas, pp 81–92

Trainor FR, Shubert LE (1974) *Scenedesmus* morphogenesis. Colony control in dilute media. J Phycol 10:28–30

Vande Berg WJ, Starr RC (1971) Structure, reproduction and differentiation in *Volvox gigas* and *Volvox powersii.* Arch Protistenkd 113:195–219

Viamontes GI, Kirk DL (1977) Cell shape changes and the mechanism of inversion in *Volvox.* J Cell Biol 75:719–730

Viamontes GI, Fochtmann LJ, Kirk DL (1979) Morphogenesis in *Volvox:* Analysis of critical variables. Cell 17:537–550

Wenzl S, Sumper M (1981) Sulfation of a cell surface glycoprotein correlates with the developmental program during embryogenesis of *Volvox carteri.* Proc Natl Acad Sci USA 78:3716–3720

14 Cellular Interaction in Plasmodial Slime Moulds

J. A. M. Schrauwen

14.1 Introduction

During this century progress in biology has been greatly facilitated by the use of model organisms which may be relatively unimportant in nature but which are good systems for the study of specific problems. The interaction between genetically different cells is a fundamental biological problem (e.g. transplantation of organs, blood transfusion, grafting of plants), and an excellent model organism for studying such problems is the myxomycete *Physarum polycephalum*, which can be grown rapidly on simple media in the laboratory; biochemical and genetical work on *P. polycephalum* has been carried out in many laboratories. Genetical work has also been carried out with a second species, *Didymium iridis*, which, however, has not been grown in pure culture on soluble media and therefore is less suitable for physiological and biochemical investigations. The life cycle and features of myxomycetes have been described extensively (ASHWORTH and DEE 1975, COLLINS and BETTERLEY 1982, DEE 1982, GOODMAN 1980, HAUGLI et al. 1980, LANE 1981, and OLIVE 1975). A brief description of the life cycle follows and is summarized in Fig. 1. The plasmodium, a mass of protoplasm containing up to several million nuclei, is not subdivided into cells and is surrounded by a plasma membrane and slime. The lack of the rigid outer wall gives the plasmodium a high flexibility, which is unique for such large cells. Starvation of a plasmodium may give rise to a sclerotium, a resting phase, which under suitable conditions will produce a plasmodium again. In the presence of light, a starving plasmodium forms sporangia. During this process the diploid nuclei undergo meiosis and haploid spores are finally released. When a spore germinates it produces a uninucleate amoeba, which may form a flagella and gives rise to a flagellate. In the absence of nutrients the amoebae or flagellates will develop into a haploid dormant stage, the thick-walled cyst. The fusion of two mating-compatible amoebae results in a zygote with a diploid nucleus. By growth and by coalescence with other zygotes a tiny plasmodium is formed. The remaining haploid amoebae may then be engulfed and digested. Finally in culture the plasmodium may grow to cover an entire Petri dish. A unique feature of a plasmodium is the shuttle protoplasmic streaming at rates of up to 1.5 mm s^{-1} (BRITZ Chap. 2.3, Vol. 7, this Series) with reversals within a minute. This shuttle streaming causes a balanced distribution of metabolic products over the whole plasmodium. When two plasmodia meet this can be followed by fusion and complete mixing or by non-fusion and growth of both plasmodia alongside each other. Fusion may be followed by a post-fusion incompatibility reaction. For *P. polycephalum*, sexual fusion

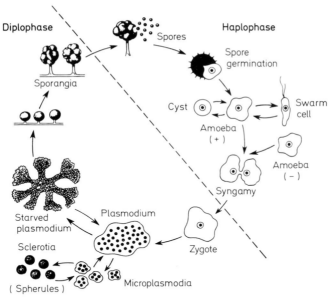

Fig. 1. Life cycle of *Physarum polycephalum*. (Hüttermann 1973)

between amoebae, somatic fusion of plasmodia and the post-fusion incompatibility reaction are controlled genetically by different sets of genes. This offers an opportunity to study interactions between cells biochemically and genetically with both the haploid and the diploid phases of a single species.

These interactions will be discussed mainly for *P. polycephalum* and to a lesser extent for *D. iridis* and other myxomycetes.

14.2 Fusion Between Amoebae

During its life cycle, myxomycetes alternate between two morphologically distinct vegetative phases, the microscopic uninucleate amoebal phase and the macroscopic plasmodial phase (Fig. 1). In the normal life cycle of the slime mould two haploid amoeba undergo cell and nuclear fusion, giving rise to a diploid multinucleate plasmodium.

In some strains the step from amoeba to plasmodium occurs without sexual interaction. This can be shown by counting the chromosomes or estimating the amount of nuclear DNA in nuclei of amoebae and plasmodia. If the plasmodial phase is initiated by sexual fusion of amoebae, the amount of DNA per nucleus is twice as much in the plasmodia as in the amoebae, whereas asexual initiation does not cause a change in the nuclear DNA content.

Dee (1960) demonstrated that a single spore of *P. polycephalum* gives a clone of amoebae, which cannot on their own give rise to plasmodia. She concluded that the fusion of amoebae is controlled by a pair of alleles at a mating-type locus, designated *mt* with alleles *mt-1* and *mt-2*. It is now known that the mating-type locus is multiallelic (DEE 1966) and at least 13 mating types of *P. polycephalum* and 12 of *D. iridis* could be demonstrated (COLLINS and BETTERLEY 1982).

So far each new diploid plasmodial isolate from nature has given amoebae showing two new mating types, based on two further alleles at the mating-type locus (CARLILE and GOODAY 1978, COLLINS 1979). Amoebae differing at the mating-type locus can fuse to form a plasmodium. The series of processes involved in the plasmodial development is, however, still largely unknown (DEE 1982).

14.2.1 *Physarum polycephalum*

Considerable attention has been given to the "amoebal plasmodial transition". Mutagenic treatment of amoebae can yield strains showing different defects in plasmodial development and lead to a better understanding of the processes involved. In 1960 DEE described the genetics of mating compatibility in a single locus, strain *mt*. Subsequently, many isolates from diverse geographic regions were studied by different groups. Their results, reviewed by CARLILE and GOODAY (1978) and COLLINS (1979) demonstrate that each isolate carries distinct *mt* alleles. At least 13 alleles are known (COLLINS 1979). So far most results of fusion experiments with mutants demonstrate that mutations affecting plasmodial development are located at or near the mating-type locus. Recent results suggest the involvement of other, unlinked loci in the development of diploid plasmodia from heterothallic amoebae. ADLER and HOLT (1975) found that cells heterozygous for the *mt* locus, especially those carrying the *mt*-h allele (which can permit haploid amoebae to develop into plasmodia by apogamy), developed more rapidly into plasmodia than cells homozygous at the *mt* locus. They suggested that the mating-type locus is a regulatory locus and that the role of the *mt* locus is to control the differentiation of zygotes into plasmodia. DEE (1978) identified in strains of *P. polycephalum* from different laboratories two alleles of a gene, *rac* (rapid crossing locus), unlinked to the *mt*-locus. YOUNGMAN et al. (1979) studied the rate and extent of plasmodium formation in mating tests involving pairs of largely isogenic amoebal strains compatible for *mt* alleles. They observed a systematic variability, plasmodia formed either rapidly and extensively or slowly and inefficiently, with a difference in rate of formation of more than a thousandfold. Their genetic experiments show the presence of a locus that affects amoebal cell fusion. YOUNGMAN et al. (1979) designated the new compatibility locus *mat B* and renamed the original *mt* locus *mat A*. The earlier described *rac* locus (DEE 1978) seems to be comparable with *mat. B*. KIROUAC-BRUNET et al. (1981) confirmed these results, and found in each natural plasmodial isolate of *P. polycephalum* studied so far two different alleles for each *mat B* and each *mat A* locus. The presence of 13 alleles at each locus, *mat A* and *mat B*, has been demonstrated. In nature the probability of mating between amoebae increases with the presence of a large number of alleles for the *mat B* as well as for the *mat A* locus.

In an attempt to clarify the specific function of both mating-type genes Youngman et al. (1981) carried out fusion experiments in mixtures of amoebae different in respect to their mating alleles. They found that mixtures of haploid amoebae could either remain haploid or develop into stable diploid amoebae or into diploid amoebae that become plasmodia. The proportion of cells which developed to a certain stage was determined by the composition of the mating alleles in the amoebae mixture. The data of Table 1 demonstrate a high frequency of diploid amoebae or plasmodia obtained from different mixtures. Youngman et al. (1981) concluded that *mat A* and *mat B* separately regulated two discrete stages of mating. In the first stage, haploid amoebae fuse in pairs, with a specificity determined by *mat B* genotype, to form diploid zygotes. The differentiation of these zygotes into plasmodia is regulated by *mat A* and is unaffected by *mat B*. Microcinematographic studies of amoebal mixtures illustrated that cell fusion is extensive in *mat B* heterothallic mixtures, regardless of *mat A* genotypes and is very rare in *mat B* homoallelic mixtures (Holt et al. 1980). Further support for the *mat A* and *mat B* mating theory came from the analysis of mutations that affect differentiation. These results demonstrated that nearly all such mutations map very close to *mat A,* and indeed most have been inseparable from *mat A* by recombination (Adler and Holt 1977, Anderson and Dee 1977, Anderson and Holt 1981, Davidow and Holt 1977, Gorman et al. 1979, Honey et al. 1979, Shinnick and Holt 1977, Shipley and Holt 1982). As in other organisms the formation of a zygote can be brought about in different ways (Bergfeld 1977).

Table 1. The result of interaction between haploid amoebae as a function of the mating-type alleles. (Based on the results of Youngman et al. 1981)

Composition of mating-type alleles in haploid amoebae mixtures		Frequency of amoebae used	Formed plasmodia	Diploid amoebae
mat A	mat B	(%)		(%)
Unlike	Unlike	10	Many	−[a]
Like	Unlike	10[b]	None	5
Like	Like	0.1[b]	None	<0.1
Unlike	Like	0.1	Few	−[a]

[a] Diploid amoebae heterozygous for *mat A* cannot easily be isolated because they develop readily into plasmodia
[b] When *mat A* alleles alike, some fused amoebae separate without nuclear fusion and therefore do not give rise to diploid amoebae

In myxomycetes the possibilities of formation of plasmodia are:
a) Non-heterothallic or clonal formation of plasmodia
    1. Apogamic – after fusion of amoebae no karyogamy occurs and the haploid condition will be preserved. Plasmodia formed in this way should be termed haploid facultative apomicts (Collins and Betterley 1982).

2. A transition from haploid to diploid state is involved with the plasmodium formation. This could happen by chromosome replication without nuclear division or after fusion of two like amoebae followed by karyogamy. This latter is homothallic mating.

b) Heterothallic mating or formation of plasmodia by mating between different haploid clones.

A diploid plasmodium develops by a two-stage process which consists of fusion (syngamy) of amoebae carrying different *mat A* alleles and the differentiation (development) of the resulting diploid zygotes into plasmodia. In contrast, a haploid plasmodium (COOKE and DEE 1975, ADLER and HOLT 1975, MOHBERG 1977) will be formed without change in ploidy, and apparently without amoebal fusion (ANDERSON et al. 1976).

Clonal development of plasmodia (selfing) that seems to be apogamic is genetically regulated and is at a moderate rate (one plasmodium from every ten amoebae) in the presence of a special allele of *mat A,* designated *mat A_h* (was *mt h*). In contrast "heterothallic" alleles of *mat A* also self, but at exceedingly low frequencies (less than one plasmodium from 10^8 amoebae) (WHEALS 1970, 1973, ADLER and HOLT 1977, YOUNGMAN et al. 1977). Mutations closely linked to *mat A* that affect plasmodial development have been demonstrated (ANDERSON 1979, HONEY et al. 1979, CARLILE and GOODAY 1978) and can be subdivided into:

1. Mutants with a decreased rate of differentiation into plasmodia (COOKE and DEE 1975, DAVIDOW and HOLT 1977, ANDERSON 1979, ANDERSON and DEE 1977, WHEALS 1973), like *apt* (amoebal plasmodial transition), *npf* (nonplasmodial formers) and the strain CLd (Colonial Leicester delayed).

2. Mutants with an increased rate of differentiation into plasmodia (ADLER and HOLT 1977, SHINNICK and HOLT 1977, GORMAN et al. 1979, ANDERSON and HOLT 1981), designated. *gad* (greater asexual differentiation) in which an increase in the frequency of plasmodial formation has been demonstrated.

Plasmodial development in these strains can be a tool for a better understanding of the initiation of amoebal-plasmodial transition and can clarify the role of syngamy. For a long time the latter was considered to be a key step in the initiation of plasmodial development. GORMAN et al. (1979) studied untreated as well as mutagenized amoebal cultures which are stable for the selfing mutation and which are able to develop into plasmodia within a clone. These investigators found an extremely low frequency of spontaneous mutation to selfing, and concluded that a highly specific type of genetic alteration is involved, and that point mutations within a single gene affecting selfing are unlikely. The haploid level of DNA (ADLER and HOLT 1977, GORMAN et al. 1979; MOHBERG 1977, COOKE and DEE 1975) of plasmodial nuclei in selfing mutants suggests that these strains no longer require syngamy to initiate differentiation.

Subsequent complementation and recombination experiments were performed with CL (Colonia Leicester) strains characterized by decreased differentiation of plasmodia (WHEALS 1973, ANDERSON and DEE 1977). HONEY et al. (1979) obtained 21 differentiation mutants (designated *dif*) from CL by NMG (N-methyl–N-nitro–N-nitrosoguanidine) mutagenesis and concluded that the mutants fall into two complementation groups, which are closely linked to,

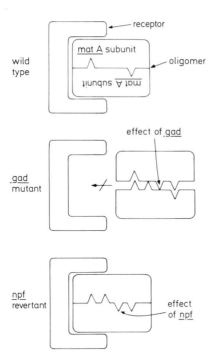

Fig. 2. A self-recognition model of *mat A* function and the effects of *gad* and *npf* mutations for fusion between amoebae of *P. polycephalum.* (ANDERSON and HOLT 1981)

or allelic at the *mat A* locus. The authors have proposed a repressor-operator model in which the *mat A* gene codes for a repressor molecule that acts at a nearby operator on the chromosome. The operator is adjacent to a structural gene, *dif*, whose expression is prevented by the interaction of the *mat A* product with the operator. Fusion of amoebae with mutants, like *npf* and revertants of *gad*, could cause a modulation of the repressor-operator hypothesis. These experiments can be performed with great efficiency between strains carrying different *mat B* alleles, without affecting the specificity of the complementation (HOLT et al. 1980, YOUNGMAN et al. 1981, ANDERSON 1979). ANDERSON and HOLT (1981) carried out fusion experiments with amoebae isolated which were revertants of selfing (*gad*) mutants and in which different alleles for *mat B* were introduced to facilitate complementation. They tested the repressor-operator model of HONEY et al. (1979) with their complementation results, and concluded that this model did not fit the facts. For the interacting system of amoebae of *P. polycephalum*, they suggested a simple model in which *mat A* and mostly *gad* and *npf* mutations are concerned with a single structural gene. In their model (Fig. 2) the structural gene makes a *mat A* product, which affects plasmodium formation by interacting at a receptor site in the cell and the specificity of the reaction is based on self-recognition. The postulated *mat A* product, a subunit of an oligomeric protein, would be active only if constructed from the subunits of a single *mat A* allele. The postulated *mat A* product would be inactive due to the inaccurate fit of allele-specific regions on unlike subunits. Experiments with strains of other origins are necessary to test this model.

The influence of the environment or the syngamy process has been studied. Clonally produced plasmodia do not form at temperatures above 28 °C with amoebal mutants of *P. polycephalum* carrying *npf A* allele, but in appropriate crosses plasmodia can be derived at such high temperatures with the *npf A* mutants (ANDERSON and DEE 1977). Zygote formation is affected by pH in a mutant carrying gene *imz* (ionic modulation of zygote formation), and SHIN-NICK et al. (1978) suggested that this is due to an amoebal surface component involved in syngamy. Comparable results with an optimum rate of mating at pH 5 were described earlier (COLLINS and TANG 1977). The presence of $MgSO_4$ in the culture medium enhanced the efficiency of mating between strains carrying different *mat A* and *mat B* alleles (SHINNICK et al. 1978, YOUNGMAN et al. 1981). Also the amoebae themselves affect the environment with extracellular products. PALLOTTA et al. (1979), SHIPLEY and HOLT (1982) demonstrated that sexual plasmodium formation is induced by products secreted by non-mating amoebae. This occurs when the non-mating amoebae are combined directly with the mating cells, as well as when the non-mating amoebae and the mating cultures are separated only by filters of 0.2 μm pore size. Similar results indicating that diffusible substances secreted by growing amoebae activated the differentiation of amoebae into plasmodia by the asexual process were described earlier (YOUNGMAN et al. 1977). Growing bacteria commonly used as a food source in the culture medium of amoebae (GOODMAN 1980), inhibit the mating process between amoebae by raising the pH of their surroundings (YOUNGMAN et al. 1981). This decrease in rate of amoebal fusion may be caused by a change of the amoebal surface (JACOBSON 1980).

14.2.2 *Didymium iridus*

As with *Physarum polycephalum,* interaction between amoebae of *Didymium iridis* has been a subject of research in several laboratories and the results have been discussed in extensive reviews (CARLILE and GOODAY 1978, COLLINS 1979, COLLINS and BETTERLEY 1982). Plasmodium initiation of *D. iridis* will therefore be surveyed briefly with special attention to the differences between *D. iridis* and *P. polycephalum*.

The number of plasmodial isolates of *P. polycephalum* studied are limited (about 8) but *D. iridis* has been sampled so extensively that 43 amoebal isolates have been obtained (COLLINS and BETTERLEY 1982). As for other myxomycetes, plasmodium initiation in *D. iridis* can be divided into two classes: *heterothallic* and *nonheterothallic*. For *D. iridis* a heterothallic mating has been shown to be based on a one-locus, multiple allelic system (COLLINS and BETTERLEY 1982). The mating characteristics for the non-heterothallic isolates may be either homothallic or apomictic. The studies of interaction between amoebae of *D. iridis* have focussed on its usefulness in understanding speciation and evolutionary relationships in the myxomycetes, whereas with *P. polycephalum* studies on the genetical fine structure of the mating locus have been performed with a view to elucidating physiological mechanisms. The results with *D. iridis* demonstrated that 12 amoebal isolate formed plasmodia only when mated and hence were heterothallic, and 31 isolated were non-heterothallic (COLLINS and BETTER-LEY 1982). As in *P. polycephalum* (for strains carrying *mat* A_h), in *D. iridis* some amoeba strains have been shown to produce plasmodia in clones as well as in appropriate crosses. Clonally developed plasmodia possess haploid nuclei (THERRIEN and YEMMA 1975, THERRIEN et al. 1977) and in their spore progeny only one mating type is present (COLLINS and LING 1968), which suggests that clonally formed plasmodia have developed apogamically, and it seems that true meiosis does not occur prior to the formation of spores (COLLINS and BETTERLEY 1982). The ploidal state of myxoflagellates and plasmodia can be determined from the DNA level of their nuclei. The results of many DNA measurements (THERRIEN and YEMMA 1974, THERRIEN et al. 1977, COLLINS 1980, MULLEAVY and COLLINS 1979, 1981) suggest a conversion of $n \rightarrow 2n$ for the sexual cycle, either heterothallic or homothallic and a constant level of $n \rightarrow n$ or $2n \rightarrow 2n$ for the apomictic non sexual cycle. Studies (COLLINS 1980) with subpopulations of *D. iridis* demonstrated that the DNA level in spores displayed in line A corresponded to n, haploid, and in line B to 2n, diploid. The heterothallic line A was spontaneously derived from the apomictic line B. These findings demonstrate that non-heterothallic apomictic isolates spontaneously can shift from a non-sexual to a sexual life-cycle mode. An explanation for this phenomenon, perhaps a result of mutation as in *P. polycephalum* (DEE personal communication) has not been found, but the occurrence of apomictic strains of *D. iridis,* ploidal level of 2n throughout, can be explained by the absence of meiosis during the development of the spore. Studies of the influence of the environment on sexual mating of amoebae of *D. iridis* are rather difficult to perform and thus the evidence is poor. Treatment of amoebae with proteases, prior to and at various times after mixing (SHIPLEY and ROSS 1978), indicate that the induction of the fusion process by a diffusible inducer is controlled by a protease-sensitive process, but the recognition and fusion of amoebae are governed by a mechanism insensitive to protease. SHIPLEY and HOLT (1980) demonstrated that exponentially growing amoebae become competent to fuse only after the cell density reaches a critical value and that a dense culture of amoebae can induce fusion competence in a sparse clonal culture when the two cultures are separated by a filter (0.2 µM pores). Plasmodium initiation in other slime moulds has not been studied so extensively, buth the results that have been obtained are reviewed by CARLILE and GOODAY (1978) and COLLINS (1979).

14.2.3 Conclusions Concerning Interactions Between Amoebae of Myxomycetes

In the conversion of the tiny uninucleate amoebae into a plasmodium that contains millions of nuclei and may cover many square centimeters many factors are involved.

First of all there is the environmental component which may make the amoebae capable of fusion. The influence of pH, presence of divalent cations, bacterial food source and chemical composition of the culture medium are external factors which have been demonstrated to affect the fusion of amoebae of many myxomycetes (COLLINS 1982). One "internal" factor in *D. iridis* as well as in *P. polycephalum* is a substance which induces rapid fusion, and is produced by a dense exponentially growing population of amoebae (SHIPLEY and HOLT 1980, 1982, YOUNGMAN et al. 1977). Whether this substance is a factor that mediates cell–cell recognition, as "golaptins" do in *Dictyostelium discoideum* and *Polyspondylium pallidum* (HARRISON and CHESTERON 1980), is unknown.

Genetic studies in *P. polycephalum* have demonstrated the function of two loci, *mat A* and *mat B,* in controlling plasmodium formation from amoebae. For the sexual process two discrete stages have been described. The formation of diploid zygotes after fusion of amoebae in pairs is determined by the *mat B* genotype and the differentiation of zygotes into plasmodia is controlled by *mat A* and unaffected by *mat B* (YOUNGMAN et al. 1981). In *D. iridis* the plasmodial development from amoebae, either apomictic or sexual, appears to be under control of one locus, the mating locus. The manner in which genetical factor(s) exert their effect is unknown, but for *P. polycephalum* a model has been proposed (Fig. 2).

14.3 Somatic Incompatibility Between Plasmodia

There is a striking difference between the research on the fusion between amoebae and between plasmodia. The work of the former topic is more genetical, whereas that on the latter is also ultrastructural and biochemical. After the germination of spores from a myxomycete a dense amoebal population develops, which gives rise to many small plasmodia. These fuse with each other to yield large plasmodia. When two large plasmodia of different origin come into close proximity the events that follow depend on the individual genetic composition of each plasmodium. Related studies on somatic incompatibility reactions will now be discussed. Earlier reviews are those of CARLILE (1973), CARLILE and GOODAY (1978), and COLLINS (1979).

14.3.1 *Physarum polycephalum*

If two plasmodia come into contact they may fail to fuse, either due to a genetical difference or to a difference in age even if they have an identical genetical composition. The former reaction is somatic fusion incompatibility (CARLILE and GOODAY 1978). The latter may be caused by the thick slime layer, which is developed by an old plasmodium, so that young and old plasmodia cannot come in close proximity. If fusion of plasmodia of different strains does occur, a series of processes may result in a lethal reaction which may destroy the fused plasmodium, partly or completely, a few hours later. The events which lead to the elimination of one or both plasmodia is known as the post-fusion incompatibility reaction. These two forms of somatic incompatibility were already known in 1967 (CARLILE and DEE).

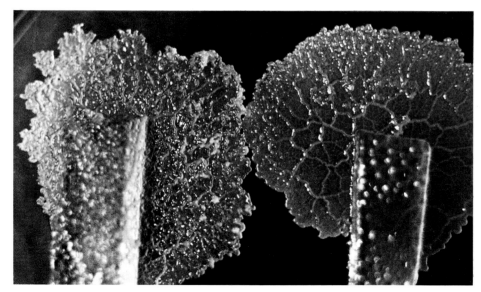

Fig. 3. Plasmodia of strains 15 and 29 of *P. polycephalum* grown out on agar medium in a 9-cm Petri dish met each other

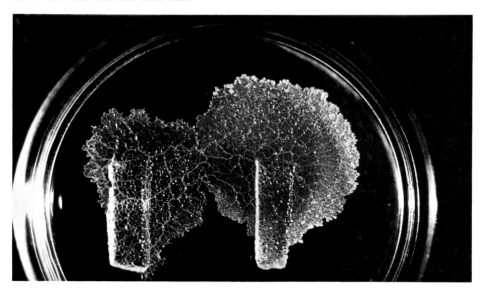

Fig. 4. About 30 min after Fig. 3. The two plasmodia are not fused completely but prominent veins have connected and changed their original position

The results of genetical studies have shown that the fusion of plasmodia is controlled by many loci with a pair of alleles at each locus, and at some loci one allele may be dominant. Studies on the progeny of Wisconsin 1 plasmodia revealed the presence of two fusion loci, now designated *fus A* and *fus B*

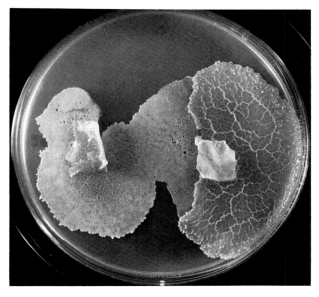

Fig. 5. Two plasmodia of the same strains 17 h after fusion. The plasmodium on the *left* (strain 29) is dead and the surviving plasmodium has just begun to re-invade. (CARLILE 1972)

(COOKE and DEE 1975); the fusion alleles are designated *fus* A_1 and A_2 and *fus* B_1 and B_2 respectively. The two alleles at the *fus* A locus are co-dominant, but B_2 is dominant to B_1. Mating of amoebae from the Colonia and Indiana isolates demonstrated a third fusion locus, *fus* C (ADLER and HOLT 1974). For plasmodial fusion to occur phenotypic identity with respect to fusion loci is required. The presence of 13 fusion loci with dominant and recessive alleles was demonstrated in other strains (COLLINS 1979). JEFFERY and RUSCH (1974) claimed that fusion incompatibility could be bypassed by gentle homogenization of mixtures of microplasmodia. This, presumably by damaging the plasma membrane, allows fusion of plasmodial fragments and regeneration of heterokaryotic plasmodia. However, fusion between two incompatible plasmodia from another strain resulted in a lethal reaction, independent of damage of the membranes by enzymatic treatment prior to fusion (SCHRAUWEN 1979). So far it seems likely that fusion compatibility is dependent upon identity with respect to some aspects of the membrane surface, but biochemical evidence is lacking. Following plasmodial fusion, an incompatibility reaction may occur due to genetical differences between the two plasmodia at certain other loci. ADLER and HOLT (1974) demonstrated that for the Colonia and Indiana strains several alleles are involved with the lethal effects. CARLILE (1976), studying the progeny of strain i × 1029, itself derived from Wis 1 plasmodia, demonstrated three loci, designated *let A, let B* and *let C,* with phenotypic differences which led to extensive lethal reactions a few hours after fusion. A dominant (*let A*) and a recessive allele (*let A* or *let a*) occurred at each locus, and lethal reactions occurred in a plasmodium that was homozygous recessive with respect to a *let* locus and was damaged by a plasmodium having one or two dominant alleles at the *let* locus. Thus a Wisconsin strain carrying *let A/let A* will fuse harmlessly with a strain carrying *let A/let a,* but both strains can kill a strain carrying *let a/let a.* Physiological experiments were performed with two closely related strains, 15 and 29, both derived from the Wisconsin strain, in which 15 acts as the killer and 29 as the sensitive strain. The events visible with the naked eye and with a light microscope which occur after fusion between both strains have been described in detail (CARLILE 1972) and are demonstrated in Figs. 3, 4 and 5.

Fusion of killer and sensitive strain is an absolute requirement for the lethal reaction to occur in the sensitive strain. Contact only of both plasmodia, intact or not, has no effect at all (SCHRAUWEN 1979).

For the post-fusion somatic incompatibility reaction in *P. polycephalum* two kinds of reactions can be distinguished: a very specific reaction that is directed only at sensitive nuclei, and an unspecific reaction which may be the result of the disorganization in the cell caused by a large number of disrupted sensitive nuclei.

Between the moment of fusion of plasmodia of strains 15 and 29 and the lethal reaction several processes occur in succession. There is an irreversible spreading of the influence of the killer strain throughout the whole plasmodium of the sensitive strain within 1 h. This is followed by the synthesis of special RNA and protein respectively 2.5 and 4 h after fusion; and the series of events ends with the lysis of the fused plasmodium (Table 2). Also the involvement of the DNA of both strains was proved experimentally (SCHRAUWEN 1981).

High resolution autoradiographic studies after fusion of another pair of strains demonstrate step-wise ultrastructural changes (LANE and CARLILE 1979). In the first 1.5 h after fusion there are apparently no changes. After that time a very specific assault is performed on the chromatin structure of the sensitive nuclei, which results in a reduction of size of the nucleolus of the sensitive strain and chromatin condensation. This is followed by disintegration of the nucleolus and the elimination of the sensitive nuclei at 3–5 h after fusion. Mean while the killer nuclei change little in appearance, and the number and size of rough endoplasmic reticulum increases. Unspecific damage to the plasmodium between 4–5 h after fusion is followed by a slow recovery of the surviving part of the plasmodium during the next few days (Table 2). After that time the fused plasmodium behaves as the killer plasmodium. This conversion of a sensitive plasmodium into a killer plasmodium as a result of the events that occur after fusion of plasmodia of both strains can be obtained in the absence of the unspecific lysis reaction (CARLILE 1972). This opportunity facilitates bio-chemical studies and has been used to follow the disappearance of sensitive nuclei after fusion of plasmodia of strain 15 and 29 by the labelling of DNA. The results showed no change in the amount of sensitive and killer nuclei for at least 8 h and a dramatic decrease in DNA from sensitive nuclei 9–11 h after fusion of plasmodia of strain 15 and 29 in ratio 1:10 (SCHRAUWEN 1981) (Table 2).

Table 2. Changes after fusion of a sensitive and a killer plasmodium. (Events are described in detail by Lane and Carlile 1980, Schrauwen 1979)

Time after fusion	Ultrastructural[a]	Biochemical[b]
0–1.5 h	No visible change	Irreversible outward spreading of the influence of the killer strain
2–2.5 h	Membranes of sensitive and killer nuclei develop small blebs Increase in amount of rough endoplasmatic reticulum	Synthesis of specific DNA
3–4 h		Synthesis of specific proteins
3–5 h	Alterations in sensitive nuclei, starting with a reduction in the size of the nucleolus to disruption Fusion of nuclei, both homologeous (from same strains) and heterologeous (from different strains)	
4–5 h	Unspecific plasmodial damage	
8–11 h		Disappearance of DNA from sensitive nuclei
12–16 h	Increase in size of remaining nuclei (11 μm) Protoplasmic streaming and movement are resumed Far fewer nuclei are present	
2–5 d	The plasmodium behaves as killer and nuclear sizes are normal (3–5 μm)	

[a] From Lane and Carlile (1979),   using strains i × A1029 (killer) and 29.02 × 29.19 (sensitive)
[b] From Schrauwen (1979, 1981), using strains i × A1015 (killer) and i × A1029 (sensitive)

14.3.2 *Didymium iridis*

Extensive work on somatic fusion of the plasmodia *Didymium iridis* has been carried out by O.R. Collins, J. Clark, H. and M. Ling, C.D. Therrien and I.C. Upadhyaya and has been reviewed by Carlile and Gooday (1978), and Collins (1979). The results of their studies demonstrate that the incompatibility reaction between plasmodia of *D. iridis* is controlled by a polygenic system with dominant alleles at many loci. The existence of at least 13 loci could be proved. Two plasmodia must be phenotypically identical at all these somatic incompatibility loci in order to fuse and to produce a heterokaryon. Thus difference of one allele at any of the large number of loci can prevent heterokaryon formation. The results of Ling and Ling (1974) suggest that these imcompatibility loci can be divided into two classes, one that controls membrane fusion and the other involved in the post-fusion reaction. The number of fusion (*Fus*) and post-fusion loci which can be distinguished vary from strain to strain. For the strains Pan-1 and Hon-1 seven *Fus* loci have been distinguished (Ling and Clark 1981). The post-fusion imcompatibility loci are designated, from their appearance, as clear-zone (*Cz*) loci. Ling and Clark (1981) recognized six *Cz* loci. Fusion between plasmodia of *D. iridis* is only possible if both plasmodia have the same fusion loci, however, this fusion terminates quickly if the two plasmodia differ at any of their clear-zone loci. The post-fusion reaction can develop within seconds of fusion. This zone of intermixture loses pigment to become a clear zone, undergoes coagulation and generally dies. The size of the intermixed zone and the speed of the reaction is determined by the *Cz* genes, each of which has a characteristic strength. The question arises whether the *Cz* loci act not only in "cytotoxic" post-fusion reactions but also in the instances of apparent non-fusion. The latter reactions would then in fact be very fast cytotoxic reactions following transient fusion. Clark (1980a and b) performed many fusion experiments between isolates of a Honduras strain and Panama–Honduras hybrids. Thirteen different plasmodial incompatibility phenotypes were prepared from eight haploid myxamoebal clones of the Hon-1 strain. The results of the fusion experiments between plasmodia of these strains show that one locus (designed G) is mainly a fusion controller (Clark 1980a). In the model, the presumed competitive survivor of each of these multiple-clone crosses could be predicted by comparing the relative cytotoxities of the potential phenotypes. Comparable fusion experiments between plasmodia from Pan-1–Hon-1 hybrids indicated that plasmodial-incompatibility phenotypes control the survival of plasmodia in crosses and demonstrated the existence of different genes which are fusion controllers (Clark 1980b). These results seem to indicate that not all seven fusion loci described by Ling and Ling (1974) are cytotoxic. Further studies may clarify the precise function of the *fus* and *Cz* loci to which Ling and Clark (1981) have given a unified nomenclature.

Somatic incompatibility studies other than in *P. polycephalum* and *D. iridis* are few and are described by Carlile and Gooday (1978) and Collins (1979).

14.3.3 Conclusions Concerning Somatic Incompatibility

Intolerance of genetically non-identical tissue is very common in nature. This phenomenon is regularly encountered, for instance with grafting in plants and the transplantation of organs in man. Unless cells of higher organisms are treated with specific agents, somatic fusion between these cells is very difficult to perform. Somatic fusion experiments between plasmodia of the slime mould may be helpful for a better understanding in the processes which are involved before and after fusion of two different cells. So far the results obtained from fusion between plasmodia of *P. polycephalum, P. cinereum, D. iridis* and *Badhamia utricularis* demonstrate that the capacity for plasmodial fusion is controlled by a polygenic system with dominant alleles at many loci. Somatic incompatibility in myxomycetes can take two forms, fusion and post-fusion incompatibility.

1. Fusion Incompatibility. Fusion between two plasmodia occurs only when they are phenotypically identical for all fusion loci. The kind of interaction processes which distinguish between fusion from non-fusion are not completely understood, but the outer surface of the cell, membrane and slime, plays an important role.

2. Post-Fusion Incompatibility. The results of fusion between two plasmodia phenotypically identical for all fusion loci, but with differences for the post-fusion loci, designated as *let* or *kil* for *P. polycephalum* (CARLILE and GOODAY 1978) and *Cz* for *D. iridis* (LING and LING 1974), is a heterokaryotic plasmodium, which, however, reverts to a single homokaryon within 24 h (COLLINS 1979, CARLILE 1976). In this conversion from a heterokaryotic to a homokaryotic plasmodium one can distinguish two phases, suggested by the results of biochemical experiments and submicroscopical observations (SCHRAUWEN 1979, 1981, LANE and CARLILE 1979).

a) Specific phase – recognition of the alien nuclei. After fusion of a killer and a sensitive plasmodium, a spreading of the killer strain follows throughout the whole plasmodium. In the subsequent reactions, the involvement of the DNA of both plasmodia, RNA and protein synthesis can be demonstrated. The process concludes with the elimination of sensitive nuclei and the acquisition of the properties of the killer plasmodium by the fused plasmodium.

b) Non-specific phase. Beside the processes which lead to the elimination of the sensitive nuclei in the specific phase, there may be the occurrence of damage to the whole or part of the fused plasmodium. This damage, designated the cytotoxic reaction for *D. iridis* and lethal or killer reaction for *P. polycephalum,* may arise within 1 min for the former or after about 5 h for the latter. Fusion experiments with other strains may show different time lags between fusion and damage of the heterokaryon formed. Dead areas did appear in a heterokaryon of fused hybrids of *P. polycephalum* 30 min after fusion (CARLILE personal communication). Both myxomycetes suffer similar damage in the disruption of membranes and organelles followed by the loss of cytoplasm. One difference is that the cytotoxic reaction is strictly local, while the lethal reaction is widespread, sometimes over the whole plasmodium.

Acknowledgements. I am grateful to R. ANDERSON, J. DEE and H.F. LINSKENS for their helpful comments, and to colleagues who provided me with reprints and preprints. I am indebted to M.J. CARLILE for his invaluable advice.

References

Adler PN, Holt CE (1974) Genetic analysis in the Colonia strain of *Physarum polycephalum:* heterothallic strains that mate with and are partially isogenic to the Colonia strain. Genetics 78:1051–1062

Adler PN, Holt CE (1975) Mating type and the differentiated state in *Physarum polycephalum.* Dev Biol Suppl 43:240–253

Adler PN, Holt CE (1977) Mutations affecting the differentiated state in *Physarum polycephalum*. Genetics 87:401–420

Anderson RW (1979) Complementation of amoebal-plasmodial transition mutants in *Physarum polycephalum*. Genetics 91:409–419

Anderson RW, Dee J (1977) Isolation and analysis of amoebal plasmodial transition mutants in the myxomycete *Physarum polycephalum*. Genet Res Camb 29:21–34

Anderson RW, Holt CE (1981) Revertants of selfing (*gad*) mutants in *Physarum polycephalum*. Dev Genet 2:253–267

Anderson RW, Cooke DJ, Dee J (1976) Apogamic development of plasmodia in the myxomycete *Physarum polycephalum*: A cinematographic analysis. Protoplasma 89:29–40

Ashworth JH, Dee J (1975) The biology of slime moulds. Arnold, London

Bergfeld R (1977) Sexualität bei Pflanzen. Ulmer, Stuttgart

Carlile MJ (1972) The lethal interaction following plasmodial fusion between two strains of the mycomycete *Physarum polycephalum*. J Gen Microbiol 71:581–590

Carlile MJ (1973) Cell fusion and somatic incompatibility in myxomycetes. Ber Dtsch Bot Ges 86:123–139

Carlile MJ (1976) The genetic basis of the incompatibility reaction following plasmodial fusion between different strains of the myxomycete *Physarum polycephalum*. J Gen Microbiol 93:371–376

Carlile MJ, Dee J (1967) Plasmodial fusion and lethal interaction between strains in a myxomycete. Nature 215:832–834

Carlile MJ, Gooday GW (1978) Cell fusion in myxomycetes and fungi. In: Poste G, Nicholson GL (eds) Membrane fusion. Cell Surf Rev Vol 5:219–265

Clark J (1980a) Competition between plasmodial incompartibility phenotypes of the Myxomycete *Didymium iridis*. I. Paired plasmodia. Mycologia 72:312–321

Clark J (1980b) Competition between plasmodial incompatibility phenotypes of the Myxomycete *Didymium iridis*. II. Multiple clone crosses. Mycologia 72:512–522

Collins OR (1979) Myxomycete biosystematics: some recent developments and future research opportunities. Bot Rev 45:146–201

Collins OR (1980) Apomictic-heterothallic conversion in a myxomycete, *Didymium iridis*. Mycologia 72:1109–1116

Collins OR, Betterley DA (1982) *Didymium iridis* in past and future research. In: Aldrich CH, Daniel JW (eds) Cell biology of *Physarum* and *Didymium* Vol 1. Academic Press, London New York, pp 25–53

Collins OR, Ling H (1968) Clonally produced plasmodia in heterothallic isolates of *Didymium iridis*. Mycologia 60:858–858

Collins OR, Tang HC (1977) New mating types in *Physarum polycephalum*. Mycologia 69:421–423

Cooke DJ, Dee J (1975) Methods for the isolation and analysis of plasmodial mutants in *Physarum polycephalum*. Genet Res 24:175–187

Davidow LS, Holt CE (1977) Mutants with decreased differentiation to plasmodia in *Physarum polycephalum*. Molec Gen Genet 155:291–300

Dee J (1960) A mating-type system in a acellular slime mold. Nature 185:780–781

Dee J (1966) Multiple alleles and other factors affecting plasmodium formation in the true slime mould *Physarum polycephalum*. J Protozool 13:610–616

Dee J (1978) A gene unlinked to mating-type affecting crossing between strains of *Physarum polycephalum*. Genet Res Camb 31:85–92

Dee J (1982) Genetics of *Physarum polycephalum*. In: Aldrich CH, Daniel JW (eds) Cell biology of *Physarum* and *Didymium* Vol 1. Academic Press, London New York, pp 211–251

Goodman EM (1980) *Physarum polycephalum*: A review of a model system using a structure-function approach. In: Bourne GH, Danielli JF, Jeon KW (eds) International review of cytology. 63:1–51

Gorman GA, Dove WF, Shaibe E (1979) Mutations effecting the initiation of plasmodial development in *Physarum polycephalum*. Dev Genet 1:47–60

Harrison FL, Chesterton CJ (1980) Factors mediating cell–cell recognition and adhesion. Galaptins, a recently discovered class of bridging molecules. FEBS Lett 122:157–165

Haugli FB, Cooke D, Sudbery P (1980) The genetic approach in the analysis of the biology of *Physarum polycephalum*. In: Dove WF, Rusch HP (ed) Growth and differentiation in *Physarum polycephalum*. Univ Press, Princeton, pp 129–156

Holt CE, Hütterman A, Heunert HH, Galle HK (1980) Role of mating specificity genes in *Physarum polycephalum*. (Institut für Wissenschaftlichen Film, 3400 Göttingen, FRG) Film B 1337. Eur J Cell Biol 22:C 940

Honey NK, Poulter RTM, Teale DM (1979) Genetic regulation of differentiation in *Physarum polycephalum*. Genet Res Camb 34:131–142

Hüttermann A (1973) *Physarum polycephalum* Object research in cell biology. Ber Dtsch Bot Ges 86:1–3

Jacobson DN (1980) Locomotion of *Physarum polycephalum* amoebae is guided by a short-range interaction with *E. coli*. Exp Cell Res 125:441–452

Jeffery WR, Rusch HP (1974) Induction of somatic fusion and heterokaryosis in two incompatible strains of *Physarum polycephalum*. Dev Biol Suppl 39:331–335

Kirouac-Brunet J, Masson S, Pallotta D (1981) Multiple allelism at the *mat B* locus in *Physarum polycephalum*. Can J Genet Cytol 23:9–16

Lane EB (1981) Somatic incompatibility in fungi and myxomycetes. In: Guel K (ed) The fungal nucleus. Academic Press, London New York

Lane EB, Carlile MJ (1979) Post-fusion somatic incompatibility in plasmodia of *Physarum polycephalum*. J Cell Sci 35:339–354

Ling H, Clark J (1981) Somatic cell incompatibility in *Didymium iridis*: locus identification and function. Am J Bot 68:1191–1199

Ling H, Ling M (1974) Genetic control of somatic fusion in a myxomycete. Heredity 32:95–104

Mohberg J (1977) Nuclear DNA content and chromosome numbers throughout the life cycle of the Colonia strain of the myxomycete. J Cell Sci 24:95–108

Mulleavy P, Collins OR (1979) Development of apogamic amoebae from heterothallic lines of a myxomycete, *Didymium iridis*. Am J Bot 66:1067–1073

Mulleavy P, Collins OR (1981) Isolation of CIPC-induced and spontaneously produced diploid myxomoebae in a myxomycete, *Didymium iridis*: a study of mating type heterozygotes. Mycologia 73:62–77

Olive LS (1975) The Mycetozoans. Academic Press, London New York

Pallotta DJ, Youngman PJ, Shinnick TM, Holt CE (1979) Kinetics of mating in *Physarum polycephalum*. Mycologia 71:68–84

Schrauwen JAM (1979) Post-fusion incompatibility in *Physarum polycephalum*. The requirement of the novo synthesized high molecular weight compounds. Arch Microbiol 122:1–7

Schrauwen JAM (1981) Post-fusion incompatibility in *Physarum polycephalum*. The involvement of DNA. Arch Microbiol 129:257–260

Shinnick TM, Holt CE (1977) Analysis of an *mt*-linked mutation affecting asexual plasmodium formation in *Physarum*. J Bacteriol 131:247–250

Shinnick TM, Pallotta DJ, Jones-Brown YR, Youngman PJ, Holt CE (1978) A gene, *imz*, affecting the pH sensitivity of zygote formation in *Physarum polycephalum*. Curr Microbiol 1:163–166

Shipley GL, Holt CE (1980) Fusion competence: an induceable state necessary for zygote formation in the myxomycete *Didymium iridis*. Eur J Cell Biol 22:M 669

Shipley GL, Holt CE (1982) Cell fusion competence and its induction in *Physarum polycephalum* and *Didymium iridis*. Dev Biol 90:110–117

Shipley GL, Ross JK (1978) The effect of proteases on plasmodium formation in *Didymium iridis*. Cell Diff 7:21–32

Therrien CD, Yemma JJ (1974) Comparative measurements of nuclear DNA in a heterothallic and a self-fertile isolate of the myxomycete, *Didymium iridis*. Am J Bot 61:400–404

Therrien CD, Yemma JJ (1975) Nuclear DNA content and ploidy values in clonally developed plasmodia of the myxomycete *Didymium iridis*. Caryologia 28:313–320

Therrien CD, Bell WR, Collins OR (1977) Nuclear DNA content of myxamoebae and plasmodia in six non-heterothallic isolates of a myxomycete *Didymium iridis*. Am J Bot 64:286–291

Wheals AE (1970) A homothallic strain of the myxomycete *Physarum polycephalum*. Genetics 66:623–633

Wheals AE (1973) Developmental mutants in a homothallic strain of *Physarum polycephalum*. Genet Res Camb 29:21–34

Youngman PJ, Adler PN, Shinnick TM, Holt CE (1977) An extracellular inducer of asexual plasmodium formation in *Physarum polycephalum*. Proc Natl Acad Sci USA 74:1120–1124

Youngman PJ, Pallotta DJ, Hosler B, Struhl G, Holt CE (1979) A new mating compatibility locus in *Physarum polycephalum*. Genetics 91:683–693

Youngman PJ, Anderson RW, Holt CE (1981) Two multiallelic mating compatibility loci separately regulate zygote formation and zygote differentiation in the myxomycete *Physarum polycephalum*. Genetics 97:513–530

15 Cell Interactions in the Cellular Slime Moulds

T. M. KONIJN and P. J. M. VAN HAASTERT

15.1 Introduction

Higher plants and the cellular slime moulds have in common that they depend mainly on their hormonal system for regulation of growth and development. The hormone or hormone-like substances are in both groups limited to a few compounds, which may bind to cell surface or intracellular receptors.

In the case of the cellular slime mould *Dictyostelium discoideum*, the hormone-like molecule cyclic AMP binds to cell-surface receptors and triggers several responses. Some occur rapidly, such as an increase of cyclic GMP, calcium fluxes, and chemotaxis, while others occur more slowly, such as the increase of cyclic AMP inside the cell. Evidence accumulates that the chemo-attractant cyclic AMP accelerates differentiation by gene activation, possibly via an increase of an internal cyclic nucleotide.

Dictyostelium is characterized by its simple and synchronous development and the relative ease of cultivating large quantities of cells within a relatively short time period. It proceeds through its life cycle within 1–2 days. Spores germinate in the soil, and the released cells feed on bacteria. When the food supply is exhausted, cells come together by chemotaxis to common collecting points (Fig. 1). The cell aggregates differentiate, after a facultative migration period, into a fruiting structure consisting of stalk cells and spores.

This review will be concerned with cell interactions all through the life cycle, with emphasis on the changes that occur during cell aggregation.

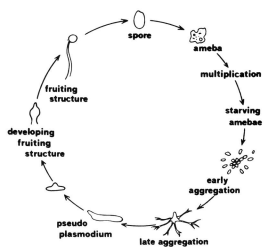

Fig. 1. Life cycle of *D. discoideum*

15.2 Spore Germination

Spore germination, like most other cell processes in the cellular slime moulds, has been studied most thoroughly in *Dictyostelium discoideum*. The spore dormancy is broken in the presence of bacteria, which release activators to promote spore germination (Hashimoto et al. 1976, Dahlberg and Cotter 1978). A heat shock will also induce spores to germinate (see Cotter 1981). After activation, spores swell, split, and a single cell emerges from one spore case. The energy required for spore germination has to be derived from internal storage materials. Protein synthesis is a prerequisite during this early stage of the life cycle (Cotter and Raper 1970). (For a thorough discussion of spore formation and germination we refer the reader to Hohl 1976).

In the absence of bacterial activators spore germination is prevented by endogenous inhibitors. The inhibitor of spore germination in *D. discoideum* has been identified as discadenine, 3-(3-amino-3-carboxypropyl)-N^6-Δ^2-isopentenyladenine (Abe et al. 1976). Its presence in other species such as *D. mucoroides* and *D. purpureum* suggests that the inhibitor is not species-specific (Taya et al. 1980). Primitive species, such as *D. minutum,* seem to lack this compound, and are not inhibited by it. *D. minutum* also lacks discadenine synthetase, the enzyme which catalyzes the synthesis of the inhibitor by transfer of the 3-amino-3-carboxypropyl moiety of S-adenosylmethionine to N^6-Δ^2-isopentenyladenine (i^6Ade). In *D. discoideum* i^6Ade is derived from i^6AMP, and 5′-AMP is the precursor of this compound (Taya et al. 1978).

Interestingly, discadenine also has cytokinin activity, as might be expected from its structure (Nomura et al. 1977). Other hormones that function in higher organisms may also be effective, or actually present, in simple eukaryotes like the cellular slime moulds. Acetylcholine has a possible role in the spacing of aggregation centers in *Polysphondylium violaceum* (Clark 1977), and progesterone inhibits cell aggregation by blocking the development to aggregation competence (Brachet and Klein 1977). Since plant hormones occur in animals and animal hormones in plants, we should be aware of the possibility that molecular origins of hormonal communication in higher organisms arose from unicellular eukaryotes or prokaryotes (see Vol. 9, this Series, Le Roith et al. 1980, Geuns 1982).

15.3 Growth Phase

The amoebae feed by phagocytosis; they engulf about 1000 bacteria between cell divisions. Their natural habitat is the soil; however, good growth occurs in the presence of bacteria on a solid agar surface or in suspension culture. The latter technique has the advantage of providing a homogeneous environment which allows synchronization of the cells. Isotopic labelling experiments are preferably carried out axenically in shaken liquid cultures to prevent loss of label to contaminating bacteria. Franke and Kessin (1977) reported a totally

defined minimal medium for the axenic strain Ax3. The capacity to grow axenically is linked to two recessive mutations on the second and third linkage group
(WILLIAMS et al. 1974).

The axenically grown cells have a much slower doubling time, ca. 10 h instead of 3 h, and contain more glycogen and protein than cells grown on bacteria
(ASHWORTH and WATTS 1970). Discrepancies in the literature may often be
due to different growth conditions, and accordingly results obtained with axenically grown strains should be interpreted cautiously. They cannot be indiscriminately extrapolated to the cells grown with bacteria on a solid medium, and
vice versa.

Cells may differ in their chromosome number. Besides haploid strains of
D. discoideum with 7 chromosomes and diploid strains with 14 chromosomes,
aneuploid cells with intermediate chromosome numbers may occur (BRODY and
WILLIAMS 1974). More information on the genetic analysis of development in
Dictyostelium has been reviewed recently by LOOMIS (1980).

Cells grown in shaken culture reach their nutrients effortlessly and take
them up by phagocytosis or pinocytosis. The search for food under natural
conditions is more laborious. Encountering their prey by accident might be
too risky to allow survival. Cells are in fact oriented towards bacteria by chemotaxis. Bacterial products not only facilitate spore germination but also attract
their predators. The best-known chemotactic factor is the vitamin, folic acid
(PAN et al. 1972). It is aspecific in its chemotactic activity; all *Dictyostelium*
and *Polysphondylium* species are attracted by folic acid, although at different
threshold concentrations. Outside the cellular slime moulds no chemotactic activity has been shown for this compound. Cell-surface-bound deaminases keep
the folic acid level around the cell low, and this maintains a sufficiently steep
gradient of the attractant. Extracellular deaminases also break down folic acid
(PAN and WURSTER 1978, KAKEBEEKE et al. 1980, BERNSTEIN et al. 1981, WUR-
STER et al. 1981). Apparently, folic acid, besides being an attractant, is also
a prerequisite for growth, since cells in the defined medium of FRANKE and
KESSIN (1977) without folic acid stopped growth within one to two doublings.

Besides being attracted, amoebae may also be repelled (SAMUEL 1961). KEAT-
ING and BONNER (1977) showed with a deep-shallow agar technique that *D.
discoideum* and *D. purpureum* repelled neighbouring cells of the same species;
the negative response within *D. mucoroides* was ambiguous and repellents secreted by *P. violaceum* cells were not demonstrated. KAKEBEEKE et al. (1979)
extended the study of negative chemotaxis by direct measurement of repulsion
in several slime mould species (Fig. 2). Only vegetative amoebae secreted repellents, while sensitivity to these compounds was shown both in the vegetative
and in the aggregative stage. Cross-reactions of various species made it clear
that there are different repellents. Since fresh amoebal supernatants inactivate
repellents, inactivating enzymes are probably present. Purified repellents could
function as tools to investigate which molecular pathways positive and negative
chemotaxis have in common.

An essential component of cell movement is the actomyosin complex. The
study of directed movement induced by chemo-attractants might contribute
to the elucidation of the filamental interaction. Microfilaments are primarily

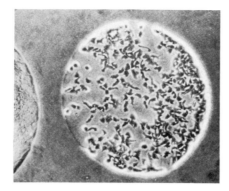

Fig. 2. *D. minutum*, V$_3$, response 30 min after deposition of a high density cell drop of *D. minutum*, V$_3$ to the left of the low density population. × 70. (Kakebeeke et al. 1979)

situated in the cell cortex, where they are associated with the cytoplasmic surface of the plasma membrane (Clarke et al. 1975). The myosin consists of heavy and light chains. Phosphorylation of the heavy chains inhibits the self-assembly into thick filaments (Kuczmarski and Spudich 1980), and may effect the contractile elements of the cell by changing the state of myosin polymerization. Malchow and co-workers (1981) showed that dephosphorylated myosin heavy chains transiently accumulated during chemotactic stimulation. Calcium inhibits phosphorylation of myosin in vitro and seems to act via calmodulin.

Microtubules are more abundant in the nuclear region than in the cortex. They radiate from microtubule-organizing centers (Cappuccinelli et al. 1981), and their location may be visualized by immunofluorescence. The microtubule-organizing centres disappear at early mitosis. Some cells, especially in shaken cultures, do not divide after nuclear division; further nuclear divisions in these multinucleated cells appear to be synchronized. The tubulin of *D. discoideum* is evidently nearly identical to that of other organisms. The molecular aspects of cell motility in simple organisms have been reviewed recently by Cappuccinelli (1980).

15.4 Social Phase

Starvation and its consequences for development have been studied extensively in *D. discoideum,* the species with the longest interphase between the growth phase and cell aggregation. Depending on environmental conditions such as light, temperature and cell density, the interphase may last 6 to 12 h (Konijn and Raper 1965, Konijn 1968). Shortly after the onset of food shortage, amoebae develop large endocytic cups, and digestive vacuoles are transformed into autophagic vacuoles (de Chastellier and Ryter 1980), which degenerate in the prespore cells while getting larger in the prestalk cells (Yamamoto et al. 1981). Dimond and co-workers (1981, Burns et al. 1981) studied the regulation and secretion of lysosomal enzymes in starving cells; they found that the lysosomal system is functionally and structurally heterogeneous. Starving *D. minutum* cells drastically reduce their size, and the excess of plasma membrane is engulfed as small vesicles that are transported to autophagic vacuoles (SCHAAP et al. 1981). This break-down of structural components may be necessary to supply the cells with energy to aggregate and differentiate into a fruiting body.

A few hours before aggregation wave-like motions pass through the whole cell population of *D. discoideum*. The interval between two waves shortens from 10 to a few minutes during the latter part of the interphase. The wave-like motions become limited to the aggregation territories, and are the visible expression of a propagated cyclic AMP signal (TOMCHIK and DEVREOTES 1981). Each territory contains one or a few cells that secrete cyclic AMP autonomously in a pulsatile manner. These cells trigger their neighbours to move to higher concentrations of attractant and also to release cyclic AMP (GERISCH and WICK 1975, SHAFFER 1975). Such a relay of signalling cells allows peripheral cells to move to the centre. These discontinuous movements create the inward waves of migrating cells that are clearly visible on time-lapse films. When repetitive pulses of cyclic AMP are applied to amoebae, cell differentiation is accelerated (DARMON et al. 1975, GERISCH et al. 1975c).

To respond to the chemotactic signals the cell needs specific receptors at its surface. The number of receptors increases drastically shortly before aggregation. The increase in receptors can be accelerated by supplying artificial pulses of cyclic AMP to an amoebae suspension (Table 1). These pulses enhance adenylate cyclase and phophodiesterase activity; the latter enzyme converts cyclic AMP to 5′-AMP. The amoebae also secrete in this stage an inhibitor of the extracellularly secreted phosphodiesterase (RIEDEL and GERISCH 1971), which changes the K_m of this enzyme from 10 µM to 2 mM (KESSIN et al. 1979). The plasmamembrane-bound phosphodiesterase is not affected by this inhibitor. The internally produced cyclic AMP may be hydrolyzed, secreted (DINAUER et al.

Table 1. Processes that are accelerated by pulses of cyclic AMP

Appearance of cyclic AMP receptors (KLEIN et al. 1977)

Appearance of adenylate cyclase (ROOS and GERISCH 1976)

Appearance of membrane-bound phosphodiesterase (KLEIN 1975)

Repression of appearance of the inhibitor of extracellular phosphodiesterase (KLEIN and DARMON 1977)

Appearance of contact sites A (GERISCH et al. 1975a)

Synthesis of mRNA (LANDFEAR and LODISH 1980)

Entrance of calcium (WICK et al. 1978)

Accumulation of cyclic GMP (MATO et al. 1977a, c, WURSTER et al. 1977)

Methylation of proteins (MATO and MARIN-CAO 1979)

Methylation of phospholipids (ALEMANY et al. 1980)

Accumulation of dephosphorylated heavy chain myosin (MALCHOW et al. 1981)

Excretion of protons (MALCHOW et al. 1978)

1980), or bound to a protein of about 40,000 molecular weight (Cooper et al. 1980, Leichtling et al. 1981).

Cells continuously exposed to a certain cyclic AMP concentration become insensitive to the attractant (Devreotes and Steck 1979). Another effect of a constant cyclic AMP level is to reduce the number of binding sites (Juliani and Klein 1981). The periodically secreted cyclic AMP can be inactivated by phosphodiesterase in the interval between two pulses; this allows the cells to respond to the new gradient of the next cyclic AMP pulse when the cells are sensitive again (Devreotes and Steck 1979). More information on the intercellular communication and intracellular regulation during aggregation has been reviewed recently by Darmon and Brachet (1979), Loomis (1979), Mato and Konijn (1979) and Robertson and Grutsch (1981).

The formation of cell aggregates depends on cell–cell adhesion, as well as on chemotaxis. Amoebae of species that are attracted to identical acrasins enter the same aggregate and separate after a few hours to construct individual fruiting bodies (Raper and Tom 1941, Konijn 1975). Species that are responsive to different acrasins do not mix, and enter separate aggregates. Since the chemotactic mechanism cannot operate in shaken cultures, clumping of aggregation-competent cells can only occur by cell adhesion. Garrod et al. (1978) mixed *Dictyostelium* and *Polysphondylium* species that required different acrasins in a shaken culture and noticed that cells clumped indiscriminately. Apparently cell–cell adhesion is a less specific process than chemotaxis.

Various glycoproteins on the cell surface of aggregating cells may contribute to the increase in cell–cell adhesion. So-called contact sites (cs) and receptors for discoidins are thoroughly investigated cell-surface glycoproteins. Two kinds of contact sites, A and B, can be distinguished (Beug et al. 1973). Contact sites B are present on vegetative and aggregative cells. They are sensitive to ethylenediaminetetraacetic acid (EDTA) and support side-to-side adhesion in

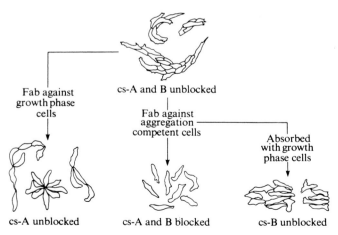

Fig. 3. Immunological analysis of the cell adhesion system in *D. discoideum*. (Müller and Gerisch 1978)

aggregating streams. Contact sites A arise later in development and are involved in the end-to-end adhesion of aggregating cells (Fig. 3). Contact sites A are insensitive to EDTA (see MÜLLER and GERISCH 1978). Univalent antibody fragments (Fab) against contact sites A and B block both end-to-end and side-to-side contacts without affect on the shape and motility of the cells. Application of specific antibodies shows that contact sites of *D. discoideum* do not cross-react with those of *P. pallidum*. This specificity might explain the sorting-out of aggregation-competent cells that clump in mixed suspensions and separate later (GARROD et al. 1978). Also the appearance of contact sites A is stimulated by pulses of cyclic AMP (Table 1) applied to interphase cells (GERISCH et al. 1975a). The total number of contact sites per cell is about 3×10^5 (see GERISCH 1980). Purification of contact sites A revealed that they are concanavalin A-binding glycoproteins with a molecular weight of 80,000 (MÜLLER and GERISCH 1978). They can neutralize the blocking activity of Fab directed against contact sites A. Contact sites do not bind discoidin; they may bind to other contact sites A or to yet unidentified receptors (GERISCH et al. 1980). Contact sites A constitute less than 1% of all the concanavalin A-binding receptors that are present on the cell surface (EITLE and GERISCH 1977).

Discoidins are lectins which can be divided into subspecies, discoidin I and II. Discoidin I is the predominant lectin. Although evidence for the involvement of discoidins in cell–cell adhesion is not yet conclusive, mutant cells which form altered discoidin I appear to be non-cohesive (RAY et al. 1979). Discoidin I and II consist of four identical subunits with a molecular weight of 26,000 each for discoidin I and 24,000 for discoidin II (FRAZIER et al. 1975). *P. pallidum* has its own lectin, called pallidin (ROSEN et al. 1975) and the lectin purpurin, which exists as seven isolectins, has been isolated from *D. purpureum* (BARONDES 1981). Cell-surface receptors for discoidin I were detected by REITHERMAN et al. (1975). BREUER and SIU (1981) isolated 11 discoidin I-binding proteins from *D. discoideum*, NC$_4$, by gel electrophoresis; the molecular weight of these various proteins ranged from 95,000 to 15,000. The synthesis of three of them depended on the developmental stage. A glycoprotein with a molecular weight of 31,000 increased eight to ten times during aggregation and bound predominantly to discoidin I; this glycoprotein may be involved in the lectin–ligand coupling resulting in an increased cell–cell adhesion. The difference in cell surface macromolecules in the wild-type strain and the axenically grown A $\times$ 3 strain is exemplified by the presence of only three discoidin I-binding proteins in the A $\times$ 3 strain. For more information on the adhesive mechanism in the cellular slime moulds reference may be made to the extensive review by ROSEN and BARONDES (1978). The latest developments, suggesting new candidates for cohesion of aggregative cells, such as a glycoprotein with a molecular weight of 150,000, which is a cell-surface protein particularly abundant where cells make contacts (GELTOSKY et al. 1980) and a cell-surface protein with a molecular weight of 95,000 (STEINEMANN and PARISH 1980), are summarized by BARONDES (1981), GARROD and NICOL (1981) and GERISCH et al. (1980).

The developmental changes in aggregative amoebae depend on gene activation. A trigger for gene activation seems to be the appearance of cell–cell contacts. Post-aggregative cells synthesize 2500 mRNA sequences that were absent before cell aggregation. These cell–cell contact-dependent sequences comprise about 30% of the total mRNA (CHUNG et al. 1981, MANGIAROTTI et al. 1982). Also the synthesis of these new mRNA's is stimulated by cyclic AMP.

Lack of space prohibits a more extensive discussion of cell aggregation in other *Dictyostelium* species. An exception should be made for *D. minutum*. This primitive species was first described by RAPER (1941), who recognized that its peculiar features do not justify a place in the *D. mucoroides* complex. Stream formation as seen in the larger species is lacking. The amoebae do not move in a pulsatile manner to the centre, and no relay of the chemotactic signal has been shown (GERISCH 1968). The aggregation centre originates from one single cell (GERISCH 1966), which stops moving, rounds up and starts to attract neighbouring cells. Such aggregation-initiating cells are only observed in cyclic AMP-insensitive species, and were reported first for *Polysphondylium* (SHAFFER 1961). Only after cell aggregates of *D. minutum* disintegrate are secondary streams formed.

An alternative pathway of development is the formation of macrocysts (BLASKOVICS and RAPER 1957). These structures originate from small aggregates consisting of about a hundred cells. Such clumps of cells, surrounded by a cellulose wall, form the macrocysts. The interesting feature in the early development of this multicellular assembly is the fusion of two cells to form a zygote, which engulfs the other cells in the macrocyst (CLARK et al. 1973, ERDOS et al. 1973). It divides into many haploid cells which emerge at the time of germination of the macrocyst. Whether amoebae follow the sexual or the asexual cycle depends on the species, strain, and the environmental conditions. Light, humidity, pH, composition of the medium, and volatile substances (FILOSA 1979, CHANG and RAPER 1981) condition the cells to become macrocysts. Unfortunately for the genetic approach, germination, even under the most favourable conditions, takes more than a week. Especially in *D. discoideum,* the best-known species at the molecular level, macrocysts germinate only after long periods.

The details of the alternative pathway of macrocyst development have been described recently by O'DAY and LEWIS (1981).

15.5 Pattern Formation

At the end of aggregation, the finger-like structure or standing slug of *D. discoideum* may fall down and migrate as a free-moving pseudoplasmodium, or migrating slug, for some hours. No true plasmodium, as in the acellular slime moulds, is formed; all cells preserve their individuality. They number from less than 100 to more than 100,000 in one slug. Slugs of most other species constantly construct stalk cells, even when they migrate over the surface of the substrate. Slugs are phototactic (RAPER 1940), thermotactic (BONNER et al. 1950) even along temperature gradients as small as 0.05 °C cm^{-1}, and chemotactic (FISHER et al. 1981). Light, ions and dryness shorten the migration period. The advantage of an accelerated culmination at lower humidity is the high resistance of spores against desiccation. The sheath surrounding the slug may also give some protection against dehydration. It is composed of proteins, cellulose fibres, a polysaccharide-containing N-acetyl glucosamine, glucose and mannose (GERISCH et al. 1969, HOHL and JEHLI 1973). Light may guide the slug to the soil surface where spore dispersal may be favoured (RAPER 1940). The response to light is mediated by a pigment, phototaxin, which is identified as a haem protein. This molecule has a molecular weight of 240,000 and is localized in the mitochondria (POFF et al. 1974). The photo-oxidation is a reversible process, Recent evidence suggests that even single cells are photoresponsive (HÄDER and POFF 1979). The photoreaction of the slug may be based on the production of a transverse gradient of an extracellular signal molecule after differential illumination (FISHER et al. 1981). This signal molecule would act as a repellent at one side of the tip, and in response the slug turns. The active or slug-turning factor (STF) has a low molecular weight.

Slugs are convenient multicellular structures to study cell sorting and differentiation. Cells differentiate along the anterior–posterior axis. The anterior third of the slug is the prestalk region, and gives rise to stalk cells. In the posterior region cells are situated that are predestined to become spores. Prestalk and prespore cells differ structurally and biochemically. Prespore cells of *D. discoideum* contain specific prespore vacuoles (HOHL and HAMAMOTO 1969, MAEDA and TAKEUCHI 1969), and the enzyme UDP galactose-polysaccharide transferase (NEWELL et al. 1969). Prestalk cells are characterized by higher cyclic AMP (BRENNER 1977), trehalase (JEFFERSON and RUTHERFORD 1976) and phosphodiesterase levels (TSANG and BRADBURY 1981). After labelling cells with ^{35}S-methionine and separation of the polypeptides by gel electrophoresis ALTON and BRENNER (1979) found that 57 of the ca. 500 polypeptides showed regional variations in labelling in different regions of the slug. Despite these differences, amputated anterior and posterior regions produce normal, although smaller, fruiting bodies (RAPER 1940, SAMPSON 1976).

For an understanding of pattern formation it is important to know whether cells differentiate before sorting out or differentiate as *consequence* of the position they take in the organism. To decide which of the two possiblities is most plausible, differences between these two cell types should be shown at early developmental stages. Cells in the prespore and prestalk region can be distinguished by use of a fluorescent antibody that specifically stains prespore cells (TAKEUCHI 1963). With this technique the earliest prespore cells were shown to be present in late aggregates shortly before the formation of a tip on the hemispherical mound of cells. SCHAAP et al. (1982) demonstrated in an electron microscopic study of *D. discoideum* that differentiation between prespore and prestalk cells is already evident during early aggregation. Some aggregative cells become more electron-dense than other cells (Fig. 4). The electron-dense cells are smaller, and are initially spread all over the aggregate. Later, the electron-dense cells occupy the posterior two thirds of the slug, together with some electron-lucent cells interspersed between them (Fig. 5). Electron-dense cells are almost completely absent in the prestalk region. In a migrating slug the pattern of electron-lucent and electron-dense cells is similar to the prespore-prestalk pattern as revealed by other techniques. Prespore and prestalk cells, therefore,

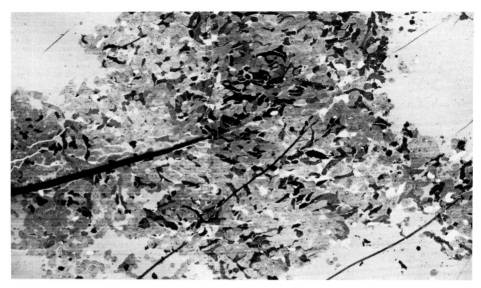

Fig. 4. The random distribution of electron-lucent and electron-dense cells in an aggregation centre with streams. × 440. (Schaap et al. 1982)

Fig. 5. The anterior part of the slug becomes almost devoid of electron-dense cells. In the posterior part electron-lucent and electron-dense cells are intermixed. × 215. (Schaap et al. 1982)

apparently are not necessarily induced by the position that they occupy in the slug. Tasaka and Takeuchi (1981) reached the same conclusion by mixing ^{3}H-labelled cells and unlabelled cells, which were grown differently. Prespore cells could be distinguished long before the labelled and unlabelled cells sorted out.

Sorting out alone cannot account for the pattern of cell differentiation (Mac Williams and Bonner 1979, Gregg et al. 1981). More information on the intriguing problem of cell interaction during sorting out in *Dictyostelium,* on the role of cyclic AMP in this process (see Durston and Vork 1979), and on cell contacts and the evidence that differentiation-inducing factor (DIF) contributes to prestalk cell formation is available in a recent review by Gross et al. (1981).

15.6 Culmination

Sorting out of the amoebae and the final differentiation require cell communication. Cell interactions via gap junctions seem to be lacking in the cellular slime moulds (JOHNSON et al. 1977). Presumably intercellular communication depends largely in the slug stage also on hormone-like signals that act via cell surface receptors. The tip of the slug may release signals that trigger the termination of migration (DURSTON and VORK 1979, SMITH and WILLIAMS 1980). Cyclic AMP is a good candidate for such a signal (RUBIN 1976). Another signal molecule that suppresses cyclic AMP production is NH_3 (SUSSMAN and SCHINDLER 1978). At the end of migration the slug rounds up, and cells that were situated at the front vacuolate and become stalk cells. The prestalk cells excrete material for a cellulose sheath surrounding the stalk. They enter the cylinder at the top and prespore cells are moved upward when the stalk lengthens (RAPER and FENNELL 1952, GEORGE et al. 1972). The cellulose wall of the stalk cells gives them a plant-cell-like appearance. When the stalk nearly reaches its final size the prespore cells become encapsulated by walls consisting of three layers, of which the thick middle layer consists mainly of cellulose (HOHL and HAMAMOTO 1969). The prespore vacuoles contribute to the spore formation by excretion of material for the distinct layers of the cell wall, and by discharge of slime in which the spores will be embedded. A peculiarity of *Dictyostelium discoideum* is that the stalk rests on a disc consisting of cells that are derived from the tail of the slug. The spores may be spread by other soil organisms or rain and germinate in a bacteria-rich environment.

15.7 Acrasins

The early work on acrasins and the identification of cyclic AMP as the chemoattractant during cell aggregation in *D. discoideum* have been summarized elsewhere (KONIJN 1975). The two most commonly used bio-assays for attractants are the cellophane square assay and the small population assay. In the cellophane square assay amoebae are allowed to settle on small cellophane squares which are transferred to small Petri dishes containing plain agar or agar mixed with the test substrate (BONNER et al. 1966). In both conditions cells cross the edge of the cellophane membrane. Within a certain period amoebae on the agar containing chemo-attractants move farther away from the edge than amoebae on the control agar.

The small population assay requires a hydrophobic agar surface on which small drops (0.1 µl) of an amoebal suspension are deposited. The amoebae move freely within the drop boundaries and do not cross the edge of the drop (KONIJN 1970). Another small drop containing a chemoattractant is placed close to the amoebal population. Chemo-attractants diffuse through the agar and trigger the cells in the neighbouring drop to move to the source of attractant (Fig. 6). Purification of wash-water of bacteria this way led to the identification

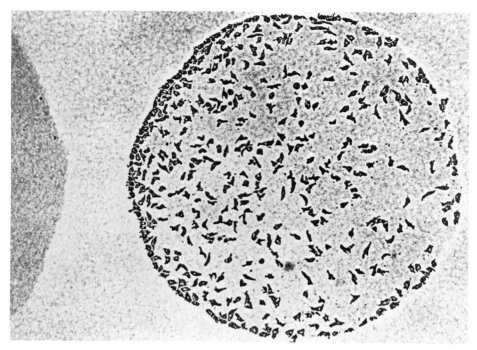

Fig. 6. Drops (0.1 μl) containing 3×10^{-13} g cyclic AMP were deposited three times at 5-min intervals to the left of the responding drop. Amoebae within this drop have been attracted but have not crossed the boundary. × 285. (KONIJN 1970)

of cyclic AMP as the acrasin of *D. discoideum* (KONIJN et al. 1967, 1969, BARK-LEY 1969). Other large species of *Dictyostelium* were also found to be sensitive to this cyclic nucleotide (KONIJN 1972). The threshold concentration of cyclic AMP is at least 100-fold lower for aggregative cells than for vegetative cells (BONNER et al. 1969).

Acrasins of *Polysphondylium, D. lacteum* and *D. minutum* have been partially purified and characterized (WURSTER et al. 1976, MATO et al. 1977b, KAKEBEEKE et al. 1978). The acrasin of *P. violaceum* appears to be a peptide which attracts aggregative but not vegetative cells, and is specific for the genus (WURSTER et al. 1976). KAKEBEEKE et al. (1978) also showed specificity for the partly purified acrasin of *D. minutum*. Other species possibly have a higher threshold for this compound, since they only respond if sufficiently high concentrations are available. The chemo-attractants of aggregative *Polysphondylium violaceum* and *D. minutum* cells are small molecules and they or their derivatives are widespread in nature. Yeast extract, bacterial products, urine and milk are good sources for attracting compound(s).

Purification of the chemotactically active yeast extract did not lead to the identification of the acrasin of *Dictyostelium minutum* (KAKEBEEKE et al. 1978) or *D. lacteum* (MATO et al. 1977b). Complicating factors are the lability of the acrasin and the presence of several active compounds, perhaps natural analogues of the acrasin, some of which were chemotactically active at low concentrations. Therefore, no other choice was left than the laborious task to isolate the acrasin from the amoebae themselves.

Secretion products of aggregating *D. lacteum* amoebae were separated from the cells and their degrading enzymes through a dialysis membrane. Only aggregating cells secreted the attractants in sufficiently large amounts to give a positive response in the small population assay. The dialyzed secretion products were analyzed by high performance liquid chromatography using different mobile phase compositions. Despite the low yield of 0.1 nM it was possible to show that the acrasin of *D. lacteum* is a pterin (VAN HAASTERT et al. 1982a). The evidence is based on UV spectra, competition with 6-methyl-pterin for the acrasinase, the antagonizing effect of 6-aminopterin and the chemical conversion of the acrasin to a pterin. Highly active derivatives, such as D-*erythro*neopterin, D-*erythro*biopterin, L-*threo*neopterin and L-*threo*biopterin can be used as a signal molecule to trigger aggregative cells to react similarly as to their acrasin during cell aggregation. The acrasinase has been shown to be a pterin deaminase.

PAN et al. (1975) postulated that pterins and folates may be food-seeking devices. The identification of this second acrasin of the cellular slime moulds as a pterin makes it clear that food detection and cell aggregation rely on similar signal compounds for their directed movement, and that these two processes are related in the evolutionary sense. We assume that the more primitive species such as *D. lacteum* "socialize" in the aggregative stage by applying the same reaction mechanism with which they detect their food source. The lack of pulsations, relay and formation of streams in this species may simplify the study of the chemotactic reaction mechanism.

15.8 Signal–Receptor Interaction

The chemotactic signal can be detected by measuring a spatial gradient, as has been proposed by ZIGMOND for leukocytes (1974), or by observing a temporal gradient (see KOSHLAND 1980), as is the case in bacterial chemotaxis. In a spatial gradient, cells compare concentrations of chemo-attractants at two or more locations of the cell, and in temporal gradients the cell measures the chemoattractant concentration at one point and after moving a certain distance measures the concentration again and compares the new level with the previous one. Evidence for an amoeboid reaction to a spatial gradient has been presented by MATO et al. (1975) and GERISCH et al. (1975b) proposed a temporal gradient as attracting mechanism for amoebae. Possibly both systems operate together in the analysis of the chemotactic signal. The large number of cyclic AMP derivatives facilitated the study of the relationship between structure and chemotactic activity in *D. discoideum* (KONIJN 1972, 1973, KONIJN and JASTORFF 1973, MATO and KONIJN 1977). Based on the chemotactic activity of about 50 derivatives, the hypothesis has been proposed that cyclic AMP binds to the receptor by its hydrogen bonds at the 6-amino, 7-nitrogen, and 3'-oxygen atom (Fig. 7). Also an ionic bond with the negatively charged phosphate could contribute

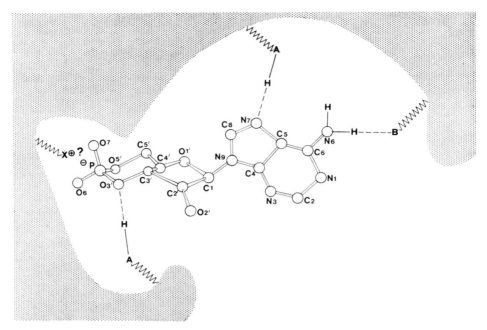

Fig. 7. A model for the cyclic AMP-receptor interaction in *D. discoideum*. (Mato et al. 1978 a)

to the ligand–receptor interaction (Mato et al. 1978a). The receptor binds to the anti-conformation of the base. Some non-cyclic AMP derivatives are also chemotactically active (Mato and Konijn 1977). The base of cyclic AMP binds to the receptor by interaction of its π-electrons and a corresponding acceptor at the active site of the receptor. This model agrees in some aspects with that proposed for the cyclic AMP-receptor interaction in higher organisms (Jastorff et al. 1979); both models have the hydrogen bond at the 3′-oxygen and the hydrophobic interactions between the aromatic base and an acceptor at the active site of the receptor in common.

The binding of cyclic AMP by aggregative cells of *D. discoideum* is reversible and saturable. Each cell has 10^5 to 10^6 receptors at its surface and the dissociation constant is 10^{-8} to 10^{-7} M (Malchow and Gerisch 1974, Green and Newell 1975, Henderson 1975, Mato and Konijn 1975). Non-linear Scatchard plots for cyclic AMP binding suggest more than one cyclic AMP receptor or the regulation of binding by negative cooperativity (Green and Newell 1975). Ligand specificity and the kinetics of association and dissociation on isolated plasma membranes are similar to those on intact cells (Klein 1981). However, these membrane preparations only contain high affinity cyclic AMP receptors and the binding is not oscillatory as was suggested by earlier reports (King and Frazier 1977, Klein et al. 1977). The binding of cyclic AMP is increased by folic acid (Kawai 1980), suggesting that the attractant during the vegetative stage enables cells to respond to the attractant of the aggregative stage.

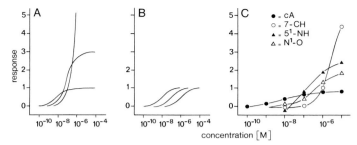

Fig. 8. Theoretical dose-response curves of a rate receptor (**A**) and an occupation receptor (**B**) for a ligand with $K_d = 10^{-8}$ M and ligand derivatives with $K_d = 10^{-7}$ M and 10^{-6} M. (**C**) Phosphodiesterase induction by cyclic AMP and cyclic AMP derivatives. • Cyclic AMP; △ N^1-oxide-adenosine 3′,5′-monophosphate; ▲ 5′-deoxy-5′ aminoadenosine 3′,5′-monophosphate; ○ 7-deazaadenosine 3′,5′-monophosphate. (After VAN HAASTERT et al. 1981)

A basic question concerning the detection of cAMP is how the chemotactic signal is transduced by the cell-surface receptor. Recently, indirect information was obtained from experiments in which the induction of phosphodiesterase activity by a set of cyclic AMP derivatives was measured (VAN HAASTERT et al. 1981). We observed that derivatives that were missing a hydrogen-bond interaction with the receptor showed higher threshold concentrations to induce phosphodiesterase activity. Cells exposed to higher concentrations of such derivatives produce more phosphodiesterase activity than cells exposed to high concentrations of cyclic AMP or a cyclic AMP derivative with a lower threshold concentration (Fig. 8). Differential degradation by phosphodiesterase seems not to be a satisfactory explanation of these results. A reasonable hypothesis is that derivatives that are lacking at hydrogen-bond interaction with the receptor fit less tightly in the receptor than cyclic AMP itself, and these derivatives dissociate faster from the receptor than cAMP. If the response of the receptor depends on the frequency of cyclic AMP–receptor combinations (rate receptor; PATON 1961), and not on the fraction of receptors occupied with cyclic AMP (occupation receptor), then a fast-dissociating cyclic AMP derivative can produce more response than cyclic AMP, which dissociates more slowly (VAN HAASTERT 1980). The close agreement of theoretical and experimental curves (Fig. 8) suggests that the signal for phosphodiesterase induction in *D. discoideum* enters the cell via a rate receptor. The phosphodiesterase induction is discussed more extensively in the next section.

15.9 Cyclic GMP as Mediator

The sequence of events after chemo-attractants bind to their receptors is poorly defined. Pseudopods are protruding at about 5 to 10 s after addition of cyclic AMP (GERISCH et al. 1975c). Cyclic AMP induces a rapid six- to tenfold transient stimulation of guanylate cyclase in a suspension of aggregative *D. discoi-*

deum cells (Mato and Malchow 1978). Cyclic GMP levels rise within 2 s, reach a maximal concentration at about 10 s, and recover prestimulated levels at about 30 s after stimulation with cyclic AMP (Mato et al. 1977a, Wurster et al. 1977). Intracellular cyclic GMP is hydrolyzed by a cyclic GMP specific phosphodiesterase (Dicou and Brachet 1980). The involvement of cyclic GMP in the chemotactic pathway is suggested by the following: (1) all attractants, such as cyclic AMP, folic acid, pterin and purified unidentified acrasins induce in the appropriate species a comparable cyclic GMP accumulation; (2) all attractants induce a cyclic GMP response at physiological concentrations; (3) mutants with altered cyclic GMP metabolism have an altered chemotactic response (Mato et al. 1977d, Ross and Newell 1981); (4) the specificity of cyclic AMP analogues for chemotaxis is similar to the specificity for the cyclic GMP response (Mato et al. 1977c).

Cyclic GMP may bind to intracellular proteins (Mato et al. 1978b, Rahmsdorf and Gerisch 1978). We do not know where cyclic GMP or the cyclic GMP-binding proteins affect the cell. A possible pathway is the stimulation of phospholipid methylation, which occurs by cyclic GMP in homogenates of *D. discoideum* cells (Alemany et al. 1980). An alternative pathway would be the induction of phosphodiesterase activity (Van Haastert et al. 1982b) or chemotaxis.

Immunohistochemical observations support the hypothesis that cyclic GMP may function inside the nucleus. Cyclic GMP has been localized particularly in the nucleus of aggregative cells of *P. violaceum* (Pan and Wedner 1979) and also in post-vegetative *D. discoideum* cells (Mato and Steiner 1980). Cyclic GMP has been localized in the nucleus in rat tissues (Ong et al. 1975) and in polytene *Drosophila melanogaster* chromosomes (Spruill et al. 1978).

15.10 Conclusions

Hormones or hormone-like signals may have originated very early during evolution and presumably constitute a conserved communication system. Since plant hormones are present in animals and vice versa, it should be not surprising that certain hormonal signals or their precursors are found in unicellular organisms, even in prokaryotes. The signal–receptor interactions may be followed by ion fluxes, changes in the cyclic nucleotide levels or increased enzyme activity. Hormone-like substances such as chemo-attractants also trigger transmethylation reactions in leukocytes and amoebae.

In the last 10 years of slime mould research several molecular events have been identified as consequence of the cyclic AMP–cell surface receptor interaction. This is not the place to tie the different processes more or less speculatively together. Instead, we will set out a scheme showing the appearance and/or disappearance of some key elements (Fig. 9) and their threshold concentrations (Fig. 10) (see Van Haastert and Konijn 1982 for further details). The threshold concentrations should be interpreted with caution since the reacting systems may be compartmentalized.

Fig. 9. Time dependency of several responses of aggregative *D. discoideum* cells to cAMP. (VAN HAASTERT and KONIJN 1982)

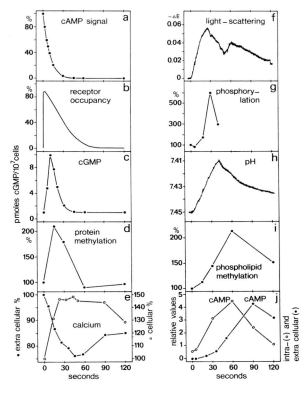

Fig. 10. Concentration dependency of several responses of aggregative *D. discoideum* cells to cAMP. The concentration of cAMP which produces a half-maximal response is indicated by an *arrow*. *Solid arrow* indicates that half-maximal response is taken from complete dose-response curves; *dashed arrow* is an estimated value. (VAN HAASTERT and KONIJN 1982)

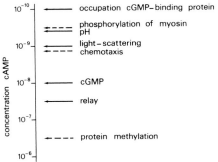

In the coming years these, and perhaps other, key elements will no doubt be located in time and space to fit this information into an integrated molecular network.

References

Abe H, Uchiyama M, Tanaka Y, Satio H (1976) Structure of discadine, a spore germination inhibitor from the cellular slime mold, *Dictyostelium discoideum*. Tetrahedron Lett 3807–3810

Alemany S, Gil MG, Mato JM (1980) Regulation by guanosine 3′,5′-cyclic monophosphate of phospholipid methylation during chemotaxis in *Dictyostelium discoideum*. Proc Natl Acad Sci USA 77:6996–6999

Alton TH, Brenner M (1979) Comparison of proteins synthesized by anterior and posterior regions of *Dictyostelium discoideum* pseudoplasmodia. Dev Biol 71:1–7

Ashworth JM, Watts DJ (1970) Metabolism of the cellular slime mould *Dictyostelium discoideum* grown in axenic culture. Biochem J 119:175–182

Barkley DS (1969) Adenosine-3′,5′-phosphate: identificiation as acrasin in a species of cellular slime mold. Science 165:1133–1134

Barondes SH (1981) Lectins: their multiple endogenous cellular functions. Ann Rev Biochem 50:207–231

Bernstein RL, Rossier C, Van Driel R, Brunner M, Gerisch G (1981) Folate deaminase and cyclic-AMP phosphodiesterase in *Dictyostelium discoideum*: their regulation by extracellular cyclic AMP and folic acid. Cell Diff 10:79–86

Beug H, Katz FE, Stein A, Gerisch G (1973) Quantitation of membrane sites in aggregation *Dictyostelium* cells by use of tritiated univalent antibody. Proc Natl Acad Sci USA 70:3150–3154

Blaskovics J, Raper KB (1957) Encystment stages of *Dictyostelium*. Biol Bull 113:58–88

Bonner JT, Clarke WW, Neely CL, Slifkin MK (1950) The orientation to light and the extremely sensitive orientation to temperature gradients in the slime mold *D. discoideum*. J Cell Comp Physiol 36:149–158

Bonner JT, Kelso AP, Gillmor RG (1966) A new approach to the problem of aggregation in the cellular slime molds. Biol Bull 130:28–42

Bonner JT, Barkley DS, Hall EM, Konijn TM, Mason JW, O'Keefe G, Wolfe PB (1969) Acrasin, acrasinase, and the sensitivity to acrasin in *Dictyostelium discoideum*. Dev Biol 20:72–87

Brachet P, Klein C (1977) Cell responsiveness to cAMP during the aggregation phase of *Dictyostelium discoideum*. Differentiation 8:1–8

Brenner M (1977) Cyclic AMP gradient in migrating pseudoplasmodia of the cellular slime mold *Dictyostelium discoideum*. J Biol Chem 252:4073–4077

Breuer W, Siu CH (1981) Identification of endogenous binding proteins for the lectin-discoidin-I in *Dictyostelium discoideum*. Proc Natl Acad Sci USA 78:2115–2119

Brody T, Williams KL (1974) Cytological analysis of the parasexual cycle in *Dictyostelium discoideum*. J Gen Microbiol 82:371–383

Burns RA, Livi GP, Dimond RL (1981) Regulation and secretion of early developmentally controlled enzymes during axenic growth in *Dictyostelium discoideum*. Dev Biol 84:407–416

Cappuccinelli P (1980) Microtubuli. Chapman and Hall, London

Cappuccinelli P, Unger E, Rubino S (1981) Immunofluorescence of microtubular structures during the cell cycle of *Dicyostelium discoideum*. J Gen Microbiol 124:207–211

Chang MT, Raper KB (1981) Mating types and macrocyst formation in *Dictyostelium rosarium*. J Bacteriol 147:1049–1053

Chastellier de C, Ryter A (1980) Characteristic ultrastructural transformations upon starvation of *Dictyostelium discoideum* and their relations with aggregation. Study of wild-type amoebae and aggregation mutants. Biol Cellulaire 38:121–128

Chung S, Landfear SM, Blumberg DD, Cohen NS, Lodish HF (1981) Synthesis and stability of developmentally regulated *Dictyostelium* mRNAs are affected by cell–cell contact and cAMP. Cell 24:785–797

Clark MA (1977) Possible role for acetylcholine in aggregation centre spacing in *Polysphondylium violaceum*. Nature 266:170–172

Clark MA, Francis D, Eisenberg R (1973) Mating types in cellular slime molds. Biochem Biophys Res Commun 52:672–678

Clarke M, Schatten G, Mazia D, Spudich JA (1975) Visualization of actin fibers associated with the cell membrane in amoebae of *Dictyostelium discoideum*. Proc Natl Acad Sci USA 72:1758–1762

Cooper S, Chambers DA, Scanion S (1980) Identification and characterization of the adenosine 3′,5′-monophosphate binding proteins appearing during the development of *Dictyostelium discoideum*. Biochim Biophys Acta 629:235–242

Cotter DA (1981) Spore activation. In: Turian D, Hohl HR (eds) Fungal spore: morphogenetic controls. Academic Press, London, pp 385–412

Cotter DA, Raper KB (1970) Spore germination in *Dictyostelium discoideum*: Trehalase and the requirement for protein synthesis. Dev Biol 22:112–128

Dahlberg KR, Cotter DA (1978) Activators of *Dictyostelium discoideum* spore germination released by bacteria. Microbios Lett 9:139–146

Darmon M, Brachet P (1979) Aggregation of *Dictyostelium discoideum* amoebae. In: Hazelbauer GL (ed) Receptors and recognition. Ser B Chapman and Hall, London, 5:101–132

Darmon M, Brachet P, Pereira da Silva LH (1975) Chemotactic signals induce cell differentiation in *Dictyostelium discoideum*. Proc Natl Acad Sci USA 72:3163–3166

Devreotes PN, Steck TL (1979) Cyclic 3′,5′-AMP relay in *Dictyostelium discoideum* II. Requirements for the initiation and termination of the response. J Cell Biol 80:300–309

Dicou EL, Brachet PA (1980) A separate phosphodiesterase for the hydrolysis of cyclic guanosine 3′,5′-monophosphate in growing *Dictyostelium discoideum* amoebae. Eur J Biochem 109:507–514

Dimond RL, Burns RA, Jordan KB (1981) Secretion of lysosomal enzymes in the cellular slime mold *Dictyostelium discoideum*. J Biol Chem 256:6565–6572

Dinauer MC, MacKay SA, Devreotes PN (1980) Cyclic 3′,5′-AMP relay in *Dictyostelium discoideum*. III. The relationship of cAMP synthesis and secretion during the cAMP signaling response. J Cell Biol 86:537–544

Durston AJ, Vork F (1979) A cinematographical study of the development of vitally stained *Dictyostelium discoideum*. J Cell Sci 36:261–279

Eitle E, Gerisch G (1977) Implication of developmentally regulated concanavalin A binding proteins of *Dictyostelium* in cell adhesion and cyclic AMP regulation. Cell Differ 6:339–346

Erdos GW, Raper KB, Vogen LK (1973) Mating types and macrocyst formation in *Dictyostelium discoideum*. Proc Natl Acad Sci USA 70:1828–1830

Filosa MF (1979) Macrocyst formation in the cellular slime mold *Dictyostelium mucoroides*: involvement of light and volatile morphogenetic substance(s). J Exp Zool 207:491–495

Fisher PR, Smith E, Williams KL (1981) An extracellular chemical signal controlling phototactic behavior by *D. discoideum* slugs. Cell 23:799–807

Franke J, Kessin R (1977) A defined minimal medium for axenic strains of *Dictyostelium discoideum*. Proc Natl Acad Sci USA 74:2157–2161

Frazier WA, Rosen SD, Reitherman RW, Barondes SH (1975) Purification and comparison of the developmentally regulated lectins from *Dictyostelium discoideum*: discoidin I and II. J Biol Chem 250:7714–7721

Garrod DR, Nicol A (1981) Cell behaviour and molecular mechanisms of cell–cell adhesion. Biol Rev 56:199–242

Garrod DR, Swan AP, Nicol A, Forman D (1978) Cellular recognition in slime mould development. Soc Exp Biol Symp 32:173–202

Geltosky JE, Birdwell CR, Weseman J, Lerner RA (1980) A glycoprotein involved in aggregation of *D. discoideum* is distributed on the cell surface in a nonrandom fashion favoring cell junctions. Cell 21:339–345

George RP, Hohl HR, Raper KB (1972) Ultrastructural development of stalk-producing cells in *Dictyostelium discoideum*, a cellular slime mold. J Gen Microbiol 70:477–489

Gerisch G (1968) Die Bildung des Zellverbandes bei *Dictyostelium minutum*. II. Analyse der Zentrengründung an Hand von Filmaufnahmen. Roux' Arch 157:174–189

Gerisch G (1968) Cell aggregation and differentiation in *Dictyostelium*. Curr Topics Dev Biol 3:157–197

Gerisch G (1980) Univalent antibody fragments as tools for the analysis of cell interactions in *Dictyostelium*. Curr Topics Dev Biol 14:243–270

Gerisch G, Wick U (1975) Intracellular oscillations and release of cyclic AMP from *Dictyostelium* cells. Biochem Biophys Res Commun 65:364–370

Gerisch G, Malchow D, Wilhelms H, Lüderitz O (1969) Artspezifität polysaccharidhaltiger Zellmembran-Antigene von *Dictyostelium discoideum*. Eur J Biochem 9:229–236

Gerisch G, Fromm H, Huesgen A, Wick U (1975a) Control of cell-contact sites by cyclic AMP pulses in differentiating *Dictyostelium* cells. Nature 255:547–549

Gerisch G, Hülser D, Malchow D, Wick U (1975b) Cell communication by periodic cyclic-AMP pulses. Phil Trans Roy Soc Lond B 272:181–192

Gerisch G, Malchow D, Huesgen A, Nanjundiah V, Roos W, Wick U (1975c) Cyclic-AMP reception and cell recognition in *Dictyostelium discoideum*. In: McMahon D, Fox CF (eds) ICN-UCLA Symposia on molecular and cellular biology. Dev Biol Vol 2 Benjamin, Menlo Park Cal, pp 76–88

Gerisch G, Krelle H, Bozzaro S, Eitle E, Guggenheim R (1980) Analysis of cell adhesion in *Dictyostelium* and *Polysphondylium* by the use of Fab. In: Curtis ASG, Pitts JD (eds) Cell adhesion and motility. Cambridge Univ Press, Cambridge London New York Melbourne, pp 293–307

Geuns JML (1982) Plant steroid hormones – what are they and what do they do? TIBS 7:7–9

Green AA, Newell PC (1975) Evidence for the existence of two types of cAMP-binding sites in aggregating cells of *Dictyostelium discoideum*. Cell 6:129–136

Gregg J, Jimenez H, Davis RW (1981) A transitional phase in cell differentiation of *Dictyostelium*. Adv Cyclic Nucl Res 14:705

Gross JD, Town CD, Brookman JJ, Jermyn KA, Peacey ML, Kay RR (1981) Cell patterning in *Dictyostelium*. Phil Trans Roy Soc Lond 295:497–508

Häder DP, Poff KL (1979) Light-induced accumulations of *Dictyostelium discoideum* amoebae. Photochem Photobiol 29:1157–1162

Hashimoto Y, Tanaka Y, Yamada T (1976) Spore germination promoter of *Dictyostelium discoideum* excreted by *Aerobacter aerogenes*. J Cell Sci 21:261–271

Henderson EJ (1975) The cyclic adenosine 3′,5′-monophosphate receptor of *Dictyostelium discoideum*. Binding characteristics of aggregation-competent cells and variation of binding levels during the life cycle. J Biol Chem 250:4730–4736

Hohl HR (1976) Myxomycetes. In: Weber DJ, Hess WM (eds) The fungal spore: form and function. Wiley, Chichester, pp 463–500

Hohl HR, Hamamoto ST (1969) Ultrastructure of spore differentiation in *Dictyostelium*: the prespore vacuole. J Ultrastruct Res 26:442–453

Hohl HR, Jehli J (1973) The presence of cellulose microfibrils in the proteinaceous slime track of *Dictyostelium discoideum*. Arch Microbiol 92:179–187

Jastorff B, Hoppe J, Morr M (1979) A model for the chemical interactions of adenosine 3′,5′-monophosphate with the R subunit of protein kinase type I. Refinement of the cyclic phosphate binding moiety of protein kinase type I. Eur J Biochem 101:555–561

Jefferson BL, Rutherford CL (1976) A stalk-specific localization of trehalase activity in *Dictyostelium discoideum*. Exp Cell Res 103:127–134

Johnson G, Johnson R, Miller M, Borysenko J, Revel JP (1977) Do cellular slime molds form intercellular junctions? Science 197:1300

Juliani MH, Klein C (1981) Photoaffinity labeling of the cell surface adenosine 3′,5′-monophosphate receptor of *Dictyostelium discoideum* and its modification in down-regulated cells. J Biol Chem 256:613–619

Kakebeeke PTJ, Mato JM, Konijn TM (1978) Purification and preliminary characterization of an aggregation-sensitive chemoattractant of *Dictyostelium minutum*. J Bacteriol 133:403–405

Kakebeeke PIJ, De Wit RJW, Kohtz SD, Konijn TM (1979) Negative chemotaxis in *Dictyostelium* and *Polysphondylium*. Exp Cell Res 124:429–433

Kakebeeke PIJ, De Wit RJW, Konijn TM (1980) Folic acid deaminase activity during development in *Dictyostelium discoideum*. J Bacteriol 143:307–312

Kawai S (1980) Folic acid increases the cAMP binding activity of *Dictyostelium discoideum* cells. FEBS Lett 109:27–30

Keating MT, Bonner JT (1977) Negative chemotaxis in cellular slime molds. J Bacteriol 130:144–147

Kessin RH, Orlow SJ, Shapiro RI, Franke J (1979) Binding of inhibitor alters kinetic and physical properties of extracellular cyclic AMP phosphodiesterase from *Dictyostelium discoideum*. Proc Natl Acad Sci USA 76:5450–5454

King AC, Frazier WA (1977) Reciprocal periodicity in cyclic AMP binding and phosphorylation of differentiating *Dictyostelium discoideum* cells. Biochem Biophys Res Commun 78:1093–1099

Klein C (1975) Induction of phosphodiesterase by cyclic adenosine 3′,5′-monophosphate in differentiating *Dictoystelium discoideum* amoebae. J Biol Chem 250:7134–7138

Klein C (1981) Binding of adenosine 3′,5′-monophosphate to plasma membranes of *Dictyostelium discoideum* amoebae. J Biol Chem 256:10050–10053

Klein C, Darmon M (1977) Effects of cyclic AMP pulses on adenylate cyclase and the phosphodiesterase inhibitor of *D. discoideum*. Nature 268:76–78

Klein C, Brachet P, Darmon M (1977) Periodic changes in adenylate cyclase and cAMP receptors in *Dictyostelium discoideum*. FEBS Lett 76:145–147

Konijn TM (1968) Chemotaxis in the cellular slime molds. II. The effect of density. Biol Bull 134:298–304

Konijn TM (1970) Microbiological assay of cyclic 3′,5′-AMP. Experientia 26:367–369

Konijn TM (1972) Cyclic AMP as a first messenger. Adv Cyclic Nucl Res 1:17–31

Konijn TM (1973) The chemotactic effect of cyclic nucleotides with substitutions in the base ring. FEBS Lett 34:263–266

Konijn TM (1975) Chemotaxis in the cellular slime moulds. In: Carlile MJ (ed) Primitive sensory and communication systems. Academic Press, London New York, pp 101–153

Konijn TM, Jastorff B (1973) The chemotactic effect of 5′-amido analogues of adenosine 3′,5′-monophosphate in the cellular slime moulds. Biochim Biophys Acta 304:774–780

Konijn TM, Raper KB (1965) The influence of light on the time of cell aggregation in the Dictyosteliaceae. Biol Bull 128:392–400

Konijn TM, Van de Meene JGC, Bonner JT, Barkley DS (1967) The acrasin activity of adenosine 3′,5′-cyclic phosphate. Proc Natl Acad Sci USA 58:1152–1154

Konijn TM, Chang YY, Bonner JT (1969) Synthesis of cyclic AMP in *Dictyostelium discoideum* and *Polysphondylium pallidum*. Nature 224:1211–1212

Koshland DE (1980) Bacterial chemotaxis as a model behavioral system. Raven Press, New York

Kuczmarski ER, Spudich J (1980) Regulation of myosin self-assembly: phosphorylation of *Dictyostelium* heavy chain inhibits formation of thick filaments. Proc Natl Acad Sci USA 77:7292–7296

Landfear SM, Lodish HF (1980) A role for cyclic AMP in expression of developmentally regulated genes in *Dictyostelium discoideum*. Proc Natl Acad Sci USA 77:1044–1048

Leichtling BH, Spitz E, Rickenberg HV (1981) A cAMP-binding protein from *Dictyostelium discoideum* regulates mammalian protein kinase. Biochem Biophys Res Commun 100:515–522

Le Roith D, Shiloach J, Roth J, Lesniak MA (1980) Evolutionary origins of vertebrate hormones: substances similar to mammalian insulins are native to unicellular eukaryotes. Proc Natl Acad Sci USA 77:6184–6188

Loomis WF (1979) Biochemistry of aggregation in *Dictyostelium*. Dev Biol 70:1–12

Loomis WF (1980) Genetic analysis of development in *Dictyostelium*. In: Leighton T, Loomis WF (eds) The molecular genetics of development. Academic Press, London New York, pp 179–212

MacWilliams HK, Bonner JT (1979) The prestalk-prespore pattern in cellular slime molds. Differentiation 14:1–22

Maeda Y, Takeuchi I (1969) Cell differentiation and fine structures in the development of the cellular slime molds. Dev Growth Differ 11:232–245

Malchow D, Gerisch G (1974) Short-term binding and hydrolysis of cyclic 3′,5′-adenosine monophosphate by aggregating *Dictyostelium* cells. Proc Natl Acad Sci USA 71:2423–2427

Malchow D, Nanjundiah V, Gerisch G (1978) pH oscillations in cell suspensions of *Dictyostelium discoideum*: their relation to cyclic AMP signals. J Cell Sci 30:319–330

Malchow D, Böhme R, Rahmsdorf HJ (1981) Regulation of myosin heavy chain phosphorylation during the chemotactic response of *Dictyostelium* cells. Eur J Biochem 117:213–218

Mangiarotti G, Lefebvre P, Lodish HF (1982) Differences in the stability of developmentally regulated mRNAs in aggregated and disaggregated *Dictyostelium discoideum* cells. Dev Biol 89:82–91

Mato JM, Konijn TM (1975) Chemotaxis and binding of cyclic AMP in cellular slime molds. Biochim Biophys Acta 385:173–179

Mato JM, Konijn TM (1977) The chemotactic activity of cyclic AMP and AMP derivatives with substitutions in the phosphate moiety in *Dictyostelium discoideum*. FEBS Lett 75:173–176

Mato JM, Konijn TM (1979) Chemosensory transduction in *Dictyostelium discoideum*. In: Levandowsky M, Hutner SH (eds) Bioch and Physiol of Protozoa 2nd edn. Academic Press, London New York, pp 181–219

Mato JM, Malchow D (1978) Guanylate cyclase activation in response to chemotactic stimulation in *Dictyostelium discoideum*. FEBS Lett 90:119–122

Mato JM, Marin-Cao D (1979) Protein and phospholipid methylation during chemotaxis in *Dictyostelium discoideum* and its relation to calcium movements. Proc Natl Acad Sci USA 76:6106–6109

Mato JM, Steiner AL (1980) Immunohistochemical localization of cyclic AMP, cyclic GMP and calmodulin in *Dictyostelium discoideum*. Cell Biol Int Rep 4:641–648

Mato JM, Losada A, Nanjundiah V, Konijn TM (1975) Signal input for a chemotactic response in the cellular slime mold *Dictyostelium discoideum*. Proc Natl Acad Sci USA 72:4991–4993

Mato JM, Krens FA, Van Haastert PJM, Konijn TM (1977a) Cyclic AMP-dependent cyclic GMP accumulation in *Dictyostelium discoideum*. Proc Natl Acad Sci USA 74:2348–2351

Mato JM, Van Haastert PJM, Krens FA, Konijn TM (1977b) An acrasin-like attractant from yeast extract specific for *Dictyostelium lacteum*. Dev Biol 57:450–453

Mato JM, Van Haastert PJM, Krens FA, Rhijnsburger EH, Dobbe FCPM, Konijn TM (1977c) Cyclic AMP and folic acid-mediated cyclic GMP accumulation in *Dictyostelium discoideum*. FEBS Lett 79:331–336

Mato JM, Krens FA, Van Haastert PJM, Konijn TM (1977d) Unified control of chemotaxis and cAMP mediated cGMP accumulation by cAMP in *Dictyostelium discoideum*. Biochem Biophys Res Commun 77:399–402

Mato JM, Jastorff B, Morr M, Konijn TM (1978a) A model for cAMP chemoreceptor interaction in *Dictyostelium discoideum*. Biochim Biophys Acta 544:309–314

Mato JM, Woelders H, Van Haastert PJM, Konijn TM (1978b) Cyclic GMP binding activity in *Dictyostelium discoideum*. FEBS Lett 90:261–264

Müller K, Gerisch G (1978) A specific glycoprotein as the target site of adhesion blocking Fab in aggregating *Dictyostelium* cells. Nature 274:445–449

Newell PC, Telser A, Sussman M (1969) Alternative developmental pathways determined by environmental conditions in the cellular slime mold *Dictyostelium discoideum*. J Bacteriol 100:763–768

Nomura T, Tanaka Y, Abe H, Uchiyama M (1977) Cytokinin activity of discadine: a spore germination inhibitor of *Dictyostelium discoideum*. Phytochemistry 16:1819

O'Day DH, Lewis KE (1981) Pheromonal interactions during mating in *Dictyostelium*. In: O'Day DH, Horgen PA (eds) Sexual interactions in eukaryotic microbes. Academic Press, London New York, pp 199–224

Ong SH, Whitley TH, Stowe NW, Steiner AL (1975) Immunohistochemical localization of 3′,5′-cyclic AMP and 3′,5′-cyclic GMP in rat liver, intestine, and testis. Proc Natl Acad Sci USA 72:2022–2026

Pan P, Wedner HJ (1979) Immunohistochemical localization of cyclic GMP in aggregating *Polysphondylium violaceum*. Differentiation 241:1–6

Pan P, Wurster B (1978) Inactivation of the chemoattractant folic acid by cellular slime molds and identification of the reaction product. J Bacteriol 136:955–959

Pan P, Hall EM, Bonner JT (1972) Folic acid as second chemotactic substance in the cellular slime moulds. Nat New Biol 237:181–182

Pan P, Hall EM, Bonner JT (1975) Determination of the active portion of the folic acid molecule in cellular slime mold chemotaxis. J Bacteriol 122:185–191

Paton WDM (1961) A theory of drug action based on the rate of drug-receptor combination. Proc R Soc 154:21–69

Poff KL, Loomis WF, Butler WL (1974) Isolation and purification of the photoreceptor pigment associated with phototaxis in *Dictyostelium discoideum*. J Biol Chem 249:2164–2167

Rahmsdorf HJ, Gerisch G (1978) Specific binding proteins for cyclic AMP and cyclic GMP in *Dictyostelium discoideum*. Cell Differ 7:249–258

Raper KB (1940) Pseudoplasmodium formation and organization in *Dictyostelium discoideum*. J Elisha Mitchell Sci Soc 56:241–282

Raper KB (1941) *Dictyostelium minutum,* a second new species of slime mold from decaying forest leaves. Mycologia 33:633–649

Raper KB, Fennell DI (1952) Stalk formation in *Dictyostelium*. Bull Torrey Bot Club 79:25–51

Raper KB, Thom C (1941) Interspecific mixtures in the *Dictyosteliaceae*. Am J Bot 28:69–78

Ray J, Shinnick Th, Lerner R (1979) A mutation altering the function of a carbohydrate binding protein blocks cell–cell cohesion in developing *Dictyostelium discoideum*. Nature 279:215–221

Reitherman RW, Rosen SD, Frazier WA, Barondes SH (1975) Cell surface species-specific high affinity receptors for discoidin: developmental regulation in *Dictyostelium discoideum*. Proc Natl Acad Sci USA 72:3541–3545

Riedel V, Gerisch G (1971) Regulation of extracellular cyclic-AMP-phosphodiesterase activity during development of *Dictyostelium discoideum*. Biochem Biophys Res Commun 42:119–123

Robertson ADJ, Grutsch JF (1981) Aggregation in *Dictyostelium discoideum*. Cell 24:603–611

Roos W, Gerisch G (1976) Receptor-mediated adenylate cyclase activation in *Dictyostelium discoideum*. FEBS Lett 68:170–172

Rosen SD, Barondes SH (1978) Cell adhesion in the cellular slime moulds. In: Garrod DR (ed) Specificity of embryological interactions. Chapmann and Hall, London, pp 235–264

Rosen SD, Reitherman RW, Barondes SH (1975) Distinct lectin activities from six species of cellular slime molds. Exp Cell Res 95:159–166

Ross FM, Newell PC (1981) Streamers: chemotactic mutants of *Dictyostelium discoideum* with altered cyclic GMP metabolism. J Gen Microbiol 127:339–350

Rubin J (1976) The signal from fruiting body and conus tips of *Dictyostelium discoideum*. J Embryol Exp Morphol 36:261–271

Sampson J (1976) Cell patterning in migrating slugs of *Dictyostelium discoideum*. J Embryol Exp Morphol 36:663–668

Samuel EW (1961) Orientation and rate of locomotion of individual amoebas in the life cycle of the cellular slime mold *Dictyostelium mucoroides*. Dev Biol 3:317–336

Schaap P, Van der Molen L, Konijn TM (1981) Development of the simple cellular slime mold *Dictyostelium minutum*. Dev Biol 85:171–179

Schaap P, Van der Molen L, Konijn TM (1982) Early recognition of prespore differentiation in the development of *Dictyostelium discoideum* and its significance for models on pattern formation. Differentiation 22:1–5

Shaffer BM (1961) The cells founding aggregation centres in the slime mould *Polysphondylium violaceum*. J Exp Biol 38:833–849

Shaffer BM (1975) Secretion of cyclic AMP induced by cyclic AMP in the cellular slime mould *Dictyostelium discoideum*. Nature 255:549–552

Smith E, Williams KL (1980) Evidence for tip control of the slug/fruit switch in slugs of *Dictyostelium discoideum*. J Embryol Exp Morphol 57:233–240

Spruill WA, Hurwitz DR, Lucchesi JC, Steiner A (1978) Assocation of cyclic GMP with gene expression of polytene chromosomes of Drosophila melanogaster. Proc Natl Acad Sci USA 75:1480–1484

Steinemann C, Parish RW (1980) Evidence that a developmentally regulated glycoprotein is target of adhesion-blocking Fab in reaggregating *Dictyostelium*. Nature 286:621–623

Sussman M, Schindler J (1978) A possible mechanism of morphogenetic regulation in *Dictyostelium discoideum*. Differentiation 10:1–5

Takeuchi I (1963) Immunochemical and immunohistochemical studies on the development of the cellular slime mold *Dictyostelium mucoroides*. Dev Biol 8:1–26

Tasaka M, Takeuchi I (1981) Role of cell-sorting in pattern formation in *Dictyostelium discoideum*. Differentiation 18:191–196

Taya Y, Tanaka Y, Nishimura S (1978) 5'-AMP is a direct precursor of cytokinin in *Dictyostelium discoideum*. Nature 271:545–547

Taya Y, Yamada T, Nishimura S (1980) Correlation between acrasins and spore germination inhibitors in cellular slime molds. J Bacteriol 143:715–719

Tomchik KJ, Devreotes PN (1981) Adenosine 3',5'-monophosphte waves in *Dictyostelium discoideum*: a demonstration by isotope dilution-fluorography. Science 212:443–446

Tsang A, Bradbury JM (1981) Separation and properties of prestalk and prespore cells of *Dictyostelium discoideum*. Exp Cell Res 132:433–441

Van Haastert PJM (1980) Distinction between the rate theory and the occupation theory of signal transduction by receptor activation. Neth J Zool 30:473–493

Van Haastert PJM, Konijn TM (1982) Signal transduction in the cellular slime molds. Mol Cell Endocr 26:1–17

Van Haastert PJM, Van der Meer RC, Konijn TM (1981) The rate of association of cyclic AMP to its chemotactic receptor induces phosphodiesterase activity in *Dictyostelium discoideum*. J Bacteriol 147:170–175

Van Haastert PJM, De Wit RJW, Grijpma Y, Konijn TM (1982a) Identification of a pterin as the acrasin of the cellular slime mold *Dictyostelium lacteum*. Proc Natl Acad Si USA 79:6270–6274

Van Haastert PJM, Pasveer FJ, Van der Meer RC, Van der Heijden PR, Van Walsum H, Konijn TM (1982b) Evidence for a messenger function of cyclic GMP during phosphodiesterase induction in *Dictyostelium discoideum*. J Bacteriol 152:232–238

Wick U, Malchow D, Gerisch G (1978) Cyclic AMP stimulated calcium influx into aggregating cells of *Dictyostelium discoideum*. Cell Biol Int Rep 2:71–79

Williams KL, Kessin RH, Newell PC (1974) Genetics of growth in axenic medium of the cellular slime mould *Dictyostelium discoideum*. Nature 247:142–143

Wurster B, Pan P, Tyan GC, Bonner JT (1976) Preliminary characterization of the acrasin of the cellular slime mold *Polysphondylium violaceum*. Proc Natl Acad Sci USA 73:795–799

Wurster B, Schubiger K, Wick U, Gerisch G (1977) Cyclic GMP in *Dictyostelium discoideum*. FEBS Lett 76:141–144

Wurster B, Bek F, Butz U (1981) Folic acid and pterin deaminases in *Dictyostelium discoideum*: kinetic properties and regulation by folic acid, pterin, and adenosine 3',5'-phosphate. J Bacteriol 148:183–192

Yamamoto A, Maeda Y, Takeuchi I (1981) Development of an autophagic system in differentiation cells of the cellular slime mold *Dictyostelium discoideum*. Protoplasma 108:55–69

Zigmond SH (1974) Mechanisms of sensing chemical gradients by polymorphonuclear leukocytes. Nature 249:450–452

16 Sexual Interactions
in the Lower Filamentous Fungi

H. Van den Ende

16.1 Introduction

Sexual development involves the formation of cells competent to fuse, the elaboration of a mechanism to bring such cells into physical contact, and finally the cellular fusion process. An important feature is that two different cells, organs or whole thalli are involved, which implies that sexual development in both partners must be well coordinated in time and space by some sort of communication. Particularly this aspect of sex has been extensively studied in some of the lower filamentous fungi, due to the fact that extracellular metabolites, acting as messengers and designated as pheromones (cf. Kochert 1978), have been amenable to chemical analysis. This has enabled us to study the biosynthesis and action of these substances in a more accurate way and so to gain a better understanding of sexual interaction processes. In this chapter, an account will be presented of the knowledge acquired in the last 10 years. The indication of some major gaps in that knowledge will hopefully stimulate further studies.

16.2 Chytridiomycetes

An outstanding case in this group of fungi is the genus *Allomyces*. In the haploid gametophytic phase, gametangial development is induced by starvation conditions. In a fairly synchronous manner, pairs of cells are delimited by the formation of septa in the apical hyphal zone. One of the cells is determined to become a male and the other to become a female gametangium. Whether the apical cell becomes male or female is dependent on the species concerned. The mechanism underlying this example of phenotypic sex determination has received considerable attention (Emerson and Wilson 1954, Ronne and Olson 1976, Morrison 1977, Nielsen and Olson 1982, Olson et al. 1982) but remains a challenge in developmental biology.

The gametes, produced by the gametangia, are cell-wall-less and uniflagellate. There are several morphological and behavioural differences between male and female gametes. The latter are larger, contain more mitochondria and gamma particles, and are less motile. They strongly attract the very motile male gametes. Sexual fusion results in biflagellate zygotes that swim for a considerable time before settling down. Pommerville and Fuller (1976) and Pommerville (1982) have provided evidence that there is a specific region of the plasma

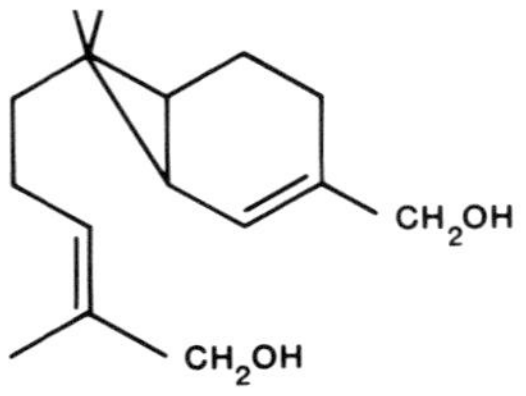

Fig. 1. Structure of sirenin

membrane at the anterior end of male gametes, which is involved in plasmogamy. Whether there is a mechanism to avoid polyspermy, is unknown.

The attraction of male gametes by female gametes is accomplished by the
secretion of a metabolite called sirenin. This substance was isolated from large
cultures of strongly female aneuploid hybrids of *Allomyces macrogynus* and
A. arbusculus (MACHLIS 1958, 1973, MACHLIS et al. 1966). Male gametes of
wild-type *A. macrogynus* respond vigorously to sirenin, much better than those
of *A. arbusculus* (MACHLIS 1968). The species specificity of sirenin has not been
further investigated.

The identification of sirenin was carried out by MACHLIS et al. (1968) and
NUTTING et al. (1968). Its structure, shown in Fig. 1, was confirmed by chemical
synthesis. Only the L-enantiomer exhibited activity (PLATTNER and RAPOPORT
1971, MACHLIS 1973). Optimal activity was found in the range of 10^{-10}–10^{-5} M
(CARLILE and MACHLIS 1965).

POMMERVILLE (1978) has studied the behavioural response of male gametes
towards sirenin by dark-field microscopy. These cells swim in short, curved
paths, interrupted by jerks, which result in a change of direction. On mixing
male and female gametes, or on introducing a sirenin source in a suspension
of male gametes, this behaviour changes dramatically. When swimming up a
gradient of sirenin, the cells move in long, spiralled paths, without the frequent
jerks observed in random patterns. In contrast, swimming parallel to, or down,
a sirenin gradient results in frequent directional changes. This pattern continues
until the cell is again moving up the gradient. The net effect of this behaviour
is an efficient approach of male gametes towards female gametes, or any other
point source of sirenin.

The mode of action of sirenin is not known. In behavioural respect there
are similarities with bacterial chemotaxis and phototaxis of, for instance, *Euglena* (MACNAB 1978, BEAN 1979). Thus, it is probable that a concentration
gradient of sirenin is sensed as a function of time, because *Allomyces* gametes,
like bacteria with respect to amino acids, show a behavioural response when
they are transferred from one uniform sirenin concentration to another (POM-
MERVILLE 1981).

An additional effect of sirenin on male gametes, noted by POMMERVILLE
(1978) is the induction of agglutination. Male cells clump together and are
sometimes attracted en masse towards a sirenin source. Homotypic agglutination
caused by an attractant has also been observed in other species with motile
gametes, such as *Fucus* (MÜLLER 1968). Whether this phenomenon has any
natural significance is unknown. Equally puzzling is the finding that male gametes produce a substance which antagonizes the attracting action of sirenin
(KLAPPER and KLAPPER 1977).

16.3 Oomycetes

Sexual differentiation in *Achlya,* as in all Saprolegniales, concerns the elaboration of antheridial and oogonial initials on the same or different, but proximal, thalli. Antheridial initials are chemotropically attracted to oogonial initials, and on contact, antheridia and oogonia are produced by delimitation of the apices by a cross-wall. Meiosis takes place in the gametangia, implying that the vegetative mycelium is diploid (BARKSDALE 1966). The egg cells, subsequently produced in the oogonia, are fertilized by fertilization tubules arising from the antheridia. The older literature on sexual interactions in *Achlya* has been reviewed by RAPER (1967).

As RAPER pointed out, the system has a number of remarkable features. In the first place, any one heterothallic, self-sterile, strain does not exhibit a predetermined sexuality, although the different strains can be arranged in order of increasing femaleness. If two strains interact sexually, then the least female strain takes on the male role. Thus the sex expressed by an individual is partially determined by the partner (RAPER 1960). The genetics of this pattern of sexuality is obscure (cf. also SHERWOOD 1969). A second feature is that sexual development follows a sequential pattern which suggests that stage-characteristic pheromones are involved. This has been confirmed by subsequent studies: the pheromone *antheridiol,* produced constitutively by strongly female mycelium (BARKSDALE and LASURE 1973), induces antheridial initials in male-acting mycelium which, in response, produces the oogoniols, inducing the female to form oogonial initials. Antheridial hyphae are attracted towards a source of antheridiol (BARKSDALE 1963), as well as to oogonial initials, which implies that oogonial initials are a relatively rich source of antheridiol. However, the involvement of other pheromones is not excluded (for a discussion, see VAN DEN ENDE 1976). The consecutive production and action of these two pheromones explains a great deal of the sequential pattern of sexual morphogenesis in *Achlya.*

Antheridiol (Fig. 2) was isolated by McMORRIS and BARKSDALE (1967) from *Achlya bisexualis,* and characterized by ARSENAULT et al. (1968) and synthesized by EDWARDS et al. (1969). The oogoniols were isolated from a bisexual strain of *Achlya heterosexualis* and characterized by McMORRIS et al. (1975) and PREUS and McMORRIS (1979). They consist of a mixture of closely related compounds (Fig. 3), of which those with unsaturated side chain proved to be

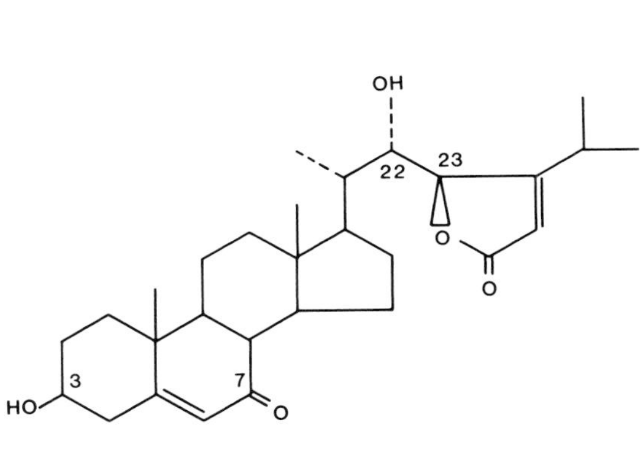

Fig. 2. Structure of antheridiol

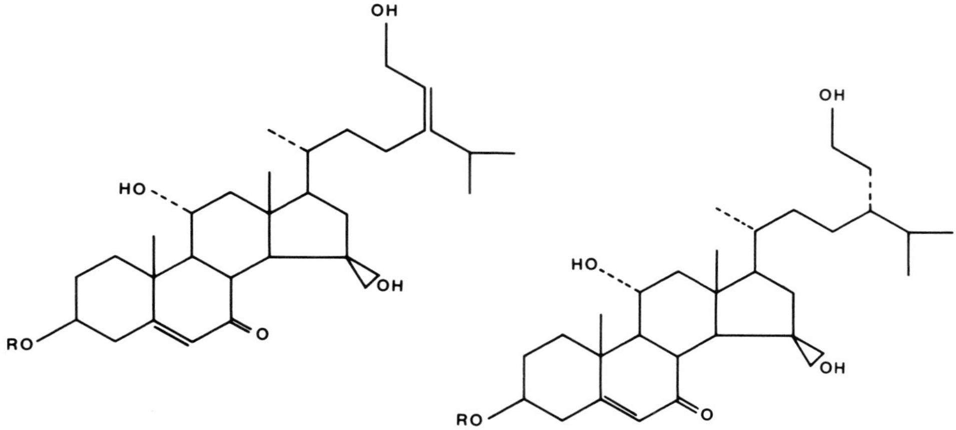

Fig. 3. Structure of the oogoniols. R = H, $(CH_3)_2CHCO$, or CH_3CH_2CO

the most active. The mixture of biologically active molecules was qualitatively similar in preparations derived from either the bisexual *A. heterosexualis* or the strongly male *A. ambisexualis* A87, which indicates that the hydrogenation of the dehydro-oogoniols to the much less active oogoniols is not the result of their oogonia-inducing action, and thus seems to be a characteristic of the male.

Several authors have demonstrated that fucosterol (I) is a precursor of these pheromones. POPPLESTONE and UNRAU (1974) found by labelling studies that the compounds I, II and III (Fig. 4) are incorporated into antheridiol (IV). Therefore, the first step in the modification of the fucosterol side chain is the dehydrogenation at C-22, C-23, which is then followed by oxidation at C-29, presumably via the sequence $-CH_3$, $-CH_2OH$, $-CHO$, $-COOH$. On the other hand, in the synthesis of oogoniol the oxidation at C-29 is the first step, for BARROW and McMORRIS (1982) have shown that 29-hydroxyfucosterol (V) as well as 29-oxofucosterol (VI) can act as precursors of the oogoniols. Reduction of the C-24, C-28 double bond may take place after the functional groups at C-7, C-11 and C-15 have been introduced, because the oogoniols (VIII) and the dehydro-oogoniols (VII) all had equal specific radioactivities after incorporation of radioactive fucosterol. There is some similarity between the Mucorales (see below) and *Achlya* with regards to pheromone biosynthesis, in that a common precursor is metabolized via two different sex-specific reaction sequences. Moreover, they are analogous in that one metabolic pathway is stimulated by the end product of the other pathway, since antheridiol stimulates oogoniol production in strong males (McMORRIS and WHITE 1977). It would be interesting to see whether antheridiol production is stimulated by the oogoniols, and thus a reciprocally stimulating action would be displayed as found in, far example *Mucor mucedo* (WERKMAN and VAN DEN ENDE 1974).

In male strains, antheridiol has still another action, namely the induction of an enzyme system which converts antheridiol to one or two biologically

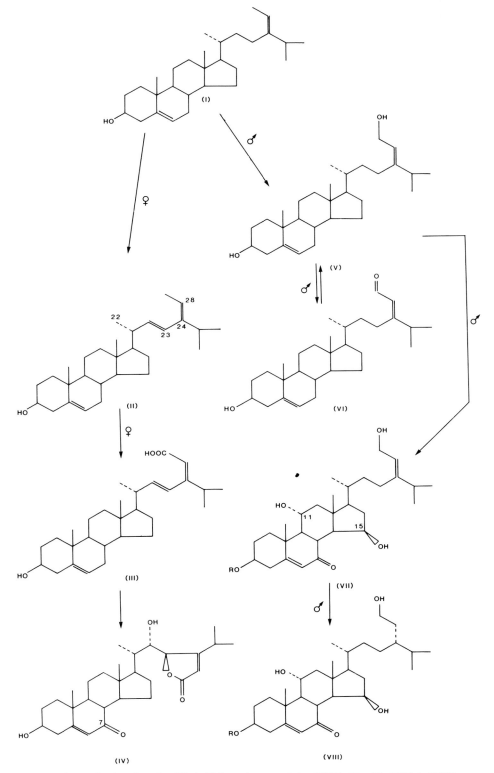

Fig. 4. Biosynthesis of antheridiol (IV) and oogoniols (VIII). R = H, (CH₃)₂CHO, or CH₃CH₂CO

inactive metabolites (Musgrave and Nieuwenhuis 1975, Musgrave et al. 1978). In several homothallic strains this enzyme system appeared to be present constitutively. It is an attractive possibility that it concerns enzyme activities involved in the oogoniol pathway, since a correlation between the rate of antheridiol metabolism and the readiness of the treated mycelium to act as a male was apparent. However, the inactivation of antheridiol could also be functional, with regards to the chemotropic growth of antheridial initials up an antheridiol gradient. By metabolizing antheridiol, male hyphae would decrease its concentration locally and so make a steeper concentration gradient, which would be easier to detect.

Several other metabolic effects of antheridiol on male mycelium have been noted. One of these is the induction of cellulase synthesis, which might be instrumental in the production of antheridial initials (Mullins and Ellis 1974, Nolan and Bal 1974). It also enhances the production of ribosomal and poly(A)-enriched RNA in vivo (Silver and Horgen 1974, Timberlake 1976), as well as in preparations of isolated chromatin (Sutherland and Horgen 1977). Horgen (1977) demonstrated that the transcriptional activity of chromatin, isolated from non-induced male cultures of *Achlya ambisexualis,* was enhanced after preincubation with a mixture of antheridiol and a cytoplasmic fraction of male mycelium, rather than with antheridiol or the cytoplasmic fraction alone. However, mRNAs, specific for antheridiol-treated mycelium, have not been detected (Rozek and Timberlake 1980, Gwynne and Brandhurst 1980).

In the related Peronosporales, and particularly in species of *Pythium* and *Phytophthora,* the sexual system is reminiscent of that in Achlya. In heterothallic species of *Phytophthora,* two "compatibility types" A_1 and A_2 are distinguished, which produce sex organs and oospores when mated (e.g. Galindo and Gallebly 1960). Many instances of interspecific mating have been described, involving these compatibility types, which suggests that pheromonal interactions take place with low species specificity (Brasier 1972). However, Brasier (1972) also confirmed an old observation, described by Kniep (1928), that during interaction between two different mycelia, selfing occurred as well. Ko (1978) demonstrated that pairing compatible mycelia of different species on opposite sides of polycarbonate membranes resulted in the production of oospores, the membranes not being penetrated by hyphae. Induction of oospore formation also occurred when heterothallic A_1 of A_2 isolates of *Phytophthora parasitica* were paired with typical homothallic strains (i.e. strains which produce oospores without the proximity of a mycelium with different sexual characteristics, Ko 1980). The stimulation of sexual reproduction in potentially homothallic species is reminiscent of results with some homothallic Mucorales (Schipper 1971, Werkman and Van den Ende 1974, Werkman et al. 1977), obtained with sex-specific pheromones. The above mentioned results infer that such agents are also operative in *Phytophthora* (cf. also Ko and Kunimoto 1981).

Species of *Pythium* and *Phytophthora* are unable to synthesize sterols, but they need them for the development of asexual and sexual reproductive organs. The effectiveness of a number of sterols was investigated by Elliott (1972).

Fucosterol, as well as β-sitosterol, stigmasterol and 5 Δ-avenasterol, showed highest activity, followed by cholesterol, ergosterol and cholestanol. Fucosterol and cholesterol are major sterols in the related Saprolegniales, but whether they serve as precursors for agents involved in sexual interaction, or rather have a structural function, is unknown (MCCORKINDALE et al. 1969). NES et al. (1980) suggested that the sterol requirement might be related to the parasitic nature of these fungi. In any case, antheridiol is inactive in *Phytophthora*.

16.4 Zygomycetes

Sexual reproduction in heterothallic Mucorales comprises (1) the elaboration of specialized hyphae ("zygophores") on vegetative mycelia of opposite mating type, when they are proximal to each other; (2) the chemotropic growth of zygophores towards those of the opposite mating type; (3) the formation of progametangia in both interacting zygophores, and (4) the hydrolysis of the adjoining cell wall and fusion of the two gametangial protoplasts. In homothallic species, two interacting hyphae originate from one thallus. Sexual morphogenesis has been described in detail by e.g. ZYCHA et al. (1969), GOODAY (1973) and O'DONNELL et al. (1978).

Incomplete sexual interaction may occur between different species, which suggests that, as in *Achlya*, common pheromones are involved (BLAKESLEE and CARTLEDGE 1927). In recent studies this has been established for several heterothallic and homothallic species. In most physiological studies, however, *Mucor mucedo* has served as model system, because it has easily recognizable zygophores.

It has been demonstrated by a number of groups (reviewed by BU'LOCK 1976, SUTTER 1977, VAN DEN ENDE 1978, JONES et al. 1981) that the trisporates as well as trisporate precursors, are active as zygophore-inducing pheromones. Figure 5 shows the structures of the metabolites of major importance as far

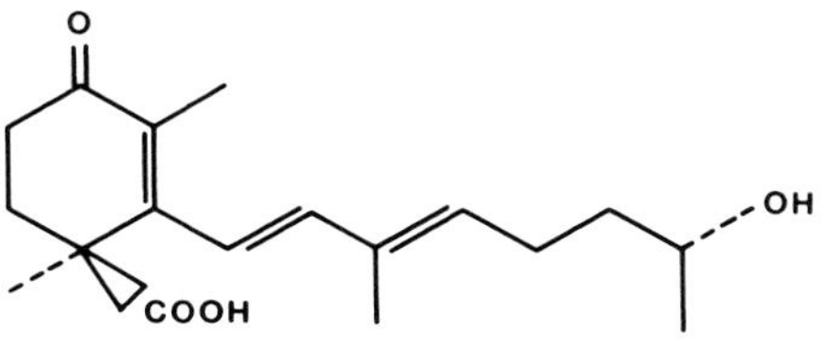

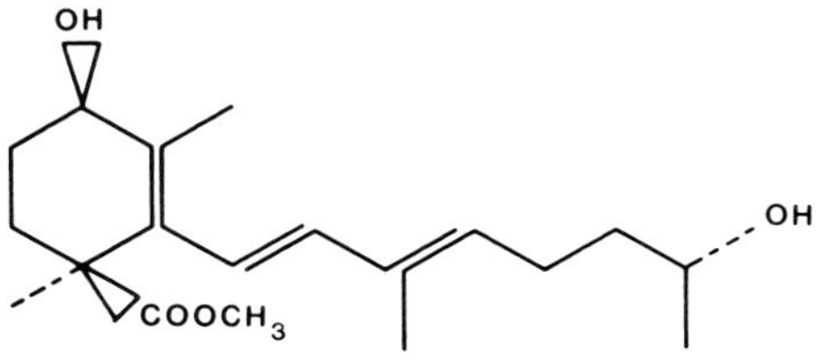

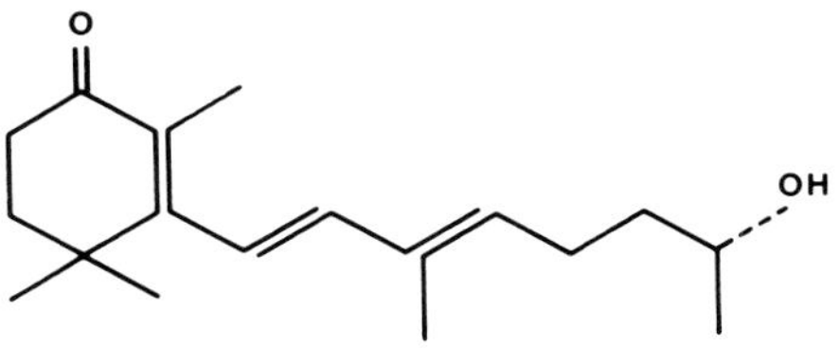

Fig. 5. Structure of the C series of trisporic acid (I), methyl 4-dihydrotrisporate (II) and trisporin (III). In the B series, a carbonyl group is present at C-13

as they have been isolated from culture media. The structures have been confirmed by chemical synthesis (e.g. Prisbylla et al. 1979).

The trisporates (I), which are the most abundant pheromones in, for example, *Blakeslea trispora* and *M. mucedo,* are synthesized by a pathway in which both mating types participate (Fig. 6). A mycelium of plus mating type (mt^+) can only accomplish a limited part of the pathway which results in the formation of the methyl 4-dihydrotrisporates (IV). The end products in the mt^- characteristic part of the pathway are the trisporins (V). When cultures of both mating types are mixed, they act complementarily to produce the trisporates. Both pathways could start with the 4-dihydrotrisporin(s) (VI), but there is no evidence to support this. Whatever its structure, it is probably a degradation product of cyclic carotenoids (Bu'Lock et al. 1974). A characteristic of mt^+ mycelium is the capacity to oxidize the pro-S-methyl group at C-1 (probably via $-CH_2OH$ and $-COOH$ intermediates, as in the biosynthesis of antheridiol, Bechtold et al. 1972). Typical for mt^- mycelia is the ability to oxidize the hydroxyl group at C-4; the dehydrogenase involved has been partially characterized; likewise an esterase, involved in the hydrolysis of the carboxy-methyl group at C-1, has been found in mt^- mycelium. Experimental support for the trisporate pathway has been presented by Sutter et al. (1973), Bu'Lock et al. (1974), Nieuwenhuis and Van den Ende (1975) and Werkman (1976) for a number of heterothallic systems.

As has been emphasized by Bu'Lock et al. (1973) and Sutter and Whitaker (1981 a, b), sexual morphogenesis and also the synthesis of enzymes, responsible for the trisporate biosynthetic pathway in heterothallic Mucorales can be envisaged as being governed by alleles of a mating-type locus. Consequently, the occurrence of specific mt^+ and mt^- metabolites might reflect the differential (de)repression of genes, coding for the appropriate enzymes. However, if this system is operating, it is leaky, particularly in mt^+ strains, for dehydrogenase activity (oxidizing methyl 4-dihydrotrisporates) has been found not only in mt^- but also in mt^+ mycelia of *Mucor mucedo, M. hiemalis* and *Phycomyces blakesleeanus* (Van der Does, personal communication). A similar lack of mating-type specificity has been noted for the hydrolysis of methyl trisporates (Bu'Lock et al. 1976). In addition, Sutter et al. (1973) found small amounts of trisporates in mt^+ cultures of *B. trispora*. Thus, pheromones with zygophore-inducing action in mt^+ strains (see below) may be produced by mt^+ mycelium, due to the incomplete repression of the responsible enzyme(s). For that reason, the inactive metabolite trisporone (Fig. 7) found in relatively large amounts in mt^+ cultures of *B. trispora,* has been considered as the product of a "detoxification" process (Sutter and Whitaker 1981 a, b).

Werkman and Van den Ende (1974) observed that the trisporates have a promoting action on their own biosynthesis; probable the synthesis of enzymes involved in the biosynthetic sequence is stimulated, since inhibitors of protein synthesis effectively block this stimulation. This positive feed-back could explain the quite substantial production of trisporates in several species.

One of the most remarkable facts of this system is the mating-type specificity of the zygophore-inducing action of the trisporates and their intermediates. Table 1 shows the absolute specificity of methyl 4-dihydrotrisporates (VI) and

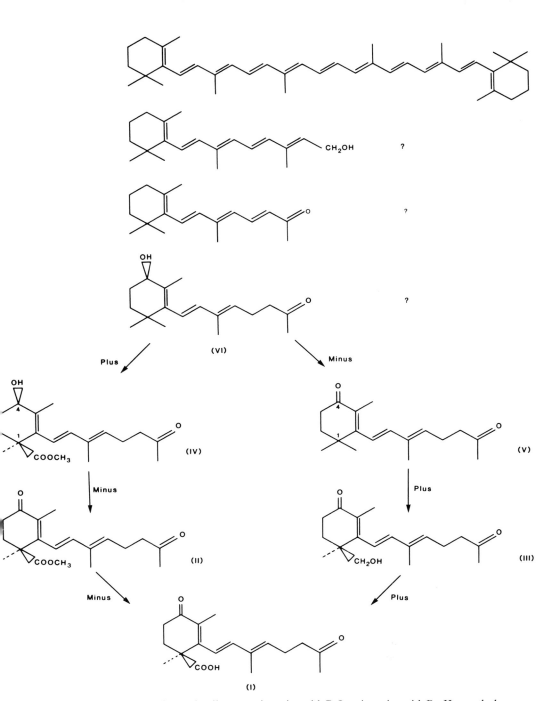

Fig. 6. Biosynthetic pathway leading to trisporic acid B I: trisporic acid B; II: methyl trisporate B; III: trisporol B; IV: methyl 4-dihydrotrisporate B; V: trisporin B; VI: 4-dihydrotrisporin B. The biosynthesis of trisporic acid C is assumed to proceed in a similar way (replace $-COCH_3$ at C12 by $-CHOHCH_3$), although interconversion between intermediates of the B- and C-series also might be possible (VAN DEN ENDE 1978)

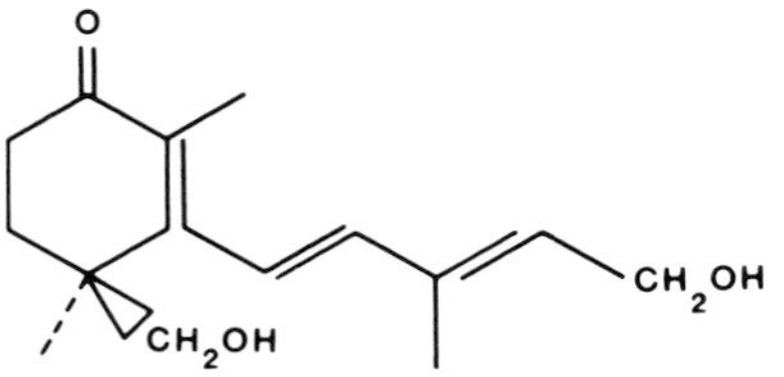

Fig. 7. Structure of trisporone

Table 1. Relative zygophore-stimulating activities of trisporic acid derivatives. (Sutter and Whitaker 1981a)

Compound	Activities[a]			
	P. blakesleeanus		M. mucedo	
	mt^+	mt^-	mt^+	mt^-
Methyl 4-dihydrotrisporate B	>320	1	>8,900	4
Methyl 4-dihydrotrisporate C	>640	4	>17,900	75
Methyl 4-dihydrotrisporate C[b]	>6,330	11	>177,000	75
Methyl trisporate B	310	3	290	4
Methyl trisporate C	2,100	21	19,200	39
Trisporic acid B	90	90	2	1
Trisporic acid C	360	360	40	10

[a] The numbers indicate the lowest amount of a compound giving a definite positive response in bioassays for zygophores (*M. mucedo*) and enlarged zygophores (*P. blakesleeanus*), relative to the compound which was active in the lowest absolute quantity. 1 equals 27 pmol in *M. mucedo* and 745 pmol in *P. blakesleeanus*

[b] Prepared by chemical modification of trisporates

methyl trisporates (II) for mt^-, while the trisporates show activity on both mating types. Comparable data from Bu'Lock et al. (1976) and Nieuwenhuis and Van den Ende (1975) show that the trisporins (V) and the trisporols (VII) are active only on mt^+. This specificity is most easily explained by assuming that the precursors derive their activity by mating-type-specific transformation to the trisporates. This would imply that the trisporates are the intracellular agents, controlling sexual morphogenesis. This is supported by results with *P. blakesleeanus,* which is quite insensitive to extracellular trisporates (Sutter 1975; cf. Table 1). Therefore the trisporate precursors, in particular the methyl 4-dihydrotrisporates (VI) and the trisporins (V), respectively, should be considered as the pheromones involved in intercellular communication (Sutter et al. 1973). This is corroborated by the observation that these relatively volatile agents can also be exchanged via the atmosphere (Burgeff 1924, Hepden and Hawker 1961, Mesland et al. 1974). It is clear that only investigations of the primary mode of action at the molecular level can support such assessments.

A second action of the trisporates is the stimulation of terpenoid production (reviewed by Gooday 1978). In *M. mucedo,* this is most conspicuous at the zones where both mating types interact, and where a massive accumulation of carotenoids takes place, but it has also been observed in other species, for example in *P. blakesleeanus* (SUTTER 1977, SCHIPPER 1978). Also the production of sterols, prenols and ubiquinones is stimulated by the trisporates (THOMAS and GOODWIN 1967, BU'LOCK and OSAGIE 1973). The mechanism is unknown. This enhanced isoprenoid biosynthesis could be connected with the fact that sporopollenin (a carotenoid polymer) is a major constituent of the zygospore cell wall (GOODAY et al. 1973). The above-mentioned stimulation of trisporate biosynthesis by trisporates could also be partly a function of enhanced isoprenoid production.

Zygophores grow towards each other (PLEMPEL 1962, GOODAY 1973) in response to sex-specific gaseous stimuli, emanating from both partners. However, it is uncertain whether this response is due to oriented tip growth, or rather to bending of the zygophores. In any case, zygophore orientation is accompanied by an increase in growth rate (MESLAND et al. 1974). The suggestion put forward by these authors, that trisporate precursors are involved as attractive agents, awaits confirmation.

There is evidence that trisporate derivatives are also operative in homothallic Mucorales. Incomplete mating between homo- and heterothallic species has been noted (BURGEFF 1924); some homothallics produce zygospores only in the presence of a heterothallic mycelium or after application of trisporate derivatives. Some of these derivatives have also been isolated from homothallic species (WERKMAN and VAN DEN ENDE 1974). It is worth mentioning in this respect that dehydrogenase activity, converting methyl 4-dihydrotrisporates to methyl trisporates, was found to be localized at specific sites in homothallic mycelium. The same sites displayed a mt^- character in interspecific matings with heterothallic species (WERKMAN 1976). An example is presented in Fig. 8.

The problem of how cells of different mating type recognize each other upon cell-to-cell contact and what triggers the fusion of the respective cell walls and protoplasts, remains unsolved. Studies of the cell surface have so far not revealed mating type-specific determinants (JONES and GOODAY 1978).

As mentioned above, it has been inferred that the genetic basis of sex in heterothallic Mucorales resides in a regulatory *mt* locus with two functional alleles, mt^+ and mt^-. Whether these alleles govern trisporate metabolism, as described above, has not been demonstrated by genetical analysis but it seems a reasonable assumption in view of the parallel biosynthetic patterns in the respective types of various species. Recent genetic studies with *P. blakesleeanus* have shown that the zygotic progeny, with mating type as marker, can essentially be interpreted in Mendelian terms (ESLAVA et al. 1975, CERDA-OLMEDO 1975, ALVAREZ et al. 1980). In these studies it appeared that in the zygospore of this fungus, out of many pairs of nuclei, only one, or very few fuse to form diploid nuclei heterozygous for mating type, which give rise to haploid progeny via meiosis, while all the other parental nuclei are destroyed. This confirms an earlier assertion of BURGEFF (1928). Thus, the germ spores (in this species normally homokaryotic) arising from the zygospore, contain mitotic descendants of one or a few meiotic divisions. However, other reported phenomena tend to complicate this picture. Immature zygospores can be induced to undergo precocious germination, but then the results with respect to the genotype of the progeny are not quite consistent with those obtained with mature zygospores

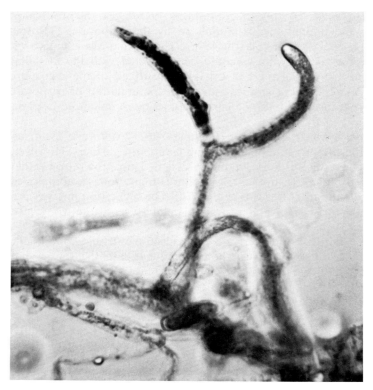

Fig. 8. Young zygophores of the homothallic *Zygorhynchus moelleri*. Glutaraldehyde-fixed mycelium was incubated with methyl 4-dihydrotrisporates, NADP and p-nitro-blue tetrazolium chloride. Dehydrogenase activity was manifested by the appearance of blue crystals, due to the formation of NADPH. (WERKMAN 1976)

(GAUGER 1977, GAUGER et al. 1980). Such immature zygospores give rise to considerable percentages of bisexual heterokaryotic mycelia, which either segregate readily into homokaryons (*Rhizopus stolonifer*, GAUGER 1977), or are quite stable and can be maintained via asexual sportes (*P. blakesleeanus*, HEISENBERG and CERDA-OLMEDO 1968). Heterokaryons of *P. blakesleeanus* have some typical features which are reminiscent of metabolic and morphological effects of sex pheromones: there is increased carotenoid production, sporangiophore formation is reduced and characteristic "pseudophores", cork-screwed, suspensor-like structures are formed. However, no reports regarding sex-pheromone production in heterokaryons is available. It should be stressed that such heterokaryons might be useful for the study of the regulation of mating-type expression. A similar point can be made concerning the diploid or aneuploid strains, heterozygous for mating type, which produce azygospores (i.e. zygospores borne on one suspensor, GAUGER 1975, SCHIPPER 1976).

No information is available with regards to the genetic basis of sex in homothallic forms. A particular part of a single hypha may attain the phenotypical characteristics of either mt^+ of mt^-, even with respect to trisporate biosynthesis (cf. Fig. 7). Since interconversion between mt^+ and mt^- characteristics is easily observable (BURGEFF 1924), it may be assumed that also at the gene level switching of mating type expression occurs in individual nuclei. Its regulation could be analogous to the expression of the mt locus in *Saccharomyces cerevisiae* (KLAR et al. 1981) and *Schizosaccharomyces pombe* (EGEL 1981, cf. SCHIPPER and STALPERT 1980). The fact that mating type conversion has been induced in *M. pusillus* by gamma irradiation, with concomitant changes in sex pheromone metabolism (NIELSEN 1978, BU'LOCK and HARDY 1979), should encourage further studies in that line.

16.5 Concluding Remarks

It must be emphasized that the systems described are a most incomplete cross-section of the wealth of sexual phenomena, encountered in the lower filamentous fungi. Many problems, aptly formulated by students of sex since the beginning of the century, remain unsolved, and the extreme diversity of this group of thallophytes prevents generous extrapolations from being made.

With the application of modern analytical techniques, much information has recently been acquired, but invariably it has aroused new questions, which are of considerable developmental interest. What is the mode of action of the pheromones which display such a remarkable specificity? How do they direct morphogenetic processes at the cellular and supracellular level? Clearly the input of still more biochemical know-how is required to obtain the answers. In yet another way, these studies are dependent on progress made in genetics. Generally, the molecular basis of sex determination is poorly understood, particularly in the fungi described in this chapter. A better understanding requires the production of sexual mutants and heterokaryons, the development of methods to germinate sexual spores and the application of the methods used in molecular genetics. This multidisciplinary approach is needed to solve the problems which are fundamental to sexual physiology in these fungi.

Acknowledgements. The author wishes to thank Dr. Alan Musgrave for detailed and helpful comments, and Dr. TC McMorris for providing unpublished data.

References

Alvarez MJ, Palaez MJ, Eslava AP (1980) Recombination between ten markers in *Phycomyces*. Mol Gen Genet 179:447–452

Arsenault GP, Biemann K, Barksdale AW, McMorris TC (1968) The structure of autheridiol, a sex hormone in *Achlya bisexualis*. J Am Chem Soc 90:5635–5636

Barksdale AW (1963) The role of Hormone A during sexual conjugation in *Achlya ambisexualis*. Mycologia 55:627–632

Barksdale AW (1966) Segregation of sex in the progeny of selfed heterozygote of *Achlya bisexualis*. Mycologia 58:802–804

Barksdale AW, Lasure LL (1973) Induction of gametangial phenotypes in *Achlya*. Bull Torrey Bot Club 100:199–202

Barksdale AW, McMorris TC, Seshadri R, Arunachalam T, Edwards JE, Sundeen J, Green DM (1974) Response of *Achlya ambisexualis* E87 to the hormone antheridiol and certain other steroids. J Gen Microbiol 82:295–299

Barrow SE, McMorris TC (1982) Studies on the biosynthesis of the oogoniols. Lipids 17:383–389

Bean B (1979) Chemotaxis in unicellular eukaryotes. In: Haupt W, Feinleib ME (eds) Encyclopedia of Plant Physiology. New Ser Vol 7. Springer, Berlin Heidelberg New York, pp 335–354

Bechtold MM, Welwicke CV, Cornai K, Gaylor GL (1972) Investigation of the component reactions of oxidative sterol demethylation. J Biol Chem 247:7650–7656

Blakeslee AF, Cartledge JL (1927) Sexual dimorphism in Mucorales. II. Interspecific reactions. Bot Gaz 84:51–58

Brasier CM (1972) Observations on the sexual mechanism in *Phytophthora palmivora* and related species. Trans Br Mycol Soc 58:237–251

Bu'Lock JD (1976) Hormones in fungi. In: Smith JE, Berry DR (eds) The filamentous fungi Vol 2. Arnold, London, pp 345–368

Bu'Lock JD, Hardy TM (1979) Mating reactions between sexually altered strains of *Mucor pusillus*. Exp Mycol 3:194–196

Bu'Lock JD, Osagie AU (1973) Prenols and ubiquinones in single-strain and muted cultures of *Blakeslea trispora*. J Gen Microbiol 76:77–83

Bu'Lock JD, Jones BE, Quarrie SA, Winskill N (1973) The biochemical basis of sexuality in Mucorales. Naturwissenschaften 60:550–551

Bu'Lock JD, Jones BE, Taylor D, Winskill N, Quarrie SA (1974) Sex hormones in the Mucorales. The incorporation of C20 and C18 precursors into trisporic acids. J Gen Microbiol 80:301–306

Bu'Lock JD, Jones BE, Winskill N (1976) The apocarotenoid system of sex hormones and prohormones in Mucorales. Pure Appl Chem 47:191–202

Burgeff H (1924) Untersuchungen über Sexualität und Parasitismus bei Mucorineen. Bot Abh 4:1–135

Burgeff H (1928) Variabilität, Vererbung und Mutation bei *Phycomyces blakesleeanus* Bgff. Z Indukt Abstamm Vererbungsl 49:26–94

Carlile MJ, Machlis L (1965) The response of male gametes of Allomyces to the sexual hormone sirenin. Am J Bot 52:478–483

Cerda-Olmedo E (1975) The genetics of *Phycomyces blakesleeanus*. Gen Res Cambridge 25:285–296

Edwards JA, Mills JS, Sundeen J, Fried JH (1969) The synthesis of the fungal sex hormone antheridiol. J Am Chem Soc 91:1248–1249

Egel R (1981) Mating-type switching and mitotic crossingover at the mating-type locus in fission yeast. Cold Spring Harbor Symp Quantitat Biol Vol XLV, pp 1003–1007

Elliott CG (1972) Sterols and the production of oospores by *Phytophthora cactorum*. J Gen Microbiol 72:321–327

Emerson R, Wilson CM (1954) Interspecific hybrids and the cytogenetics and cytotaxonomy of Euallomyces. Mycologia 46:396–434

Eslava AP, Alvarez MJ, Burke PV, Delbrück M (1975) Genetic recombination in sexual crosses of Phycomyces. Genetics 80:445–462

Galindo AJ, Gallebly ME (1960) The nature of sexuality in *Phytophthora infestans*. Phytopathology 50:123–128

Gauger W (1975) Further studies on sexuality in azygospores strains of *Mucor* hiemalis. Trans Brit Mycol Soc 64:113–118

Gauger W (1977) Meiotic gene segregation in *Rhizopus stolonifer*. J Gen Microbiol 101:211–217

Gauger W, Pelaez MJ, Alvarez MJ, Eslava AP (1980) Mating type heterokaryons in *Phycomyces blakesleeanus*. Exp Mycol 4:56–64

Gooday GW (1973) Differentiation in the Mucorales. Symp Soc Gen Microbiol 23:269–294

Gooday GW (1978) Functions of trisporic acid. Phil Trans Royal Soc Lond B 284:509–520

Gooday GW, Fawcett P, Green D, Shaw J (1973) The formation of fungal sporopollemin in the zygospore wall of *Mucor mucedo:* a role for the sexual carotenogenesis in the Mucorales. J Gen Microbiol 74:233–239

Gwynne DJ, Brandhurst BP (1980) Antheridiol-induced differentiation of *Achlya* in the absence of detectable synthesis of new proteins. Exp Mycol 4:251–259

Heisenberg M, Cerda-Olmedo E (1968) Segregation of heterokaryons in the asexual cycle of *Phycomyces*. Mol Gen Genet 102:187–195

Hepden PM, Hawker LE (1961) A volatile substance controlling early stages of zygospore formation in *Rhizopus sexualis*. J Gen Microbiol 24:155–164

Horgen PA (1977) Cytosol-hormone stimulation of transcription in the aquatic fungus *Achlya ambisexualis*. Biochem Biophys Res Commun 75:1022–1028

Jones BE, Gooday GW (1978) An immunofluorescent investigation of the zygospore surface of Mucorales. FEMS Microbiol Lett 4:181–184

Jones BE, Williamson JP, Gooday GW (1981) Sex pheromones in Mucor. In: O'Day DH, Horgen PA (eds) Sexual interactions in eukaryotic microbes. Academic Press, London New York, pp 179–198

Klapper BF, Klapper MH (1977) A natural inhibitor of sexual attraction in the water mold *Allomyces*. Exp Mycol 1:352–355

Klar AJS, Strathern JN, Broach JR, Hicks JB (1981) Regulation of transcription in expressed and unexpressed mating type cassettes of yeast. Nature 289:239–244

Kniep H (1928) Die Sexualität der niederen Pflanzen. Fischer, Jena

Ko WH (1978) Heterothallic Phytophthora Evidence for hormonal regulation of sexual reproduction. J Gen Microbiol 207:15–18

Ko WH (1980) Hormonal regulation of sexual reproduction in *Phytophthora*. J Gen Microbiol 116:459–463

Ko WH, Kunimoto RK (1981) Hormone production and reception among different isolates of *Phytophthora parasitica* and *Phytophthora palmivora*. Mycologia 73:440–444

Kochert G (1978) Sexual pheromones in algae and fungi. Annu Rev Plant Phys 29:461–486

Machlis L (1958) Evidence for a sexual hormone in *Allomyces*. Physiol Plant 11:181–192

Machlis L (1968) The response of wild-type male gametes of *Allomyces* to sirenin. Plant Physiol 43:1319–1320

Machlis L (1973) The chemotactic activity of various sirenins and analogues of sirenin by the sperm of *Allomyces*. Plant Physiol 52:527–531

Machlis L, Nutting WH, William MW, Rapoport H (1966) Production, isolation and characterization of sirenin. Biochemistry 5:2147–2152

Machlis L, Nutting WH, Rapoport H (1968) The structure of sirenin. J Am Chem Soc 90:1674–1676

MacNab RM (1978) Bacterial motility and chemotaxis. The molecular biology of a behavioral system. CRC Crit Rev Biochem 5:291–341

McCorkindale NJ, Hutchinson GA, Pursey BA, Scott WT, Wheeler R (1969) A comparison of the types of sterol found in species of the Saprolegniales and Leptomitales with those found in some other Phycomycetes. Phytochemistry 8:861–867

McMorris TC, Barksdale AW (1967) Isolation of a sex hormone from the water mould *Achlya bisexualis*. Nature 215:320–321

McMorris TC, White RH (1977) Biosynthesis of the oogoniols, steroidal sex hormones of *Achlya:* the role of fucosterol. Phytochemistry 16:359–362

McMorris TC, Seshadri R, Weihe GR, Arsenault GP, Barksdale AW (1975) Structures of oogoniol-1, -2 and -3, steroidal sex hormones of the water would *Achlya*. J Am Chem Soc 97:2544–2545

Mesland DAM, Huisman JG, van den Ende H (1974) Volatile sexual hormones in *Mucor mucedo*. J Gen Microbiol 80:111–117

Morrison PJ (1977) Gametangial development in *Allomyces macrogynus*. II. Evidence against mitochondrial involvement in sexual differentiation. Arch Microbiol 113:173–179

Müller DG (1968) Versuche zur Charakterisierung eines Sexuallockstoffes bei der Braunalge *Ectocarpus siliculosus*. I. Methoden, Isolierung und gaschromatographischer Nachweis. Planta 81:160–168

Mullins JT, Ellis EA (1974) Sexual morphogenesis in *Achlya:* ultrastructural basis for the hormonal induction of antheridial hyphae. Proc Natl Acad Sci USA 71:1347–1350

Musgrave A, Nieuwenhuis D (1975) Metabolism of radioactive antheridiol by *Achlya* species. Arch Microbiol 105:313–317

Musgrave A, Ero L, Scheffer R (1978) The selfinduced metabolism of antheridiol in water moulds. Acta Bot Neerl 27:397–404

Nes WD, Patterson GW, Bean GA (1980) Effect of steric and nuclear changes in steroids and triterpenoids on sexual reproduction in *Phythophthora cactorum*. Plant Physiol 66:1008–1011

Nielsen RJ (1978) Sexual mutants of a heterothallic *Mucor* species, *Mucor pusillus*. Exp. Mycol 2:193–197

Nielsen TAB, Olson LW (1982) Nuclear control of sexual differentiation in *Allomyces macrogynus*. Mycologia 74:303–312

Nieuwenhuis M, van den Ende H (1975) Sex specificity of hormone synthesis in *Mucor mucedo*. Arch Microbiol 102:167–169

Nolan RA, Bal AK (1974) Cellulase localization in hyphae of *Achlya ambisexualis*. J Bacteriol 117:840–843

Nutting WH, Rapoport H, Machlis L (1968) The structure of sirenin. J Am Chem Soc 90:6434–6438

O'Donnell KL, Flegler SL, Hooper GR (1978) Zygosporangium and zygospore formation in *Phycomyces nitens*. Can J Bot 56:91–100

Olson LW, Nielsen TAB, Heldt-Hansen HP, Grant NG (1982) Maleness, its inheritance and control in the aquatic phycomycete *Allomyces macrogynus*. Trans Br Mycol Soc 78:331–336

Plattner JJ, Rapoport H (1971) The synthesis of D- and L-sirenin and their absolute configurations. J Am Chem Soc 93:1758–1761

Plempel M (1962) Die zygotropische Reaktion bei Mucorineen III. Planta 58:509–520

Pommerville J (1978) Analyses of gamete and zygote motility in *Allomyces*. Exp Cell Res 113:166–172

Pommerville J (1981) The role of sexual pheromones in *Allomyces*. In: O'Day DH, Horgen PA (eds) Sexual interactions in eukaryotic microbes. Academic Press, London New York, pp 53–72

Pommerville J (1982) Morphology and physiology of gamete mating and gamete fusion in the fungus *Allomyces*. J Cell Soc 53:193–209

Pommerville J, Fuller MSL (1976) The cytology of the gametes and fertilization of *Allomyces macrogynus*. Arch Microbiol 109:21–30

Popplestone CR, Unrau AM (1974) Studies on the biosynthesis of antheridiol. Can J Chem 52:462–468

Preus MW, McMorris TC (1979) The configuration at C-24 in oogoniol (24R-3B, 11, 15B, 29-tetrahydroxystigmast-5-en 7-one) and identification of 24(28)-dehydroooogeniols as hormones in *Achlya*. J Am Chem Soc 101:3066–3071

Prisbylla MP, Takabe K, White JD (1979) Stereospecific synthesis of (±)-trisporol B, a prohormone of *Blakeslea trispora* and a facile synthesis of (±) trisporic acids. J Am Chem Soc 101:762–763

Raper JR (1960) The control of sex in fungi. Am J Bot 47:794–808

Raper JR (1967) The role of specific secretions in the induction and development of sexual organs and in the determination of sexual affinity. In: Ruhland W (ed) Handbuch zu Pflanzenphysiologie vol 18. Springer, Berlin Göttingen Heidelberg, pp 214–237

Ronne M, Olson LW (1976) Isolation of male strains of the aquatic phycomycete *Allomyces macrogynus*. Hereditas 83:191–202

Rozek CE, Timberlake WE (1980) Absence of evidence for changes in messenger RNA populations during steroid hormone-induced cell differentiation in *Achlya*. Exp Mycol 4:34–47

Schipper MAA (1971) Induction of zygospore production in *Mucor saximontensis*, an agamic strain of *Zychorhynchus moelleri*. Trans Br Mycol Soc 56:157–159

Schipper MAA (1976) Induced azygospore formation in *Mucor* (*Rhizomucor*) *pusillus* by *Absidia corymbifera*. Antonie van Leeuwenhoek, J Microbiol Serol 42:141–144

Schipper MAA (1978) On certain species of *Mucor* with a key to all accepted species. Stud Mycol (Centraalbureau voor Schimmelcultures, Baarn, Holland) 17:1–52

Schipper MAA, Stalpers JA (1980) Various aspects of the mating system in Mucorales. Persoonia 11:39–63

Sherwood WA (1969) Sexual reactions between clonal subcultures of a strain of *Dictyuchys monosporus*. Mycologia 61:251–263

Silver JC, Horgen PA (1974) Hormonal regulation of presumptive mRNA in the fungus *Achlya ambisexualis*, Nature 294:252–254

Sutherland RB, Horgen PA (1977) Effects of the steroid sex hormone, antheridiol, on the initiation of RNA synthesis in the simple eukaryote *Achlya ambisexualis*. J Biol Chem 252:8812–8820

Sutter RP (1975) Mutations affecting sexual development in *Phycomyces blakesleeanus*. Proc Natl Acad Sci USA 72:127–130

Sutter RP (1977) Regulation of the first stage of sexual development in *Phycomyces blakesleeanus* and in other mucoraceous fungi. In: O'Day DH, Horgen PA (eds) Eukaryotic microbes as model developmental systems. Dekker, New York, pp 251–272

Sutter RP, Whitaker JP (1981a) Zygophore stimulating precursors (pheromones) of trisporic acids active in (−)-*Phycomyces blakesleeanus*. J Biol Chem 256:2334–2351

Sutter RP, Whitaker JP (1981b) Sex pheromone metabolism in *Blakeslea trispora*. Naturwissenschaften 68:147–148

Sutter RP, Capage DA, Harrison TL, Keen WA (1973) Trisporic and biosynthesis in separate plus and minus cultures of *Blakeslea trispora*: identification by *Mucor* assay of two mating-type-specific components. J Bacteriol 114:1074–1082

Thomas DM, Goodwin TW (1967) Studies on carotenogenesis in *Blakeslea trispora*. Phytochemistry 6:355–360

Timberlake W (1976) Alterations in RNA and protein synthesis associated with steroid hormone-induced sexual morphogenesis in the water mould *Achlya*. Dev Biol 51:202–214

Van den Ende H (1976) Sexual interactions in plants. Academic Press, London New York

Van den Ende H (1978) Sexual morphogenesis in the Phycomycetes. In: Smith JE, Berry DR (eds) The filamentous fungi Vol 3. Arnold, London, pp 257–274

Werkman BA (1976) Localization and partial characterization of a sex-specific enzyme in homothallic and heterothallic Mucorales. Arch Microbiol 109:209–213

Werkman BA, Smits HL, Van den Ende H (1977) Lipid content during sexual development in the homothallic Mucor *Zygorhynchus moelleri*. Phytochemistry 16:1511–1514

Werkman BA, Van den Ende H (1974) Trisporic acid synthesis in homothallic and heterothallic Mucorales. J Gen Microbiol 82:273–278

Zycha H, Siepmann R, Linnemann G (1969) Mucorales. Cramer, Uhre

17 Barrage Formation in Fungi

K. Esser and F. Meinhardt

17.1 Introduction

If fungal hyphae of different mycelia grow towards each other, in general four main types of interaction occur, as may easily be demonstrated on agar media:

1. *Mutual intermingling, normal contact*

   After approach, the hyphae intermingle in the zone of contact and show (with few exceptions, e.g. Oomycetales) numerous hyphal fusions via anastomosis formation. After a time the border between the two mycelia becomes hardly recognizable (Fig. 1a, b, c, d).

2. *Fusion*

   After contact between the hyphal tips, one mycelium overgrowth the other. In general this does not inhibit hyphal fusions.

3. *Inhibition*

   When coming into contact with each other an inhibition zone free of hyphae is formed between the two mycelia. This phenomenon may be caused by unilateral action or by mutual interaction, due to the excretion and diffusion of inhibitory substances.

4. *Mutual intermingling and inhibition (barrage formation)*

   When the two mycelia *grow into each other* and intermingle, an antagonistic reaction takes place. In contrast to the inhibition by diffusible substances, the barrage reaction requires cytoplasmic contact via hyphal fusions. The phenotype of the barrages differs with respect to both species and mode of genetic control (Fig. 1). However, all barrages known so far have in common that, although hyphal fusions occur, no nuclear exchange takes place and in most cases the two types of mycelia form abnormal and often lethal hyphal fusions. The hyphal tips may branch profusely. A clear line of contact

---

Fig. 1a–e. Compilation of mycelial interactions in fungi. **a** *Schizophyllum commune. Above* normal contact in a compatible combination $(A \neq B \neq)$; *below* formation of a sexual barrage in a hemicompatible combination with common B-factors $(A \neq B =)$. **b** *Podospora anserina.* Barrage formation between different geographical races. Sexual reproduction is not affected. Perithecia are produced in any combination between different mating types. **c** Hyphal morphology of *Podospora anserina. Above* hyphae in the contact zone of an intrarace combination; *below* hyphae within the barrage. (Adapted from Blaich and Esser 1971). **d** Intra- and interspecific interactions between monokaryons of *Polyporus ciliatus* (*cil*) and *Polyporus brumalis* (*bru*) showing normal contact *left,* barrage *bottom,* and border line *top* and *right.* All monokaryons are compatible in mating type. (Hoffmann and Esser 1978). **e** Cross-section of a log colonized by wood-destroying basidiomycetes. *Dark zones* indicate the aversion lines of mycelia

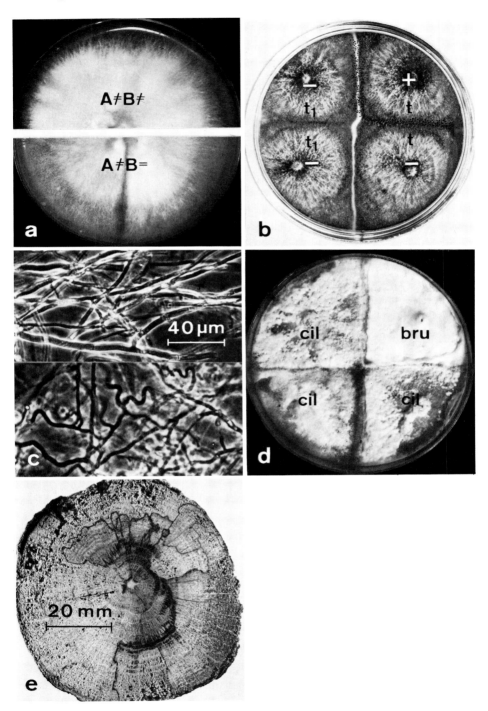

appears with increasing age of the culture. Depending on the species, the barrage may be white or pigmented (Fig. 1 a, b, c, d).

The phenomenon of barrage formation was observed decades ago by many mycologists (e. g. Reinhardt 1892, Cayley 1923, 1931, Nakata 1925). Vandendries (1932) probably first used the term barrage. Most certainly older descriptions of barrages, not under this name, comprehended also mutual or unilaterial mycelial respulsion. In this paper we want to separate the latter phenomenon from the concept of barrage, because this kind of mycelial inhibition may be due to the actions of antibiotics, especially when occurring between different races or species. Therefore we use the term barrage only when the macroscopic observation of a zone of aversion occurring between two mycelia is brought about by hyphal contact, as revealed by microscopic observations.

As already indicated above, barrage formation is not correlated with a unique phenotype, nor is it controlled by a common genetic mechanism. Therefore in examples of barrages in subsequent sections, we shall use this versatility as a basis for classification. In this review we attempt no complete overview of the literature, but aim to give the reader not familiar with fungal genetics or mycology an idea of the nature of this peculiar phenomenon, often observed in experiments in different fields of research.

General references are given by Porter 1924, Porter and Carter 1938, Rizet 1952, Esser and Kuenen 1967, Burnett 1968, Carlile and Gooday 1978, Rayner and Todd 1979.

17.2 Intraspecific Barrages

Within a single species, barrage formation is frequently correlated with the hereditary traits which control the breeding system (references, Esser 1971).

17.2.1 Barrages Associated with Homogenic Incompatibility

In fungi, as in many higher plants, the advantages of dioecism (bisexuality) as a means of decreasing inbreeding are obtained by sexual incompatibility.

This means that a plant or a fungus, while having the capacity of a hermaphrodite, is not capable of self-fertilization, but depends for zygote formation on another hermaphrodite partner of a different mating type. The genetic control of this breeding system has been very well studied in fungi and depends in general on the inability of genetically like gametes (or nuclei with gamete function) to fertilize each other. This system is therefore called homogenic incompatibility. Two genetic mechanism are known: the bipolar and the tetrapolar; they have in common that karyogamy only takes place when the two mating types are heterogeneous at all loci concerned (Esser 1976, Raper 1966).

Since sexual barrage was mainly observed in fungi with tetrapolar mechanism, we shall here ignore the bipolar mechanism.

Tetrapolar incompatibility is controlled by two (in general unlinked) factors, A and B, both complex genetic traits, often having many allelic configurations. The monokaryotic mycelia originating from the spores of a single fruiting body belong to four different mating types (tetrapolar), which may be abbreviated as A_xB_x, A_yB_y, A_xB_y, A_yB_x. Sexual compatibility and therewith fruiting body formation occurs only if the mating types are heterogeneous for both factors, e.g. $A_xB_x \times A_yB_y$ or abbreviated $A \neq B \neq$. In all other combinations no fruiting takes place. However, if there is only heterogeneity for one factor $A \neq B =$ or $A = B \neq$, the two mating types only fuse and exchange nuclei, but stable dikaryons, the prerequisite for fruiting, are never established.

With respect to barrage formation, the combination $A \neq B =$ is of interest, because between the two mating types, e.g. $A_xB_x \times A_yB_x$, a zone of mutual aversion of the intermingling hyphae may be seen. This is exactly the phenomenon for which the term "barrage sexuel" was introduced by VANDENDRIES (1932). Subsequentially this barrage phenomenon was found in many other tetrapolar basidiomycetes (VANDENDRIES and BRODIE 1933a, b) and it has been studied in detail in *Schizophyllum commune*, (Fig. 1a) one of the genetically best-analyzed fungi, by RAPER and his numerous associates (for references see RAPER 1966).

The occurrence of barrage allows identification of macroscopic combinations with different common A and B factors and so is of practical value for mycologists, but the biological significance still remains obscure. It is not even known whether the barrage is caused by the heterogeneity of the A factors or the homogeneity of the B factors. This view is supported by the fact that this "sexual barrage", in contrast to what was earlier suggested by RAPER, does not in general occur in tetrapolar basidiomycetes.

However, in some fungi the existence of a sexual barrage may not be immediately evident. On the one hand, it may be hidden under masses of aerial hyphae, as was recently found for another type of barrage in *Neurospora crassa* (see Sect. 17.2.2.3). On the other hand additional genetic traits are known which appear to be epistatic and "cover" the $A \neq B =$ action. This results in barrage formation in almost any combination of different mating types (MEINHARDT and LESLIE 1982).

17.2.2 Barrages Associated with Heterogenic Incompatibility

In contrast to the homogenic system, heterogenic incompatibility occurs only between different races of the same species. It may affect both the sexual and the parasexual cycle. In the latter case it prevents the hyphal fusion which is the prerequisite for mitotic recombination. Since barrage formation may also be under chromosomal and extrachromosomal control we shall exemplify all these modifications using the coprophilous ascomycete *Podospora anserina*. Barrage formation in *Podospora*, elucidated by the fundamental work of RIZET (1952) is one of the best-understood examples of barrage formation, due to the detailed experimentation to which it has been subjected.

17.2.2.1 *Podospora anserina*

The mating competence of *Podospora* mycelia originating from a uninucleated ascospore is controlled by a bipolar incompatibility system. The two mating types both being hermaphrodite are self-incompatible and designated as + and −. No multiple alleles are known (Rizet and Engelmann 1949).

In studying various races of different geographic origin, which showed no recognizable macroscopic differences, Rizet (1952) found that in all combinations between strains of different races a barrage occurred irrespective of the mating type. As may be seen from Fig. 1b, this barrage is macroscopically characterized by a sharp white zone between the two dark pigmented (melaninic) mating partners. A microscopical examination revealed that in this zone the hyphae, after having attempted to perform anastomoses, become curled, blown-up and degenerate (Fig. 1c, d). The barrage line eventually consists of paralyzed dead cells.

In many cases barrage formation does not affect fruiting, for perithecia are formed on both sides of the "iron curtain". This peculiar phenomenon is explained by the fact that the trichogynes of the +female sex organs (protoperithecia) only fuse with the −male gametes (spermatia), and vice versa. Spermatia, however, are never formed in the barrage zone.

The trichogynes pass the barrage unharmed because no anastomoses take place. Obviously the fusion of the tip of the trichogyne with a spermatium does not bring about the incompatibility reaction occurring between two vegetative hyphae, although the reason for this remains obscure. The fact that the barrage separates the fruiting bodies of the two reciprocal crosses between + and − mating types is an excellent tool in genetic analysis, as will be seen later.

Concerning genetic control, in the simplest case heterogeneity at a single locus is sufficient to cause a barrage (Rizet 1953, see Fig. 1, b). However, in nature the geographical races mostly differ by more than one barrage gene, which all act additively to cause a barrage (Rizet and Esser 1953, Esser 1956, 1959a, b).

In addition to this rather simple mechanism of heterogenic incompatibility which concerns only the vegetative phase, another more complicated mechanism was detected which interferes with sexual compatibility. In these cases fruiting is suppressed either on one side of the barrage (semi-incompatibility) or on both sides (complete incompatibility). Semi-incompatibility requires heterogeneity at two, and complete incompatibility at four loci (since heterogenic incompatibility is not the aim of this review, we refer for more details to Esser and Blaich 1973).

Apart from this kind of barrage formation which is exclusively under chromosomal control, Rizet in his first general paper (1952) showed that barrage formation between the two strains S and s also depends on extrachromosomal traits.

These strains S and s are not different geographical races; indeed the strain s was found to occur spontaneously from the race S. This may explain why this form of nucleo-cytoplasmic control of the barrage phenomenon is unique in *Podospora,* because, as numerous analyses have shown (Esser 1971, Bernet et al. 1973) that all other examples of barrage formation between races are exclusively under chromosomal control. When the S/s barrage phenomenon was first described 30 years ago, it caused excitement among geneticists, because at that time the scientific world was beginning to understand the phenomenon of extrachromosomal inheritance on a quantitative level and its various relationships with chromosomal inheritance. The spirit of that time may be found in Ephrussi's booklet published in 1953. The essential data concerning the

S/s barrage subsequently obtained by the comprehensive investigations of BEISSON-SCHECROUN (1962) may be summarized as follows:

The strains *S* and *s* show an allelic chromosomal difference. In addition, both strains carry in the cytoplasms factors which are believed to be produced under the influence of the *S* and *s* allele respectively. These cytoplasmic determinants behave antagonistically and therefore cause the formation of a barrage. Genetic and microsurgical experiments have revealed that the *s* particles, at least, are infective, because they may be transferred from *s* strains into those of the same genotype which have themselves lost the ability to produce the *s* particles.

Later attempts in our laboratory using modern techniques to identify fungal viruses or plasmids have failed (KÜCK personal communication), and furthermore no differences in the mitochondrial DNA's of *S* and *s* were found; the genetic basis of this peculiar barrage phenomenon remains therefore still unclear.

17.2.2.2 Wood-Destroying Basidiomycetes

After it became clear that in *Podospora* barrage formation is the macroscopic phenotypic expression of heterogenic incompatibility between geographical races, and that heterogenic incompatibility is a very general, basic phenomenon in biology (for see references see ESSER and BLAICH 1973), it was obviously important to extend the investigations of barrage formation to other fungi. Wood-destroying fungi are easily accessible, because fallen twigs, trees and stumps are covered with fruiting bodies, especially of white rot fungi.

In a preliminary study SCHERER (1973) analyzed 21 wild isolates of *Ganoderma applanatum* and 26 of *Trametes confragosa* collected from a forest of 40 ha. Barrage formation was found within the isolates of both species. It was very interesting to observe that isolates obtained from the fruiting bodies which covered a single fallen tree belonged to different barrage groups. This shows that races reacting by barrage formation may grow in close proximity to each other, and may infect the same tree or trunk. Since it was not possible to obtain fruiting bodies under laboratory conditions, either from *G. applanatum* or from *T. confragosa,* no genetic data became available.

SCHERER's observations were later confirmed and extended in *Coriolus versicolor* (RAYNER and TODD 1977, TODD and RAYNER 1978) *Stereum hirsutum* (COATES et al. 1981) and *Piptoporus betulinus* (ADAMS et al. 1981). On the polished disc of stumps colonized by one or the other of these fungi, the decay regions were found to be separated by narrow black interaction zones (Fig. 1e). After testing the mycelia obtained from the corresponding areas, it became evident that these regions were infected by different geographical races, and that the black lines corresponded to the barrage which these mycelia showed in the laboratory as a narrow dark line.

More examples, some of them quite recent, describing barrages, demarcation lines, or whatever expressions are used for this phenomenon of intraspecific antagonisms in wood-decaying basidiomycetes, have been quoted by ESSER and BLAICH (1973), and further references given by ADAMS and ROTH 1967, BARRAT and USCULPIC 1971, GOLDSTEIN and GILBERTSON 1981, KEMP 1970, and LI 1981.

Unfortunately no data are available yet which would reveal the nature of the genetic control of barrage formation. However, it is most likely that, as in *Podospora*, heterogeneity is the cause. This assumption is confirmed by a more detailed analysis of barrage formation earlier investigated in some species of another white rot fungus, *Polyporus brumalis* (Hoffmann and Esser 1978). The breeding system in *Polyporus* is controlled by the tetrapolar mechanism of homogenic incompatibility. In intra-race mating all compatible combinations ($A \neq B \neq$) are fertile, and no barrage correlated with the AB factors is observed. However, in crosses between different races on agar medium a barrage zone about 1–2 mm wide, free of aerial hyphae and of reduced hyphal density, is macroscopically visible.

In comparison to the $A \neq B \neq$ combinations reacting with normal contact, the barrage-forming crosses are characterized by the following events (Fig. 1 d):
1. Hyphal fusion is delayed up to 7 days, regardless of the combinations of mating-type factors. The hyphae frequently grow past each other and avoid anastomosis.
2. Nuclear migration is mainly unilateral in that after hyphal fusions have occurred, only one of the mating types is invaded by the nuclei of the other.
3. Two weeks after the first contact, a gradual lysis of hyphae may be seen in the barrage zone.

As a consequence of effects 1 and 2 there is a delay in fruiting: 7–9 days on the "unhindered" side and 21–23 days on the other. After this time the culture medium is often too exhausted to support the formation of normal-shaped fruiting bodies, or even of fruiting bodies at all.

If one compares this type of barrage with that in *Podospora*, some similarity becomes obvious. In both cases fruiting is not suppressed, but it may be restricted, and the intermingling, fused hyphae may degenerate, showing the mutual aversion caused by heterogeneity.

This analogy becomes even more evident if the genetic control of barrage formation, as revealed in *P. ciliatus*, is considered. Barrage formation is started by a specific interaction of three non-linked genes (bi^+/bi = barrage initiation, $bf I_1/bf I_2$ and $bf II_1/bf II_2$ = barrage formation) in a way characteristic for mechanisms of heterogenic incompatibility. Barrage requires the presence of the allele bi^+ in at least one mating type, additional to a heterogeneity for both bf-genes.

17.2.2.3 Comparable Phenomena in Other Fungi

In *Neurospora crassa*, genetically one of the most thoroughly investigated fungi, Beadle and Coonradt as early as 1944 showed that opposite mating types (called A and a in this fungus) never form a heterokaryon. That vigorous growth and stable heterokaryons are only formed between like mating types was subsequentially accepted by most neurosporologists as being a property peculiar to the red bread mould. Later, however Garnjobst, William and Wilson (for references see Esser and Blaich) became aware that this phenomenon is under the control of other genes, not identical with the mating-type alleles. They identified two non-linked loci C/c and D/d controlling heterokaryon formation in such a manner that any heterogeneity at either locus provoked an incompatibility reaction leading to a destruction of the hyphal parts which had fused. Since sexual compatibility of the opposite mating was not at all inhibited, this reaction is exactly identical to that taking place in the barrage zone of *Podospora*. The question why this physiological and genetical barrage cannot be identified as a "morphological" barrage is easy to answer. In contrast to the colonial growth of *Podospora, N. crassa* shows very vigorous growth with masses of aerial hyphae which never form a dense mycelial mat. In addition there are no black mycelial pigments in *Neurospora* to contrast with the white barrage zone. This view is supported by the observation that if *Podospora* is grown on a medium poor in carbohydrate no melanin is produced, and the barrage may only be identified with difficulty.

In this connection it was no surprise to find a communication of GRIFFITHS and RIECK (1981), who described for the first time barrage formation in *Neurospora crassa* which in its phenotype closely resembles the barrage of *Podospora*. These authors also state that this phenomenon had been overlooked for years in *N. crassa* due to the rich growth of aerial hyphae, and to the specific mating methods via conidia. Using conidialess mutants (with reduced aerial hyphae) for pair-wise crosses on Petri dishes, in some combinations normal contacts (one line of perithecia) were observed and in others, barrage (two lines of perithecia separated by a distinct clear zone). Unfortunately no genetic data are available yet.

A second example reminiscent of barrage formation is the incompatibility reaction observed between diploid plasmodia in the true slime moulds, which leads to a macroscopically visible line of mutual destruction. In the two well-studied objects *Didymium iridis* and *Physarum polycephalum,* this interaction was shown to be caused by heteroallelic interactions, which as in *Podospora* and *Neurospora,* do not necessarily inhibit the sexual compatibility of different mating types (for references see CARLILE 1973, 1976, CARLILE and GOODAY 1978, CLARK and COLLINS 1979).

In the previously quoted review (BLAICH and ESSER 1973), other examples of mutual aversion in fungi, are listed which, according to their microscopical morphological reactions and their genetic control so far as this is known, very closely resemble the examples of more evident barrage formations described in this chapter. Comparable cases may also be found in other reviews under the heading vegetative or cytoplasmic incompatibility.

17.3 Interspecific Barrages

The dark interaction lines (Fig. 1e) found in trees or wood stumps infected by wood-decaying basidiomycetes may not always be attributable to barrage formation between different races. There is much evidence (for references, see RAYNER and TODD 1979) that the demarcation lines are the result of interspecific reactions.

The genus *Polyporus* may be used again as an example. During the course of the studies of intraspecific barrage formation mentioned in the previous chapter (Hoffmann and Esser 1978), the mating reactions between 26 isolates belonging to the species. *P. brumalis, P. ciliatus* and *P. arcularius* were studied. Apart from the frequently occurring intraspecific barrages, in all inter-species combinations a distinct zone of aversion was found (see Fig. 1d). This zone was called border line (Fig. 1d) because, in contrast to the barrage there is no hyphal anastomosis, but rather a mutual repulsion in the area of contact which is often associated with pigment production.

All efforts to obtain a plasmogamy via protoplast fusion by matings between dicaryons and monocaryons (dimon matings), or by complementation of auxotrophs failed. Therefore a genetic investigation of this kind of barrage formation is not possible.

Reports of comparable interactions in interspecific combinations of *Pleurotus* (Anderson et al. 1973, Bresinsky et al. 1977) and *Auricularia* (Duncan and MacDonald 1967) indicate its general occurrence.

Apart from providing examples of yet another phenomenon of interaction, these studies are of great interest for taxonomy, because they support the genetic species concept which considers the fertility between different isolates as a major species criterion. Especially in *Polyporus,* in which the specification follows morphological criteria such as size of hymenial pores, an unequivocal accordance of the inter-fertility with species limits was found. Another investigation along these lines is of interest. Anderson and Ullrich (1979) studied inter-strain matings of mycelia obtained from 97 fruiting bodies of *Armillaria mellea,* obtained from all over the United States. They found bipolar incompatibility as the basic system, but with ten intersterility groups. These were called biological species because they are not clearly distinguishable by morphological criteria. It would be certainly interesting to know whether segregation within *A. mellea* is also accompanied by a barrage or a comparable mutual aversion; unfortunately, however, there is a lack of data referring to this. It is imaginable that in fungi barrage formation linked with sterility may eventually be accepted by the taxonomists as a species criterion (see Boidin 1980).

17.4 Intergeneric Barrages

In reviewing the literature, one finds many reports of mutual aversion between fungi. However, according to the definition of the barrage we have to exclude those which appear only as inhibition haloes such as those provoked by antibiotic-forming fungi. We are well aware that in giving the following few examples it is difficult to recognize a general principle, but a general "attack–defense mechanism" while fighting for the utilization of a common substrate may be involved, especially with plant pathogens.

Mutual antagonistic effects of mycelia belonging to different genera were described extensively by Rayner and Todd (1979). These effects occur in wood as well as on agar media. The authors list several different species mostly belonging to the wood-rotting basidiomycetes, which act as mutual antagonists after hyphal contact. It is not easy to say whether the formation of coloured contact zones or the development of clear zones between different colonies fulfill the criteria of a barrage, but some of their observations might support this kind of interpretation.

There is one type of antagonism among fungi called hyphal interference (first described by IKEDIUGWU and WEBSTER 1970a, b) which is apparently different from the formation of a barrage. Hyphal interference is always unilateral, i.e. hyphae of one of the involved mycelia are killed after getting in contact with the hyphae of the other mycelium, for example the antagonistic ability of *Peniophora gigantea* against *Heterobasidion annosum* (IKEDIUGWU et al. 1970)

Mutual antagonism between *Alternaria* and *Fusarium* was described by TAHA (1953), but because of the lack of microscopical and physiological data it still remains obscure whether this kind of antagonism results in the formation of a barrage.

17.5 Biological Significance of Barrage Formation

Despite the still scattered phenomena of barrage formation which have also not all been equally thoroughly analyzed, the biological significance of a barrage is becoming somewhat more evident.

At first sight the formation of a barrage may appear to have different biological meanings; compare for example the sexual barrages with the parasexual barrages, or the intraspecific with the intergeneric. More intensive reflection suggests that all the different types of barrage infact support one general biological aim: the conservation of the species as a breeding unit in evolution. What does this mean? Intraspecific sexual barrages support the outbreeding effects of the homogenic incompatibility mechanisms. Barrages in the heterogenic system enhance the separation of geographical races, and promote their transformation into new species. Barrages occurring between species and genera protect the species from external genetic material and ensure that the species remains the smallest breeding unit. Last but not least, all barrages avoid the transduction of infective material such as fungal viruses from one partner to the other.

References

Adams DH, Roth LF (1967) Demarcation lines in paired cultures of *Fomes cajanderi* as a basis for detecting genetically distinct mycelia. Can J Bot 45:1583–1589
Adams TJH, Todd NK, Rayner ADM (1981) Antagonism between dikaryons of *Piptoporus betulinus*. Trans Br Mycol Soc 76:510–513
Anderson JB, Ullrich RC (1979) Biological species of *Armillaria mellea* in North America. Mycologia 71:402–414
Anderson NA, Wang SS, Schwandt JW (1973) The *Pleurotus ostreatus-sapidus* species complex. Mycologia 65:28–35

Barrett DK, Uscuplic M (1971) The field distribution of interacting strains of *Polyporus schweinitzii* and their origin. New Phytol 70:581–598

Beadle GW, Coonradt VL (1944) Heterocaryosis in *Neurospora crassa.* Genetics 29:291–308

Beisson-Schecroun J (1962) Incompatibilité cellulaire et interaction nucléoplasmiques dans les phénomènes de "barrage" chez le *Podospora anserina.* Ann Génét 4:4–50

Bernet J, Begueret J, Labarere J (1973) Incompatibility in the fungus *Podospora anserina.* Mol Gen Genet 124:35–50

Blaich R, Esser K (1971) The incompatibility relationships between geographical races of *Podospora anserina.* V. Biochemical characterization of heterogenic incompatibility on cellular level. Mol Gen Genet 111:265–272

Boidin J (1980) La notion d'espèce. III. Le critère d'interfertilité ou intercompatibilité: résultats et problèmes. Bull Soc Mycol France 96:43–57

Bresinsky A, Hilber O, Molitoris HP (1977) The genus *Pleurotus* as on aid for understanding the concept of species in basidiomycetes. In: Clémençon H (ed) Herbette Symposium Lausanne 1976. Lehre, pp 229–258

Burnett JH (1968) Fundamentals of mycology. Arnold, London

Carlile MJ (1973) Cell fusion and somatic incompatibility in myxomycetes. Ber Dtsch Bot Ges 86:125–139

Carlile MJ (1976) The genetic basis of the incompatibility reaction following plasmodial fusion between different strains of the myxomycete *Physarum polycephalum.* J Gen Microbiol 93:371–376

Carlile MJ, Gooday GW (1978) Cell fusion in myxomycetes and fungi. In: Poste G, Nicolson GL (eds) Membrane fusion. Elsevier/North-Holland Biomedical Press, Amsterdam New York, pp 219–265

Caylay DM (1923) The phenomenon of mutual aversion between monospore mycelia of the same fungus (*Diaporthe perniciosa* March.) with a discussion on sex heterothallism in fungi. J Genet 13:353

Caylay DM (1931) The inheritance of the capacity for showing mutual aversion between monospore mycelia of *Diaporthe perniciosa.* J Genet 24:1–63

Clark J, Collins OR (1979) Plasmodial incompatibility in the myxomycete *Didymium nigripes.* Mycologia 70:1249–1253

Coates D, Rayner ADM, Todd NK (1981) Mating behaviour, mycelial antagonism and the establishment of individuals in *Stereum hirsutum.* Trans Br Mycol Soc 76:41–51

Duncan EG, MacDonald JA (1967) Micro-evolution in *Auricularia auricula.* Mycologia 59:803–818

Ephrussi B (1953) Nucleo-cytoplasmic relations in microorganisms. Clarendon, Oxford

Esser K (1956) Die Incompatibilitätsbeziehungen zwischen geographischen Rassen von *Podospora anserina* (Ces). Rehm. I. Genetische Analyse der Semi-Incompatibilität. Z Indukt Abst Vererbl 87:595–624

Esser K (1959a) Die Incompatibilitätsbeziehungen zwischen geographischen Rassen von *Podospora anserina* (Ces). Rehm. II. Die Wirkungsweise der Semi-Incompatibilitäts-Gene. Z Vererbungsl 90:29–52

Esser K (1959b) Die Incompatibilitätsbeziehungen zwischen geographischen Rassen von *Podospora anserina* (Ces). Rehm. III. Untersuchungen zur Genphysiologie der Barragebildung und der Semi-Incompatibilität. Z Vererbungsl 90:445–456

Esser K (1971) Breeding systems in fungi and their significance for genetic recombination. Mol Gen Genet 110:86–100

Esser K (1976) Genetic factors to be considered in maintaining living plant collections. In: Simmons JB, Beyer RI, Brandham PE, Lucas GL, Parry VTH (eds) Conservation of threatened plants. Plenum, New York London, pp 185–198

Esser K, Blaich R (1973) Heterogenic incompatibility in plants and animals. Adv Genet 17:107–152

Esser K, Kuenen R (1967) Genetics of fungi. Springer, Berlin Heidelberg New York

Goldstein D, Gilbertson RL (1981) Cultural morphology and sexuality of *Inonotus arizonicus.* Mycologica 73:167–180

Griffiths AJF, Rieck A (1981) Perithecial distribution pattern in standard and variant strains of *Neurospora crassa.* Can J Bot 59:2610–2617

Hoffmann P, Esser K (1978) Genetics of speciation in the basidiomycetous genus *Polyporus*. Theor Appl. Genet 53:273–282

Ikediugwu FEO, Webster J (1970a) Antagonism between *Coprinus heptemerus* and other coprophilous fungi. Trans Br Mycol Soc 54:181–204

Ikediugwu FEO, Webster J (1970b) Hyphal interference in a range of coprophilous fungi. Trans Br Mycol Soc 54:205–210

Ikediugwu FEO, Dennis C, Webster J (1970) Hyphal interference by *Peniophora gigantea* against *Heterobasidion annosum*. Trans Br Mycol Soc 54:307–309

Kemp RFO (1970) Interspecific sterility in *Coprinus bisporus, C. congregatus* and other basidiomycetes. Trans Br Mycol Soc 54:488–489

Li CY (1981) Phenoloxidase and peroxidase activities in zone lines of *Phellinus weirii*. Mycologia 73:811–820

Meinhardt F, Leslie JF (1982) Mating types of *Agrocybe aegerita*. Curr Genet 5:65–68

Nakata K (1925) Studies on *Sclerotium rolfsii*. I. The phenomenon of aversion and its relation to the biologic forms of the fungus. II. The possible cause of the phenomenon of aversion in the fungus and morphological features of the phenomenon. Bull Sci Fakult Terkulturen Kjusa Imp Univ 1:4

Porter CL (1924) Concerning the characters of certain fungi as exhibited by their growth in the presence of other fungi. Am J Bot 11:168–188

Porter CL, Carter JC (1938) Competition among fungi. Bot Rev 4:165–182

Raper JR (1966) Genetics of sexuality in higher fungi. Ronald, New York

Rayner ADM, Todd NK (1977) Interspecific antagonism in natural populations of wood-decaying basidiomycetes. J Gen Microbiol 103:85–80

Rayner ADM, Todd NK (1979) Population and community structure and dynamics of fungi in decaying wood. Adv Bot Res 7:334–417

Reinhardt OM (1892) Das Wachstum der Pilzenhyphen. Jahrb Wiss Bot 23:479

Rizet G (1952) Les phénomènes de barrage chez *Podospora anserina*. I. Analyse génétique des barrages entre souches S et s. Rev Cytol Biol Végét 13:51–92

Rizet G (1953) Sur la multiplicité des mécanismes génétiques conduisant à des barrages chez *Podospora anserina*. C R Acad Sci Paris 237:666–668

Rizet G, Engelmann G (1949) Contribution à l'étude génétique d'un ascomycète tétra-sporé: *Podospora anserina* (Ces.) Rehm. Rev Cytol Biol Végét 11:202–304

Rizet G, Esser K (1953) Sur des phénomènes d'incompatibilité entre souches d'origines différentes chez *Podospora anserina*. C R Acad Sci Paris 237:760–761

Scherer P (1973) Heterogenische Incompatibilität und ihre Bedeutung für die Evolution, unter Berücksichtigung eigener Experimente an höheren Pilzen. Staatsexamensarbeit Bochum

Taha EEM (1953) Über Wechselbeziehungen zwischen Schimmelpilzen. I. Der Einfluß verschiedener physiologischer Faktoren auf die Wechselbeziehungen zwischen *Alternaria tenuis* und *Fusarium semitectum*. Arch Mikrobiol 19:45–51

Todd NK, Rayner ADM (1978) Genetic structure of a natural population of *Coriolus versicolor* (L. ex Fr.) quél. Genet Res 32:55–65

Vandendries R (1932) La tétrapolarité sexuelle de *Pleurotus columbinus*. Démonstration photographique d'un tableau de croisement. Cellule 41:267–279

Vandendries R, Brodie HJ (1933a) Nouvelles investigations dans le domaine de la sexualité des basidiomycètes et étude expérimentale des barrages sexuels. Cellule 42:165–209

Vandendries R, Brodie HJ (1933b) La tétrapolarité et l'étude expérimentale des barrages sexuels chez les basidiomycetes. Bull Acad R Belg Cl Sci 19:120–125

18 Physiological Interactions Between the Partners of the Lichen Symbiosis [1]

M. Galun and P. Bubrick

"Two are better than one; because they have a good reward for their labour. For if they fall, the one will lift up his fellow: but woe to him that is alone when he falleth; for he hath not another to help him up. Again, if two lie together, then they have heat: but how can one be warm alone? And if one prevail against him, two shall withstand him; and a three-fold cord is not quickly broken."

Ecclesiastes, Chapter 4
Verses 9–12: (King James' Version)

18.1 Introduction

Lichens are a symbiotic association comprised of two unrelated organisms, a fungus and an alga; usually one type of alga (the phycobiont) combines with one type of fungus (the mycobiont). When fully integrated they form a new biological entity with no resemblance to either of the components. There are to date about 20,000 lichen species known, which account for about 25% of all the fungi known in the world (lichens are incorporated in fungal classification).

Lichens exhibit great diversity of shape, colour and growth-form, and grow on a wide range of organic and inorganic substrates in almost all the environments of the world. Some lichens are crust-like and adhere closely to their substrate. In others the thallus is foliose with a leafy, dorsiventral plant body (Fig. 1) or fruticose, shrub-like or filamentous (Fig. 2). Their sizes range from 1–2 mm up to about 5 m.

---

1 Vol. 12C of this Encyclopedia (Physiological Plant Ecology III, Lange OL, Nobel PS, Osmond CB, Ziegler H, eds.) contains another chapter on lichens: Matthes U, Feige CB: Ecophysiology of lichen symbioses (pp. 423–467)

Abbreviations. ABP = algal binding protein; ADH = alanine dehydrogenase; CB = coomassie brilliant blue R 250; Con A = concanavalin A; CT = cultured alga from *Xanthoria parietina*; DNP = 2,4-dinitrophenol; FCCP = fluorocarbonylcyanidephenylhydrazone; FITC = fluorescein isothiocyanate; FT = freshly isolated alga from *Xanthoria parietina*; GAT = glutamate-aspartate-aminotransferase; GDH = glutamate dehydrogenase; GlcNac = N-acetylglucosamine; GOGAT = glutamate synthase; GS = glutamine synthetase; RIA = radioimmunoassay; RR = ruthenium red; WGA = wheat germ agglutinin

Fig. 1. A foliose lichen–*Xanthoria parietina* on bark (× 0.5)

Fig. 2. A fruticose lichen–*Teloschistes lacunosus* growing on loess (× 0.5)

18.1.1 The Lichen Components

The dual nature of lichens was recognized in 1869 by the botanist SCHWENDENER. According to SCHWENDENER "the algae are the nutrition-serving slaves of the dominating master, the fungus. The two become intimately associated and form new plants with individual characters". Before and also several years later, scientists contradicting the SCHWENDENER theory claimed that the coloured cells were reproductive bodies of the fungus, calling them gonidia. While studying the composite nature of lichens, DE BARY (1879) coined the term symbiosis, and defined symbiosis as "the living together of dissimilar organisms".

Table 1. Algal genera found in the lichen symbiosis[a]

Eukaryotic algae	Cyanobacteria[b]
Asterochloris (Tschermak-Woess 1980a)	*Anacystis* (Wetmore 1970)
Cephaleuros (A)	*Calothrix* (A)
Chlorella (A, Tschermak-Woess 1978b)	*Chroococcus* (A)
Chlorosarcina (A)	*Dichothrix* (A)
Coccobotrys (A)	*Gloeocapsa* (A)
Coccomyxa (A)	*Hyella* (A)
Dictyochloropsis (Tschermak-Woess 1978c, 1980b)	*Hyphomorpha* (Henssen 1981)
Dilabifilum (Tschermak-Woess 1976)	*Nostoc* (A)
Elliptochloris (Tschermak-Woess 1980c)	*Scytonema* (A)
Gloeocystis (A)	*Stigonema* (A)
Heterococcus (A, Parra and Redon 1977)	
Hyalococcus (A)	
Leptosira (A, Tschermak-Woess and Poelt 1976)	
Myrmecia (A)	
Nannochloris (Tschermak-Woess 1981)	
Phycopeltis (A)	
Protococcus (A)	
Pseudochlorella (A)	
Pseudopleurococcus (A, Tschermak-Woess 1970)	
Pseudotrebouxia (Archibald 1975)	
Stichococcus (A)	
Trebouxia[c] (Archibald 1975, Hildreth and Ahmadjian 1981)	
Trentepohlia (A)	
Trochiscia (A)	

[a] Genera followed by (A) are taken from Ahmadjian (1967a). In applicable cases, more recent literature is provided
[b] The following cyanobacterial genera listed by Ahmadjian (1967a) have been excluded: *Anabaena, Aphanocapsa, Gloeothece, Mastigocoleus* and *Microcystis*
[c] *Trebouxia* species listed by Ahmadjian (1967a) have been redefined by Archibald (1975)

18.1.1.1 The Algal Partner

Literature records species from 34 different chlorophycean genera that have been found in the lichen symbiosis. The genera range from unicellular to hetero-trichous forms. The most common phycobionts belong to the genera *Trebouxia* (Chlorococcales) and *Pseudotrebouxia* (Chlorosarcinales). These two genera have been monographed by Archibald (1975) and Hildreth and Ahmadjian (1981). Approximately 10 genera of cyanobacterial (blue-green) phycobionts have been recorded from lichens. They are either single cells or trichome-forming species (Table 1). Only one known phycobiont belongs to the Xanthophyceae.

18.1.1.2 The Fungal Partner

The majority of fungi integrated in lichen symbiosis belong to the Ascomycetes. There are very few basidiolichens, about 20 species, according to recent ac-counts. Naturally, therefore, almost the entire literature on lichens refers to ascolichens.

Techniques for isolating the partners from the intact lichen thallus and culturing methods have been developed by AHMADJIAN (1967b, 1973a). Fungal spore germination and the initial stages of hyphal growth in culture are relatively fast, at a rate comparable with free-living fungi. Thereafter, the growth of the fungal isolates in culture is extremely slow. It takes several months for a cultured mycobiont to form a colony of 1–2 cm in diameter. The colonies are usually compact, elevated and hard.

This sluggish growth of the cultured mycobiont is one of several reasons which render the experimental approach in lichen investigations problematic. Many attempts to improve the in vitro growth of the isolated mycobionts (mostly unpublished), have not been successful.

18.1.2 The Lichen Thallus

Inside the lichen thallus the two partners form a well-organized structure. In all but the most primitive lichens, the algal cells are confined to a distinct layer between two specialized fungal tissues, the cortical layer and a medulla of loosely interwoven hyphae. These two layers are interconnected by hyphae passing between the algal cells (Fig. 3). There is between 4 and 10 times less chlorophyll per volume or per weight in the lichen thallus in comparison to the chlorophyll-containing cells in leaves of higher plants. There are, therefore, much lower rates of net CO_2 assimilation, 10–24 mg CO_2 dm^{-2} h^{-1} in some higher plants as compared to 0.44–3.8 mg CO_2 dm^{-2} h^{-1} in the lichen *Hypogymnia physodes* (RICHARDSON 1973).

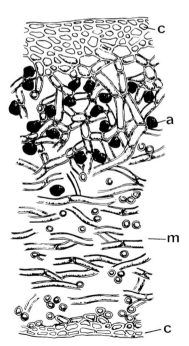

Fig. 3. Schematic section through a foliose lichen thallus, *c* cortex; *a* algal layer; *m* medulla

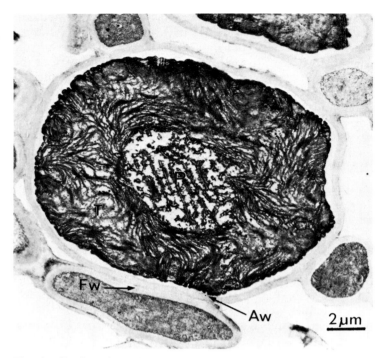

Fig. 4. Section through a mature algal cell of *Caloplaca erythrocarpa* with adjacent fungal hyphae, *Aw* algal wall; *P* pyrenoid; *T* thylakoids. (GALUN et al. 1971 a)

The fungus constitutes the major part of the thallus, and is morphologically the dominant partner which determines the size and shape of the lichen. Physical contact and the physiological interaction between the partners is established within the algal layer. According to COLLINS and FARRAR (1978), in *Xanthoria parietina,* the alga occupies only 7% of the thallus volume, the fungus 43% and the remainder is divided between air spaces (18%) and extracellular matrix (34%).

Most lichens are two-membered symbioses. However, certain lichens are three-membered, the fungus being associated with two phycobionts, a green alga and a cyanobacterium. The green (primary) phycobiont is found in a well-defined zone between the cortex and medulla. The cyanobacterium is restricted to a specialized, delimited part of the thallus called a cephalodium. This structure, which comprises approximately 2–6% of the thallus dry weight, may be located in the thallus medulla, or on the upper or lower surface of the thallus. The most common are external cephalodia which vary considerably in their morphology, ranging from small discs to branched bush-like structures.

18.1.2.1 Fungus–Alga Relation

The fungus–alga relation in lichens is an ectotrophic association. In most lichen species with green phycobionts the components are in simple proximity to one

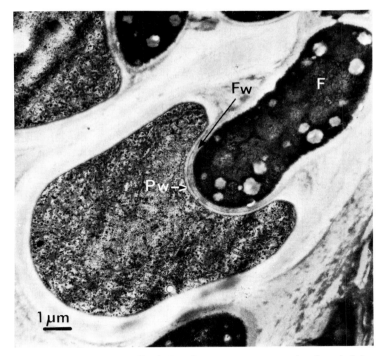

Fig. 5. Section through a mature algal cell of *Gonohymenia mesopotamica* invaded by fungal hypha, *F* fungus; *Fw* fungal wall; *Pw* phycobiont cell wall. (PARAN et al. 1971)

another or with wall-to-wall contact (Fig. 4). Even when haustorial penetration occurs, the percentage of invaded algal cells is small (summarized by COLLINS and FARRAR 1978). It appears that fungal penetration into the algal partner is not an obligatory phenomenon for the mutual relationship between the symbionts of lichens (GALUN et al. 1971a, COLLINS and FARRAR 1978). GALUN et al. (1970a, 1971b) have substantiated light microscope observations by PLESSL (1963) on the physical contact between the symbionts, by using electron microscopy. These authors concluded that there is a correlation between the developmental differentiation of the lichen and the intimacy of the symbionts. In primitive lichens, all phycobiont cells from all phases of the life cycle (i.e. young, mature, scenescing, and decaying cells) could be invaded by intracellular haustoria (GALUN et al. 1970b). In more differentiated types, there was no invasion of young and healthy algal cells. Penetrations appeared only in cells in which the disintegration of the chloroplast had begun and the cells had reached the mature or senescing stage. Finally, in lichens with a foliose or fruticose differentiation, haustoria were found only in already decaying or entirely distorted cells. PEVELING (1968) has suggested that the degeneration of the algal protoplast was caused by the penetration of fungal haustoria, rather than the invasion occurring when the protoplast was already distorted.

With the advancement of differentiation, the number of algal cells penetrated by haustoria also decreases. Parallel with this advancement, the relation of the symbionts gradually changes from a "host–parasite" contact to a looser association of merely close proximity.

Fungal penetrations have rarely been observed in the phycobiont *Nostoc,* which is the more common partner of cyanolichens (Bousfield and Peat 1976, Peat 1968, Peveling 1969 a). On the other hand, light and electron microscope studies on other cyanolichens have revealed pronounced fungal intrusions in a number of both filamentous and unicellular cyanobacterial lichen components (Boissière 1977, Paran et al. 1971, Bubrick 1978, Marton and Galun 1976, Marton 1977). The fungus indents the cyanobacterial cell wall, causing it to invaginate, usually deep into the cell and often into each cell (Fig. 5). Both partners' cell walls remain intact, except for the outer sheath of the cyanobacterium, which is no longer apparent.

18.2 Alga–Fungus Interactions

Both partners of the lichen symbiosis undergo a series of physiological, biochemical and quite complex and dramatic morphological modifications. These presumably facilitate the interaction between the symbionts, being essential for the successful differentiation and for maintaining the stability of the association. All transformation processes known so far are reversible; the components when liberated from the intact lichen regain their free-living characteristics.

18.2.1 Carbohydrate Metabolism

A very basic character of the lichen symbiosis is the supply of carbohydrates by the alga to the fungus. The carbohydrate metabolism of the phycobiont in the intact lichen is strikingly different from that of the alga separated from the lichen.

18.2.1.1 Carbohydrate Efflux

The type of sugar released is determined by the alga. In lichens containing cyanobacteria the carbohydrate released by the phycobiont and transferred to the fungus is glucose. In lichens containing green algae, the carbohydrate released by the phycobiont and transferred to the fungus is a polyol–ribitol, erythritol or sorbitol (Table 2).

The nature of the compound that passes from the alga to the fungus has been determined mainly with aid of the "inhibition technique" devised by Drew and Smith (1967). This technique is based on the competition between labelled and unlabelled compounds. Lichen discs are allowed to photosynthesize in a solution of $NaH^{14}CO_3$ and a high concentration of the unlabelled counterpart of the transferred compound. The unlabelled compound is taken up by the fungus, whereas the radioactive compound produced by photosynthesis enters the medium and can then be measured by chromatography or radioautography.

Table 2. Nature of the mobile carbohydrate in 42 lichen species. (SMITH 1980)

Symbiont	Algal genus	Number of lichens investigated		Mobile carbohydrate
		Genera	Species	
Green alga	*Trentepohlia*	4	6	Erythritol (C_4)
Green algae	*Trebouxia*	10	13	Ribitol (C_5)
	Myrmecia	2	4	
	Coccomyxa	3	3	
Green algae	*Hyalococcus*	1	2	Sorbitol (C_6)
	Stichococcus	2	2	
Cyanobacteria	*Nostoc*	5	10	Glucose (C_6)
	Calothrix	1	1	
	Scytonema	1	1	

Massive and selective carbohydrate efflux from algal cells is characteristic of the symbiotic situation and occurs only in the intact lichen. Free-living *Nostoc* does not release glucose into the medium; the bulk of the photosynthetates are used for growth and storage. Green algal isolates either do not produce any polyols in culture or produce them to a limited extent. Moreover, after isolation from the thallus, the release of sugar declines until it completely ceases, with a subsequent increase of ^{14}C incorporation into the ethanol-insoluble fraction (Table 3) (SMITH 1974).

In three-membered symbioses, carbon is supplied to the fungus almost solely from the green phycobiont. ENGLUND (1977) has shown that *Nostoc* from the three-membered *Peltigera aphthosa* supplied only 2–3% of its photosynthetically fixed carbon to the fungus. Carbon fixed by the cephalodial phycobiont has been detected in polyglucoside granules (RAI et al. 1981c), cyanobacterial lipids (FEIGE 1976a) and water-soluble glucose polymers (FEIGE 1976b). The reason for this altered pattern of carbon efflux in three-membered lichens is partially explained by the elevated heterocyst frequencies in the cyanobacterial biont (see Sect. 18.2.2.2.1). It is also possible that the fungus selectivity modifies the carbon metabolism of only the green phycobiont, while not affecting carbon metabolism of the cyanobacterial symbiont.

The changes in the algal carbohydrate metabolism are presumably induced by the fungus, but how is a question not yet solved.

HILL (1972) incubated discs of *Peltigera polydactyla,* a species which contains *Nostoc* and releases glucose, in $H^{14}CO_3$ solutions in the light for 10 min, and then studied the distribution of this pulse in media without radioactivity. His findings suggested that the alga does not release glucose directly, but first forms a glucan pool outside its plasmalemma. This glucan is then hydrolyzed by an extracellular enzyme produced by the fungus. The evidence for HILL's theory is the rapid turnover of ^{14}C in the ethanol-insoluble fraction during transfer of ^{14}C to the fungus. This theory is attractive, but unfortunately has not found support from evidence of other studies. SMITH (1975) found that

Table 3. Algae or lichens incubated in distilled water containing $NaH^{14}CO_3$ for 3 h at room temperature. Algae took 30 min to 1 h to isolate from lichens. (Smith 1974)

Condition of alga	*Nostoc* from *Peltigera canina*	*Hyalococcus* from *Dermatocarpon miniatum*	*Coccomyxa* from *Peltigera aphthosa*	*Trebouxia* from *Xanthoria aureola*
% Fixed ^{14}C released from alga				
In lichen	60	55	65	40
0 h isolate	15.3	26.1	23.1	8.0
24 h isolate	7.0	6.2	3.6	1.0
Culture	4.0	1.3	1.4	2.5
% Fixed ^{14}C incorporated into ethanol-insoluble				
In lichen	9.0	2.0	21.0	1.0
0 h isolate	27.8	29.6	56.2	35.3
24 h isolate	50.7	40.2	62.8	53.2
Culture	46.0	50.3	50.8–72.3	57.9

	As % ^{14}C							
	re-leased	in cells	re-leased	in cells	re-leased	in cells	re-leased	in cells
"Mobile" carbohydrate								
0 h isolate	22.0	Trace	91.0	73.0	85.0	49.0	85.0	83.5
24 h isolate	0	0	15.0	75.7	0.9	5.1	20.0	58.5
Culture	0	0	0	Trace	0	Trace	17.1	32.8
Identity of carbohydrate								
	Glucose		Sorbitol		Ribitol		Ribitol	

Release from algae in lichens calculated from ^{14}C accumulation in fungal polyols

enzyme inhibitors did not change the process. Using autoradiography, Peveling and Hill (1974) have later shown that the glucan pool was located inside rather than outside the algal cells. Peveling (1976) suggested that storage bodies which she found in the algal cells are degraded by enzymes of lomasomes or lysosomes adjacent to the storage bodies.

18.2.1.2 Transfer of Carbohydrates

Using the "inhibition technique" Smith and Hill (1972) measured 0.7–1.1 mg g^{-1} dry weight of thallus per hour glucose transferred from the phycobiont to the mycobiont of *Peltigera polydactyla*. Richardson (1973) distinguished between fast, intermediate, slow, and very slow carbohydrate movement between the symbionts and recorded the rate of movement of fixed ^{14}C from the algal layer to the medulla of various lichens. Smith (1980) and Tapper (1981) have expressed reservations about data on the rate of movement of photosynthates from alga to fungus. Most of the measurements are based on

Table 4. Proportion of ^{14}C contained in the symbionts of *Cladonia convoluta*. (Tapper 1981)

	Medullary hyphae	Algal cells	Cortex	Total grain number
30-min pulse only				
Untreated control	36.54	29.93	33.53	745.0
30-min pulse followed by 1.25 h chase				
Untreated control	57.76	10.07	32.18	811.9
DNP 10^{-4} M	44.98	27.00	28.02	771.8
DNP 10^{-3} M	23.35	29.60	47.06	547.8
FCCP 5×10^{-5} M	45.75	21.55	32.70	851.9
Valinomycin 10 µg ml^{-1}	54.84	26.51	18.65	788.5
Arsenate 3.3×10^{-2} M	46.62	24.87	28.51	770.6

Figures show percentage of ^{14}C label over different parts of sections
DNP = 2,4 dinitrophenol.
FCCP = Fluorocarbonylcyanidephenylhydrazon

experiments using the "inhibition technique" which does not suffice for quantitative calculations. By using quantitative autoradiography of whole thalli, Tapper (1981) could measure directly the amount of fixed carbon released by the algal symbiont. He showed that 70% of the photosynthetically fixed ^{14}C by *Trebouxia* of *Cladonia convoluta* was released by the alga during a 30-min pulse, and an additional 20% during a subsequent 1.25 h chase. FCCP, DNP and arsenate inhibited release of ^{14}C from the algal symbiont (Table 4). These substances are known as inhibitors of active transport systems. Tapper therefore assumed that release of carbohydrate by *Trebouxia* was an active process, though the photosynthesis process may, at least to some extent, be affected by these substances as well. Tapper also found that the fixed carbon was transferred from *Trebouxia* to both the medulla and the cortex. During the pulse, both tissues received similar amounts of ^{14}C. During chase, the amount of ^{14}C increased only in the medulla. Tapper's report is the first on any physiological activity in the cortex, which was regarded as an inert layer protecting the algal cells underneath; the cortical tissue has in previous investigations simply not been taken into consideration (see Richardson 1973).

In contrast to a previous theory that carbohydrate efflux was the result of fungal-induced changes in the algal membrane potentials (Smith 1974), a more recent theory postulates that transfer from the alga is accomplished by a carrier-mediated, facilitated diffusion process (Collins and Farrar 1978, Smith 1975, 1980). This process does not depend on close contact between large surface areas of the symbionts. From ultrastructural studies it is clear that photosynthetates released by the phycobiont and transferred to the mycobiont must cross the plasma membranes, cell walls, and extracellular matrices of both partners. According to Collins and Farrer (1978), 20% of the algal cell wall is in direct contact with fungal cell wall; a further 30–35% is in contact with extracellular matrix. These authors believe that the extracellular matrix renders excreted sugar available to a larger fungal surface area than would otherwise be possible.

Peveling (1969a) has observed convolutions in the fungal plasmalemma and suggested that the enlarged surface area facilitates metabolite uptake, in a manner similar to "transfer cells" in higher plants (Gunning and Pate 1969). In a later study, Peveling (1973b) found numerous vesicles in the sheath of *Calothrix* from the lichen *Lichina pygmaea* and in the matrix between the symbionts. She suggested that these vesicles may be involved in a mechanism for promoting transport between the symbionts.

18.2.1.3 Uptake and Conversion of Transferred Carbohydrate by the Mycobiont

A great part of the sugar that reaches the fungus, whether glucose or ribitol, is immediately converted into mannitol. When samples of the *Nostoc*-containing lichen *Peltigera polydactyla* were incubated in the light in solutions of $NaH^{14}CO_2$, 60% of the radioactivity was incorporated into mannitol and about 10% in insoluble compounds (presumably polysaccharide) during the first 45 min of the experiment (Smith 1974).

Similarly, ribitol released by *Trebouxia* and by *Coccomyxa* phycobionts was also converted in the mycobiont into mannitol. Arabitol was formed as an intermediate during the conversion of ribitol to mannitol (Richardson and Smith 1968, Feige 1970). Erythritol released by *Trentepohlia* accumulated partially as erythritol; sorbitol released by *Hyalococcus* accumulated as sorbitol or volemitol (Feige 1975).

The conversion into mannitol is a one-way process; mannitol that accumulates in the fungus is not available to the alga. It seems logical to assume that such a "source–sink" system would allow for a continuous flow from the donor to the recipient. However, Chambers et al. (1976) have shown that the massive efflux from the phycobiont continues even when the fungal uptake is inhibited by digitonin. As of yet, we cannot explain the mechanism underlying the specific sugar production and release, neither do we know the pathway by which ribitol or glucose are converted into mannitol.

The phycobiont appears to be an altruistic partner. Farrar (in Smith 1980) estimated that 80–90% of the carbon fixed by photosynthesis in *Xanthoria parietina* is transferred to the fungus. However, not all this carbon is nutritional or used for the synthesis of structural components. Periodical drying and wetting are an integral part of the life of lichens. During dry periods, metabolic activity comes to a standstill, or at least is not measurable by conventional techniques (Lange 1969). They have to survive drought and re-establish metabolic activity after every desiccation. The surplus of carbohydrates produced and transferred to the fungus is essential for survival in this fluctuating water economy. The high concentrations of sugar alcohol in the mycobiont's cytoplasm are thought to act as a physiological buffer by replacing water removed from the hydration shells of protein (Smith 1979, Cowan et al. 1979).

The rewetting of a dry and "dormant" lichen thallus also causes a substantial loss of CO_2. By rewetting dry *Peltigera polydactyla* discs, Smith and Moles-

WORTH (1973) found that while basic respiration accounted for 121 µl CO_2, 115 µl CO_2 were wasted by "wetting burst" and 114 µl by re-saturation respiration.

Mannitol is apparently not metabolized by free-living fungi, but is a specific metabolite of lichenized fungi. GALUN et al. (1976) showed that the cultured mycobionts isolated from *X. parietina, Tornabenia intricata* and *Sarcogyne* sp. incorporated ³H-mannitol, ³H-mannose and ³H-ribitol (the latter was tested only on the *X. parietina* mycobiont), whereas free-living fungi (*Trichoderma viride* and *Phytophthora citrophthora*) did not incorporate these compounds.

18.2.2 Nitrogen Metabolism

There are two main sources of nitrogen available to lichens: nitrogen (organic or inorganic), which is absorbed from the environment, and/or atmospheric N_2, which can be reduced by lichens with cyanobacterial phycobionts. Nitrogen assimilated by lichens may, under various environmental circumstances, be made available to the surrounding environment. This input, due to such factors as fire, mineralization, leaching or decomposition, may be considerable, or even crucial to, the nitrogen economy of the specific environment (CRITTENDEN and KERSHAW 1978, FORMAN 1975, FORMAN and DOUDEN 1977, MILLBANK 1978, 1981, PIKE 1978, ROGERS et al. 1966).

18.2.2.1 Nitrogen and Amino Acid Analysis of Whole Thalli

Numerous studies have reported on the total nitrogen content of intact lichen thalli (ENGLUND 1977, GOAS and BERNARD 1967, GREEN et al. 1980, HITCH and MILLBANK 1975b, HITCH and STEWART 1973, MILLBANK 1972, MILLBANK and KERSHAW 1969, SCOTT 1956, see MILLBANK and KERSHAW 1973 for a partial summary of available data). In general, lichens with cyanobacterial phycobionts have significantly higher nitrogen contents than do lichens with green phycobionts. For example, GREEN et al. (1980), in examining the nitrogen content of 44 members of the lichen genera *Pseudocyphellaria* and *Sticta,* found that the mean nitrogen content (as % dry weight of thallus) of cyanolichens was 3.4%; in lichens with green phycobionts it was only 0.5%. The nitrogen content of cephalodia is similar to that of the thallus of cyanolichens, although ENGLUND (1977) reported a somewhat higher nitrogen content in the cephalodia of *Peltigera aphthosa.*

The soluble amino acids of lichen thalli have been reported by BERNARD and GOAS (1968), BERNARD and LAHRER (1971), GOAS and BERNARD (1967), JÄGER and WEIGEL (1978) and RAMAKRISHNAN and SUBRAMANIAN (1964a, b, 1965, 1966a, b). High levels of glutamine and glutamic acid were noted (GOAS and BERNARD 1967, JÄGER and WEIGEL 1978). The latter authors suggested that glutamine may function as a nitrogen pool in three green-phycobiont lichens, and that the major portion of inorganic nitrogen taken up was incorporated into amino acid metabolism via glutamic acid.

Bound amines, such as methylamine, dimethylamine, trimethylamine (BER-
NARD and GOAS 1968, BERNARD and LAHRER 1971), putrescine, cadaverine, sper-
midine, agmatine and spermine (JÄGER and WEIGEL 1978) have also been reported.

18.2.2.2 Nitrogen Fixation

Lichens Which Fix Atmospheric Nitrogen. Nitrogen fixation has been demonstrated only in lichens with cyanobacterial phycobionts. A survey of some lichen
genera in which fixation has been shown either by means of $^{15}N_2$ incorporation
or acetylene reduction is given in Table 5.

The cyanobacterial phycobionts of the lichens listed in Table 5 are all filamentous and heterocystous (*Nostoc, Calothrix, Stigonema*). Unpublished data
(HOCHMAN, BUBRICK and GALUN) suggest that some lichens with unicellular
cyanophycobionts may reduce acetylene. Of interest was the finding of acetylene
reduction in the lichen, *Heppia* spp. The phycobiont of *Heppia* is *Scytonema*
which, in the free-living state, is filamentous and heterocystous. However, in
symbiosis, it is reduced to a unicellular form completely devoid of heterocysts
(MARTON and GALUN 1976). If confirmed, this suggests that nitrogenase is
expressed in the vegetative cells of this phycobiont.

There are a number of modifications in the cyanobacterial phycobionts
of cephalodia which are not seen in two-membered symbioses. The most conspicuous of these is the increased heterocyst frequency of the cephalodial phycobiont
(ENGLUND 1977, GREEN et al. 1980, GRIFFITHS et al. 1972, HITCH and MILLBANK
1975a, b, HUSS-DANELL 1979, STEWART and ROWELL 1977). Whereas the average
range of heterocyst frequency in two-membered symbioses is 2%–5% of the
vegetative cells (which is comparable to the frequency in free-living cyanobac-

Table 5. Representative genera of lichens tested for nitrogen fixation

Lichen	Phycobiont	Method[a]	Reference
Collema	*Nostoc*	C_2H_2	HITCH and MILLBANK (1975b)
Ephebe	*Stigonema*	C_2H_2	HITCH and MILLBANK (1975b)
Leptogium	*Nostoc*	$^{15}N_2$	BOND and SCOTT (1955)
Lichina	*Calothrix*	$^{15}N_2$	STEWART (1970)
Lobaria[b]	*Nostoc*	$^{15}N_2$	MILLBANK and KERSHAW (1970)
Massalongia	*Nostoc*	C_2H_2	HITCH and MILLBANK (1975b)
Nephroma[c]	*Nostoc*	C_2H_2	HITCH and MILLBANK (1975b)
Pannaria	*Nostoc*	C_2H_2	HITCH and MILLBANK (1975b)
Peltigera[c]	*Nostoc*	C_2H_2	HITCH and MILLBANK (1975b)
Placynthium	*Dichothrix*	C_2H_2	HITCH and MILLBANK (1975b)
Polychidium	*Nostoc*(?)	C_2H_2	HITCH and MILLBANK (1975b)
Pseudocyphellaria	*Nostoc*	C_2H_2	GREEN et al. (1980)
Stereocaulon[b]	*Nostoc/Stigonema*	C_2H_2	HUSS-DANELL (1977)
Sticta[c]	*Nostoc*	C_2H_2	GREEN et al. (1980)

[a] Nitrogen-fixation tested by $^{15}N_2$ incorporation or acetylene reduction (C_2H_2)
[b] With cephalodia
[c] With cephalodia and/or *Nostoc* as primary phycobiont

teria; STEWART and ROWELL 1977), heterocysts of cephalodial phycobionts may range from 15%–30% of the vegetative cells.

In three-membered symbioses, heterocyst frequencies were found to vary between different parts of the thallus. HITCH and MILLBANK (1975b) reported that in three lichens[2] heterocyst frequencies were 2–10 times higher in cyanobacteria adjacent (0–2 mm) to the green algal layer than when remote (5–10 mm) from the primary phycobiont. ENGLUND (1977) has demonstrated that heterocyst frequency is higher (22%) in cephalodia on the more mature thallus parts, and lower (14%) in cephalodia at the actively growing apex of *Peltigera aphthosa*.

Rate of Nitrogen Fixation. Numerous authors have reported on the rates of nitrogen fixation in lichens maintained in the laboratory and in the field (references listed in Table 5 are applicable). Rates of acetylene reduction on the order of 3 nmol ethylene evolved h^{-1} mg^{-1} dry wt are typical (ENGLUND 1977, HENRIKSSON and SIMU 1971, KALLIO et al. 1972, KERSHAW 1974, MILLBANK 1972), and compare favourably with the rates of free-living cyanobacteria (ENGLUND and MEYERSON 1974, HITCH and STEWART 1973, MILLBANK 1974a).

A major problem has been to relate the observed rates of acetylene reduction to the actual rates of nitrogen fixed. An attempt to find suitable conversion factors for lichens has recently been undertaken by MILLBANK (1981) and MILLBANK and OLSEN (1981), using a controlled environmental growth chamber. Data, based on the ratio of acetylene reduced during a 1-h period/$^{15}N_2$ fixed on a 14-day basis, yielded conversion factors of 6.5–12.9 in *Peltigera membranacea*, and 7.1–10.1 in *Lobaria pulmonaria*. Factors as high as 22 were found for *Peltigera polydactyla*. Factors were shown to vary considerably between seasons as well as between collection sites; a major departure from the theoretical value of 3 was found in all cases.

Variation in the rates of acetylene reduction in different parts of the thallus of the two-membered *Peltigera canina* have been noted by HITCH and STEWART (1973). Acetylene reduction was highest in the actively growing tips of vegetative thalli, decreasing toward the older thallus parts. These authors also reported that acetylene reduction per unit weight of fruiting thalli was only 20% that of purely vegetative thalli. Since there were no differences in the algal content (heterocyst frequencies were not determined), the authors suggested that the algae were not all in the same metabolic state.

In the three-membered symbiosis, *Peltigera aphthosa*, ENGLUND (1977) reported that cephalodia closest to the actively growing apices reduced acetylene at rates lower than cephalodia in the mature thallus areas. In this case, there was a positive correlation between heterocyst frequency and rate of acetylene reduction.

The rate of acetylene reduction of excised cephalodia is reduced as compared to that of cephalodia still attached to intact thalli (see ENGLUND 1977). HUSS-DANELL (1979) reported that excised cephalodia retained an average of 59% of the activity of cephalodia connected to the thallus. He also reported a reduc-

---

2 The three lichens used in the study were *Solorina spongiosa, Sticta dufourii/S. canariensis* and *Dendriscocaulon umhausense/Lobaria amplissima*

tion of activity when thalli were cut into discs without rupturing the connection between cephalodium and thallus.

Transfer and Partition of Fixed Nitrogen. The nature of the products liberated by nitrogen-fixing cyanobacterial phycobionts was first reported by Millbank (1974b). Using several three-membered lichens, he noted a wide range of amino acids and peptides, which were similar to those liberated by the cultured *Nostoc* phycobionts. From this he concluded that, in contrast to carbon metabolism (Sect. 18.2.1), the basic pattern of nitrogen metabolism in symbiotic cyanobacteria did not undergo marked alterations. Subsequent studies have shown that NH_4^+ is a major product liberated from cyanobacterial phycobionts, and its liberation is the result of a highly altered nitrogen metabolism in the lichenized cyanobacterium (Englund 1977, Rai et al. 1980, 1981a, 1981b, Stewart and Rogers 1978, Stewart and Rowell 1977, Stewart et al. 1980). The bulk of the liberated combined nitrogen is assimilated by the fungal component with only minor amounts made available to *Nostoc* in two-membered symbioses, or to *Coccomyxa* and *Nostoc* in three-membered symbioses (Kershaw and Millbank 1970, Millbank and Kershaw 1969, Rai et al. 1981b).

Assimilation of Fixed Nitrogen. In order to simplify the following discussion, some salient points relevant to nitrogen assimilation in both prokaryotes and eukaryotes are listed below.
1. The primary route of NH_4^+ assimilation in cyanobacteria is via the glutamine synthetase-glutamate synthase (GOGAT) pathway (cf. Lea and Miflin 1979, Miflin and Lea 1982), with other pathways being of secondary importance under N_2-fixing conditions.
2. Glutamate dehydrogenase (GDH) is the primary NH_4^+ assimilating enzyme in certain eukaryotes, including fungi (also see Bernard and Goas 1969). The enzymes alanine dehydrogenase (ADH) and glutamate-aspartate-aminotransferase (GAT) are also found in both prokaryotes and eukaryotes.
3. High levels of combined nitrogen, especially of NH_4^+, inhibit nitrogen fixation and heterocyst production in free-living cyanobacteria.
4. Evidence suggests that GS or a product of its activity, and not NH_4^+ alone, is involved in the regulation of nitrogenase activity and heterocyst production in free-living cyanobacteria.

For a more detailed description and background, the reader is referred to Gallon (1980), Stewart (1974, 1980) and Stewart et al. (1980), and references therein.

Activities of Some Important NH_4^+-Assimilating Enzymes in Symbiosis. It has been demonstrated that several enzymes important in NH_4^+ assimilation undergo modification in symbiosis (Rai et al. 1980, 1981a, 1981b, Sampaio et al. 1979, Stewart and Rodgers 1978, Stewart and Rowell 1977, Stewart et al. 1980). In Table 6, the activities of various NH_4^+-assimilating enzymes from different parts of the three-membered *Peltigera aphthosa* are presented. In the cultured *Nostoc,* GS activity was high while GDH activity was negligible, whereas in the cultured *Coccomyxa* GS and GDH activities were appreciable and GAT

Table 6. Activities of NH_4^+-assimilating enzymes in thalli, cultured and symbiont *Nostoc* and *Coccomyxa* (After RAI et al. 1980)

	Glutamine synthetase	NADPH-dependent glutamate dehydrogenase	Alanine dehydrogenase	Glutamate-aspartate-aminotransferase
Whole thallus	28	90	9.7	58
Thallus without cephalodia	18	35	10	47
Excised cephalodia	2	398	8.5	20
Lichen medulla (fungus)	ND	24	15	37
Cultured *Nostoc*	60	2	11	25
Cultured *Coccomyxa*	37	22	3	520

Activities expressed as nmol product formed min^{-1} mg^{-1} protein, ND not detected

activity high. In symbiosis, low GS activities were associated with *Nostoc* in the cephalodia, while high levels of GDH were associated with the cephalodial fungus. The high GDH levels in cephalodial fungi have also been demonstrated by histochemical techniques (STEWART and ROWELL 1977). Similar altered levels of GS and GDH activity have been reported in the two-membered *P. canina*; in this case, high GDH activity was found in fungal hyphae closest to the *Nostoc* cells (STEWART and ROWELL 1977).

Although it was originally reported that the amounts of GS protein were similar in cultured and symbiotic cyanobacteria from two- and three-membered symbioses (SAMPAIO et al. 1979), it was subsequently shown that the amount of GS protein in symbiotic *Nostoc* in the two-membered *P. canina* was less than 5% of the protein present in the cultured phycobiont (see STEWART et al. 1980).

It is interesting to note that while the activity of GOGAT is reduced in the *Nostoc* phycobiont of the three-membered *P. aphthosa* as compared to the cultured *Nostoc,* the activity of GOGAT is unaltered in the *Coccomyxa* phycobiont (RAI et al. 1981a). Hence, in symbiosis there is an apparent preferential inhibiton of cyanobacterial GOGAT.

The alterations in the level of NH_4^+-assimilating enzymes may partially explain the observation that NH_4^+ is the major product liberated from N_2-fixing cyanobacteria in symbiosis, and the observation that while exogenously supplied NH_4^+ completely inhibits nitrogenase activity in free-living cyanobacteria, it has no effect on N_2-fixing excised cephalodia (RAI et al. 1980, STEWART and ROWELL 1977). Thus, when GS activity is low and GDH activity negligible, and alternative pathways of NH_4^+ assimilation are of secondary importance in cyanobacteria, NH_4^+ produced as a result of N_2 fixation is liberated extracellularly by the symbiotic cyanobacterium. The normal regulation of GS activity by NH_4^+ is not operative as GS is essentially not available to be regulated.

STEWART et al. (1980) brought up the point that reduced GS levels may, in part, contribute to the high heterocyst frequencies in three-membered symbioses, as glutamate or a related product may be involved in the regulation of heterocyst development. However, and as STEWART et al. (1980) rightly recognize, this cannot account for the "normal" heterocyst frequencies found in two-membered symbioses. In the latter case, phycobionts must supply both fixed carbon and nitrogen to the fungus. These phycobionts probably could not survive the double demand with up to 30% of their cells converted to non-CO_2-fixing heterocysts. In three-membered symbioses, the possibility that CO_2 fixed by the green phycobiont is cycled to the cyanobacterial symbiont (hence allowing their carbon demand to be fulfilled) should be further investigated. The demonstration by Rai et al. (1981c) that cephalodial cyanobacteria store considerable polyglucoside in the light (presumably because of the lack of carbon demand from the alga) which it can mobilize in the dark to support substantial nitrogenase activity may also be relevant to this problem.

Fate of Newly Assimilated NH_4^+. Rai et al. (1981b) have studied the incorporation and subsequent metabolism of liberated NH_4^+ in the three-membered *Peltigera aphthosa.* Excised cephalodia were incubated for 30 min in the presence of $^{15}N_2$ and the distribution of labelled nitrogenous compounds determined. After 30 min, labelling was greatest in glutamate, with significant labelling found in NH_4^+, glutamine, alanine and aspartate. The highest initial labelling was into NH_4^+ which subsequently levelled off (12 min) while incorporation into glutamine and glutamate increased steadily over the 30-min period. Pulse-chase experiments suggested that alanine was labelled secondarily as a result of aminotransferase activity of fungal origin (also see Bernard and Goas 1969). It was concluded that the bulk of NH_4^+ assimilated for onward transport to the main thallus was channelled via fungal GDH in the cephalodia. The glutamate formed was converted to alanine via aminotransferase activity, and it appears that alanine may be the main form of combined nitrogen exported to the remainder of the thallus (also see Cowan et al. 1979, Stewart et al. 1980).

18.3 Other Effects of Symbiosis

18.3.1 Effects of Symbiosis on the Algal Partner

18.3.1.1 Cyanobacterial Phycobionts

Symbiosis has remarkable effects on the morphology and developmental cycles of cyanobacterial phycobionts. In certain lichens, such as *Ephebe lanata,* the *Stigonema* phycobiont is essentially unaltered in morphology. In many *Collema* species, the phycobiont is easily recognized as *Nostoc,* although complex developmental cycles, such as described by Lazaroff (1973) in free-living *Nostoc,* are not apparent in symbiosis. In other cases, the true nature of the phycobiont is completely obscured. In *Heppia echinulata,* the *Scytonema* phycobiont, normally filamentous and heterocystous with abundant false branchings, is reduced to a unicellular form devoid of heterocysts (Marton and Galun 1976). Similar dramatic alterations in the *Dichotrix* phycobiont of *Placynthium nigrum* have been observed by Geitler (1934).

Alterations have also been observed in unicellular phycobionts. BUBRICK (1978) has observed that, whereas many phycobionts from the Lichinaceae and Heppiaceae appear to be *Gloeocapsa* in the lichenized state, upon isolation and culture a wide diversity of forms and developmental cycles were apparent.

The subcellular structure of cyanobacteria is basically unaltered in symbiosis (BOISSIÈRE 1972, 1982, BOUSFIELD and PEAT 1976, BUTLER and ALLSOPP 1972, PARAN et al. 1971, PEAT 1968, PEVELING 1969b, 1973a, 1974, SPECTOR and JENSEN 1977). Subcellular structures such as the cell envelope and sheath, cyanophycin granules, polyphosphate granules, polyhedral bodies, thylakoids etc. are present in both lichenized and free-living forms. Differences in the size/distribution of these components have been noted, but it is difficult to evaluate the significance of these observations. The broad pattern of cell division is also unaltered. Polyglucoside bodies, previously thought to be absent in lichenized phycobionts (PEAT 1968), have been subsequently demonstrated with the aid of specific cytochemical methods (BOISSIÈRE 1972).

The presence of neutral phosphatase in the plasmalemma of symbiotic *Nostoc* in *Peltigera canina* and its absence from free-living *Nostoc* has been noted (BOISSIÈRE 1973).

18.3.1.2 Eukaryotic Phycobionts

As with the cyanobacteria, lichenization results in conspicuous modifications in the morphology and reproductive cycles of eukaryotic phycobionts. For example, the phycobionts of both *Vezdaea aestivalis* and *Verrucaria maura* appear to be unicellular in the lichen. Upon culturing, they develop into the branched filamentous structures of *Leptosire* (TSCHERMAK-WOESS and POELT 1976) and *Heterococcus* (PARRA and REDON 1977) respectively. Such dramatic modifications do not take place in the two most common phycobionts, *Trebouxia* and *Pseudotrebouxia.*

The "normal" reproductive pathways of phycobionts are usually modified when lichenized. Many phycobionts listed in Table 1 produce abundant zoospores when free-living or in culture, whereas zoospores are rarely seen in the lichen. In a study of algal zoosporogenesis in the lichen *Parmelia caperata,* SLOCUM et al. (1980) observed that in the early stages of zoosporogenesis within the lichen, flagella and basal apparatuses were formed in the naked zoospores. Later, flagella were absorbed and presumably the basal structures disappeared; potential zoospores formed walls and were transformed into aplanospores. TSCHERMAK-WOESS (1978a) has noted that *Trebouxia,* when in fortuitous contact with fungal hyphae or when free-living, produced abundant zoospores. *Trebouxia* encircled by fungal hyphae multiplied only by aplanospores. Nothing is known about the factors responsible for this effect, although it appears that more than just physical constraint of the algal cell is involved.

Light and electron microscopic observations on numerous lichens have shown that the normal complement of subcellular organelles is present in symbiosis (CHERVIN et al. 1968, FISHER and LANG 1971a, b, GALUN et al. 1970b, JACOBS and AHMADJIAN 1971, PEVELING 1973a). Changes that have been observed upon culturing include the appearance of polyphosphate bodies, an increase in the amount of starch surrounding the pyrenoid, and a decrease in the size and number of pyrenoglobuli (Ahmadjian 1973 b).

18.3.1.3 Cell-Wall Modifications

Several modifications in the phycobiont cell wall as a result of lichenization have been noted. In *Trebouxia* from *Cladonia* species, a gelatinous sheath surrounding cultured phycobiont cells is apparently absent in symbiosis (Ahmadjian 1973 b). Similar observations have been made on *Chlorella* phycobionts (Tschermak-Woess 1978 b). Ultrastructural studies on *Trebouxia* from *Cladonia cristatella* have shown that an outer fibrillar coat on the cultured phycobiont cell wall was not present in symbiosis (Jacobs and Ahmadjian 1971). These authors also noted that the cell wall of cultured *Trebouxia* was thinner than that of the lichenized alga.

A detailed study of the cell walls of *Coccomyxa* and *Myrmecia* phycobionts has been made by Honegger and Brunner (1981). Acid hydrolysis of isolated cell walls of freshly isolated and cultured *Coccomyxa* from *Peltigera leucophlebia* yielded the sugars rhamnose, xylose, mannose, arabinose, glucose and galactose. Interestingly, no qualitative differences in the sugar composition between symbiotic and cultured *Coccomyxa* cell walls were noted. Sporopollenin was also detected in the symbiotic and cultured *Coccomyxa* cell walls, as well as in cultured *Myrmecia* cell walls. Contrary to the report of König and Peveling (1980), sporopollenin was not detected in *Trebouxia* cell walls.

Studies in our laboratory have shown that there are a number of differences in the cell-wall properties between freshly isolated and cultured phycobionts from the lichen *Xanthoria parietina,* as well as from several other lichens. We have found that algae rapidly isolated from the lichen and cleaned of debris, hyphae, etc. are much more suitable for study than phycobionts actually in the lichen. Numerous studies have convinced us that the cell-wall properties of these freshly isolated algae are comparable to algae in situ. Whenever possible, we have checked our results using freeze-microtome sections. In all cases to date, reactions observed on freshly isolated or in situ algae have been identical.

In experiments which are not monitored visually, it is essential to remove all fungal debris from freshly isolated preparations. Methods for the isolation of cyanobacterial phycobionts (e.g. Millbank and Kershaw 1969) have proved to be very efficient. However, published methods for cleaning *Trebouxia* phycobionts have not been generally applicable. Ascaso (1980) suggested the use of $CsCl_2$ and KI step gradients for the purification of phycobionts from *Parmelia conspersa* and *Lasallia pustulata*. Upon using these methods, as well as self-generated Percoll gradients, Bubrick, Sonesson and Galun (unpublished) found that when optimized, $CsCl_2$ step gradients remove nearly all fungal debris. Percoll, although somewhat less efficient than $CsCl_2$, has the advantage that morbid cells and cell walls are separated from living cells. Unpublished observations by R. Ronen in our laboratory have shown that $CsCl_2$ and Percoll gradients have not noticeable effect on O_2 evolution in freshly isolated algae. Algae purified on KI gradients showed variable responses; with longer exposures to KI, O_2 evolution was inhibited.

Table 7. ABP-binding and cytochemical staining of cell walls of freshly isolated and cultured phycobionts. (BUBRICK and GALUN 1980a)

Phycobiont from	ABP					
	Xanthoria-protein		Coomassie blue		Ruthenium red	
	Fresh	Culture	Fresh	Culture	Fresh	Culture
Xanthoria parietina	−	+ +	−	+ +	±	+ +
Caloplaca aurantia	−	+ +	±	+ +	−	+ +
C. citrina	−	+ +	−	+ +	−	+
Cladonia convoluta	−	−	−	−	−	−
Ramalina duriaei	−	−	−	−	±	±
R. pollinaria	−	−	±	−	−	−

− Staining indistinct to invisible; ± staining faint; + staining clearly visible; + + staining intense

BUBRICK and GALUN (1980b) have shown that there are differences in the binding of fluorescently labelled concanavalin A (FITC-Con A) to the cell walls of freshly isolated phycobionts from several lichens. With the exception of the *Cladonia convoluta* phycobionts, all cultured phycobionts bound FITC-Con A, whereas freshly isolated phycobionts did not. This was not due to the absence or removal of the freshly isolated phycobiont cell wall; walls were clearly visible in the light microscope, and were stained with the cell-wall brightener, calcofluor white. Cell-free supernatants from the ground lichens did not prevent FITC-Con A binding to cultured phycobionts (unpublished). FITC-Con A binding sites reappeared on the freshly isolated algal cell walls after 7–21 days in culture. In many cases, binding was apparent prior to the first cellular division; released daughter cells always bound FITC-Con A. Con A binding to the cultured phycobiont of *Hypogymnia physoides,* and to *Pseudotrebouxia corticola,* has been reported by ROBENEK et al. (1982).

These results have recently been extended to include other lectins (GALUN, MALKI and BUBRICK, unpublished). Using a sensitive spectrofluorometric assay, results of FITC-Con A binding have been confirmed. In addition, we have found that several other FITC-lectins (e.g. wheat germ agglutinin, castor bean lectin and others) bind the cultured, but not the freshly isolated, phycobiont from *X. parietina.*

Cytochemical examinations of several freshly isolated-cultured phycobiont pairs also demonstrated differences between the two cell types (BUBRICK and GALUN 1980a). Cell walls of the cultured *X. parietina* phycobiont stained with coomassie blue (CB) and ruthenium red (RR); neither dye stained cell walls of freshly isolated algae. Differences were also detected between other freshly isolated-cultured pairs. Both CB and RR stained cultured walls of algae from *Caloplaca citrina* and *C. aurantia,* but not walls from freshly isolated phycobionts; CB and RR also did not stain the cell walls from cultured or freshly isolated phycobionts from *Ramalina duriaei, R. pollinaria* or *Cladonia convoluta* (Table 7).

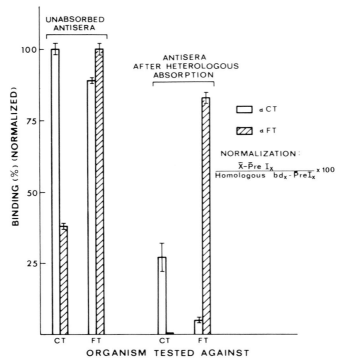

Fig. 6. Shared and unique antigen(s) on cultured and freshly isolated phycobionts from *X. parietina*. Normalized data. (Bubrick et al. 1982)

Similarities, as well as differences, have been detected between symbiotic and cultured phycobionts from *X. parietina* with the aid of a solid phase radioim-munoassay (RIA) (Bubrick et al. 1982). In this study antibodies raised against whole cells were reacted with whole cells so that immunogens detected were located primarily in and on the cell wall. Results from a RIA are presented, in a normalized form, in Fig. 6. Antisera (α) raised against cultured (CT) or freshly isolated (FT) phycobionts show significant cross-reaction, suggesting that they possess shared immunogens. Heterologous absorptions followed by homologous reactions (e.g. αCT absorbed with FT and reacted with CT) clearly demonstrate that each algal type possesses at least one unique immunogen. Stated somewhat differently, symbiosis-specific and culture-specific antigens can be detected.

Unpublished experiments show that antiserum to the cultured *X. parietina* phycobiont can be used to detect this alga in nature, as well as to distinguish between a variety of cultured *Trebouxia/Pseudotrebouxia* strains in culture.

Mechanisms of Cell-Wall Modification. The modifications in phycobiont cell-wall structure and molecular architecture are undoubtedly of importance to the success of the symbiosis. At present, nothing is known about how such modifications take place, although several possibilities may be considered. First, the fungus may actively modify the algal cell wall through extracellular or cell-wall-bound enzyme activity. Some lichen fungi are known to utilize cellulose as a carbon source (AHMADJIAN 1977), presumably through the action of extra-cellular cellulases. Although this has not been tested, it is likely that certain lichen fungi also produce enzymes capable of hydrolysing pectins, hemicellulose, cell-wall-associated proteins and other cell-wall macromolecules. In addition, several authors (e.g. WEBBER and WEBBER 1970) have suggested that haustorial penetration of algal cell walls was by enzymatic rather than purely physical means. This hypothesis suggests that the fungus must not only produce enzymes at the sites of mutual contact between the symbionts, but also secrete them into the intercellular matrix of the thallus. In *X. parietina,* the entire phycobiont cell wall is modified; the fungus, however, is in contact with only 20% of the cell-wall surface (COLLINS and FARRAR 1978). Such a mechanism of modification might be viewed as uneconomical from the metabolic point of view.

A more likely hypothesis is that the fungus regulates, either directly or indirectly, some aspect of algal cell-wall biosynthesis. Due to the massive export of carbon from the alga to the fungus (Sect. 18.2.1.1) phycobiont cells may lack sufficient carbon skeletons for all but the most essential cell-wall polymers. Alternatively, fungal products may interfere with or inhibit algal cell-wall associated enzymes involved in wall biosynthesis. Finally, the fungus may regulate the intracellular environment of the alga so that one or more metabolic pathways leading to cell-wall biosynthesis are repressed. Such mechanisms appear to underlie the regulation of both nitrogen (Sect. 18.2.2.2.5) and probably carbon export from the alga to the fungus.

18.3.2 Effect of Symbiosis on the Fungal Partner

There is evidence for a number of metabolic and structural differences in the mycobiont of the intact thallus in comparison to the isolated one grown in culture. Some studies have also compared cultured mycobionts to free-living fungi.

18.3.2.1 Concentric Bodies

Characteristic to the symbiotic situation is the presence of organelles, designated "concentric bodies" (PEVELING 1969a), in the cytoplasm of the lichenized fungus. GRIFFITHS and GREENWOOD reported in 1972 on 43 lichen species containing concentric bodies; in fact there are many more. Almost every ultrathin section observed yields concentric bodies. They may thus be considered of general occurrence in lichens. The only exceptions so far known are two species of the cyanolichen *Gonohymenia* (PARAN et al. 1971), and one aquatic species *Hydrothyra venosa* (JACOBS and AHMADJIAN 1973). Concentric bodies usually appear in clusters, situated in a matrix which appears in the electron microscope to have a different consistency from that of the surrounding cytoplasm, although there is no membrane separating these regions. Each concentric body has a

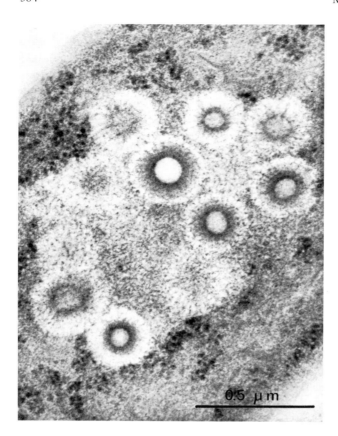

Fig. 7. Concentric bodies

0.5 µm

translucent central core surrounded by the dense material of the organelle. The latter is delineated from the core by a structure resembling a unit membrane. The dense material is subdivided into two or more distinct regions. The outer zone is usually surrounded by radiating substructures and sometimes also by an outer "halo" (Fig. 7). Using proteolytic enzymes, Galun et al. (1974) have shown that the concentric bodies are in part proteinaceous. According to Bois-sière (1982) they do not contain any polysaccharidic substance.

The genesis and function of these organelles are still a puzzle. They are completely absent from the cultured mycobiont (Jacobs and Ahmadjian 1971, Galun unpublished), but have been detected in about ten non-lichenized asco-mycetes (see Samuelson and Bezerra 1977).

18.3.2.2 Secondary Metabolic Products – "Lichen Substances"

"Lichen substances" are synthesized as the result of symbiotic conditions. They are extracellular products, crystallizing on the hyphal cell walls, usually insoluble in water, and can be extracted only with organic solvents. They amount to between 0.1–10% of the dry weight of the thallus, sometimes up to 30%. One species may contain one to several different substances.

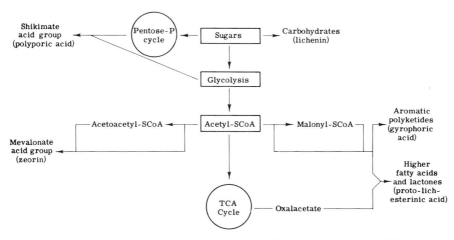

Fig. 8. Carbon metabolism leading to the formation of the major groups of lichen substances with a typical representative given in parentheses. (MOSBACH 1973)

These secondary metabolites are very heterogenous and belong to a variety of chemical groups (CULBERSON 1969, MOSBACH 1973).

A. Acetate–polymalonate group
   1. Higher fatty acids and lactons
   2. Aromatic polyketides
       depsides
       depsidones
       dibenzoquinons
       dibenzofurans
       chromones
       xanthones
       naphtoquinons
       anthrones
       anthraquinones
B. Mevalonate group
C. Shikimate group
       diketo piperazines
       terpenylquinons
       pulvic acid derivatives.

The aromatic polyketides are the most common lichen products. Depsides, depsidones, dibenzoquinons and dibenzofurans are found exclusively in lichens. The other aromatic polyketides are not unique to lichens and have been detected also in non-lichenized fungi and in higher plants. The pathway leading to the formation of these groups, according to MOSBACH (1973), is presented in Fig. 8.

In most cases orsellinic acid and its homologues are the principle structural units. The two principle mechanisms by which these monomers are connected are: (1) esterification to form depsides (Fig. 9a) (2) oxidative coupling to form

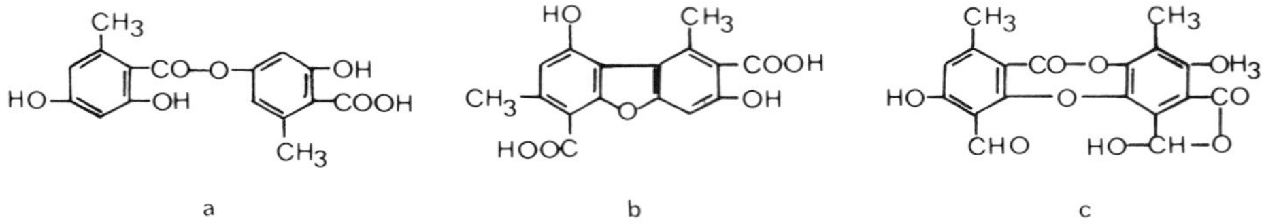

Fig. 9. Lichen substances: **a** depside: lecanoric acid; **b** dibenzofuran: pannaric acid; **c** depsidone: norstictic acid

dibenzofurans (Fig. 9b) and dibenzoquinons. The synthesis of depsidones (Fig. 9c) requires both linkages.

The chemical structure of about 220 secondary metabolites found in lichens has been established (Huneck 1973), and every year several more are chemically characterized.

Information on whether or not isolated cultured mycobionts synthesize lichen substances is rather limited, perhaps because negative results are unfortunately rarely published. Culberson and Ahmadjian (1980) could not detect characteristic lichen substances in cultures of the mycobiont of *Cladonia cristatella,* neither could Renner and Gerstner (1982) find such substances in cultures of the mycobiont isolated from *Baeomyces rufus.* There is only one case reported on the production of a typical lichen depside by the mycobiont of *Cladonia crispata* (Ejiri and Shibata 1975).

We may assume that the synthesis of lichen substances is a combined effort of both symbionts, but that the nature of the substance produced is apparently determined by the fungus. In in vitro synthesis experiments (Culberson and Ahmadjian 1980), combinations of the mycobiont from *C. cristatella* with a variety of phycobionts from other lichens always yielded the substance (barbatic acid) characteristic to *C. cristatella;* in no case were lichen substances, characteristic of the lichen from which the alga was derived, detected.

Culberson and Ahmadjian (1980) hypothesize on the synthesis of the specific lichen substances that the alga prevents decarboxylation of the phenolic acid precursor (orsellinic acid) by the fungus, which would lead to the production of typical fungal products, rather than lichen products. Fungal acetate-polymalonate, in combination with the carbon source supplied by the alga, form polyketide precursors. Those produce typical lichen substances, by esterification only when decarboxylase is inhibited by the phycobiont.

The biological role of these metabolites is not known; their production is apparently not required to maintain the symbiotic state, since there are lichens that do not produce such compounds. As mentioned before, the phycobionts produce and release more carbohydrates than necessary for the slow growth of the lichens. It is therefore conceivable that part of this surplus is channelled into the formation of secondary metabolites.

Lichens have a history over many centuries as a source of useful medicinal and chemical products, attributable to the lichen substances. There are many amusing legends on use of lichens, very interestingly presented by Richardson (1975). Certain lichen products are still an important commodity in the perfume and soap industry and as dyes for textiles.

18.3.2.3 The Cell Wall of the Mycobionts

GALUN et al. (1976) have shown that N-^{3}H-acetylglucosamine (GlcNac) was incorporated into the hyphae of cultured mycobionts and the pattern of incorporation revealed by microradioautography was the same as the pattern of fluorescently labelled wheat germ agglutinin (FITC-WGA) binding to these hyphae. Similarly, FITC-WGA and ^{3}H-GlcNac labelled the hyphal tips, young hyphal walls and septae of the free-living fungus *Trichoderma viride,* but did not react with *Phytophthora citrophthora.* These results indicated that chitin is an important component of mycobiont hyphal walls and that the spatial distribution and apical synthesis of chitin in mycobionts' hyphae is similar to that of free-living fungi having chitin-glucan hyphal walls. The mycobionts tested were from taxonomically widely separated species. It was therefore assumed that chitin is a major component of many, if not all cultured mycobionts. Binding experiments with other lectins indicated that there are obvious differences among the mycobionts in respect to hyphal wall components.

18.4 Interactions Between Lichen Symbionts

18.4.1 Selectivity and Recognition: General Considerations

The terms specificity and selectivity, which have been used to describe cell behaviour and cell–cell adhesion in animal cells (GARROD and NICOL 1981), can also be applied to symbiotic associations. Specificity describes cell–cell interactions in which absolute exclusivity is expressed; symbiotic partners associate *only* with one another, and no other combinations between them are possible. In contrast, selectivity describes a situation where organisms interact *preferentially* with one another; if the host organism is presented with a choice between symbionts (or vice versa), it will preferentially select one potential symbiont over the other. This type of selective behaviour has been demonstrated by PROVASOLI et al. (1968) in the marine flatworm, *Convoluta roscoffensis.* In this study, established but non-preferred algal symbionts were rapidly displaced when preferred symbionts were made available to the animal.

An overriding factor governing selectivity will be the biological fitness and efficiency of the combination. For example, it is well-documented (TRENCH 1979) that many marine invertebrates can be infected with non-preferred algal symbionts. However, the fitness, in terms of algal numbers present, persistence, and growth of the host, is much lower than for animals infected with preferred symbionts.

The processes which lead to selectivity are complex. Organisms must pass through a series of events which culminate in the most preferred combination between symbionts. The term "recognition" can be used to describe "the set of phenomena resulting in the expression of specificity or selectivity in associations between symbionts" (SMITH 1981).

Selectivity is most probably expressed during various stages in the development of a symbiosis. In the *Rhizobium*-clover symbiosis, it is thought that binding of *R. trifolii* to clover root hairs, mediated by the lectin trifoliin A, is a highly selective recognition event (DAZZO 1981, DAZZO and HUBBELL 1975); however, it is by no means the sole selective event (BHUVANESWARI 1981, DAZZO 1981). In addition, an organism's ability to pass through one selective event does not necessarily mean that it will pass through subsequent events. Latex particles and symbiotic *Chlorella* are phagocytized by the Florida strain of *Hydra viridis* by the same selective phagocytic mechanism (MCNEIL 1981). However, only symbiotic *Chlorella* are incorporated into *Hydra;* latex particles are eventually excluded.

An essential property of a selective recognition event is that it must elicit a response and this response leads to subsequent phases of selectivity. In the case of the *R. trifolii*-clover symbiosis, both preferred and non-preferred rhizobia may attach to root hairs. However, only preferred attachment elicits a response which, in this case, is fixation and stabilization of the bacterium to the root hair (DAZZO 1981).

Finally, variations in the selectivity and recognition mechanisms between symbioses are prevalent. Even within the same symbiosis variations have been observed. In the Florida strain of the *Hydra-Chlorella* symbiosis, the initial recognition event appears to be during the contact phase between symbionts (POOL and MUSCATINE 1980). In the European strain, there are no obvious recognition events at contact, but only later during stages of integration (JOLLEY and SMITH 1978).

18.4.2 Selectivity in Lichens

Lichens may be regarded as a highly selective symbiosis. Some evidence in support of this *general* view stems from the following observations:
1. Among the 1600 known genera of algae (BOLD 1973), very few are potential phycobionts.
2. Free-living algae and lichens often colonize the same habitats. These algae, which are often present in high numbers and growing around, under, or on lichen thalli, are rarely, if ever, incorporated into the lichen.
3. Even if incorporated, there are no reports of algae other than preferred phycobionts persisting in lichen thalli (SMITH 1981).
4. Phycobionts, especially *Trebouxia* and *Pseudotrebouxia,* are infrequently encountered free-living in nature. When they are, they are interdispersed among the myriad of free-living algae found in the same area. Yet these algae are the only ones with which lichen fungi can associate.
5. Of the approximately 34 known genera of phycobionts (Table 1) and 20,000 species of lichens, it has been estimated (AHMADJIAN 1982) that 50–70% of these lichen fungi associate with only one of two potential phycobionts, *Trebouxia* or *Pseudotrebouxia.*

There is no general evidence against the selective nature of the lichen symbiosis. The major question revolves around the degree of selectivity. It must be noted that lichens are a heterogenous assemblage of organisms which are derived polyphyletically from various orders of non-lichenized ascomycetes and, to a much lesser extent, basidiomycetes (HAWKSWORTH 1973). It is also possible that lichenized associations arose at different times in different groups of ascomycetous fungi (JAMES and HENSSEN 1976). Hence, we should expect that the degree of selectivity, as well as the mechanisms by which it is achieved, will vary to some extent between phylogenetically distinct groups of lichens.

There have been two basic approaches to the study of selectivity in lichens; these are the identification of phycobionts, and the study of resynthesis in vitro and in situ.

18.4.2.1 Identification of Phycobionts

The introduction and use of standardized techniques for the taxonomic study of lichen algae have, as yet, only been applied to the algal genera *Trebouxia* and *Pseudotrebouxia* (ARCHIBALD 1975, 1977, HILDRETH and AHMADJIAN 1981). To date, the phycobiont of 22 species of *Cladonia* has been identified as *Trebouxia* (AHMADJIAN 1982, BUBRICK unpublished observations). Only *Trebouxia* has been isolated from *Stereocaulon* species (4) (ARCHIBALD 1975, HILDRETH and AHMADJIAN 1981). *Pseudotrebouxia* has been identified from all tested species of *Buellia* (5), *Lecanora* (7) and *Ramalina* (6) (ARCHIBALD 1975, HILDRETH and AHMADJIAN 1981, BUBRICK, unpublished observations). A preliminary survey of the lichen family Teloschistaceae (3 genera, 8 species) indicate that the phycobiont is *Pseudotrebouxia* (BUBRICK unpublished observation).

On the other hand, several lichen species are associated with more than one species of algae; identical species of algae have also been identified from different genera of lichens (AHMADJIAN 1982). The report that the same lichen thallus may have several strains of phycobiont (WANG-YANG and AHMADJIAN 1972) should be reinvestigated.

In the cases of *Trebouxia* and *Pseudotrebouxia*, taxononomic studies imply that the majority of the lichens which harbour these symbionts are selective at the algal generic level. It is difficult to evaluate the distribution of the remaining phycobionts in Table 1. However, among the more widely distributed phycobionts (e.g. *Nostoc, Scytonema, Trentepohlia, Stichococcus, Coccomyxa*), it appears that lichens are also selective at the generic level.

18.4.2.2 Resynthesis in Vitro

AHMADJIAN and JACOBS (1981) combined the cultured bionts of *Cladonia cristatella* and inoculated the mixture on mica strips soaked in nutrient solution, or on the soil surface in clay flower pots (AHMADJIAN et al. 1980). They also combined the *C. cristatella* mycobiont with a number of other algae, as well as combining cultured mycobionts and phycobionts from several other lichen species. The mycobiont of *C. cristatella* in combination with its preferred algal partner formed thallus squamules "identical in morphology to the natural squamules of this lichen" (AHMADJIAN et al. 1980). The *C. cristatella* mycobionts also formed squamules with other species of *Trebouxia* (4). It did not form squamules with other *Trebouxia* (5 species), *Pseudotrebouxia* (12 species), other phycobionts (3 species) or other species of free-living algae (12 species); in these cases only soredia were formed (Ahmadjian and Jacobs 1981). Soredia consist of several algal cells tied by fungal hyphae which serve as vegetative diaspores on the natural thallus.

It appears that soredial clusters can be formed in almost every combination; even glass beads of the proper size became encircled by the fungus in the same manner as algal cells (Ahmadjian and Jacobs 1981). From the soredial stage, perhaps only preferred combinations continue development whereas non-preferred combinations do not develop further. It seems that there are critical events expressed during this stage which signal whether or not lichenized development will continue.

Soredial clusters are also commonly found in nature, either as interim stages during thallus development (Jahns et al. 1979), or soredial clusters as such which do not undergo further development (Galun unpublished observation). There is some evidence that soredial clusters are formed during the initial stages of the in situ resynthesis of *Xanthoria parietina* (Bubrick, Garty and Galun unpublished). It would appear that the formation of soredial clusters during lichen resynthesis is a basic developmental stage.

The production of characteristic secondary lichen substances is apparently induced at an early stage of interaction, determined by the mycobiont and, at least to a certain extent, independent of the degree of selectivity of the fungal–algal combination. Artificially synthesized *C. cristatella* squamules contained barbatic acid (a depside), the major secondary product of natural *C. cristatella*, as well as small quantities of some other minor chemical constituents of this lichen. The same lichen substances were already present in the soredial stages of *C. cristatella*, as well as in the soredia of combinations between the *C. cristatella* mycobiont and *Trebouxias* from other lichens. Usnic acid, another lichen substance in natural *C. cristatella*, could not be detected either in the squamules or on the soredia of preferred or non-preferred combinations (Culberson and Ahmadjian 1980).

The assumption that carbohydrate efflux and transport would be induced at an early stage of interaction, and is a key requirement for a stable association has not yet been supported by experimental evidence (Smith 1981). Hill and Ahmadjian (1972) found that 4 months after initiation of a synthesis between the symbionts of *C. cristatella*, only a small proportion of photosynthetically fixed ^{14}C moved from the alga to the fungus.

The available evidence from in vitro resyntheses further support the belief that lichen fungi are selective at the generic level of phycobionts.

18.4.2.3 Selectivity at the Contact Phase

The study of selectivity and the mechanisms by which it is achieved in lichens has been hampered by the inability to resynthesize lichens routinely in vitro. In those cases where resynthesis is possible, it is a protracted process (1–6 months) and hence is not suited to many types of biochemical analyses. However, some evidence has been found that selective events take place during the contact phase of resynthesis.

Lectins. Lectins are thought to mediate highly selective recognition events during the contact phase in the *Rhizobium*–legume symbiosis (DAZZO 1981, PAAU et al. 1981). Lectins have also been isolated from numerous lichens (BARRETT and HOWE 1968, BERNHEIMER and FARKAS 1953, ESTOLA and VARTIA 1955, HOWE and BARRETT 1970, LOCKHART et al. 1978, PETIT 1982). To date, only the lectin from *Parmelia michauxiana* has been purified and characterized (HOWE and BARRETT 1970).

Lectins are common in the lichen genus *Peltigera* (BARRETT and HOWE 1968, LOCKHART et al. 1978, PETIT 1982). LOCKHART et al. (1978) found that the lectins from *P. canina* and *P. polydactyla* (both two-membered) were associated exclusively with the mycobiont. Fluorescently labelled lectin from *P. polydactyla* bound to the cultured *Nostoc* phycobiont; bound lectin could be eluted from the alga by D-galactose. Labelled lectin also bound significantly to the cyanobacterium *Gloeocapsa alpicola;* this cyanobacterium is not a known lichen phycobiont (AHMADJIAN 1967a).

PETIT (1982) has isolated a lectin from the two-membered *P. horizontalis*. It was demonstrated that fluorescently labelled lectin bound to the cultured *Nostoc* from this lichen, as well as to cultured *Nostocs* from two other *Peltigera* species. Labelled lectin did not bind to any of the freshly isolated phycobionts; it also did not bind to cultured *Scytonema* or *Trebouxia* phycobionts from other lichens. Bound lectin could be eluted from cells with D-galactose. However, galactose did not inhibit haemagglutination of erythrocytes. The lectin binding was found to be heat-stable (105 °C, 10 min) but sensitive to pH; similar properties were reported for the *Parmelia michauxiana* lectin (HOWE and BARRETT 1970).

Lectins have also been isolated from *P. rufescens, P. aphthosa* and *P. leucophlebia* (BUBRICK and GALUN unpublished). Fluorescently labelled lectins from the three-membered *P. aphthosa* and *P. leucophlebia* bound to their respective cultured *Nostoc* phycobionts (from the cephalodia). However, we could not demonstrate binding to the cultured primary phycobiont *Coccomyxa* from either lichen.

As of yet, there is little evidence to suggest a *general* role for lectins in contact phase selectivity between lichen symbionts. However, available evidence does point to the possibility that lectins may mediate selective events in *Nostoc*-containing lichens.

Algal-binding Protein (ABP). BUBRICK and GALUN (1980a) have isolated a crude protein fraction from the lichen *Xanthoria parietina,* a component(s) of which bound to the cell wall of the cultured, but not to the freshly isolated phycobiont. This component(s) has also been isolated from the mycobiont grown in vitro (BUBRICK et al. 1981b). ABP bound selectively to cell walls of cultured phycobionts isolated from lichens of the same taxonomic family (Teloschistaceae). It did not bind to any of the freshly isolated phycobionts, or to cultured phycobionts from lichens of other taxonomic families. Binding was correlated to cytochemical properties of the cultured phycobiont cell wall. Cell walls which bound ABP were also visibly stained by both coomassie blue and ruthenium red (Table 7).

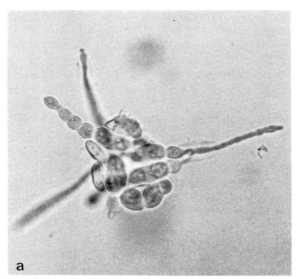

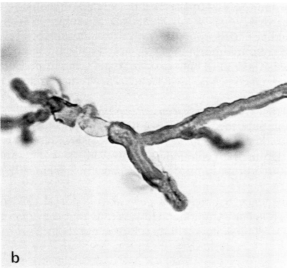

Fig. 10a, b. Visualization of ABP component(s) on the *X. parietina* mycobiont cultured in vitro. **a** Untreated mycobionts. **b** Mycobiont treated with α-ABP and visualized with peroxidase labelled goat α-rabbit serum (× 1000). (BUBRICK et al. 1981a)

Components of ABP were localized in the intact thallus of *X. parietina* and the cultured mycobiont with the aid of an indirect immunoperoxidase assay (BUBRICK et al. 1981a). Components were localized in the cortical hyphae of the intact lichen, as well as in the cell walls of the mycobiont cultured in vitro (Fig. 10). Thus, ABP is not dependent upon symbiosis for its synthesis.

The demonstration of ABP in the cell wall of the cultured mycobiont and its ability to bind selectively to cultured phycobionts suggest that ABP may have a role in selective-contact events in this lichen. ABP may serve as an initial discriminating mechanism between potential phycobionts by selectively

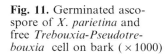

Fig. 11. Germinated asco-spore of *X. parietina* and free *Trebouxia-Pseudotrebouxia* cell on bark (× 1000)

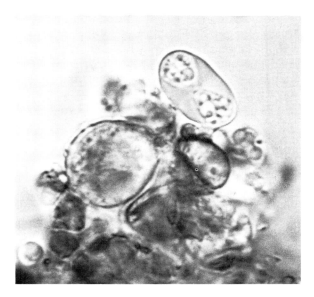

binding potential phycobionts to fungal hyphae. In a competitive resynthesis experiment between the mycobiont of *X. parietina* and two phycobionts (the *Pseudotrebouxia* from this lichen and *Coccomyxa* from *Peltigera aphthosa*), only *Pseudotrebouxia* was firmly attached to fungal hyphae; *Coccomyxa* cells in contact with hyphae were easily dislodged with a gentle stream of water (BUBRICK unpublished). As yet the degree of selectivity conferred by ABP is unknown. Under natural conditions it may prove to be very selective in the location of potential phycobionts, or in discrimination against the myriad of free-living algae found in the same area.

18.4.2.4 Resynthesis in Situ

As discussed previously (Sect. 18.4.2), the degree of selectivity between lichen partners in nature is high. However, there is little information on how this selectivity is achieved. The process of resynthesis has been infrequently observed, although the potential for resynthesis between lichen symbionts clearly exists. The majority of phycobionts listed in Table 1 can be found free-living in nature (AHMADJIAN 1967a). The two most common phycobionts, *Trebouxia* and *Pseudotrebouxia,* have also been observed completely free from fungal hyphae (BU-BRICK et al. 1981b, TSCHERMAK-WOESS 1978a), and these two phycobionts are probably more widespread in nature than has previously been assumed. The presence of free-living mycobionts is more difficult to detect. Many lichens produce, and eject, hundreds (if not thousands) of spores into the environment. Numerous germinated spores of *X. parietina* have been observed, often in close proximity to free-living *Trebouxia-Pseudotrebouxia* (Fig. 11). Later stages of resynthesis have also been seen (BUBRICK, GARTY and GALUN unpublished).

In the case of cephalodia formation in three-membered associations, a high degree of selectivity is observed. Of the variety of algae which come into contact with the lichen surfaces, only the preferred cyanobacterial phycobiont is incorporated into the thallus, and induces the formation of cephalodia (Brodo and Richardson 1978, Jahns 1972, James and Henssen 1976, Jordan 1970, Jordan and Rickson 1971).

18.5 Concluding Remark

Over the 100 years since the dual nature of lichens was recognized, we have gained considerable insight into this unique group of organisms. Many aspects of the physiological interaction between lichen symbionts have been recognized and studied. However, the mechanisms which underlie these interactions have yet to be uncovered. Experimental research has been hindered not only by the slow growth of lichens in nature, but also by the inability to manipulate intact lichens, or their separated bionts, in the laboratory. We hope that today's technological sophistication, combined with the increasing interest in cell–cell interactions, will encourage further experimental research on the lichen symbiosis.

References

Ahmadjian V (1967a) A guide to the algae occurring as lichen symbionts: isolation, culture, cultural physiology, and identification. Phycologia 6:127–160
Ahmadjian V (1967b) The lichen symbiosis. Blaisdell, Waltham
Ahmadjian V (1973a) Methods of isolating and culturing lichen symbionts and thalli. In: Ahmadjian V, Hale ME (eds) The lichens. Academic Press, London New York, pp 653–659
Ahmadjian V (1973b) Resynthesis of lichens. In: Ahmadjian V, Hale ME (eds) The lichens. Academic Press, London New York, pp 565–579
Ahmadjian V (1977) Qualitative requirements and utilization of nutrients: lichens. In: Rechcigl M Jr (ed) CRC handbook series in nutrition and food, Sec D/Vol 1. CRC Press, Cleveland, pp 203–215
Ahmadjian V (1982) Algal/fungal symbioses. In: Round F, Chapman DJ (eds) Progress in phycological research, Vol 1. Elsevier/North-Holland Biomedical Press, Amsterdam New York, pp 179–233
Ahmadjian V, Jacobs JB (1981) Relationship between fungus and alga in the lichen *Cladonia cristatella* Tuck. Nature 289:169–172
Ahmadjian V, Russell LA, Hildreth KC (1980) Artificial reestablishment of lichens I. Morphological interactions between the phycobiont of different lichens and the mycobionts of *Cladonia cristatella* and *Lecanora chrysoleuca*. Mycologia 72:73–89
Archibald PA (1975) *Trebouxia* de Puymaly (Chlorophyceae, Chlorococcales) and *Pseudotrebouxia* gen. nov. (Chlorophyceae, Chlorosarcinales). Phycologia 14:125–137
Archibald PA (1977) Physiological characteristics of *Trebouxia* (Chlorophyceae, Chlorococcales) and *Pseudotrebouxia* (Chlorophyceae, Chlorosarcinales). Phycologia 16:295–300
Ascaso C (1980) A rapid method for the quantitative isolation of green algae from lichens. Ann Bot 45:483

Barrett JT, Howe ML (1968) Hemagglutination and hemolysis by lichen extracts. Appl Microbiol 16:1137–1139

DeBary A (1879) Die Erscheinung der Symbiose. Trübner, Straßburg

Bernard T, Goas G (1968) Contribution à l'étude du métabolisme azoté des lichens: charactérisation et dosages des méthylamines de quelques espèces de la famille des Strictacées. CR Acad Sci 267:622–624

Bernard T, Goas G (1969) Contribution à l'étude du métabolisme azoté des lichens. Mise en évidence de quelques transaminases; activité de la glutamate oxalacétate transaminase dans cinq espèces de la famille des Stictacées. CR Acad Sci 269:1657–1659

Bernard T, Lahrer F (1971) Contribution à l'étude du métabolisme azoté des lichens: rôle de la glycine ^{14}C-2 dans la formation des methylamines chez Lobaria laetevirens Zahlbr. CR Acad Sci 272:568–571

Bernheimer AW, Farkas ME (1953) Hemagglutinins among higher fungi. J Immunol 70:197–198

Bhuvaneswari TV (1981) Recognition mechanisms and infection process in legumes. Econ Bot 35:204–223

Boissière JC (1977) Présence d'haustoriums chez le lichen Lichina pygmaea (Light) A.C. Ag. et role de la paroi des gonidies dans le contact entre les symbiontes des lichens à cyanophytes. Rev Bryol Lichenol 43:176–182

Boissière M-C (1972) Mise en évidence cytochimique en microscopie électronique de polyglucosides de réserve chez des Nostoc libes et lichénisés. CR Acad Sci 274:2643–2646

Boissière M-C (1973) Activité phosphatasique neutre chez le phycobionte de Peltigera canina comparée à celle d'un Nostoc libre. CR Acad Sci 277:16949–1651

Boissière M-C (1982) Cytochemical ultrastructure of Peltigera canina: Some features related to its symbiosis. Lichenologist 14:1–27

Bold HC (1973) Morphology of plants, 3rd edn. Harper and Row, New York

Bond G, Scott GD (1955) An examination of some symbiotic systems for fixation of nitrogen. Ann Bot 19:67–77

Bousfield J, Peat A (1976) The ultrastructure of Collema tenax, with particular reference to microtubule-like inclusions and vesicle production by the phycobiont. New Phytol 76:121–128

Brodo IM, Richardson DHS (1978) Chimeroid associations in the genus Peltigera. Lichenologist 10:157–170

Bubrick P (1978) Studies on the phycobionts of desert cyanolichens. MSc Thesis. Florida State Univ, Tallahassee

Bubrick P, Galun M (1980a) Proteins from the lichen Xanthoria parietina which bind to phycobiont cell walls. Correlation between binding patterns and cell-wall cytochemistry. Protoplasma 104:167–173

Bubrick P, Galun M (1980b) Symbiosis in lichens: differences in cell wall properties of freshly isolated and cultured phycobionts. FEMS Microbiol Lett 7:311–313

Bubrick P, Galun M, Frensdorff A (1981a) Proteins from the lichen Xanthoria parietina which bind to phycobiont cell walls. Localization in the intact lichen and cultured mycobiont. Protoplasma 105:207–211

Bubrick P, Frensdorff A, Galun M (1981b) Differences in the cell wall properties between freshly isolated and cultured phycobionts from the lichen, Xanthoria parietina. XIII Int Bot Congr, Sydney

Bubrick P, Galun M, Ben-Yaacov M, Frensdorff A (1982) Antigenic similarities and differences between symbiotic and cultured phycobionts from the lichen, Xanthoria parietina. FEMS Microbiol Lett 13:435–348

Butler RD, Allsopp A (1972) Ultrastructural investigations in the Stigonemataceae (Cyanophyta). Arch Mikrobiol 82:283–299

Chambers S, Morris M, Smith DC (1976) Lichen physiology XV. The effect of digitonin and other treatments on biotrophic transport of glucose from alga to fungus in Peltigera polydactyla. New Phytol 76:485–500

Chervin RE, Baker GE, Hohl HR (1968) The ultrastructure of phycobiont and mycobiont in two species of Usnea. Can J Bot 46:241–245

Collins CR, Farrar JF (1978) Structural resistance to mass transfer in the lichen *Xanthoria parietina*. New Phytol 81:71–83

Cowan DA, Green TGA, Wilson AT (1979) Lichen metabolism 1. The use of tritium-labelled water in studies of anhydrobiotic metabolism in *Ramalina celastri* and *Peltigera polydactyla*. New Phytol 82:489–503

Crittenden PD, Kershaw KA (1978) Discovering the role of lichens in the nitrogen cycle in boreal-arctic ecosystems. Bryologist 81:258–267

Culberson CF (1969) Chemical and botanical guide to lichen products. Univ of North Carolina Press, Chapel Hill

Culberson CF, Ahmadjian V (1980) Artificial reestablishment of lichens II. Secondary products of resynthesized *Cladonia cristatella* and *Lecanora chrysoleuca*. Mycologia 72:90–109

Dazzo FB (1981) Bacterial attachment as related to cellular recognition in the *Rhizobium*-legume symbiosis. J Supramol Struct Cell Biochem 16:29–41

Dazzo FB, Hubbell DH (1975) Cross-reactive antigens and lectins as determinants of symbiotic specificity in the *Rhizobium*-clover association. Appl Microbiol 30:1017–1033

Drew EA, Smith DC (1967) Studies on the physiology of lichens. VIII. Movement of glucose from alga to fungus during photosynthesis in the thallus of *Peltigera polydactyla*. New Phytol 66:389–400

Ejiri H, Shibata S (1975) Squamatic acid from the mycobiont of *Cladonia crispata*. Phytochemistry 14:2505

Englund B (1977) The physiology of the lichen *Peltigera aphthosa*, with special reference to the blue-green phycobiont (*Nostoc* sp). Physiol Plant 41:298–304

Englund B, Meyerson H (1974) In situ measurement of nitrogen fixation at low temperatures. Oikos 25:283–287

Estola E, Vartia KO (1955) Phytoagglutinins in lichens. Ann Med Exp Biol Fenn 33:392–395

Feige GB (1970) Untersuchungen zur Stoffwechselphysiologie der Flechten unter Verwendung radioaktiver Isotope. Ber Dtsch Ges NF 4:35–44

Feige GB (1975) Untersuchungen zur Ökologie und Physiologie der marinen Blaualgenflechte *Lichina pygmaea*. III. Einige Aspekte der photosynthetischen C-Fixierung unter osmoregulatorischen Bedingungen. Z Pflanzenphysiol 77:1–15

Feige GB (1976a) Untersuchungen zur Physiologie der Cephalodien der Flechte *Peltigera aphthosa* (L.) Willd. I. Die photosynthetische ^{14}C-Markierung der Lipidfraktion. Z Pflanzenphysiol 80:377–385

Feige GB (1976b) Untersuchungen zur Physiologie der Cephalodien der Flechte *Peltigera aphthosa* (L.) Willd. II. Das photosynthetische ^{14}C-Markierungsmuster und der Kohlenhydrattransfer zwischen Phycobiont und Mycobiont. Z Pflanzenphysiol 80:386–394

Fisher KA, Lang NJ (1971a) Comparative ultrastructure of cultured species of *Trebouxia*. J Phycol 7:155–165

Fisher KA, Lang NJ (1971b) Ultrastructure of the pyrenoid of *Trebouxia* in *Ramalina menziesii* Tuck. J Phycol 7:25–27

Forman RTT (1975) Canopy lichens with blue-green algae: a nitrogen source in a Colombian rain forest. Ecology 56:1176–1184

Forman RTT, Dowden DL (1977) Nitrogen-fixing lichen roles from desert to alpine in the Sangre de Cristo Mountains, New Mexcio. Bryologist 80:561–570

Gallon JR (1980) Nitrogen fixation by photoautotrophs. In: Stewart WDP, Gallon JR (eds) Nitrogen fixation. Academic Press, London New York, pp 197–238

Galun M, Paran N, Ben-Shaul Y (1970a) The fungus–alga association in the Lecanoraceae: an ultrastructural study. New Phytol 69:599–603

Galun M, Paran N, Ben-Shaul Y (1970b) Structural modifications of the phycobiont in the lichen thallus. Protoplasma 69:85–96

Galun M, Ben-Shaul Y, Paran N (1971a) Fungus–alga association in lichens of the Teloschistaceae: an ultrastructural study. New Phytol 70:837–839

Galun M, Ben-Shaul Y, Paran N (1971b) The fungus–alga association in the Lecideaceae: an ultrastructural study. New Phytol 70:483–485

Galun M, Behr L, Ben-Shaul Y (1974) Evidence for protein content in concentric bodies of lichenized fungi. J Microscop 19:193–196

Galun M, Braun A, Frensdorff A, Galun E (1976) Hyphal walls of isolated lichen fungi-autoradiographic localization of precursor incorporation and binding of fluorescein-conjugated lectin. Arch Microbiol 108:9–16

Garrod DR, Nicol A (1981) Cell behavior and molecular mechanisms of cell–cell adhesion. Biol Rev 56:199–242

Geitler L (1934) Beiträge zur Kenntis der Flechtensymbiose IV-V. Arch Protistenkd 82:51–85

Goas G, Bernard T (1967) Contribution à l'étude du metabolisme azoté des lichens: les différentes formes d'azote de quelques espèces de la famille des Stictacées. CR Acad Sci 265:1187–1190

Green TGA, Horstmann J, Bonnett H, Wilins A, Silvester WB (1980) Nitrogen fixation by members of the Stictaceae (lichenes) of New Zealand. New Phytol 84:339–348

Griffiths HB, Greenwood AD (1972) The concentric bodies of lichenized fungi. Arch Mikrobiol 87:285–302

Griffiths HB, Greenwood AD, Millbank JW (1972) The frequency of heterocysts in the *Nostoc* phycobiont of the lichen *Peltigera canina* Willd. New Phytol 71:11–13

Gunning BES, Pate JS (1969) Transfer cells – plant cells with wall ingrowths specialised in relation to short distance transport of solutes, their occurrence, structure and development. Protoplasma 68:107–133

Hawksworth DL (1973) Some advances in the study of lichens since the time of E.M. Holmes. Bot J Linn Soc 67:3–31

Henriksson E, Simu B (1971) Nitrogen fixation by lichens. Oikos 22:119–121

Henssen A (1981) *Hyphomorpha* als Phycobiont in Flechten. Plant Syst Evol 137:139–143

Hildreth KC, Ahmadjian V (1981) A study of *Trebouxia* and *Pseudotrebouxia* isolates from different lichens. Lichenologist 13:65–86

Hill DJ (1972) The movement of carbohydrate from the alga to the fungus in the lichen *Peltigera polydactyla*. New Phytol 71:31–39

Hill DJ, Ahmadjian V (1972) Relationship between carbohydrate movement and the symbiosis in lichens with green algae. Planta 103:267–277

Hitch CJB, Milbank JW (1975a) Nitrogen metabolism in lichens VI. The blue-green phycobiont content, heterocyst frequency and nitrogenase activity in *Peltigera* species. New Phytol 74:473–476

Hitch CJB, Milbank JW (1975b) Nitrogen metabolism in lichens VII. Nitrogenase activity and heterocyst frequency in lichens with blue-green phycobionts. New Phytol 75:239–244

Hitch CJB, Stewart WDP (1973) Nitrogen fixation by lichens in Scotland. New Phytol 72:509–524

Honegger R, Brunner U (1981) Sporopollenin in the cell walls of *Coccomyxa* and *Myrmecia* phycobionts of various lichens: an ultrastructural and chemical investigation. Can J Bot 59:2713–2734

Howe ML, Barrett JT (1970) Studies on a hemagglutinin from the lichen *Parmelia michauxiana*. Biochim Biophys Acta 215:97–104

Huneck S (1973) Nature of lichen substances. In: Ahmadjian V, Hale ME (eds) The lichens. Academic Press, London New York, pp 495–522

Huss-Danell K (1977) Nitrogen fixation by *Stereocaulon paschale* under field conditions. Can J Bot 55:585–592

Huss-Danell K (1979) The cephalodia and their nitrogenase activity in the lichen *Stereocaulon paschale*. Z Pflanzenphysiol 95:431–440

Jacobs JB, Ahmadjian V (1971) The ultrastructure of lichens. II. *Cladonia cristatella*: the lichen and its isolated symbionts. J Phycology 7:71–82

Jacobs JB, Ahmadjian V (1973) The ultrastructure of lichens. V. *Hydrothyra venosa* a fresh water lichen. New Phytol 72:155–160

Jäger H-J, Weigel H-J (1978) Amino acid metabolism in lichens. Bryologist 81:107–113

Jahns HM (1972) Die Entwicklung von Flechten-Cephalodien aus *Stigonema*-Algen. Ber Dtsch Bot Ges 85:615–622

Jahns HM, Mollenhauer D, Jenninger M, Schönborg D (1979) Die Neubesiedlung von Baumrinde durch Flechten I. Nat Mus 109:40–51

James PW, Henssen A (1976) The morphological and taxonomic significance of cephalodia. In: Brown DH, Hawksworth DL, Bailey RH (eds) Lichenology: progress and problems. Academic Press, London New York, pp 27–77

Jolley E, Smith DC (1978) The green *Hydra* symbiosis I. Isolation, culture and characteristics of the *Chlorella* symbiont of "European" *Hydra viridis*. New Phytol 81:637–645

Jordan WP (1970) The internal cephalodia of the genus *Lobaria*. Bryologist 73:669–681

Jordan WP, Rickson FR (1971) Cyanophyte cephalodia in the lichen genus *Nephroma*. Am J Bot 58:562–568

Kallio P, Suhonen S, Kallio H (1972) The ecology of nitrogen fixation in *Nephroma arcticum* and *Solorina crocea*. Rep Kevo Subarct Res Stn 9:7–14

Kershaw KA (1974) Dependence of the level of nitrogenase activity on the water content of the thallus in *Peltigera canina, P. evansiana, P. polydactyla*, and *P. praetextata*. Can J Bot 52:1423–1427

Kershaw KA, Millbank JW (1970) Nitrogen metabolism in lichens. II. The partition of cephalodial-fixed nitrogen between the mycobiont and phycobionts of *Peltigera aphthosa*. New Phytol 69:75–79

König J, Peveling E (1980) Vorkommen von Sporopollenin in der Zellwand des Phycobionten *Trebouxia*. Z Pflanzenphysiol 98:459–464

Lange OL (1969) Die funktionellen Anpassungen der Flechten an die ökologischen Bedingungen arider Gebiete. Ber Dtsch Bot Ges 82:3–22

Lazaroff N (1973) Photomorphogenesis and nostocacean development. In: Carr NG, Whitton BA (eds) The biology of blue-green algae. Univ California Press, Berkeley Los Angeles, pp 279–319

Lea PJ, Miflin BJ (1979) Photosynthetic ammonia fixation. In: Gibbs M, Latzko E (eds). Encyclopedia of plant physiology, vol. 6. Springer, Berlin Heidelberg New York, pp 445–456

Lockhart CM, Rowell P, Stewart WDP (1978) Phytohaemagglutinins from the nitrogen-fixing lichens *Peltigera canina* and *P. polydactyla*. FEMS Microbiol Lett 3:127–130

Marton K (1977) The cyanophilous lichen population of the Arava Valley and the Judean Desert, their means of propagation and culturing under controlled conditions. Ph D Thesis, Tel-Aviv Univ, Tel Aviv

Marton K, Galun M (1976) In vitro dissociation and reassociation of the symbionts in the lichen *Heppia echinulata*. Protoplasma 87:135–143

McNeil PL (1981) Mechanisms of nutritive endocytosis. I. Phagocytic versatility and cellular recognition in *Chlorohydra* digestive cells, a scanning electron microscope study. J Cell Sci 49:311–339

Miflin BJ, Lea PJ (1982) Ammonia assimilation and amino acid metabolism. In: Boulter D, Parthier B (eds) Encyclopedia of plant physiology, vol. 14A. Springer, Berlin Heidelberg New York, pp 5–64

Millbank JW (1972) Nitrogen metabolism in lichens. IV. The nitrogenase activity of the *Nostoc* phycobiont in *Peltigera canina*. New Phytol 71:1–10

Millbank JW (1974a) Associations with blue-green algae. In: Quispel A (ed) The biology of nitrogen fixation. Elsevier North Holland, Amsterdam New York, pp 238–264

Millbank JW (1974b) Nitrogen metabolism in lichens V. The forms of nitrogen released by the blue-green phycobiont in *Peltigera* spp. New Phytol 73:1171–1181

Millbank JW (1978) The contribution of nitrogen-fixing lichens to the nitrogen status of the environment. Ecol Bull 26:260–265

Millbank JW (1981) The assessment of nitrogen fixation and throughout by lichens. I. The use of a controlled environment chamber to relate acetylene reduction estimates to nitrogen fixation. New Phytol 89:647–655

Millbank JW, Kershaw KA (1969) Nitrogen metabolism in lichens. I. Nitrogen fixation in the cephalodia of *Peltigera aphthosa*. New Phytol 68:721–729

Millbank JW, Kershaw KA (1970) Nitrogen metabolism in lichens. III. Nitrogen fixation by internal cephalodia in *Lobaria pulmonaria*. New Phytol 69:595–597

Millbank JW, Kershaw KA (1973) Nitrogen metabolism. In: Ahmadjian V, Hale ME (eds) The lichens. Academic Press, London New York, pp 289–307

Millbank JW, Olsen JD (1981) The assessment of nitrogen fixation and throughput by lichens II. Construction of an enclosed growth chamber for the use of $^{15}N_2$. New Phytol 89:657–665

Mosbach K (1973) Biosynthesis of lichen substances. In: Ahmadjian V, Hale ME (eds) The lichens. Academic Press, London New York, pp 523–546

Paau AS, Leps WT, Brill WJ (1981) Agglutinin from alfalfa necessary for binding and nodulation by *Rhizobium meliloti*. Science 213:1513–1515

Paran N, Ben-Shaul Y, Galun M (1971) Fine structure of the blue-green phycobiont and its relation to the mycobiont in two *Gonohymenia* lichens. Arch Microbiol 76:103–113

Parra OO, Redon J (1977) Isolation of *Heterococcus caespitosus* Vischer, phycobiont of *Verrucaria maura* Wahlenb. Bol Soc Biol Concepcion 51:219–224

Peat A (1968) Fine structure of the vegetative thallus of the lichen, *Peltigera polydactyla*. Arch Mikrobiol 61:212–222

Petit P (1982) Phytolectins from the nitrogen-fixing lichen *Peltigera horizontalis*: the binding pattern of the primary protein extract. New Phytol 91:705–710

Peveling E (1968) Elektronenoptische Untersuchungen an Flechten. 1. Strukturveränderungen der Algenzellen von *Lecanora muralis* (Schreber) Rabenh. [= *Placodium saxicolum* (Nyl.) sec. Klem.] beim Eindringen von Pilzhyphen. Z Pflanzenphysiol 59:172–183

Peveling E (1969a) Electroneoptische Untersuchungen an Flechten. III. Cytologische Differenzierungen der Pilzzellen im Zusammenhang mit ihrer symbiotischen Lebensweise. Z Pflanzenphysiol 61:151–164

Peveling E (1969b) Elektronenoptische Untersuchungen an Flechten. IV. Die Feinstruktur einiger Flechten mit Cyanophyceen-Phycobionten. Protoplasma 68:209–222

Peveling E (1973a) Fine structure. In: Ahmadjian V, Hale ME (eds) The lichens. Academic Press, London New York, pp 147–182

Peveling E (1973b) Vesicles in the phycobiont sheath as possible transfer structures between the symbionts in the lichen *Lichina pygmaea*. New Phytol 72:343–345

Peveling E (1974) Biogenesis of cell organelles during the differentiation of the lichen thallus. Port Acta Biol Sér A 14:357–368

Peveling E (1976) Investigations into the ultrastructure of lichens. In: Brown DH, Hawksworth DL, Bailey RH (eds) Lichenology: progress and problems. Academic Press, London New York, pp 17–26

Peveling E, Hill DJ (1974) The localization of an insoluble intermediate in glucose production in the lichen *Peltigera polydactyla*. New Phytol 73:767–769

Pike LH (1978) The importance of epiphytic lichens in mineral cycling. Bryologist 81:247–257

Plessl A (1963) Über die Beziehungen von Haustorientypus und Organisationshöhe bei Flechten. Oesterr Bot Z 110:194–269

Pool RR Jr., Muscatine L (1980) Phagocytic recognition and the establishment of the *Hydra viridis–Chlorella* symbiosis. In: Schwemmler W, Schenk HEA (eds) Endocytobiology, endosymbiosis and cell biology, Vol 1. de Gruyter, Berlin, pp 223–238

Provasoli L, Yamasu T, Manton I (1968) Experiments on the resynthesis of symbiosis in *Convoluta roscoffensis* with different flagellate cultures. J Mar Biol Assoc UK 48:465–479

Rai AN, Rowell P, Stewart WDP (1980) NH_4^+ assimilation and nitrogenase regulation in the lichen *Peltigera aphthosa* Willd. New Phytol 85:545–555

Rai AN, Rowell P, Stewart WDP (1981a) Glutamate synthase activity in symbiotic cyanobacteria. J Gen Microbiol 126:515–518

Rai AN, Rowell P, Stewart WDP (1981b) $^{15}N_2$ incorporation and metabolism in the lichen *Peltigera aphthosa* Willd. Planta 152:544–552

Rai AN, Rowell P, Stewart WDP (1981c) Nitrogenase activity and dark CO_2 fixation in the lichen *Peltigera aphthosa* Wild. Planta 151:256–264

Ramakrishnan S, Subramanian SS (1964a) Amino acids of *Rocella montagnei* and *Parmelia tinctorium*. Indian J Chem 2:467

Ramakrishnan S, Subramanian SS (1964b) Amino acids of *Peltigera canina*. Curr Sci 33:522–523

Ramakrishnan S, Subramanian SS (1965) Amino acids of *Cladonia rangiferina, Cl. gracilis,* and *Lobaria isidiosa.* Curr Sci 34:345–347

Ramakrishnan S, Subramanian SS (1966a) Amino acids of *Lobaria subisidiosa, Umbilicaria pustulata, Parmelia nepalensis,* and *Ramalina sinensis.* Curr Sci 35:124

Ramakrishnan S, Subramanian SS (1966b) Amino acids of *Dermatocarpon moulinsii.* Curr Sci 35:284

Renner B, Gerstner E (1982) Stoffwechselunterschiede zwischen dem lichenisierten und dem isolierten Mycosymbionten von *Baeomyces rufus* (Huds.) Rebent. Z Pflanzenphysiol 107:47–57

Richardson DHS (1973) Photosynthesis and carbohydrate movement. In: Ahmadjian V, Hale ME (eds) The lichens. Academic Press, London New York, pp 249–288

Richardson DHS (1975) The vanishing lichens. David & Charles Douglas, Vancouver

Richardson DHS, Smith DC (1968) Lichen physiology IX. Carbohydrate movement from the *Trebouxia* symbiont of *Xanthoria aureola* to the fungus. New Phytol 67:61–68

Robenek H, Marx M, Peveling E (1982) Gold-labelled concanavalin A-binding sites at the cell surface of two phycobionts visualized by deep-etching. Z Pflanzenphysiol 106:63–68

Rogers RW, Lange RT, Nicholas DJD (1966) Nitrogen fixation by lichens of arid soil crusts. Nature 209:96–97

Sampaio MJAM, Rai AN, Rowell P, Stewart WDP (1979) Occurrence, synthesis and activity of glutamine synthetase in N_2-fixing lichens. FEMS Microbiol Lett 6:107–110

Samuelson DA, Bezerra J (1977) Concentric bodies in two species of the Loculoascomycetes. Can J Microb 23:1485–1488

Schwendener S (1869) Die Algentypen der Flechtengonidien. Universitätsbuchdruckerei C Schultze, Basel

Scott GD (1956) Further investigations of some lichens for fixation of nitrogen. New Phytol 55:111–116

Slocum RD, Ahmadjian V, Hildreth KC (1980) Zoosporogenesis in *Trebouxia gelatinosa:* ultrastructural potential for zoospore release and implications for the lichen association. Lichenologist 12:173–187

Smith DC (1974) Transport from symbiotic algae and symbiotic chloroplasts to host cells. Symp Soc Exp Biol 28:485–520

Smith DC (1975) Symbiosis and the biology of lichenised fungi. Symp Soc Exp Biol 29:373–405

Smith DC (1979) Is a lichen a good model for biological interactions in nutritional-limited environments? In: Shilo M (ed) Strategies of microbial life in extreme environments. Dahlem Konferenzen, Berlin, pp 291–303

Smith DC (1980) Mechanisms of nutrient movement between lichen symbionts. In: Cook CB, Pappas PW, Rudolph ED (eds) Cellular interactions in symbiosis and parasitism. Ohio State Univ Press, Columbus, pp 197–227

Smith DC (1981) The role of nutrient exchange in recognition between symbionts. Ber Dtsch Bot Ges 94:517–528

Smith DC, Hill DJ (1972) Lichen physiology. XII. The "inhibition technique". New Phytol 71:15–30

Smith DC, Molesworth S (1973) Lichen physiology. XIII. Effects of rewetting dry lichens. New Phytol 72:525–533

Spector DL, Jensen TE (1977) Fine structure of *Leptogium cyanescens* and its cultured phycobiont *Nostoc commune.* Bryologist 80:445–460

Stewart WDP (1970) Algal fixation of atmospheric nitrogen. Plant Soil 32:555–588

Stewart WDP (1974) Blue-green algae. In: Quispel A (ed) The biology of nitrogen fixation. Elsevier, North Holland, Amsterdam New York, pp 202–237

Stewart WDP (1980) Some aspects of structure and function in N_2-fixing cyanobacteria. Annu Rev Microbiol 34:497–536

Stewart WDP, Rodgers GA (1978) Studies on the symbiotic blue-green algae of *Anthoceros, Blasia* and *Peltigera.* Ecol Bull 26:247–259

Stewart WDP, Rowell P (1977) Modifications of nitrogen-fixing algae in lichen symbioses. Nature 265:371–372

Stewart WDP, Rowell P, Rai AN (1980) Symbiotic nitrogen-fixing cyanobacteria. In: Stewart WDP, Gallon JR (eds) Nitrogen fixation. Academic Press, London New York, pp 239–277

Tapper R (1981) Direct measurement of translocation of carbohydrate in the lichen *Cladonia convoluta*, by quantitative autoradiography. New Phytol 89:429–437

Trench RK (1979) The cell biology of plant–animal symbiosis. Annu Rev Plant Physiol 30:485–531

Tschermak-Woess E (1970) Über wenig bekannte und neue Flechtengonidien. V. Der Phycobiont von *Verrucaria aquatilis* und die Fortpflanzung von *Pseudopleurococcus arthropyreniae*. Oesterr Bot Z 118:433–455

Tschermak-Woess E (1976) Algal taxonomy and the taxonomy of lichens: the phycobiont of *Verrucaria adriatica*. In: Brown DH, Hawksworth DL, Bailey RN (eds) Lichenology, progress and problems. Academic Press, London New York, pp 89–105

Tschermak-Woess E (1978a) *Myrmecia reticulata* as a phycobiont and freeliving – freeliving *Trebouxia* – the problem of *Stenocybe septata*. Lichenologist 10:69–79

Tschermak-Woess E (1978b) Über den *Chlorella*-Phycobionten von *Trapelia coarctata*. Plant Syst Evol 130:253–263

Tschermak-Woess E (1978c) Über die Phycobionten der Sektion *Cystophora* von *Chaenotheca*, insbesondere *Dictyochloropsis splendida* und *Trebouxia simplex*, spec. nova. Plant Syst Evol 129:185–208

Tschermak-Woess E (1980a) *Asterochloris phycobiontica* gen. et spec., nov., der Phycobiont der Flechte *Varicellaria carneonivea*. Plant Syst Evol 135:279–294

Tschermak-Woess E (1980b) *Chaenothecopsis consociata* – kein parasitischer oder parasymbiontischer Pilz, sondern lichenisiert mit *Dictyochloropsis symbiontica*, sp. nov. Plant Syst Evol 136:287–306

Tschermak-Woess E (1980c) *Elliptochloris bilobata* gen. et sp. nov., der Phycobiont von *Catolechia wahlenbergii*. Plant Syst Evol 136:63–72

Tschermak-Woess E (1981) Zur Kenntnis der Phycobionten von *Lobaria linita* und *Normandina pulchella*. Nova Hedwigia 35:63–73

Tschermak-Woess E, Poelt J (1976) *Vezdaea*, a peculiar lichen genus, and its phycobiont. In: Brown DH, Hawskworth DL, Bailey RH (eds) Lichenology, progress and problems. Academic Press, London New York, pp 89–105

Wang-Yang J-R, Ahmadjian V (1972) A morphological study of the algal symbionts of *Cladonia rangiferina* (L) Web and *Parmelia caperata* (L) Ach. Taiwania 17:170–181

Webber MM, Webber PJ (1970) Ultrastructure of lichen haustoria: symbiosis in *Parmelia sulcata*. Can J Bot 48:1521–1524

Wetmore CM (1970) The lichen family Heppiaceae in North America. Ann Mo Bot Gard 57:158–209

19 Mating Systems and Sexual Interactions in Yeast

N. YANAGISHIMA

19.1 Introduction

Ascosporogenous yeasts such as *Saccharomyces cerevisiae, Hansenula wingei* and *Schizosaccharomyces pombe* are unicellular but eukaryotic both functionally and structurally. Taking advantage of these characteristics, these yeasts have been used as model systems for studies on regulation mechanism in eukaryotic cells. Sexual differentiation and sexual interactions have been studied intensively, because in these phenomena we can find examples of many biologically important problems concerning cell-type differentiation, cell–cell recognition and pheromonal regulation. In Fig. 1, the life cycle of *S. cerevisiae* is shown, with the alternation of nuclear phases and sexual differentiation. *H. wingei* has the same type of life cycle as *S. cerevisiae*. The life cycle of *Schiz. pombe* is different from that of *S. cerevisiae* in that spore formation takes place in zygotes without diploid vegetative growth. Unless otherwise specified, this article deals with experimental results from *S. cerevisiae* because the most important results have been obtained with this yeast so far.

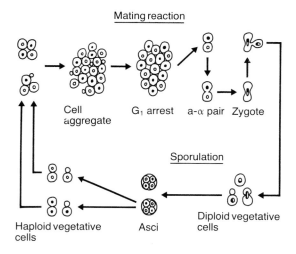

Mating reaction

Cell aggregate G_1 arrest a-α pair Zygote

Sporulation

Haploid vegetative cells Asci Diploid vegetative cells

Haploid cell

⊙ a cell   ◉ Diploid cell
◎ α cell

Fig. 1. Life cycle and mating process of *S. cerevisiae*

19.2 Sexual Differentiation

In *S. cerevisiae*, *H. wingei* and *Schiz. pombe* haploid strains are classified into two opposite mating types, *a* and *α*, *5* (*a*) and *21* (*α*), and *h*⁺ and *h*⁻, respectively. The mating reaction takes place only between the opposite mating types. Mating caused by mixing vegetative cells of opposite mating types (mass mating) was found first in *S. cerevisiae* by LINDEGREN and LINDEGREN (1943). Mass mating has been generally used for the study on mating reaction. Since it is difficult to obtain stable *a* or *α* cells in homothallic strains, as will be described later (see Sect. 2.1), heterothallic strains, in which *a* or *α* cells grow stably, are used for mass mating.

19.2.1 Genetic Control

The two mating types of *S. cerevisiae*, *a* and *α* are controlled by the mating type alleles, *MATa* and *MATα*, respectively. It has been proposed from results of studies on interspecific or intergeneric sexual interactions (FUJIMURA and YANAGISHIMA 1981, FUJIMURA et al. 1982, YANAGISHIMA and FUJIMURA 1981), that ascosporogenous yeasts generally, if not always, have two mating types, each of which corresponds to the *a* or *α* mating type in *S. cerevisiae*. The mating type locus (*MAT* locus) has regulatory function on the *a*- and *α*-specific genes which are not linked to the *MAT* locus and encode mating-type-specific functions (STRATHERN et al. 1981). Not only haploid cells (*MATa* or *MATα*) but also diploid cells homozygous in the *MAT* locus (*MATa/MATa* or *MATα/MATα*) do not sporulate. However, heterozygous diploid cells (*MATa/MATα*) sporulate efficiently, indicating that the *MAT* locus also controls the ability to sporulate. Hence, the *MAT* locus regulate both the cell functions, mating and sporulation which includes meiosis (for a review see HERSKOWITZ and OSHIMA 1981).

The structure of the *MAT* locus has been studied genetically mainly by using mutants defective in mating ability (MACKAY and MANNEY 1974a, b, STRATHERN et al. 1981). The results are as follows: *MATa* has one functional unit *MATa1*, of which *mata1* mutations have no effect on mating ability, but have an inhibitory effect on sporulation ability. On the other hand, *MATα* has the two complementation groups, *MATα1* and *MATα2*. Mutations in *MATα1*, *matα1* types, prevent production of the α pheromone (α-factor peptidyl pheromone) and have no mating ability. Mutations in *MATα2*, *matα2* types result in simultaneous expression of *a*- and *α*-specific functions (SPRAGUE Jr and HERSKOWITZ 1981, TOHOYAMA and YANAGISHIMA 1982), which may cause loss of mating ability.

Parallel with the genetic studies, the molecular structure of the *MAT* locus has been studied (HICKS et al. 1979, NASMYTH and TATCHELL 1980, NASMYTH et al. 1981a, STRATHERN et al. 1980) (Fig. 2). The mating type alleles, *MATa* and *MATα* have the specific sequences, Y*a* and Y*α* regions, consisting of about 650 and 750 base pairs, respectively, in addition to common regions (NASMYTH

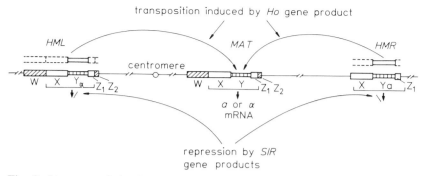

Fig. 2. Structure of the *MAT, HML* and *HMR* loci, and mechanism of mating type interconversion. *MAT, HML,* and *HMR* are located on chromosome III. Region Y is an *a*-specific (Y*a*) or α-specific (Yα) sequence. Regions X and Z1 are present at all the three loci. Regions W and Z2 are absent only at *HMR*. Mating type interconversion occurs by transposition of a copy of Y*a* (or Yα) at the *HM* loci to Yα (or Y*a*) at the *MAT* locus by the action of *HO* gene. *MAT* genes at the *HML* and *HMR* loci are repressed by *SIR* gene products. (Nasmyth et al. 1981a)

and Tatchell 1980, Nasmyth et al. 1981a, Strathern et al. 1980) (Fig. 2). *MAT*α produces two kinds of transcripts, supporting the above notion that *MAT*α has two complementation groups (Klar et al. 1981, Nasmyth et al. 1981b).

Strathern et al. (1981) have proposed the α1-α2 hypothesis for genetic control of mating type. The hypothesis, shown schematically in Fig. 3, is as follows. *MAT*α1 is a positive regulator of the α-specific genes and *MAT*α2 is a negative regulator of the *a*-specific genes, of which the function is constitutive. Hence, in α cells, the α-specific genes are activated by the *MAT*α1 product and the *a*-specific genes are repressed by the *MAT*α2 product, resulting in expression of the α mating type (α-specific characters). In *a* cells, the α-specific genes are repressed by the absence of *MAT*α1 and the *a*-specific genes are active because of absence of *MAT*α2, resulting in expression of *a* mating type (*a*-specific characters). Since the *MAT*α2 product represses the *a*-specific genes and gene products of *MAT*α1 and *MAT*α2 repress *MAT*α1, resulting in no activation of the α-specific genes, *MATa/MAT*α diploids have no mating ability. In *MATa/MAT*α diploid cells, gene products of both *MAT*α1 and *MAT*α2 cooperatively activate sporulation-specific genes, probably by repressing negative regulators of these genes, thus giving sporulation ability.

The mating types, *a* and α interconvert with high frequency in homothallic strains of *S. cerevisiae* carrying the *HO* gene (Oshima and Takano 1971, Takano and Oshima 1970, Hicks and Herskowitz 1976a).

The results of genetic analyses allowed Hicks et al. (1977a) to propose the "cassette" model, which suggests that the interconversion occurs by transposition of a copy of silent genetic information of *MATa* or *MAT*α at the *HML* and *HMR* loci to the *MAT* locus where the information is expressed (for a review see Herskowitz and Oshima 1981) (Fig. 2). The transposition occurs by the action of *HO* gene located on chromosome IV (G. Kawasaki personal

Fig. 3a–c. The α1-α2 hypothesis. *Straight arrows* indicate positive control and *T bars* negative control. *Wavy arrows* indicate expression. (STRATHERN et al. 1981)

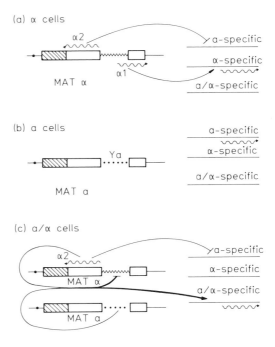

communication). The silent information which normally is repressed by the action of *SIR* genes (Fig. 2) (HABER and GEORGE 1979, KLAR et al. 1979, 1981, NASMYTH et al. 1981a, b) is located on the "right" (*HMRa* and *HMRα* genes having *MATa* and *MATα* information respectively) or "left" (*HMLa* and *HMLα* having *MATa* and *MATα* information respectively) distal part of chromosome III on which the *MAT* locus is located near the centromere (HARASHIMA and OSHIMA 1976, KLAR et al. 1980). Ordinary laboratory homothallic strains carry *HMLα* and *HMRa,* and these strains switch mating type much more frequently than do variant homothallic strains carrying *HMLa* and *HMRα* (RINE et al. 1981). The model of the transposition of silent information to the *MAT* locus is consistent with results of experiments at DNA level (HICKS et al. 1979, KLAR et al. 1981, NASMYTH and TATCHELL 1980, NASMYTH et al. 1981a, b, STRATHERN et al. 1980). The *HO* gene is repressed in the diploid cells heterozygous at the *MAT* locus (HICKS et al. 1977b).

19.2.2 Molecular Basis

The sexuality of *a* and *α* cells is expressed through the mating-type-specific substances of which functions are encoded by the *a*- or *α*-specific genes and under the control of the *MAT* locus. Among the mating-type-specifically active substances, agglutination substances responsible for sexual cell agglutination, sex pheromones regulating mating process and pheromone-binding substances

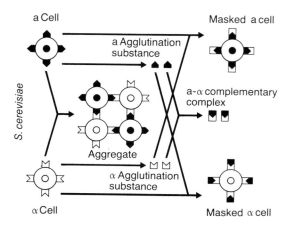

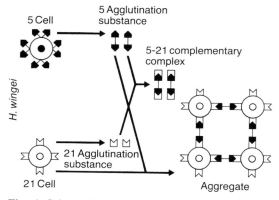

Fig. 4. Schematic explanation of sexual agglutination and action of agglutination substance. For simplification, *5* agglutination substance in *H. wingei* is supposed to be divalent. The complementary complexes are biologically inactive. The masked cells have no sexual agglutinability. *5* and *21* cells form cell aggregates by complementary binding between *5* and *21* agglutination substances on the cell surface, as in the case of *a* and *α* cells

involved in the action of the *α* pheromone have been isolated and studied both biochemically and physiologically (for reviews see MANNEY et al. 1981, THORNER 1981, YANAGISHIMA 1978 and YANAGISHIMA and YOSHIDA 1981 a).

19.2.2.1 Agglutination Substances

The molecular basis of sexual agglutination was studied systematically first with *H. wingei.* Soon after the discovery of the phenomenon of sexual agglutination in *H. wingei* by WICKERHAM (1956), BROCK (1958 a, b) began to study the molecular basis. The sex-specific substance responsible for the sexual cell agglutination (agglutination factor or agglutinin) will be called agglutination

Table 1. Agglutination substances from *S. cerevisiae* and *H. wingei*

	Mating type of agglutination substance			
	H. wingei		*S. cerevisiae*	
	5 (a)	*21 (α)*	*a*	*α*
Biological activity	Selfagglutination of *21* cells. Formation of complementary complex with *21* agglutination substance	Inactivation of *5* agglutination substance by formation of complementary complex	Masking of agglutinability of *α* cells. Formation of complementary complex with *α* agglutination substance	Masking of agglutinability of *a* cells. Formation of complementary complex with *a* agglutination substance
Binding activity	Multivalent	Univalent	Univalent	Univalent
Molecular weight	$15{,}000$–10^8 (heterogenous)	About 27,000	About 23,000	About 130,000
Subunit	2–7?	?	2	?
Chemical nature	Glycoprotein	Glycoprotein	Glycoprotein	Glycoprotein
Sugar content (%)	50–96	5	61	47
pI	?	3.8	About 4.5	About 4.3

Modified from YANAGISHIMA and YOSHIDA (1981b). Constructed based on the results of TAYLOR (1964) CRANDALL and BROCK (1968b), YEN and BALLOU (1973, 1974) BURKE et al. (1980), YOSHIDA et al. (1976), HAGIYA et al. (1977), K. YOSHIDA (personal communication), HAGIYA (1980)

substance in this chapter. TAYLOR (1964, 1965) succeeded in solubilizing agglutination substance from wall of *5* cells by enzymatic digestion (*5* agglutination substance), and tried to purify it. BROCK (1965) isolated *5* agglutination substances with relatively low molecular weights from the cytoplasm of *5* cells by disruption and proposed that the agglutination substance with the lowest molecular weight is the ultimate unit of the *5* agglutination substance. YEN and BALLOU (1973, 1974) isolated a *5* agglutination substance from subtilisin digest of *5* cells, and proposed two possible structures for it, based on their own results of and those of TAYLOR and ORTON (1971). *5* Agglutination substance is a mannanprotein having multivalent action on type *21* cells, causing cell agglutination (Fig. 4) (TAYLOR 1964). *21* Agglutination substance was purified first using trypsin digest of *21* cells as the starting material. This substance was shown to be a glycoprotein, which masks the biological activity of *5* agglutination substance but causes no cell agglutination itself in *5* cells (CRANDALL and BROCK 1968b). Hence the *21* agglutination substance is univalent (Fig. 4).

BURKE et al. (1980) also purified from trypsin or zymolyase digest of *21* cells an agglutination substance, and chemically characterized it. In Table 1 the chemical characters and biological activity of 5 and 21 agglutination substances are summarized, based on the above results.

In *S. cerevisiae,* Shimoda et al. (1975) succeeded in solubilizing *a* and α agglutination substances from cell-wall fractions with glusulase. The solubilized *a* and α agglutination substances were glycoproteins having molecular weights of about 1×10^6 (Shimoda and Yanagishima 1975). These substances had univalent action, masking sexual agglutinability of cells of the opposite mating type without causing agglutination (Fig. 4). We tried to obtain agglutination substances with much lower molecular weights for analytical studies. Yoshida et al. (1976) and Hagiya et al. (1977) solubilized the substances by autoclaving *a* or α cells for 3–5 min at a pressure of 1 kg cm^{-2}. The agglutination substances extracted by this method came almost exclusively from the cell wall (Hagiya et al. 1977). The autoclave method is applicable to ascosporogenous yeasts other than *S. cerevisiae,* although some modifications seem to be necessary, depending on yeast species (M. Yamaguchi, personal communication, K. Yoshida, personal communication, K. Yoshida et al. in preparation). The chemical character and biological activity of *a* and α agglutination substances extracted by the autoclave method are summarized in Table 1.

Yamaguchi et al. (1982a) isolated an α agglutination substance from cytoplasm, using cell extract obtained by disrupting α cells from which wall agglutination substance had been removed by the autoclave method as starting material. The molecular weight and carbohydrate content of α agglutination substance from cytoplasm are higher than those of the substance from the cell wall. Both of the substances are univalent and no difference between the two was found in sensitivity to proteolytic enzymes and certain chemical modifications (Hagiya 1980, Yamaguchi et al. 1982a). Recently, Yamaguchi et al. (1982b) succeeded in isolating cytoplasmic *a* agglutination substance, which is univalent, from α-pheromone-treated *a* cells and showed that in the case of *a* agglutination substance also molecular weight of cytoplasmic agglutination substance is higher than that of wall agglutination substance. Complementary binding between opposite mating type agglutination substances in vitro has been demonstrated in *H. wingei* (Crandall et al. 1974) and *S. cerevisiae* (Yoshida et al. 1976), confirming that the solubilized agglutination substances are really responsible for the sexual cell agglutination.

No active agglutination substances were detected in diploid cells heterozygous at the mating type locus in either *H. wingei* (Crandall and Brock 1968a) or *S. cerevisiae* (Hagiya et al. 1977, Tohoyama et al. 1979). However, in *H. wingei,* addition of vanadium salts caused production of 5 agglutination substance and addition of chelating agents caused production of 21 agglutination substance, even in diploid cells (Crandall and Caulton 1973). These results suggest that mutual repression of the mating-type-specific genes responsible for synthesis of the agglutination substances occurs in diploid cells.

19.2.2.2 Sex Pheromones

Levi (1956) found that α cells produce a sex pheromone which caused cell elongation in *a* cells. Later, Stötzler et al. (1976) succeeded in chemically char-

S. cerevisiae

α1   H−Trp−His−Trp−Leu−Gln−Leu−Lys−Pro−Gly−Gln−Pro−Met−Tyr−OH

α2   H−His−Trp−Leu−Gln−Leu−Lys−Pro−Gly−Gln−Pro−Met−Tyr−OH

α3 and α4 are methionine sulfoxides of α1 and α2, respectively

S. kluyveri

X−His−Trp−Leu−Ser−Phe−Ser−Lys−Gly−Glx−Pro−Met(O)−Tyr−OH

Fig. 5. Amino acid sequences of α pheromones in *S. cerevisiae* and *S. kluyveri*. *Met (0)* methionine sulphoxide residue. *X* unknown residue. *Glx* glutamic acid or glutamine residue. Judging from biological activity, *Met (0)* in the *S. kluyveri* pheromone may possibly be *Met* (FUJIMURA et al. 1982)

acterizing this pheromone, called α factor (α pheromone in this article) (Fig. 5). The α pheromone arrests the cell cycle at the G_1 phase, and causes formation of large, elongated, pear-shaped cells ("shmoo"-shaped cells) specifically in *a* cells. The presence of the pheromone produced by *a* cells (*a* factor or *a* pheromone), which inhibits DNA synthesis of α cells, was indicated first by SHIMODA and YANAGISHIMA (1973). Later, WILKINSON and PRINGLE (1974) confirmed the presence of this pheromone, and finally, BETZ and DUNTZE (1979) isolated and partially characterized it. Both *a* and α pheromones in *S. cerevisiae* are peptidyl substances.

SAKAI and YANAGISHIMA (1972) found that α cells secrete a pheromone inducing sexual agglutinability of *a* cells. The pheromone which we purified on the basis of agglutinability-inducing action on *a* cells and the pheromone purified on the basis of G_1-arresting action on *a* cells were shown to be the same (SAKURAI et al. 1976, STÖTZLER et al. 1976). Later studies confirmed this conclusion (BETZ et al. 1978). Thus, the α pheromone of *S. cerevisiae* has dual action, agglutinability-inducing and G_1-arresting. *a* Cells were shown to secrete *a* pheromone which induces sexual agglutinability of α cells (YANAGISHIMA et al. 1976). BETZ and DUNTZE (1979) showed conclusively that purified *a* pheromone also has dual action, agglutinability-inducing and G_1-arresting, on α cells.

Sex pheromones which induce sexual agglutinability and/or G_1 arrest were found also in *Hansenula wingei, S. kluyveri, S. globosus* and *Pichia heedii* (FUJIMURA and YANAGISHIMA 1981, FUJIMURA et al. 1982, McCULLOUGH and HERSKOWITZ 1979, N. YANAGISHIMA unpublished result, YANAGISHIMA and FUJIMURA 1981). In Fig. 6, induction of sexual agglutinability by opposite mating type pheromone in *H. wingei* is shown. The α pheromone of *S. kluyveri* has been chemically characterized (SAKURAI et al. 1980, SATO et al. 1981). It is different in amino acid sequence from the α pheromone of *S. cerevisiae* (Fig. 5), but shows both agglutinability-inducing and G_1-arresting actions on *a* cells of *S. cerevisiae* (FUJIMURA et al. 1982).

19.2.2.3 Binding Substances

Not only *a* cells but also cell-free culture medium of *a* cells inactivate α pheromone, and a peptidase(s) was shown to be mostly responsible for the inactivation

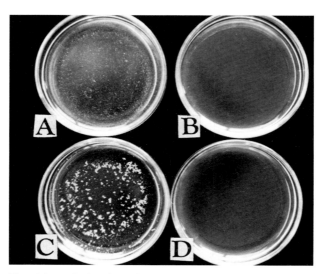

Fig. 6 A–D. Induction of sexual agglutinability by opposite mating type pheromone in *H. wingei*. 5 (*a*) cells and *21* (α) cells treated with the culture medium of the same or the opposite mating type were mixed in the presence of cycloheximide and shaken. **A** 5 + X *21*−. **B** 5− X *21*+. **C** 5+ X *21*+. **D** 5− X *21*−. + and − indicate the cells treated with the culture medium of the opposite mating type and those with the same mating type, respectively

(Ciejek and Thorner 1979, Chan 1977, Finkelstein and Strausberg 1979, Hicks and Herskowitz 1976b, Mannes and Edelman 1978, Yanagishima et al. 1977). We found that a substance(s) inactivating the α pheromone is secreted from *a* cells, and the inactivation of the α pheromone by this substance is reversible, indicating that the inactivation is through binding (Shimizu et al. 1977, Yanagishima et al. 1977). This substance(s) is extractable with hot water from *a* cells and Fujimura et al. (1980) purified a substance from this extract which inactivates the α pheromone by binding to it. The isolated substance is proteinaceous and has a molecular weight of about 60,000. The relationship between the binding substance(s) and α pheromone-inactivating enzyme(s) has not been made clear.

19.3 Sexual Interactions

19.3.1 The Process of Mating Reaction

When cells of *a* and α mating types of *S. cerevisiae* are mixed, the mating reaction begins with sexual agglutination (Fig. 1). In sexual cell aggregates the ratio of *a* to α cells is usually one to one, regardless of the ratio at the time

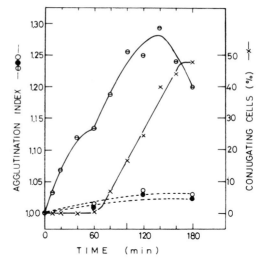

Fig. 7. Time course of mating reaction of *S. cerevisiae*. Agglutination index; mixed culture of *a* and α cells ——⊖——, separate culture of *a*-cells ---○--- and that of α-cells ---●---. Percent conjugating cells in the mixed culture ——×——. For the measurement of agglutination index, absorbance at 530 nm (A_{530}) of cell suspension was measured before and after sonic treatment which dispersed agglutinating cells completely. Agglutination index = A_{530} after sonication/A_{530} before sonication. For the theoretical basis of the index, see YOSHIDA and YANAGISHIMA (1978). (OSUMI et al. 1974)

of mixing (KAWANABE et al. 1979). The aggregates consist of cells at various stages of the cell cycle at the early stage of the mating reaction. In the aggregates, G_1-arrest may occur by the action of opposite mating type pheromones, resulting in the accumulation of unbudded cells in the aggregates. Then follows the formation of cell pairs between *a* and α cells as the initial step of zygote formation. Only unbudded cells participate in zygote formation (HARTWELL 1973). Since non-agglutinable mutants hardly form zygotes (Y. MATSUSHIMA et al., unpublished result) and zygotes are formed almost exclusively in cell aggregates (KAWANABE et al. 1979), sexual cell agglutination seems to be prerequisite for the formation of zygotes, although the intensity of the agglutination is not necessarily correlated with the frequency of zygote formation (KAWANABE et al. 1978).

Cell-wall-bound β-glucanase is responsible, at least partly, for the wall degradation which causes morphological changes and fusion of paired cells (SHIMODA 1973). The nuclei of the paired cells come near the contacting region of paired cells and fuse at the spindle pole bodies (BYERS and GOETSCH 1975), and this is probably followed by breakdown and reassociation of the nuclear membranes after fusion of paired cells (OSUMI et al. 1974). The time course of the mating reaction and zygote formation are shown in Figs. 7 and 8, respectively. During the mating reaction we can detect many physiological and biochemical changes, such as induction and enhancement of sexual agglutinability, inhibition of DNA synthesis with little inhibition of protein and RNA syntheses (SHIMODA and

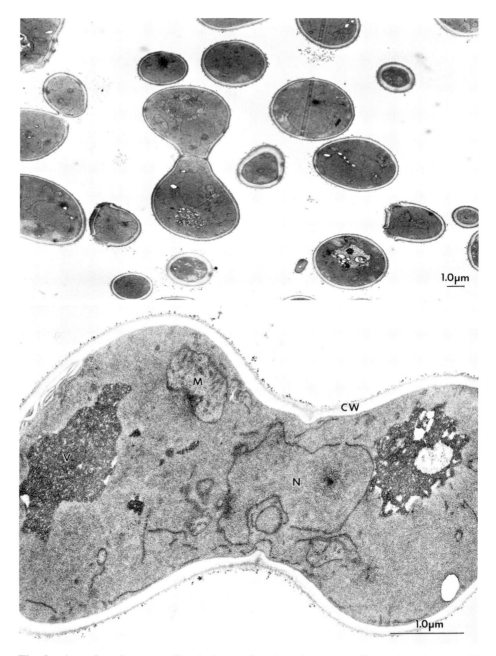

Fig. 8. A conjugating *a–α* cell pair (*upper figure*) and a zygote (*lower figure*). *CW* cell wall; *M* mitochondrion; *N* nucleus; *V* vacuole. (OSUMI et al. 1974)

Fig. 9. Increase in cell number, DNA, RNA and protein during mating reaction in *S. cerevisiae*. Each value for the separate culture represents the average of *a* and α separate cultures. *Closed circles* separate cultures; *open circles* mixed culture of *a* and α cells. (SHIMODA and YANAGISHIMA 1973)

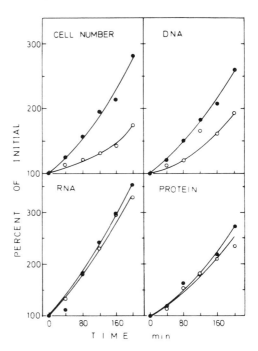

YANAGISHIMA 1973), and enhancement of cellular autolytic activity (SHIMODA and YANAGISHIMA 1972). In Fig. 9, changes in DNA, RNA and protein syntheses during the mating reaction are shown. The inhibition of cell division was observed both in *a* and α cells (SHIMODA and YANAGISHIMA 1973).

The mechanism of transition from sexual agglutination to the formation of the *a*–α cell pairs which are to be zygotes is unknown. However, ethyl N-phenylcarbamate (EPC), which binds to tubulin to disturb the formation of microtubules in plants (MIZUNO et al. 1981), is known to inhibit the transition without inhibiting the induction and enhancement of sexual agglutinability (YANAGISHIMA and HASEGAWA 1980, HASEGAWA and YANAGISHIMA 1981). If EPC is added to a mating mixture of *a* and α cells, sexual agglutination progresses without zygote formation, resulting in a highly agglutinating cell suspension. This result implies the possibility that microtubules play an important role in the transition.

2-Deoxyglucose hardly inhibits progress of sexual agglutination at a concentration which inhibits zygote formation completely. Since 2-deoxyglucose inhibits the synthesis of mannan and glucan much more significantly than protein synthesis, the inhibitory action on the formation of zygotes is thought to be due to the inhibition of the synthesis of cell wall materials necessary for *a*–α cell pair formation (SHIMODA and YANAGISHIMA 1974). Cycloheximide inhibits both enhancement of sexual agglutinability and formation of zygotes, indicating that both processes require protein synthesis (SHIMIODA 1973). Fermentable sugars are required for both processes (SHIMODA and YANAGISHIMA 1974).

19.3.2 Control of Mating Processs

19.3.2.1 Control of Sexual Agglutinability

The ability of cells to agglutinate with cells of the opposite mating type is controlled by genetic and physiological factors. There are two types of strains in haploid *S. cerevisiae;* one is inducible, and the other constitutive for sexual agglutinability. The inducible strains produce active sexual agglutination substances only in response to sex pheromones of the opposite mating type, and the constitutive strains do so without stimuli from the opposite mating type (SAKAI and YANAGISHIMA 1972. YANAGISHIMA and NAKAGAWA 1980) (Fig. 10). We have found two types of inducible strains in both *a* and α mating types; one has dominant inducible agglutinability, and the other has recessive inducible agglutinability (YANAGISHIMA and NAKAGAWA 1980). The strains having recessive inducible agglutinability are temperature-sensitive, being constitutive at 22 °C, inducible at 28 °C and non-agglutinable at 37 °C, while the strains having dominant inducible agglutinability are inducible at all the temperatures tested so far (YANAGISHIMA and NAKAGAWA 1980). We propose that the strains having dominant inducible agglutinability are wild-type with a pheromonal regulation system. When cultured at 35–38 °C (DOI and YOSHIMURA 1977, TOHOYAMA et al. 1979) or in the absence of fermentable sugars (YANAGISHIMA et al. 1976), even constitutive cells having the ability to produce agglutination substances without sex pheromones of the opposite mating type produced no detectable amount of active agglutination substance, and show no sexual agglutinability. Addition of Triton X-100 (N. YANAGISHIMA unpublished result) or imidazole (NAKAGAWA and YANAGISHIMA 1980) brings about the same results. The physiologically repressed sexual agglutinability mentioned above is reversed in a short time when the sex pheromone of the opposite mating type is added. Even in the strains producing active agglutination substances constitutively, sex pheromones of the opposite mating type enhances the production, especially in the case of *a* cells (SHIMODA et al. 1976).

19.3.2.2 Mechanism of Agglutinability Induction

The induction of sexual agglutinability by sex pheromone of the opposite mating type is accompanied by the appearance of the active agglutination substance on the cell surface in both *a* and α inducible cells where the respective agglutination substance is hardly detected before the induction, and the appearance is inhibited by cycloheximide which inhibits the induction (SHIMODA et al. 1976, TOHOYAMA et al. 1979, YANAGISHIMA and NAKAGAWA 1980), indicating that the induction of sexual agglutinability is brought about through production of the active agglutination substances.

   The induction of sexual agglutinability by the α pheromone has been studied mainly by using recessive inducible *a* cells. The induction is inhibited by osmotic shock, imidazole, Triton X-100 and high temperature (37°–38 °C), suggesting the involvement of the cell membrane in the induction mechanism (NISHI and YANAGISHIMA 1982, YANAGISHIMA and NAKAGAWA 1980, NAKAGAWA and YANAGISHIMA 1980). In this connection, it is noteworthy that the production of *a*

Fig. 10. Inducible and constitutive cells concerning sexual agglutinability. The superscripts, i and c, given to the mating type symbols, *a* and α, designate the inducible and constitutive types, respectively

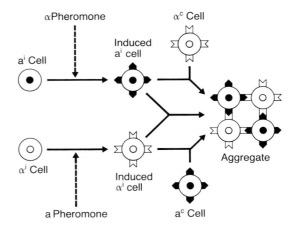

agglutination substance is always associated with the formation of invaginations of cell membrane in *a* cells (NAKAGAWA et al. 1983).

The α pheromone binds to inducible *a* cells before induction (YANAGISHIMA et al. 1977, SHIMIZU et al. 1977). The bound α pheromone induces sexual agglutinability even in the absence of external α pheromone (YANAGISHIMA and INABA 1981). When the binding of α pheromone to *a* cells is inhibited by 5M urea, no induction occurs (T. SHIMIZU unpublished result), indicating that the binding is the first step of α pheromone action in inducing sexual agglutinability. The binding needs no metabolic activity but the induction needs both carbon and nitrogen sources, and physiological temperature (SHIMODA et al. 1976, YANAGISHIMA and INABA 1981). Chemical alteration of α pheromone by *a* cells was suggested to be necessary for the G_1-arresting action (MANNES and EDELMAN 1978) and agglutinability induction (NISHI and YANAGISHIMA 1982). However, recently mutants defective in α-pheromone-inactivating action were shown to be highly sensitive to G_1-arresting action of α pheromone (CHAN and OTTE 1983, SPRAGUE Jr and HERSKOWITZ 1981). Hence there may be two kinds of alterations of α pheromone, one necessary for the α pheromone action and the other not. Anyway, we cannot draw any conclusion from the results obtained so far. The induction of sexual agglutinability by α pheromone has a 10-min lag period which is temperature-sensitive, the restrictive and permissive temperature for this period being 38° and 28 °C, respectively (NISHI and YANAGISHIMA 1982). If cycloheximide (10^{-4} M) was added at 10, 20 and 30 min after the start of treatment with α pheromone, complete, partial and no inhibition of the induction was detected, respectively. It seems that not only for the completion of the temperature-sensitive period (lag period) but also for the progress of the induction at initial stage, protein synthesis is required (NISHI and YANAGISHIMA 1982).

The action of *a* pheromone in inducing sexual agglutinability of α cells, which requires both nitrogen and energy sources, is inhibited by cycloheximide (YANAGISHIMA and NAKAGAWA 1980). Inactivation by, or binding to, α cells of *a* pheromone was hardly detected (N. YANAGISHIMA unpublished result). At the moment we know little about the action mechanism of *a* pheromone.

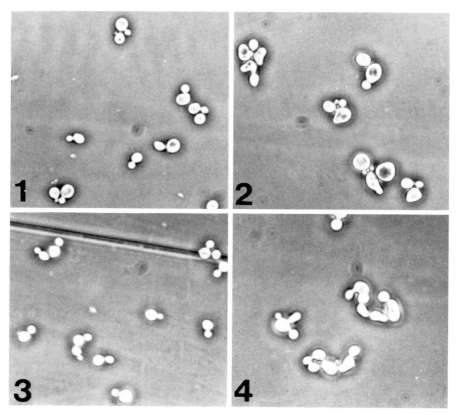

Fig. 11. α-Pheromone-induced "shmoo" in *S. cerevisiae* and *S. kluyveri*. *S. cerevisiae;*
1 Control, **2** "Shmoo". *S. kluyveri;* **3** Control, **4** "Shmoo". Each strain was cultured
with or without addition of respective α pheromone for 4 h. (By courtesy of Mr. Fujimura, Nagoya Univ.)

19.3.2.3 G₁ Arrest Caused by Sex Pheromones

The α pheromone inhibits initiation of DNA synthesis with little inhibition
of RNA and protein syntheses in *a* cells (Throm and Duntze 1970), resulting
in the formation of "shmoo" (Duntze et al. 1970), accompanied by changes
in the chemical nature of the cell wall especially near the elongating tip (Lipke
and Ballou 1980, Tkacz and MacKay 1979) (Fig. 11). The α pheromone
arrests the cell cycle near or at the step which *CDC*28 gene controls (Hereford
and Hartwell 1974). The α pheromone concentration necessary for the induction of G₁ arrest is about 10^3 times higher than that necessary for the induction
of sexual agglutinability (Shimoda et al. 1978). The agglutination induction by
α pheromone (0.1 μg ml⁻¹) began after a 10-min lag period, and the time till
maximum induction was about 1 h. However, only slight inhibition of DNA
synthesis was detected even after 1 h treatment with the α pheromone at 10
times higher concentration (Shimoda et al. 1978). Agglutinability is induced

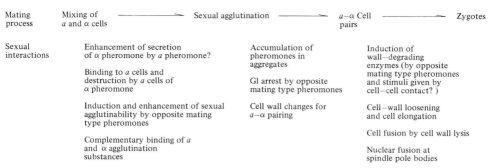

Fig. 12. Sexual interactions during mating reaction (mass mating) in *S. cerevisiae*

independently of the cell cycle (N. YANAGISHIMA unpublished result), but DNA synthesis is inhibited only in cells at the stage of G_1 (HEREFORD and HARTWELL 1974). The G_1 arrest caused by the α pheromone is reversed by washing the arrested cells with α-pheromone-free medium (BÜCKING-THROM et al. 1973). The α pheromone absorbed by *a* cells induced sexual agglutinability without causing G_1 arrest (YANAGISHIMA and INABA 1981). The *a* pheromone causes G_1 arrest and inhibition of DNA synthesis in α cells (SHIMODA and YANAGISHIMA 1973, WILKINS and PRINGLE 1974), but few further studies on the *a* pheromone action have been performed.

From the above results, it seems that during the mating reaction, cell agglutinability is at first induced or enhanced by opposite mating type pheromones, resulting in the formation of cell aggregates, followed by G_1 arrest of the cells in cell aggregates in which high concentrations of the sex pheromones can be maintained at least locally.

19.3.2.4 Biochemical Changes Caused by α Pheromone

α Pheromone inhibits activity of membrane-bound adenylate cyclase and addition of cyclic AMP shortens the period of G_1 arrest caused by the α pheromone (LIAO and THORNER 1980). These findings suggest that α pheromone causes G_1 arrest by changing cellular cyclic AMP level. α Pheromone has been shown to cause the modification of a ^{32}P-labelled protein, an alteration of molecular weight (FINKELSTEIN and MCALISTER 1981). These findings may give a clue to biochemical analysis of α pheromone action. In Fig. 12, the sexual interactions which occur or may occur when *a* and α cells of *S. cerevisiae* are mixed are summarized.

19.3.3 Interspecific and Intergeneric Sexual Interactions

The ascosporogenous yeasts seem to show sexual agglutination in general (YOSHIDA and YANAGISHIMA 1978). Hence, by making clear the interspecific or intergeneric sexual agglutination, together with the chemical nature of isolated agglutination substances responsible for the agglutination, we may be able to trace phylogenetic relationships among the yeasts.

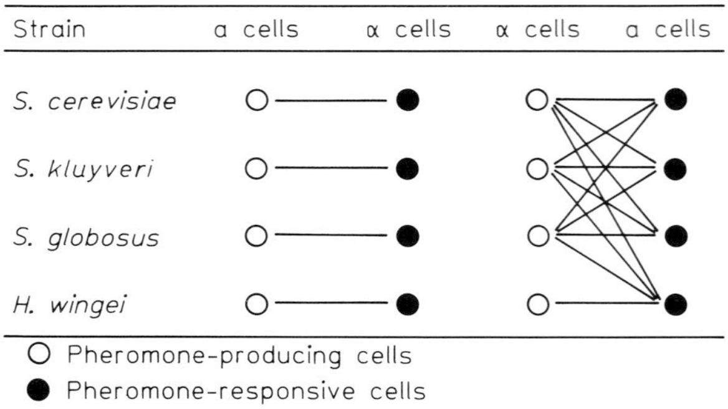

Strain	a cells	α cells	α cells	a cells
S. cerevisiae	○———————●	○⤭———————●
S. kluyveri	○———————●	○⤭———————●
S. globosus	○———————●	○⤭———————●
H. wingei	○———————●	○———————●

○ Pheromone-producing cells
● Pheromone-responsive cells

Fig. 13. Pheromonal interactions among ascosporogenous yeasts. Pheromone action was detected by agglutinability induction and/or "shmoo" induction. (Based on the results of FUJIMURA and YANAGISHIMA 1981, and YANAGISHIMA and FUJIMURA 1981)

We are performing comparative studies on sex pheromones in ascosporogenous yeasts among which neither sexual agglutination nor mating reaction occurs (FUJIMURA and YANAGISHIMA 1981, YANAGISHIMA and FUJIMURA 1981). In Fig. 13, the results obtained so far are summarized. α Pheromones of *S. cerevisiae*, *S. kluyveri* and *S. globosus* induced "shmoo" in *a* cells of these yeasts except that the α pheromone of *S. globosus* hardly induced "shmoo" in *a* cells of *S. cerevisiae*. Contrary to α pheromones, *a* pheromones of these yeasts induced "shmoo" only in α cells of the same species, with the exception of *S. cerevisiae*, of which *a* pheromone hardly induced "shmoo" even in α cells of the same species in our system. α Pheromones of these yeasts induced sexual agglutinability of recessive inducible *a* cells in *S. cerevisiae* and 5 (*a*) cells of *H. wingei*. However, no interspecific agglutinability-inducing action of *a* pheromones was observed so far. Hence it is concluded that α pheromones are effective interspecifically, although the action is exactly mating-type-specific. However, the action of *a* pheromones is not only mating-type-specific, but also species-specific. It is important that even among yeast species in which no conjugation occurs pheromonal interactions occur. Thus, communication through sex pheromones must play an important role for the maintenance and changes of population in the ecological system as a whole. We already know the chemical structures of α pheromones of *S. cerevisiae* and *S. kluyveri* (SAKURAI et al. 1976, SATO et al. 1981, STÖTZLER et al. 1976) (Fig. 5). By comparing chemical structures of sex pheromones in ascosporogenous yeasts, together with their actions, it will be possible to open up a new approach to phylogenetic relationship among the yeasts.

19.4 Conclusion

As described above, the phenomenon of mating reaction in yeast involves various problems from molecular to ecological or phylogenetic levels, which are interrelated. Hence it is necessary to take a wide view of the phenomenon, even when we treat only one of the problems in the mating reaction. Only thus will we be able to reach a biologically significant goal.

Acknowledgement. A part of author's research described in this article was supported by Grants-in-Aid from the Ministry of Education, Science and Culture, Japan.

References

Betz R, Duntze W (1979) Purification and partial characterization of *a*-factor, a mating hormone produced by mating-type-*a* cells from *Saccharomyces cerevisiae*. Eur J Biochem 95:469–475

Betz R, Duntze W, Manney TR (1978) Mating-factor-mediated sexual agglutination in *Saccharomyces cerevisiae*. FEMS Microbiol Lett 4:107–110

Brock TD (1958a) Mating reaction in the yeast *Hansenula wingei*. Preliminary observations and quantitation. J Bacteriol 75:697–701

Brock TD (1958b) Protein as a specific cell-surface component in the mating reaction of *Hansenula wingei*. J Bacteriol 76:334–335

Brock TD (1965) The purification and characterization of an intracellular sex-specific mannan protein from yeast. Proc Natl Acad Sci USA 54:1104–1112

Bücking-Throm E, Duntze W, Hartwell LH, Manney TR (1973) Reversible arrest of haploid yeast cells at the initiation of DNA synthesis by a diffusible sex factor. Exp Cell Res 76:99–110

Burke D, Mendonca-Previato L, Ballou CE (1980) Cell–cell recognition in yeast: Purification of *Hansenula wingei* 21-cell sexual agglutination factor and comparison of the factors from three genera. Proc Natl Acad Sci USA 77:318–322

Byers B, Goetsch L (1975) Behavior of spindles and spindle plaques in the cell cycle and conjugation of *Saccharomyces cerevisiae*. J Bacteriol 124:511–523

Chan RK (1977) Recovery of *Saccharomyces cerevisiae* mating type *a* cells from G_1 arrest by α factor. J Bacteriol 130:766–774

Chan RK, Otte CA (1982) Physiological characterization of *Saccharomyces cerevisiae* mutants supersensitive to G_1 arrest by *a*-factor and α-factor pheromones. Mol Cell Biol 2:21–29

Ciejek E, Thorner J (1979) Recovery of *S. cerevisiae a* cells from G_1 arrest by α-factor pheromone requires endopeptidase action. Cell 18:623–635

Crandall M, Brock TD (1968a) Mutual repression of haploid genes in diploid yeast. Nature 219:533–534

Crandall MA, Brock TD (1968b) Molecular basis of mating in the yeast *Hansenula wingei*. Bacteriol Rev 32:139–163

Crandall M, Caulton JH (1973) Induction of glycoprotein mating factors in diploid yeast of *Hansenula wingei* by vanadium salts or chelating agents. Exp Cell Res 82:159–167

Crandall M, Lawrence ML, Saunders RM (1974) Molecular complementarity of yeast glycoprotein mating factors. Proc Natl Acad Sci USA 71:26–29

Doi S, Yoshimura M (1977) Temperature-sensitive loss of sexual agglutinability in *Saccharomyces cerevisiae*. Arch Microbiol 114:287–288

Duntze W, MacKay V, Manney TR (1970) *Saccharomyces cerevisiae*: A diffusible sex factor. Science 168:1472–1473

Finkelstein DB, McAlister L (1981) α-Factor-mediated modification of a ^{32}P-labelled protein by *MATa* cells of *Saccharomyces cerevisiae*. J Biol Chem 256:2561–2566

Finkelstein DB, Strausberg S (1979) Metabolism of α factor by *a* mating type cells of *Saccharomyces cerevisiae*. J Biol Chem 254:796–803

Fujimura H, Yanagishima N (1981) Sex pheromones in *Saccharomyces globosus* in relation to sex pheromones in *Saccharomyces cerevisiae* and *Saccharomyces kluyveri*. Proc 46th Ann Meet Bot Soc Jpn p 173 (in Japan.)

Fujimura H, Yoshida K, Yanagishima N (1980) α Pheromone-binding substance in *Saccharomyces cerevisiae*. Proc 45th Ann Meet Bot Soc Jpn p 78 (in Japan.)

Fujimura H, Yanagishima N, Sakurai A, Kitada C, Fujino M, Banno I (1982) Sex pheromone of α mating type in the yeast *Saccharomyces kluyveri* and its synthetic analogues in relation to sex pheromones in *Saccharomyces cerevisiae* and *Hansenula wingei*. Arch Microbiol 132:225–229

Haber JE, George JP (1979) A mutation that permits the expression of normally silent copies of mating-type information in *Saccharomyces cerevisiae*. Genetics 93:13–35

Hagiya M (1980) Studies on the mechanism of sexual agglutination in the yeast *Saccharomyces cerevisiae*. Ph D Thesis Nagoya Univ, Nagoya

Hagiya M, Yoshida K, Yanagishima N (1977) The release of sex-specific substances responsible for sexual agglutination from haploid cells of *Saccharomyces cerevisiae*. Exp Cell Res 104:263–272

Harashima S, Oshima Y (1976) Mapping of the homothallic genes, *HMα* and *HMa* in *Saccharomyces yeasts*. Genetics 84:437–451

Hartwell LH (1973) Synchronization of haploid yeast cell cycles, a prelude to conjugation. Exp Cell Res 76:111–117

Hasegawa S, Yanagishima N (1981) Effect of phenylurethane on the action of the α pheromone in *Saccharomyces cerevisiae*. Proc 46th Ann Meet Bot Soc Jpn p 172 (in Japan.)

Hereford LM, Hartwell LH (1974) Sequential gene function in the initiation of *Saccharomyces cerevisiae* DNA synthesis. J Mol Biol 84:445–461

Herskowitz I, Oshima Y (1981) Control of cell type in *Saccharomyces cerevisiae*: Mating type and mating type interconversion. In: Strathern JN, Jones EW, Broach J (eds) Molecular biology of the yeast *Saccharomyces*. Life cycle and inheritance. Cold Spring Harbor Lab, New York, pp 181–209

Hicks JB, Herskowitz I (1976a) Interconversion of yeast mating types I. Direct observation of the action of homothallism (*HO*) gene. Genetics 83:245–258

Hicks JB, Herskowitz I (1976b) Evidence for a new diffusible element of mating pheromones in yeast. Nature 260:246–248

Hicks JB, Strathern JN, Herskowitz I (1977a) The cassette model of mating type interconversion. In: Bukhar A, Shapiro J, Adhya S (eds) DNA insertion elements, plasmids and episomes. Cold Spring Harbor Lab, New York, pp 457–462

Hicks JB, Strathern JN, Herskowitz I (1977b) Interconversion of yeast mating type III. Action of the homothallism (*HO*) gene in cells homozygous for the mating type locus. Genetics 85:395–405

Hicks JB, Strathern JN, Klar JS (1979) Transposable mating type genes in *Saccharomyces cerevisiae*. Nature 282:478–483

Kawanabe Y, Hagiya M, Yoshida K, Yanagishima N (1978) Effect of concanavalin A on the mating reaction of *Saccharomyces cerevisiae*. Plant Cell Physiol 19:1207–1216

Kawanabe Y, Yoshida K, Yanagishima N (1979) Sexual cell agglutination in relation to the formation of zygotes in *Saccharomyces cerevisiae*. Plant Cell Physiol 20:423–433

Klar AJS, Fogel S, MacLeod M (1979) *Mar*1-A regulator of the *HMa* and *HMα* loci in *Saccharomyces cerevisiae*. Genetics 93:37–50

Klar AJS, MacIndoo J, Hicks JB, Strathern JN (1980) Precise mapping of the homothallism genes, *HML* and *HMR* in yeast *Saccharomyces cerevisiae*. Genetics 96:315–320

Klar AJS, Strathern JN, Broach JR, Hicks JB (1981) Regulation of transcription in expressed and unexpressed mating type cassettes of yeast. Nature 289:239–244

Levi JD (1956) Mating reaction in yeast. Nature 177:753–754

Liao H, Throner J (1980) Yeast mating pheromone α-factor inhibits adenylate cyclase. Proc Natl Acad Sci USA 77:1898–1902

Lindegren CC, Lindegren G (1943) A new method for hybridizing yeast. Proc Natl Acad Sci USA 29:306–308

Lipke PN, Ballou CE (1980) Altered immunochemical reactivity of *Saccharomyces cerevisiae* a cells after α-factor-induced morphogenesis. J Bacteriol 141:1170–1177

MacKay VL, Manney TR (1974a) Mutations affecting sexual conjugation and related processes in *Saccharomyces cerevisiae* I. Isolation and phenotypic characterization of nonmating mutants. Genetics 76:255–271

MacKay V, Manney TR (1974b) Mutations affecting sexual conjugation and related processes in *Saccharomyces cerevisiae* II. Genetic analysis of nonmating mutants. Genetics 76:273–288

Mannes PF, Edelman GM (1978) Inactivation and chemical alteration of mating factor α by cells and spheroplasts of yeast. Proc Natl Acad Sci USA 75:1304–1308

Manney TR, Duntze W, Betz R (1981) The isolation, characterization, and physiological effects of the *Saccharomyces cerevisiae* sex pheromones. In: O'Day DH, Horgen PA (eds) Sexual interactions in eukaryotic microbes. Academic Press, London New York, pp 21–51

McCullough J, Herskowitz I (1979) Mating pheromones of *Saccharomyces kluyveri*: Pheromone interactions between *Saccharomyces kluyveri* and *Saccharomyces cerevisiae*. J Bacteriol 138:146–154

Mizuno K, Koyama M, Shibaoka H (1981) Isolation of plant tubulin from azuki bean epicotyls by ethyl N-phenylcarbamate-Sepharose affinity chromatography. J Biochem 89:329–332

Nakagawa Y, Yanagishima N (1980) Inducible genes concerning sexual agglutinability and their physiological activity. Proc 45th Ann Meet Bot Soc Jpn p 79 (in Japan.)

Nakagawa Y, Tanaka K, Yanagishima N (1983) Occurrence of plasma membrane invagination associated with sexual agglutination ability in the yeast *Saccharomyces cerevisiae*. Molec Gen Genet 189:211–214

Nasmyth KA, Tatchell K (1980) The structure of transposable yeast mating type loci. Cell 19:753–764

Nasmyth KA, Tatchell K, Hall BD, Astell C, Smith M (1981a) Physical analysis of mating-type loci in *Saccharomyces cerevisiae*. Cold Spring Harbor Symp Quant Biol 45:961–981

Nasmyth KA, Tatchell K, Hall BD, Astell C, Smith M (1981b) A position effect in the control of transposition at yeast mating type loci. Nature 289:244–250

Nishi K, Yanagishima N (1982) Temperature dependency of induction of sexual agglutinability by α pheromone in the yeast *Saccharomyces cerevisiae*. Arch Microbiol 132:236–240

Oshima Y, Takano I (1971) Mating types in *Saccharomyces*: Their convertibility and homothallism. Genetics 67:327–335

Osumi M, Shimoda C, Yanagishima N (1974) Mating reaction in *Saccharomyces cerevisiae* V. Changes in the fine structure during the mating reaction. Arch Microbiol 97:27–38

Rine J, Jensen R, Hagen D, Blair L, Herskowitz I (1981) Pattern of switching and fate of replaced cassette in yeast mating type interconversion. Cold Spring Harbor Symp Quant Biol 45:951–960

Sakai K, Yanagishima N (1972) Mating reaction in *Saccharomyces cerevisiae* II. Hormonal regulation of agglutinability of α type cells. Arch Mikrobiol 84:191–198

Sakurai A, Tamura S, Yanagishima N, Shimoda C (1976) Structure of the peptidyl factor inducing sexual agglutination in *Saccharomyces cerevisiae*. Agric Biol Chem 40:1057–1058

Sakurai A, Sato Y, Park KH, Takahashi N, Yanagishima N, Banno I (1980) Isolation and chemical characterization of a mating pheromone produced by *Saccharomyces kluyveri*. Agric Biol Chem 44:1451–1453

Sato Y, Sakurai A, Takahashi N, Hong Y-M, Shimonishi Y, Kitada C, Fujino M, Yanagishima N, Banno I (1981) Amino acid sequence of α_k substance, a mating pheromone of *Saccharomyces kluyveri*. Agric Biol Chem 45:1531–1533

Shimizu T, Yoshida K, Yanagishima N (1977) Specific binding substance for the agglutinability-inducing pheromone, α substance-I in *Saccharomyces cerevisiae*. Proc Ann Meet Jpn Soc Plant Physol p 77 (in Japan.)

Shimoda C (1973) Physiological and biochemical studies on mating reaction in *Saccharomyces cerevisiae*. Ph D Thesis Osaka City Univ, Osaka

Shimoda C, Yanagishima N (1972) Mating reaction in *Saccharomyces cerevisiae* III. Changes in autolytic activity. Arch Mikrobiol 85:310–318

Shimoda C, Yanagishima N (1973) Mating reaction in *Saccharomyces cerevisiae* IV. Retardation of deoxyribonucleic acid synthesis. Physiol Plant 29:54–59

Shimoda C, Yanagishima N (1974) Mating reaction in *Saccharomyces cerevisiae* VI. Effect of 2-deoxyglucose on conjugation. Plant Cell Physiol 15:767–778

Shimoda C, Yanagishima N (1975) Mating reaction in *Saccharomyces cerevisiae* VIII. Mating-type-specific substances responsible for sexual cell agglutination. Antonie van Leeuwenhoek 41:521–532

Shimoda C, Kitano S, Yanagishima N (1975) Mating reaction in *Saccharomyces cerevisiae* VII. Effect of proteolytic enzymes on sexual agglutinability and isolation of crude sex-specific substances responsible for sexual cell agglutination. Antonie van Leeuwenhoek 41:513–519

Shimoda C, Yanagishima N, Sakurai A, Tamura S (1976) Mating reaction in *Saccharomyces cerevisiae* IX. Regulation of sexual agglutinability of *a* type cells by a sex factor produced by α type cells. Arch Microbiol 108:27–33

Shimoda C, Yanagishima N, Sakurai A, Tamura S (1978) Induction of sexual agglutinability of *a* mating type cells as the primary action of the peptidyl sex factor from α mating type cells in *Saccharomyces cerevisiae*. Plant Cell Physiol 19:513–517

Sprague Jr GF, Herskowitz I (1981) Control of yeast cell type by the mating locus I. Identification and control of expression of the *a*-specific gene, *BAR*1. J Mol Biol 153:305–321

Stötzler D, Kiltz H-H, Duntze W (1976) Primary structure of α-factor peptides from *Saccharomyces cerevisiae*. Eur J Biochem 69:397–400

Strathern JN, Spatola E, McGill C, Hicks JB (1980) Structure and organization of transposable mating type cassettes in *Saccharomyces* yeasts. Proc Natl Acad Sci USA 77:2839–2843

Strathern JN, Hicks J, Herskowitz I (1981) Control of cell type in yeast by the mating type locus, the α1-α2 hypothesis. J Mol Biol 147:357–372

Takano I, Oshima Y (1970) Mutational nature of an allele-specific conversion of the mating type by the homothallic gene HO_α in *Saccharomyces*. Genetics 65:421–427

Taylor NW (1964) Specific, soluble factor involved in sexual agglutination of the yeast *Hansenula wingei*. J Bacteriol 87:863–866

Taylor NW (1965) Purification of sexual agglutination factor from the yeast *Hansenula wingei* by chromatography and gradient sedimentation. Arch Biochem Biophys 111:181–186

Taylor NW, Orton WL (1971) Cooperation among the active binding sites in the sex-specific agglutinin from the yeast, *Hansenula wingei*. Biochemistry 10:2043–2049

Thorner J (1981) Pheromonal regulation of development in *Saccharomyces cerevisiae*. In: Strathern JN, Jones EW, Broach J (eds) Molecular biology of the yeast *Saccharomyces*. Life cycle and inheritance. Cold Spring Harbor, New York, pp 143–180

Throm E, Duntze W (1970) Mating-type-dependent inhibition of deoxyribonucleic acid synthesis in *Saccharomyces cerevisiae*. J Bacteriol 104:1388–1390

Tkacz JS, Mackay VL (1979) Sexual conjugation in yeast. Cell surface changes in response to the action of mating hormones. J Cell Biol 80:326–333

Tohoyama H, Yanagishima N (1982) Control of the production of the sexual agglutination substances by the mating type locus in *Saccharomyces cerevisiae*: Simultaneous expression of specific genes for *a* and α agglutination substances in *mat* α2 mutant cells. Molec Gen Genet 186:322–327

Tohoyama H, Hagiya M, Yoshida K, Yanagishima N (1979) Regulation of the production of the agglutination substances responsible for sexual agglutination in *Saccharomyces cerevisiae*: Changes associated with conjugation and temperature shift. Molec Gen Genet 174:269–280

Wickerham LJ (1956) Influence of agglutination on zygote formation in *Hansenula wingei*, a new species of yeast. Compt Rend Trav Lab Carlsberg Ser Physiol 26:423–443

Wilkinson LE, Pringle JR (1974) Transient G_1 arrest of *S. cerevisiae* cells of mating type α by a factor produced by cells of mating type *a*. Expt Cell Res 89:175–187

Yamaguchi M, Yoshida K, Yanagishima N (1982a) Isolation and partial characterization of cytoplasmic α agglutination substance in the yeast *Saccharomyces cerevisiae* FEBS Lett 139:125–129

Yamaguchi M, Yoshida K, Yanagishima N (1982b) Purification and characterization of cytoplasmic *a* agglutination substance in *Saccharomyces cerevisiae*. Proc 47th Ann Meet Bot Soc Jpn, p 171 (in Japan.)

Yanagishima N (1978) Sexual cell agglutination in *Saccharomyces cerevisiae*: Sexual cell recognition and its regulation. Bot Mag Tokyo Special Issue 1:61–81

Yanagishima N, Fujimura H (1981) Sex pheromones of the yeast *Hansenula wingei* and their relationship to sex pheromones in *Saccharomyces cerevisiae* and *Saccharomyces kluyveri*. Arch Microbiol 129:281–284

Yanagishima N, Hasegawa S (1980) Differential inhibition of induction of sexual agglutinability and zygote formation in *Saccharomyces cerevisiae*. Proc Ann Meet Jpn Soc Plant Physiol p 208 (in Japan.)

Yanagishima N, Inaba R (1981) Induction of sexual agglutinability by the absorbed peptidyl factor in *a* mating type cells of *Saccharomyces cerevisiae*. Plant Cell Physiol 22:317–321

Yanagishima N, Nakagawa Y (1980) Mutants inducible for sexual agglutinability in *Saccharomyces cerevisiae*. Molec Gen Genet 178:241–251

Yanagishima N, Yoshida K (1981a) Sexual interactions in *Saccharomyces cerevisiae* with special reference to the regulation of sexual agglutinability. In: O'Day H, Horgen PA (eds) Sexual interactions in eukaryotic microbes. Academic Press, London New York, pp 261–295

Yanagishima N, Yoshida K (1981b) Sexuality and conjugation. In: Yanagishima N, Oshima Y, Osumi M (eds) Atlas of yeast. Kodanshya, Tokyo, pp 85–97 (in Japan.)

Yanagishima N, Yoshida K, Hagiya M, Kawanabe Y, Sakurai A, Tamura S (1976) Regulation of sexual agglutinability in *Saccharomyces cerevisiae of a* and α types by sex-specific factors produced by their respective opposite mating types. Plant Cell Physiol 17:439–450

Yanagishima N, Shimizu T, Yoshida K, Sakurai A, Tamura S (1977) Physiological detection of a binding substance for the agglutinability-inducing pheromone, α substance-I in *Saccharomyces cerevisiae*. Plant Cell Physiol 18:1182–1192

Yen PH, Ballou CE (1973) Composition of a specific intercellular agglutination factor. J Biol Chem 248:8316–8318

Yen PH, Ballou CE (1974) Partial characterization of the sexual agglutination factor from *Hansenula wingei* Y-2340 type 5 cells. Biochemistry 13:2428–2437

Yoshida K, Yanagishima N (1978) Intra- and intergeneric mating behaviour of ascosporogenous yeasts I. Quantitative analysis of sexual agglutination. Plant Cell Physiol 19:1519–1533

Yoshida K, Hagiya M, Yanagishima N (1976) Isolation and purification of the sexual agglutination substance of mating type *a* cells in *Saccharomyces cerevisiae*. Biochem Biophys Res Commun 71:1085–1094

20 Cellular Interactions During Early Differentiation

L. STANGE

20.1 Introduction

The invention of structures and mechanisms allowing cellular interactions has been a prerequisite for the evolution of multicellular organisms. In multicellular organisms, by means of the process of differentiation, division of labour between the parts becomes possible; some cells continue replication of genetic material, while others specialize in new functions enabling the multicellular organism to conquer new environments. In the plant kingdom this evolutionary trend is most clearly documented in the origin of species adapted to conditions on the land surface.

Cooperation to reach higher achievements requires precise modes of communication between the units. The structural base for communication has evolved in plants and animals in different ways, in close relationship to other specific structural features. In contrast to animals (cf. KARKINEN-JÄÄSKELÄINEN et al. 1977), plants are distinguished by the tight spatial order of their cells due to the formation of rigid cell walls; the cells are connected by plasmodesmata, protoplasmic threads with an elaborate ultrastructure, which penetrate through the intervening cell wall and provide a symplastic route for cell-to-cell transport of low molecular weight materials (GUNNING 1976, ROBARDS 1976, GUNNING and ROBARDS 1976). In addition, intercellular exchange of "messages" is possible by diffusion through the apoplast. At still another level, plants have evolved special vascular tissues for long-distance transport of nutrients and correlative signals (cf. RAVEN 1977).

Multicellular organisms arise from single cells in a continuous sequence of cell divisions. In plants, during cytokinesis coupled to nuclear division, plasmodesmata are formed at sites where fusion of cell plate vesicles is prevented by the presence of strands of endoplasmic reticulum and the plasmalemma which lines the plasmodesma is derived from fusion of the membranes of the cell plate vesicles (JONES 1976). Thus, during development, coupled to the number of cell divisions, the network for intercellular communication becomes increasingly complex. It is very probable that the structural possibilities for intercellular information exchange are used by the developing plant from the very beginning to reap the advantages of division of labour. The development of channels for transport, and of transport functions, can be visualized as essential components of the overall process of differentiation.

Abbreviations. IAA indole-3-acetic acid; 2,4-D 2,4-dichlorophenoxyacetic acid; PCIB p-chlorophenoxyisobutyric acid.

Differentiation is the process by which, during development, differences arise between the parts of an individuum (cf. definition of the term in STANGE 1965, HESLOP-HARRISON 1967). Although the results of differentiation are most striking and, as such, sometimes taken for "differentiation", the critical directive events of the process have already taken place during the early phases of development. Within the scope of this chapter, the discussion will be limited to a few selected aspects of early differentiation. No attempt can be made to achieve a complete coverage of even these aspects, and the aim will be rather to focus on some general features of early differentiation in which the involvement of cellular interactions is suggested, and to emphasize, on the cellular level, the close relationship between growth and differentiation.

The term interaction is often used rather loosely for effects of one part of a system on another. In a strict sense, however, interaction means reciprocal effects. In a complex self-regulating system these are especially feed-back operations. Keeping this in mind, one has to realize that, despite theoretical postulations, so far experimental research has contributed only fragmentary evidence concerning the functioning and nature of cellular interactions during differentiation.

Earlier work and general aspects related to this subject have been discussed by BÜNNING 1953, 1956, 1965, SINNOTT 1960, BLOCH 1965, LANG 1965, 1966, 1973, WARDLAW 1968.

20.2 Cell Cycle and Cell Differentiation

A synergetic system composed of a high number of units can exhibit properties not exhibited by the single unit alone (HAKEN 1980); however, the nature of these properties is determined by the properties of the individual unit. The essential properties of the cell relevant in this context are its inherited genetic information, its tendency to growth caused by the capacity for self-replication of DNA, and its ability for continued gene activity when growth functions are arrested. The expression of these properties of the nucleus depends upon the surrounding cytoplasm and cellular environment (cf. BRACHET and LANG 1965, MATHER 1965, WHALEY 1965). When all necessary environmental requirements are given, the independent cell will pass continuously through the cell reproduction cycle, the sequence of a phase of synthetic activity and a phase of partition of augmented material to new cell individuals.

The cell reproduction cycle has been the subject of extensive studies during the last decades (cf. MITCHISON 1971, 1973, PRESCOTT 1976, BRYANT 1976, NAGL 1976a, YEOMAN and AITCHISON 1976, PARDEE et al. 1978, HOCHHAUSER et al. 1981, JOHN 1981). The elucidation of the molecular mechanisms governing cell cycle progression is of extreme importance for an understanding of molecular regulation of cell differentiation. Since the pioneer experiments of HOWARD and PELC (1951, 1953) on the cell cycle events in root tips of *Vicia faba,* it has been established in numerous experiments with a wide variety of organisms that the sequence of events in the cell cycle can be divided into four distinct phases, (1) the interval between mitosis and DNA synthesis (G_1-phase), (2) the phase of DNA synthesis (S-phase), (3) the interval between DNA synthesis and mitosis (G_2-phase), and (4) the partitioning phase (mitosis, usually followed by cell division). Much effort has been concentrated on studies of processes occurring during G_1- and G_2-phases which are required for the cell cycle progression, and on the analysis of their temporal order. The general picture, recently supported by experiments with G_1-specific mutants, is that there are several dependent pathways of events. In each pathway the completion of early events is required for the later ones to occur; at some point, however, the pathways must converge before further progression becomes possible (cf. HOCH-HAUSER et al. 1981). Referring to expression of gene activity in these dependent pathways, it has been stated that "progression through the cell cycle is based on a program of causally connected sequential transcriptions and translations of a set of cell-cycle genes" (PRESCOTT 1976).

It has been shown that two separate pathways, the nuclear DNA-division cycle and the cytoplasmic growth cycle (MITCHISON 1971) are usually coordinated, but can be uncoupled under special developmental conditions. An independent control of DNA synthesis and mitosis opens the possibility that cells, unable to complete the events of a mitotic cycle, continue to replicate DNA within so-called endocycles, resulting in endopolyploid nuclei or, in the case of inhibited separation of the endochromosomes, in nuclei with polytene chromosomes, both with a high capacity for uninterrupted RNA synthesis (NAGL 1976a, 1978, 1981).

The duration of cell cycles within one organism can vary over a large range depending on the developmental stage, on the type and position of the cell (cf. CLOWES 1976, LYNDON 1976, BARLOW 1976), as well as on environmental conditions (e.g. GOULD 1977, GOULD et al. 1974). Furthermore, for comparable cells of different species, different cycle durations have been reported (e.g. TORREY 1972, LYNDON 1976). In most cases variation in total cycle time can be attributed to an extension or reduction of the G_1-phase, and the importance of this phase is seen in providing a period for control of proliferation in multicellular organisms (HOCHHAUSER et al. 1981, YEOMAN and AITCHISON 1976). In experiments on the effect of different temperatures on the cell cycle in plant cell suspension cultures, the G_1-phase was distinguished from the other phases by a very low temperature coefficient, characteristic of physical processes such as diffusion (GOULD 1977). This observation might point to a rate-limiting control of the G_1-phase by intercellular transport processes.

During development of multicellular plants, initiation of cell differentiation is expressed most strikingly by the suppression of divisional functions in those cells which, in symplastic contact with meristematic cells, take over specialized functions for the total organism. Differentiation can, however, already be initiated within a population of dividing cells (see Sects. 3 and 5). As can be concluded from regeneration experiments in which specialized cells, after isolation, take up divisional functions again (see Sect. 3), the blockage of the cell cycle in situ must have been caused by cellular interactions. Central questions with respect to early differentiation, therefore, are: what causes the arrest of the cell cycle during differentiation, and which events in the cell cycle are affected? In many cases specialized cells have the G_1-content of DNA and are, therefore, arrested in the G_1-phase. It is, however, still a matter for discussion whether there are one or multiple control points in the G_1-phase at which is decided whether a cell continues or ceases proliferation (cf. HOCHHAUSER et al. 1981). Arrest of the cell cycle in the G_2-phase has also been reported (e.g. D'AMATO 1972, VAN'T HOF 1973). It has been shown that cell arrest in G_2 in root and shoot meristems in *Pisum* could be promoted by a factor from the cotyledons (EVANS and VAN'T HOF 1974). The biochemical nature of the restricting events is still unknown. VAN'T HOF and coworkers (WEBSTER and VAN'T HOF 1969, VAN'T HOF and ROST 1972) have advanced the hypothesis that control points represent peaks of energy requirement. By carbohydrate starvation of root meristems, which resulted in reduced protein synthesis, two principal control points of the cell cycle were found, one in the G_1- and one in the G_2-phase (cf. VAN'T HOF and KOVACS 1972, VAN'T HOF 1973). In synchronized plant cell cultures, peaks in succinate dehydrogenase activity and respiration rate coincide with G_1/S and G_2/mitosis transitions (KING et al. 1974).

In excised root tips of *Pisum sativum,* excess IAA in the culture medium results in cell arrest in G_1 and G_2 in a ratio similar to that obtained in the absence of sucrose (SCADENG and MACLEOD 1977). Selective inhibition of cell-cycle stages in root meristems of *Pisum* resp. *Allium* by IAA, kinetin and abscisic acid has been reported (VAN'T HOF 1968, NAGL 1972). In this context, recent results on differential synthesis of auxin and of cytokinins during the cell cycle are of great interest. These results could be achieved in work with synchronized cell cultures of tobacco (NISHINARI and YAMAKI 1976, NISHINARI and SYONO 1980a, b) and of English sycamore (HALL and ELLIOTT 1982, ROBINSON et al. 1982). Deprivation of auxin from the culture medium led to arrest in G_1, and later addition of auxin to partial synchronization of carrot cell culture (NISHI et al. 1977).

While, as a consequence of restriction, cells become "quiescent" with respect to their growth functions, the metabolic activities of the cell continue, at least partly, along new pathways controlled by the expression of hitherto inactive genes responsible for specialized functions. The diversion of cell activity from the G_1-phase to a phase which is not a part of the proliferation cycle, but a state in which the cell can perform specific biochemical functions, has been signified by the term G_0-state (LAJTHA 1963, cf. LAJTHA 1979, PARDEE et al. 1978, HOCHHAUSER et al. 1981).

As briefly mentioned above, the strongest evidence for the existence of cellular interactions during early differentiation is derived from numerous experiments in which specialized cells, either as single cells or as tissue or organ fragments, were isolated from the mother plant, or at least from the growth centres, by interruption of intercellular symplastic connections. In many cases, these isolated cells, or some of them, react to the interruption of previous intercellular relationships by activation of growth and division functions and then regenerating new multicellular plants (cf. BÜNNING 1955, STANGE 1964, DORE 1965, MÜLLER-STOLL 1965, see Sects. 3 and 4.2). Their behaviour does not only reveal their inherent totipotency, but also shows that the stability of the differentiated state in undisturbed development and the concomitant blockage of the cell cycle must be founded on intercellular exchange of information between the specialized cells and the growth centres via symplastic connections (cf. CARR 1976).

20.3 Re-Differentiation in Organ and Tissue Fragments

Regeneration experiments in general and especially work with cell, tissue and organ cultures open the possibility of studying effects of external factors, and of selected substances which are known to play a role in development, on the course of growth activation and of re-differentiation in the explants (cf. STREET 1977, EVERETT et al. 1978). Great success in controlling morphogenesis in proliferating explants has been achieved (RAGHAVAN 1976, REINERT and BAJAJ 1977, TRAN THANH VAN 1981). Regeneration experiments and work with cell and tissue culture have given indications of the nature of cellular interactions during differentiation in the developing multicellular plant. While the relevant aspects of in vitro differentiation of embryoids will be discussed in Section 4.2, observations on re-differentiation between similar mature cells in a tissue or organ fragment during activation of divisional functions will be considered here. The advantage of this type of experimental approach is the possibility of choosing defined and simplified multicellular systems composed of only few cells of a selected type in which to study their behaviour after isolation.

In many cases it can be observed that in organ fragments and larger tissue fragments some, but not all, cells take up divisional functions and regenerate. The regenerating cells are not restricted to the wound border of the fragment, as would be expected if only correlations between cells in direct physical contact were interrupted. Destruction of some symplastic connections by cutting a fragment results in disturbance of the previous organization of the isolated part, which had been incorporated in physical and chemical gradients and, probably, in directed solute transports within the original total plant.

In extensive research TRAN THANH VAN and coworkers have studied the organogenetic capacities of organ and tissue fragments of several higher plants (TRAN THANH VAN et al. 1974, TRAN THANH VAN 1981). The type of organ regenerated is strongly dependent on the auxin/cytokinin ratio in the culture medium, as confirmed in numerous cases since the report of SKOOG and MILLER (1957). In organ fragments composed of several tissues, de novo formation of shoot and root meristems is confined to certain tissues, but those tissues which do not regenerate within a complex fragment, can do so when isolated. Interestingly, while in stem segments of *Torenia fournieri* buds are formed in the epidermal cells, excised epidermis grown alone is incapable of proliferating, but can form buds when replaced on the subepidermal tissue either directly or after intercalation of a thin agar layer (CHLYAH 1974a).

The onset of cellular divisions and their propagation in the epidermal layer of stem segments of *Torenia fournieri* has been studied in great detail (CHLYAH 1974b). After isolation, an increase in the volume and staining capacity of the nucleus and nucleolus is observed in certain groups of cells within 24 to

36 h. One of the cells then starts to divide about 48 h after isolation. Soon after, division of some adjacent cells is observed and a "cell division centre" is formed, in which one of the cells divides more rapidly than the others; division activity then spreads out to neighbouring cells. The number of divisions in each original cell decreases with the distance from the central cell(s), thus exhibiting a radial gradient in cell-cycle duration. Among the numerous division centres initiated, a few form meristem primordia and only a reduced number of these become buds. These observations are interpreted by assuming competition among cell-division sites. The division centres are not distributed at random, but along a gradient of increasing density in the apical-basal direction and at a distance from underlying vascular tissue. A more uniform distribution of division centres can be obtained by addition of auxin to the medium. It is assumed that the endogenous auxin content is one of the factors controlling cell division and meristem distribution during bud formation (CHLYAH et al. 1975).

These studies have been extended in the same material to investigate the relationship between initiation of DNA synthesis and cell division in epidermal cells of stem segments. During the first 48 h after isolation, about 20% of the analyzed epidermal cells entered the S-phase of the cell cycle. These cells were not synchronized, and it is assumed that at the time of excision they were not in the same stage of G_1-phase. Most, but not all, of the cells that have synthesized DNA divide until the 5th day after excision, after which time cell division in the still undivided epidermal cells becomes very rare. Competition among the epidermal cells for nutrient or hormonal factors as the base for the different behaviour of the cells also with respect to DNA synthesis is assumed (CHLYAH 1978).

Similar re-differentiation processes during the formation of meristematic centres in tissue fragments excised from the unistratose thallus of the liverwort *Riella* have been described (STANGE 1957, 1964). After isolation, the first reaction observed is an increase in nucleolar size in all cells of a broad polar zone of the fragment. A parallel increase in rate of RNA synthesis has been demonstrated (SCHULZ 1971). The variance between the enlarging nucleoli in adjacent cells increases with time after isolation of the tissue fragment; only some of the cells synthesize DNA and divide. Together with adjacent cells they can form meristematic centres (STANGE and KLEINKAUF 1968). The number of dividing cells in the autotrophic tissue increases with increasing light intensity and with the duration of the light period (KARSTEN 1967). Following a blockage of DNA synthesis by treatment with 5-fluorodeoxyuridine, the nucleoli increase in size to a threefold value of those in untreated fragments, and increase in nucleolar size spreads out over the fragment (GROTHA and STANGE 1969); no difference in the rate of synthesis of any species of RNA could be detected (GROTHA 1973). These results suggest that DNA synthesis plays a central role in directing intercellular correlations during re-differentiation.

Normally, soon after onset of nuclear division some of the fragment cells with a specially large nucleolus, but still with the G_1 DNA content, grow out to form long unicellular rhizoids. Their number can be drastically increased by adding auxin to the culture medium (STANGE 1957, 1958), a phenomenon

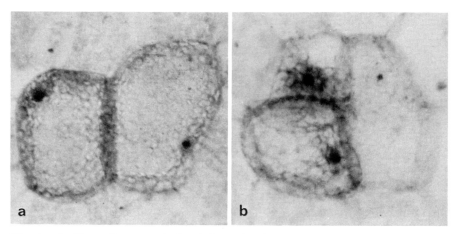

Fig. 1 a, b. Two-cell system isolated from the thallus of *Riella helicophylla* **a** 5 d after isolation, **b** 7 d after isolation. Staining with gallocyanine. (By courtesy of I. Eisenbeiser)

probably comparable to the formation of unicellular hairs in leaf fragments of *Begonia rex* under the influence of auxin (Tran Thanh Van 1974). The formation of rhizoids is suppressed and the number of dividing cells and meristematic centres is strongly increased when the culture medium contains PCIB (Grotha 1976). PCIB is a substance to which anti-auxin properties have been attributed (Burström 1950, McRae and Bonner 1953, Hertel et al. 1969); an increase in the activity of IAA-oxidase in the presence of PCIB has been reported (Frenkel and Haard 1973). The effect of PCIB on re-differentiation in thallus fragments of *Riella* has been interpreted as showing that, normally, auxin levels arising in the neighbourhood of auxin-producing meristematic centres delay and finally block the cell cycle in the respective cells (Grotha 1976).

The most simplified system in which to study cellular interactions during re-differentiation is a system consisting of only two cells. From the thallus of *Riella*, Lehmann (1966) isolated systems of two cells connected to each other by plasmodesmata. When two similar mature cells are isolated, the nucleoli in both cells enlarge (Fig. 1 a), but with a different rate (Eisenbeiser unpublished results), in most cases only one cell divides (Fig. 1 b) and regenerates a multicellular plant. If, however, the cell in which the nucleus had just divided is killed, the other cell divides within a very short time interval (Fig. 2), demonstrating that it had been arrested during its activation of divisional functions by the neighbouring cell. When single cells were compared with groups of two, six or more cells, the speed of regeneration was enhanced with increasing number of cells in the isolated group.

From these observations and experiments on meristematic activity in *Riella* (see Sect. 6.3), Stange (1977) has proposed a hypothesis on cellular interactions during differentiation of adjacent cells (Fig. 3). This hypothesis refers to results on discontinuous auxin synthesis during the cell cycle (see Sect. 2), and takes

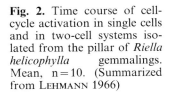
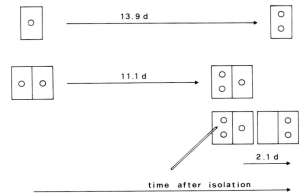

Fig. 2. Time course of cell-cycle activation in single cells and in two-cell systems isolated from the pillar of *Riella helicophylla* gemmalings. Mean, n = 10. (Summarized from LEHMANN 1966)

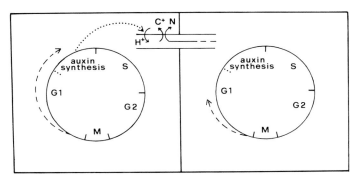

Fig. 3. Hypothesis on cellular interactions during differentiation of two adjacent cells. C^+ cation; N non-electrolyte. (Cf. STANGE 1977)

into consideration recent concepts of polar metabolite transport (cf. KOMOR and TANNER 1974, TANNER et al. 1977) and the role of auxin in membrane transport (cf. HAGER et al. 1971, 1980, RAY 1977, see Sect. 5.1). If one of the cells in a two-cell system progresses through the cycle slightly faster than the neighbouring cell, it enters the phase of auxin production earlier than the other cell. By its effect on the boundary membrane the auxin then induces unidirectional transport of substances from the slower cycling cell to the auxin-producing cell. This import speeds up the cycle in the receiving cell while the cycle in the exporting cell is slowed down.

The cycle will even be arrested as soon as the amount of substances exported equals the net production of the cell. During continuation of cell division, directed transport of substances required for growth could be perpetuated to a larger field of cells. The general idea is that the transformation of genetic information into a time sequence of metabolic events within the cell is used for communication between neighbouring cells. The importance of the temporal organization of individual cells, together with intercellular functional coupling, for the appearance of ordered heterogeneity in a tissue has been emphasized by GOODWIN and COHEN (1969).

20.4 Differentiation in Embryos and Embryoids

To study cellular interactions during differentiation, it is advantageous to start at the onset of development, when the number of cells in the system can still be comprehended, i.e. to focus on embryogenesis or comparable developmental phases, such as the formation of embryoids in tissue culture.

20.4.1 Embryos

Informative research on the biophysical, biochemical and cytological basis of cell polarity and on the processes associated with differentiation of two-celled embryos has been carried out with the free-floating eggs and zygotes of the *Fucales*. Methods and results of this work has been considered in Chapter 22, this Volume. The pattern of polar development and differentiation in embryos of spermatophytes is similar to that of the *Fucales* (cf. SINNOTT 1960, WARDLAW 1968, QUATRANO 1978), in spite of the fact that the polar axis of the egg or zygote is determined by influences coming from the mother plant.

A comprehensive survey of early embryogenesis in plants has documented that all plant embryos have essential features of early differentiation in common (WARDLAW 1965a, b, 1968, RAGHAVAN 1976). The most elementary step of differentiation consists in the determination of a polar axis, often established already in the egg cell and later in development expressed as different capacities of certain regions for meristematic activity (BÜNNING 1956). The division of the zygote is generally an unequal division, in which the cell plate is perpendicular to the long axis of the zygote. The smaller apical (terminal) cell and the larger basal (micropylar) cell are different, both structurally and chemically. These differences are critical to the future development of the embryo, which has been traced in detailed microscopical and ultrastructural analysis in a number of plants (e.g. SCHULZ and JENSEN 1968a, b, HASKELL and POSTLETH-WAIT 1971, NORSTOG 1972, SINGH and MOGENSEN 1975, NAGL and KÜHNER 1976). Differences between the cells with respect to stainability for RNA and protein, to the number and diversification of plastids, mitochondria and dictyosomes, the appearance of the endoplasmic reticulum and the occurrence of polysomes, starch, lipid bodies and vacuoles have been described. In most species investigated, the cells of the embryo, including the suspensor, have been shown to be connected by plasmodesmata, but there are no plasmodesmata in the walls separating the embryo from the embryo sac. The basal cells of the suspensor of most species studied so far develop wall ingrowths (e.g. SCHULZ and JENSEN 1969, NEWCOMB 1973, NEWCOMB and FOWKE 1974, NAGL 1976b, YEUNG and CLUTTER 1978) and can be interpreted, therefore, as transfer cells (PATE and GUNNING 1972).

As has been revealed by quantitative analyses with regard to cell number, cell size and distribution of cell divisions (POLLOCK and JENSEN 1964), and as can be deduced from many published pictures of early embryo development, at least in those species with an elaborated embryo suspensor, cells at the two poles of the embryo are clearly distinguished in early stages of embryo development by different rates of cell division and cell enlargement, the embryo proper being characterized by a higher division rate and strongly reduced cell size. Local increase in mitotic activity in the distal portion of the globular embryo, probably due to shorter cell-cycle duration, accounts for the formation of the cotyledon primordia. Unequal distribution of mitoses and cell elongation in connection with procambium formation accompanies the development to the torpedo stage. The factors which determine which cells divide most frequently are unknown. JENSEN (1976) points out that cell growth in the embryo is guided by an intercellular system of organization and control, and that subtle differences, such as the position of a cell in the embryo and its relationship to centres of hormone production, will determine its pattern of differentiation.

As to the differentiation of the embryo suspensor, directed intercellular metabolite transport within the embryo is strongly suggested. Suspensor cells in many species are characterized by a high degree of endopolyploidy or polyteny as a result of continued DNA synthesis in endo-cycles after cessation of cell division (NAGL 1974, 1978, 1976c), allowing these cells to carry out rapid RNA and protein synthesis. In *Phaseolus coccineus* the amount of chromosome endo-reduplication increases progressively from the cells adjoining the embryo proper to the giant micropylar cells (NAGL 1974). The rate of RNA synthesis in the embryo proper increases concurrently with the increase in cell number, while in the suspensor this rate is highest during the heart stage and then declines (WALBOT et al. 1972). Expressed per unit DNA, polytene suspensor cells are more efficient in synthesizing RNA than the diploid cells of the embryo proper (CLUTTER et al. 1974). These peculiarities, as well as ultrastructural details of suspensor cells (SCHULZ and JENSEN 1969, SCHNEPF and NAGL 1970, NEWCOMB 1973, NEWCOMB and FOWKE 1974, SIMONCIOLI 1974, NAGL 1976d), suggest an important nutritive role not only with respect to absorption and transport of nutrients from the surrounding tissue of the ovule to the developing embryo (cf. YEUNG 1980), but also with respect to active production and secretion of substances needed by the embryo proper.

This assumption has been confirmed by in vitro culture of young embryos with and without suspensor, completely isolated from the endosperm and the integuments (CORSI 1972, CIONINI et al. 1976, YEUNG and SUSSEX 1979). In heart-shaped embryos, removal of the suspensor reduces embryo development, this reduction being greater with younger embryos. Gibberellic acid (GA_3) concentrations of 10^{-8} to 10^{-6} M can replace the suspensor in heart-shaped and early cotyledonary embryos. In embryos of the heart stage the gibberellin-like activity in the suspensor was 30 times greater than in the embryo proper; a drastic decrease in the level of gibberellin-like substances occurs in suspensors of cotyledonary embryos, while the level in the embryo proper increases 10 times in comparison with the heart stage (ALPI et al. 1975). These observations suggest a transport of gibberellins from the suspensor to the embryo proper during defined developmental phases. The gibberellin present in the suspensor of the heart-shaped embryo of *Phaseolus coccineus* has been identified as gibberellin A_1 (ALPI et al. 1979). In a cell-free system of *Phaseolus coccineus* suspensors, the enzymatic potential for the biosynthesis of GA_5, GA_1 and GA_8 was revealed (CECCARELLI et al. 1981). There is also evidence for the suspensor playing an essential role in supplying specific forms of cytokinin to the growing heart-shaped embryo until it acquires autonomy for cytokinin synthesis during the stage of cotyledon development (LORENZI et al. 1978). Determination of IAA in early embryos of *Tropaeolum* revealed significantly higher amounts in the suspensor than in the embryo proper (PRZYBYLLOK and NAGL 1977). It is, however, unknown whether IAA is actually synthesized within the suspensor cells. When the embryo proper has enlarged enough to take over its nourishment by itself, autolysis of the suspensor starts at the basal end and progresses to the embryonal end, the lysed material being utilized by the embryo during the phase of maturation (NAGL 1976b).

Polar transport of auxin has been demonstrated in mature embryos of *Pinus lambertiana* as soon as they were excised from the seed and hydrated (GREENWOOD and GOLDSMITH 1970). Strongly polarized basipetal auxin transport has also been shown to occur in hypocotyls of embryos from unripe seeds of *Phaseolus vulgaris* and *Acer pseudoplatanus*. It is assumed that the early globular embryo is apolar and that the initiation of polarized transport of auxin towards the point of attachment of the suspensor determines the morphological polarity of the embryo. Procambial cells are probably responsible for at least part of the observed polar transport (FRY and WANGERMANN 1976).

It is of interest that BENNICI et al. (1976) report callus formation from the suspensor cells of *Phaseolus coccineus* embryos cultured in a medium free of growth regulators but rich in sucrose. This result reveals that suspensor cells are still capable of divisional functions, and that the differentiated state is dependent upon a precisely adjusted balance between nutritive and regulatory substances. As to an understanding of cellular interactions during differentiation in embryogenesis, central questions remain unanswered, e.g.: how is directed transport of nutrients and of growth regulators generated, and can it possibly be traced back to the first critical division of the zygote?

20.4.2 Embryoids

The formation of embryoids or "somatic embryos" in cell, tissue and organ culture is in many respects comparable to embryogenesis. The control of embryoid formation is very important for practical purposes such as breeding and propagation, and a vast literature exists on the specific culture conditions allowing embryoid development in a great number of plant species (cf. RAGHAVAN 1976, REINERT and BAJAJ 1977). In the study of fundamental problems of cell differentiation, wide use is made of cell and tissue cultures in a search for critical controls which can turn cultured cells from unorganized proliferation to the formation of specialized cells, tissues and organs (cf. STREET 1977, EVERETT et al. 1978). With respect to the operation of cellular interactions during differentiation of embryoids, studies of associated cellular events and the dependency of polarization on intrinsic conditions, which can be influenced by external control, are of interest.

Detailed information is available on cytological and histological changes during embryoid formation in a number of species (STEWARD et al. 1958a, b, STEWARD 1958, REINERT 1959, SUSSEX 1972, LANG and KOHLENBACH 1975, REUTHER 1977, cf. RAGHAVAN 1976) and on the ultrastructure of embryoids (HALPERIN and JENSEN 1967, KONAR et al. 1972, STREET and WITHERS 1974). Formation of embryoids can follow several ontogenetic patterns, which can correspond very closely to, or may deviate from, normal embryogeny from the zygote. Embryoids can be initiated from a single cell or from several cells, at the surface or from the interior of coherent tissues in culture. In many cases it has been shown that auxin is required in high concentrations to give rise to disorganized so-called embryogenic clumps or nodules (STREET and WITHERS 1974, KOHLENBACH 1977). Depending on the size of these clumps, they consist of small, densely cytoplasmic cells at the surface and large vacuolated cells in the centre, or they are entirely composed of densely cytoplasmic cells. If auxin is reduced or omitted in the subculture medium, bipolar embryoids can develop from single superficial cells of the embryogenic clumps.

In suspension and callus cultures of carrot, 2,4-D has proved to be the most effective auxin for the maintenance of active growth and cell division. In such cultures, embryoid formation is routinely promoted by transfer to a medium lacking 2,4-D, in which the rate of cell division accelerates, the cell size decreases and globular embryoids are formed. After some days of growth in auxin-free medium the establishment of a bipolar axis is revealed by a sharp distinction between meristematic cells and non-dividing cells in a separate "suspensor-like" region of the developing embryoid. The most striking change in cell ultrastructure during development of embryoids involves the appearance of microtubules, which are rarely observed in cells growing in the presence of auxin (HALPERIN and JENSEN 1967). When 2,4-D in the culture medium was replaced by 3,5-dichlorophenoxyacetic acid or PCIB, two aryloxyalkanecarboxylic acids to which anti-auxin properties have been attributed (FAWCETT et al. 1955, see Sect. 3), the number of embryos was significantly enhanced over those recorded during subculture in auxin-free medium (CHANDRA et al. 1978). In interpreting these results, differences in the action of native and synthetic auxins have to be taken into account (cf. LAM and STREET 1977). In callus cultures derived from unfertilized ovules of *Citrus sinensis* and independent in their growth of exogenous auxin and cytokinin, addition of IAA to the medium inhibited the formation of embryoids, while addition of inhibitors of auxin synthesis, 5-hydroxynitrobenzylbromide or 7-aza-indole, markedly stimulated their development (KOCHBA and SPIEGEL-ROY 1977).

These results suggest that a defined endogenous auxin level is one of the main factors controlling formation of embryoids. Probably, exogenous application of auxin inhibits the establishment of the required endogenous balance between auxin and other essential factors and obliterates a differential spatial distribution of auxin necessary for polarization.

20.5 Differentiation Within and at the Border of Apical Meristems

Plants continuously differentiate new organized tissues at their shoot and root apices which are easily accessible to experiment. Also, at the shoot apex, morphogenesis is initiated by differentiation of a pattern of meristematic centres and formation of leaf primordia. A vast literature exists dealing with descriptive and experimental investigations of shoot and root meristems and the onset of differentiation in the apical regions (cf. WARDLAW 1965c, d, 1968, CLOWES 1961a, 1972, 1976, TORREY 1965, 1972, 1976, LYNDON 1976).

20.5.1 Root Apical Meristems

It became clear from numerous studies of cell population kinetics, especially in root apices, that an apical meristem is not a homogenous population of dividing cells, but that considerable differences between cells already exist within a meristem, at least with respect to cell-cycle duration and cell size (cf. WEBSTER and MACLEOD 1980, GREEN 1976). The most striking example is the existence of the quiescent centre in root meristems (CLOWES 1954, 1956, 1961a), where the cells rarely or never divide, are distinguished by their smaller nucleoli and lower RNA content of the cytoplasm and have very prolonged cell cycle times compared to the other cells of the apex (e.g. CLOWES 1961b, 1975, TORREY 1972, PHILLIPS and TORREY 1972). It has been pointed out that the quiescent centre cell population itself is not homogenous, but shows an approximately radial gradient of decreasing cell cycle times outward from the centre (e.g. PHILLIPS and TORREY 1971). Other subpopulations within the root meristem can be distinguished with respect to cycle time, the shortest cycle time usually being found in the cap initials. Differences in proliferative activity are present in the stele, the cortex and the epidermis and along the axis of the root tip, and arise mainly by variation of the length of G_1-phase (CLOWES 1968, 1975, TORREY 1972, BARLOW and MACDONALD 1973).

There is experimental evidence that these differences in proliferation are controlled from within the meristem by cellular interactions. In primary roots of *Zea mays* the entire cap can be removed or the distal half of the cap can be cut off; in both cases the result is a stimulation of DNA synthesis, mitosis and cell division in the quiescent centre, which regenerates a new set of cap initials, eventually producing a new cap, and then reverts to mitotic quiescence (CLOWES 1972, BARLOW 1974). Apparently, different cell types interact with one another with regard to cell cycle control. It has been assumed that there is competition between the regions of the meristem for some substance essential for cell growth or division (CLOWES 1972).

For growth regulators known to be required for cell growth and division (cf. PÄTAU et al. 1957), much evidence indicates that auxins are transported within roots from root base to root apex and that the transport is polarized (cf. TORREY 1976, JACOBS 1977). In the root tip of *Zea mays* the amount of cytokinin is higher in the quiescent centre and the proximal meristem region than in the root cap; synthesis of cytokinins in the root tip is strongly suggested (TORREY 1976). Taking into account these results, BARLOW (1976, cf. TORREY 1972) has proposed a hypothesis about the behaviour of root meristems: The quiescent centre may be the source of a gradient of a substance (perhaps cytokinin) which, as long as its concentration is at an appropriate level, triggers mitosis and cytokinesis. A second gradient is established by a substance (perhaps auxin) which moves towards the quiescent centre from the maturing cells of the root apex; its concentration is appropriate to maintain nuclear DNA synthesis in cells of the meristem. The relative concentrations of the two substances may regulate the rates of mitosis, cell growth and the switch from mitosis to endomitosis. Supply to the root apex of carbohydrates synthesized in the shoot and essential for cell-cycle progression would contribute to modification of cell behaviour (VAN'T HOF 1973, TORREY 1972). Thus, the behaviour of each cell would be a function of its position within the apex, "positional information" being available from the quiescent centre as a "boundary region" (cf. WOLPERT 1969, 1971, BARLOW 1976).

A detailed description of early differentiation in the diminutive root of the water fern *Azolla* is the result of extensive microscopical and ultrastructural analysis carried out by GUNNING and coworkers (GUNNING et al. 1978, cf. GUNNING 1982). This system has the advantage that it provides a record of its own development in the form of a sequence of merophytes that reflects the past meristematic activity of the apical cell and the way in which the other types of cell division occur in time and space. The descriptive term "formative division" has been used to refer to divisions that give rise in the most apical zone of the root to the initial cells of cell files with a different destiny, and the term "proliferative division" to refer to divisions that increase the number of cells per file. The orderly progression through sequences of divisions with determinate microtubuli and cell-wall formation is interpreted as demonstrating an ability of cells in the *Azolla* meristem to sense the developmental stage reached by their older and younger neighbours, and to programme or be programmed accordingly. Cell-cycle durations vary widely with cell type and position in the root apex of *Azolla*; variation over a tenfold range of durations is reported for certain proliferative division cycles. The root of *Azolla* has a limited lifespan. There is a progressive diminution of the frequency of plasmodesmata which are found in the transverse cell walls formed by the apical cell and by later-produced merophytes before the apical cell becomes inactive. Electrical coupling is strongly correlated with the number of plasmodesmata between the coupled cells (OVERALL and GUNNING 1982). These results suggest that symplastic connections are important in providing nutrients and substrates, or in some form of morphogenetic signalling required for the activity of the apical cell (GUNNING et al. 1978, GUNNING 1978). CARR (1976) has summarized the evidence on the significance of plasmodesmata as the structural basis of cellular interactions in the regulation of growth and development.

The question why cells stop dividing at the border of a meristem has not yet been answered. Experimental data have been interpreted to show that the end of cycling is preceded by a drop in the fraction of cells in the cycle ("growth fraction") (Clowes 1975), or by variation in cycle times (cf. Webster and Mac-Leod 1980) caused by gradients of unknown nature. Cell expansion or cell elongation occurs in meristematic cells in connection with their biosynthetic activity, and continues for some time when cell division has ceased at the border of a meristem (cf. Green 1976). It has been suggested that the degree of coordination between nuclear and cell growth regulates meristematic activity. If the rate of cell growth is too fast, possibly some factor required for mitosis cannot attain the critical concentration at which it is effective (Barlow 1977).

In this connection the two ways in which auxin can promote growth could be relevant: it causes loosening and extension of the cell wall following activation of a membrane-bound ATPase which initiates or stimulates a proton pump (Hager et al. 1971, cf. Cleland 1977, 1980), and it enhances synthesis of rRNA and mRNA (cf. Key 1969, Jacobsen 1977, Bevan and Northcote 1981). While the first effect can be sufficient to promote cell expansion, both effects, probably, are operating in the cycling cell. Differential sensitivity of these processes with respect to auxin concentration, together with their dependence on other regulating factors, would allow differential "answering" of neighbouring cells to minute changes of conditions. Detailed knowledge on the location and timing of auxin synthesis at the cellular level is very fragmentary (cf. Sect. 2). Emergence of polar auxin transport is, probably, a crucial component of the general phenomenon of polarization (cf. Bentrup Chap. 22, this Vol.; Van Steveninck 1976). Therefore, the mechanism of polar auxin transport (Rubery and Sheldrake 1974, Raven 1975, cf. Goldsmith 1977, Rubery 1980, 1981, Bentrup Chap. 22, this Vol.) has to be included into considerations on cellular interactions during early differentiation.

20.5.2 Shoot Apical Meristems

The shoot apex is less accessible to experimentation and analysis than the root apex; it has, however, been the subject of intense research on orderly pattern formation of growth centres (cf. Wardlaw 1968). For analysis at the cellular level, information is now available on the rates and planes of division in the shoot apex and how these change with time in relation to the processes of morphogenesis (cf. Lyndon 1973, 1976, 1982, Halperin 1978). Direct measurements of the rates of division in different regions of the shoot apex of a number of species revealed that the rate of division in the cells at the summit of the apex was about half or a third of that at the base of the apical dome, the region of leaf initiation; a faster rate of cell division at the site of initiation of procambial strands was also recorded (Denne 1966, Lyndon 1970, 1976). Estimates of the length of the phases of the cell cycle in each region of the shoot apical meristem of *Pisum* suggest that in the cells at the summit the whole interphase is slowed down, G_1- and G_2-phases, however, to a greater extent than S-phase (Lyndon 1973). The importance of variations in the directions of growth for changes in shape during initiation of primordia are stressed by observations on the planes of division at intervals throughout a single plastochron. Essentially no periclinal divisions could be found in the apical dome in the first half of the plastochron, whereas one-third of all mitotic figures would give periclinal divisions during the second half of the plastochron. At the point of initiation of the new primordia, local differences in division rate are also reported (LYNDON 1976).

Scanty information is available for young leaves regarding cessation of cell division at the border of the shoot apex. During its development the leaf primordium is at first dependent on a supply of metabolites from older leaves (cf. DALE 1976, MOORBY 1977). In cucumber plants, during this period of import the rate of cell division in the developing leaf could be increased by exposure of the plants to higher light intensity (WILSON 1966). Cessation of cell division coincides with the transition of the young leaf from an importing organ to one exporting photosynthetic products (cf. DALE 1976).

The demonstration of morphogenetic independence of shoot apical meristems excised and grown in vitro has contributed crucially to the concept of an autonomous meristem capable of self-organization (cf. WARDLAW 1968). The size of the excised apical region and the number of leaf primordia on it are of critical importance for continued growth and morphogenesis. In several herbaceous angiosperm species, excised meristem domes less than 0.1 mm tall developed to complete plants in a nutrient medium that contained IAA as the only growth regulator (SMITH and MURASHIGE 1970). In experiments with *Dianthus caryophyllus,* apical meristem dome explants developed at a much higher percentage into complete plants when the two growth regulators kinetin and IAA were added to the nutrient medium (SHABDE and MURASHIGE 1977). It is emphasized by these authors that the perpetuation of growth and organogenesis in shoot apices is clearly regulated by the same chemical mechanism as observed earlier in tobacco callus cultures by SKOOG and MILLER (1957). No exogenous growth substances were required if the explant consisted of the apical meristem together with two or more pairs of primordial leaves. This result suggests that the primordial and expanding leaves are the sources of the growth regulators required by the apical dome, and that subtle cellular interactions exist within the shoot apex.

20.5.3 Meristems in Lower Plants

Thallophytes and bryophytes, with their lower level of organization, usually have a simpler meristem structure than higher plants. The unistratose liverwort *Riella* offers the advantage of an easy analysis of meristematic activity and its spatial distribution. During early development, the one intercalary meristem of the gemma differentiates by subdivision into two lateral and, later, apical meristems. In these meristems a gradient of the duration of the cell cycle exists; the shortest cycles occur in the marginal region, and the polar differentiation between cycling and non-cycling cells is preceded by a prolongation of the cycle duration near the border of the meristem. This polarization of the meristem is abolished under the influence of the anti-auxin PCIB, the cycle in the marginal cells is delayed, and no subdivision of the original meristem occurs. Treatment with PCIB reduces cell expansion and starch degradation. During the light period of the culture, the cell cycle is reversibly arrested in the G_1-phase (STANGE 1977, 1979). In young prothallia of *Athyrium filix-femina,* an influence of PCIB on the rates of cell division and cell elongation has also been reported (BÄHRE 1976, 1977). From these results it has been assumed that auxin plays a critical role in the regulation of the cycle duration in adjacent cells within the meristem and in the transition from the cycling to the non-cycling conditions. Recently, it has been shown that transfer of young *Riella* plants to higher light intensity reduces the duration of the cell cycle in the two meristems, while in plants grown in lower light intensity, in parallel to prolonged cycle time, only one meristem develops from the intercalary zone of the gemma (STANGE 1982). These observations have contributed to the formulation of the hypothesis proposed in Fig. 3. In the liverwort *Marchantia* the presence of auxin (IAA) has been demonstrated (SCHNEIDER et al. 1967) and polar basipetal transport of IAA occurs in the midrib tissue (MARAVOLO 1976).

Also in mosses, the operation of cellular interactions during differentiation has been demonstrated. While whole protonemata of *Funaria hygrometrica* and other mosses spontaneously produce buds by a change of the ratio between the rate of cell division and the rate of cell enlargement in single cells of caulonema filaments, isolated caulonema filaments only do so when cytokinin is added to the culture medium. Caulonema differentiation precedes bud formation and is stimulated by auxin. Cytokinines increase the number of buds in protonemata, while simultaneous addition of auxin reduces or, in high concentrations, inhibits bud formation (cf. BOPP 1965, 1968, 1980).

20.6 Onset of Vascular Differentiation

The formation of tissues which specialize in solute transport takes place in close spatial relationship to the growing apical regions and is an expression of the integration of growth and differentiation in early development of a multicellular plant. The function of these tissues is a central component in division of labour (see Sect. 1). The structural properties of the terminal stages of vascular cells and the cellular continuity along pattern-forming files have contributed to make xylem and phloem cells among the most extensively studied cell types in research on cellular differentiation (cf. ROBERTS 1976, JACOBS 1979, SHININGER 1979). Earlier work on differentiation of vascular tissues, especially with respect to the phenomena of "homeogenetic induction" and pattern formation has been discussed by LANG (1965, 1966, 1973) and BLOCH (1965).

Vascular differentiation starts with formation of the procambium, which becomes recognizable at the base of apical meristems by the local increase of longitudinal cell division and cell elongation resulting in files of prosenchymatic cells. Procambium differentiation can be seen in the shoot apex at the base of young primordia at the same times as their lateral outgrowth. The young primordia are required for procambium formation. When they are removed, the differentiation of procambial cells is prevented (YOUNG 1954), and instead the respective meristematic cells develop vacuoles and become isodiametric parenchyma cells. With respect to procambium formation, the primordium could not be replaced by a local supply of auxin, although IAA inhibited parenchymatization of meristematic cells. YOUNG has assumed the existence in primordia of an unknown factor "desmin" being involved in procambium differentiation.

During initiation of primordia, procambium is reported to differentiate acropetally or basipetally (cf. SHININGER 1979, HALPERIN 1978). It has been argued that the cytologically observed "direction of differentiation" cannot be used as indication that vascular tissues determine leaf primordia or vice versa, because recognizable signs of differentiation depend on the maturity of the cells which itself varies along the file (SACHS 1981, cf. SHININGER 1979). This is one of the reasons that make experiments on primary differentiation of vascular tissue difficult. The discussion in this section, therefore, focusses on some results of recent experimental work on the induction of regenerative differentiation of vascular tissue in shoots after wounding, and on those conclusions which are relevant to the existence and nature of the cellular interactions involved.

Important progress in the analysis of cellular interactions responsible for differentiation of vascular tissue has been achieved in the experiments of JACOBS (1952), in which he showed that externally applied auxin could replace the effects of leaves and the terminal bud on regeneration of xylem strands around a wound (cf. Fig. 4b–d). IAA also substitutes for the "sieve-tube stimulus" normally produced in leaves (LA MOTTE and JACOBS 1963, THOMPSON and JACOBS 1966). These observations show that auxin is a limiting factor by which leaves control vascular differentiation. A number of contributions by SACHS to the experimental approach on the role of auxin in vascular differentiation are summarized in Fig. 4 (cf. SACHS 1981). By applying auxin directly to the parenchyma in the vicinity of a wound, an orientated differentiation of xylem is produced which follows the known polarity of auxin transport and is directed towards the existing vascular strands (SACHS 1968, 1969; Fig. 4e). The development of contact between new and preexisting strands is inhibited when the latter are connected with young leaves or loaded with auxin from an exogenous source. When the differentiation of vascular tissue in the basipetal direction is prevented, vessels are formed in the reverse direction (Fig. 4f). When two sources of auxin are placed one above the other, the xylem strand induced by the upper source is diverted away from the vicinity of the lower source (SACHS 1974; Fig. 4 h to be compared with Fig. 4g). This diversion even occurs if the lower source is applied 2 days after the upper one, showing that long-term presence of auxin is required. When a redirection of vascular differentiation is impossible, an auxin source induces a strand passing straight through another, even stronger, source (Fig. 4i). Under these conditions, the directional differentiation of the vessels cannot be explained by a gradient of auxin concentration, but as a response to polar transport of auxin through the differentiating cells (SACHS 1974).

The organization of discrete strands of procambium and vascular tissues shows that relations between cells along the axis of the strand are different from relations in other directions. Since vascular tissues are capable of regenera-

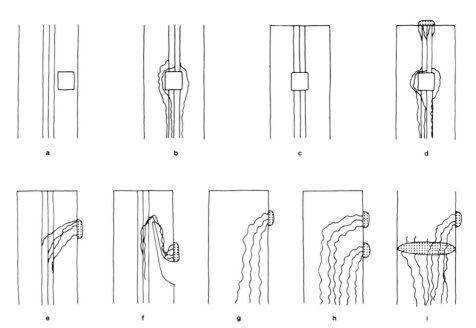

Fig. 4a–i. Vascular differentiation from parenchyma of wounded pea seedlings. All experiments were on stems in contact with cotyledons and roots. Half of each stem was removed by a longitudinal cut, exposing a wounded surface on which the experiments were performed. *Straight lines* existing vascular strands; *wavy lines* new vascular tissues; *dots* location of sources of auxin (1% IAA in lanolin). **a–d** Vascular regeneration around a wound **a** no damage to vascular tissue; **b–d** vascular tissue damaged and shoot tip present, removed or replaced by auxin, respectively. **e–i** Induction of vascular differentiation by auxin **e** auxin applied laterally; **f** differentiation directly towards the existing vascular tissue inhibited by a cut; **g** compared to **h**, **i** application of two auxin sources. Further explanation in the text. (Selected and redrawn from SACHS 1981, Fig. 9, with kind permission of the author)

tion, these relations must involve oriented interactions. SACHS (1978, 1981) points out that a flux of signals could become canalized in discrete channels of cell files when the process of differentiation itself increases the ability of the cells to transport the signals for differentiation (SACHS 1975). This would mean a control by a positive feedback: differentiation depending on a flux of signals and the flux depending on differentiation. This idea of a positive feedback is supported by and extended to the fact that leaf development both induces vascular differentiation and requires the import of nutrients and, possibly, growth regulators, which are transported through the vascular tissue from other parts of the plant. In a wider context, shoots produce auxin, which is transported basipetally and can control the initiation of root apices, which produce cytokinin necessary for shoot development. To this extent control of vascular differentiation is seen as part of a general positive feedback between shoot and root.

20.7 Concluding Remarks

In this chapter early differentiation at the cellular level has been discussed with regard to the spatial and temporal organization of the developing multicellular plant. The close relationship between growth and differentiation has been described in several cases as due to competition between cells. Here, order arises from a growth centre, a single cell or a group of cells, where cell division rate and, consequently, the rate of DNA synthesis are highest. This synthetic activity depends on the availability of building blocks being supplied by other cells arrested in their growth functions. Accordingly, competition between cells is based on sink-source relationships.

The concept of sink-source relationships is usually applied to the relationships between parts of an adult plant (cf. MOORBY 1977, WAREING 1979). It stresses the import and export characteristics of the component parts and the required connecting transport processes. The regulation of these relationships by "hormone-directed transport" has been pointed out (cf. PATRICK 1976, WAREING 1977, HEROLD 1980, ZERONI and HALL 1980), however, the way in which it is achieved is still unresolved, stimulation of sink activity or direct action of growth regulators on the transport system both being discussed as possibilities.

Sink-source relationships must be established during development and probably become realized during the onset of development between adjacent cells. Some results presented in this chapter suggest that hormone-directed transport might play a role during early differentiation. However, to substantiate this idea, much more information is required on the subtle mechanisms underlying early directive events in differentiation, such as the emergence of cell asymmetry, the determination of planes of cell division and of axis of cell elongation, the polarization of cells for auxin transport, and the spatial separation of capacities for growth-regulator synthesis.

Acknowledgement. The author is indebted to Dr. Mark Stitt for critical reading of the manuscript and linguistic help.

References

Alpi A, Lorenzi R, Cionini PG, Bennici A, D'Amato F (1979) Identification of gibberellin A$_1$ in the embryo suspensor of *Phaseolus coccineus*. Planta 147:225–228

Alpi A, Tognoni F, D'Amato F (1975) Growth regulator levels in embryo and suspensor of *Phaseolus coccineus* at two stages of development. Planta 127:153–162

Bähre R (1976) Zur Regulation des Protonemawachstums von *Athyrium filix-femina* (L.) Roth. III. Wirkung von Auxin (IAA) und Antiauxin (PCIB) unter verschiedenen Lichtbedingungen. Z Pflanzenphysiol 77:323–335

Bähre R (1977) Zur Regulation des Protonemawachstums von *Athyrium filix-femina* (L.) Roth. IV. Wirkung cholinerger Substanzen in Gegenwart eines Antiauxins (PCIB). Z Pflanzenphysiol 81:278–282

Barlow PW (1974) Regeneration of the cap of primary roots of *Zea mays*. New Phytol 73:937–954

Barlow PW (1976) Towards an understanding of the behaviour of root meristems. J Theoret Biol 57:433–451

Barlow PW (1977) An experimental study of cell and nuclear growth and their relation to cell diversification within a plant tissue. Differentiation 8:153–157

Barlow PW, MacDonald PDM (1973) An analysis of the mitotic cell cycle in the root meristem of Zea mays. Proc R Soc Lond Ser B 183:385–398

Bennici A, Cionini PG, D'Amato F (1976) Callus formation from the suspensor of Phaseolus coccineus in hormone-free medium: a cytological and DNA cytophotometric study. Protoplasma 89:251–261

Bevan M, Northcote DH (1981) Some rapid effects of synthetic auxins on mRNA levels in cultured plant cells. Planta 152:32–35

Bloch R (1965) Histological foundations of differentiation and development in plants. In: Ruhland W (ed) Encyclopedia of plant physiology, Vol XV/1. Springer, Berlin Heidelberg New York, pp 146–188

Bopp M (1965) Entwicklungsphysiologie der Moose. In: Ruhland W (ed) Encyclopedia of plant physiology, Vol XV/1. Springer, Berlin Heidelberg New York, pp 802–843

Bopp M (1968) Control of differentiation in fern allies and bryophytes. Annu Rev Plant Physiol 19:361–380

Bopp M (1980) The hormonal regulation of morphogenesis in mosses. In: Skoog F (ed) Plant growth substances. Springer, Berlin Heidelberg New York, pp 351–361

Brachet J, Lang A (1965) The role of the nucleus and the nucleo-cytoplasmic interactions in morphogenesis. In: Ruhland W (ed) Encyclopedia of plant physiology, Vol XV/1. Springer, Berlin Heidelberg New York, pp 1–40

Bryant JA (1976) The cell cycle. In: Bryant JA (ed) Molecular aspects of gene expression in plants. Academic Press, London New York, pp 177–216

Bünning E (1953) Entwicklungs- und Bewegungsphysiologie der Pflanze, 3rd edn. Springer, Berlin Heidelberg New York

Bünning E (1955) Regenerationen bei Pflanzen. In: Büchner F, Letterer E, Roulet F (eds) Handbuch der allgemeinen Pathologie, Vol 6. Springer, Berlin Heidelberg New York, pp 383–404

Bünning E (1956) General processes of differentiation. In: Milthorpe FL (ed) The growth of leaves. Butterworths, London, pp 18–30

Bünning E (1965) Die Entstehung von Mustern in der Entwicklung von Pflanzen. In: Ruhland W (ed) Encyclopedia of plant physiology, Vol XV/1. Springer, Berlin Heidelberg New York, pp 383–408

Burström H (1950) Studies on growth and metabolism of roots. IV. Positive and negative auxin effects on cell elongation. Physiol Plant 3:277–292

Carr DJ (1976) Plasmodesmata in growth and development. In: Gunning BES, Robards AW (eds) Intercellular communication in plants: studies on plasmodesmata. Springer, Berlin Heidelberg New York, pp 243–289

Ceccarelli N, Lorenzi R, Alpi A (1981) Gibberellin biosynthesis in Phaseolus coccineus suspensor. Z Pflanzenphysiol 102:37–44

Chandra N, Lam TH, Street HE (1978) The effects of selected aryloxyalkanecarboxylic acids on the growth and embryogenesis of a suspension culture of carrot (Daucus carota L.). Z Pflanzenphysiol 86:55–60

Chlyah H (1974a) Inter-tissue correlations in organ fragments: organogenetic capacity of tissues excised from stem segments of Torenia fournieri Lind cultured separately in vivo. Plant Physiol 54:341–384

Chlyah H (1974b) Formation and propagation of cell division centers in the epidermal layer of internodal segments of Torenia fournieri grown in vitro. Simultaneous surface observations of all the epidermal cells. Can J Bot 52:867–872

Chlyah H (1978) Intercellular correlations: Relation between DNA synthesis and cell division in early stages of in vitro bud neoformation. Plant Physiol 62:482–485

Chlyah H, Tran Thanh Van M, Demarly Y (1975) Distribution pattern of cell division centers on the epidermis of stem segments of Torenia fournieri during de novo bud formation. Plant Physiol 56:28–33

Cionini C, Bennici A, Alpi A, D'Amato F (1976) Suspensor gibberellin and in vitro development of Phaseolus coccineus embryos. Planta 131:115–117

Cleland RE (1977) The control of cell enlargement. In: Integration of activity in the higher plant. Symp Soc Exp Biol 31:101–115

Cleland RE (1980) Auxin and H^+-excretion: the state of our knowledge. In: Skoog F (ed) Plant growth substances 1979. Springer, Berlin Heidelberg New York, pp 71–78

Clowes FAL (1954) The promeristem and the minimal constructional centre in grass root apices. New Phytol 53:108–116

Clowes FAL (1956) Nucleic acids in root apical meristems of *Zea*. New Phytol 55:29–34

Clowes FAL (1961a) Apical meristems. Blackwell, Oxford

Clowes FAL (1961b) Duration of the mitotic cycle in a meristem. J Exp Bot 12:283–293

Clowes FAL (1968) The DNA content of the cells of the quiescent centre and root cap of Zea mays. New Phytol 67:631–639

Clowes FAL (1972) The control of cell proliferation within root meristems. In: Miller MW, Kuehnert CC (eds) The dynamics of meristem cell populations. Plenum, New York London, pp 133–143

Clowes FAL (1975) The cessation of mitosis at the margins of a root meristem. New Phytol 74:263–271

Clowes FAL (1976) The root apex. In: Yeoman MM (ed) Cell division in higher plants. Academic Press, London New York, pp 254–284

Clutter M, Brady T, Walbot V, Sussex I (1974) Macromolecular synthesis during plant embryogeny. Cellular rates of RNA synthesis in diploid and polytene cells in bean embryos. J Cell Biol 63:1097–1102

Corsi G (1972) The suspensor of *Eruca sativa* Miller (Cruciferae) during embryogenesis in vitro. Giorn Bot Ital 106:41–54

Dale JE (1976) Cell division in leaves. In: Yeoman MM (ed) Cell division in higher plants. Academic Press, London New York, pp 315–345

D'Amato F (1972) Morphogenetic aspects of the development of meristems in seed embryo. In: Miller MW, Kuehnert CC (eds) The dynamics of meristem cell populations. Plenum, New York London, pp 149–163

Denne MP (1966) Morphological changes in the shoot apex of *Trifolium repens* L. I. Changes in the vegetative apex during the plastochron. N Z J Bot 4:300–314

Dore J (1965) Physiology of regeneration in cormophytes. In: Ruhland W (ed) Encyclopedia of plant physiology, Vol XV/2. Springer, Berlin Heidelberg New York, pp 1–91

Evans LS, Van't Hof J (1974) Promotion of cell arrest in G_2 in root and shoot meristems in *Pisum* by a factor from the cotyledons. Exp Cell Res 87:259–264

Everett NP, Wang TL, Street HE (1978) Hormone regulation of cell growth and development in vitro. In: Thorpe TA (ed) Frontiers of plant tissue culture 1978. Proc 4th Int Congr Plant Tissue Cell Cult. Univ of Calgary, Calgary, pp 307–316

Fawcett CH, Wain RL, Wightman F (1955) Studies on plant growth regulating substances. VIII. The growth-promoting activity of certain aryloxy- and arylthio-alkanecarboxylic acids. Ann Appl Biol 43:342–354

Frenkel C, Haard NF (1973) Initiation of ripening in bartlett pear with an antiauxin α(p-chlorophenoxy) isobutyric acid. Plant Physiol 52:380–384

Fry SC, Wangermann E (1976) Polar transport of auxin through embryos. New Phytol 77:313–317

Goldsmith MHM (1977) The polar transport of auxin. Annu Rev Plant Physiol 28:439–478

Goodwin B, Cohen MH (1969) A phase-shift model for the spatial and temporal organization of developing systems. J Theoret Biol 25:49–107

Gould AR (1977) Temperature response of the cell cycle of *Haplopappus gracilis* in suspension culture and its significance to the G_1 transition probability model. Planta 137:29–36

Gould AR, Bayliss MW, Street HE (1974) Studies on the growth in culture of plant cells. XVII. Analysis of the cell cycle of asynchronously dividing *Acer pseudoplatanus* L. Cells in suspension culture. J Exp Bot 25:468–478

Green PB (1976) Growth and cell pattern formation on an axis: critique of concepts, terminology, and modes of study. Bot Gaz 137:187–202

Greenwood MS, Goldsmith MHM (1970) Polar transport and accumulation of indole-3-acetic acid during root regeneration by *Pinus lambertiana* embryos. Planta 95:297–313

Grotha R (1973) Über die Wirkung von FdUrd auf den Nukleinsäure-Stoffwechsel von Zellen isolierter Gewebefragmente aus dem Thallus von *Riella helicophylla*. Planta 115:147–160

Grotha R (1976) Der Einfluß des Antiauxins p-Chlorphenoxyisobuttersäure auf die Bildung meristematischer Zentren bei der Regeneration isolierter Gewebefragmente von *Riella helicophylla* (Bory et Mont.) Mont. Planta 129:235–238

Grotha R, Stange L (1969) Ausmaß und räumliche Verteilung der nucleolären RNS-Synthese in Gewebefragmenten von *Riella* nach Blockierung der DNS-Synthese. Planta 86:324–333

Gunning BES (1976) Introduction to plasmodesmata. In: Gunning BES, Robards AW (eds) Intercellular communication in plants: studies on plasmodesmata. Springer, Berlin Heidelberg New York, pp 1–13

Gunning BES (1978) Age-related and origin-related control of the number of plasmodesmata in cell walls of developing *Azolla* roots. Planta 143:181–190

Gunning BES (1982) The root of the water fern *Azolla*: cellular basis of development and multiple roles for cortical microtubules. Symp Soc Dev Biol 40:379–421

Gunning BES, Robards AW (1976) Plasmodesmata: current knowledge and outstanding problems. In: Gunning BES, Robards AW (eds) Intercellular communication in plants: studies on plasmodesmata. Springer, Berlin Heidelberg New York, pp 297–311

Gunning BES, Hughes JE, Hardham AR (1978) Formative and proliferative cell divisions, cell differentiation, and developmental changes in the meristem of *Azolla* roots. Planta 143:121–144

Hager A, Menzel H, Kraus A (1971) Experiments and hypothesis concerning the primary action of auxin in elongation growth. Planta 100:47–75

Hager A, Frenzel R, Laible D (1980) ATP-dependent proton transport into vesicles of microsomal membranes of *Zea mays* coleoptiles. Z. Naturforsch 35c:783–793

Haken H (1980) Lines of developments of synergetics. In: Haken H (ed) Dynamics of synergetic systems. Springer, Berlin Heidelberg New York, pp 2–19

Hall JF, Elliott MC (1982) The regulation of growth of English sycamore cells. 11th Int Conf Plant Growth Substances, Aberystwyth Wales (abstracts), p 10

Halperin W (1978) Organogenesis at the shoot apex. Annu Rev Plant Physiol 29:239–262

Halperin W, Jensen WA (1967) Ultrastructural changes during growth and embryogenesis in carrot cell cultures. J Ultrastruct Res 18:428–443

Haskell DA, Postlethwait SN (1971) Structure and histogenesis of the embryo of *Acer saccharinum*. I. Embryo sac and proembryo. Am J Bot 58:595–603

Herold A (1980) Regulation of photosynthesis by sink activity – the missing link. New Phytol 86:131–144

Hertel R, Evans ML, Leopold AC, Sell AM (1969) The specificity of the auxin transport system. Planta 85:238–249

Heslop-Harrison J (1967) Differentiation. Annu Rev Plant Physiol 18:325–348

Hochhauser SJ, Stein JL, Stein GS (1981) Gene expression and cell cycle regulation. Int Rev Cytol 71:95–243

Howard A, Pelc SR (1951) Nuclear incorporation of P[32] as demonstrated by autoradiographs. J Exp Cell Res 2:178–187

Howard A, Pelc SR (1953) Synthesis of deoxyribonucleic acid in normal irradiated cells and its relation to chromosome breakage. Heredity 6 (Suppl):261–273

Jacobs WP (1952) The role of auxin in differentiation of xylem around a wound. Am J Bot 39:301–309

Jacobs WP (1977) Regulation of development by the differential polarity of various hormones as well as by effects of one hormone on the polarity of another. In: Schütte HR, Gross D (eds) Regulation of developmental processes in plants. Fischer, Jena, pp 361–380

Jacobs WP (1979) Plant hormones and plant development. Cambridge University Press, Cambridge London New York Melbourne

Jacobsen JV (1977) Regulation of ribonucleic acid metabolism by plant hormones. Annu Rev Plant Physiol 28:537–564

Jensen WA (1976) The role of cell division in angiosperm embryology. In: Yeoman

MM (ed) Cell division in higher plants. Academic Press, London New York, pp 391–405

John PCL (ed) (1981) The cell cycle. Cambridge University Press, Cambridge London New York Melbourne

Jones MGK (1976) The origin and development of plasmodesmata. In: Gunning BES, Robards AW (eds) Intercellular communication in plants: studies on plasmodesmata. Springer, Berlin Heidelberg New York, pp 81–105

Karkinen-Jääskeläinen M, Saxen L, Weiss L (eds) (1977) Cell interactions in differentiation. Academic Press, London New York

Karsten I (1967) Über den Einfluß des Lichts auf die Embryonalisierung differenzierter Zellen von *Riella helicophylla*. Z Pflanzenphysiol 56:305–324

Key JL (1969) Hormones and nucleic acid metabolism. Annu Rev Plant Physiol 20:449–474

King PJ, Cox BJ, Fowler MW, Street HE (1974) Metabolic events in synchronised cell cultures of *Acer pseudoplatanus* L. Planta 117:109–122

Kochba J, Spiegel-Roy P (1977) The effects of auxins, cytokinins and inhibitors on embryogenesis in habituated ovular callus of the "Shamouti" orange (*Citrus sinensis*). Z Pflanzenphysiol 81:283–288

Kohlenbach HW (1977) Regulation of embryogenesis in vitro. In: Schütte HR, Gross D (eds) Regulation of developmental processes in plants. Fischer, Jena, pp 236–251

Komor E, Tanner W (1974) The hexose-proton symport system of *Chlorella vulgaris*. Specificity, stoichiometry and energetics of sugar-induced proton uptake. Eur J Biochem 44:219–233

Konar RN, Thomas E, Street HE (1972) Origin and structure of embryoids arising from epidermal cells of the stem of *Ranunculus sceleratus* L. J Cell Sci 11:77–93

Lajtha LG (1963) On the concept of the cell cycle. J Cell Comp Physiol 62 (Suppl 1):143–145

Lajtha LG (1979) Stem cell concepts. Differentiation 14:23–34

Lam TH, Street HE (1977) The effects of selected aryloxyalkanecarboxylic acids on the growth and levels of soluble phenols in cultured cells of *Rosa damascena*. Z Pflanzenphysiol 84:121–128

La Motte CE, Jacobs WP (1963) A role of auxin in phloem regeneration in *Coleus* internodes. Dev Biol 8:80–98

Lang A (1965) Progressiveness and contagiousness in plant differentiation and development. In: Ruhland W (ed) Encyclopedia of plant physiology, Vol XV/1. Springer, Berlin Heidelberg New York, pp 409–423

Lang A (1966) Intercellular-regulation in plants. In: Locke M (ed) Major problems in developmental biology. Academic Press, London New York, pp 251–287

Lang A (1973) Inductive phenomena in plant development. In: Basic mechanisms in plant morphogenesis. Brookhaven Symp Biol 25:129–144

Lang H, Kohlenbach HW (1975) Morphogenese in Kulturen isolierter Mesophyllzellen von *Macleaya cordata*. In: Mohan Ram HY, Shah JJ, Shah CK (eds) Form, structure and function of plants. Prof BM Johri, Commemoration Volume, Sarita Prakashan Meerut/Indien, pp 125–133

Lehmann H (1966) Über die Regenerationsleistungen isolierter Einzelzellen und kleiner Zellverbände von *Riella helicophylla*. Planta 71:240–256

Lorenzi R, Bennici A, Cionini PG, Alpi A, D'Amato F (1978) Embryo-suspensor relations in *Phaseolus coccineus*: Cytokinins during seed development. Planta 143:59–62

Lyndon RF (1970) Rates of cell division in the shoot apical meristem of *Pisum*. Ann Bot (Lond) 34:1–17

Lyndon RF (1973) The cell cycle in the shoot apex. In: Balls M, Billett FS (eds) The cell cycle in development and differentiation. Cambridge Univ Press, Cambridge London, pp 167–183

Lyndon RF (1976) The shoot apex. In: Yeoman MM (ed) Cell division in higher plants. Academic Press, London New York, pp 285–314

Lyndon RF (1982) Changes in polarity of growth during leaf initiation in the pea, *Pisum sativum* L. Ann Bot (Lond) 49:281–290

Mather K (1965) Genes and cytoplasm in development. In: Ruhland W (ed) Encyclopedia of plant physiology, Vol XV/1. Springer, Berlin Heidelberg New York, pp 41–67

McRae DH, Bonner J (1953) Chemical structure and antiauxin activity. Physiol Plant 6:485–510

Maravolo NC (1976) Polarity and localization of auxin movement in the hepatic *Marchantia polymorpha*. Am J Bot 63:526–531

Mitchison JM (1971) The biology of the cell cycle. Cambridge University Press, Cambridge London New York Melbourne

Mitchison JM (1973) Differentiation in the cell cycle. In: Balls M, Billett FS (eds) The cell cycle in development and differentiation. Cambridge University Press, Cambridge London New York Melbourne, pp 1–11

Moorby J (1977) Integration and regulation of translocation within the whole plant. In: Jennings DH (ed) Integration of activity in the higher plant. Symp Soc Exp Biol 31:425–454

Müller-Stoll WR (1965) Regeneration bei niederen Pflanzen (in physiologischer Betrachtung). In: Ruhland W (ed) Encyclopedia of plant physiology, Vol XV/2. Springer, Berlin Heidelberg New York, pp 92–155

Nagl W (1972) Selective inhibition of cell cycle stages in the *Allium* root meristem by colchicine and growth regulators. Am J Bot 59:346–359

Nagl W (1974) The *Phaseolus* suspensor and its polytene chromosomes. Z Pflanzenphysiol 73:1–44

Nagl W (1976a) Zellkern und Zellzyklen. Ulmer, Stuttgart

Nagl W (1976b) Ultrastructural and developmental aspects of autolysis in embryo-suspensors. Ber Dtsch Bot Ges 89:301–311

Nagl W (1976c) Early embryogenesis in *Tropaeolum majus* L.: evolution of DNA content and polyteny in the suspensor. Plant Sci Lett 7:1–8

Nagl W (1976d) Early embryogenesis in *Tropaeolum majus* L.: Ultrastructure of the embryo-suspensor. Biochem Physiol Pflanz 170:253–260

Nagl W (1978) Endopolyploidy and polyteny in differentiation and evolution. Elsevier/North-Holland, Amsterdam New York Oxford

Nagl W (1981) Polytene chromosomes of plants. Int Rev Cytol 73:21–53

Nagl W, Kühner S (1976) Early embryogenesis in *Tropaeolum majus* L.: Diversification of plastids. Planta 133:15–19

Newcomb W (1973) The development of the embryo sac of sunflower *Helianthus annuus* after fertilization. Can J Bot 51:879–890

Newcomb W, Fowke LC (1974) Stellaria media embryogenesis: The development and ultrastructure of the suspensor. Can J Bot 52:607–614

Nishi A, Kato K, Takahashi M, Yoshida R (1977) Partial synchronization of carrot cell culture by auxin deprivation. Physiol Plant 39:9–12

Nishinari N, Syono K (1980a) Changes in endogenous cytokinin levels in partially synchronized cultured tobacco cells. Plant Physiol 65:437–441

Nishinari N, Syono K (1980b) Identification of cytokinins associated with mitosis in synchronously cultured tobacco cells. Plant Cell Physiol 21:383–393

Nishinari N, Yamaki T (1976) Relationship between cell division and endogenous auxin in synchronously-cultured tobacco cells. Bot Mag (Tokyo) 89:73–81

Norstog K (1972) Early development of the barley embryo: Fine structure. Am J Bot 59:123–132

Overall RL, Gunning BES (1982) Intercellular communication in *Azolla* roots. II. Electrical coupling. Protoplasma 111:151–160

Pätau K, Das NK, Skoog F (1957) Induction of DNA synthesis by kinetin and indoleacetic acid in excised tobacco pith tissue. Physiol Plant 10:949–966

Pardee AB, Dubrow R, Hamlin JL, Kletzien RF (1978) Animal cell cycle. Annu Rev Biochem 47:715–750

Pate JS, Gunning BES (1972) Transfer cells. Annu Rev Plant Physiol 23:173–196

Patrick JW (1976) Hormone-directed transport of metabolites. In: Wardlaw IF, Passioura JB (eds) Transport and transfer processes in plants. Academic Press, London New York, pp 433–446

Phillips HL, Torrey JG (1971) The quiescent center in cultured roots of *Convolvulus arvensis* L. Am J Bot 58:665–671

Phillips HL, Torrey JG (1972) Duration of cell cycles in cultured roots of *Convolvulus*. Am J Bot 59:183–188

Pollock EG, Jensen WA (1964) Cell development during early embryogenesis in *Capsella* and *Gossypium*. Am J Bot 51:915–921

Prescott DM (1976) Reproduction of eucaryotic cells. Academic Press, London New York

Przybyllok T, Nagl W (1977) Auxin concentration in the embryo and suspensors of *Tropaeolum majus*, as determined by mass fragmentation (single ion detection). Z Pflanzenphysiol 84:463–465

Quatrano RS (1978) Development of cell polarity. Annu Rev Plant Physiol 29:487–510

Raghavan V (1976) Experimental embryogenesis in vascular plants. Academic Press, London New York

Raven JA (1975) Transport of indoleacetic acid in plant cells in relation to pH and electrical potential gradients, and its significance for polar IAA transport. New Phytol 74:163–172

Raven JA (1977) The evolution of vascular land plants in relation to supracellular transport processes. Adv Bot Res 5:153–219

Ray PM (1977) Auxin-binding sites of maize coleoptiles are localized on membranes of the endoplasmic reticulum. Plant Physiol 59:594–599

Reinert J (1959) Über die Kontrolle der Morphogenese und die Induktion von Adventivembryonen an Gewebekulturen aus Karotten. Planta 53:318–333

Reinert J, Bajaj YPS (eds) (1977) Applied and fundamental aspects of plant cell, tissue and organ culture. Springer, Berlin Heidelberg New York

Reuther G (1977) Embryoide Differenzierungsmuster im Kallus der Gattungen *Iris* und *Asparagus*. Ber Dtsch Bot Ges 90:417–438

Robards AW (1976) Plasmodesmata in higher plants. In: Gunning BES, Robards AW (eds) Intercellular communication in plants: studies on plasmodesmata. Springer, Berlin Heidelberg New York, pp 15–57

Roberts LW (1976) Cytodifferentiation in plants. Cambridge Univ Press, Cambridge

Robinson GM, Hall JF, Moloney MM, Barker RDJ, Elliott MC (1982) Regulation of mitosis in English sycamore cells. 11th Int Conf Plant Growth Substances, Aberystwyth Wales (abstracts), p 73

Rubery PH (1980) The mechanism of transmembrane auxin transport and its relation to the chemiosmotic hypothesis of the polar transport of auxin. In: Skoog F (ed) Plant growth substances 1979. Springer, Berlin Heidelberg New York, pp 50–60

Rubery PH (1981) Auxin receptors. Annu Rev Plant Physiol 32:569–596

Rubery PH, Sheldrake AR (1974) Carrier-mediated auxin transport. Planta 118:101–121

Sachs T (1968) On the determination of the pattern of vascular tissues. Ann Bot (Lond) 32:781–790

Sachs T (1969) Polarity and the induction of organized vascular tissues. Ann Bot (Lond) 33:263–275

Sachs T (1974) The induction of vessel differentiation by auxin. In: Proc 8th Int Conf Plant Growth Substances. Hirokawa, Tokyo, pp 900–906

Sachs T (1975) The induction of transport channels by auxin. Planta 127:201–206

Sachs T (1978) Patterned differentiation in plants. Differentiation 11:65–73

Sachs T (1981) The control of the patterned differentiation of vascular tissues. In: Woolhouse HW (ed) Advances in botanical research, Vol 9, Academic Press, London New York, pp 151–262

Scadeng DWF, Macleod RD (1977) The effect of indol-3yl-acetic acid concentration on cell arrest in interphase in the apical meristem of excised roots of *Pisum sativum* L. Cytobiologie 15:49–57

Schneider MJ, Troxler RF, Voth PD (1967) Occurrence of indoleacetic acid in the bryophytes. Bot Gaz 128:174–179

Schnepf E, Nagl W (1970) Über einige Strukturbesonderheiten der Suspensorzellen von *Phaseolus vulgaris*. Protoplasma 69:133–143

Schulz P, Jensen WA (1969) *Capsella* embryogenesis: The suspensor and the basal cell. Protoplasma 67:139–163

Schulz R (1971) Untersuchungen über den Nukleinsäure-Stoffwechsel während der Entdifferenzierung von Zellen aus dem Thallus von *Riella helicophylla*. Z Pflanzenphysiol 64:335–349

Schulz R, Jensen WA (1968a) *Capsella* embryogenesis: The egg, zygote and young embryo. Am J Bot 55:807–819

Schulz R, Jensen WA (1968b) *Capsella* embryogenesis: The early embryo. J Ultrastruct Res 22:376–392

Shabde M, Murashige T (1977) Hormonal requirements of excised *Dianthus caryophyllus* L. shoot apical meristem in vitro. Am J Bot 64:443–448

Shininger TL (1979) The control of vascular development. Annu Rev Plant Physiol 30:313–337

Simoncioli C (1974) Ultrastructural characteristics of *Diplotaxis erucoides* (L.) D.C. suspensor. Giorn Bot Ital 108:175–189

Singh AP, Mogensen HL (1975) Fine structure of the zygote and early embryo in *Quercus gambelii*. Am J Bot 62:105–115

Sinnott EW (1960) Plant morphogenesis. McGraw Hill, New York

Skoog F, Miller CO (1957) Chemical regulation of growth and organ formation in plant tissues cultured in vitro. Symp Soc Exp Biol 11:118–131

Smith RH, Murashige T (1970) In vitro development of isolated shoot apical meristems of angiosperms. Am J Bot 57:562–568

Stange L (1957) Untersuchungen über Umstimmungs- und Differenzierungsvorgänge in regenerierenden Zellen des Lebermooses *Riella*. Z Bot 45:197–244

Stange L (1958) Weitere Untersuchungen über den Einfluß von Indolylessigsäure auf die Vorgänge in regenerierenden Zellen des Lebermooses *Riella*. Z Bot 46:199–208

Stange L (1964) Regeneration in lower plants. In: Abercrombie M, Brachet J (eds) Advances in morphogenesis, Vol 4. Academic Press, London New York, pp 111–153

Stange L (1965) Plant cell differentiation. Annu Rev Plant Physiol 16:119–140

Stange L (1977) Meristem differentiation in *Riella helicophylla* (Bory et Mont.) Mont. under the influence of auxin or antiauxin. Planta 135:289–295

Stange L (1979) Reversible blockage of the cell cycle in the meristem of *Riella helicophylla* (Bory et Mont.) Mont. by p-chlorophenoxyisobutyric acid (PCIB). Planta 145:347–350

Stange L (1982) Influence of light intensity on meristem structure and activity in *Riella helicophylla* (Bory et Mont.) Mont. J Hattori Bot Lab 53:249–254

Stange L, Kleinkauf H (1968) Der Zeitpunkt der DNS-Synthese bei der Embryonalisierung von Zellen des Lebermooses *Riella* in bezug auf RNS-Synthese und Kernteilung. Planta 80:280–287

Steveninck RFM van (1976) Effect of hormones and related substances on ion transport. In: Lüttge U, Pitman MG (eds) Transport in plants II part B tissues and organs. Encyclopedia of plant physiology, new ser, Vol 2. Springer, Berlin Heidelberg New York, pp 307–342

Steward FC (1958) Growth and organized development of cultured cells. III. Interpretations of the growth from free cell to carrot plant. Am J Bot 45:709–713

Steward FC, Mapes MO, Smith MS (1958a) Growth and organized development of cultured cells. I. Growth and division of freely suspended cells. Am J Bot 45:693–703

Steward FC, Mapes MO, Mears K (1958b) Growth and organized development of cultured cells. II. Organization in cultures grown from freely suspended cells. Am J Bot 45:705–708

Street HE (1977) Differentiation in cell and tissue cultures – regulation at the molecular level. In: Schütte HR, Gross D (eds) Regulation of developmental processes in plants. Fischer, Jena, pp 192–218

Street HE, Withers LA (1974) Anatomy of embryogenesis in culture. In: Street HE (ed) Tissue culture and plant science. Academic Press, London New York, pp 71–100

Sussex IM (1972) Somatic embryos in long-term carrot tissue cultures: Histology, cytology, and development. Phytomorphology 22:50–59

Tanner W, Komor E, Fenzl F, Decker M (1977) Sugar proton cotransport systems. In: Marré E, Ciferri O (eds) Regulation of cell membrane activities in plants. Elsevier/North-Holland Biomedical Press, Amsterdam New York, pp 79–90

Thompson NP, Jacobs WP (1966) Polarity of IAA effect on sieve tube and xylem regeneration in *Coleus* and tomato stems. Plant Physiol 41:673–682

Torrey JG (1965) Physiological bases of organization and development in the root. In: Ruhland W (ed) Encyclopedia of plant physiology, Vol XV/1. Springer, Berlin Heidelberg New York, pp 1256–1327

Torrey JG (1972) On the initiation of organization in the root apex. In: Miller MW, Kuehnert CC (eds) The dynamics of meristem cell populations. Plenum, New York London, pp 1–13

Torrey JG (1976) Root hormones and plant growth. Annu Rev Plant Physiol 27:435–459

Tran Thanh Van KM (1981) Control of morphogenesis in in vitro cultures. Annu Rev Plant Physiol 32:291–311

Tran Thanh Van KM, Chlyah H, Chlyah A (1974) Regulation of organogenesis in thin layers of epidermal and subepidermal cells. In: Street HE (ed) Tissue culture and plant science. Academic Press, London New York, pp 101–139

Van't Hof J (1968) The action of IAA and kinetin on the mitotic cycle of proliferative and stationary phase excised root meristems. Exp Cell Res 51:167–176

Van't Hof J (1973) The regulation of cell division in higher plants. In: Basic mechanism in plant morphogenesis. Brookhaven Symp Biol 25:152–165

Van't Hof J, Kovacs CJ (1972) Mitotic cycle regulation in the meristem of cultured roots: the principle control point hypothesis. In: Miller MW, Kuehnert CC (eds) The dynamics of meristem cell populations. Plenum, New York London, pp 15–32

Van't Hof J, Rost TL (1972) Cell proliferation in complex tissues: the control of the mitotic cycle of cell populations in the cultured root meristem of sunflower (*Helianthus*). Am J Bot 59:769–774

Walbot V, Brady T, Clutter M, Sussex IM (1972) Macromolecular synthesis during plant embryogeny: Rates of RNA synthesis in *Phaseolus coccineus* embryos and suspensors. Dev Biol 29:104–111

Wardlaw CW (1965a) General physiological problems of embryogenesis in plants. In: Ruhland W (ed) Encyclopedia of plant physiology, Vol XV/1. Springer, Berlin Heidelberg New York, pp 424–442

Wardlaw CW (1965b) Physiology of embryonic development in cormophytes. In: Ruhland W (ed) Encyclopedia of plant physiology, Vol XV/1. Springer, Berlin Heidelberg New York, pp 844–965

Wardlaw CW (1965c) The morphogenetic rôle of apical meristems: fundamental aspects (illustrated by means of the shoot apical meristem). In: Ruhland W (ed) Encyclopedia of plant physiology, Vol XV/1. Springer, Berlin Heidelberg New York, pp 443–451

Wardlaw CW (1965d) The organization of the shoot apex. In: Ruhland W (ed) Encyclopedia of plant physiology, Vol XV/1. Springer, Berlin Heidelberg New York, pp 966–1076

Wardlaw CW (1968) Morphogenesis in plants. Methuen, London

Wareing PF (1977) Growth substances and integration in the whole plant. In: Jennings DH (ed) Integration of activity in the higher plant. Symp Soc Exp Biol 31:337–365

Wareing PF (1979) Plant development and crop yield. In: Marcelle R, Clijsters H, Poucke M van (eds) Photosynthesis and plant development. Junk, The Hague, pp 1–17

Webster PL, Macleod RD (1980) Characteristics of root apical meristem cell population kinetics: A review of analyses and concepts. Environ Exp Bot 20:335–358

Webster PL, Van't Hof J (1969) Dependence on energy and aerobic metabolism of initiation of DNA synthesis and mitosis by G_1 and G_2 cells. Exp Cell Res 55:88–94

Whaley WG (1965) The interaction of genotype and environment in plant development. In: Ruhland W (ed) Encyclopedia of plant physiology, Vol XV/1. Springer, Berlin Heidelberg New York, pp 74–99

Wilson GL (1966) Studies on the expansion of the leaf surface. V. Cell division and expansion in a developing leaf as influenced by light and upper leaves. J Exp Bot 17:440–451

Wolpert L (1969) Positional information and the spatial pattern of cellular differentiation. J Theoret Biol 25:1–47
Wolpert L (1971) Positional information and pattern formation. Curr Topics Dev Biol 6:183–224
Yeoman MM, Aitchison PA (1976) Molecular events of the cell cycle: a preparation for division. In: Yeoman MM (ed) Cell division in higher plants. Academic Press, London New York, pp 111–133
Yeung EC (1980) Embryogeny of *Phaseolus*: the role of the suspensor. Z Pflanzenphysiol 96:17–28
Yeung EC, Clutter ME (1978) Embryogeny of *Phaseolus coccineus*: growth and micro-anatomy. Protoplasma 94:19–40
Yeung EC, Sussex IM (1979) Embryogeny of *Phaseolus coccineus*: the suspensor and the growth of the embryo-proper in vitro. Z Pflanzenphysiol 91:423–433
Young BS (1954) The effects of leaf primordia on differentiation in the stem. New Phytol 53:445–460
Zeroni M, Hall MA (1980) Molecular effects of hormone treatment on tissue. In: MacMillan J (ed) Hormonal regulation of development I. Molecular aspects of plant hormones. Encyclopedia of plant physiology new ser, Vol 9. Springer, Berlin Heidelberg New York, pp 511–586

21 Cellular Recognition Systems in Grafting

M. M. YEOMAN

21.1 Introduction

It is over 50 years since KOSTOFF (1928) first promulgated his theory of acquired immunity in plants. Subsequently in a series of papers (KOSTOFF 1929, 1930, 1931), he claimed that his experimental data were consistent with the conclusion that a higher plant may acquire immunity and that this may be induced by grafting. In a later publication WHITAKER and CHESTER (1933) repeated many of KOSTOFF's experiments and showed convincingly, using the "precipitin" technique, that there was no indication that as a result of grafting there was an acquired immunity of stock to scion or of scion to stock. However, these authors were careful to point out that they did not claim to have shown that antibody formation in plants was not possible. It is now generally accepted that antibodies as such are not formed in plants, but there are certain phenomena which occur in plants in which a recognition event appears to take place (HESLOP-HARRISON 1978, NOGGLE 1979) and this in turn suggests the presence of molecules, probably proteins, which possess a recognition function not dissimilar from the antigen–antibody reaction.

Some interactions between the cells of higher plants have been interpreted as a cell–cell recognition phenomenon. Two good examples are, first, the interaction between the pollen grain and the stigma surface, particularly in self-incompatibility (HESLOP-HARRISON et al. 1975), and, second, for which the evidence is less secure, graft formation, which concerns the cellular interactions which take place during the formation of a graft union.

21.2 Inter-Relationships Between the Wound Reaction and Graft Formation

The formation of any graft involves the assembly of two or more pieces of an organ or tissue. This provides an occasion in which the constituent cells of the donor and the recipient are brought into direct opposition. It is in the nature of grafting that, apart from approach grafts, the two interacting surfaces are initially open wounds caused by the explantation process, and therefore many of the early cellular events which lead to the formation of a successful or unsuccessful graft union are part of a general wound reaction (see LIPETZ 1970). However, graft formation cannot be explained solely in terms of a wound reaction (YEOMAN and BROWN 1976), and there are important basic differences between wounding and grafting. Even casual observation of the histological events which accompany the development of a successful graft and the events which take place at a wounded surface exposed to the air make this clear, particularly with respect to the differentiation of vascular elements. This is further supported by the results presented in YEOMAN et al. (1978), which show that some proteins are preferentially synthesized in response to wounding, but that when a graft is forming, additional proteins are also preferentially synthesized. It may be noted, however, that MOORE and WALKER (1981 a) have argued that incompatibility in grafting may be the result of a sustained wound reaction which generates enough scar tissue to provide a physical barrier between stock and scion, thus preventing the joining together of the vascular elements on either side of the union.

21.3 Evidence for Recognition from Investigations into Graft Compatibility and Graft Incompatibility

So far, the best evidence for the presence of a cellular recognition system in grafting in higher plants has come from studies of incompatible combinations (ROBERTS 1949, ROGERS and BEAKBANE 1957, YEOMAN et al. 1978, MOORE and WALKER 1981 a). However, a major difficulty in approaching the phenomenon of incompatibility in grafting lies in the definition of the term "incompatible". YEOMAN et al. (1978) have used a restricted definition in which they have placed emphasis on the union of the vascular elements of stock and scion as the critical event in the formation of a compatible graft. Subsequently such grafted plants survive, grow, flower and fruit under normal conditions. In contrast, incompatible combinations do not display vascular continuity and can only survive in a very humid environment for a limited period (YEOMAN and BROWN 1976). In addition, when the graft-breaking weight (ROBERTS and BROWN 1961) is plotted against time of development, the shape of the curve obtained is characteristic of whether the graft is compatible or incompatible. MOORE and WALKER (1981 a) have stated, "an 'incompatible' graft is not synonymous with an 'unsuccessful' graft", even compatible combinations may fail for a variety of reasons including, infection, desiccation, over-watering, bruising due to careless handling, or the mechanical separation of the surfaces of stock and scion. It can be, and has been, argued that anatomical differences between stock and scion, the presence of a virus, marked differences in growth rate leading eventually to mechanical problems, metabolic differences, and many other causes may result in an unsuccessful union, to which effects many have ascribed the term incompatibility. Also, in horticultural practice a graft may survive for many years and then break for mechanical reasons, and this is widely accepted as indicating incompatibility (GARNER 1970). Clearly, within the general use of the term graft incompatibility we must be more precise and use the term in a narrow or restricted sense (YEOMAN et al. 1978). In the studies in this laboratory within the family Solanaceae, it is the failure to achieve vascular continuity which appears to be the critical event determining whether a graft is compatible or not, and therefore it is logical to focus attention on the surfaces of regenerating cells at the interface of stock and scion, paying particular attention to the development of putative xylem and phloem initials.

A systematic approach to this problem demands that the histological observations are assembled on a precise developmental framework. Here the use of mechanical strength as a measure of graft development, first described by ROBERTS and BROWN (1961) and later refined by LINDSAY et al. (1974), has proved to be decisive. It is impossible within a large population of grafted plants to be certain of the stage of development attained by a particular individual by an examination of its external morphology. Graft strength provides a means, although partially destructive, of providing a developmental framework within which the precise order of the structural events at the interface may be ascertained. Without such a means interpretation becomes difficult and unreliable. Other workers (e.g. STODDARD and MCCULLY 1980, MOORE 1982) have now realized the usefulness and importance of this technique in establishing a temporal order of events.

21.3.1 Incompatibility and Graft Rejection

Studies on graft incompatibility inevitably tend to focus on the phenomenon of graft rejection. There is thus a danger that the striking histological and morphological events that accompany rejection are over-emphasized and thereby obscure the less spectacular events which precede rejection. If there is a primary recognition event, it must occur long before rejection takes place. An excellent parallel exists which emphasizes this point. A pollen grain which is incompatible with the stigma on which it alights may, depending on its nature, fail to germinate, or may produce a short pollen tube which quickly stops growing. This prevention of germination or the retardation of growth of the tube is accompanied in some species by a striking event, the rapid deposition of large amounts of callose, but this is the result of a less visually spectacular event which has taken place earlier between the interacting cell surfaces (HESLOP-HARRISON 1975). Thus the possibility exists that although the early cell interactions which mediate cell recognition in grafting may be similar throughout the higher plants, the events which are consequent on recognition and which accomplish rejection may appear to be quite different. Although the nature of these cell interactions is far from completely understood, the histological and morphological events which follow are now much clearer, but as might be expected, they differ between grafts. This is especially true of the process of graft rejection. MOORE and WALKER (1981a) have shown that the inability of a *Sedum telephoides* scion to graft onto a stock of *Solanum pennellii* is due to the formation of a substantial necrotic layer between the surfaces of stock and scion which physically prevents any fusion of the two opposing components. Here the metabolic differences between a C_4 and a C_3 plant could contribute towards the massive rejection. In complete contrast observations in this laboratory over a number of years with incompatible species within a single family, the Solanaceae, show no massive necrotic layer. A simple explanation of this difference may be that the more taxonomically distant the graft partners, the greater may be the rejection response, and conversely that the closer the relationship, the more successful the graft. The available evidence, although fragmentary, would support this view. KLOZ (1971) has shown that the ability of his so-called graft-hybrids to form an integrated unit, at least within the Leguminosae, can be used as an indicator of taxonomic similarity or diversity. In this investigation he showed that a variety of leguminous species could be grafted successfully onto *Phaseolus vulgaris* stocks, but that some species grew more rapidly than others. These observations, in conjunction with serological evidence, were interpreted as showing the taxonomic relationship of the scion to the stock. In a series of investigations in this laboratory we have shown that the rate of graft-take and the subsequent development of the mechanical strength of the graft union between compatible heterografts in the Solanaceae varies, and appears to be related to the taxonomic relationship of the stock and scion. FUJII and NITO (1972) have shown that callus raised from the cambia of a number of woody species fused together successfully only when the two species were compatible in a normal graft situation, and that, generally, the closer the taxonomic relationship, the stronger the union. From this it would appear that considerable differences in graft-take might be expected between diverse species. The subsequent histological pattern of development might be expected to reflect these differences. However, such differences as are observed could still be consistent with a single basic mechanism of cell recognition.

21.3.2 How May Incompatibility Be Overcome?

The successful combination of two incompatible individuals may be achieved in certain instances by the use of "inter-stock" which is compatible with both stock and scion. This principle is exploited in standard horticultural practice (GARNER 1970). Various explanations have been given to account for the success of this procedure. MOSSE (1962), who worked on graft incompatibility in fruit trees, has suggested that the inter-stock traps the compounds responsible for the incompatibility response, and refers to this as localized incompatibility. He also correlated this effect with grafts that have a line of callus cells with little or no differentiated elements at the graft interface. The nature of these incompatibility substances is not known. However, they must be compounds with a fairly high tissue specificity, and this would appear to eliminate normal metabolites and toxins. MOSSE (1962) also argues that in cases where incompatibility is not overcome by the use of an inter-stock this is due to "translocated incompatibility" and is correlated with cell necrosis at the graft interface.

The inability to graft could be due either to very high or very low levels of diffusible growth substances. For example the "localized incompatibility" of MOSSE (1962) could be due to a suboptimal level of auxin at the interface, and cell necrosis might be caused by a supra-optimal level of one of the growth hormones. Here the observations of SHIMOMURA and FUZIHARA (1977) are relevant. They have shown that the application of auxin or related growth regulators can assist graft-take in autografts of some cacti. In instances where the insertion of an inter-stock assists grafting, it could be that the presence of the bridge corrects for the deficiency or excess of a particular growth substance. An inter-stock could also provide a means of bridging the anatomical differences between stock and scion which it is alleged prevents or slows down graft formation.

In an attempt to resolve the role of cellular interactions and growth regulators in graft formation in both compatible and incompatible grafts, PARKINSON and YEOMAN (1982) have established a method in which excised internodes from a range of Solanaceous species can be grafted in culture. The successful grafts exhibit characteristics similar to those formed in whole plants of the same species with respect to the attainment of mechanical strength and vascular continuity. However, the process is slower, taking about twice the time. Auxin applied apically is always necessary, underlining the role of growth substances at some stage of graft formation. The incompatibility responses previously reported (YEOMAN and BROWN 1976) between *Nicandra physaloides* and *Lycopersicon esculentum* also occur in culture, although in a few instances – rather less than 10% of grafts – occasional vascular elements differentiate across the union. However, these are unable to translocate solutes between stock and scion. In these instances an inter-stock may be inserted between the incompatible stock and scion, resulting, as with the whole plant, in a successful graft union. The inter-stock can be a very thin slice of tissue (ca. 2 mm), and this would obviously reduce the filtering and source properties of the bridge. While it is not possible to eliminate filtering as a factor, the finding does make it more probable that the important function of the bridge is to provide a compatible surface. This emphasizes the importance of the surface and the activities which take place between opposing cells at the union. For example, it would seem unlikely that such a thin bridge could restore a balance of growth regulators between stock and scion, although it could still filter out "incompatibility factors".

Conversely an incompatible bridge of *Nicandra* can be used to separate a tomato stock and scion in culture. Here, as expected, a graft is not formed. No vascular continuity is observed apart from the occasional non-functioning connection. As with the whole plant graft, there is, of course, considerable vascular differentiation within 1 mm of the unions in both the tomato stock and scion, and in the *Nicandra* bridge tissue. Indeed the total number of wound vessel members (WVM's) is similar to that in a tomato bridge between a tomato stock and scion. The difference is that there are many functional connections in the tomato-tomato-tomato (TTT), and only an occasional non-functional one in the tomato-*Nicandra*-tomato (TNT). Even where connections exist in TNT there are only a few WVM's in each connection. Inversion of the tomato bridge between stock and scion (TLT) produces a situation almost indistinguishable from TNT. Vascular connections with a few WVM's are formed only occasionally and are non-functional. However, as with TTT and TNT the total number of WVM's formed within 1 mm of the tissue interfaces is similar. The only difference between TTT and TLT concerns the production of vascular strands induced in the stock which do not cross the interface. An explanation of the failure of an inverted bridge to produce a graft would appear to be related to polarity. Even a thin slice of tissue when inverted would presumably affect the transport of auxin from scion to stock, thus preventing the formation of trans-graft vascular strands. It should be noted that the total number of WVM's is not, however, affected by the inversion of the bridge, but only the formation of vascular connections. Experiments currently in progress in which IAA is added at the graft interfaces of the inverted bridge and in which TIBA is applied to prevent IAA transport in an upright bridge should help to resolve this problem. However, the possibility remains that what is being observed is an intrinsic polarity of the cell acquired at the previous division (YEOMAN and BROWN 1976). Again these observations place emphasis on the importance of the tissue interface and the interactions between cells from opposing sides of the graft union.

21.4 Cellular Events at the Stock–Scion Interface

21.4.1 Initial Adhesion of Opposing Pith Cells

There seems little doubt that initial adhesion between the stock and scion of compatible and incompatible species (YEOMAN and BROWN 1976, YEOMAN et al. 1978, MOORE and WALKER 1981a and b) is promoted by the production of pectinaceous material within the cells on opposing sides of the graft. These materials become incorporated into a common wall complex. YEOMAN et al. (1978) have described the complex as a sandwich of walls reinforced with a deposition of pectin. In the light microscope it appears as a dark line which increases in width as grafting proceeds. Sections of this complex when viewed with the electron microscope show clearly the sandwich and how it increases in width (Fig. 1). The increase in thickness can be measured from sections, and these data are presented in Fig. 2. From this it can be seen that there is an increase of approximately three times over a 4-day period. MOORE (1981) considers this complex to be due to the polymerisation of wall precursors secreted into the wall space, a common wound response in plants (AIST 1976).

21.4.1.1 Specificity of Cellular Interactions

The initial adhesion response appears to be non-specific, and occurs in compatible and non-compatible grafts (YEOMAN et al. 1978, MOORE 1981). Indeed MOORE and WALKER (1981a) have demonstrated that internodes of *Sedum telephoides* and *Solanum pennellii* will "graft" weakly to a piece of dead wood. This fairly weak "non-specific" adhesion between incompatible species within the Solanaceae and between such diverse types as *Cucumis sativus* and *Phaseolus coccineus* has been quantified in this laboratory using the breaking weight technique and shown to be of the order of 100–200 g. Some very recent observations (PARKINSON 1981, unpublished results) on explanted grafted internodes in culture using the technique described by PARKINSON and YEOMAN (1982) have shown that "grafts" can be constructed between *Lycopersicon esculentum* and

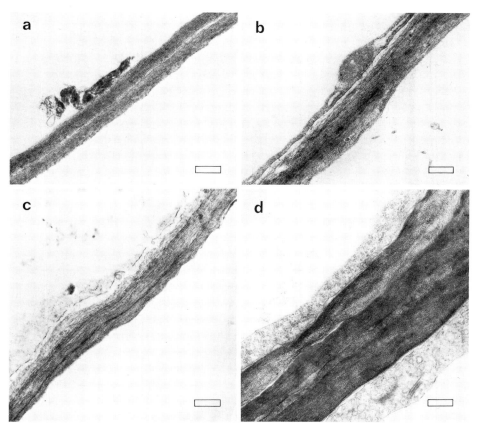

Fig. 1 a–d. Electron micrographs of vertical sections through the pith region of the graft union of tomato autografts, at intervals after assemblage of the stock and scion. **a** *At day one* a sandwich of two walls can be distinguished with a thin dividing zone, probably pectin. **b** *At day two* the thickness of the cell-wall complex (two walls) has increased. **c** *At day three* the sandwich is still increasing in thickness (probably two walls). **d** *At day four* the sandwich, which in this case is made up of four walls because of the infolding of the two collapsed walls of the cut cells (see YEOMAN et al. 1978), has increased considerably in thickness due to massive pectin deposition. Note the presence of Golgi bodies in close proximity to the wall complex. The magnification is the same in all four micrographs. The Bar 0.5 μm

Nicandra physaloides, an incompatible pair of species in which the mechanical strength can exceed 500 g without the formation of any vascular connections. This has been regarded as a non-specific ("superglue") effect. The non-specificity of this initial adhesion strongly suggests that cellular recognition is not involved at this stage.

21.4.1.2 Nature and Mechanism of Secretion of the Adhesive

The exact chemical nature of the adhesive is still a matter for speculation. However, there is general agreement (LINDSAY et al. 1974, YEOMAN and BROWN

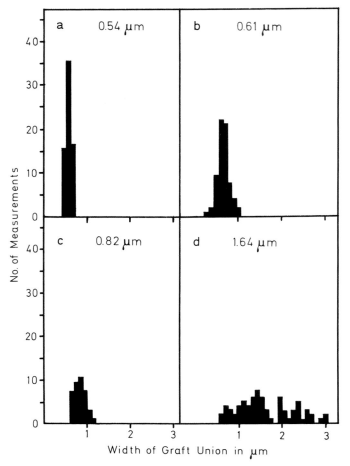

Fig. 2. A comparison of the thickness of the graft union of tomato autografts in the pith region at 1, 2, 3 and 4 days after graft assemblage. All measurements were made from electron micrographs at the junction of two pith cells and include two walls (see Fig. 1 a, b, c). Every effort was made to exclude oblique sections and only to include sections at right angles. Each "histogram" is made up of between 40 and 70 individual measurements. The spread of values at Day *4* is very large and reflects the difficulty of making accurate determinations at this advanced stage of development

1976, Yeoman et al. 1978, Moore 1981) that it is pectinaceous. Jeffree (1980 unpublished observations) has shown that during the early stages of graft formation the cell wall complex in the pith region gives a strong positive reaction for pectin when stained with hydroxylamine and ferric chloride (Reeve 1979). Also the observations by Moore and Walker (1981 b) and in this laboratory have shown a marked increase in the number of Golgi bodies in the cells adjacent to the graft union; this would support the view that polysaccharide, probably pectic, materials are being produced and deposited within the cell-wall sandwich.

21.4.1.3 Significance of the Attainment of Vascular Continuity

Although scions may survive for some time without the attainment of vascular continuity (MUZIK 1958, *Vanilla* orchid; BRADFORD and SITTON 1929, apples and pears) and even in some instances for several years (HERRERO 1951, fruit trees), it is normally a pre-requisite for survival. MOORE (1981) argues that these grafts are "unsuccessful" rather than "incompatible", suggesting that the separation of the stock and scion does not result directly from cellular and tissue interactions. However, the persistence of a region of callus through which the vascular elements fail to differentiate, enabling the joining together of the conducting systems of stock and scion, suggests the operation of a recognition process which determines cellular incompatibility (YEOMAN et al. 1978). Indeed later in the same article MOORE (1981) points out that "the differentiation of pro-cambium between the stock and scion requires some form of cellular communication" and he continues to invoke "some means of cellular communication and/or recognition between the two cells involved". Clearly a more precise morphological/histological framework is required before further progress can be made.

21.4.2 Proliferation at the Graft Interface and the Establishment of Cellular Contact

A reasonably precise developmental framework has been established in this laboratory using tomato autografts (JEFFREE and YEOMAN 1983), the essential features of which are summarized in Fig. 3. After initial cohesion is established between the pith regions of stock and scion, cell division is induced in the outer regions of the stem, with the greatest activity around the vascular bundles, in the vascular cambium and in the outer cortex (chlorenchyma), and epidermis. This activity is more pronounced in the scion than in the stock. Within 48 h the new cells formed enlarge markedly, and protrude into the space between stock and scion formed by the shrinkage of cells in the outer regions of the stem. These new cells, which form what may be described as a callus, grow towards one another and eventually make contact. The cells forming the initial bridge have extremely thin walls (ca. 300 nm) the outer surfaces of which are smooth and free from any projections (Fig. 4a). As the cells approach, nodules appear on the outer surfaces of the wall, mainly over the apical portion (Fig. 4b). These nodules appear to enlarge and become more flattened as they develop, and they probably consist largely of pectinaceous material. The coalescence of this pectinaceous layer forms the putative middle lamella. A particularly good illustration of how this may be achieved is seen in Fig. 4c, which shows a callus cell (arrowed) in a broken 4-day graft which presumably has been in contact with a similar cell from the stock (not shown). The area of contact with the cell from the stock can be clearly seen, since the nodules appear to have fused to form a continuous layer at the junction between the cell pair. Where the cells have not been in contact the nodules have retained their shape. DAVIES and LEWIS (1981) have reported pectic protuberances from callus cells

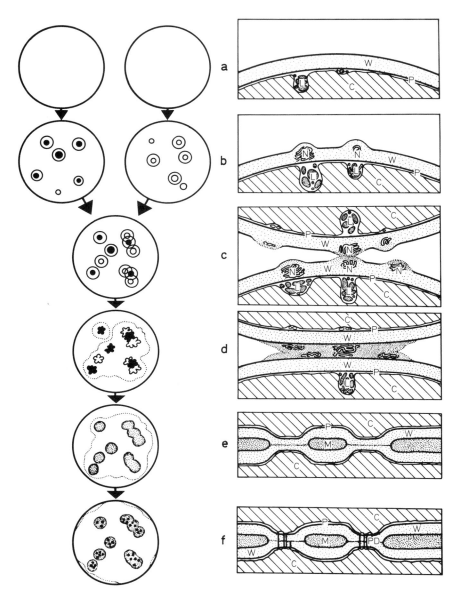

Fig. 3 a–f. A diagrammatic representation (*in surface view and in section*) of the events which take place when cells proliferating from the outer regions of the surfaces of stock and scion of a tomato autograft approach meet and begin to form a graft union (see Figs. 4 and 5). **a** The opposing cells approach one another; the walls are smooth and free of projections (Fig. 4a). Lomasomes are at this stage closely associated with the plasmalemma. **b** Before the cells touch, pectinaceous nodular structures begin to appear at their surfaces (Fig. 4b). These nodules, which can be seen in section in Fig. 4d, possess an internal membranous structure. **c** As the cells touch, the nodules become associated with the opposite cell wall or an opposing nodule. **d** The pressure generated by the growth of the opposing cells and the tissues beneath them spreads out the pectina-

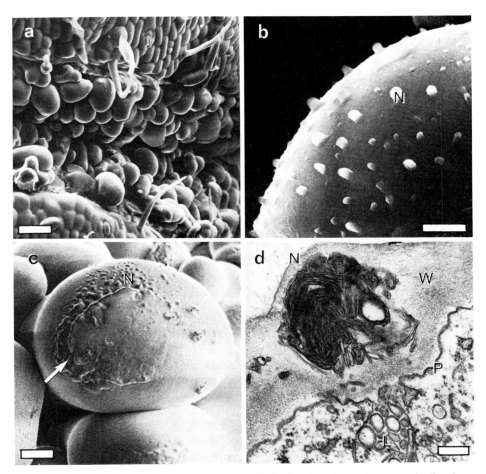

Fig. 4 a–d. Electron micrographs (SEM and TEM) showing the early stages in development of a tomato autograft. **a** A general view of the intact developing union in a 4 day graft, as seen from the outside (SEM cold stage). Bar 100 μm. **b** Callus cell in an intact 4-day graft (stock) showing pectinaceous nodules covering the distal portion of the cell surface (SEM cold stage). Bar 10 μm. **c** Callus cell in a broken 4-day graft (scion) showing flattened nodules forming a continuous sheet over the distal portion of the cell, due to contact with a similar cell from the surface of the stock (SEM cold stage). Bar 20 μm. **d** A section (TEM) through a nodular structure in the cell wall of a 6-day graft showing an internal membranous structure. A lomasome is situated directly below the nodule in the cytoplasm. Bar 0.2 μm

ceous nodules to form a putative middle lamella (Fig. 4c) and liberates the internal membranous structure (Fig. 4d). **e** It is proposed (see text) than an enzyme complex is associated with these membranous structures and removes the middle lamella in localized areas (Fig. 5a, b). **f** Subsequent to the removal of the pectinaceous material of the middle lamella the cellulosic walls are pushed together by the pressure which has built up within the cells, subsequently plasmodesmata form across the pair of cellulosic walls (Fig. 5a, b) of carrot and these formed bridges between closely packed cells. Similar structures were reported earlier by CARLQUIST (1956) in Compositae.

21.4.3 Erosion of the Middle Lamella

The pectinaceous layer (putative middle lamella) subsequently becomes eroded locally, bringing the cell walls on each side closer together, and presenting a picture of thick and thin walls (Fig. 5a, b). The thinning process appears to remove the pectic materials from the sandwich, leaving the original cellulosic wall intact. However, the wall bilayer, which now consists of two walls in intimate contact, appears to be reduced by up to half the original combined thickness of the two cell walls, probably as a consequence of the removal of pectins from the wall itself. At this stage a new phase of development begins, the formation of plasmodesmata within the regions of the wall which have been reduced in thickness.

21.4.4 Formation of Plasmodesmata

In Fig. 5a, b many plasmodesmata-like structures can be seen (arrowed) traversing the graft union in a 6-day-old tomato autograft. Some of these appear unbranched, at least in the plane of the section, while others are branched, and may follow a tortuous path with short deviations into the middle lamella. A compound plasmodesmum can be seen connecting the two cells across the graft interface in Fig. 5b. There is little evidence to suggest whether plasmodesmata arise simultaneously from both sides of the cell-wall complex, or commence at one side and proceed to the other. Some of the plasmodesmata (Fig. 5a) appear to penetrate only partially through the wall, although this could again be due to the plane of the section. However, the branched structures observed are consistent with the hypothetical scheme for secondary formation of plasmodesmata described by JONES (1976). It is clear that the frequency of plasmodesmata which are always within thinned areas of wall is very high. Recent unpublished observations by JEFFREE (1982) have shown that at no stage of graft development was there any evidence of attempts by cells to produce thinned areas of wall, plasmodesmata or holes when confronted with inert membranes whether porous or impermeable. In cells grown against Millipore membrane filters (Fig. 5d) the cell walls often penetrate the open mesh of the membrane structure, forming lobes and protuberances in intimate contact with the inert materials. Apart from these morphological abnormalities, and the overall general flattening of the cells, no unusual features were observed in the cell walls.

21.4.5 Development of Interconnecting Pits and the Differentiation of Cell Units

The formation of holes or pores between cells opposing one another across the graft union also occurs (Fig. 5c) but at a much lower frequency than thinned

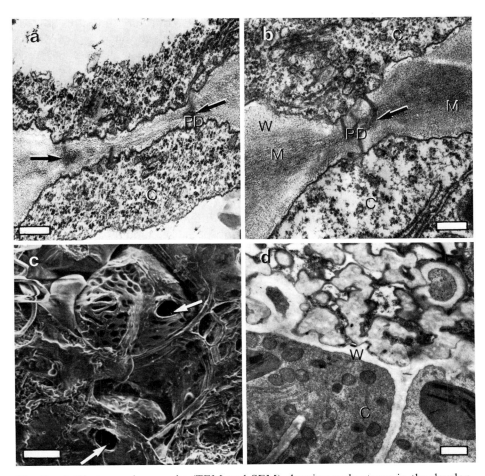

Fig. 5 a–d. Electron micrographs (TEM and SEM) showing early stages in the development of a tomato autograft. **a** Plasmodesmata-like structures traversing the graft union in a 6-day graft. Some of the plasmodesmata appear unbranched, while others are branched and may follow a tortuous path with short deviations into the middle lamella. The plasmodesmata are all within the thinned area. Bar 0.2 μm. **b** A compound plasmodesma can be seen connecting two cells across the graft interface. Bar 0.2 μm. **c** SEM of a freeze fractured 4-day graft showing wound vessel members (WVM's) exposed at the graft union. Note the circular apertures between cells (*arrowed*). Bar 30 μm. **d** TEM of a 5-day graft in which the stock and scion were separted at graft formation with a Millipore membrane filter (0.45 μm porosity). The cells in contact with the Millipore have become flattened and penetrated the surface. Pores and plasmodesmata are absent from the contact surfaces. *W* cell wall; *C* cytoplasm; *L* lomasome; *N* nodule; *P* plasmalemma; *PD* plasmodesmata; *M* middle lamella. Bar 10 μm

areas of wall with plasmodesmata. These holes are the equivalent of perforation plates which are commonly observed in the xylem of angiosperms (ESAU 1977, BENAYOUN et al. 1981). MOORE and WALKER (1981 b) have not observed perforation plates in sections across the graft union of autografted *Sedum telephoides*.

This is consistent with the low frequencies observed in our studies on Solanaceous species, and it is only by the use of scanning electron microscopy that these holes can be easily detected. From the recent studies of JEFFREE and YEOMAN (1982) it would appear probable that although perforation plates may be formed this only occurs to a limited extent and therefore the major channel of conduction of water and mineral salts must be across the pitted areas. This represents a slight modification of the view presented by YEOMAN et al. (1978), in which it was assumed that the pores were present at a much higher frequency and represented the major conduction channels.

It is clear that at present the information available is inadequate to attempt to interpret all the major events which occur between the initial contact of opposing cells and the final linkage of differentiated wound vessels. There is, for example, no direct way of determining whether a callus cell seen at an early stage of development is destined to differentiate into a wound vessel, or into a phloem sieve tube, or to remain unspecialized, and it cannot be determined solely from the examination of EM specimens whether wound vessels are involved, in their early stages, in a communication system across the graft union. Here the picture is incomplete and much remains to be done; however, it is probable that the formation of perforation plates or their equivalent is too infrequent and occurs too late in the developmental sequence to play a role in cell–cell recognition.

21.5 Cellular and Molecular Basis of a Recognition System in Grafting

The observation that the walls of opposing cells at the graft union develop thinned areas with plasmodesmatal connections and holes between them suggests the existence of some type of sensing mechanism, capable of mobilizing appropriate enzyme systems. Also, thinning of cell walls and the formation of plasmodesmata are not observed when a cell is grown up against an inert surface. The cells are apparently able to "sense" when they are in contact with another cell, despite the fact that the membranes of each cell are not exposed at the surface, do not come into contact with each other, and are separated by a pair of cell walls which are continuous at the time contact takes place.

21.5.1 Location of the Sensing Mechanism

The evidence available, although fragmentary, would support the view that some exchange of material takes place when the cells which proliferate from the opposite sides of the graft union come into physical contact. The exchange of material must be facilitated by the thinness of the confronting cell walls which provide only a minimal impedance to the movement of large molecules because of the low density and relatively open mesh of the primary walls. The short distance between adjacent plasmalemmas will also enable rapid transfer. Also because of the pressures generated as the cells grow in opposition and the plastic nature of these new walls, the cells become extensively flattened, providing a very large contact area through which ready exchange of materials can occur. However, the deposition of pectinaceous material into the cell-wall complex while binding the cells together might impede any intercellular exchange. Such an impedance will be short-lived in a compatible graft because the newly formed middle lamella is quickly modified to reinstate a line of communication. The local removal of the pectin brings the thin primary cellulosic walls, now low in pectin, into intimate contact and it is at this stage that the formation of plasmodesmata is initiated. Both the formation of thinned areas of the wall and the formation of plasmodesmata require the presence of an opposing cell, for neither of these events takes place when the cells are grown up against an inert barrier. From this it would appear that a response to contact with a surface mediated by the sensing of pressure at a point on the cell surface does not provide an explanation. The obvious alternative is that the response might be due to the sensing by one of the cells of the presence of a diffusible molecule produced by the other, or might be due to the exchange of such diffusible molecules. Such molecules might be small (sugars etc.) or large (e.g. proteins), but they must be such as to pass readily across the primary wall. The formation of holes between opposing cells does occur, albeit infrequently, and probably at a later developmental stage than plasmodesmata formation. As yet we are unsure of whether the formation of these holes or pores precedes or follows the destruction of the protoplast in differentiating wound vessel members. If it occurred while the protoplast was intact, which is unlikely, then direct contact between plasmalemmas through the hole would take place. However, this point remains to be resolved so that no firm conclusions can be drawn at this time. This leaves the following likely possibilities for the exchange of material and/or information in temporal sequence:

1. At the point of contact of the two opposing cells immediately before massive pectin deposition.
2. Across the thinned wall areas immediately after the local removal of pectinaceous material but before the formation of plasmodesmata.
3. Via plasmodesmata formed de novo.

The simple fact that erosion of the middle lamella and the formation of plasmodesmata follow as a result of the initial contact of the opposing cell surfaces strongly suggests that the sensing mechanism operates immediately cell contact is made, that it is an exchange, and that is located in the wall or in the plasmalemma. The subsequent exchanges that occur are likely to have been set in train by this initial contact.

21.5.2 Nature of the Sensing Mechanism

As has been suggested in an earlier paper (YEOMAN et al. 1978), a recognition molecule would presumably be a protein, a glycoprotein, a glycolipid or a glycopeptide. An obvious candidate for a recognition role in plants is a lectin, and this class of molecule has already been implicated in the recognition reaction that occurs between pollen and stigma (HESLOP-HARRISON et al. 1975). Some lectins (BOYD and SHAPLEIGH 1954) are superficially similar to vertebrate immunoglobulins in being di- or multivalent and able to recognize specific saccharides or saccharide sequences. The interaction of lectins with the surfaces of animal cells has been studied intensively (NICOLSON 1974) and it is well-known that they can bring about major changes in cell-surface architecture and initiate important cellular events. A protein with these properties is an obvious possible candidate for a recognition role in the formation of grafts in higher plants. The lectin could either be a soluble messenger which travels across the graft (YEOMAN and BROWN 1976) conveying information by binding to a glycoprotein receptor on the plasmalemma; or the lectin itself could be a component of the plasmalemma, able to recognize and specifically bind to a diffusible carbohydrate-containing messenger substance. So far attempts in this laboratory to link lectins with the graft process have failed. Immunofluorescence studies on the distribution of the *Datura* lectin have shown that the lectin is found closely associated with the cell and organelle membranes in all cells of the plant (KILPATRICK et al. 1979, JEFFREE and YEOMAN 1981) and is therefore in a suitable location to engage in recognition. However, lectin distribution at the graft union, observed by immunofluorescence, was found to be undifferentiated in respect to adjacent tissues, implying that no special increase or reduction of lectin concentration occurs in response to grafting (JEFFREE 1981, unpublished results).

Attempts to affect the development of autografts of *Datura stramonium* by the application of the lectin's saccharide inhibitor (chitobiose or chitotriose, see KILPATRICK and YEOMAN 1978) have not been successful, but this may be entirely due to the difficulty of maintaining the sugar within the graft union. The sugar may be rapidly removed by translocation and/or degradation, and experiments of this sort give inconclusive results. However, here the new techniques of producing grafts in culture (PARKINSON and YEOMAN 1982) and epidermal grafts either on intact plants or on cultured internodes (HOLDEN 1981, unpublished results) will provide a better approach. Clearly the involvement of a lectin in grafting compatibility is a possibility worthy of continued experimental investigation, especially in view of the results of KNOX et al. (1976), who have shown that glycoproteins (lectin receptors?) are present on the stigma surface of *Gladiolus gandavensis* and when these receptors were blocked by treating the stigma surface with concanavalin A, pollen tubes failed to penetrate the cuticle, preventing pollination.

However, there are now other candidates for the role of recognition molecules in grafting. Recently, ALBERSHEIM et al. (1980) have suggested that fragments of plant cell walls may act as informational molecules in phytoalexin induction. RYAN et al. (1981) have proposed that informational (hormonal) systems involving specific cell-wall fragments may be a regulatory feature of many plant genes, particularly those involved in natural plant protection. The erosion and subsequent fragmentation of the cell wall could provide a broad spectrum of chemical signals, effective either near to or at a distance from the site of wounding, which cause specific genes to respond to a variety of environmental stresses. The relevance of this discovery to the grafting situation is obvious, and it could prove important in future research strategy complementing or to replace the ideas concerning the involvement of lectins.

21.5.3 A Tentative Model

In the conclusions to this article I think it is justified and perhaps helpful to offer a tentative model which is consistent with the available facts on graft formation in higher plants. It has, like all models, distinct limitations, but it can be tested and hopefully improved as more facts become available. The sequence of events which culminates in the formation of a successful graft union begins with two wounded surfaces from which cells proliferate. In herbaceous plants many of the constituent tissues are stimulated to proliferate, while in woody plants the major sources of viable cells are the vascular cambium and the medullary rays. This has led to certain misunderstandings of the role of the vascular cambium in graft formation. In all instances however, it is callus cells which grow out from the wounded surfaces. These cells are at first smooth and then, before contact is made, their surfaces become covered with pectic nodules which have a complex ultrastructure (JEFFREE and YEOMAN 1983). It is presumed that these vesicular structures encapsulate membrane bound cell-wall-degrading enzymes kept apart from their substrates (see Fig. 4d). As the cells make contact and continue to expand the pectic material is spread out, forming a putative middle lamella and the packaged cell-wall-degrading enzymes are released but still localized within a membranous structure. It seems likely that eventually the pressure generated between the opposing cell surfaces will burst the nodules bringing together enzyme(s) and substrate(s), after the partial formation of the middle lamella. It is envisaged that contributions of enzyme(s) will be made from both surfaces but that in all probability each enzyme package will face a pectinaceous layer contributed by the other cell. It is at this point that massive erosion of the pectinaceous material will occur, producing the thinned areas which are characteristic of the graft union (compare with break-down of xylem vessel end walls, BENAYOUN et al. 1981). There is, however, an inconsistency requiring explanation, namely that if a cell is grown up against an inert surface, no obvious erosion of the deposited pectin is apparent. Here there are several possibilities, of which two offer an explanation consistent with the facts. The first is that the pressure generated when the nodule is pressed against an inert surface is insufficient to burst the vesicles and release the enzyme complex. This seems possible particularly for a regenerating surface grown against a Millipore filter where the cells penetrate the membrane structure (see Fig. 5d) but is less probable for grafts grown up against silicone rubber or polythene. A second, perhaps less satisfactory, explanation is that the nodules with vesicles do actually burst and the enzyme complex is released and begins to attack the substrate. At an early stage, however, the immediate products of the enzyme reaction(s) accumulate at the site of enzyme action because they cannot diffuse quickly through the inert surface and inhibit further degradation of the substrate. Certainly diffusion of the products of the reactions would be impeded through silicone rubber or polythene, but probably not through the Millipore membrane filter. It is, of course, possible that both explanations are valid and that the surface against which the cells are grown determines the course of events, e.g. the pectic nodules are pushed into the Millipore and fail to burst. It is envisaged that as a result of the action of the enzymes, wall fragments (oligosaccharides) will be released, and that it is these molecules which initiate (compatible) or do not initiate (incompatible) the formation of plasmodesmata. This is consistent with the proposal of Ryan et al. (1981) that these cell wall fragments can provide a broad spectrum of chemical signals which can initiate developmental sequences, perhaps by gene activation. Clearly the nature of the wall fragments will be determined by the composition of the wall and could prescribe specificity. It is likely that the "wall fragments" in the compatible graft will be recognized as self, irrespective of whether they originated from cells of the stock or scion and grafting will therefore proceed. In the incompatible combinations the "wall fragments" will be recognized as non-self by both sides and grafting will fail. In other words the recognition of self/self is by default, to use a computing analogy (i.e. the system reports nothing unusual) and recognition of self/non-self is by a positive response, the generation of an unfavourable complex of the receptor with recognition molecule. The formation of plasmodesmata completes intercellular communication and restores short-range signalling. The subsequent differentiation of these cells follows, depending on the position of the cells, availability of hormones etc. It is also probable that once compatibility has been established, the rate of proliferation will be slowed down, thereby reducing the possibility of a large callus mass between stock and scion which is frequently an indication of incompatibility (Yeoman and Brown 1976, Yeoman et al. 1978, Moore and Walker 1981 a).

Acknowledgements. The author wishes to express his thanks to Dr. C. E. Jeffree, Mark Holden and Mike Parkinson for their many frank and full discussions during the preparation of this article. I also wish to thank Mrs. E. Raeburn for typing the manuscript and Andrew Thompson and Bill Foster for technical and photographic assistance. I am also indebted to the Agricultural Research Council for a grant in support of this research.

References

Aist JR (1976) Papillae and related wound plugs of plant cells. Ann Rev Phytopathol 14:145–163

Albersheim P, Darvill AG, McNeil M, Valent BS, Hahn MG, Lyon G, Sharp AE, Desjardins MW, Spellman LM, Ross BK, Robertson BK, Åman P, Franzén L-E (1980) Structure and function of complex carbohydrates active in regulating plant–microbe interactions. Pure Appl Chem 53:79–88

Benayoun J, Catesson AM, Czaninski Y (1981) A cytochemical study of differentiation and breakdown of vessel end walls. Ann Bot 47:687–698

Boyd WC, Shapleigh E (1954) Specific precipitating activity of plant agglutinins (lectins). Science 119:419

Bradford FC, Sitton BG (1929) Defective graft unions in the apple and the pear. Tech Bull Agr Expt Stn Michigan State Coll 99:106

Carlquists S (1956) On the occurrence of inter-cellular pectic warts in Compositae. Am J Bot 43:425–429

Davies WP, Lewis BG (1981) Development of pectic projections on the surface of wound callus cells of *Daucus carota* L. Ann Bot 47:409–414

Esau K (1977) Anatomy of seed plants. Wiley and Sons, New York

Fujii T, Nito N (1972) Studies on the compatibility and grafting of fruit trees. I Callus fusion between rootstock and scion. J JPN Soc Hortic Sci 41:1–10

Garner RJ (1970) The grafter's handbook, 3rd edn. Faber and Faber, London

Herrero J (1951) Studies of compatible and incompatible graft combinations with special reference to hardy fruit trees. J Hortic Sci 26:186–237

Heslop-Harrison J (1975) Incompatibility and the pollen–stigma interaction. Annu Rev Plant Physiol 26:403–425

Heslop-Harrison J (1978) Cellular recognition systems in plants. Institute of Biology Studies in Biology No 100, Arnold, London

Heslop-Harrison J, Heslop-Harrison Y, Barber J (1975) The stigma surface in incompatibility responses. Proc R Soc Lond Ser B 188:282–297

Jeffree CE, Yeoman MM (1981) A study of the intracellular and intercellular distribution of the *Datura stramonium* lectin using an immunofluorescent technique. New Phytol 87:463–471

Jeffree CE, Yeoman MM (1983) Development of intercellular connections between opposing cells in a graft union. New Phytol 93:491–509

Jones MGK (1976) The origin and development of plasmodesmata. In: Gunning BES, Robards AW (eds) Intercellular communication in plants; studies on plasmodesmata. Springer, Berlin Heidelberg New York, pp 81–105

Kilpatrick DC, Yeoman MM (1978) Purification of the lectin from *Datura stramonium*. Biochem J 175:1151–1153

Kilpatrick DC, Yeoman MM, Gould AR (1979) Tissue and sub-cellular distribution of the lectin from *Datura stramonium* (thornapple). Biochem J 184:215–219

Kloz J (1971) Serology of the Leguminosae. In: Harborne JB, Boulter D and Turner BL (eds) Chemotaxonomy of the Leguminosae. Academic Press, London New York, pp 309–365

Knox RB, Clarke A, Harrison S, Smith P, Marchalonis JJ (1976) Cell recognition in plants: Determinants of the stigma surface and their pollen interactions. Proc Natl Acad Sci USA 73:2788–2792

Kostoff D (1928) Induced immunity in plants. Proc Natl Acad Sci USA 14:236–237

Kostoff D (1929) Acquired immunity in plants. Genetics 14:37–77

Kostoff D (1930) Induced immunity in plants by grafting. Congrès Int Microbiol Paris 1:642

Kostoff D (1931) Studies on the acquired immunity in plants induced by grafting. Z Immunität 74:339–346

Lindsay DW, Yeoman MM, Brown R (1974) An analysis of the development of the graft union in *Lycopersicon esculentum*. Ann Bot 38:639–646

Lipetz J (1970) Wound-healing in higher plants. Int Rev Cytol 27:1–28

Moore R (1981) Graft compatibility and incompatibility in higher plants. Dev Comp Immunol 5:377–389

Moore R (1982) Graft formation in *Kalanchoë blossfeldiana*. J Exp Bot 33:533–540

Moore R, Walker DB (1981a) Studies on vegetative compatibility–incompatibility in higher plants. II. A structural study of an incompatible heterograft between *Sedum telephoides* (Crassulaceae) and *Solanum pennellii* (Solanaceae). Am J Bot 68:831–842

Moore R, Walker D (1981b) Studies on vegetative compatibility-incompatibility in higher plants. I. A structural study of a compatible autograft in *Sedum telephoides* (Crassulaceae). Am J Bot 68:820–830

Mosse B (1962) Graft-incompatibility in fruit trees. Tech Common No 28 Comm Bur Hortic and Plant Crops, East Malling, England

Muzik TJ (1958) Role of parenchyma cells in graft union in *Vanilla* orchid. Science 127:82

Nicolson GL (1974) Interaction of lectins with animal cell surfaces. Int Rev Cytol 39:89–190

Noggle GR (1979) Recognition systems in plants. What's New in Plant Physiol 10:5–8

Parkinson M, Yeoman MM (1982) Graft formation in cultured, explanted internodes. New Phytol 91:711–719

Reeve RM (1979) A specific hydroxylamine–ferric chloride reation for histochemical localisation of pectin. Stain Technol 34:209–211

Roberts RH (1949) Theoretical aspects of graftage. Bot Rev 15:423–463

Roberts JR, Brown R (1961) The development of the graft union J Exp Bot 12:294–302

Rogers WS, Beakbane AB (1957) Stock and scion relations. Annu Rev Plant Physiol 8:217–236

Ryan CA, Bishop P, Pearce G, Darvill AG, MnNeil M, Albersheim P (1981) A sycamore cell wall polysaccharide and a chemically related tomato leaf polysaccharide possess similar proteinase inhibitor-inducing activities. Plant Physiol 68:616–618

Shimomura T, Fuzihara K (1977) Physiological study of graft union formation in cactus II Role of auxin on vascular connections between stock and scion. J Jpn Soc Hortic Sci 45:397–406

Stoddard FL, McCully ME (1980) The effects of excision of stock and scion organs on the formation of the graft union in *Coleus:* a histological study. Bot Gaz 141:401–412

Whitaker TW, Chester KS (1933) Studies on the precipitin reaction in plants. IV The question of acquired reactions due to grafting. Am J Bot 20:297–308

Yeoman MM, Brown R (1976) Implications of the formation of the graft union for organisation in the intact plant. Ann Bot 40:1265–1276

Yeoman MM, Kilpatrick DC, Miedzybrodzka MB, Gould AR (1978) Cellular interactions during graft formation in plants, a recognition phenomenon? Symp Soc Exp Biol 32:139–160

22 Cellular Polarity [1]

F.-W. BENTRUP

22.1 Introduction

„Unter die allgemeinste naturwissenschaftliche Aufgabe, allen Wechsel der Erscheinungen auf Bewegung zurückzuführen, und nach mathematischen Gesetzen aus Grundkräften der Anziehung und Abstossung zu erklären, fällt auch die Construction des Bildungstriebes" (SCHLEIDEN 1845).

This virtual call for a biophysical approach to identify the fundamental morphogenetic mechanisms seems particularly justified regarding cell polarity in plant morphology and physiology. In fact, this conclusion was already drawn by H. VÖCHTING in 1878, who introduced the term polarity into botanical literature and started to subject polar properties of the living plant to rigorous experimentation. He did so with explicit reference to this quotation. Later, resuming the ideas of VÖCHTING (1918), BÜNNING (1958) emphasized that polarity in plants is established on the cellular level as a structural asymmetry of the protoplast.

There are several predecessors to this article. In the Encyclopedia 1st Series (Vol. XV/1) 1965, BLOCH, BÜNNING, and VON WETTSTEIN treated various aspects of polarity and pattern formation in plants rather comprehensively. In Vol. 7 of the New Series, WEISENSEEL (1979) reviewed subsequent progress in the field. The reader should also consult recent treatments with special reference to electrical control of cell polarity by JAFFE and NUCCITELLI (1977), by JAFFE (1979), and by WEISENSEEL and KICHERER (1981); concerning primarily structural, biochemical, and cytological aspects, the reviews of QUATRANO (1978), and SIEVERS and SCHNEPF (1981) are recommended. Like most of the cited literature, this article deals predominantly with the single, mostly lower plant cell; significant progress is still largely confined to this level, where the *Fucus* egg continues to be the leader in the field since the discovery of its photopolarizability by ROSENVINGE (1888).

With regard to the general topic of this volume, an attempt is made to outline the present state of our knowledge and thinking about the origin, structural and molecular base of cell polarity. The sensory aspect of how cell polarity is controlled by environmental and endogeneous factors will be treated only as far as it helps to understand cell polarization.

Also the diversity of polar phenomena in cell structure and functions will not be considered per se. The reader is encouraged to recall this diversity by inspection of the quoted literature, also including intercellular and intracellular pattern formation (see BÜNNING 1958, 1965, BLOCH 1965).

---

1 This article is dedicated to Professor WOLFGANG HAUPT who in 1960 introduced the author into this field of research

22.2 Rise of Cell Polarity in Single Plant Cells

22.2.1 Segregation and Redistribution of Cellular Components: Observations on the Fucus Zygote

The fucacean zygote is one of the very few instances of a virtually apolar, i.e. radially symmetrical, germ cell in plants. The process of irreversible polarization of the zygote has been studied on the ultrastructural level by QUATRANO and coworkers, who found segregation of intracellular components, in particular, migration of vesicles with fibrillar contents and of osmiophilic bodies to the presumptive rhizoid pole (QUATRANO 1972, BRAWLEY et al. 1976). In eggs of *Fucus vesiculosus,* fertilization elicits an increase of RNA and protein synthesis (KOEHLER and LINSKENS 1967, PETERSON and TORREY 1968); in fact, the increase of protein synthesis 2 h after fertilization is accompanied by formation of a

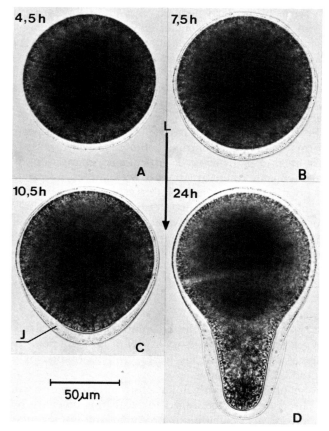

Fig. 1 A–D. Photomicrographs of a zygote of *Pelvetia fastigiata* developing under continuous illumination by white light (*arrow, L*) at 15 °C. Secretion of a transparent jelly (J) is visualized by resin beads (0.1–2 µm) added at **A** 4.5 h, **B** 7.5 h, **C** 10.5 h, and again **D** at 24 h after fertilization. (After SCHRÖTER 1978)

heavy polysomal fraction probably containing mRNA unmasked upon fertilization (LINSKENS 1969). Whereas these products appear too early to be asymmetrically distributed in the *Fucus serratus* zygote which is photopolarizable only 6–10 h later (BENTRUP unpublished), transport of cytoplasmic RNA towards the rhizoid pole has been reported for *Fucus evanescens* by NAKAZAWA (1966).

The earliest evidence for cell polarization is most interesting with respect to putative trigger mechanisms. Therefore, the recent finding of SCHRÖTER (1978) that jelly secretion by zygotes of the fucacean *Pelvetia fastigiata* precedes cytoplasmic segregation is especially intriguing. Figure 1 shows that asymmetrical jelly secretion clearly preceeds rhizoid formation, i.e. localized growth on the shaded side of unilaterally illuminated zygotes. Secretion occurs already 4 h after fertilization and, as SCHRÖTER states, so far this is the earliest polarization event observable on the cytological level. It enables the zygote to adhere to its natural substratum, the marine rock washed by the surf. In the absence of sulfate, zygotes do not become adhesive because they do not polymerize fucoidin; that is, extracellular sulphation of the polysaccharide fucan is blocked (CRAYTON et al. 1974, NOVOTNY and FORMAN 1974). Jelly secretion is a localized process, whereas cell-wall formation by synthesis of alginic acid and cellulose is for obvious reasons not localized (NOVOTNY and FORMAN 1975, QUATRANO and STEVENS 1976).

Figure 1 C demonstrates a further early polarization event, namely a similarly localized clearing of the cortical cytoplasm, which has been described by NUCCITELLI (1978). It is caused by vesicle accumulation and might be involved in the localized jelly secretion (SCHRÖTER 1978). A localized fusion of vesicles with the plasmalemma therefore seems to be the preliminary step in the causal chain of events during cell polarization. The observation of KÜSTER (1906) that this presumptive rhizoid region preferentially plasmolyzes provides circumstantial evidence for a loose association of plasmalemma and cell wall exactly in this region.

22.2.2 Electrophysiology and Morphogenesis: Transcellular Electric Fields and Ionic Currents

It is an old idea that the stationary cortical cytoplasm is the site where cell polarity is determined and structurally fixed. Several authors have pointed out that spatial asymmetries in the activity of the plasmalemma will create electrical and chemical transcellular gradients. LUND studied transcellular electrical fields extensively from the early 1920's (see his monograph of 1947). SPEK (1930) and WENT (1932) proposed that transcytoplasmic electrophoresis might be a potent mechanism to stratify the cytoplasm. JAFFE (1966a) elegantly demonstrated that a battery of developing *Fucus* zygotes aligned in a glass tube drives a steady electric current which is clearly related to cell polarity: a current of about 100 pA (taking as positive ionic charge) enters the presumptive rhizoid pole. Meanwhile, analysis of the existence of such transcellular electric currents has been advanced by a technique developed in JAFFE's laboratory (JAFFE and NUCCITELLI 1974). Essentially, a platinum-black-filled glass microelectrode of

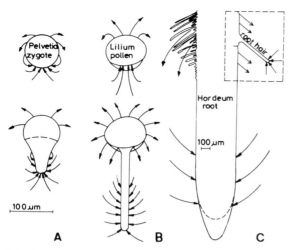

Fig. 2 A–C. Examples of natural ionic currents generated by growing plant cells and organs; after WEISENSEEL and KICHERER (1981). *Arrowed lines* indicate the directions of net flow of positive charge, i.e. a cation flux and/or a counter flux of anions. The relative density of the lines indicates current density, A m^{-2}. (For sake of clarity only part of the current loops, i.e. only outside and near the cell surface, has been depicted.) For the employed technique see text. **A** *Pelvetia fastigiata:* developing zygote and two-celled embryo (cf. Fig. 1) according to NUCCITELLI and JAFFE (1975) and NUCCITELLI (1978). **B** *Lilium longiflorum:* pollen grain before and after germination, according to WEISENSEEL et al. (1975). **C** *Hordeum:* root and root hair (*inset*) according to WEISENSEEL et al. (1979)

10–30 µm diameter vibrates at some 200 Hz, picking up at this frequency by capacitative coupling the voltage of two points with respect to a remote reference electrode. The amplitude of the vibration gives the distance of the two points (30 µm), and the voltage drop across this distance, $\Delta\psi$, V m^{-1}, may be converted into a current density, i, A m^{-2}, by using the medium's ohmic conductivity ρ, S m^{-1}, and Ohm's law: $i = \rho \cdot \Delta\psi$. (The electrical noise is very low due to the use of a lock-in amplifier tuned to the frequency of electrode vibration.) JAFFE and coworkers thus could show that already 30 min after fertilization *Pelvetia* zygotes generate a transcytoplasmic current which enters the presumptive rhizoid pole and leaves the zygote at its opposite thallus pole (Fig. 2 A). The current density is highest at the rhizoid pole and finally reaches values of 10 mA m^{-2}.

Similarly, in lily pollen a longitudinal electric current arises during germination entering the pollen grain at the site of tube outgrowth (Fig. 2 B). It finally reaches the remarkable density of 0.6 A m^{-2}, causing the cell to expend about 1% of its turnover of ATP to drive this current (WEISENSEEL et al. 1975). Figure 2 C shows current loops generated by the tissues as well as by individual trichoblasts of barley roots. In regenerating anucleate stalk segments of *Acetabularia mediterranea,* a longitudinal current accompanies cell growth localized at one end of the segment (NOVAK and BENTRUP 1972); experimental control of the current revealed that only a minute fraction is essential for growth (i.e. incipient

Fig. 3. Synopsis of two developmental events occurring in two different fucacean species: membrane potential (*left ordinate*) in *Fucus serratus* (BENTRUP 1970) and in *Pelvetia fastigiata* (WEISENSEEL and JAFFE 1972), and orientation of cell polarity by light (*right ordinate*) according to FEUCHT and BENTRUP (1972) and JAFFE (1968), respectively

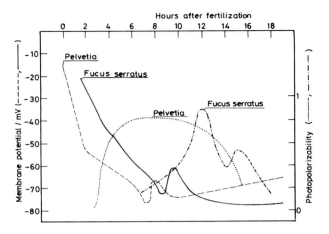

cap regeneration), namely, less than 0.1% of the average current of 10^{-7} A leaving the growing end.

Two questions immediately arise from these observations. Firstly, where is the electromotive force located which drives these currents? Secondly, which ions carry the current – that is, add up to the observed *net* current – across the cytoplasmic compartment(s), and the membrane of the growing cell as well as of the other cell regions? (For the sake of clarity, in Fig. 2 only the sections of the current loops outside and near the cell surface have been depicted.)

Basically, a transcytoplasmic charge displacement or current may arise from a fixed-charge gradient in the cytoplasm and/or differences in membrane potential between growing and non-growing regions of the cell. In the fucacean zygote the membrane potential develops in a unique time-course which is shown for *Fucus serratus* and *Pelvetia fastigiata* in Fig. 3. Upon fertilization, and strictly depending upon this event, the plasmalemma of the zygote develops a high selectivity for K^+ over Na^+ (BENTRUP 1970, WEISENSEEL and JAFFE 1972); roughly speaking, predominantly K^+ channels are installed, while the permeability of the plasmalemma to Cl^-, Na^+ and Ca^{2+} remains low, as is usually found in the plant cell membrane (cf. BENTRUP 1979). On the other hand, fertilization does not significantly change the electromechanical properties of the plasmalemma in *Fucus serratus* (GAUGER and BENTRUP 1979).

In *Pelvetia* the Ca^{2+} permeability seems to depend upon the membrane potential (CHEN and JAFFE 1978), and upon the Ca^{2+} concentration: at 1 mM Ca^{2+}, it is 60 times larger than at the normal Ca^{2+} concentration of sea water, 10 mM Ca^{2+} (ROBINSON 1977). Thus a spontaneous (during the rise of the membrane potential, Fig. 3) or triggered *local* change of the Ca^{2+} permeability would be a plausible mechanism to initiate a transcellular current. Indeed, ROBINSON and JAFFE (1975) have demonstrated that net amounts of Ca^{2+} leak in at the presumptive rhizoid pole, and are pumped out at the opposite, thallus pole, respectively. This result is reproduced in Fig. 4; it also shows that the net Ca^{2+} movement ceases (flux ratio becomes unity) when the cell polarity

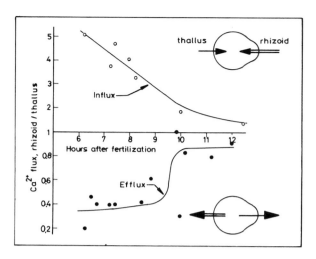

Fig. 4. Net influx and efflux, respectively, of $^{45}Ca^{2+}$ measured as function of time after fertilization on the presumptive rhizoid and thallus hemispheres of zygote populations of *Pelvetia fastigiata* (Fucaceae). *Ordinate* indicates the flux ratio for the two hemispheres. The eggs had been arranged on a specially designed nickel screen, and their rhizoid pole aligned by unilateral illumination. (After ROBINSON and JAFFE 1975)

is fixed at about 12 h after fertilization (cf. time-course of photopolarizability in Fig. 3). JAFFE et al. (1974) have argued that a localized increase of the Ca^{2+} permeability accomplishes both; it depolarizes the membrane at the rhizoid pole, and creates a cytoplasmic field due to binding of Ca^{2+} to anionic fixed charges in the cytoplasm. The field, in turn, may guide and drive electrophoresis of sufficiently densely charged and large particles (e.g. vesicles), so that their electrophoresis may prevail over their back-diffusion. (For a discussion on the numerical level see JAFFE et al. 1974.) Regarding cell polarization this hypothesis is especially attractive, because a local Ca^{2+} entry tends to amplify its transcytoplasmic driving force, and thus could provide a positive feed-back loop for normal cell polarization. Incidentally, cell polarization may fail, so that zygotes develop several rhizoides under special conditions like plasmolysis (KÜSTER 1906) or upon illumination by plane polarized light (JAFFE 1956).

The putative key role of Ca^{2+} is even more convincingly established by the absolute Ca^{2+} requirement for pollen germination and pollen-tube growth (BREWBAKER and KWACK 1963). Specifically, the current through the lily pollen tube (Fig. 2B) requires the presence of at least 10^{-4} M Ca^{2+} and K^+ in the medium; most of the growth current leaks in at the growing pollen tube tip as K^+ and presumably is pumped out by the grain as H^+ (WEISENSEEL and JAFFE 1976); Ca^{2+} constitutes only a minor but significant fraction of the current entering the tube tip. Selective accumulation of Ca^{2+} in the pollen tube tip has been demonstrated through 45-calcium autoradiography by L.A. JAFFE et al. (1975), and tetracycline fluorescence by REISS and HERTH (1978).

Superimposed upon the steady electrical currents, conspicuous transient depolarizations have been recorded on regenerating stalk segments of *Acetabularia mediterranea* (NOVAK and BENTRUP 1972), on *Pelvetia* zygotes (NUCCITELLI and JAFFE 1975) and on *Lilium* pollen (WEISENSEEL et al. 1975). The significance on these current pulses of a frequency in the order of 1 min seems unsettled; in *Acetabularia* they resemble action potentials associated with turgor regulation (WENDLER et al. 1983).

22.3 Polarity Induction

22.3.1 Application of Ionic and Electrical Gradients

It is plausible from the preceding paragraph that application of transcellular gradients in membrane potential and/or ion concentration should also orient cell growth. Indeed, evidence for this supposition has a long record, since LUND (1923) discovered that zygotes of *Fucus inflatus* may be polarized by an applied electric field. A current record includes examples from green, brown and red algae; the required field strength for a significant response is in the order of one millivolt per cell diameter (see Table 6 in JAFFE and NUCCITELLI 1977). Recent work on *Fucus, Equisetum* and *Funaria* spores has shown that growth is not consistently oriented to the same pole, i.e. anode of the applied field, as would be expected from the previous paragraph (BENTRUP 1968, PENG and JAFFE 1976, CHEN and JAFFE 1979). Curve C in Fig. 5 illustrates the dependence of the field effect upon the K^+ concentration of the medium. Plausible, on the other hand, seems the similarity between this electric field effect and the response towards a K^+ concentration gradient (Fig. 5 B), because both stimuli will modify the membrane potential. Figure 5 A shows that *Fucus* eggs may be polarized by a pH-gradient, although the membrane potential does not notably vary with the external pH.

Of particular interest, of course, is the action of a Ca^{2+} concentration gradient. It has been established by use of the calcium ionophore A 23187: *Fucus*

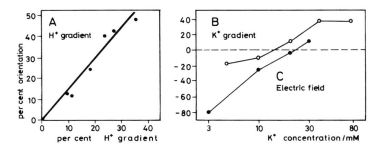

Fig. 5 A, B. Induction of cell polarity in *Fucus* zygotes by different chemical and electrical transcellular gradients. Positive values of orientation indicate that the given percentage of a *Fucus* egg population developed the rhizoid pole toward the higher concentration (H^+, K^+) and the negative electrode of the applied constant electric field, respectively. **A** Response of *Fucus furcatus* zygotes exposed for 16 h to a pH-gradient. In the pH-range between 7.6 and 8.3, the eggs faced at opposite cell poles sea water of two pH-values, pH_1 and pH_2; the transcellular ΔpH difference (gradient) varied from 0.047 to 0.187 units, and the relative gradient, $\Delta pH/pH_1$, varied as given in the abscissa. (After BENTRUP et al. 1967.) **B** Response of *Fucus furcatus* zygotes exposed for 16 h to a K^+ concentration gradient. In this case, the relative gradient, $\Delta K^+/K_1^+$, was 37% throughout, whereas the abscissa indicates the mean K^+ concentration near the egg surface, i.e. $0.5 \Delta K^+$, over the tested range of K^+ from 5 to 80 mM. (Natural sea water contains 10 mM K^+.) **C** Response of *Fucus serratus* zygotes exposed for 6 h to a constant electric field of 15 mV per egg, i.e. of a field strength of 200 V m^{-1}, as a function of the K^+ concentration of the sea water. **A, B** after BENTRUP et al. (1967), **C** after BENTRUP (1968)

zygotes develop their rhizoids toward the higher Ca^{2+} concentration (ROBINSON and CONE 1980); a similar behaviour was reported for spore germination in *Funaria* by CHEN and JAFFE (1979). Furthermore, rhizoid development in the fungus *Blastocladiella* is triggered by A 23187 and Ca^{2+} (VAN BRUNT and HAROLD 1980). The ionophore triggers cell growth also in *Neurospora* (REISSIG 1977); A 23187-dependent cell-wall synthesis by incorporation of Golgi vesicles has been inferred for caulonema cells of *Funaria* by SCHMIEDEL and SCHNEPF (1980), and demonstrated in germinating lily pollen by REISS and HERTH (1978). Thus the capability of Ca^{2+} to initiate localized growth in plant cells seems fairly well established.

22.3.2 Cell Polarization by Light

Since the discovery that unilateral illumination will strongly polarize *Equisetum* spores (STAHL 1885) and *Fucus* zygotes (ROSENVINGE 1888), no other aspect of polarity has been studied as thoroughly as its induction by light. In this article we will discuss it only in so far as it may help to understand the polarization process. The other, sensory aspect falls into the category of blue-light responses in plants. Briefly, dichroic photoreceptors lie near the cell periphery, the main photon transition moment being oriented either parallel (*Fucus, Equisetum, Botrytis*) or perpendicular to the cell surface (*Osmunda*); chemically, these photoreceptors seem to be flavoproteins (cryptochrome). For work on *Fucus* see JAFFE (1956), BENTRUP (1963, 1964), on *Equisetum* see HAUPT (1957), and BENTRUP (1963, 1964), on *Botrytis* and *Osmunda* see JAFFE and ETZOLD (1962). Our knowledge concerning the sensory steps *following* light absorption, however, still is very poor, as is generally true in photosensory transduction in plants (cf. BENTRUP 1979, SHROPSHIRE 1979).

The time-course of photopolarizability has been determined repeatedly in order to detect that stage prior to germination where the cell polarity is irreversibly fixed, although not yet apparent cytologically. In the Fucaceae the time of photopolarizability is surprisingly variable between the different species. Whereas the time when, say, 50% of a zygote population has visibly formed a rhizoid protuberance generally ranges between 15 and 20 h after fertilization, the time of photopolarizability falls between 0.5 and 4 h in *Cystoseira barbata* (KNAPP 1931), and may be as late as 8 to 18 h in *Fucus serratus* (BENTRUP 1963). Figure 3 illustrates the difference between *Pelvetia* and *Fucus serratus*, and reveals that this difference is independent of the differentiation of the plasmalemma toward K^+ selectivity. For better comparison it should be mentioned that the photosensitive period of the individual zygote lasts for about 1 h (FEUCHT and BENTRUP 1972). On the other hand, among the several polarizing factors, light seems to be the most effective one, and the other stimuli are effective only prior to or simultaneous with light. This applies to polarization by the site of sperm entry in *Cystoseira* (KNAPP 1931), to the period of susceptibility to electric fields in *Fucus serratus* (BENTRUP 1968) and *Cystoseira* (BENTRUP unpublished), and to polarization by laminar flow in *Fucus furcatus* (BENTRUP and JAFFE 1968). Altogether polarity seems to be fixed irreversibly when photopolarizability ceases (HAUPT 1957, has demonstrated this on *Equisetum* spores). In this respect photopolarization is well suited to test the significance of the transcytoplasmic current outlined in the preceding paragraph. In *Pelvetia*, where initiation of the current and photopolarization coincide, NUCCITELLI (1978) was able to establish that the current consistently enters unilaterally illuminated eggs within the shaded side, and, even more convincingly, a change of the light direction within the period of photopolarizability was followed by a corresponding change of the site of current entry.

Finally, we return to SCHRÖTER's experiment (Fig. 1). This author found that two consecutive illuminations from different directions will generate two regions of enforced jelly secretion. Besides the efficiency of applied electrochemical gradients (Fig. 5) the strict correlation between photopolarization and the early electrical polarization events strongly support JAFFE's hypothesis that cell polarization involves a transcytoplasmic regenerative flow of electrically charged particles. It would be interesting to test whether this relationship can be established as well in *Fucus serratus* where the period of photopolarizability is extremely late compared to the development of the membrane potential (Fig. 3) and to the period of electric polarizability (2 to 10 h after fertilization; BENTRUP 1968).

22.3.3 Cell Polarization by Signal Molecules

ROSENVINGE (1888) and KNIEP (1907) had observed that *Fucus* eggs show polarization by neighbour zygotes (*group effect*), or a piece of thallus; JAFFE (1968) found germination of *Fucus* eggs towards all of a dozen of tested marine plants, six brown algae, five red algae, and a higher plant, *Phyllospadix*. The plant material obviously emits one or several signal substances stable enough to diffuse even some millimeters. In analogy to the other treated polarizing factors, electrochemical or light absorption gradients, it seems reasonable to assume transcellular concentration gradients of the emitted signal molecules to be the effective stimulus. Experimentally, a concentration gradient may be established by a laminar flow across the cell enhancing the signal molecule concentration at the cell's lee pole and depleting it at the cell's luff pole. Quantitatively, JAFFE (1965) has worked out a relationship between the medium's streaming velocity and the diffusion coefficient of the emitted stuff. Using this rheotropic analysis, MÜLLER and JAFFE (1965) found that conidia of *Botrytis cinerea* will germinate downstream as is expected from emission of a growth stimulator and the known autotropism (group effect) of this fungus. The diffusion ceofficient of the order of 10^{-6} cm^2 s^{-1} indicates that the signal molecule is macromolecular. In a subsequent study (JAFFE 1966b) postulated the release of both growth stimulator and inhibitor substances by *Botrytis* conidia, because the response changed its sign from positive to the negative group effect, when the medium was equilibrated with air enriched with 0.3 or 3% CO_2, instead of the normal 0.03%. The chemical composition of the signal molecules is unknown.

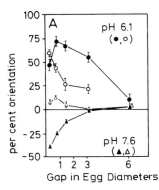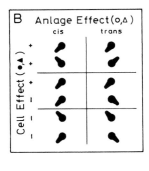

Fig. 6 A, B. Mutual polarization of *Fucus furcatus* zygotes lying in pairs. **A** As a function of distance between the eggs (*abscissa*) the angular orientation of the rhizoid pole of each egg within a pair has been measured at pH 6.1 and 7.6 as indicated. **B** Rhizoid orientation categories used in **A**: The rhizoid formation of each egg in a total of 8008 pairs was classified whether it occurred towards (plus values) or away from (minus values) the other egg (cell effect, *solid symbols*), and whether it occurred more specifically toward or away from the other cell's rhizoid pole (anlage effect, *open symbols,* cis orientation = positive values). (After JAFFE and NEUSCHELER 1969)

Returning to the *Fucus* zygote, we notice that its group effect is pH-dependent; the positive group effect prevails at pH < 7.6, its counterpart at pH > 7.6, optimally at pH 8.4 (WHITHAKER and LOWRANCE 1940, JAFFE 1968), again suggesting the existence of two antagonistic effector molecules. It was not surprising, therefore, that *Fucus* zygotes display a pH-dependent rheotropic response, namely, downstream rhizoid orientation at pH 6.5 and upstream orientation at pH 7.1 and 8.1 (BENTRUP and JAFFE 1968). The analysis remained unsatisfactory, however, because the downstream response only partly matched the quantitative flow model (JAFFE 1965). Progress came from a detailed study by JAFFE and NEUSCHELER (1969), who investigated the group effect under the conditions of pH, washing pretreatment of the eggs, and a flow across zygote pairs interfering with the zygotes diffusional communication. For the understanding of their conclusions regarding the cell polarization mechanism, their detailed evaluation of the group effect must be presented. Figure 6 B shows that the response of each zygote in a pair is categorized, as to whether it forms its rhizoid toward its partner (positive cell effect) or away from it (negative cell effect), as well as whether both rhizoids will be formed in the cis-arrangement, that is, toward the partner's rhizoid "anlage", or in the trans-arrangement (positive and negative anlage effect, respectively). Figure 6 A shows that the cell effect is positive at pH 6.1 and negative at pH 7.6. The pH effect is modified by a pretreatment of the zygotes: thorough washing of the freshly shed eggs with sea water enforces the positive cell effect, whereas the strongest negative cell effect is displayed by unwashed eggs. Figure 6 A further illustrates that the positive cell effect operates over a bigger distance, and solely exhibits a clear anlage effect. Assuming that the zygote development of the average pair is asynchronous, JAFFE and NEUSCHELER concluded from the existence of the anlage effect that rhizin, the hypothetical growth stimulator substance, is emitted from the rhizoid anlage of the faster developing zygote and causes the slower egg to form its rhizoid anlage to that of its partner, thus giving rise to the observed cis-arrangement of the pair. By contrast, the growth inhibitor, antirhizin, presumably is emitted synchronously and uniformly by the whole zygote, because there is no anlage effect, and a low percentage of the negative cell effect (Fig. 6 A), as may be expected for empirical reasons from cells interacting through a steady-state diffusion gradient of a stable substance. Vice versa, the authors conclude that rhizin is an unstable molecule, and a link in the polarization process; a feed-back loop between emission and stimulation might cause the observed high degree of growth orientation.

What is the chemical nature of rhizin and antirhizin? Du Buy and Olson (1937) had argued that the positive group effect is mediated by indoleacetic acid (IAA). Jaffe (1955) ruled out interaction through H^+, CO_2 or O_2 and later favoured the idea that rhizin is IAA (Jaffe 1968). Indeed, Jaffe and Neu-scheler (1969) derived from cross-flow experiments that rhizin has a diffusion coefficient of 10^{-5} cm^2 s^{-1} which is consistent with a small molecule like IAA.

On the other hand, IAA is relatively stable, and a high background concentration of this auxin should drown any cell-born concentration gradient, hence extinguishing the positive group effect. However, Klemke and Bentrup (1973) observed in well-buffered sea water no effect of three auxins tested, IAA, α- and β-naphtylacetic acid, upon the group effect in *Fucus serratus*. They found that adenosine 5'-monophosphate (AMP) and, to a lesser degree, 3',5'-cyclic monophosphate (cAMP) partially behaved like rhizin, or might modulate the secretion or action of rhizin. The observation of many investigators that the positive group effect does not disappear in large *Fucus* egg populations (i.e. 2000 eggs; Jaffe 1968), suggests that rhizin in fact decays quickly. One could think of a low molecular weight decomposable carbohydrate secreted by the vesicles involved in localized cell wall extension and synthesis.

Finally, it should be added that *Fucus* eggs release another signal molecule, a volatile, lipid-soluble sperm-attractant octatriene compound. In *Fucus serratus* the most effective species is the 1, 3, 5-octatriene isomer *fucuserraten* (Müller and Jaenicke 1971); it also attracts spermatozoids of *F. vesiculosus*, but apparently does not lead to cross-fertilization (Müller and Seferiadis 1977).

22.4 Cellular Polarity in Higher Plants

The variety of mechanisms supposedly involved in the establishment of cell polarity in lower plants basically suffices to describe also the polar phenomena and their integration within a tissue or organ of a higher plant. Few examples might illustrate this rationale. Firstly, unequal cell division occurring, for instance, in pollen or stoma development a priori requires no other mechanisms than observed on *Fucus* zygote or *Equisetum* spore development; the analogy also includes the known variety of orienting stimuli, where chemical gradients imposed upon a cell by the surrounding tissue are particularly likely. For a recent treatment of sporogenesis and pollen-grain development see Buchen and Sievers (1981). The group effect of *Fucus* zygotes or *Botrytis* conidia provides a simple model of how intercellular communication might operate in a tissue; namely, emission of and response to only two antagonistic growth substances with different life-times and few additional characteristics, as outlined for the hypothetical signal substances rhizin and antirhizin, might suffice to create more-dimensional growth patterns like, for instance, the stoma distribution in a leaf epidermis (cf. Bünning 1965).

Polar transport of auxin seems to occur through transcellular membrane transport rather than by way of the symplasm; plasmodesmata are not required for auxin movement down the *Avena* coleoptile (Cande and Ray 1976). Polar membrane transport of the weak acid IAAH is adequately described by a carrier-mediated increase of the plasmalemma permeability in the basal region of each cell in the tissue for the dissociated form, IAA^-, whereas the neutral auxin molecule, IAAH, permeates the membrane well over the whole cell surface (Rubery and Sheldrake 1974, Raven 1975, 1979). Raven (1975) measured a permeability ratio P_{IAAH}/P_{IAA^-} of about 10^{-3} and pointed out that a positive feed-back loop is conceivable between auxin transport and the presumably auxin-controlled proton pump, because auxin transport is driven by the trans-membrane protonmotive force generated by this pump.

Longitudinal transcellular electric fields could easily develop from this asymmetrical membrane transport adding up across the tissue. Indeed, broad experimental evidence exists for electric fields generated by higher plant tissues including the *Avena* coleoptile (see Scott 1967). Thus the hypothesis is supported that tissue or organ polarity originates from the aligned polarity of the individual cells in that material.

22.5 Concluding Discussion

Since all matter of the living protoplast is subjected to metabolic turnover, no individual macromolecule or compound molecular structure can ensure a permanent structural cell polarity. The cell wall as the permanent structure next to and synthesized by the protoplast can be envisaged to be the primary manifestation of a cellular polarity. Further support comes from the earlier observation of Küster (1906) and Kniep (1907) on *Fucus* that detachment of the protoplast from the cell wall by means of plasmolysis will disturb the polarization process and will lead to multiple rhizoid formation. An analogous effect of plasmolysis on *Equisetum* spore development was observed by Nakazawa and Takahashi (1981). In *Bryopsis* cell polarity is lost if protoplast fragments are experimentally separated from the cell wall (Nakazawa 1975). On the other hand, in filaments of *Cladophora*, plasmolysis will abolish only the polarity of the filament, whereas the cellular polarity is retained (Miehe 1905). Generally, complete removal of the cell wall during protoplast preparation from any material yields protoplasts which are virtually spherical and reveal no asymmetry under the light or electron microscope regarding, for instance, organelle distribution. A more natural case is the *Fucus* egg released from the oogonium as a naked turgorless cell. Loss and reversal of cell polarity has been a matter of continuous debate in the past (cf. BÜNNING 1958). Rather convincingly it has been claimed by HÄMMERLING (1955) for cap regeneration by anucleate stalk segments of *Acetabularia* (see below).

The protoplast surface, that is, the plasmalemma lends itself favourably to control and initiate the structural polarization of a cell. The highly ordered molecular organization of the plasmalemma features an asymmetry due to the specific orientation of the intrinsic proteins within the membrane. BÜNNING (1958) already pointed out that a membrane asymmetry would be favourable for the establishment of cell polarity. In fact, an initial, completely labile and episodic polarity may be envisaged as a local rearrangement of some membrane protein domains functioning, for instance, as transmembrane pathway ("channel") or sensory transducer in photo- or chemoreception. These domains are known to have a good lateral mobility within the lipid core of the membrane. If they carry a net charge, they might easily be redistributed, that is, for instance locally accumulated under the influence of a transcellular electric field generated by spontaneous local depolarization or by neighbour cells in a tissue (JAFFE 1977).

Evidence exists that the concanavalin A receptor on the embryonic muscle cell membrane is displaced by fields of only 1 mV per cell (POO and ROBINSON 1977). Field strengths of this order suffice to polarize plants cells, as mentioned above.

A reversible, purely functional differentiation of the plasmalemma into macroscopic transport patches is known from photosynthesizing *Chara* cells. Upon illumination the internodal cell develops alternating acid and alkaline regions (Fig. 7). At the acid regions (A) ATP-fueled bicarbonate import (or proton export) occurs and requires the presence of Ca^{2+} (LUCAS and DAINTY 1977a). The alkaline bands (B) are generated by passive OH^- efflux produced by photosynthetic carbon fixation. The spatial differentiation optimizes bicarbonate uptake: the high pH, due to the disposal of OH^-, converts HCO_3^- into CO_3^{2-} and thus would limit the availability of bicarbonate to the pump if both transport processes would share the same site. Since both processes are electrogenic, electric currents similar to those in Fig. 2 circulate between the sites A and B (WALKER and SMITH 1977, LUCAS and NUCCITELLI 1980). Inhibition of protoplasmic streaming by cytochalasin B reversibly changes the OH^- efflux pattern from the discrete bands into a network of numerous small but still localized efflux sites; LUCAS and DAINTY (1977b) argue that this localization in fact arises from localized activation of uniformly distributed microscopic efflux sites in the plasmalemma; thus localization would not even require a physical rearrangement of the transport proteins within the membrane, but would call for some control mechanism of unspecified location. In two other green algae, functional patterns of membrane transport may be related to both cell growth and photosynthesis. In *Acetabularia* apex regeneration by anucleate posterior stalk segments kept in the dark for 5 days occurs virtually independently of the segment's former cell polarity (HÄMMERLING 1955). Upon illumination a gradient of photo-

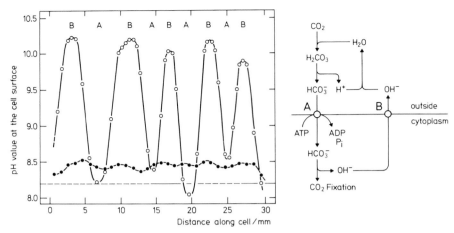

Fig. 7. Profile of extracellular pH along the surface of an internodial cell of *Chara corallina* (o) after 1 h of illumination (20 W m^{-2}), and (●) after a subsequent period of 30 min darkness and another hour of illumination. Additionally, the K$^+$ concentration was increased from 0.2 to 10 mM. *Dashed line* indicates the background pH of the artificial pond water. The electrogenic membrane transport mechanisms putatively operating at the alternating membrane patches, *A* and *B*, are given in the scheme on the right. (After Lucas and Dainty 1977a, from Bentrup 1980)

synthetic activity and incipient cap regeneration arises at the end of highest activity (Hämmerling and Hämmerling 1959). The transcellular electric current arising also during regeneration (see above) presumably is due to the higher activity of the ATP-driven chloride pump hyperpolarizing the regenerating end (Novak and Bentrup 1972).

Tubes of the siphonaceous *Vaucheria sessilis* locally irradiated with blue light start to grow from this patch (Kataoka 1975); the growth activity is preceded by a similarly localized reticulation of cortical fibers consisting of actin filaments and subsequent chloroplast aggregation (Blatt et al. 1980). Associated with these cytological events a hyperpolarizing *outward* current arises from this patch (similar to the current leaving the regenerating end of the *Acetabularia* segment); after 10 to 130 min, however, the current changes its sign to inward, and then is followed by the localized growth (Blatt et al. 1981, Weisenseel and Kicherer 1981).

In this context it seems useful to recall that chloroplast migration is the earliest event observed hitherto during polarization of the *Equisetum* spore (Nienburg 1924), and, secondly, that spore polarization depends upon photosynthesis (Haupt 1957).

Altogether these examples might serve to suggest just one possible evolutionary origin of cell polarity: A fundamental cellular process such as photosynthetic carbon fixation is optimized or accidentally followed by a spatial differentiation of the membrane transport involved. The resulting inevitable transcellular electric currents eventually provide a simple mechanism for growth localization irreversibly polarizing the cell. The assembly of the light-transducing purple membrane in *Halobacterium halobium* (SUMPER et al. 1976) suggests that membrane differentiation does indeed have a long history.

References

Bentrup FW (1963) Vergleichende Untersuchungen zur Polaritätsinduktion durch das Licht an der *Equisetum*spore und der *Fucus*zygote. Planta 59:472–491

Bentrup FW (1964) Zur Frage eines Photoinaktivierungs-Effektes bei der Polaritätsinduktion in *Equisetum*sporen und *Fucus*zygoten. Planta 63:356–365

Bentrup FW (1968) Die Morphogenese pflanzlicher Zellen im elektrischen Feld. Z Pflanzenphysiol 59:309–339

Bentrup FW (1970) Elektrophysiologische Untersuchungen am Ei von *Fucus serratus:* Das Membranpotential. Planta 94:319–332

Bentrup FW (1979) Reception and transduction of electrical and mechanical stimuli. In: Haupt W, Feinleib ME (eds) Encyclopedia of plant physiology. Springer, Berlin Heidelberg New York, Vol 7, pp 42–70

Bentrup FW (1980) Electrogenic membrane transport in plants. A review. Biophys Struct Mech 6:175–189

Bentrup FW, Jaffe LF (1968) Analyzing the "Group Effect": Rheotropic responses of developing *Fucus* eggs. Protoplasma 65:25–35

Bentrup FW, Sandan T, Jaffe LF (1967) Induction of polarity in *Fucus* eggs by potassium ion gradients. Protoplasma 64:254–266

Blatt M, Wessells N, Briggs W (1980) Actin and cortical fiber reticulation in the siphonaceous alga *Vaucheria sessilis*. Planta 147:363–375

Blatt M, Weisenseel MH, Haupt W (1981) A light-dependent current associated with chloroplast aggregation in the alga *Vaucheria sessilis*. Planta 152:513–526

Bloch R (1965) Polarity and gradients in plants: a survey. In: Ruhland W (ed) Encyclopedia of plant physiology. Springer, Berlin Göttingen Heidelberg Vol XV/1, pp 234–274

Brawley SH, Wetherbee R, Quatrano RS (1976) Fine-structural studies of the gametes and embryo of *Fucus vesiculosus* L. (Phaeophyta). II. The cytoplasm of the eggs and young zygote. J Cell Sci 20:255–271

Brewbaker JL, Kwack BH (1963) The essential role of calcium ion in pollen germination and pollen tube growth. Am J Bot 50:859–865

Buchen B, Sievers A (1981) Sporogenesis and pollen grain formation. In: Kiermayer O (ed) Cytomorphogenesis in plants, cell biology monographs. Springer, Berlin Heidelberg New York, Vol 8, pp 349–376

Bünning E (1958) Polarität und inäquale Teilung des pflanzlichen Protoplasten. Protoplasmatologia VIII/9a:1–86

Bünning E (1965) Die Entstehung von Mustern in der Entwicklung von Pflanzen. In: Ruhland W (ed) Encyclopedia of plant physiology. Springer, Berlin Göttingen Heidelberg, Vol XV/1, pp 383–408

Cande WZ, Ray PM (1976) Nature of cell-to-cell transfer of auxin in polar transport. Planta 129:43–52

Chen TH, Jaffe LF (1978) Effects of membrane potential on calcium fluxes of *Pelvetia* eggs. Planta 140:63–67

Chen TH, Jaffe LF (1979) Forced calcium entry and polarized growth of *Funaria* spores. Planta 144:401–406

Crayton MA, Wilson E, Quatrano RS (1974) Sulfation of fucoidan in *Fucus* embryo. II. Separation from initiation of polar growth. Dev Biol Suppl 39:164–167

Du Buy HG, Olson RA (1937) The presence of growth regulators during the early development of *Fucus*. Am J Bot 24:609–611

Feucht U, Bentrup FW (1972) Über die photosensible Phase der Polaritätsinduktion bei *Equisetum*sporen und *Fucus*zygoten. Z Pflanzenphysiol 66:233–242

Gauger B, Bentrup FW (1979) A study of dielectric membrane breakdown in the *Fucus* egg. J Membr Biol 48:249–264

Hämmerling J (1955) Neuere Versuche über Polarität bei *Acetabularia*. Biol Zentralbl 74:545–554

Hämmerling J, Hämmerling Chr (1959) Über Bildung und Ausgleich des Polaritätsgefälles bei *Acetabularia*. Planta 53:522–531

Haupt W (1957) Die Induktion der Polarität bei der Spore von *Equisetum*. Planta 49:61–90

Haupt W (1960) Die Entstehung der Polarität in pflanzlichen Keimzellen, insbesondere die Induktion durch Licht. Ergeb Biol 25:1–32

Jaffe LA, Weisenseel MH, Jaffe LF (1975) Calcium accumulations within the growing tips of pollen tubes. J Cell Biol 67:488–492

Jaffe LF (1955) Do *Fucus* eggs interact through a CO_2-gradient? Proc Natl Acad Sci USA 41:267–270

Jaffe LF (1956) Effect of polarized light on polarity of *Fucus*. Science 123:1081–1082

Jaffe LF (1958) Tropistic responses of zygotes of the Fucaceae to polarized light. Exp Cell Res 15:282–299

Jaffe LF (1965) On the concentration gradient across a spherical source washed by slow flow. Biophys J 5:201–210

Jaffe LF (1966a) Electrical currents through the developing *Fucus* egg. Proc Natl Acad Sci USA 56:1102–1109

Jaffe LF (1966b) On autotropism in *Botrytis:* Measurement technique and control by CO_2. Plant Physiol 41:303–306

Jaffe LF (1968) Localization in the developing *Fucus* egg and the general role of localizing currents. Adv Morphogen 7:295–328

Jaffe LF (1977) Electrophoresis along cell membranes. Nature 265:600–602

Jaffe LF (1979) Control of development by ionic currents. In: Cone RA, Dowling JE (eds) Membrane transduction mechanisms. Raven Press, New York, pp 199–231

Jaffe LF, Etzold H (1962) Orientation and locus of tropic photoreceptor molecules in spores of *Botrytis* and *Osmunda*. J Cell Biol 13:13–31

Jaffe LF, Neuscheler W (1969) On the mutual polarization of nearby pairs of fucaceous eggs. Dev Biol Suppl 19:549–565

Jaffe LF, Nuccitelli R (1974) An ultrasensitive vibrating probe for measuring extracellular currents. J Cell Biol 63:614–628

Jaffe LF, Nuccitelli R (1977) Electrical controls of development. Annu Rev Biophys Bioeng 6:445–476

Jaffe LF, Robinson KR, Nuccitelli R (1974) Local cation entry and self-electrophoresis as an intracellular localization mechanism. Ann NY Acad Sci 238:372–389

Kataoka H (1975) Phototropism in *Vaucheria geminata* II. The mechanism of bending and branching. Plant Cell Physiol 16:439–448

Klemke I, Bentrup FW (1973) Zur Frage nach den Signalstoffen beim Gruppeneffekt keimender *Fucus*zygoten. Plant Sci Lett 1:315–320

Knapp E (1931) Entwicklungsphysiologische Untersuchungen an Fucaceen-Eiern. I. Zur Kenntnis der Polarität an den Eiern von *Cystosira barbara*. Planta 14:731–751

Kniep H (1907) Beiträge zur Keimungsphysiologie und -biologie von *Fucus*. Jahrb Wiss Bot 44:635–724

Koehler LD, Linskens HF (1967) Incorporation of protein and RNA precursors into fertilized *Fucus* eggs. Protoplasma 64:209–212

Küster E (1906) Normale und abnormale Keimungen bei *Fucus*. Ber Dtsch Bot Ges 24:522–528

Linskens HF (1969) Changes in the polysomal pattern after fertilization in *Fucus* eggs. Planta 85:175–182

Lucas WJ, Dainty J (1977a) HCO_3^- influx across the plasmalemma of *Chara corallina*. Divalent cation requirement. Plant Physiol 60:862–867

Lucas WJ, Dainty J (1977b) Spatial distribution of functional OH^- carriers along the characean internodal cell: Determined by the effect of cytochalasin B on $H^{14}CO_3^-$ assimilation. J Membr Biol 32:75–92

Lucas WJ, Nuccitelli R (1980) HCO_3^- and OH^- transport across the plasmalemma of *Chara*. Spatial resolution obtained using extracellular vibrating probe. Planta 150:120–131

Lund EJ (1923) Electrical control of organic polarity in the egg of *Fucus*. Bot Gaz 76:288–301

Lund EJ (1947) Bioelectric fields and growth. Univ Texas Press, Austin

Miehe H (1905) Wachstum, Regeneration und Polarität isolierter Zellen. Ber Dtsch Bot Ges 23:257–264

Müller DG, Jaenicke L (1971) Fucoserraten, the female sex attractant of *Fucus serratus* L. (Phaeophyta). FEBS Lett 30:137–139

Müller DG, Jaffe LF (1965) A quantitative study of cellular rheotropism. Biophys J 5:317–335

Müller DG, Seferiadis K (1977) Specificity of sexual chemotaxis in *Fucus serratus* and *Fucus vesiculosus* (Phaeophyceae) Z Pflanzenphysiol 84:85–94

Nakazawa S (1966) Regional concentration of cytoplasmic RNA in *Fucus* eggs in relation to polarity. Naturwissenschaften 53:138

Nakazawa S (1975) Regeneration polarity in *Bryopsis*. Bull Jpn Soc Phycol 23:139–143

Nakazawa S, Takahashi T (1981) Induction of the secondary rhizoid by plasmolysis in *Equisetum* sporelings. Bull Yamagata Univ 10:207–212

Nienburg W (1924) Die Wirkung des Lichtes auf die Keimung der *Equisetum*-Spore. Ber Dtsch Bot Ges 42:95–99

Novak B, Bentrup FW (1972) An electrophysiological study of regeneration in *Acetabularia mediterranea*. Planta 108:227–244

Novotny AM, Forman M (1974) The relationship between changes in cell-wall composition and the establishment of polarity in *Fucus* embryos. Dev Biol Suppl 40:162–173

Novotny AM, Forman M (1975) The composition and development of cell walls of *Fucus* embryos. Planta 112:67–78

Nuccitelli R (1978) Ooplasmic segregation and secretion in the *Pelvetia* egg is accompanied by a membrane-generated electrical current. Dev Biol Suppl 62:13–33

Nuccitelli R, Jaffe LF (1975) The pulse current pattern generated by developing fucoid eggs. J Cell Biol 64:636–643

Peng HG, Jaffe LF (1976) Polarization of fucoid eggs by steady electrical fields. Dev Biol Suppl 53:277–284

Peterson DM, Torrey JG (1968) Amino acid incorporation in developing *Fucus* embryos. Plant Physiol 43:941–947

Poo MM, Robinson KR (1977) Electrophoresis of concanavalin A receptors along embryonic muscle cell membrane. Nature 265:602–605

Quatrano RS (1972) An ultrastructural study of the determined site of rhizoid formation in *Fucus* zygotes. Exp Cell Res 70:1–12

Quatrano RS (1978) Development of cell polarity. Annu Rev Plant Physiol 29:487–510

Quatrano RS, Stevens PT (1976) Cell-wall assembly in *Fucus* zygotes. I. Characterization of the polysaccharide components. Plant Physiol 58:224–231

Raven JA (1975) Transport of indoleacetic acid in plant cells in relation to pH and electrical potential gradients, and its significance for polar IAA transport. New Phytol 74:163–172

Raven JA (1979) The possible role of membrane electrophoresis in the polar transport of IAA and other solutes in plant tissues. New Phytol 82:285–291

Reiss HD, Herth W (1978) Visualization of the Ca^{2+}-gradient in growing pollen tubes of *Lilium longiflorum* with chlorotetracycline fluorescence. Protoplasma 97:373–377

Reissig JL (1977) The divalent cation ionophore A 23187 induces branching in *Neurospora*. J Cell Biol 75:30a

Robinson KR (1977) Reduced external calcium or sodium stimulates calcium influx in *Pelvetia* eggs. Planta 136:153–156

Robinson KR, Cone R (1980) Polarization of fucoid eggs by a calcium ionophore gradient. Science 207:77–78

Robinson KR, Jaffe LF (1975) Polarizing fucoid eggs drive a calcium current through themselves. Science 187:70–72

Rosenvinge LK (1888) Undersögelser öfver ydre faktorers inflydelse paa organdannelsen hos planterne. Diss Kopenhagen

Rubery PH, Sheldrake AR (1974) Carrier-mediated auxin transport. Planta 118:101–121

Schleiden MJ (1845) Grundzüge der wissenschaftlichen Botanik, 2nd edn. Leipzig

Schmiedel G, Schnepf E (1980) Polarity and growth of caulonema tip cells of the moss *Funaria hygrometrica*. Planta 147:405–413

Schröter K (1978) Asymmetrical jelly secretion of zygotes of *Pelvetia* and *Fucus:* An early polarization event. Planta 140:69–73

Scott BIH (1967) Electric fields in plants. Annu Rev Plant Physiol 18:409–418

Shropshire W (1979) Stimulus perception. In: Haupt W, Feinleib ME (eds) Encyclopedia of plant physiology. Springer, Berlin Heidelberg New York, Vol 7, pp 10–41

Sievers A, Schnepf E (1981) Morphogenesis and polarity of tubular cells with tip growth. In: Kiermayer O (ed) Cytomorphogenesis in plants, cell biology monographs. Springer, Berlin Heidelberg New York, Vol 8, pp 265–299

Spek J (1930) Zustandsänderung der Plasmakolloide bei Befruchtung und Entwicklung des *Nereis*-Eies. Protoplasma 9:370–425

Stahl E (1885) Einfluß der Beleuchtungsrichtung auf die Teilung der *Equisetum*-Sporen. Ber Dtsch Bot Ges 3:334–340

Sumper M, Reitmeier H, Oesterhelt D (1976) Zur Biosynthese der Purpurmembran von Halobakterien. Angew Chem 88:203–210

Van Brunt J, Harold FM (1980) Ionic control of germination of *Blastocladiella emersonii* zoospores. J Bacteriol 141:735–744

Vöchting H (1878) Über Organbildung im Pflanzenreich. Physiologische Untersuchungen über Wachsthumsursachen und Lebenseinheiten. Pt I, Cohen, Bonn

Vöchting H (1918) Untersuchungen zur experimentellen Anatomie und Pathologie des Pflanzenkörpers. II. Die Polarität der Gewächse. Tübingen

Walker NA, Smith FA (1977) Circulating electric currents between acid and alkaline zones associated with HCO_3^- assimilation in *Chara*. J Exp Bot 28:1190–1206

Weisenseel MH (1979) Induction of polarity. In: Haupt W, Feinleib ME (eds) Encyclopedia of plant physiology. Springer, Berlin Heidelberg New York, Vol 7, pp 485–505

Weisenseel MH, Jaffe LF (1972) Membrane potential, and impedance of developing fucoid eggs. Dev Biol Suppl 27:555–574

Weisenseel MH, Jaffe LF (1976) The major growth current through lily pollen tubes enters as K^+ and leaves as H^+. Planta 133:1–7

Weisenseel MH, Kicherer RM (1981) Ionic currents as control mechanism in cytomorphogenesis. In: Kiermayer O (ed) Cytomorphogenesis in plants, cell biology monographs. Springer, Berlin Heidelberg New York, Vol 8, pp 379–399

Weisenseel MH, Nuccitelli R, Jaffe LF (1975) Large electrical currents traverse growing pollen tubes. J Cell Biol 66:556–567

Weisenseel MH, Dorn A, Jaffe LF (1979) Natural currents traverse growing roots and root hairs of barley (*Hordeum vulgare* L.) Plant Physiol 64:512–518

Wendler S, Zimmermannn U, Bentrup FW (1983) Relationship between cell turgor pressure, electrical membrane potential, and chloride efflux in *Acetabularia mediterranea*. J Membr Biol 72:75–84

Went FW (1932) Eine botanische Polaritätstheorie. Jahrb Wiss Bot 76:528–554

Wettstein D von (1965) Die Induktion und experimentelle Beeinflussung der Polarität bei Pflanzen. In: Ruhland W (ed) Encyclopedia of plant physiology. Springer, Berlin Heidelberg New York, Vol XV, pp 275–330

Whitaker DM, Lowrance EW (1940) The effect of alkalinity upon mutual influences determining the developmental axis in *Fucus* eggs. Biol Bull 78:407–411

23 Fusion of Somatic Cells

T. NAGATA

23.1 Introduction

At the onset it should be noticed that the contents of this chapter are confined to the fusion of protoplasts; normal somatic plant cells with thick cell walls cannot fuse with each other and only naked protoplasts can do so. Nowadays fusion of protoplasts has attracted attention as the first step towards somatic hybridization, but the phenomenon is also interesting from the view of plant physiology. The first report of the fusion of protoplasts was by KÜSTER (1909, 1910). He observed the fusion of protoplasts which were isolated from the epidermis of tulip bulbs according to the mechanical method of KLERCKER (1892). Later MICHEL (1937) observed the fusion between protoplasts from different species and tissues. These protoplasts were also isolated mechanically from the tissues plasmolyzed previously with 0.5 M KNO_3; however, when sucrose was employed as a plasmolyticum, fusion was not observed. In these pioneering studies the work of MICHEL is especially important, because fusion could be induced not only between the protoplasts from the same tissue, but also between protoplasts from different tissues and different plants. However, these interesting observations went unnoticed until the report of POWER et al. (1970). This neglect may be due to the fact that at the time of MICHEL the technique of plant cell culture had not yet been established and the yield of protoplasts from tissues was extremely restricted because of the mechanical method utilized. The new possibility for the development of protoplast research started from the report of COCKING (1960), who isolated protoplasts from root tissues of tomato by a self-made crude cellulase preparation from a fungus, *Myrothecium verrucaria*. This possibility was substantiated, and further progress was accelerated by the finding of TAKEBE et al. (1968) that the commercially available cell-wall-degrading enzymes produced in Japan could liberate biologically active protoplasts from plant tissues. The use of protoplasts as an experimental material, has contributed remarkably to the elucidation of many problems whose resolution had previously been hampered by the presence of thick cell walls.

In this chapter recent progress in the fusion of protoplasts is reviewed by considering the interaction between cell surfaces of protoplasts. Genetical aspects such as somatic hybridization will not be included.

23.2 Preparation of Protoplasts

Since the application of commercially available enzymes (TAKEBE et al. 1968), protoplasts have been isolated from various plants and tissues. Since several reviews covering this subject are available (COCKING 1972, CONSTABEL 1975, BAJAJ 1977, ERIKSSON et al. 1978), the author will mention only two frequently utilized methods of preparing protoplasts from tobacco mesophyll and from suspension culture cells according to his more recent protocols using newly developed enzymes.

23.2.1 Preparation from Tobacco Mesophyll

Leaf mesophyll is one of the favourite materials for preparing protoplasts, because protoplasts from this source retain physiological functions at high biological activity, regenerate cell walls, and divide to form colonies at high frequency. Regeneration of whole plants could be shown in these colonies (NAGATA and TAKEBE 1971, TAKEBE et al. 1971).

After the lower epidermis was peeled from fully expanded leaves of tobacco (*Nicotiana tabacum* L. cv. Xanthi, Xanthi nc, Samsun etc.), they were cut into small pieces and infiltrated in vacuo with an enzyme solution consisting of 0.1% Pectolyase Y23 (Seishin Pharmaceutical Co. Ltd., Nihonbashi-Koamicho, Tokyo, Japan) (NAGATA and ISHII 1979), 0.5% potassium dextran sulphate (average degree of polymerization 3.5, Meito Sangyo Co. Ltd., Nagoya, Japan) and 0.6 M D-mannitol, pH being adjusted to 5.8. The enzymic digestion is carried out by shaking the tissues at 120 reciprocal excursions min^{-1} at 25 °C. The fraction released during the initial 5 min incubation is discarded because it consists mostly of cell debris and spongy parenchyma cells. After the renewal of the enzyme solution, incubation is continued for 30–45 min. The cells released, consisting mostly of palisade parenchyma cells, are collected by centrifugation at 100 g for 2 min. They are then washed once with 0.6 M D-mannitol and further converted to protoplasts by incubation at 30 °C for 20–30 min with a solution of 1% cellulase Onozuka RS (Yakult Pharmaceutical Ind. Co. Ltd., Takarazuka, Hyogo, Japan) (NAGATA et al. 1981) and 0.6 M D-mannitol being adjusted to pH 5.2. This procedure can be easily carried out aseptically (NAGATA and TAKEBE 1970). In this procedure, separation of cells and their conversion to protoplasts are carried out successively, and so the method is called "a two-step procedure". A one-step procedure is also possible when enzyme treatment is carried out with an enzyme mixture of Pectolyase Y23 and cellulase Onozuka RS (NAGATA and ISHII 1979, see also COCKING 1972).

23.2.2 Preparation from Tissue Culture Cells

Tissue culture cells, especially cells grown in suspension culture, are suitable for cell biological and biochemical studies because they provide sterile biologically active cells, and large-scale preparation is also easily possible. However, they have very often aberrant chromosome numbers and tend to lose the ability to differentiate during long-term propagation in vitro.

Tobacco BY-2 cells from *Nicotiana tabacum* L. cv. BY-2 are subcultured weekly in modified liquid medium of LINSMAIER and SKOOG (1965): KH_2PO_4 370 mg l^{-1}, thiamine HCl 1 mg l^{-1}, supplemented with 0.2 mg l^{-1}2,4-dichlorophenoxyacetic acid (2,4-D) and 3% sucrose, pH 5.8. Log-stage cells are washed once with 0.4 M D-mannitol and incubated at 30 °C with occasional gentle swirling in an enzyme mixture of 0.1% Pectolyase Y23, 1% cellulase Onozuka RS, 0.4 M D-mannitol at a pH adjusted to 5.5. After 50–60 min of incubation the cells are converted to protoplasts. In this procedure it is necessary to monitor the removal of cell walls by staining with a fluorescent brightener, Calcofluor White ST (American Cyanamid Co. Ltd., WAYNE, N.J., U.S.A) (NAGATA and TAKEBE 1970), because spherical cells sometimes retained thin walls after digestion with cellulase Onozuka R10 or Driselase (Kyowa Hakko Kogyo Co. Ltd., Tokyo, Japan). The protoplasts are collected by centrifugation at 100 g for 2 min. This protocol is essentially according to NAGATA et al. (1981).

23.3 Methods of Protoplast Fusion

23.3.1 Spontaneous Fusion

Spontaneously fused protoplasts are frequently observed, especially when they are isolated from young tissues according to the one-step procedure. WITHERS and COCKING (1972) studied this process by electron microscopy and concluded that spontaneous fusion occurred as a passive phenomenon by the expansion of plasmodesmata, which interconnect cells in plant tissues. In contrast, liliaceous meiotic protoplasts, isolated by a brief incubation with enzymes, tended to fuse by collision only, when observed within 30 min after isolation (ITOH 1973). The cell membranes of these protoplasts were supposed to be in an unstable condition, and temporarily susceptible to spontaneous fusion.

The fusion of protoplasts in these cases is caused by the special condition of the materials and is not experimentally controllable. This phenomenon is not, therefore, considered further.

23.3.2 Sodium Nitrate Method

POWER et al. (1970) observed the intra- and interspecific fusion of protoplasts from monocotyledonous plants in the presence of 0.25 M $NaNO_3$. In the presence of KNO_3, which was originally employed as a plasmolyticum by MICHEL (1937), fusion was induced as a rare event. POTRYKUS (1971) applied this method for the fusion between protoplasts from flower petals of two different species of *Torenia,* and observed interspecific fusion. Subsequently CARLSON et al. (1972) reported using this method that they obtained the fusion between protoplasts from leaves of *Nicotiana glauca* and *N. langsdorffii*. Further, they added that they regenerated whole plants from these fusion products. Thus this work played a role in gathering attention on the subject of protoplast fusion. However, the sharp criticism aimed at this work has not been resolved (MELCHERS 1977). This method was found to be applicable only to protoplasts from young tissues, and the results were found not to be reproducible. Furthermore, protoplasts deteriorated during this procedure. The technique was, therefore, replaced by other, more reliable methods.

23.3.3 High pH–High Ca^{2+} Method

KELLER and MELCHERS (1973) observed fusion of protoplasts from tobacco mesophyll when they were incubated at 37 °C for 30–40 min with 50 mM glycine-NaOH buffer (pH 10.5) containing 50 mM CaCl$_2$ and 0.4 M D-mannitol. The fusion percentage was 20–50%. Further, MELCHERS and LABIB (1974) employed this method for the fusion of protoplasts from mesophyll of chlorophyll deficient two light-sensitive haploid mutant tobaccos, sublethal and virescent, and obtained genetically complemented whole plants after selection under the high light intensity. This method was most suitable for obtaining fusion products and their regenerants from mesophyll protoplasts (POWER et al. 1980). It was observed later that some protoplasts, especially from tissue culture cells, were not easily susceptible to this treatment. It was soon found, however, that this medium increased the fusion frequency of protoplasts when it was applied as a washing medium to remove polyethylene glycol (see Sect. 23.3.4). On the other hand, it is interesting that cell fusion between plant protoplasts and amphibian cells was induced by this procedure with higher efficiency than others when the amphibian cells were pretreated with a protease (WARD et al. 1979).

23.3.4 Polyethylene Glycol Method

KAO and MICHAYLUK (1974) found strong adhesion of protoplasts after the addition of 33% (wt/wt) polyethylene glycol (PEG) 1540 or 4000, and subsequent removal of PEG by washing with a culture medium could induce the fusion between protoplasts from *Vicia hajastana* and *Pisum sativum* and between protoplasts from *Glycine max* and *Hordeum vulgare*. Independently, WALLIN et al. (1974) reported essentially the same results in protoplasts from cultured cells of *Daucus carota*. In this method a suitable range of molecular weight of PEG was around 1540–6000, and the presence of CaCl$_2$ and KH$_2$PO$_4$ increased the fusion frequency. Subsequent washing with high pH–high Ca^{2+} medium to remove PEG could significantly increase the fusion frequency, and 35% fusion was observed between *Pisum sativum* and *Glycine max* protoplasts (KAO et al. 1974). Because of the high reproducibility of this method it has been successfully utilized for obtaining fusion of protoplasts of any origin, and it was further applied to the fusion of animal cells and microbial protoplasts (PONTECORVO et al. 1977, PEBERDY 1979, HOPWOOD 1981). This method opened a new era of somatic hybridization of plants. As there are several reviews covering this subject (COCKING 1977, SCHIEDER and VASIL 1980), the present author will confine himself to stating that MELCHERS et al. (1978) regenerated whole plants from the fusion products between dihaploid cells of *Solanum tuberosum* and chlorophyll deficient light-sensitive mutants of *Lycopersicon esculentum,* which are sexually incompatible. Further, SCHIEDER (1978) regenerated whole plants from the fusion products between *Datura innoxia* and *D. discolor* and between *D. innoxia* and *D. stramonium* and GLEBA and HOFFMANN (1980) reported the somatic hybrids between *Arabidopsis thaliana* and *Brassica campestris.*

By means of this procedure, plant protoplasts and animal cells formed somatic hybrid cells (JONES et al. 1976, DUDITS et al. 1976) and fusion between plant and algal protoplasts was also reported (FOWKE et al. 1981). However, in the combination between *Vinca rosea* protoplasts and *Agrobacterium tumefaciens* spheroplasts, treatment with PEG was found to induce the uptake of the spheroplasts into protoplasts rather than fusion (HASEZAWA et al. 1981). This response might be determined by difference in size of the partners.

23.3.5 Polyvinyl Alcohol Method

NAGATA (1978) found the adhesion and fusion of tobacco mesophyll protoplasts by adding more than 20% (wt/wt) polyvinyl alcohol (PVA, average degree of polymerization 500) and after removing the PVA by washing with the high pH–high Ca^{2+} buffer of KELLER and MELCHERS (1973). Fusion was observed with similar frequency as with PEG. Although the adhesive force of this chemical was slightly weaker than that of PEG, it was less harmful to protoplasts. Further, he found that PVA of lower degree of polymerization (DP 200 or 300) was even more effective in inducing fusion. Although the effect of PVA on the surface of protoplasts is less understood than that of PEG, both chemicals have similar characteristics in that both are weak surfactants and have high solubility in water. The chemical formulae are shown in Fig. 1.

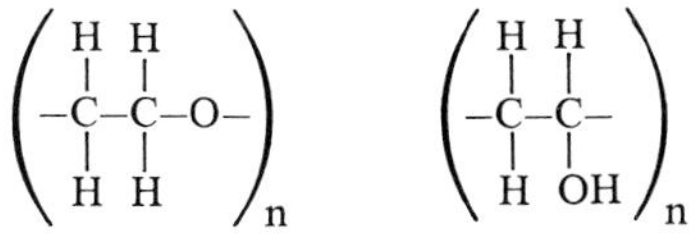

$$\left(\begin{matrix} H & H \\ | & | \\ -C-C-O- \\ | & | \\ H & H \end{matrix}\right)_n \qquad \left(\begin{matrix} H & H \\ | & | \\ -C-C- \\ | & | \\ H & OH \end{matrix}\right)_n$$

Fig. 1. Chemical formulae of polyethylene glycol and polyvinyl alcohol

Polyethylene Glycol   Polyvinyl Alcohol

23.3.6 Electrical Pulse Method

ZIMMERMANN and SCHEURICH (1981) reported on a completely physical method of protoplast fusion. When protoplasts from mesophyll cells of *Vicia faba* were placed in a highly inhomogenous alternating electric field (sine wave, 200 V cm^{-1} at maximum field strength, 0.5 M Hz, 0.2 mm electrode distance), they formed aggregates of 2 to 3 protoplasts on the electrodes or bridges con-

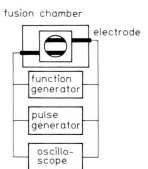

fusion chamber

electrode

function generator

pulse generator

oscillo-scope

Fig. 2. Schematic presentation of an apparatus for electrically induced fusion. Two electrodes are glued to a slide glass in parallel and can be seen under a microscope. The frequency and amplitude of the dielectrophoretic voltage are controlled by a function generator. A pulse generator is connected in parallel to elicit the breakdown pulse. The applied voltages are recorded on an oscilloscope

sisting of 5 to 6 protoplasts between the electrodes. The schematic presentation of this apparatus is shown in Fig. 2. This phenomenon has been known as dielectrophoresis. Additional application of a single field pulse (square wave, 750 Vcm^{-1}, 50 µs) to these aggregates induced fusion between protoplasts. Both dielectrophoretic voltage, amplitude and pulse frequency should be appropriately chosen for materials utilized. The characteristics of this procedure is that extensive fusion can be synchronously induced. One prerequisite is that the conductivity of the medium should be lower than 10^{-4} S cm^{-1}. Thus, a simple non-electrolytic solution such as 0.5 M D-mannitol, which gives conductivity of around 1.4×10^{-5} S cm^{-1}, is suitable. This procedure was also successfully applied to the fusion of other plant protoplasts, animal cells and liposomes (Zimmermann et al. 1981).

23.4 Mechanism of Protoplast Fusion

The methods of protoplast fusion shown in the preceding section were developed independently from each other and it is difficult to find a common explanation for them. However, if there is such an explanation, efforts to find it would contribute to an understanding of the mechanism of fusion, and develop further new fusion methods. There are thus far no published discussions of this subject, but there are several suggestive reviews concerning this matter dealing with the fusion of animal cells (Poste and Allison 1973, Knutton and Pasternak 1979). Thus the common features of the fusion methods will be represented in reference to some of these discussions.

Fusion of protoplasts can be divided into two processes, namely the adhesion and the subsequent membrane fusion, although swelling of the fusion products was counted as a third step in animal cells (Knutton and Pasternak 1979). In order to understand the adhesion of cells, it is necessary to consider their surface characteristics (Jones 1975). Thus in the following discussion the surface characteristics of protoplasts are first depicted and then adhesion and membrane fusion are described.

23.4.1 Surface Characteristics of Protoplasts

Adhesion of protoplasts is rarely observed, when protoplasts are suspended in a non-electrolytic osmoticum such as D-mannitol or sucrose without cations or with very low concentration of cations. This phenomenon can be explained according to the DLVO theory which has been originally developed for the understanding of the interaction of hydrophobic colloid particles. The DLVO theory was independently developed in the 1940's by Derjaguin and Landau in the USSR, and by Verwey and Overbeek in the Netherlands (JONES 1975).

CURTIS (1960) applied this theory to the understanding of the cell–cell interaction of animal cells. He succeeded in explaining why adhesion of animal cells was observed with the separation of 10–20 nm from the secondary minimum of the potential energy curve of cell–cell interaction. According to the DLVO theory the potential energy of cell–cell interaction is expressed as the summation of the potential energy of electrostatic repulsive forces and that of Van der Waals forces. However, as the Van der Waals forces which originate from the molecular attractive force of the cell surface are constant, most of the potential energy of the cell–cell interaction is governed by electrostatic forces. Thus it is predicted from this theoretical treatment that cell–cell interaction is controlled mostly by the charged state of the cell surface.

The charged state of the cell surface can be experimentally measured by cell electrophoresis. As is shown schematically in Fig. 3, when protoplasts are put in a capillary in an electric field, they move according to surface charges. From the velocity of the protoplasts in this electric field the electrophoretic mobility (U) can be calculated. Then according to Smoluchowski's equation [Eq. (1)] the ζ-potential can be calculated (BANGHAM et al. 1958).

$$\zeta = \frac{4\pi\eta U}{\varepsilon} \tag{1}$$

$\zeta = \zeta$-potential (mV); $U =$ electrophoretic mobility ($\mu \cdot s^{-1} V^{-1}$ cm); $\varepsilon =$ dielectric constant; $\eta =$ viscosity (poise).

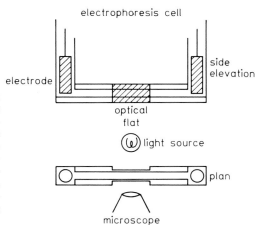

Fig. 3. Schematic presentation of a cell electrophoresis apparatus. Protoplasts are put into the capillary under an electric field. The velocity of protoplasts is measured with a microscope at the optical flat taking into account the electroosmosis of the capillary. From this velocity the electrophoretic mobility and then the ζ-potential are calculated according to Smoluchowski's equation [Eq. (1)]

Table 1. ζ-potential of protoplasts from several origins[a]

Source of protoplasts	ζ-potential
Nicotiana tabacum (2n)	−25 to −35 mV
Nicotiana tabacum (n)	−25 mV
Petunia hybrida (2n)	−30 mV
Brassica rapa (2n)	−23 mV
Vigna unguiculata (2n)	−10 to −15 mV

[a] ζ-potential of protoplasts from several origins was calculated according to the Smoluchowski's equation [Eq. (1)] from the electrophoretic mobility (U) measured in the apparatus shown in Fig. 3

In the usual apparatus, as is shown schematically in Fig. 3, the velocity of protoplasts is measured with a microscope, but recently laser Doppler spectroscopy has also been employed in an optical system which facilitated the rapid measurement of the velocity (Smith et al. 1976).

Nagata and Melchers (1978) measured the ζ-potential of protoplasts from mesophyll of several plants. As is shown in Table 1, these values were in the range of −20 to −30 mV. Grout et al. (1973) reported a similar value for tomato fruit protoplasts. The ζ-potential measured by cell electrophoresis is theoretically equal to the surface potential, if the potential value is not too high, as is the case of living cells (Jones 1975). Based on these values, the potential energy of the interaction between two protoplasts was calculated (Ohshima, Gotoh, Nagata, unpublished results). These calculations used the computer programme of the equation of Ohshima (1975), which was modified from the original DLVO theory assuming that the surface charge density is constant, which reflects the characteristics of living cells better than the original assumption for colloidal particles. Actually the results of calculation showed that the adhesion of protoplasts is improbable in the usual non-electrolytic osmoticum.

In this context care should be taken that the surface potential or ζ-potential should not be confused with the membrane potential. Even in a recent review (Galun 1981) such confusion was observed. Surface potential originates from the fixed charges of the surface, while membrane potential originates from Donnan membrane equilibrium. Therefore, both are completely different (Jones 1975).

From the value for the ζ-potential, Nagata and Melchers (1978) calculated the surface charge density (σ) of protoplasts according to the Gouy-Chapman equation [Eq. (2)]. When tobacco mesophyll protoplasts had a ζ-potential of −30 mV, σ was calculated to be 2×10^3 electrostatic units cm^{-2}, which corresponds to the value of animal cells (Haydon and Seaman 1962).

$$\sigma = \left(\frac{2\,k\,T\,n\,\varepsilon}{\pi}\right)^{1/2} \sin h \left(\frac{v e\,\Psi}{2\,k\,T}\right) \qquad (2)$$

σ = charge density (esu cm^{-2}); Ψ = surface potential ($=\zeta$) (mV); k = Boltzmann constant; v = valency of ion; T = temperature (K); n = ion concentration.

However, this negative charge of tobacco protoplasts did not originate from sialic acid, which is responsible for the negative charges of animal cells, because α-neuraminidase, which can remove sialic acid and thus reduce the ζ-potential of animal cells (EYLER 1962), did not affect the surface charge of these protoplasts. On the other hand, acid phosphate reduced the surface charges of protoplasts most significantly, which suggests that the surface charges of these protoplasts mostly originates from phosphatase side chains in the cell membrane. Further, this phenomenon can readily explain why plant protoplasts were not caused to adhere and fuse by Sendai virus (HVJ), as was observed by BAWA and TORREY (1971). The fusion of animal cells by Sendai virus requires the presence of sialic acid on the cell surface (KNUTTON and PASTERNAK 1979).

23.4.2 Adhesion of Protoplasts

23.4.2.1 High pH–High Ca^{2+} Method

Although protoplasts do not aggregate under usual conditions, KELLER and MELCHERS (1973) showed that in a high pH–high Ca^{2+} medium a high concentration of Ca^{2+} contributed to the aggregation of protoplasts. NAGATA and MELCHERS (1978) measured the decrease in ζ-potential as a function of the increase in Ca^{2+} concentration as is shown in Table 2. Protoplasts aggregated in the range of 50–100 mM Ca^{2+} which is used for fusion, and in this range the ζ-potential became nearly 0 mV. According to the DLVO theory, this phenomenon can be explained as follows. The decrease in ζ-potential is brought about by neutralization of surface charges, which weakens the repulsive forces of the negative charges of protoplast surface. The weakened repulsive forces can be overcome by the attractive Van der Waals forces, and aggregation will then ensue. GROUT et al. (1973) reported that the aggregation of protoplasts in the high concentrations of $NaNO_3$ was accompanied by a decrease in ζ-potential.

On the other hand, as was shown by WILKINS et al. (1962), high concentration of various cations brought about a decrease in ζ-potential accompanied by aggregation of leucocytes, and this phenomenon was observed also in plant

Table 2. ζ-potential of protoplasts in the presence of calcium [a]

$CaCl_2$	ζ-potential	Aggregation
0 mM	−28 mV	−
1 mM	−25 mV	−
10 mM	− 9 mV	−
100 mM	0 mV	+ + +

[a] ζ-potential of protoplasts from tobacco mesophyll was calculated in the presence of various concentrations of calcium (NAGATA and MELCHERS 1978)

protoplasts (unpublished results), but only Ca^{2+} was found to be effective for fusion of protoplasts. Other cations could not be replaced by Ca^{2+}.

23.4.2.2 Polyethylene Glycol Method

Maggio et al. (1976) observed a decrease of surface potential of an artificial monolayer of phosphatidylcholine and phosphatidylethanolamine with increasing concentrations of PEG. They discussed this decrease in surface potential as closely related to membrane fusion. So far there are no available data for the change in ζ-potential of protoplasts caused by PEG, because the viscosity at higher concentration of PEG prevented measurement of the ζ-potential by means of cell electrophoresis. Nevertheless, it is highly probable that the decrease of ζ-potential of protoplasts is brought about by the higher concentration of PEG, as was observed in the case of artificial membranes. The decrease in ζ-potential could cause the aggregation of protoplasts according to the mechanism discussed for the Ca^{2+}-mediated aggregation. Thus the DLVO theory should also help to understand the aggregation of protoplasts by PEG. In this context, needless to say, it is urgently necessary to develop a new experimental method which enables one to measure the ζ-potential (surface potential) of protoplasts in the presence of higher concentration of PEG.

However, there is some difference between the aggregation of protoplasts by PEG and that by cations. Aggregation by salts is essentially reversible, while that induced by PEG is irreversible. This behavior may be due to the fact that in higher concentrations of PEG, cell surfaces became extremely dehydrated, which brought the two opposing cell membranes into close contact. Blow et al. (1978) showed that in the higher concentrations of PEG animal cells were in a dehydrated condition, which was measured as the absence of free water by scanning calorimetry.

23.4.2.3 Electrically Induced Aggregation

The aggregation of protoplasts in an alternating inhomogeneous electric field reported by Zimmermann and Scheurich (1981) was explained by the authors (Zimmermann et al. 1981) as follows. As shown schematically in Fig. 4, the negative charges on the protoplast surface were "masked" or neutralized in an alternating electric field and at the same time dipoles were induced in protoplasts by a non-uniform electric field. The induced dipoles brought the protoplasts together and these aggregates moved to the higher electric field. The small aggregates or bridges between the two electrodes were formed depending on the protoplast density. The electrically induced aggregation was also reversible after the removal of the electric field.

23.4.2.4 Lectin-Induced Aggregation

Glimelius et al. (1974) observed that aggregation of protoplasts from suspension culture cells of *Daucus carota* was induced by concanavalin A (Con A). Further Larkin (1978) examined the action of 11 different lectins on protoplasts,

Fig. 4. Schematic presentation of the formation of protoplast chains during dielectrophoresis. The surface
charges of protoplasts are neutralized by an alternating
electric field and protoplasts adhere to each other because of the dipoles induced by an inhomogeneous electric field

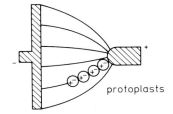

but only Con A, soybean (*Glycine max*) lectin, castor bean (*Ricinus communis*)
lectin, and peanut (*Arachis hypogaea*) lectin induced aggregation of protoplasts
from four species of *Nicotiana, Petunia hybrida, Vicia faba, Daucus carota,
Triticum aestivum, Avena sativa, Zea mays* and *Bromus inermis*. This aggregation
was counteracted by methyl-α-D-glucoside and/or methyl-β-galactoside and thus
it was specific aggregation. However, these lectins could not differentiate protoplasts from different species. In contrast, in animal cells such differentiation
was studied extensively and is important in tumour biology (INBAR and SACHS
1969) and enabled the cell–cell adhesion in special combination (RUTISHAUSER
and SACHS 1975).

Although the aggregation of protoplasts induced by lectins did not result
in fusion, this binding of lectins to protoplasts was utilized for reducing loss
of fusion products (GLIMELIUS et al. 1978). In the fusion of protoplasts by PEG
procedure, significant amounts of protoplasts may be lost during the washing
procedure; however, pretreatment of the surface of Petri dishes with Con A
decreases the detachment of protoplast from the surface of dishes and increased
the yield of fusion products.

BURGESS and LINSTEAD (1976) ascertained that Con A bound specifically
to cell membranes of protoplasts from cultured cells of *Vitis vinifera* and of
tobacco mesophyll. They used electron microscopy and Con A conjugated with
colloidal gold. The bridging was also inhibited by α-methyl mannoside.

23.4.3 Fusion of Protoplasts

In studies of the fusion of animal cells it was shown by POSTE and ALLISON
(1973) and by KNUTTON and PASTERNAK (1979) that cell fusion following cell
adhesion is a phenomenon which is closely related to a phase separation as
a result of the increase in fluidity of the cell membrane. Cell fusion is a general
phenomenon which can be induced in animal cells, plant protoplasts and also
in microbial protoplasts.

23.4.3.1 High pH–High Ca^{2+} Method

In this condition cell fusion was induced at 37 °C and not at lower temperature.
The presence of such a critical temperature is a typical characteristic of phase
separation of the cell membrane as a result of an increase in membrane fluidity.
TRÄUBLE and EIBL (1974) showed that in artificial lipid membranes the fluidity

of membranes increased at high pH. Accordingly, protoplast membrane should be more fluid at high pH. All other conditions utilized in high pH–high Ca^{2+}-mediated fusion also have been shown to be related to the increase in membrane fluidity. In this procedure protoplasts suspended in 0.7 M D-mannitol were transferred to fusion medium employing 0.4 M D-mannitol. BOROCHOV and BOROCHOV (1979) measured the fluorescence polarization of a protoplast labelled with 1,6-diphenyl-1,3,5-hexatriene in a single-cell microviscosimeter (Elscint, Israel) and observed that the decrease in osmotic pressure was accompanied by increased membrane fluidity.

Furthermore, BOSS and MOTT (1980) determined that the addition of Ca^{2+} contributed to the increase in membrane fluidity in electron spin resonance (ESR) spectra as determined by using 5-nitroxy stearic acid as a spin-label probe, although Ca^{2+} contributed to the aggregation of protoplasts (see preceding section). Thus, all the conditions in the high pH–high Ca^{2+} medium concertedly induce the increase in fluidity of the membrane and bring about phase separation of the membrane to fusion. This effect of Ca^{2+} could not be replaced by other cations.

23.4.3.2 Polyethylene Glycol Method

Fusion of protoplast mediated by PEG was not observed in the presence of PEG, but was observed during the removal of the PEG by washing with high pH–high Ca^{2+} medium (KAO et al. 1974). BOSS and MOTT (1980), using 5-nitroxy stearic acid as a spin-label probe, observed that no increase in membrane fluidity was observed in the presence of PEG alone. Addition of Ca^{2+} to the PEG solution increased the membrane fluidity sufficient for phase separation and fusion.

ROBENEK and PEVELING (1978) used freeze-etched cell membrane and observed the localization of intramembraneous particles during the removal of PEG by treating with washing medium. Similar results were also reported by ROBINSON et al. (1979) during fusion of animal cells by PEG. Membrane fusion was predominantly observed by electron microscopy in the particle-free area, which was presumed to be the lipid region resulting from phase-separation. In addition to this observation ROBINSON et al. (1979) observed also that PEG treatment did not disrupt the cytoskeletal elements such as microtubules and microfilaments, while in virus and Ca^{2+}-induced fusion of hen erythrocytes the partition of cytoskeletal elements to cell fusion was suggested.

23.4.3.3 Electrical Pulse Method

Electrical pulse-induced fusion of protoplasts was explained by ZIMMERMANN et al. (1981) as follows: A field pulse of short duration brought about dielectric breakdown and formed holes for a short while in the opposing two membranes, as is shown schematically in Fig. 5. In this condition exchange of lipid molecules should be possible at the holes. During the resealing process, lipid bridges are formed between the opposing membranes and then lead to form the short radia of curvature at the holes and to expose to high surface tensions. These bridges

Fig. 5. Schematic presentation of dielectric breakdown in opposing two protoplasts. Electrical breakdown occurs in the membrane areas of the zone of contact which is orientated vertically with respect to the field line

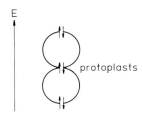

progressed to membrane fusion, as membrane fusion is energetically more favoured than separate resealing of the two membranes. The process following the lipid exchanges at holes reflects the fluid mosaic characteristics of the cell membrane and might be closely related to membrane fluidity.

23.5 Liposomes as Models of Cell Fusion

WILSCHUT et al. (1980) observed the fusion of liposomes made of phosphatidyl-serine by the addition of Ca^{2+}. As one liposome contained $Tb(citrate)_3^{6-}$ and the other contained dipicolinic acid (DPA), the aggregation could be monitored by light scattering and the fusion process by the fluorescence spectra of the complex of Tb/DPA. It was ascertained that treatment of Ca^{2+} could induce the aggregation and subsequent fusion of these liposomes. This effect of Ca^{2+} could not be duplicated by other cations including Mg^{2+}. These results are very important for understanding the fusion of negatively charged protoplasts.

The fusion of protoplasts could be induced solely by liposomes. NAGATA et al. (1979) reported on the fusion of protoplasts induced by liposomes consisting of a positively charged synthetic phospholipid of 1,2-O-dipentadecyl-methylidene-glycerol-3-phosphoryl-(N-ethylamino)-ethanolamine. The positive charge contributed to the aggregation of the protoplasts, and the dioxolane ring of this lipid probably disturbed the membrane structure and induced fusion.

NAGATA et al. (1981) also studied the interaction of protoplasts with liposomes. The liposomes consisting of phosphatidylserine and cholesterol encapsulating tobacco mosaic virus (TMV) RNA were added to protoplasts of cultured tobacco cells isolated according to the procedure shown in Section 23-2. After this the mixture was treated with PEG or PVA, these substances were removed by washing with high pH–high Ca^{2+} medium, and these cells were cultured for 24 h. The functional expression of TMV-RNA in protoplasts was monitored by staining with the fluorsescent antibody against TMV. Using this procedure, more than 60% of the protoplasts were infected. However, the introduction of liposomes into protoplasts was shown by electron microscopy to be carried out by the process of endocytosis, not by fusion between liposomes and cell membranes, although fusion between liposomes and protoplasts had been expected because of the use of PEG or PVA (NAGATA 1983). This phenomenon may be related to the interaction of protoplasts with bacterial spheroplasts

(Hasezawa et al. 1981). The endocytosis of liposomes or bacterial spheroplasts by plant protoplasts might be energetically more favoured than fusion between them.

23.6 Conclusion

In this chapter several empirically developed fusion methods of protoplasts have been arranged in an attempt to deduce common features and to understand the mechanism of protoplast fusion. Several important points need to be emphasized. As has been shown, the fusion process could be subdivided into two processes, adhesion and subsequent membrane fusion. In the initial adhesion it is very useful to introduce the DLVO theory. As is theoretically deduced in all of these fusion methods, control of surface chages of protoplasts plays an important role for aggregation. In the membrane fusion the increase of membrane fluidity has been shown to be the most important factor for fusion. Even in the pure physical method of electrical pulse-induced fusion, the latter process obeyed the general features of the fluid mosaic model of membrane.

Nevertheless, the present status of the research on this subject is still in its infancy and there is still not enough information for full understanding. Thus it is urgently necessary to introduce new methods and, no less important, fresh ideas and concepts. Such efforts should result in the better understanding of protoplast fusion.

Finally it should be stressed that protoplast fusion offers many new and attractive research opportunities in the fields of plant physiology and cell biology.

References

Bajaj YPS (1977) Protoplast isolation, culture and somatic hybridization. In: Reinert J, Bajaj YPS (eds) Applied and fundamental aspects of plant cell, tissue, and organ culture. Springer, Berlin Heidelberg New York, pp 467–577

Bangham AD, Flemans R, Heard DH, Seaman GVF (1958) An apparatus for microelectrophoresis of small particles. Nature 182:642–644

Bawa SB, Torrey JG (1971) "Budding" and nuclear division in cultured protoplasts of corn, convolvulus, and onion. Bot Gaz 132:240–245

Blow AMJ, Botham GM, Fisher D, Goodall AH, Tilcock CPS, Lucy JA (1978) Water and calcium in cell fusion induced by poly(ethylene glycol). FEBS Lett 94:305–310

Borochov A, Borochov H (1979) Increase in membrane fluidity in liposomes and plant protoplasts upon osmotic swelling. Biochim Biophys Acta 550:546–549

Boss WF, Mott RL (1980) Effect of divalent cations and polyethylene glycol on the membrane fluidity of protoplasts. Plant Physiol 66:835–837

Burgess J, Linstead PJ (1976) Ultrastructural studies of the binding of concanavalin A to the plasmalemma of higher plant protoplasts. Planta 130:73–79

Carlson PS, Smith HH, Dearing RD (1972) Parasexual interspecific plant hybridization. Proc Natl Acad Sci USA 69:2292–2294

Cocking EC (1960) A method for the isolation of plant protoplasts and vacuoles. Nature 187:962–963

Cocking EC (1972) Plant cell protoplasts–isolation and development. Annu Rev Plant Physiol 23:29–50

Cocking EC (1977) Selection and somatic hybridization. In: Thorpe TA (ed) Frontiers of plant tissue culture 1978. The Bookstore, Univ Calgary, Calgary pp 151–158

Constabel F (1975) Protoplast isolation. In: Gamborg OL, Wetter LR (eds) Plant tissue culture methods. National Research Council of Canada, Ottawa, pp 11–21

Curtis ASG (1960) Cell contacts: Some physical considerations. Am Nat 94:37–56

Dudits D, Rasko I, Jadlaczky GY, Lima-de-Faria A (1976) Fusion of human cells with carrot protoplasts induced by polyethylene glycol. Hereditas 82:121–124

Eriksson T, Glimelius K, Wallin A (1978) Protoplast isolation, cultivation and development. In: Thorpe TA (ed) Frontiers of plant tissue culture 1978. The Bookstore, Univ Calgary, Calgary, pp 131–139

Eyler EH (1962) The contribution of sialic acid to the surface charge of the erythrocyte. J Biol Chem 237:1992–2000

Fowke LC, Marchant HJ, Gresshoff PM (1981) Fusion of protoplasts from carrot cell cultures and the green alga *Stigeoclonium*. Can J Bot 59:1021–1025

Galun E (1981) Plant protoplasts as physiological tools. Annu Rev Plant Physiol 32:237–266

Gleba YY, Hoffmann F (1980) "Arabidobrassica": A novel plant obtained by protoplast fusion. Planta 149:112–117

Glimelius K, Wallin A, Eriksson T (1974) Agglutination effects of concanavalin A on isolated protoplasts of *Daucus carota*. Physiol Plant 31:225–230

Glimelius K, Wallin A, Eriksson T (1978) Concanavalin A improves the polyethylene glycol method for fusing plant protoplasts. Physiol Plant 44:92–96

Grout BWW, Willison JHM, Cocking EC (1973) Interactions at the surface of plant cell protoplasts; An electrophoretic and freeze-etch study. J Bioenerget 4:586–611

Hasezawa S, Nagata T, Syono K (1981) Transformation of *Vinca* protoplasts mediated by *Agrobacterium* spheroplasts. Mol Gen Genet 182:206–210

Haydon DA, Seaman GVF (1962) An estimation of the surface ionogenic groups of the human erythrocyte and of *Escherichia coli*. Proc R Soc Lond Ser B 156:533–549

Hopwood HA (1981) Genetic studies with bacterial protoplasts. Annu Rev Microbiol 35:237–272

Inbar M, Sachs L (1969) Interaction of the carbohydrate-binding protein concanavalin A with normal and transformed cells. Proc Natl Acad Sci USA 63:1418–1425

Itoh M (1973) Studies on the behavior of meiotic protoplasts. II. Induction of a high fusion frequency in protoplasts from liliaceous plants. Plant Cell Physiol 14:865–872

Jones MN (1975) Biological interfaces. Elsevier, North/Holland, Amsterdam Oxford New York

Jones CW, Mastrangelo IA, Smith HH, Liu HZ (1976) Interkingdom fusion between human (HeLa) cells and tobacco hybrid (GGLL) protoplasts. Science 193:401–403

Kao KN, Michayluk MR (1974) A method for high-frequency intergeneric fusion of plant protoplasts. Planta 115:355–367

Kao KN, Constabel F, Michayluk MR, Gamborg OL (1974) Plant protoplasts fusion and growth of intergeneric hybrid cells. Planta 120:215–227

Keller WA, Melchers G (1973) The effect of high pH and calcium on tobacco leaf protoplast fusion. Z Naturforsch 28c:737–741

Klercker J af (1892) Eine Methode zur Isolierung lebender Protoplasten. Oefversigt Kongl Vetens-Akad Foerhandlingar 49:463–474

Knutton S, Pasternak CA (1979) The mechanism of cell–cell fusion. Trends Biochem Sci 4:220–223

Küster E (1909) Über die Verschmelzung nackter Protoplasten. Ber Dtsch Bot Ges 27:589–598

Küster E (1910) Eine Methode zur Gewinnung abnorm großer Protoplasten. Arch Entw Mechan 30:351–355

Larkin PJ (1978) Plant protoplast agglutination by lectins. Plant Physiol 61:626–629

Linsmaier EM, Skoog F (1965) Organic growth factor requirements of tobacco tissue cultures. Physiol Plant 18:100–127

Maggio B, Ahkong QF, Lucy JA (1976) Poly(ethylene glycol), surface potential and cell fusion. Biochem J 158:647–650

Melchers G (1977) Plant hybrids by fusion of protoplasts. In: Beers Jr RF, Bassett EG (eds) Recombinant molecules: Impact on science and society. Raven Press, New York, pp 209–227

Melchers G, Labib G (1974) Somatic hybridization of plants by fusion of protoplasts. I. Selection of light-resistant hybrids of "haploid" light sensitive varieties of tobacco. Mol Gen Genet 135:277–294

Melchers G, Sacristàn MD, Holder AA (1978) Somatic hybrid plants of potato and tomato regenerated from fused protoplasts. Carlsberg Res Comm 43:203–218

Michel W (1937) Über die experimentelle Fusion pflanzlicher Protoplasten. Arch Exp Zellforsch 20:230–252

Nagata T (1978) A novel cell-fusion method of protoplasts by polyvinyl alcohol. Naturwissenschaften 65:263–264

Nagata T (1983) Interaction of liposomes and protoplasts as a model system of protoplast fusion. In: Beers Jr RF, Bassett EG (eds) Cell fusion: Gene transfer and transformation. Raven Press, New York, pp 217–226

Nagata T, Ishii S (1979) A rapid method for isolation of mesophyll protoplasts. Can J Bot 57:1820–1823

Nagata T, Melchers G (1978) Surface charges of protoplasts and their significance in cell–cell interaction. Planta 142:235–238

Nagata T, Takebe I (1970) Cell wall regeneration and cell division in isolated tobacco mesophyll protoplasts. Planta 92:12–20

Nagata T, Takebe I (1971) Plating of isolated tobacco mesophyll protoplasts on agar medium. Planta 99:12–20

Nagata T, Eibl H, Melchers G (1979) Fusion of plant protoplasts by a positively charged synthetic phospholipid. Z Naturforsch 34c:460–462

Nagata T, Okada K, Takebe I, Matsui C (1981) Delivery of tobacco mosaic virus RNA into plant protoplasts mediated by reverse-phase evaporation vesicles (liposomes). Mol Gen Genet 184:161–165

Ohshima H (1975) Diffuse double layer interaction between two spherical particles with constant surface charge density in an electrolyte solution. Colloid Polymer Sci 263:158–163

Peberdy JF (1979) Fungal protoplasts: Isolation, reversion, and fusion. Annu Rev Microbiol 33:21–39

Pontecorvo G, Riddle PN, Hales A (1977) Time and mode of fusion of human fibroblasts treated with polyethylene glycol (PEG). Nature 265:257–258

Poste G, Allison AC (1973) Membrane fusion. Biochim Biophys Acta 300:421–465

Potrykus I (1971) Intra and interspecific fusion of protoplasts from petals of *Torenia beillonii* and *Torenia fournieri*. Nat New Biol 231:57–58

Power JB, Cummins SE, Cocking EC (1970) Fusion of isolated protoplasts. Nature 225:1016–1018

Power JB, Berry SF, Chapman JV, Cocking EC (1980) Somatic hybridization of sexually incompatible petunias: *Petunia parodii, Petunia parviflora*. Theor Appl Genet 54:1–4

Robenek H, Peveling E (1978) Beobachtungen am Plasmalemma während der Fusion isolierter Protoplasten von *Skimmia japonica* Thunb. Ber Dtsch Bot Ges 91:351–359

Robinson JM, Roos DS, Davidson RL, Karnovsky MJ (1979) Membrane alterations and other morphological features associated with polyethylene glycol-induced cell fusion. J Cell Sci 40:63–75

Rutishauser U, Sachs L (1975) Cell-to-cell binding induced by different lectins. J Cell Biol 65:247–257

Schieder O (1978) Somatic hybrids of *Datura innoxia* Mill. + *Datura discolor* Bernh. and of *Datura innoxia* Mill. + *Datura stramonium* L. var *tabula* L. Mol Gen Genet 162:113–119

Schieder O, Vasil IK (1980) Protoplast fusion and somatic hybridization. In: Vasil I (ed) Recent advances in plant cell and tissue culture. Int Rev Cytol Suppl 11 B, Academic Press, London New York, pp 21–46

Smith BA, Ware BR, Weiner RS (1976) Electrophoretic distributions of human peripheral blood mononuclear white cells from normal subjects and from patients with acute lymphocytic leukemia. Proc Natl Acad Sci USA 73:2388–2391

Takebe I, Labib G, Melchers G (1971) Regeneration of whole plants from isolated mesophyll protoplasts of tobacco. Naturwissenschaften 58:318–320

Takebe I, Otsuki Y, Aoki S (1968) Isolation of tobacco mesophyll cells in intact and active state. Plant Cell Physiol 9:115–124

Träuble H, Eibl H (1974) Electrostatic effects on lipid phase transitions: Membrane structure and ionic environment. Proc Natl Acad Sci USA 71:214–219

Wallin A, Glimelius K, Eriksson T (1974) The induction and fusion of *Daucus carota* protoplasts by polyethylene glycol. Z Pflanzenphysiol 74:64–80

Ward M, Davey MR, Mathias RJ, Cocking EC, Clothier RH, Balls M, Lucy JA (1979) Effects of pH, Ca^{2+}, temperature and protease pretreatment on interkingdom fusion. Somat Cell Genet 5:529–536

Wilkins DJ, Ottewill RH, Bangham AD (1962) On the flocculation of sheep leucocytes: I. Electrophoretic studies. J Theoret Biol 2:165–175

Wilschut J, Düzgünes N, Fraley R, Papahadjopoulos D (1980) Studies on the mechanism of membrane fusion: Kinetics of calcium ion-induced fusion of phosphatidylserine vesicles followed by a new assay for mixing of aqueous vesicle contents. Biochemistry 19:6011–6021

Withers LA, Cocking EC (1972) Fine-structural studies on spontaneous and induced fusion of higher plant protoplasts. J Cell Sci 11:59–75

Zimmermann U, Scheurich P (1981) High frequency fusion of plant protoplasts by electric fields. Planta 151:26–32

Zimmermann U, Scheurich P, Pilwat G, Benz R (1981) Cells with manipulated functions: New perspectives for cell biology, medicine, and technology. Angew Chem 20:325–344

24 Pollen–Pistil Interactions

R. B. Knox

24.1 Introduction

Pollination is a process that is vital for the survival not only of flowering plants, but also for mankind. Yet human efforts to manipulate the reproduction of crop plants have been largely empirical. Today, research into plant reproduction is not given high priority. Yet it provides the means of ensuring successful pollination, in order to maximize fertilization and hence crop production. The pollination process is susceptible to modification to permit the introduction of desirable genes, for example, for disease resistance, from wild species into crop plants. The potential applications are endless, yet all depend on a precise knowledge of the events of pollination (see reviews by STANLEY and LINSKENS 1974, FRÄNKEL and GALUN 1977, CLARKE and KNOX 1978).

Pollination begins when pollen which carries the male gametes or their progenitor cell is shed from the anthers. Both the flowers and the animal vectors which transport the pollen to the pistil, the site of the female gametes, are usually specifically adapted to ensure efficient pollen transfer (see review by FAEGRI and VAN DER PIJL 1979). The shape of the pollen, its organization as a single or composite structure, surface pattern, odour and colour, may be precise adaptations for a particular animal vector or mode of pollination, and are usually species-specific features.

The pollen grain is constructed essentially like a single cell, but it interacts with a complex multicellular organ – the pistil. Pollen is deposited first on the stigma, whose cells are coated with an adhesive for pollen attachment. The arrival of the pollen sets in train a series of interactions which lead to discrimination by the pistil of self from foreign pollen, and in many cases, of pollen from the same plant, from pollen from other individuals of the same species. This remarkable series of recognition events is determined by surface interactions. These are likely to be chemical and of the key-in-lock type. The interaction results in the exchange of signals between pollen and stigma cells. Receipt and translation of the signals must occur in both partners since the interactions are mutual.

Acceptance leads to pollen germination and tube growth to the ovule; rejection leads to inhibition of the interaction.

The life cycle involves alternation of sporophytic and gametophytic generations (see reviews by HESLOP-HARRISON 1972, 1979c, JENSEN 1974). The sporophytes are the diploid vegetative plants which produce haploid spores. These differentiate into the gametophytes which bear the gametes. The male gametes are produced by the microspores which undergo cell division to form the pollen grain with its sperm transport organ, the pollen tube. The megaspores divide to form the embryo sac which includes the female gamete, the egg cell, and the central cell. These cells of the embryo sac are essential for double fertilization, that is characteristic of angiosperm reproduction. The embryo sac lies within the ovule, which is part of the ovary, the parental sporophytic tissue of the pistil.

The pistil is tripartite, with the stigma acting as a specific pollen receptor organ, overlying the style which provides the guiding tissue to the ovary. The pollen – pistil interactions therefore basically involve interactions between the male gametophyte, the pollen grain or pollen tube on one hand, and the sporophyte which forms the female tissues on the other. However, interactions with the female gametophyte must occur at the interface of the egg cell and central cell within the embryo sac when fertilization takes place. In gymnosperms the pollen grains make direct contact with the outer envelopes of the ovule, which serves as a "stigmatic" surface.

The purpose of this review is to give a coherent picture of the cell biology and physiology of pollination, especially as they concern the mutual interactions that are essential for successful pollination. Although the pollen – stigma recognition processes cannot yet be explained in molecular terms, the available information on the molecules that are sited on the surface of the pollen grain and stigma is presented, together with the models proposed for their interactions.

This story is largely based on the pioneering studies of three botanists – J. HESLOP-HARRISON, now of Aberystwyth, whose work on pollen biology and the physiology of pollen–pistil interactions has spanned two continents; D. LEWIS of London whose work has provided evidence for recognition genes and their products in pollen grains and models for their interactions; and H.F. LINSKENS of Nijmegen whose work has provided a biochemical basis for understanding pollen–pistil interactions.

It is hoped that the information brought together here will first help elucidate the mechanisms of the remarkable co-adaptations between pollen, pollinating animals and the stigma. Secondly, it may stimulate new efforts to explore the molecular basis of pollen–pistil recognition and responses and their genetic control.

24.2 Cell Biology of Pollen–Pistil Interactions

24.2.1 The Pollen Grain

The pollen grain is the carrier of the male gametes or their progenitor cell in higher plants. It represents one of the simplest isolated cellular systems, since it is bicellular or tricellular at maturity, and is the product of two rounds of cell division (as opposed to five in the gymnosperms). The pollen grain is an example of an exposed plant cell which has a wall that is not enveloped in a cuticle; instead it is surrounded by a complex exine.

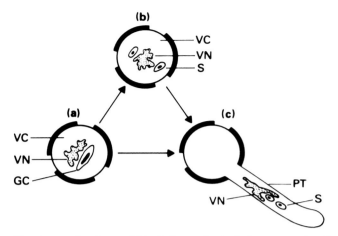

Fig. 1 a–c. Cytology of bicellular and tricellular types of pollen. **a** bicellular type is shed with vegetative nucleus (*VN*), and generative cell (*GC*). After germination, the generative cell divides to form the pair of sperm cells (*S*) as shown in **c**. **b** tricellular type, containing pair of sperm cells, resulting from division of the generative cell **a** during pollen development. (Dexheimer 1970)

Pollen grains exhibit a wide diversity in form and structure, suggesting that they have evolved by remarkable processes of adaptation – on the one hand, to the environment during pollen dispersal by wind or water currents, and on the other hand, by mutual co-adaptation with specific animal vectors. In all these cases, pollen structure has been influenced in a way that accounts for its species or group-specific patterns of wall sculpturing, and the composite nature of certain types of pollen.

Pollen grains vary considerably in size, e.g. from 3.5 μm in diameter in *Myosotis* to about 300 μm in *Hibiscus* and *Citrullus* (see Erdtman 1952). These grains are spherical. However, some types are elongate and filiform, up to 5000 μm in length, e.g. the seagrass *Amphibolis* (Ducker and Knox 1976). Other types are composite, with tetrads as small as 6.5 μm in *Mimosa* (Guinet 1981). Pollen mass varies widely, e.g. a grass pollen grain is 22×10^{-9} g (Stanley and Linskens 1974).

In higher plants, two types of pollen grain are found:
1. Bicellular type. In gymnosperms and in two-thirds of flowering plant families, the pollen grain comprises two cells at maturity, that is, a vegetative cell which is concerned with tube growth and metabolism, and a generative cell which divides within the pollen tube to produce two sperm cells (Fig. 1).
2. Tricellular type. In the remainder of families the generative cell divides precociously so that all three cells are present within the pollen grain at maturity (Fig. 1). The sperm cells are sited wholly within the cytoplasm of the vegetative cell, separated not by walls but by their own plasma membrane and that of the host cell (Larson 1965). The sperm cells have a limited cytoplasm containing Golgi bodies, mitochondria, endoplasmic reticulum, polysomes, and plastids in some species, but all show reduced internal structure com-

pared with their counterparts in the vegetative cell (see review by KNOX 1983).

When we consider that the pair of sperm cells or their progenitor, the generative cell, are the vehicles of male heredity in plants, surprisingly little is known about both their structure and their specificity. There are reports that the sperm cells are surrounded by a cell wall, while in other systems they are bounded by their own plasma membrane and that of the vegetative cell (see review by KNOX 1983). Knowledge at present is restricted to ultrastructural studies, which have revealed striking differences in the heritable organelle content of the pair of sperm cells in *Plumbago* (RUSSELL 1979, 1980).

Because of polarized division of the generative cell, one sperm cell has high numbers of mitochondria, while the other has high numbers of plastids (RUSSELL and CASS 1983). In *Plumbago*, the sperm cells are linked to each other, and by a connective to the vegetative nucleus (RUSSELL and CASS 1981). In *Brassica,* the sperm cells of mature viable pollen, are bounded only by plasma membranes, which take the form of long processes or pseudopodia (DUMAS et al. 1983). These link the sperm cells together in a tail:tail configuration. One of the sperm cells is associated by its pseudopodia with the vegetative nucleus. This sperm cell has the highest content of mitochondria of the pair, and plastids are absent, presumably maternally inherited. While only these two cases have been studied in detail, DUMAS et al. (1983) have proposed that since the nuclear and cytoplasmic elements of heredity are linked together in the mature pollen, the resultant unit, termed the male germ line unit, may function in directing fertilization.

There is a need for cytochemical studies at both light and electron microscope levels to characterize the cell surface of the sperm cells. At the same time, little is known of their role in fertilization (see review by JENSEN 1974, 1983). It is widely reported that the cytoplasm of sperm cells is left behind when fusion with the egg cell occurs; however, in two cases, ultrastructural studies have revealed that male cytoplasm may fuse with the egg cytoplasm, giving biparental transmission of cellular organelles. This has been shown for *Oenothera erythrosepala* (MEYER and STUBBE 1974) and for *Plumbago zeylanica* (RUSSELL 1980).

The state of the plasma membrane of mature, dormant pollen grains has been called into question by HESLOP-HARRISON (1979a, b). Lipid micelles were detected at the potential site of a plasma membrane in mature dry grass pollen in ultrastructural studies of glutaraldehyde-fixed anthers. The usual double lipid bilayer of the membrane was detected in developing grains and grains fixed within 5 min or more following hydration at pollination. Such changes are reflected in the altered physiology, especially water relations, of the pollen grains. Most pollens resemble seeds in their low water content (15–35%) at maturity (STANLEY and LINSKENS 1974).

Cytoskeletal elements, including microtubules and microfilaments, have not been found associated with the plasma membrane of ungerminated pollen grains, although microtubules have been described in the generative cell of several species, including *Endymion* (BURGESS 1970) and *Haemanthus* (SANGER and JACKSON 1971), where they are considered to give the generative cell its elongate shape. In germinating pollen, motility of the generative cell to ensure its transfer to the pollen tube may be conferred through microfilament complexes within the tube cytoplasm (FRANKE et al. 1972, CRESTI et al. 1976b). F-actin-containing elements have been demonstrated in pollen tubes and protoplasts of *Amaryllis belladonna* (CONDEELIS 1974).

The surface of a pollen grain, the wall which houses the component cells, is intricately sculptured and ornamented, showing species- or group-specific features. It consists of the outer patterned layer, the exine, which is composed of sporopollenin; and an inner largely unpatterned layer, the intine, composed of polysaccharides. Their structure and composition will now be briefly outlined.

24.2.1.1 Pollen-Wall Structure and Compartments

The exine is the patterned layer and comprises wall units enveloped in the resistant polymer, sporopollenin. The exine covers the intine except at the germinal apertures, where it may be absent, reduced in thickness or represented by a cap or operculum. The exine commonly comprises two layers, which in some cases show chemical differences. There are few structures which have provoked greater confusion in terminology than the pollen-wall and its stratigraphy. In this review, the terminology of ERDTMAN (1952) has been used (terms used by FAEGRI and IVERSEN 1975 are given in parentheses in the following paragraph).

1. The outer layer, the sexine (ectexine) is patterned, containing a foot layer supporting rod-like bacula (columellae) (Fig. 2A). The bacula may support a roof or tectum that may be perforated by micropores. The tectum is often ornamented with spines or other protuberances. In pilate exines, the tectum is absent and the tops of the rods may be thickened into swollen heads. The sexine is generally enveloped in a sticky pigmented layer, the pollen coat (Fig. 2B).
2. The inner nexine (endexine) is non-sculptured. It may become detached from the sexine, especially in Compositae, forming large cavea (HESLOP-HARRISON 1969). The nexine may be greatly reduced or absent in certain genera.

The exine is often chemically defined by its resistance to degradation, including acetolysis (ERDTMAN 1952). The resistant component is sporopollenin, considered to be an oxidative co-polymer of carotenoids and carotenoid esters (SHAW 1971, BROOKS and SHAW 1978). The conclusion is based on chemical analysis of sporopollenin derivatives produced by ozonolysis and by potassium hydroxide fusion, but the detailed chemistry remains uncertain. Pollen exines may be solubilized by oxidizing solutions and by 2-amino-ethanol (BAILEY 1960, ROWLEY and FLYNN 1966, SOUTHWORTH 1974). They are slowly broken down by microorganisms (ELSIK 1971), and may be degraded by enzymes during germination on the stigma (GHERARDINI and HEALEY 1969, DICKINSON and LEWIS 1974).

Cytochemical studies have indicated considerable heterogeneity between the exines of different groups of angiosperms, and chemical changes also occur during pollen development. The reaction of the pollen exine of several genera to a variety of cell-wall stains is shown in Table 1. The sexine of mature pollen can usually be defined by its positive staining with lipid dyes (e.g., auramine 0,

Fig. 2. A Stratification of the pollen grain wall in radial section; **a** shows tectate type of exine; **b** pilate type. (Modified from HESLOP-HARRISON 1968). **B** Sporophytic and gametophytic domains of the pollen wall, and their origin from the microspore or anther tapetum (*tapis staminal*). Exine sporopollenin originates initially from the microspore, and later by accretions of the tapetum. Exine proteins and coat substances originate from the tapetum and accumulate in the exine arcades and on the surface, while the intine proteins originate mainly from the microspore. (ZANDONELLA et al. 1981)

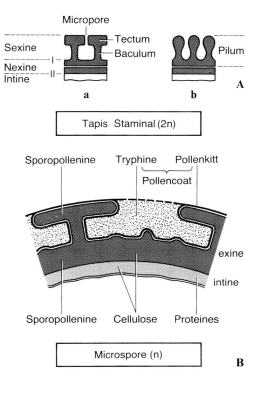

Table 1. Cytochemical reactions of the pollen grain wall to various cell-wall stains, based on data collected by KNOX (1983). Specificity of the stains is reviewed by JENSEN (1962), SOUTHWORTH (1973) and HESLOP-HARRISON (1979a)

Cytochemical technique	Specificity	Pollen-wall layer		
		Tectum	Nexine	Intine
1. Lipid-containing components				
Auramine O	Lipids, especially of cuticles	+	+ or − or absent	−
Nile blue after pyridine extraction	Specific reactive groups of lipids; anionic groups	+	+	−
Osmium tetroxide	Specific reactive groups of lipids	+	+	−
Scarlet R (Sudan IV)	Lipids	+	−	−
2. Lignin and phenolic components				
Autofluorescence	Phenolic compounds or lignin-like components	+	+ or − or absent	−
Fast blue B salt	Non-specific binding to phenolic compounds	+	absent	−

+ = positive staining reaction; − = negative reaction

Table 1 (continued)

Cytochemical technique	Specificity	Pollen-wall layer		
		Tectum	Nexine	Intine
Maule reaction	Lignin components	+ (brown)	+ (brown)	
Toluidine blue	Acidic polyanionic groups stain blue, purple; lignin and phenolics stain green; acidic polysaccharides (pectic acid) stain red or purple	+ (green) + (blue)	+ (green)	+ (red)
3. Wall polysaccharides				
Alcian blue	Acidic polysaccharides	−	−	+ or ±
Calcofluor	β-glucans	−	− or absent	+ or ±
Decol. Aniline blue	Callose	−	+ or − or absent	+ or −
Hydroxylamine Iron after methylation	Pectic acids	−	−	+
PAS	Vicinal glycol groups of carbohydrates	− or absent	+ or − or absent	+
Ruthenium Red	Acidic polysaccharides (pectic acids)	−	− or absent	+
Zinc chloro-iodide	Cellulose or hemicellulose	+ (orange)	+ (orange)	+ (blue)
4. Charged groups				
Azure B	Negatively charged groups	+	+	−
Basic fuchsin	Negatively charged groups	+	+ or − or absent	−
5. Wall proteins				
Aniline blue black	Proteins	+	+ or −	+
Coomassie blue	Proteins	+	−	+

yellow fluorescence), dyes binding to phenolic compounds (e.g., green colour with toluidine blue) and by its negative staining with PAS reaction for polysaccharides. In contrast, a positive PAS reaction may often be detected in the early sexine of microspores after release from the tetrad, indicating that changes in composition occur during development of the exine. The exine somewhat resembles lignin and suberin in some of its cytochemical reactions.

The structure of the exine and the mechanism for sporopollenin polymerization have been approached by J. Rowley and coworkers in Sweden. They have used chemical etching methods to dissect away the sporopollenin, to reveal filamentous exinous subunits, 15–40 nm in diameter, each comprising an axially-oriented branched tubule 10–15 nm in diameter (ROWLEY et al. 1980, 1981 a, b). Subunits are embedded within the sporopollenin, and, on the basis of electron staining properties, are made up of polysaccharide, protein and lipid components, ROWLEY et al. (1981 a) propose that these subunits form part of the exine glycocalyx network, occurring throughout the exine, and function as sporopollenin receptors (see review by KNOX 1983). The glycocalyx network is believed to originate at the microspore plasma membrane, and to pass through the nexine and sexine layers to the grain surface.

Recently, BARTHLOTT (1981) showed that the surface characters of plant cells may be grouped into four categories (1) cellular arrangement or pattern; (2) shape; (3) outer wall relief; (4) epicuticular secretions. The exine can be conveniently considered under these headings.

Firstly, the exine usually encloses free structures, the monads. However, in many families of angiosperms the grains may be linked in tetrads, octads or higher numbers as polyads. Tetrads include all four meiotic products from a pollen mother cell; polyads include more than one meiotic lineage. A further attribute of the exine – the presence of apertures, the designated sites for pollen tube emergence at germination – are determined within the meiotic tetrad according to the orientation of the microspore. Apertures are distinguished by their number, shape and position (WALKER 1974) and by the nature of the exine layer from which they are derived (ERDTMAN 1952; FAEGRI and IVERSEN 1975). Recently, HIDEUX and FERGUSON (1976) delineated ectoapertures from endoapertures on the basis of their origin from either the sexine (ectexine) or the nexine (endexine). Ectoapertures are well-defined slits or pores, while endoapertures are diffuse areas, uncovered or covered only with nexine. In inaperturate grains, there is no designated site for pollen-tube emergence.

Secondly, the exine controls the shape of pollen grains; most are spherical or triangular, although occasionally the grains are elongated into a tube (e.g., *Crossandra* BRUMMITT et al. 1980) or are filiform, e.g., in seagrasses of the families Cymodoceaceae, Zosteraceae or Posidoniaceae (PETTITT et al. 1981).

Thirdly, the physiological significance of the pollen grain wall lies in the structure of the exine. The exine layer may completely seal the intine and protoplast from the outer surface, or it may be perforated by micropores which permit entry and release of macromolecules (cf. HESLOP-HARRISON et al. 1973, HOWLETT et al. 1973, BELIN and ROWLEY 1971). In other grains, the exine may be greatly reduced in structure and functional significance, and is totally absent in one group, the seagrasses (DUCKER et al. 1978).

The sculpturing is group or even species-specific in its patterning. Surface ornamentations include spinules and microspinules, often with complex form apparently related to attachment to animal pollinators, rather than to the stigma surface (see review by KNOX 1982). Within the exine are arcades connected to the grain surface by micropores; these have been carefully investigated as anatomical features by palynologists, but have only recently revealed their function. They are sites for the storage of a range of proteins, glycoproteins, carbohydrates, lipids and pigments, held at the surface of the pollen grain (Figs. 2 B,3). These have a sporophytic origin in pollen of *Malvaviscus* and *Hibiscus* of the

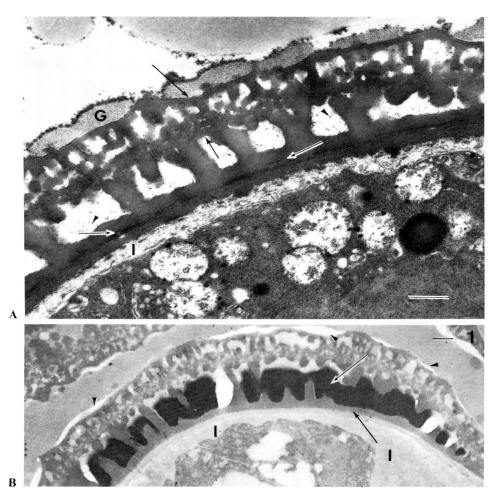

Fig. 3 A, B. Ultrastructure of the pollen wall of *Artemisia vulgaris*. **A** Exine components and layer coating the surface above a granular-fibrillar secretion. Elements of the tectum (*between black arrows*) are thickened, and lamellations can be seen within the foot layer (*white arrows*). There are many cytoplasmic evaginations within the intine zone (*I*). Fixation: glutaraldehyde-phosphotungstic acid-osmium tetroxide; post-stained with uranyl acetate and lead citrate. (Rowley and Dahl 1977). **B** Wall of grain fixed earlier in the maturation period than **A**. Globular substances are present in either or both portions of the exine arcade. Without section staining, the nexine (*above black arrow*) and exine surface coating (*arrow heads*) are low in contrast after fixation, as in **A**. Material within the exine arcade (*white arrow*) that includes lipid is darkened. (Rowley and Dahl 1977)

family Malvaceae (J. Heslop-Harrison et al. 1973) and in various Cruciferae and Compositae (see review by Knox et al. 1975). They are produced by the tapetal cells and transferred to the exine late in pollen development.

Fourthly, the oily, pigmented viscous surface coat on pollen grains, for example in *Lilium* or *Oenothera*, may confer a characteristic odour, stickiness, and colour to the pollen. The material is present in particular abundance in animal-pollinated grains where it has been termed Pollenkitt (PANKOW 1957) and in other cases tryphine (see review by ECHLIN 1971). Because of this confusion in terminology, the term pollen-coat suggested by ZANDONELLA et al. (1981) has been used in this chapter. Its adhesive properties may be strengthened by rods or strands of viscin, a material cytochemically similar to sporopollenin, especially in, for example, pollen of *Onagraceae* and *Ericaceae* which is often released in strings linked by adhesive-coated viscin threads.

The intine is the inner polysaccharide layer of the pollen wall; it is structurally and chemically related to the primary wall of other plant cells (SITTE 1953). It is the essential layer of the pollen grain, as the intine is present in all pollen types. The intine exhibits two kinds of patterning: (1) it is markedly thickened and more complex in structure at the apertures, where it forms the surface layer; (2) it has a microfibrillar appearance as viewed by transmission electron microscopy, at least in its principal layer. It is synthesized by the microspore or pollen protoplast during the vacuolate period of development (see review by KNOX 1983).

The intine has been shown to contain cellulose, hemicelluloses and pectic polymers (SITTE 1953, BOUVENG 1965 and ROLAND 1971, SOUTHWORTH 1973, HARA et al. 1977). It can be wholly solubilized by acetolysis (ERDTMAN 1966), and partly by monoethanolamine (BOUVENG 1965) or EDTA (HARA et al. 1977) treatments.

In pollen of the gymnosperm *Cryptomeria japonica* pectic polymers were extracted by treatment with ammonium oxalate or hot water (OHTANI 1955), while with 0.5% EDTA at pH 6.8, HARA et al. (1977), obtained a homogeneous pectin fraction. On hydrolysis, a fraction produced by ion exchange chromatography and GLC contained galactose, rhamnose, arabinose, xylose and uronic acid.

Hemicelluloses were extracted by BOUVENG (1965) from *Pinus mugo* pollen using monoethanolamine as solvent. Two polysaccharides were obtained:

1. An arabinogalactan with D-galactose, L-rhamnose, and D-glucuronic acid backbone, with L-arabinose at branching positions and D-galactose as terminal group (BOUVENG and LUNDSTROM 1965).
2. A linear $(1 \rightarrow 3)$-β-D-glucan chain and xylogalacturonan side chains with D-xylose in terminal position.

The pectic polymers provide the intine with a cation exchange capacity, and negative charges by means of the free carboxyl groups, generally in the uronic acid portion (KNIGHT et al. 1972).

Cytochemical studies of the intine, listed in Table 1, have provided further evidence for the presence of these three major classes of wall polysaccharides and also for the chemical hetero-morphism of the layer in many pollen types. The intine stains positively for polysaccharides. In many cases, it stains red with toluidine blue; in *Populus* pollen this metachromatic staining is absent if the section is digested with pectinase, indicating the presence of pectic polymers (ASHFORD and KNOX 1980). Certain zones of the intine also stain with a variety of protein stains (TSINGER and PETROVSKAYA-BARANOVA 1961, KNOX and HESLOP-HARRISON 1969, 1970, SOUTHWORTH 1973). J. HESLOP-HARRISON and Y. HESLOP-HARRISON (1980a) and Y. HESLOP-HARRISON and J. HESLOP-HARRISON (1982b) have shown that the intine of several types of pollen comprises cytochemically distinct layers. In grasses, for example, the outer Z-layer (Zwischenkörper) forms a lens-shaped oncus at the aperture, and is composed of pectins, staining with alcian blue and ruthenium red. At pollen germination, it forms a gel which precedes the pollen tube (see Sect. 24.2.1.3). Recently KRESS and STONE (1982) have concluded from studies of several monocot pollens that the intine contains two major layers: the outer composed of acidic polysaccharides, and the inner containing neutral polysaccharides (see Fig. 4).

The intine proteins have been shown in ultrastructural studies to be located in tubules or lamellar structures within the intine polysaccharide matrix, for example, acid phosphatase in the intine of *Crocus* pollen (KNOX and HESLOP-HARRISON 1971b). The available evidence is that these proteins are secreted from the pollen protoplast, and laid down in the intine when it is synthesized (see review by HOWLETT et al. 1979). Intine proteins have been detected by their antigenic activity, and fluorescent antibody methods have been used to investigate their routes of emission from moistened mature pollen. In *Gladiolus* and *Iris* pollen, intine proteins are released to the external surface through the exine from all around the grain (KNOX 1971); in contrast in *Malvaviscus, Hibiscus, Ambrosia,* and *Cosmos,* the intine proteins were found to reach the surface only at the germinal apertures (HESLOP-HARRISON et al. 1973, HOWLETT et al. 1973). Recently, in pollen of ryegrass *Lolium perenne,* immunocytochemical evidence suggests that during the maturation of the pollen, intine antigens pass through micropores in the exine into the arcades, and onto the external surface of the exine in mature pollen (KNOX et al. 1980, HOWLETT et al. 1981, VITHANAGE et al. 1982). In contrast, in *Lilium* pollen, components of tapetal origin have been demonstrated in both exine and intine (MIKI-HIROSIGE and NAKAMURA 1980, cited by PANDEY 1980). In pollen grains with no exine, as in various seagrasses, the intine may contain proteins within tubular inclusions (PETTITT 1980) and the presence of the marker enzyme, acid phosphatase, has been detected cytochemically (DUCKER et al. 1978). Slime, of tapetal origin, may coat the pollen grains, bearing at its surface, a layer with cytochemically detectable esterase activity, also of tapetal origin (PETTITT 1980).

24.2.1.2 The Pollen Tube

Following pollen hydration, the pollen tube emerges from one of the previously differentiated germinal apertures in the wall of aperturate grains. The tube wall in these types is continuous with the inner layer of the intine (LARSON 1965, CRANG and MILES 1969). Growth of the pollen tube occurs at the tip, shown first by ROSEN and co-workers (ROSEN et al. 1964).

Physiological and ultrastructural studies of pollen germination in vivo and in vitro have been made in *Lycopersicon peruvianum.* On the stigmas, pollen grains hydrate within 15 min, and germination begins after 3 h 30 min (PACINI and SARFATTI 1976). In contrast, in germination medium, hydration occurs in seconds, and germination follows in 45 min (CRESTI et al. 1977). The grains have three semi-spherical apertures which are extruded at hydration. On germi-

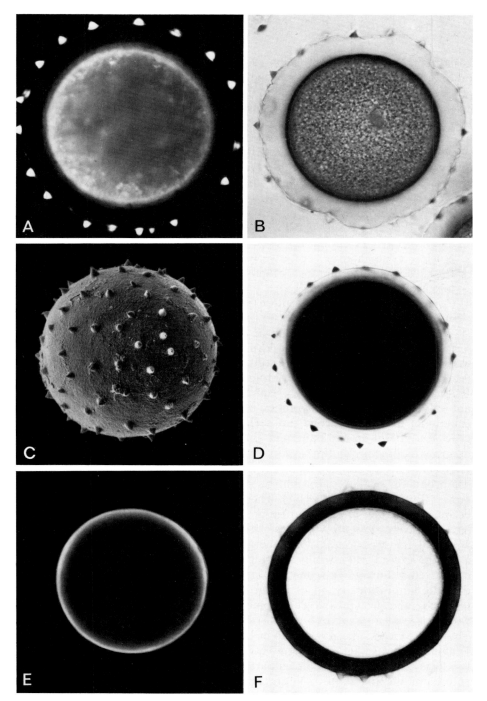

Fig. 4 A–F. Cytochemistry of the pollen wall of the tropical monocotyledon, *Canna indica* (KRESS and STONE 1982). **A** Auramine O fluorescence, showing exine spinules around grain surface. **B** Periodic acid-Schiff technique showing specific staining of inner intine layer. **C** Scanning electron micrograph of whole grain showing prominent surface exine spinules. **D** Fuchsin staining of exine spinules. **E** Calcofluor fluorescence staining of inner zone of intine. **F** Alcian blue staining at low pH of outer zone of intine. Specificity of stains is described in Table 1. (All figures × 1000 except E, × 665)

nation, "the pore opens like a port-hole door; ... the thicker portion of the pore, protruding through the corresponding part of the exine, breaks off along a large part of its edge, forming a kind of door that opens towards the outside. The portion of the edge that remains attached to the intine functions as a hinge" (CRESTI et al. 1977).

The developing tube wall of *Lycopersicon* is bi-layered, comprising an outer pecto-cellulosic wall continuous with the intine, and an inner layer of callose, detected by its fluorescence with decolorized aniline blue. Later, when the tube reaches a length of about 150 μm, the vegetative nucleus and generative cell enter into the tube. Four zones, reflecting functional specialization, have been detected by electron microscopy.

1. and 2. Apical and sub-apical zones: cytoplasm with abundant Golgi bodies and vesicles; thin fibrillar cell wall.
3. Nuclear zone: houses vegetative nucleus and elongate generative cell surrounded by two membranes, often with fibrillar material between them; callose layer in inner tube wall.
4. Zone of vacuolation and callose plug formation: very thick callosic wall ending with a callose plug.

The process of wall deposition in *Lycopersicon* tubes is similar to that described for *Petunia* by CRESTI and VAN WENT (1976). The participation of Golgi vesicles in secretion of the tube-wall polysaccharides, especially callose, has been suggested in several studies (MORRE et al. 1971, FRANKE et al. 1972, ENGELS 1974a, b, HELSPER et al. 1977). Indeed, VAN DER WOUDE et al. (1971) have not only described the ultrastructure of the vesicles but have isolated a fraction containing them and carried out a chemical analysis. In *Lilium,* pollen tubes grow at a rate of up to 12 μm min^{-1}, and produce 2,150 vesicles min^{-1} during tip growth of the pollen tube (VAN DER WOUDE and MORRE 1967). In *Tradescantia,* recent stereological studies indicate that nearly twice as many Golgi vesicles are produced, 5388 vesicles min^{-1}, although tube growth rate is only 6 μm min^{-1} (PICTON and STEER 1981). There certainly appears to be an association between the Golgi vesicles and tip growth in these types of pollen tube which have relatively slow growth rates. HESLOP-HARRISON (1979a) has shown that pollen of the grass *Secale* has tube growth rates of 72–120 μm min^{-1}, while *Zea* may exceed 240 μm min^{-1}. Golgi vesicles are infrequent and apparently do not contribute to tube metabolism. Instead, Golgi activity is confined to the late maturation period of pollen development, and the grains are geared to rapid tube growth and development. This is ensured by the presence of vesicle-derived storage carbohydrate organelles, the P (polysaccharide)-particles. These have recently been shown to be largely pectic in nature, readily solubilized in ammonium oxalate (J. HESLOP-HARRISON and Y. HESLOP-HARRISON 1982b). Products of hydrolysis of isolated particles include galactose, arabinose, glucose and rhamnose, and about 12% protein content. The P-particles of *Secale* represent 30% of reserves of mature pollen, and about 10^6 particles are present in each pollen grain.

In many types of pollen, the pollen-tube tip is preformed in the aperture. The emergence of the plug-like tip or oncus of pollen tubes has been reported in *Lychnis* (CRANG and MILES 1969) and in various Compositae: *Cosmos* (KNOX and HESLOP-HARRISON 1970) sunflower, *Helianthus* (VITHANAGE and KNOX 1977) and *Artemisia* (ROWLEY and DAHL 1977). In *Artemisia*, ultrastructural studies have demonstrated that the emerging tube is covered by a bubble of coarse fibrillar material through which the tube grows, surrounded by a modified intine-like wall. In germinating pollen of rye, *Secale cereale*, the oncus-like Zwischenkörper layer at the aperture forms a gel that swells, displacing the sporopollenin cap, and the tube emerges within the gel (HESLOP-HARRISON 1979 b).

During submarine pollination in seagrasses, the filiform pollen of *Amphibolis antarctica* has pollen grains which are inaperturate with no definable exine layer. The grains form an aperture in the pollen wall by local autolysis, when they come in contact with the stigma. A tube-like outgrowth of the plasma membrane subsequently emerges within a mucilaginous bubble (PETTITT et al. 1980). The tube wall develops soon after emergence.

In all these cases, extracellular secretions of slime, probably containing pectic carbohydrates, are implicated in pollen germination, especially in the early stages of tube emergence and growth.

24.2.2 The Pistil

The pistil of flowering plants consists of three parts: the stigma, which provides the cells for pollen reception, and germination; the style, through which the pollen tubes grow; and the ovary, which contains one or more ovules, each containing an embryo sac, the female gametophyte. This contains the target cells for the pollen tube: the egg and central cell, whose polar nuclei will participate in double fertilization, characteristic of most angiosperms. In gymnosperms, the ovary provides for both pollen reception and pollen tube growth to the archegonium, and fertilization involves a single event, fusion of sperm and egg (see Sect. 24.2.3).

The pistil is precisely adapted for its interactions with the male gametophyte, both with the pollen grain and the pollen tube. During development, the style and stigma secrete components which provide a virtual "culture medium" for pollen germination and tube growth. The concepts of the pistil have changed very little during the 20th century, for example, F.E. LLOYD (1910) observed the route of the pollen tube in the pistil of the date palm, *Phoenix dactylifera*. "The short style is traversed by a canal lined by secretory cells–pollen tube guiding tissue. The canal is continuous with a groove, which ... bifurcates as it passes downward, one groove passing on each side of the funicle ... The pollen tube is guided in its course by the differential distribution of a stimulant arising from the egg-apparatus, chiefly the synergidae, and that the whole phenomenon is chiefly a chemical one ... where different guiding tissues occur, each one involved produces its specific secretion which restimulates the pollen tube from time to time on its course". How much progress has been made in understanding pistil structure in the intervening 70 years?

Stigma and style are glandular organs, and their metabolism is geared to the temporal processes of flowering and pollination. The sequence of events

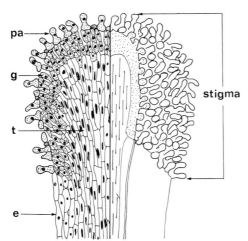

Fig. 5. Diagram of the upper part of the style of *Lycopersicon peruvianum. Left* longitudinal section; *right* superficial view of the stigmatic region. *pa* papillae; *g* glandular cells; *t* transmitting tissue; *e* epidermis. (DUMAS et al. 1978)

in the pistil leading to union of sperm and egg is the progamic stage (STANLEY and LINSKENS 1974).

Pistils show characteristic periods of development, first documented for the pistil of *Forsythia,* by DUMAS (1973a, b, 1974b):
1. Presecretory period of cell growth and elongation following pistil initiation.
2. Synthetic period; characterized by accumulation of endoplasmic reticulum and vacuoles in the cells of the style and stigma.
3. Extrusion period prior to or following flower opening, when secretion begins.

In this section, the structural and cytochemical features of the pistil will be reviewed as a prelude to biochemical analyses to be given later (Sect. 24.3).

24.2.2.1 Types of Stigma

The stigma may be looked upon as a gland, which provides the receptive cells for pollination, and may comprise a few or many modified epidermal cells (Fig. 5). These may be specialized elongate papillae composed of either unicellular or multicellular units, or in other cases, the receptive cells may be bullate or flattened in appearance. The surface characteristics of stigmas in the living state are readily observed by scanning electron microscopy without use of any pretreatment. This has made it possible to observe with remarkable clarity morphological features of stigmas, and has shown that in some species, the stigmas are covered when receptive with a copious liquid exudate, i.e., they are wet in appearance, while others have only an adhesive coating or pellicle and are "dry" (HESLOP-HARRISON 1975b).

Morphological Adaptations

The stigmas of many genera of flowering plants have been surveyed morphologically (Y. HESLOP-HARRISON and SHIVANNA 1977, Y. HESLOP-HARRISON 1981).

Table 2. Types of receptive surface present on mature angiosperm stigmas. (Y. HESLOP-HARRISON 1981)

Type of receptive surface	Category
Surface dry	
Receptive cells dispersed on a plumose stigma	D, Pl
Receptive cells concentrated in ridges, zones or heads	
Surface non-papillate	D, N
Surface distinctly papillate; papillae unicellular	D, P, U
Papillae multicellular	
Papillae uniseriate	D, P, M, Us
Papillae multiserate	D, P, M, Ms
Surface Wet	
Receptive cells papillate; secretion moderate to slight, flooding interstices	W, P
Receptive cells non-papillate; secretion usually copious	W, N

D dry; W wet; Pl plumose; N non-papillate; P papillate; U unicellular; M multicellular; Us uniseriate; Ms multiseriate

A classification of the types of receptive surface is given in Table 2. Among monocots, most families have the dry type of stigma. For example, grasses have plumose stigmas with receptive cells arranged in multiseriate branches. In other families, the receptive cells are concentrated into distinct receptive areas, which may be in the form of papillae, either unicellular (for example, *Gladiolus* and *Crocus* of the family Iridaceae) or multicellular (for example, *Philesia* of family Liliaceae). Non-papillate types occur especially among aquatic monocots, for example, in *Potamogeton, Typha,* and seagrasses. Multiseriate papillae occur in only a few families, notably in *Limnobium* (family Hydrocharitaceae). Monocots with wet-type stigmas include various sub-types, comprising those without papillae, for example, *Hedychium, Zingiber* and *Maranta* in the tropical Scitamineae, and those with low or medium papillae, for example, *Bromelia* (family Bromeliaceae) *Canna* (Cannaceae), *Costus* (Costaceae), *Anigozanthos* (Haemodoraceae), *Aloe* and *Lilium* (Liliaceae), many orchids, and several other Scitamineae, including *Musa* and *Heliconia*. The wide diversity of stigma types within the Liliaceae has been described by Y. HESLOP-HARRISON (1981).

Among dicots, there is also a wide expression of stigmatic variation (Y. HESLOP-HARRISON and SHIVANNA 1977). Dry papillate stigmas, with unicellular or multicellular papillae, occur in such families as the Cruciferae and Malvaceae. Non-papillate dry stigmas are also found, as for example, in *Populus* (Salicaceae). In the Boraginaceae, stigma papillae are often capitate with thickly cutinized heads, apparently an adaptation for pollination by solitary bumble bees which force the pollen between the papillae (Y. HESLOP-HARRISON 1981). Among the many families with wet-type stigmas are the Orchidaceae, Leguminosae,

Ericaceae, Epacridaceae and Pyrolaceae. Among dicots, the receptive cells may also be necrotic when the stigma is receptive, for example, in *Oenothera* of the family Onagraceae (DICKINSON and LAWSON 1975a).

An interesting correlation has been noted in the type of pollen associated with the various stigma types (Y. HESLOP-HARRISON and SHIVANNA 1977). In the monocots, bicellular pollen grains are found in genera with both wet and dry types of stigma, while tricellular pollen is confined to the dry stigma types. However, all wet stigmas possess bicellular grains. Among the dicots differences are not as striking, bicellular pollen occurring in both wet and dry stigma types, and tricellular pollen occurring mainly in dry stigma types.

The distinction between wet and dry types is operationally convenient, although the amount of exudate secreted in wet stigma types may vary widely during the flowering season. In some species, it may be conspicuous, in others restricted to the interstices between the papillae, or consist of little more than a film of lipid droplets over the surface of the receptive cells (Y. HESLOP-HARRISON and SHIVANNA 1977).

Cellular Adaptations

Having observed a wide variation in the morphology of the stigma surface, the nature of the wet and dry stigma surfaces needs to be examined. Do these represent two quite different types in terms of their physiology? The stigmas of more than a hundred genera have been examined by light or electron microscopy (see review by Y. HESLOP-HARRISON 1981). In this section, the cellular adaptations of stigma cells will be considered in terms of the features of both cytoplasm, cell walls, and their extracellular secretions.

Cytoplasmic Features. There are characteristic differences in the cytoplasmic organisation of the stigma cells of different genera of angiosperms (Table 3). These are mainly seen in quantitative differences in organelles important for the secretory processes. Some stigma cells contain a prevalence of endoplasmic reticulum, while others contain many Golgi bodies and vesicles (Table 3). This suggests that there may be differences in the nature of the secreted products, and also in the modes of secretion.

Stigmas with a lipophilic exudate characteristically have extensive networks of endoplasmic reticulum, and have been termed reticulate types (DUMAS 1974b), as for example, in *Petunia, Forsythia* and *Persea*. The stigmas of *Aptenia* are of this type, despite the different nature of their secretion. Both ER and Golgi bodies have been observed in other genera, for example *Lycopersicon,* and this is presumably related to the mixed secretion observed. Abundant Golgi and vesicles are characteristic of the mucilaginous secretion of *Citrus, Prunus* and *Malus* stigma cells (Table 3).

Stigmas of *Crocus* and *Gladiolus* belong to the dry type (Y. HESLOP-HARRISON and SHIVANNA 1977). Papillae of both contain abundant Golgi bodies and vesicles, often associated with the plasma membrane (Y. and J. HESLOP-HARRISON 1975b, CLARKE et al. 1980). In *Lilium* the cytoplasm of the wet stigma cells is rich in both smooth and rough endoplasmic reticulum, ribosomes, Golgi

Table 3. Ultrastructural and cytochemical features of the stigma surface in angiosperms. Stigma type categories are given in Table 2

Species	Stigma Type	Pellicle or exudate	Cuticle	Characteristic cellular organelles	Reference
1. Monocotyledons					
Amphibolis antarctica, Thalassodendron ciliatum, Zostera marina	DN	Esterase +, proteins +	Perforated by channels	—	DUCKER and KNOX 1976; PETTITT 1977, 1980
Crocus chrysanthus	DP	Esterase +	Chambered	—	HESLOP-HARRISON and HESLOP-HARRISON 1975b; Y. HESLOP-HARRISON 1977
Enhalus acoroides, Halophila decipiens, Thalassia hemprichii	DP	Esterase +, proteins +, binds Con A	Perforated by channels	—	PETTITT 1980
Gladiolus gandavensis	DP	Esterase + binds Con A, tridacnin β-glu Yariv antigen	Perforated by channels	Golgi and vesicles	KNOX et al. 1976; CLARKE et al. 1979
Hordeum bulbosum, Secale cereale	DPI	Esterase + (15–20 nm thick)	Cuticular discontinuities frequent (190–250 nm thick)	ER	HESLOP-HARRISON and HESLOP-HARRISON 1980b
Lilium longiflorum, L. regale, L. davidii	WP	—	Degraded	ER, Golgi	ROSEN 1971; VASILIEV 1970
Phalaris minor	DPI	Binds Con A	Perforated by channels	—	Y. HESLOP-HARRISON 1976
2. Dicotyledons					
Acacia iteaphylla, A. retinodes	WN	Esterase +, exudate: Coomassie Blue +, Scarlet R +, PAS + Ruthenium, Red +	Absent	ER	KENRICK and KNOX 1981 a, b

Table 3 (continued)

Species	Stigma Type	Pellicle or exudate	Cuticle	Characteristic cellular organelles	Reference
Aptenia cordifolia	WP	Exudate: Ferric sulphate + (tannins); Sudan IV −; PAS +	Detached over exudate	ER and vesicles	KRISTEN 1977; KRISTEN et al. 1979
Brassica oleracea	DP	Esterase +, digested by protease but not lipase	Perforated by channels	Golgi and vesicles	DICKINSON and LEWIS 1974; MATTSSON et al. 1974; HESLOP-HARRISON et al. 1975
Citrullus lanatus	WP	Exudate: Sudan Black +, Coomassie Blue +, Ruthenium Red −	Intact over exudate	ER	SEDGLEY and SCHOLEFIELD 1980; SEDGLEY 1981
Citrus unshui, *C. tamurana*	WN	Vesicular extracuticular exudate	Degraded	Golgi and vesicles	SHIRAISHI et al. 1976; YAMASHITA 1978
Forsythia intermedia	WP	Exudate: Sudan IV+, Argentaffin reaction +	−	ER	DUMAS 1973b, 1974a, b, 1977
Helianthus annuus	DP	Esterase +, binds Con A	Present	−	VITHANAGE and KNOX 1977
Linum pubescens	DP	Sudan IV+, Nile blue + (blue); Ruthenium Red+	± Degraded	−	DULBERGER 1974
Linum grandiflorum, *L. usitatissimum* and other spp.	D/WP	Esterase +, acid phosphatase +, protein stains +, binds Con A	Present	−	GHOSH 1981
Lycopersicon peruvianum	WP	Exudate, Sudan IV +	Degraded	ER and Golgi vesicles	DE NETTANCOURT et al. 1973; DUMAS et al. 1978
Malus domestica, *M. hupehensis*	WP	Exudate Sudan IV+; protein stains+, PAS+	Degraded	ER and Golgi	HESLOP-HARRISON 1976
Persea americana	WP	Lipid droplets and fibrillar material in exudate	Intact over exudate	ER	SEDGLEY and BUTTROSE 1978; SEDGLEY 1979

Species					References
Petunia hybrida	WP	Esterase −, acid phosphatase +, exudate, Sudan IV +, PAS +	Degraded	ER and vesicles	KONAR and LINSKENS 1966a, b; HERRERO and DICKINSON 1979, 1980
Phaseolus vulgaris	WP	Exudate: 2,6-dichloroindophenol + (red, reducing acids), p-nitro-aniline + (phenols), Bromophenol blue + (proteins), osmiophilic	−	−	WEBSTER et al. 1977; LORD and WEBSTER 1979
Populus deltoides	DN	Esterase +, protein stains +, Sudan black +, Toluidine blue + (green)	−	−	KNOX et al. 1972 and unpublished
Prunus avium, P. subhirtella	WP	Esterase +, exudate: Aeramine 0+, Coomassie Blue +, PAS +	Intact over secretion	Golgi	HESLOP-HARRISON 1976; RAFF et al. 1982
Tradescantia virginiana, T. pallida	DP	Protein stains +, PAS −	± Degraded	ER and Golgi vesicles	HERD and BEADLE 1980
Vitis vinifera	WP	Exudate: Alcian blue +, Ruthenium red +, Toluidine blue + (purple) Coomassie blue +, PAS +	Intact over secretion	−	CONSIDINE and KNOX 1979
Zephyranthes candida, Z. citrina	DP	Esterase +, acid phosphatase +, protein stains +, Sudan black +, PAS +	Present	−	GHOSH 1981

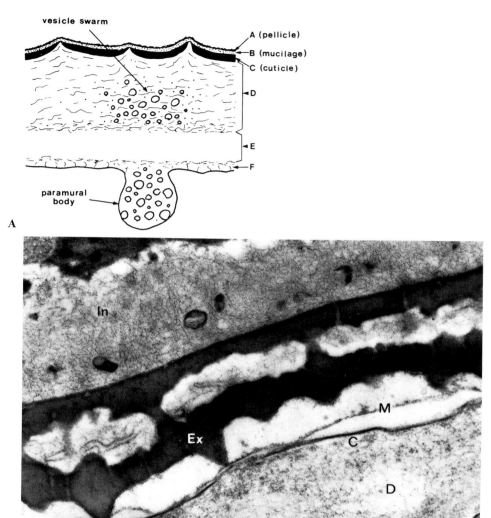

Fig. 6 A, B. Ultrastructural features of the stigma surface of grasses. **A** Diagram showing an interpretation of stigma wall structure and the mechanism of pellicle secretion. Six wall layers (*A–F*) are indicated, which differ in ultrastructural appearance, and, in addition, paramural bodies and a swarm of vesicles. (Heslop-Harrison and Heslop-Harrison 1980b). **B** Electron micrograph showing the interface between a pollen grain (*above*) and stigma surface (*below*) of the grass, *Gaudinia fragilis*. *In* intine; *Ex* exine; *M* material at interface *C* Stigma cuticle; *D* outer stigma wall. (Heslop-Harrison and Heslop-Harrison 1981a)

Table 4. Occurrence of wall thickenings of the transfer-cell type in the pistils of angiosperms

Species	Site	Reference
Ornithogalum caudatum	Stigma papilla cells	TILTON and HORNER 1980
Citrullus lanatus	Stigma papilla cells	SEDGLEY 1981
Aneilema aequinoctiale	Stima papilla cells (tip)	OWENS and HORSFIELD 1982
Lilium longiflorum	Stylar canal	ROSEN and THOMAS 1970
L. davidii	Epithelial cells	VASILIEV 1970
L. regale		
Muscari sp.	Stylar cells in contact with pollen tube	JOHRI 1981
Medicago sativa	Ovary	JOHNSON et al. 1975
Jasione montana	Wall of embryo sac at micropylar end	JOHRI 1981
Gossypium hirsutum	Filiform apparatus of synergids	JENSEN 1974
Plumbago zeylanica	Egg cell base	RUSSELL and CASS 1981 b

apparatus, secretory vesicles and mitochondria (ROSEN 1971, VASILIEV 1970). YAMADA (1965) observed "colloidal bodies" in the cytoplasm which stained positively with aniline blue and hematoxylin, like the stigma secretion. It seems likely that these are polyphenol vacuoles, and that the secretion contains some phenolic compounds. Certainly these would be stained by methylene blue, as observed by ROSEN (1971). There is a need for a re-examination of the cytochemistry of the *Lilium* stigma in the light of the biochemical characteristics of the style secretions (see Sect. 24.2.1).

Cell-Wall Structure. The cell walls of mature stigma cells prior to pollination show various adaptations to their role in the interaction with the pollen.

Paramural bodies, and vesicles in the wall have been detected in the stigmas of grasses (HESLOP-HARRISON and HESLOP-HARRISON 1981a), and are possibly involved in the secretory system (Fig. 6). Freeze-fracture studies of the *Gladiolus* stigma wall have revealed the presence of tubules radiating from the plasma membrane to the surface cuticle, which is perforated, presumably permitting secretion of the surface exudate (CLARKE et al. 1980). In *Crocus* the secretion is stored within a chambered cuticle, at the surface (Y. HESLOP-HARRISON 1977).

Wall ingrowths of the transfer cell type (GUNNING and PATE 1969) have been detected in both stigma and style secretory cells in certain species (Table 4) and may facilitate fluid secretion. In *Citrullus* stigmas the wall ingrowths contain aniline blue-fluorescent polysaccharides (SEDGLEY 1982). In this respect they resemble the filiform apparatus of the synergids of the embryo sac (WILLIAMS et al. 1982a).

Stigma-Surface Pellicle. The pellicle is defined as the outermost extracellular coating of the stigma surface in dry type stigmas, and is a hydrophilic layer, containing proteins and lipids (MATTSSON et al. 1974). It coats the papillar surface of both monocot and dicot stigmas of this type, forming an electron-dense layer, about 15–20 nm in thickness. It is separated from the cuticle by an intervening layer, shown in grasses to be mucilaginous in nature (HESLOP-HARRISON and HESLOP-HARRISON 1980a).

In monocot stigmas, for example, *Crocus* and *Gladiolus* (Table 3), the pellicle is secreted only when the stigma is receptive. The pellicle shows a positive cytochemical reaction for non-specific esterase activity, containing enzymes capable of hydrolyzing α-naphthyl acetate, detected by binding of the released α-naphthol to coloured diazonium salts or hexazotized pararosanilin at the stigma surface (Fig. 7). The esterase test provides a valuable indication of the functional sites of the stigma, since in whole pistils the reaction product is usually confined to the receptive cell surface (HESLOP-HARRISON et al. 1975a, b, SHIVANNA and SASTRI 1981). Previously, the hydrogen peroxide test had been used for this purpose (ZEISLER 1938).

What are the other characteristics of the pellicle? Firstly, the surface secretions of *Gladiolus* have an affinity for the lectin Con A, detected by binding of FITC-Con A to the stigma surface and fluorescence microscopy (Fig. 7). Controls incubated in labelled lectin in the presence of the inhibitory sugar, α-methyl mannoside, were negative (KNOX et al. 1976). Ultrastructural studies of Con A-binding to the stigma surface (KNOX and CLARKE 1979, CLARKE et al. 1980), and to the surface of *Phalaris* stigmas (Y. HESLOP-HARRISON 1976) reveal that the lectin binds to the secreted layer on the surface. In these cases, the lectin was complexed with iron mannan, or ferritin before stigma binding, in a one-step reaction in the *Gladiolus* experiments, or in *Crocus,* a three-step reaction involving first lectin binding, secondly peroxidase binding to the lectin and thirdly, cytochemical detection of the bound enzyme. PETTITT (1980) has since detected lectin binding to the surface secretion of seagrass stigmas. Specificity of binding in these electron-microscopic cytochemical studies has been difficult to demonstrate with use of inhibitory sugars because of the remarkable adhesive properties of the stigma surface components (see Sect. 24.4.1), the level of non-specific binding being as great as specific binding at the lectin concentration used (see CLARKE et al. 1980, HERD and BEADLE 1980).

Secondly, binding of the fluorescent-labelled lectin, tridacnin, specific for galactose, has been demonstrated in the surface secretion of *Gladiolus* stigmas (Fig. 7 and CLARKE et al. 1979). This follows a previous demonstration that the secretion binds the Yariv artificial carbohydrate antigen, specific for β-lectins (JERMYN and YEOW 1975). The specificity of these coloured phenylazo dyes varies according to the sugar residues bound to the surface of the molecule. The stigma surface of *Gladiolus* binds specifically to the β-glucosyl antigens, and not to α-galactosyl antigens (KNOX et al. 1976). The isolated binding component has proved to be an arabinogalactan (see Sect. 24.4).

Among dicotyledons with dry type stigmas, the pellicle is also extra-cellular. Cytochemically the layer stains for proteins, carbohydrate and often lipids, and contains proteins with esterase activity and, in a few cases, glycoproteins with affinity for the lectin concanavalin A (see Table 3). This surface layer

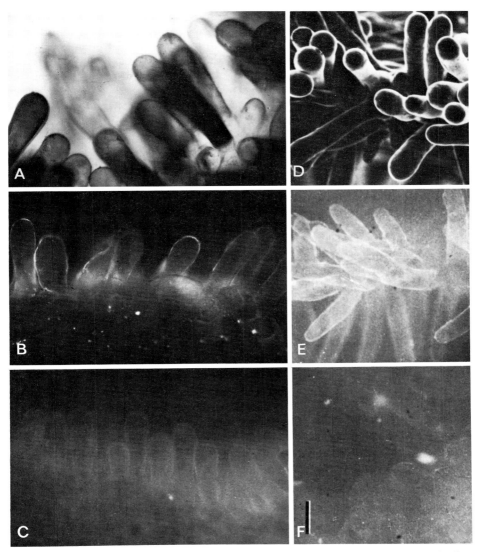

Fig. 7 A–F. Cytochemistry of the stigma surface of *Gladiolus,* which has dry type, unicellular papillae. **A** Esterase activity of pellicle. **B** Binding of FITC-labelled concanavalin A to pellicle. **C** Control stigma cells treated with FITC labelled ConA in presence of α-methyl mannoside. **D** Scanning electron micrograph of stigma papillae after freeze-drying. **E** Binding of FITC-labelled tridacnin to stigma papillae. **F** Control for **E**, stigma cells after treatment with labelled lectin in presence of 0.1 M lactose. (KNOX et al. 1976, CLARKE et al. 1979)

is solubilized by protease treatment, which led to its discovery by MATSSON et al. (1974) in stigmas of *Raphanus* and *Brassica* (Table 3).

In *Linum grandiflorum,* the stigma surface differs markedly between pin- and thrum-style types (GHOSH and SHIVANNA 1980a, b) with pin styles having a smooth dry pellicle, and thrum types have a copious exudate that wells up over the cuticle. Both secretions contain proteins, carbohydrates and lipids, and are remarkable for the presence of both esterase activity (smooth film over surface in pin stigmas) and acid phosphatase activity (reticulate pattern over surface in pin stigmas). In *Linum pubescens* a sub-cuticular exudate, probably containing lipids and pectins, may accumulate in the space between the papillar cell wall and the cuticle (DULBERGER 1974).

Recent electron microscope cytochemical studies of *Brassica* pollen have revealed that the pellicle exhibits ATPase activity (GAUDE 1982) a property it shares with the plasma-membrane. Other cytochemical markers for the pellicle included cationised ferritin and Con A.

Stigma Exudates. The nature of the exudate in wet stigmas appears to vary widely between different genera (Table 3). There are two principal types:

1. Stigmas with a lipophilic and hydrophobic surface exudate as the continuous phase, involving holocrine secretion;
2. Stigmas with a mucilaginous secretion of carbohydrates and proteins as the continuous phase, involving merocrine secretion mechanisms.

In both types, lipids and polyphenols may be present.

In the best-studied system, the pistil of *Petunia hybrida,* the exudate is of the first type; it is rich in lipids and the stigma cells undergo dissolution when the stigma is receptive so that the exudate acts as a "liquid cuticle" (KONAR and LINSKENS 1966a, b). In this "wet" stigma, the exudate actually contains little or no water, so that the term hardly applies. Pollen germination does not begin until the grains hydrate at the stigma surface, beneath the lipid exudate (KONAR and LINSKENS 1966a). The lipids in the exudate are produced in vesicles derived from the endoplasmic reticulum (KROH 1967) and are stored in the vacuoles. Most of the lipid content of the exudate is not present until anthesis, when many of the papillae degenerate (HERRERO and DICKINSON 1979). The secretion mechanism is obscure, but it seems that lipids may be secreted directly through the plasmamembrane into the stigma wall, or be passed through the tonoplast into the vacuoles for release when the protoplast degenerates, a holocrine mechanism. These concepts have been applied to exudate production in the lipidic secretion of *Forsythia* stigmas (DUMAS 1973b, 1974a, 1977) which show an eccrine secretion mechanism. The hydrophilic slime of *Aptenia* stigmas (KRISTEN 1977, KRISTEN et al. 1979) is produced first by modifications to the endomembrane system in the course of exudate secretion by a granulocrine mechanism (see Fig. 8).

The presence of this hydrophobic lipid secretion must presumably have an adaptive advantage for those genera possessing it. The lipophilic exudate may have a protective role, as a "liquid cuticle" (KONAR and LINSKENS 1966b), or serve as an attraction or reward to pollinating insects which may collect it (see LORD and WEBSTER 1979).

The stigmatic cuticle covering the exudate undergoes dissolution in several wet stigmas, including *Forsythia* (DUMAS 1973a, b; 1974), *Oenothera* (DICKINSON and LAWSON 1975a), *Lycopersicon* (CRESTI et al. 1977). The exudate remains covered by the detached cuticle when stigmas are receptive in the avocado, *Persea americana* (SEDGLEY and BUTTROSE 1978) and in *Aptenia cordifolia*

Fig. 8. Diagram of the postulated secretion transport pathways in a stigma papilla of *Aptenia cordifolia:* **1** Granulocrine secretion of the exudate (*EX*) by ER-vesicles (*ERV*) via exocytosis occurring in immature papillae. **2** Incorporation of ER-vesicles into the cell vacuole (*CV*) by a process resembling phagocytosis; the vesicle contents are mixed with the vacuole substances after membrane dissolution. **3** Holocrine excretion of a mixed exudate via degeneration of the protoplast. *PL* plasmalemma; *T* tonoplast; *CW* cell wall. (KRISTEN et al. 1979)

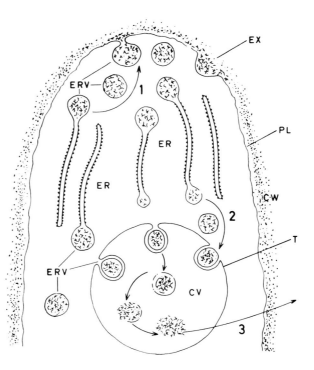

(KRISTEN et al. 1979). No cuticle is present in developing or mature stigma cells of *Acacia* (KENRICK and KNOX 1981 b).

The nature of the exudate itself varies from the osmiophilic lipids of *Petunia* and *Forsythia;* PAS-positive vesicular components which also stain for proteins in *Citrullus, Citrus, Lycopersicon,* and *Prunus;* and polyphenolic compounds detected by their cytochemical reactions in *Forsythia* and *Lycopersicon* (Table 3). Recent cytochemical studies have revealed the presence of proteins and carbohydrates as the discontinuous phase in the exudate of *Petunia* stigmas, and HERRERO and DICKINSON (1979) have detected droplets with acid phosphatase activity within the lipid matrix. Among monocots, the wet stigma of *Lilium* has been extensively studied ultrastructurally and biochemically. The multicellular stigma papillae are embedded in a mucilaginous secretion when receptive. Initially this is covered by a thin cuticle, which detaches from the papillae and later breaks down by the time the stigma is receptive. This mucilaginous secretion is continuous with that of the stylar canal. In ultrastructural appearance, the mucilage is foamy and vesicular, with electron-dense and electron-lucent structures. The microfibrillar cell walls are thick and fibrous on the outer surface, and contain numerous vesicles, apparently the secreted products. Other properties of stigmatic surfaces have been described by UWATE and LIN (1981a, b) for *Prunus.* Developing stigmas here have an affinity for the dye crystal violet, while conspicuous stigma autofluorescence is characteristic of stigmas just prior to secretion. The staining properties of crystal violet are non-specific, and the dye-binding may simply indicate the increased permeability of the cell walls. A similar observation has recently been made for stigmas of *Acacia* which are able to take up neutral red, ruthenium red, PAS and other dyes in the living state (KENRICK and KNOX 1981b). There is a need for the size of molecules capable of passing through stigma cell walls to be determined, and methods are available for primary cell walls (CARPITA et al. 1979).

A special case is the translator organ of the gynostegium of asclepiads of the Cynanchoideae (SCHNEPF et al. 1979). This organ is secreted by glandular epithelial cells of the stigmatic chamber, and is the means of attaching the pollinia to flower-visiting insects. The translator organ has two parts, a corpus and a pair of caudiculae. Both comprise an emulsion-like lipophilic substance and an enclosed hydrophilic mucilage. This slime occurs in threads or lamellae, surrounded by foamy structures. The lipophilic component originates from ER by an eccrine mode of secretion, while the mucilage may be produced by Golgi vesicles. The translator organ is produced as a result of coordinated secretory activity of the epithelial cells accompanied later by shrinking processes which give it its shape. The role of the translator in pollination is considered in Section 24.5.4.4.

A fluid secretion also characterizes pollination sites in cycads and gymnosperms (see Sect. 24.2.3).

Post-Pollination Responses. A post-pollination secretion of aqueous exudate has been observed, for example, in *Citrus* spp. (SHIRAISHI et al. 1976, YAMASHITA 1978). In stigmas of watermelon, *Citrullus lanatus,* secretions are most apparent within the first 30 min following pollination (SEDGLEY and SCHOLEFIELD 1980). In *Acacia subulata,* the wet stigma surface becomes inundated and covered by a large drop of sticky aqueous exudate within the first 2 h after pollination (KENRICK and KNOX 1981a). The drop may enlarge, and flow over the cup-like stigma and run down the style, often bearing the pollen aggregates (polyads) away with it. In this case, the secondary secretion is different in physical and chemical properties from the initial wet stigma surface secretion, and appears to resemble the guttation exudates of leaves. In both *Acacia* and *Citrullus,* the post-pollination exudate is stimulated by self- or foreign pollen, but not by wounding (KENRICK and KNOX 1981a) or by inanimate objects such as glass beads of similar size to the pollen (SEDGLEY and BLESING 1982).

In some stigmas, following pollination holocrine secretion from stigma cells may occur. The cellular contents are autolyzed and released, as the pollen tubes penetrate the stigma. Such post-pollination secretion has been observed in *Citrullus* (SEDGLEY 1982) and *Acacia* (JOBSON et al. 1983).

24.2.2.2 The Style–Pathway for Pollen Tube Growth

The style in its variety of structural and morphological features provides an added dimension for pistil variation. Three types of style have been described, based on morphological features:

1. Open style, in which the route for pollen tube growth to the ovary is through a mucilage-filled stylar canal (characteristic of many monocotyledons);

2. Closed style, in which the pollen tubes pass through the transmitting tissue to the ovary (characteristic of many dicotyledons);
3. Semi-closed style, which exhibits intermediate features (characteristic e.g., of the family *Cactaceae*).

A review of these three systems is given by DUMAS (1975), and their presence in nearly 20 genera is listed in Table 5.

The open style is characteristic of monocotyledons (e.g., *Lilium*, see review by ROSEN 1971, and *Gladiolus*, CLARKE et al. 1977). Canals filled with mucilage also occur in dicotyledons e.g. *Rhododendron* (WILLIAMS et al. 1982a). In all cases, the canal is lined with epithelial secretory cells whose function is to secrete the mucilage. In *Gladiolus*, the upper style is a stylodium, which provides a guiding tissue from the fan-shaped array of stigmatic papillae, and funnels the pollen tubes to the style. In grass stigmas, the stigma constitutes two stylodia (HESLOP-HARRISON 1982). Structural differences in the style lie in the presence or absence of a cuticle within the canal. In both *Lilium* and *Gladiolus*, the cuticle becomes detached from the walls of the secretory cells on the canal side as secretion progresses, ending up in the centre of the mucilage-filled canal (CLARKE et al. 1977). A cuticle within the style has been observed in some species with closed styles, for example sunflower, *Helianthus annuus* (VITHANAGE and KNOX 1977).

In most closed style systems, the transmitting route has been delineated on structural and cytochemical features into two zones (Fig. 5) an initial, glandular *stigmatic zone* or neck encountered by pollen tubes on passing from the stigma, which leads to the stylar *transmitting tissue* (DUMAS 1974b, DUMAS et al. 1978, HERRERO and DICKINSON 1979). The stigmatic zone may differ from the transmitting tissue in both the spherical cell shape and in content of cytoplasmic organelles. The cells of the transmitting tissue are usually elongated and fusiform in the axial direction, and have relatively thick pectocellulosic cell walls. However, both tissues may secrete a thick intercellular matrix (IM) into the intercellular spaces of the wall. In *Trifolium*, in developing pistils, the style is solid, but later a lysigenous canal is formed, complete with an entasis (swelling) towards the stigma, but the canal does not extend to the tip of the style. It is formed by dissociation of the core tissue, and becomes filled with an aqueous fluid containing sugars and glycoproteins (Y. HESLOP-HARRISON and J. HESLOP-HARRISON 1982a).

The stylar secretory cells have a similar glandular role to the stigma cells in secreting the exudate, even though there may be chemical differences between the IM and surface exudate (see Sect. 4.1). The files of cells within closed styles are usually directly connected via plasmodesmata on their transverse walls (Fig. 9) although their longitudinal walls are separated by the intercellular mucilage at maturity. In some species, for example *Petunia hybrida*, special "key junctions", ingrowths of one cell wall into the neighbouring cell in the file (Fig. 10), apparently provide direct plasmodesmatal links from one cell to its neighbour (HERRERO and DICKINSON 1979). These transmitting cells are connected directly to the papillae of the receptive surface by plasmodesmata, for example, in *Citrus* (SHIRAISHI et al. 1976, YAMASHITA 1978) and the tract terminates in the ovarian cavity (Fig. 11). The transmitting tissue thus provides not

Table 5. Characteristic features of the transmitting tissue of the style of selected angiosperms

Species	Style		Intercellular matrix (IM)	Cell wall	Characteristic cellular organelle	Reference
	Type					
1. Monocotyledons						
Gladiolus gandavensis	Open		Mucilage binds tridacnin, β-glu Yariv, antigen	Cuticle detached from epithelial cells	–	KNOX et al. 1976; CLARKE et al. 1979, 1980
Hordeum bulbosum, *Secale cereale*	Closed					HESLOP-HARRISON and HESLOP-HARRISON 1980b, 1981 a
Lilium longiflorum, *L. davidii, L. regale*	Open		Mucilage	Wall ingrowths in epithelial cells	ER and Golgi bodies	ROSEN 1964; ROSEN and THOMAS 1970; VASILIEV 1970; ROSEN 1971
2. Dicotyledons						
Acacia spp.	Closed		Esterase +, PAS +, Coomassie blue +	–	–	KENRICK and KNOX 1981b
Begonia tuberhybrida	Semi-closed		Acid phosphatase +	–	–	LECOCQ and DUMAS 1975
Brassica oleracea, *Capsella bursa-pastoris*	Closed		–	Many PD	–	SASSEN 1974
Forsythia intermedia	Closed		Sudan IV +, osmiophilic	–	ER	DUMAS 1973b, 1974a, b, 1977
Gossypium hirsutum	Closed		No IM	CW 4-layered PAS+, protein stains +, PD in walls	Golgi bodies	JENSEN and FISHER 1969

Species	Stigma	Staining / chemical reactions	Structure	ER and Golgi	References
Helianthus annuus	Closed	PAS +, Coomassie blue +	–	–	VITHANAGE and KNOX 1977
Lycopersicon peruvianum	Closed	PAS +, Sudan black –, Ruthenium red +, Protein stains +, Digested by protease and pectinase	PD in transverse walls	ER and Golgi vesicles	DE NETTANCOURT et al. 1973; CRESTI et al. 1976a; DUMAS et al. 1978
Lythrum virgatum, L. salicaria	Closed	No IM	CW 2-layered No PD	–	SASSEN 1974
Malus domestica, M. hupehensis	Closed	Protein stains +	–	ER and Golgi vesicles	HESLOP-HARRISON 1976
Nicotiana alata, N. tabacum	Closed	Peroxidase +, glucose 6 –, phosphatase +	–	ER and Golgi bodies	BELL and HICKS 1976; BREDEMEIJER 1979
Persea americana	Semi-closed	PAS +, protein stains +, Sudan black +	–	ER and Golgi bodies	SEDGLEY and BUTTROSE 1978; SEDGLEY 1979
Petunia hybrida	Closed	Esterase –, peroxidase +, Not digested by protease PAS + (esterase +, PAS +, acid phosphatase +, Coomassie blue +, Digested by protease in cells below stigma)	Key junctions in cells below stigma	ER and Golgi vesicles	VANDER et al. 1966; KONAR and LINSKENS 1966a, b; KROH 1967, 1973; SASSEN 1974; HERRERO and DICKINSON 1979, 1980
Prunus avium, P. subhirtella	Closed	Protein stains +	PD in transverse walls. Callose in outer cell walls	ER and Golgi vesicles	HESLOP-HARRISON 1976; CRESTI et al. 1978
Vitis vinifera	Closed	Alcian blue +, Coomassie blue +, PAS +	–	–	CONSIDINE and KNOX 1979

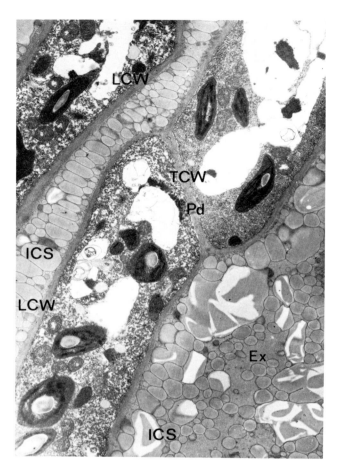

Fig. 9. Longitudinal section of the stigmatic zone of the pistil of *Lycopersicon peruvianum* at anthesis. Secretory products (*Ex*) have accumulated within enlarged intercellular spaces (*ICS*) between longitudinal cell walls (*LCW*). Cell cohesion relies solely on transverse walls (*TCW*) which contain plasmodesmata (*Pd*). × 7500. (DUMAS et al. 1978)

only a guiding tissue for the pollen tubes, but possibly also a means for the passage of electrical and other signals from the stigma surface to the ovarian cavity. Changes in electrical potential within the transmitting tissue of *Petunia* styles have been detected following cross-pollination (LINSKENS and SPANJERS 1973). The existence of pollination signals has been demonstrated by a number of experiments (see Sect. 24.2.4).

In some cases, for example, *Petunia,* the IM is present even in young stigmas, while in others, it is not secreted until maturity. In such cases, the IM occupies the intercellular spaces, the sites of the middle lamella in the presecretory styles; these spaces become greatly enlarged in mature styles. In other genera, for example *Gossypium* (cotton) the microfibrillar walls (Fig. 12) provide the route for pollen-tube growth (JENSEN and FISHER 1969). The style canal in species with an open canal is not usually filled with mucilage until the flowers open, as in *Lilium* and *Gladiolus* (CLARKE et al. 1977). In *Prunus* and *Malus* the transmitting tissue is demarcated from the style cortex by a zone of cells all around

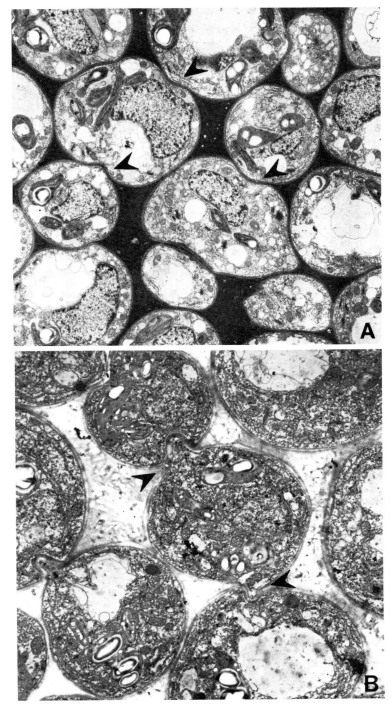

Fig. 10 A, B. Ultrastructure of the stigmatic zone transmitting cells of *Petunia hybrida*, viewed in transverse section. **A** The cells of this region are spherical and possess unusual "key" junctions (*arrows*) in electron-opaque intercellular spaces. × 3900. **B** Section after digestion. The loss of electron opacity of the intercellular spaces is striking. The "key" junctions between cells (*arrows*) are also conspicuous in this micrograph. × 6400. Glutar-aldehyde-osmium tetroxide fixation. (HERRERO and DICKINSON 1979)

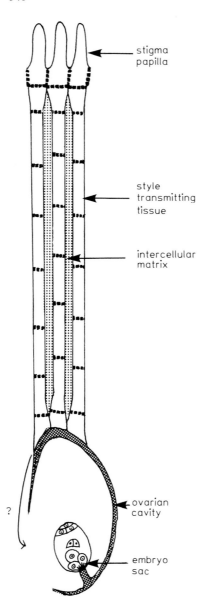

Fig. 11. Diagram of a pistil reconstructed from data of Table 2, showing the symplastic system. Plasmodesmata link the cells of the stigma papillae in the radial epidermal walls, and on the transverse wall of the style transmitting tissue. In some instances there appears to be direct plasma membrane contact between the pistil cells and transmitting cells at the ovarian cavity. Plasmodesmata are shown by a *thick broken line*. Such a system, based on isolated electronmicroscopic observations, needs to be confirmed, but it does provide for the maintenance of electrical signals from the surface of the stigma to the ovarian cavity, or for chemical signaling within the extracellular mucilage, the intercellular matrix, secreted when the style is receptive, which is the route for pollen tubes en route to the ovarian cavity. Ultrastructural studies are needed to determine whether plasmodesmatal connections continue through the placental tissue (*arrow* and *question mark*) or whether the signal travels, like the pollen tube, via the ovarian cavity

the tract whose cell walls stain intensely with decolorized aniline blue, presumably because of their callose content, although the central transmitting cells contain no callose (HESLOP-HARRISON 1976).

Both cytochemical and ultrastructural evidence implicate the cells of the stigmatic zone and transmitting tissue in the secretion of the intercellular matrix. Like the stigma cells, the style cells are characteristically rich in endoplasmic

Fig. 12. Transverse section of the transmitting tissue of the stigmatic zone of cotton, *Gossypium hirsutum*. The radial walls are thick and four layers can be identified morphologically and chemically. Plasmodesmata are lacking except in those localized portions of the wall that are considerably thinner (*arrows*), perhaps resembling the "key" junctions of *Petunia* (Fig. 10). Small vesicles (*Ve*) are present in wall layer *3*. A portion of a pollen tube (*PT*) can be seen to be in wall layer *3*. Glutaraldehyde-osmium tetroxide fixation. × 6000. (JENSEN and FISHER 1969)

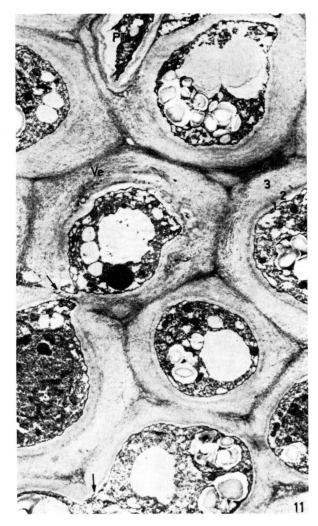

reticulum and/or Golgi bodies (Table 5). ER is abundant in species with a mainly lipid exudate, for example, in the stigmatic zone of *Petunia* and *Forsythia;* while both types of organelles are frequent in styles with secretory vesicles in the IM. A merocrine mode of secretion appears to be involved. The role of the style canal cells in secreting the mucilage is further supported by the presence of transfer cell wall ingrowths on the wall interfacing with the canal in *Lilium longiflorum* (Table 4 and ROSEN 1971).

Cytochemical tests have shown similarities and differences between the IM of styles of different genera (Table 5). The IM of *Petunia* is electron-dense, and following protease digestion the electron opacity is greatly reduced in the stigmatic zone (Fig. 10) but not in the transmitting tissue. HERRERO and DICKINSON (1979) consider that this indicates the presence of proteins in the matrix of the stigmatic zone. In both zones enzymes are active, and cytochemical tests have detected reaction products for acid phosphatase and peroxidase; in contrast, esterase has been found only in the stigmatic zone IM. In this latter feature, *Petunia* is remarkably similar to *Acacia* and *Populus* (Table 5). In *Lycopersicon,* which has a lipidic stigmatic exudate, the stylar IM contains carbohydrates and proteins (CRESTI et al. 1976a), and is partly also digested by pectinase and protease. Lipids, in addition to carbohydrate, have been detected in the IM of stylar transmitting tissue of avocado, *Persea americana* (Table 5).

It appears, then, that nearly all styles possess a mucilaginous intercellular matrix which provides a route for the pollen tubes to the ovary. In the cases tested (Table 5), this matrix contains carbohydrates and proteins as major constituents and is hydrophilic. In some cases, a similar secretion is also found at the stigma surface.

There are indications that the transmitting tissue (or stylar canal) directs pollen tube growth to the ovary, perhaps partly chemotropically (see review by ROSEN 1971). In ovules of the grass *Paspalum,* mucilaginous material fills the micropylar cavity (CHAO 1972, 1977), and this may have a role in pollen-tube attraction. In the few cases where pollen-tube penetration of the embryo sac has been observed by light or electron microscopy, the tubes enter one synergid cell, with its conspicuous filiform apparatus (transfer cell-like wall ingrowths). The sperm cells are discharged into the embryo sac, one sperm cell fusing with the egg cell and the other with the pair of polar nuclei forming the primary endosperm cell (see review by JENSEN 1974). The possibilities offered by the embryo sac of *Plumbago,* which has no synergids, have been explored by RUSSELL (1979, 1980) and RUSSELL and CASS (1981a). The pollen tube is received through transfer-cell-like ingrowths at the base of the egg (RUSSELL and CASS 1981b).

These events are associated with fertilization and have been exceedingly difficult to approach experimentally because the embryo sac lies deep within the pistil. Some success has been achieved by quite different approaches. First, attempts have been made to examine living pollinated embryo sacs in species with transparent ovules at the time when fertilization might be expected. Using this approach, ERDELSKA (1974) has recorded by light microscopy the fusion of male and female nucleoli of the living primary endosperm nuclei of the central cell in *Jasione,* and after dissection in *Galanthus.* A second approach, akin to in vitro culture, has been adopted by PREIL and KEYSER (1975), who have observed the growth of germinating pollen directly on living ovules of *Begonia,* using a mutant strain where the anthers have been transformed into styles above masses of exposed ovules. Pollen grains germinated on the ovule surfaces, and many tubes penetrated the micropyles. In vitro culture provides a third approach which has been employed by ZENKTELER (1970) and JOHRI and SHIVANNA (1974) to culture isolated ovules and pollen grains together on nutrient media. Success has been obtained in overcoming self-incompatibility and in crossing even unrelated genera, (see review by SHIVANNA 1982) providing a system for the exploration of sperm–egg interactions.

24.2.3 Mating Systems in Plants: An Evolutionary Perspective

It is appropriate at this point to consider the nature of the mutual interactions between gamete-bearing cells in plants. They represent a sequential series of encounters between self or foreign cells that may lead to successful fertilization. In algae and lower plants, sexual reproduction usually involves direct fusion or interaction between motile gametes in an aqueous environment. An elegant example is given by the green alga *Oedogonium* during sexual reproduction, here the role of mucilage as a sperm trap is displayed (Fig. 13). A mucilaginous envelope surrounds the oogonial mother cell, which swells to cover the oogon-

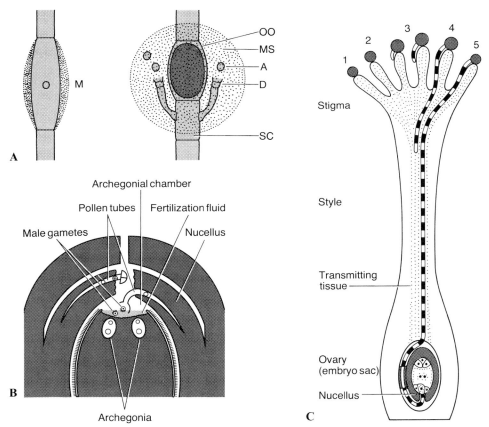

Fig. 13 A–C. Mating systems in plants. **A** Formation of a mucilaginous sperm trap around the oogonium of the green alga, *Oedogonium* (RAWITSCHER-KUNKEL and MACHLIS 1961; BONEY 1981). *O* original mother cell; *OO* oogonium; *MS* mucilage sheath; *SC* suffultory cell; *D* dwarf male plant releasing spermatozoids (*A*). **B** Pollination in cycads. Pollen grains are borne by air currents to the micropyle, where they germinate on the inner surface of the integument, releasing the motile sperm cells into the fluid bathing the archegonia. (PETTITT 1977). **C** Pollination in flowering plants. Diagram of pistil with stigma papillae, showing the sequential events following arrival of compatible pollen. (KNOX 1976). Compare with Fig. 14d

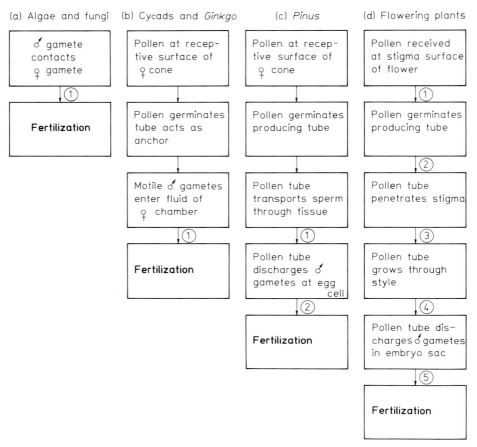

(a) Algae and fungi   (b) Cycads and *Ginkgo*   (c) *Pinus*   (d) Flowering plants

Fig. 14 a–d. A systems approach to the events of fertilization in plants. The possible steps where extracellular secretions may be involved in discrimination of foreign or self-incompatible pollen are indicated by the *circled numbers*. (CLARKE and KNOX 1980, WILLIAMS et al. 1982a)

ium after settlement and maturation of the male plants (RAWITSCHER-KUNKEL and MACHLIS 1961, and review of BONEY 1981). Fertilization occurs within the mucilage. In algae with flagellated gametes, for example *Chlamydomonas,* fusion of the male gametes involves an agglutination reaction. The tips of the flagellae of both partners adhere together to ensure direct cell-cell contact. In lower plants, fertilization is essentially a one-step event (Fig. 14).

In the cycads and lower gymnosperms, such as *Ginkgo biloba,* the maidenhair tree, we can find the first step in the mutual co-adaptations that have led to the explosive success of angiosperms into terrestrial environments. *Ginkgo* is dioecious, and pollen is borne from the anther to the female megaspores by air currents. A pollen tube is produced when the pollen germinates within the pollen chamber, but it simply acts as an anchor for the sperm sac, which ruptures, releasing the ripe motile sperm (Fig. 13 and PETTITT 1977). They swim freely in the pool of fluid bathing the egg-containing structures, the archegonia, and sexual fusion occurs within this aqueous environment.

Surface secretions form the fluid constituting the archegonial pool. The female megaspore has a patterned outer surface structurally resembling the exine wall of flowering plant pollen, and is coated with proteins and glycoproteins derived from surrounding parental tissue (PETTITT 1977). These include non-specific esterase, a Con A-binding glycoprotein and a haemagglutinin. Hydrolytic enzymes have also been detected at the germinal pores of cycad and *Ginkgo* pollen (PETTITT 1977) in similar wall sites to those occurring in *Pinus* and flowering plants (KNOX and HESLOP-HARRISON 1970), and they presumably function in pollen-tube penetration and growth.

Foreign pollen may be rejected within the archegonial pool, presumably indicating a sperm–egg interaction. In these primitive vascular plants, there is little evidence for specificity in the initial stages of pollination (Fig. 14).

Pinus and other higher gymnosperms have a fertilization system where the pollen grains are borne from the anthers to the megaspores by air currents. When receptive, the megaspores secrete a "pollination drop" of fluid within which the grains germinate, and the pollen tubes grow slowly through the nucellus to the archegonia (McWILLIAM 1958, ZIEGLER 1959). A single fertilization event is involved. The sperm cells are non-motile, and are borne to the egg within the pollen tube, the process of siphonogamy. Foreign pollen may be rejected within the nucellus in incompatible matings (Fig. 14).

24.2.4 Pollination Systems in Flowering Plants

In keeping with their place in the evolutionary scale, the angiosperms have a more advanced system for discrimination of self and non-self during pollination than the gymnosperms. The major difference is the involvement of the pollen tube and its interactions with special receptor cells of the stigma (Fig. 13), compared with the direct contact of naked sperm with the archegonium of *Ginkgo* and cycads and between the pollen tube and the nucellus of *Pinus*. The presence of the stigma and style allow for the intervention of a series of physiological barriers to pollen-tube growth that may prevent fertilisation. The mating interactions of angiosperms can be considered as a sequential series of sub-systems (Fig. 14). In this section, the interactions in self-incompatible systems where the genetic control is more precise will be considered first. Subsequently the possibilities offered by interspecific pollinations and compatible self-pollinations will be considered.

24.2.4.1 Self-Incompatibility Systems

Nearly all families of flowering plants exhibit some kind of self-incompatibility, the incapacity of plants to produce seed when self-pollinated. In such cases, although pollination does not lead to fertilization, the self-pollen may germinate and the pollen tube grow to an extent specific for each system. At some stage the subsequent events of fertilization (Fig. 14) are prevented. The reaction is controlled by one or more major gene loci (see reviews by LEWIS 1976, DE NETTANCOURT 1977).

Heteromorphic Systems

In about 24 families, self-incompatibility is manifested by a marked dimorphism or heteromorphism in the appearance, structure and size of pollen and anthers, stigma and styles in flowers of different individuals of a species (VILLEUMIER 1966). Generally unequal style and stamen length distinguish the floral morphs from each other. In addition differences are found in the shape and sculpturing of the exine of the pollen and in the stigmatic papillae, and differences may also occur in wall structure and the type of secretion (OCKENDON 1968, HEITZ et al. 1971, DULBERGER 1974). Rejection occurs following self-pollination, but not following inter-morph pollination. The system is controlled by a single di-allelic locus (LEWIS 1976). Since all pollen grains within an anther behave similarly, the mechanism involves sporophytic control. In *Linum,* inhibition following self-pollination of short-styled thrum types occurs in the stigma; in pin × pin pollinations, few grains adhere, and they either fail to hydrate (LEWIS 1942, DICKINSON and LEWIS 1974) or germinate and are then inhibited in the stigma (GHOSH 1981). LEWIS (1943) explored the differences in osmotic potential between pollen and stigma, and found that intermorph (compatible) pollinations showed a median ratio of osmotic differences between pollen and style, while self pollinations showed higher or lower ratios. These disparities may be related to differences in the stigma surface between the two types (see Sect. 24.2.2.1).

Sporophytic Systems

In most cases of sporophytic self-incompatibility, there are no morphological differences associated with S-gene expression. Pollen grains display the parental phenotype at the stigma rather than expressing their own haploid genotype. A list of the features of this incompatibility system is given in Table 6 and the systems have recently been reviewed by HESLOP-HARRISON (1978a) and DU-MAS and GAUDE (1981). In *Cosmos bipinnatus,* a single multi-allelic S-gene is involved, (CROWE 1954), while in sunflower, two major gene loci may be involved (HABURA 1957, SHCHORI 1969). Following self-pollination in both systems, pollen tubes are produced, but they are inhibited on the stigma surface. A characteristic feature of the response is that the $(1 \rightarrow 3)$-β-glucan callose is produced (detected cytochemically by its staining reactions with decolorized aniline blue) occluding both pollen tubes and stigmatic papillae (KNOX 1973). This rejection response can be detected within 15 min in *Cosmos* (HOWLETT et al. 1975). In compatible matings, much less callose is produced.

Several genera of Cruciferae possess a well-defined sporophytic self-incompatibility system whose barriers have been investigated (CHRIST 1959). In these systems, the stigma callose response to self-pollen is also elicited by pollen-wall fractions derived from the exine cavities of self but not compatible pollen (DICKINSON and LEWIS 1973a, b, HESLOP-HARRISON et al. 1974). The stigmatic site

Table 6. Summary of characteristic features of sporophytic and gametophytic self-incompatibility systems

Feature	Self-incompatibility system	
	Sporophytic	Gametophytic
Reaction of pollen on stigma	Determined by parental S-genotype. Dominance and reciprocal differences. $S_1S_2 \rightarrow S_1$ or S_2 parent pollen	Determined by pollen S-genotype. No dominance or reciprocal differences. $S_1S_2 \rightarrow S_1$ and S_2 parent pollen
Type of pollen	Tricellular	Bicellular
Site of pollen or pollen-tube arrest	Mostly stigma surface	Mostly stylar transmitting tissue
Incompatibility factors of pollen	S-specific exine proteins in Cruciferae	Diffusible S-specific pollen protein in *Oenothera*
Incompatibility factors of stigma and style	S-specific glycoprotein of *Brassica* stigmas	S-specific glycoprotein of *Prunus* styles
Rejection response	Callose produced in pollen and stigma papillae	Callose produced in pollen tube; tube tip bursts
Experimental systems	*Brassica oleracea, B. campestris, Raphanus alba, Cosmos bipinnatus, Helianthus annuus*	*Petunia hybrida, Lycopersicon peruvianum, Prunus avium, Nicotiana alata*

of incompatibility was demonstrated by SEARS (1937) in *Brassica oleracea*: removal of the stigmatic papillae, or macerating the surface before pollination led to self-fertility. The degree of pollen hydration apparently affects the incompatibility response (CARTER et al. 1975), as do high levels of CO_2 in the atmosphere (SHARMA et al. 1981). Methods of overcoming the incompatibility by manipulation of pollen or stigma surface are reviewed by CLARKE and KNOX 1978, DUMAS and GAUDE 1982a, b, and by STETTLER and AGAR, Chap. 25, this Vol.).

Gametophytic Systems

In gametophytic systems of self-incompatibility, the expression of the S-gene is determined by the genotype of the individual pollen grain at the stigma surface or in the style. This means that in a plant heterozygous for the S gene, the pollen grains will segregate equally within the anther for the two S-alleles, each grain then acting according to its own S-genotype. A list of the features of this incompatibility system is given in Table 6. The system has been recently reviewed (HESLOP-HARRISON 1978a, b, LEWIS 1980, and DUMAS and GAUDE 1981). Pollen-tube inhibition can occur at a variety of sites, from within the stigma to the embryo sac itself (LEWIS 1956, BREWBAKER 1957). Self-pollen inhibition may occur within the stigma, for example, in the grasses and *Oeno-*

thera organensis (EMERSON 1940) and *Tradescantia pallida* (HERD and BEADLE 1980). Generally, however, pollen-tube arrest occurs within the transmitting tissue of the style, as for example, in *Petunia hybrida* (STRAUB 1946), *Nicotiana alata, Lycopersicon peruvianum* (DE NETTANCOURT et al. 1973), and *Prunus avium* (CRANE and BROWN 1937). In *Lilium longiflorum* with an open style, inhibition occurs at the base of the canal (ROSEN 1971). In some species, incompatible self-pollen tubes grow as rapidly as compatible ones until arrest occurs.

In a few cases, inhibition may take place in the ovary, at the micropylar cavity or in the embryo sac, for example in *Hemerocallis citrina* (STOUT and CHANDLER 1933) and certain other monocotyledons (ARASU 1968). In *Gasteria verrucosa* (SEARS 1937) and *Borago officinalis* (CROWE 1971), the inhibition is said to be post-zygotic, taking place after fertilization in *Borago,* and after the first division of the endosperm in *Gasteria* presumably mediated by lethal genes. In cocoa, *Theobroma cacao,* COPE (1962) has summarized cytological and genetic experiments demonstrating that inhibition occurs in the embryo sac just before syngamy. In a certain proportion of the ovules, abortion occurs, even though sperm cells are released. This has been explained by postulating the presence of self-incompatibility substances at the interacting surfaces of sperm and egg (see review by DE NETTANCOURT 1977). In this case, the interaction may truly be gamete–gamete. In nearly all other cases of gametophytic self-incompatibility, the interaction is gametophytic an the male side, and sporophytic on the female side.

Arrest of self-tubes is usually accompanied by marked thickening and swelling of the tips, which become occluded with callose and in many instances appear to burst, as if in premature discharge of the male gametes (DE NETTANCOURT et al. 1973, HERRERO and DICKINSON 1980). In others, the mitotic division forming the sperm cells may be inhibited (see review by DE NETTANCOURT 1977). Some preliminary evidence on the nature of pollen-tube inhibition has been obtained. DE NETTANCOURT et al. (1973) observed an unusual concentric type of endoplasmic reticulum in the pollen-tube cytoplasm of self-tubes, that was absent in compatible tubes, and which they consider may reflect inhibition of protein synthesis. This hypothesis has been supported by CRESTI et al. (1977), who showed that *Lycopersicon peruvianum* pollen exposed to lethal doses of gamma irradiation produced pollen tubes which were arrested in the styles of compatible pistils. The inhibited tubes possessed this characteristic concentric type of endoplasmic reticulum.

24.2.4.2 Dichogamy and Other Isolating Mechanisms

Self-incompatibility is primarily a mechanism to ensure outcrossing, especially in those species with hermaphrodite flowers in which self-crosses are possible through geitonogamy or autogamy. However, some genera with self-incompatibility systems are dichogamous, the flowers having two periods of reproductive activity, once at anther dehiscence (the male phase) and when the stigma is receptive (the female phase). For example, in the avocado, *Persea americana,* the flower opens twice. The first occasion is the female phase, when pollination of the stigma results in fertilization; in contrast, on the second occasion, anther dehiscence occurs, but if the stigma receives pollen, it mostly fails to germinate and fertilization does not occur (SEDGLEY 1977, 1979). Structural changes have been defined in the stigma and style, in terms of starch depletion and callose deposition, which occur in both pollinated and unpollinated pistils, but there has been a marked absence of any evidence for cellular changes to account for the changed stigma response.

A similar phenomenon has been found in *Amyema,* an Australian genus of Loranthoid mistletoes, where certain species are markedly protandrous, and stigmas permit fewer pollen tubes to germinate and grow through the styles in the male phase than in the female phase (BERNHARDT et al. 1980). Secretion of the stigma surface pellicle on the dry-type papillae, as detected cytochemically by esterase activity, is delayed until the female phase in protandrous species such as *A. miquelii* and *A. miraculosum.* The appearance of this layer in other systems has been correlated with stigma receptivity (see Sect. 24.4.1).

Dichogamous systems of this kind offer a useful experimental system to approach the molecular basis of stigma receptivity in that they provide a temporal basis for detection of biochemical differences. However, they are important also for advancing our understanding of the chemical characteristics of the stigma surface. Dichogamy is intimately related to the behaviour of animal pollinators, which need to be attracted on at least two separate occasions to the same flower. For insects and vertebrate pollinators such as birds and bats, the lure is usually nectar secreted by the floral nectaries (see FINDLAY Chap. 18, Volume 13A and ZANDONELLA et al. 1981). Nectar comprises an aqueous solution of sucrose and fructose, with traces of free amino acids. Since the stigma is the site of pollen reception, it is likely that the exudate may function not only in cell-cell interactions, but also in vector attraction by providing even a secondary source of sugar secretions. In addition, the need for an effective adhesive to ensure that the pollen is trapped efficiently on the stigma surface must also have adaptive significance (see review by DUMAS and GAUDE 1982a).

24.2.4.3 Compatible Systems

Compatible pollination occurs in species where both self- and cross-pollinations result in successful fertilization and seed-setting. In many ways it must be looked upon as the basic system of reproduction, although the existence of self-incompatibility systems in a high proportion of angiosperm families might cast doubt on such a view. In genera with sporophytic self-incompatibility, for example, *Brassica,* self-compatibility is believed to result from the expression of compatibility genes usually suppressed in the presence of S-alleles high in the dominance series; while in other systems, compatibility may result from deletions at the S-locus (see review by DE NETTANCOURT 1977 and references therein). Whatever their ultimate genetic basis may be, compatible systems have proved of value in the study of the events of pollination with a view to establishing the genetic control and physiological features of the component subprocesses.

Direct observation allows the definition of various landmarks of pollination, through studies of the behaviour of self-, cross- and foreign pollen. Adhesion and hydration of compatible pollen is usually one of the first events occurring at touchdown on the stigma surface. In two systems, *Crocus* and *Gladiolus* (family Iridaceae), which are self-fertile and have easily accessible stigmas and anthers, the major events of pollination have been delineated. Both have dry type stigmas and the recognition events constitute a sequential series (HESLOP-HARRISON 1975b, KNOX et al. 1976) each having an option in which the interaction can be terminated (Fig. 14).

1. Foreign pollen may fail to hydrate as seen, for example, in the response of *Gladiolus* stigmas to pollen of *Gloriosa,* family Liliaceae (KNOX et al. 1976), or it may successfully pass this barrier only to be halted at a later stage. Specific attachment of compatible pollen to the stigma surface has been demonstrated in some systems, for example *Brassica* (ROGGEN 1972, ROBERTS et al. 1980, STEAD et al. 1980). Adhesion presumably precedes hydration, but these events are difficult to separate.
2. Pollen from closely related species may germinate but the pollen tubes fail to penetrate the stigma surface as seen, for example, in the response of *Gladiolus* stigmas to pollen of *Crocosmia,* another genus in the same family (KNOX et al. 1976).
3. Occasionally, foreign pollen may successfully penetrate the stigma surface and be arrested within the style (Y. HESLOP-HARRISON 1977). In some cases, even fertilization may be completed, but the embryo may fail early in differentiation because the genomes are incompatible.

LEWIS and CROWE (1958) crossed species belonging to different families, and observed in some cases that pollen tubes grew successfully through the style to the ovule. In reciprocal crosses, the successful pollen tube growth often occurred in one direction only. The site of arrest in this unilateral type of interspecific incompatibility characteristically parallels that of self-incompatibility (see review by DE NETTANCOURT 1977).

In crosses between closely related species in the same family, the responses of pollen and pistil may reflect control by the S-supergene, or closely linked genes. If pollination is inhibited, it has been attributed to *interspecific incompatibility,* and pollen–pistil recognition mechanisms are implicated. However, in foreign pollinations, or interspecific crosses between species that are not closely related, the genetic distance may be so great that the pollination process is uncoordinated, the pollen and/or pistil no longer receiving or responding to mutual signals. HOGENBOOM (1975) has termed such interactions *incongruity* rather than incompatibility. The precise operational criteria which separate the two processes now need to be defined; incompatibility is controlled by the S-supergene; incongruity is an uncontrolled process arising from the lack of co-adaptation.

Modification of the pollen or stigma surface markedly affects the pollination process. After short-term exposure to proteolytic enzymes, stigmas of *Agrostemma* or *Crocus* still have living papillae, as judged by retention of cytoplasmic streaming, but in subsequent compatible pollinations, the pollen tubes germinate but are unable to attach themselves to the papillae and fail to penetrate the stigma cuticle (Y. HESLOP-HARRISON 1977, J. and Y. HESLOP-HARRISON 1975a). Binding of the lectin Con A to the surface of *Gladiolus* and *Secale* stigmas has a similar consequence for normally compatible pollen (KNOX et al. 1976;

HESLOP-HARRISON and HESLOP-HARRISON 1981 a). Pollen-tube growth, however, appeared to be stimulated by the presence of the lectin on the stigma surface (cf. SOUTHWORTH 1975) and many grains developed more than one pollen tube. These data, together with the results of manipulated pollinations in which the pollen or stigma were treated with, for example, various organic solvents prior to pollination (see CLARKE and KNOX 1978 for review), all suggest that communication between the intact pollen and stigma surfaces is essential for discrimination of pollen from other species or genera.

What then is the basis for pistil discrimination of self-pollen on the one hand, and foreign pollen on the other? The experiments on the control of close interspecific matings by LEWIS and CROWE (1958) suggested that self-incompatible female parents often showed unilateral cross-incompatibility, while compatible species were more permissive. This view has received general support in other systems (DE NETTANCOURT 1977) and it seems likely that control may be exerted by the S-gene and other controlling gene systems of self-incompatibility. The possible relationships of the pollen–pistil recognition systems have been given by HESLOP-HARRISON (1982) based on extensive studies of grasses (Fig. 15).

24.2.4.4 Pollination–Stimulated Ovule Differentiation

The assumption implicit in Fig. 13 C is that the pistil, when the stigmas are receptive, is ready and waiting for fertilization. Certainly this is not the case in gymnosperms: in *Ginkgo* fertilization does not occur until the "fruits" fall to the ground, long after pollination; *Larix* and *Pinus*, pollen-tube-growth to the megaspore may take many months. Completion of megaspore development does not occur until after pollination. In angiosperms, fertilization usually takes minutes or hours rather than months from pollen arrival on the stigma, and necessarily, the ovary is usually fully differentiated and ovules mature at time of pollination. Exceptions do occur, however, in certain orchids, e.g., *Cymbidium* (ARDITTI 1979) and in the oak *Quercus*, and hazel nut *Corylus* (THOMPSON 1979). In hazel nut, the pollen tube forms a cyst-like structure within the style, while awaiting ovary differentiation, which is triggered by pollination.

24.3 Physiology of Interacting Cells and Tissues

24.3.1 Physiological Characteristics of Pollen

A mature, living pollen grain has a low water content, estimated to lie between 15%–35% fresh weight ($\sim 22 \times 10^{-9}$ g for a grass pollen grain) (STANLEY and LINSKENS 1974). Pollen grains are insulated from their environment by the complex wall and the aperture(s) when present. The wall may be sealed by lipids, in the form of pollencoat. The aperture structure appears to be related to the germination mechanism, comprising a pore or slit with complex infolding, or buckling or interleaving shutters made of sporopollenin plates (HESLOP-HARRISON 1979c).

Table 7. Relationships between water content and plasma membrane characteristics in pollen grains. (DUMAS and GAUDE 1983a, b)

Pollen grain	Water content	Plasma membrane of vegetative cell
Secale cereale	<20%	Porous and ineffective
	Critical level (20%)	
Brassica napus	>20%	Continuous and effective

The significance of the low water state in the physiology of the pollen grain is that it not only reduces the mass of the grain, but the metabolic activity is altered: activity of cellular water and turgor pressure is reduced; spatial relationships between endomembrane systems may be changed; macromolecules may change by removal of water or modification of the structure of vicinal water (see review by DUMAS and GAUDE 1983).

When freshly shed from the anther, a rye pollen grain has a water content of 30% to 35%, while if allowed to dry out for 2 h at 23 °C in an atmosphere of 50% RH, its water content will fall to 20% to 25% (HESLOP-HARRISON 1979b). In pollen of grasses, SHIVANNA and HESLOP-HARRISON (1981) have obtained ultrastructural and physiological evidence that the state of the plasma membrane is one of the principal factors determining germination capacity. In *Secale*, they have shown that at time of dispersal, the membranes are dissociated and do not form an osmotic barrier. Normal membrane properties are recovered when the grains are hydrated. The porous plasma membrane in the dehydrated living pollen may be a characteristic of drier types of pollen (less than 20% water) (Table 7 and DUMAS and GAUDE 1982a). In pollen types with higher than 20% water content, e.g., *Brassica*, DUMAS and GAUDE (1983a, b) show that there is an effective plasma membrane in the living dehydrated grain, which only becomes porous at death. Striking differences between the two types exist: grass pollen quickly loses viability within 1–2 h, and has no lipid seal (SHIVANNA and HESLOP-HARRISON 1981, HESLOP-HARRISON 1979a, b); *Brassica* pollen retains some viability after dehydration and storage, and has a thick lipid seal (DUMAS and GAUDE 1983).

Tricellular pollen grains occur in evolutionarily advanced families, and we may wonder what selective advantage the precocious development of the sperm cells may confer. This question has been approached by HOEKSTRA and BRUINSMA (1979), who have explored the metabolic differences between bicellular pollen, for example, *Nicotiana*, *Typha* and *Tradescantia*, and tricellular pollen of *Aster*, *Chrysanthemum* and *Cosmos*. Tricellular pollen contains fully developed mitochondria when mature (HOEKSTRA 1979), with a high rate of respiration, attaining maximum electron-transducing capacity within 2 min of hydration, thus permitting rapid germination. In comparison, bicellular types have poorly differentiated mitochondria at maturity, but become progressively more normally differentiated as germination proceeds. Respiration is low, and maximum electron-transducing capacity is not reached until 15 to 75 min in different genera. Bicellular pollen thus has a lag period whose duration is determined not by the division of the generative cell but by the level and rate of mitochondrial development (HOEKSTRA 1979).

24.3.2 Pollen Germination and Tube Growth

Pollen germination is triggered by hydration, either in vitro, or in vivo on the moist surface of the stigma. The pollen tubes may grow extremely rapidly and can be observed within 1 min in wheat pollen (CHANDRA and BHATNAGAR 1974). In vitro, successful conditions were established by BREWBAKER and KWACK (1964), and have since been adapted for germination of a wide variety of pollen types by varying the carbohydrate or boron content. In-vitro techniques for pollen germination include: (1) hanging-drop cultures (SCHRAUWEN and LINSKENS 1967); (2) suspension cultures (SHARMA et al. 1981); (3) semi-solid agar medium (BAR-SHALOM and MATTSSON 1977). The greatest difficulties have been found in germinating tricellular pollen types. In these grains, the mode of hydration is apparently the problem, and BAR-SHALOM and MATTSSON (1977) have shown that semi-solid media, simulating the "dry" stigma surface, give highest percentage germination.

The degree of hydration of the grains determines their rapidity of germination on the stigma (HESLOP-HARRISON 1979 b). A freshly released grain of rye will germinate in 100 s, while one that has been stored for 2 h, and allowed to lose nearly half its water content will take 9 min to rehydrate and germinate. The time course for rehydration on the stigma for pollen of two monocotyledons and one dicotyledon is shown in Fig. 15.

Pollen germination and tube growth provides a biological indicator of the physiological effects of antimetabolites, and also of pistil components implicated

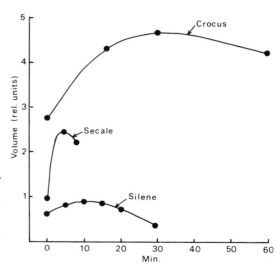

Fig. 15. Hydration of three types of pollen grain in vivo on the stigma, monitored by estimation of volume changes of the grain. Pollen tubes emerged at the point of maximum volume in each. (HESLOP-HARRISON 1979 c)

in the control of self-incompatibility. Such systems could provide a bio-assay to monitor the activity of stigma and style fractions.

Plant growth substances have variable, although usually inhibitory effects on pollen germination and tube growth in different species (LINSKENS and KROH 1970). Effects are primarily dependent on hormone concentration, but there may be interactions with other hormones (SHUKLA and TEWARI 1974). For example, IAA may stimulate these processes at low concentration, but higher concentrations are inhibitory (KWAN et al. 1969, BHANDAC and MALIK 1980) although in one case, IAA is an effective inhibitor at even low concentrations (MEHAN and MALIK 1976).

Growth substances isolated from the glandular secretions of ants are also powerful inhibitors of pollen germination and tube growth (see review by BEATTIE 1982). One compound, myrmicacin (p-hydroxydecanoic acid) isolated from the metathoracic gland of the ant, *Atta sexdens,* reduces pollen germination at very low concentrations (IWANAMI and IWADARE 1978, 1979). At 10 ppm, cytoplasmic streaming and generative cell division in emergent tubes is reduced, and inhibited altogether at higher concentrations (IWANAMI et al. 1979). The inhibitor apparently acts by blocking cell wall synthesis (IWANAMI et al. 1981). These insect secretions, although not apparently primarily evolved to affect pollen germination, played an important role in the evolution of pollination systems (BEATTIE 1982).

What are the effects of inhibitors of RNA and protein synthesis on pollen germination and tube growth? The results indicate that the two processes are independent. RNA transcription is not essential for germination, but is needed for sustained pollen-tube growth in some, but not all, species (MASCARENHAS 1975, 1978, SHIVANNA et al. 1979). Both processes, however, may require protein synthesis, determined from the effects of cycloheximide and other inhibitors of translation (Table 8). Stable mRNA has been demonstrated in one system using an in-vitro cell-free translation system (FRANKIS and MASCARENHAS 1980). One important event of pollination in bicellular pollen types is generative cell division, which requires protein synthesis in one system (Table 8). These data are important in providing an essential background against which the effects of natural stigma and style molecules can be tested.

In bicellular pollen types (see Sect. 3.1) polysome assembly begins only after hydration, whereas in tricellular types it is virtually complete in the mature grain (HOEKSTRA and BRUINSMA 1979). This is reflected in data for protein synthesis in the germinating grains; incorporation of ^{3}H-leucine into proteins varied from 0.01 pmol min^{-1} mg^{-1} in *Tradescantia* to 0.44 pmol min^{-1} mg^{-1} in *Aster.* It is supported by data for cycloheximide inhibition which show that protein synthesis is blocked in all hydrating grains, but that germination and tube emergence is blocked only in the bicellular pollen types. In tricellular types, for example, *Aster,* pollen tube emergence is completely unaffected by the inhibitor (Table 13 and HOEKSTRA and BRUINSMA 1979).

The growing pollen tubes of *Lilium longiflorum* utilize the stigma and style exudate for wall biosynthesis. The carbohydrates are broken down to monosaccharides before incorporation into the pollen tube polysaccharides (ROSENFIELD and LOEWUS 1975). Inositol and arabinose are utilized for the biosynthesis of

Table 8. Physiological effects of cycloheximide on pollen germination and tube growth in vitro

Species	Response	Reference
Pollen germination		
Aster tripolium	−	HOEKSTRA and BRUINSMA 1979
Brassica oleracea	+	FERRARI and WALLACE 1977a, b
Clivia miniata	+	FRANKE et al. 1972
Impatiens balsamina	−	SHIVANNA et al. 1974
Lilium longiflorum	+	FRANKE et al. 1972
Lilium longiflorum	−	ASCHER and DREWLOW 1970
Nicotiana alata	+	HOEKSTRA and BRUINSMA 1979
Nicotiana tabacum	+	CAPKOVA-BALATKOVA et al. 1980
Tradescantia paludosa	+	HOEKSTRA and BRUINSMA 1979
Tradescantia virginiana	+	MASCARENHAS 1975, 1978
Typha latifolia	+	HOEKSTRA and BRUINSMA 1979
Zephyranthes candida	−	GHOSH 1981
Z. citrina	+	GHOSH 1981
Pollen tube growth		
Amaryllis vitata	+	MASCARENHAS 1975, 1978
Antirrhinum majus	+	MASCARENHAS 1975, 1978
Aster tripolium	−	HOEKSTRA and BRUINSMA 1979
Impatiens balsamina	−	SHIVANNA et al. 1974
Lilium longiflorum	+	ASCHER and DREWLOW 1970
Nicotiana alata	+	HOEKSTRA and BRUINSMA 1979
Petunia hybrida	+	SONDHEIMER and LINSKENS 1974
Tradescantia paludosa	+	HOEKSTRA and BRUINSMA 1979
Trigonella foenum-graecum	+	SHIVANNA et al. 1974
Typha latifolia	+	HOEKSTRA and BRUINSMA 1979
Zephyranthes candida	+	GHOSH 1981
Z. citrina	+	GHOSH 1981
Generative cell division		
Impatiens balsamina	+	SHIVANNA et al. 1979

+ = inhibition; ± = some inhibition; − = no inhibition

xylosyl and arabinosyl units of wall polysaccharides (LOEWUS et al. 1973). The conversion pathways are illustrated in Fig. 16. Three compounds are utilized for tube wall formation: UDP-xylose, UDP-arabinose, and UDP-galacturonic acid. LOEWUS and co-workers have shown that the germinating grain obtains these compounds in two ways.

1. Initially on germination, the pollen relies on its own reserves, and arabinose is formed from hexose via the inositol oxidation pathway and/or formation and oxidation of UDP-glucose (LOEWUS et al. 1973).
2. During pollen-tube growth through the pistil canal exudate, arabinose from this source is converted to UDP-arabinose and used for wall polysaccharide formation (ROSENFIELD and LOEWUS 1975).

The nutritional dependence of *Lilium* pollen tubes on the pistil exudates was demonstrated by LOEWUS and LABARCA (1973) and has been confirmed

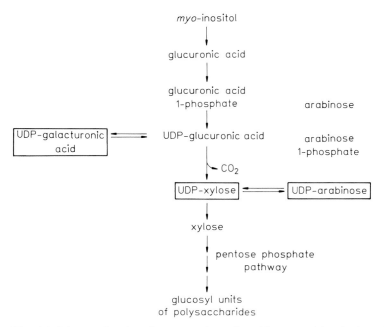

Fig. 16. Scheme showing the conversion of arabinose and inositol to the glucosyl units of pollen wall polysaccharides. The compounds utilized by lily pollen tubes for wall formation are shown in boxes. (ROSENFIELD and LOEWUS 1975)

in recent autoradiographic experiments (MIKI-HIROSIGE 1981). It had earlier been implicated in *Petunia* (LINSKENS and ESSER 1959).

The mechanism of pollen-tube inhibition in the self-incompatibility response appears to involve interference with either polysaccharide biosynthesis in the control of tube growth, or protein synthesis in the tube cytoplasm, or both. These conclusions are drawn both from physiological experiments, and from ultrastructural studies (see LEWIS 1976, DE NETTANCOURT 1977). RAFF (1981) has shown that the site of inhibition of self-tubes in the style of *Prunus avium* is genotype-specific; the extent of tube penetration is different in different cultivars. At the same time, the extent of tube growth is dependent on temperature (LEWIS 1942, RAFF 1981). In *Petunia hybrida,* the temperature dependence has been shown to relate not only to the temperature prevailing during the pollination period, but to that experienced during prior floral development (LINSKENS 1977, HERPEN and LINSKENS 1981). Similar effects of temperature on self-incompatibility have been observed by WILLIAMS and KNOX (1981) in *Lycopersicon peruvianum.* In this species, the incompatibility reaction appeared to be blocked at both low and high temperatures. Recent experiments have shown that other environmental factors, for example, concentration of atmospheric carbon dioxide, may change the expression of sporophytic self-incompatibility in *Brassica* (DHALIWAL et al. 1981). Both changes in temperature, and high concentrations of carbon dioxide may alter the recognition mechanism, and perhaps negate the onset of inhibition. Pollen tube growth is dramatically affected by various pistil extracts in experiments on the mechanism of self-incompatibility, including the isolated S-glycoprotein (see Sect. 24.5.4.2).

24.3.3 Stigma Physiology Before and After Germination of Pollen

Like other floral structures, the stigma lobes or branches exhibit movement, the closing process being sensitive to tactile stimuli, and occurring within a few seconds (see review by LINSKENS 1976b and references therein). After the initial closure, the stigma branches open again and close permanently only after pollen germination, for example, in *Mimulus* (GUTTENBERG and REIFF 1958) and *Hibiscus* (BUTTROSE et al. 1977). Mechanical stimuli do not induce permanent closure, although alcoholic extracts of pollen will do so, while heated pollen will not.

Chemotropic attractants for pollen tubes have been experimentally demonstrated in several systems, including the dicotyledons *Antirrhinum* and *Primula*, and the monocotyledons *Amaryllis*, *Clivia*, *Lilium* and *Narcissus* (see review by LINSKENS 1976b and references therein). When stigmas are placed on germination medium, the pollen tubes will grow towards the stigma, the effective distance for attraction being approx. 1500 μm.

Information on transport processes within the flower has come from the work of LINSKENS (1973b, 1974). Internal translocation of U-^{14}C-glucose, ^{14}C-labelled protein hydrolysate and ^{32}-P-phosphate occurs in flowers, mainly to the anthers but also to the ovary and style of *Petunia hybrida*, a self-incompatible system. Pollination, however, results in a redistribution of organic substances, and the style becomes a major sink for components from the anther, calyx

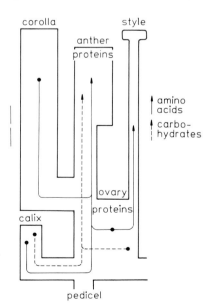

Fig. 17. Flow-diagram illustrating the routes of transport and accumulation within flowers of *Petunia hybrida* at time of anthesis. (LINSKENS 1973b)

and corolla (Fig. 17). The influx of organic substances occurs about 10 h after pollination and is especially intense after self pollination. Unpollinated styles remain in a steady state. After 18 h, the cross-pollinated (compatible) ovary becomes the major sink for materials from other floral parts, an event that does not occur in self-pollinated flowers.

After pollination, different types of electrical responses have been demonstrated (see LINSKENS 1980). No differences apparently occur in the type and quantity of growth substances present (BARENDSE et al. 1970). The rate of respiration increases within 6–14 h after pollination in *Petunia,* although substrate utilization remains unchanged. As shown by LINSKENS (1973a), the total protein content of pollinated stigmas shows quantitative differences within the first 48 h following self- or cross-pollination. Changes occur in the degree of saturation of fatty acids in the lipid pool, and differences occur in the distribution of fatty acids in the lipid pool, and differences occur in the distribution of fatty acids among the various glycosphingolipids in self- or cross-pollinated styles of *Petunia* (DELBART et al. 1980c).

The initial function of the stigma, especially in dry types, is to release water for uptake by the pollen grain. In rye, *Secale cereale,* the rate of water movement has been computed by HESLOP-HARRISON and HESLOP-HARRISON (1981a) to be initially 5.3×10^{-10} cm^3 s^{-1}, and later during pollen-tube growth, uptake increases to 7.5 to 14×10^{-11} cm^3 s^{-1}.

24.4 Molecular Aspects of Pollen–Pistil Interactions

During the pollen–stigma interaction, surface contact is initiated at several sites. Germinating pollen grains and pollen tubes on the male side interact with the stigma surface, the stigmatic zone, and the style (all parental sporophytic tissues) to determine the fate of a pollination. Ultimately contact is between the two gametophytes, the pollen tube and embryo sac. The interface is provided initially by the wall of the pollen tube on one hand, with the wall of the stigma papillae or cells of the transmitting tissue. Ultimately the contact is between the sperm cells and the egg and central cell of the embryo sac. In exceptional cases, initial pollen tube contact with the stigma surface appears to be between the plasma membrane of the tube and the stigma wall; this has been observed in the avocado, *Persea americana,* (SEDGLEY 1979), and in the seagrass *Amphibolis antarctica* (PETTITT et al. 1980). The information available on the molecular nature of these interacting surfaces will now be presented.

24.4.1 The Pollen Surface

The exine is the wall layer that first interacts with the stigma surface. In fact, the exines of most pollen grains when shed from the anther are covered by a thin osmiophilic layer, enveloping whatever materials are transferred from the tapetum for example in *Gladiolus* (KNOX 1971) and *Artemisia* (ROWLEY and DAHL 1977). The chemical nature of this layer is not known, although its affinity for the electron stain PTA at low pH suggests a high acidic polysaccharide content. The presence of an extra exinic layer on the exine surface of *Brassica* pollen has recently been demonstrated by DUMAS and GAUDE (1982a) and may be involved in interactions with the stigma (see Fig. 2B).

Short-term diffusates of the pollen grain contain a variety of proteins (STANLEY and LINSKENS 1964), glycoproteins, carbohydrates, pigments, and lipids that have been qualitatively demonstrated in a number of systems (for review see KNOX et al. 1975, HESLOP-HARRISON 1975a, HOWLETT et al. 1979).

There is little doubt that the diffusible proteins and glycoproteins of the pollen wall constitute the major source of informational molecules for the interactions with the stigma, i.e. pollen information (see DUMAS and GAUDE 1982a). Macromolecules present at the pollen surface include enzymes (TSINGER and PETROVSKAYA-BARANOVA 1961, KNOX and HESLOP-HARRISON 1969); antigens and allergens (HESLOP-HARRISON et al. 1970); lectins (DENBOROUGH 1964, GAUDE and DUMAS 1983, ZEITZ et al. 1971) and recognition factors (KNOX et al. 1972, HOWLETT et al. 1975).

A summary of the carbohydrates present in pollen as free monosaccharides and in polysaccharides is given in Table 9. If pollen is moistened in an osmotically favourable medium, wall-held components will diffuse away from the pollen grain into the medium and may be collected and analyzed as a cell-free diffusate. The extracellular origin of this material from the wall has been confirmed both by cytochemistry and by examining the pollen protoplast after moistening: the intactness of the plasma membrane is demonstrated by means of the fluorochromatic reaction (J. and Y. HESLOP-HARRISON 1970). A novel demonstration involves testing for the presence in the medium of UDP glucuronic acid pyrophosphorylase. This enzyme is cytoplasmic, and only occurs in *Lilium* pollen medium if cells are disrupted (FETT et al. 1976).

Recently, the first quantitative analysis of a pollen diffusate of *Gladiolus* showed that the major macromolecular components were protein, carbohydrate, and lipid in the ratio 10:6:0.2 (CLARKE et al. 1979). Monosaccharides present in high molecular weight components were galactose, mannose, glucose, arabinose, and rhamnose. Uronic acids were not detected. All the mannose was associated with an antigenically active glycoprotein fraction that bound to Con A-Sepharose, and the major monosaccharide of the lipid fraction was glucose. At least nine antigens have been detected in these pollen diffusates (KNOX 1971, CLARKE et al. 1979).

24.4.1.1 Extracellular Pollen Proteins

Mature pollen is a source of many proteins which appear in solution when the pollen is suspended in aqueous media. The capacity of moistened pollen to hydrolyze starch was first demonstrated by GREEN (1894). Since then, many enzymic activities have been demonstrated in pollen diffusates. The presence of hydrolases, transferases, dehydrogenases, oxidases, ligases, and lyases has been demonstrated (TSINGER and PETROVSKAYA-BARANOVA 1961, KNOX and

Table 9. Selected list of carbohydrates identified in pollen diffusates either as free sugars, polysaccharides or glycoproteins

Species	Pollen extract		References
	Free monosaccharides	Sugars present in hydrolysate of extracts	
Monocotyledons			
Phleum pratense, Dactylis glomerata	Glc, Fru, inositol 11% of pollen dry wt.	–	AUGUSTIN 1959a
Typha latifolia	Rha, Xyl, Ara, Glc, Fru + 3% di- and tri-saccharides	ND	WATANABE et al. 1961
Zea mays	Sucrose	Man, GlcA, Glc, Gal	GLADYSHEV 1962 PORTNOI and HOROVITZ 1977
Lilium longiflorum	30 µg mg^{-1} carbohydrate in buffer extract, 150 µg mg^{-1} in EDTA extract ≃ 5% total pollen carbohydrate	Hypro, Ara in pollen tube wall. Rha, Fuc. Ara, Xyl, Man, Gal, Glc in pollen tube membrane preparation	DASHEK et al. 1970; VAN DER WOUDE et al. 1971; FETT et al. 1976
Yucca aloifolia	Sucrose	ND	PORTNOI and HOROVITZ 1977
Hemerocallis fulva	Sucrose	ND	PORTNOI and HOROVITZ 1977
Gladiolus gandavensis	ND	Gal, Man, Glc, Ara, Rha, Con A-bound: Man, Gal, Glc, Not Con A-bound: Gal, Ara, Glc, Rha 2:1 Chloroform/ methanol extract Glc, Man, Gal	CLARKE et al. 1979
Thalassia hemprichii	ND	Man (59% by wt), Xyl, Gal, Glc, Ara, Rha	PETTITT 1980
Dicotyledons			
Ambrosia elatior	Ara, Gal, associated with proteins and pigments	L-Ara (37%), Hexosamine, Hexuronic acid	LEA and SEHON 1962; KING and NORMAN 1962
Cosmos bipinnatus	Ara, Glc, Gal		HOWLETT et al. 1975
Rosa damascena	Fru, Sucrose, Glc, Rha	ND	ZOLOVITCH and SECENSKA 1963
Citrus aurantium	Sucrose and reducing sugars	ND	PORTNOI and HOROVITZ 1977
Oenothera drummondii	Glc, Fru, reducing sugars	ND	PORTNOI and HOROVITZ 1977
Citrullus lanatus	Sucrose (high), Glc, Fru (traces)	ND	HAWKER et al. 1983

HESLOP-HARRISON 1969, 1970, and reviews by KNOX et al. 1975, HOWLETT et al. 1979, MALIK and SINGH 1980). The activity of only a few enzymes has been demonstrated using natural substrates; most have been detected with artificial substrates. This means that the potential function of many of these enzymes in pollination is not known.

Wall-Associated Enzymes

The capacity of pollen-wall proteins to hydrolyze a number of defined polysaccharides has been examined. Pollen diffusates of *Gladiolus* degraded the style arabinogalactan to about 30%, and effectively hydrolyzed the $(1 \rightarrow 3)$-β-glucan laminarin and carboxymethyl cellulose, indicating the presence of $(1 \rightarrow 3)$-β- and $(1 \rightarrow 4)$-β-glucan hydrolases (CLARKE et al. 1980). $(1 \rightarrow 3)$-β-glucan hydrolase activity has also been detected in pollen of other species (STANLEY and THOMAS 1967, ROSENFIELD and MATILE 1979) and a $(1 \rightarrow 4)$-β-glucan hydrolase in pear pollen diffusates (STANLEY and THOMAS 1967) and a pectinase (KROH and LOEWUS 1968). Glycosidase activity of germinating pollen of *Petunia* has been demonstrated by in vitro tests (LINSKENS et al. 1969). Recently, the relative activities of extracts of pear, *Pyrus communis,* pollen have been estimated by hydrolysis of a number of p-nitrophenyl glycosides (ROSENFIELD and MATILE 1979). Their presence at the surface is indicated by their ready diffusion into aqueous media. The nature of the pollen and pollen-tube glycosidases is of particular interest because of their potential role in hydrolysis of stigma and style polymeric carbohydrates to provide material for the growing pollen tube. However, the few studies published have been mainly based on measurements of hydrolysis of artificial substrates. Pollen diffusates hydrolyze a number of these substrates. During in-vitro pollen germination and tube growth no increase has been detected in the hydrolase activity released into the incubation medium (ROSENFIELD and MATILE 1979, SHARMA et al. 1981).

The enzyme believed to be responsible for breakdown of the stigmatic cuticle, cutinase, has been examined in some detail. Approaches have been at two levels. The first involves cytochemical observations in which dissolution of a natural substrate, the cuticle, provides an index of enzyme activity. The second involves in-vitro studies using defined substrates. This is an extremely difficult task, mainly because the precise structure of the cuticle is not established, so that it is difficult to select defined substrates for in-vitro studies with direct elevance to the in-vivo situation. It seems that the pollen of many species has the enzymic capacity to degrade stigmatic cuticles. There are apparently a few exceptions. LINSKENS and HEINEN (1962) suggested that only pollen of plants with waxy stigmatic cuticles contained the enzyme, which was absent in pollen of species which lacked such structures (i.e., with wet stigmas); for example, *Lilium* pollen. Y. HESLOP-HARRISON (1977) has shown with substrate films of cuticle that *Crocus* pollen does not degrade them, and yet the enzyme is essential for pollination.

The cuticle is composed of "cutin", a complex of polyesters containing aliphatic monomers esterified to phenolic compounds. The aliphatic compounds are predominantly in the C_{16}–C_{18} family of fatty acids, both saturated and unsaturated, which may be modified to include epoxide and diol groups;

p-coumaric acid is the major phenolic compound (SHAYK and KOLATTUKUDY 1977, see KOLATTUKUDY et al. Chap. 10, Vol. 13 B, this Series).

In pollen of nasturtium, *Tropaeolum majus,* cutinase has been shown to be synthesized during pollen maturation. The enzyme hydrolyzed the stigmatic cuticle very rapidly, suggesting it may be located in the pollen wall (SHAYK and KOLATTUKUDY 1977). The cutinase has been isolated, and shown to be a glycoprotein of molecular weight 40,000 containing 7% carbohydrate and having an isoelectric point of pH 5.5 (MAITI et al. 1979). With apple cutin as substrate, the isolated pollen enzyme is capable of catalyzing the hydrolysis of p-nitrophenyl esters of C_2–C_{18} fatty acids and has a pH optimum of 6.8. Unlike previously isolated fungal cutinases, the pollen cutinase is strongly inhibited by sulfhydryl reagents N-ethyl maleimide and p-hydroxymercuribenzoate, and has been termed an SH-cutinase. The pollen enzyme showed no antigenic cross-reactivity with fungal cutinase, nor was its activity inhibited by antisera to the fungal cutinase of *Fusarium solani pisi.* MAITI et al. (1979) were able to differentiate their cutinase fraction from a non-specific esterase in nasturtium pollen which hydrolyzed p-nitrophenyl acetate preferentially and had an isoelectric point of pH 5.6.

It is not known whether cutinases affect sporopollenin. However, degradation of the exine polymer at sites close to the germinal aperture during early stages of pollen germination has been observed for two species *Pharbitis nil* (GHERARDINI and HEALEY 1969) and *Linum grandiflorum* (DICKINSON and LEWIS 1974). Also, filiform pollen grains of the seagrass *Amphibolis antarctica,* which are inaperturate, produce a germinal aperture by autolysis of the pollen wall (PETTITT et al. 1980). While these three pieces of evidence are entirely circumstantial, and based on electron microscopic observations, the involvement of enzymic degradation would be hard to dispute.

Another enzyme, alkaline phosphatase, is concentrated mainly at the tips of the pollen tube in *Calotropis procera* (MALIK et al. 1975). A specific alkaline phosphatase (MI-I-phosphatase) has been isolated from mature pollen of *Lilium* by LOEWUS and LOEWUS (1982). The enzyme is involved in the myo-inositol oxidation pathway, in the conversion of glucose-6-P to UDP-D-glucuronate which leads to the uronosyl and pentosyl residues of plant cell wall polysaccharides, and may be involved in phytate degradation.

Acid Hydrolases and Transferases

The properties of several acid hydrolases and transferases from pollen of Japanese cycad, *Cycas revoluta,* have been investigated by HARA et al. (1970). Ribonuclease (pH optimum 5.7) in pollen extracts readily degraded RNA into nucleotide 2′, 3′-cyclic phosphates and 3′-mononucleotides. This enzyme, together with a phosphomonoesterase was not eluted in glucose or sucrose solutions, but rapidly eluted in various salt solutions, suggesting that the enzyme was located in the pollen wall. This was later confirmed by using short-term extractions, and comparing activity with those from homogenates of thoroughly washed pollen. Nearly half the ribonuclease activity proved to be located at the surface of pollen grains, together with phosphomonoesterase, acid pyrophosphatase, and invertase activities (HARA et al. 1972). One enzyme, Mg^{2+}-dependent alkaline pyrophosphatase, was found to be entirely cytoplasmic in cycad pollen. About 90% of the invertase activity of whole pollen (pH optimum 4.0) was detected in the cell wall fraction, and the enzyme hydrolyzed sucrose and raffinose specifically, so that it is a β-fructofuranoside-fructohydrolase (HARA et al. 1972).

Cytochemical evidence has been obtained to show that enzymes are held in both the exine and intine. This has been firmly established by cytochemical studies, e.g. for acid phosphatase of *Crocus* pollen by light microscopy (KNOX and HESLOP-HARRISON 1969, 1970) and transmission electron microscopy (KNOX and HESLOP-HARRISON 1971a). The enzymes may have quite different origins. The transfer of non-specific esterase activity from parental tapetal cells (on their dissolution in the mature anther) has been demonstrated cytochemically, both qualitatively and quantitatively in kale, *Brassica oleracea* (HESLOP-HARRISON et al. 1974, VITHANAGE and KNOX 1976) and for acid phosphatase in sunflower *Helianthus annuus* (VITHANAGE and KNOX 1979). These data suggest sporophytic synthesis of exine enzymes which form part of the pollen coat, although no direct biochemical or immunological evidence for such transfers has yet been obtained. This is probably because of the great difficulties in separating fractions from tapetum and pollen in the developing anthers. Such an achievement has been reported by HERDT et al. (1978) for the location of the enzymes involved in phenylpropanoid metabolism in the tapetal fraction of developing anthers.

Several enzymes have been associated with the successive appearance of pollen surface phenylpropanoid components during development of tulip anthers (WIERMANN 1979). The exine of pollen grains is compartmented, providing sites for storage of phenyl propanes, chalcones, flavonols and anthocyanins in the pollen coat. The enzymes involved in their biosynthesis are thought to be synthesized in the parental tapetal cells and to function at the surface of the exine. Of especial interest is the final flavonoid stage, when the flavonols are glycosylated. WIERMANN (1979) has evidence that this is carried out by a UDPG-O-glucosyltransferase at the surface of mature pollen.

Some pollen enzymes may be antigenically active; for example, in pollen diffusates of *Cosmos* and ragweed, antigens with esterase and acid phosphatase activity have been detected using cytochemical methods to detect enzymic activity associated with immunodiffusion precipitin bands (HOWLETT et al. 1975).

Other Pollen-Wall Proteins and Glycoproteins

Phosphatases and other hydrolases have been detected cytochemically in spore walls of a wide range of both pro- and eukaryotes, including spores of bacteria, fungi, pteridophytes, and pollen of higher plants. In bacteria, investigations suggest the enzymes may have an autolytic function (ABADIE 1979) but their function in other systems remains unknown, although the pollen enzymes are implicated in degradation of the stigma surface at pollination.

It remains an interesting fact that the major diffusible pollen proteins and glycoproteins have no known function in nature – the allergens characterized by their remarkable ability to stimulate the human immune systems to synthesize a special class of antibodies, and hence provoke the symptoms of seasonal asthma and hayfever (see review by HOWLETT and KNOX, Chap. 28, this Vol.). Together they account for 10% of the proteins of ragweed pollen. All known functions in pollen biology and pollination remain speculative, and include several possibilities: dispensable proteins; source of storage reserves available after hydrolysis for pollen-tube growth; enzymes important in pollen germination; or stigma recognition factors.

Pollen grains are found only in higher plants, and are one of the major factors in their evolutionary success in the conquest of terrestrial environments. It is an interesting fact that all allergens are largely confined to angiosperm pollens, where they occur as proteins or glycoproteins. Occasional reports of allergens in gymnosperm pollen refer to polysaccharides (e.g., *Cryptomeria* pollen, HARA et al. 1977). It is reasonable to suppose that the allergens perform some function that is unique to angiosperm pollen. Examination of Fig. 14 reveals that there are five sequential interactions between pollen and pistil in angiosperms, while there are only two in gymnosperms. The logical conclusion from this argument is that the allergens are needed for one or more of the three unique events – namely interactions with the stigma surface or style, all of which are vital for self-incompatibility responses. Perhaps allergens act as cutinases, in permitting pollen tube penetration of the stigma surface; or perhaps as recognition factors which initiate the stigma interactions.

Little is known of the recognition factors in pollen grains. An antigen specific for the S gene controlling self-incompatibility has been demonstrated in pollen of *Oenothera* (LEWIS 1952). The putative S-antigen consisted of at least 20% of total pollen protein and diffused from the pollen into isotonic medium within 30 min (MÄKINEN and LEWIS 1962), suggesting that the antigen is located in the pollen wall. Radial immunodiffusion tests using antisera to homozygous S_6 pollen of *Oenothera* provided convincing evidence of antigen specificity. Also individual pollen grains produced precipitates when sprinkled on gel containing anti-S_6 pollen sera (LEWIS et al. 1967). No reaction was obtained with pollen from S_2S_3 anthers, while 53% and 62% of pollen from S_3S_6 anthers formed precipitates in two experiments, which is close to the expectation of 50%, while pollen from homologous S_6S_6 anthers gave 81%, 91% or 100% in three experiments. Following these early demonstrations of S-antigen specificity in pollen, there has been little progress made in its isolation or characterization.

Manipulation of the pollen information, through the mentor effect (see STETTLER and AGAR, Chap. 25, this Vol.) has implicated pollen surface proteins and glycoproteins as recognition factors, e.g. in interspecific incompatibility in *Populus* (KNOX et al. 1972) and self-incompatibility in *Cosmos* (HOWLETT et al. 1975).

Lectins are bifunctional molecules with considerable potential for mediating cellular interactions. Lectinlike components have been detected in diffusates of grass pollen, by their agglutinin activity (DENBOROUGH 1964) and by their mitogenic activity (ZEITZ et al. 1971). Using a pollen-rosetting technique, the presence of agglutinins is indicated in *Brassica* and *Populus* pollen (GAUDE and DUMAS 1983). One interpretation of these studies is that both soluble and membrane-bound agglutinins may be present in pollen.

24.4.2 The Stigma Surface and Style Mucilage

24.4.2.1 Composition

Chemical analysis of exudates or homogenates of angiosperm pistils shows that they may contain a variety of macromolecules, including carbohydrates as free sugars or polysaccharides, free amino acids, proteins, glycoproteins, and proteoglycans, lipids and phenolic compounds. Data showing analyses for 16 genera of flowering plants are given in Table 10.

These chemical analyses reveal that the stigma surface secretions may be divided into two broad groups:
1. Stigmas with lipophilic secretions with very low water content
2. Stigmas with hydrophilic secretions, containing proteins and carbohydrates, usually with lipids and polyphenols.

Exudates of Monocotyledons

In Monocotyledons, stigmas with exudates belonging to the lipophilic class occur in *Strelitzia* (Table 10). MARTIN (1970) has shown that the major lipid is oleic acid, and there are esters and glycosides of various phenolic compounds present. Among other genera analyzed, *Lilium* has a hydrophilic exudate with more than 99% of the exudate comprising water, carbohydrate and protein. The major components are high molecular weight, arabinogalactans composed of galactose, arabinose, rhamnose, glucuronic and galacturonic acids (Table 10).

Soluble sugars have been found in the sticky exudate of one stigma (Table 10). In *Yucca aloifolia* D-glucose, D-fructose and sucrose are present in alcohol or aqueous extracts giving an apparent sugar concentration of 0.16 M (HOROWITZ et al. 1972). These authors noted that this is considerably less than the concentration of sugars required for pollen germination in vitro (0.3–0.5 M), and attributed the discrepancy to the presence of films of higher sugar concentration within the exudate. It is also possible that the free sugar present in pollen diffusates would combine with the stigma exudate to produce the optimum sugar concentration.

The dry stigmas of *Gladiolus* "glisten" with a surface secretion when receptive. This secretion has proved to be hydrophilic and readily extracted by buffer or detergent treatment. The *Gladiolus* extract is considered to be extracellular as the stigmatic papillae, after extraction, showed cytoplasmic streaming, indicating that intracellular components were unlikely to be present in the extracts. The high molecular weight fraction contained almost equal quantities of protein and carbohydrate with only trace amounts of lipid (Table 10; CLARKE et al. 1979). Analysis of the monosaccharide composition showed that galactose and arabinose are present with small amounts of glucose, rhamnose and mannose. The stigma surface extract contained a complex array of proteins, glycoproteins and glycolipids when analyzed by SDS-polyacrylamide gel electrophoresis. Twelve protein components were present, ranging in molecular weight from 10,000 to 60,000; seven stained for both protein and carbohydrate and are likely to be glycoprotein. A low molecular weight component (10,000) stained for carbohydrate and lipid, and is likely to be glycolipid in nature. Several glycoproteins bind to the lectin Con A (KNOX et al. 1976). Chromatography on Con A-Sepharose showed that the lectin-binding glycoprotein contained all the mannose residues. The remainder, not binding to the lectin, consisted mainly of a high molecular weight arabinogalactan or arabinogalactan protein. The arabinogalactans isolated from *Gladiolus* pistils are major components of both stigma and style – they represent about 20% of the total stigma surface preparation and 40% of the total carbohydrate of the style (CLARKE et al. 1979). The polysaccharide component of *Lilium longiflorum* stigmatic exudate also contain an arabino-$(3 \rightarrow 6)$-galactan (ASPINALL and ROSELL 1978), but this differs from the *Gladiolus* style material in having a higher arabinose content as well as rhamnose and glucuronic acid as major components (GLEESON and CLARKE 1979).

These components of the *Gladiolus* stigma and style have some of the same characteristics as animal cell surface glycoproteins (Clarke 1981). Galactose is a common constituent of both animal and plant glycoproteins; glucose is not commonly associated with secreted animal glycoproteins although intermediates of biosynthesis of asparagine-linked glycoproteins contain three glucosyl residues per mol. (see review by Clarke and Hoggart 1982 as well as Chaps. 8 and 19, Vol. 13 B, this Series).

Glucose is the major monosaccharide in the chloroform/methanol-soluble lipid fraction of the *Gladiolus* stigma surface preparation. This observation, taken in conjunction with the very low content of organically bound phosphate and the absence of both mono and digalactosyl diglycerides, suggests that it may be a glucoside such as a sterol glucoside (Clarke et al. 1979). Readily extractable lipids, together with phenolic glycosides and pigments are present in two other monocots with dry stigmas, *Zea mays* and *Iris pseudacorus* (Table 10).

We can conclude that when the stigmas of the monocots, whether of wet or dry types, are analyzed, they contain a range of surface proteins and carbohydrates. A proteoglycan, the arabinogalactan protein, has been partly defined and is a major component in two genera, *Gladiolus* with a dry stigma, and *Lilium* which has a wet stigma.

Exudates of Dicotyledons
The stigmas of dicotyledons that have been analyzed fall into either lipophilic, hydrophobic types, or hydrophilic types (Table 10).

Table 10. Selected list summarizing the data from chemical analyses of stigma surface secretions or exudates and style mucilages of flowering plants, and the micropylar fluid of gymnosperms

| Species | Pistil type | | Tissue analyzed | Carbohydrate analysis |
	Stigma[a]	Style[b]		Method
Monocotyledons				
A. *Strelitzia reginae*	C	ND	Isolated exudate	PC
B. *Zea mays*	N	C	Whole extract	Molisch, Fehling and DNS tests
C. *Iris pseudacorus*	N	C	Homogenate	ND
D. *Hemerocallis fulva*	C	ND	Whole extract	Molisch, Fehling and DNS tests
E. *Lilium longiflorum*	C	O	Isolated exudate from stigma and style canal	GF, β-glucosyl artificial antigen
F. *Gladiolus gandavensis*	N	O	Isolated buffer and detergent extracts of stigma surface and style	GLC, GF, AC, with β-glucosyl artificial antigens and lectins. Cellulose acetate and PAGE

[a] C=copious exudate, wet-type stigma; N=no detectable exudate; dry-type stigmas
[b] C=closed type style; O=open type style; ND=not determined

A great deal of work has been carried out on pistils of *Petunia hybrida,* principally by H.F. LINSKENS and coworkers at the University of Nijmegen in the Netherlands. They have determined several physical properties of the exudate and its rate of release. It is interesting that the principal fatty acids present are different from the major constituents of the monocot stigma exudates (Table 10). A most comprehensive analysis of the lipids and glycolipids has detected differences in composition before and after various types of pollination (DELBART et al. 1980a, b, c). The free sugars present in the exudate were sucrose, glucose and fructose with traces of galactose. Xylose and other saccharides were present in exudates collected 48 h after anthesis, when the exudate appeared mucilaginous. KROH (1973) analyzed the stylar intercellular matrix of *Petunia* which is extractable in warm water. Fractionation on Sephadex G-15 showed a peak containing low molecular weight acidic carbohydrates which gave glucuronic acid and galacturonic acid after hydrolysis (Table 10). The stigmatic exudate became mucilaginous in older flowers, perhaps from secretion of carbohydrates similar to those present in the stylar transmitting tissue.

Stigmas of *Forsythia* are similar to *Petunia* in the chemical composition of their exudate. The elegant studies of DUMAS (1974a, 1977) have revealed the presence of a series of fatty acids (Table 10) and several phenolic compounds and flavonoid aglycones. The stigma exudate consists of neutral lipids in a glyceridic mixture probably containing hydrocarbons and terpenes. The fatty acids identified are between C:7 and C:12 with the maximum unsaturated form being C:18 (Table 10). During the secretory process no qualitative differences are evident between the neutral lipids and fatty acids found in the stigmas of this heterostylous species in either pin or thrum forms (see Sect.

Carbohydrate analysis		Amino acid and protein analysis	
Result		Method [c]	Result
Glc. present after acid hydrolysis		ND	
No free sugars		ND	
		ND	
No free sugars		ND	
85%–90% carbohydrate polysaccharide and complex pectic polymers. G100-I: 30% Gal; 28% Ara; 12.5% Rha; 9.5% GlcA 2.5% GalpA; G200-I: Gal, Ara, Uronic acid, tr. of Xyl, Man; arabinogalactan component isolated; highly branched; 57% Gal; 26% Ara; 11% GlcA; 6% Rha; contains sequence: 0-Rhap $(1 \to 4)$-GlcpA $(1 \to 6)$-Galp		Spectrophotometry	7% protein
Arabino 3–6 galactan protein major component, and several glycoproteins identified. 23% of extract carbohydrate containing 59% Gal, 22% Ara, 11% Glc, 1% Man and 7% Rha		Biuret PAGE	20% protein

[c] AC = affinity chromatography; GF = gel filtration; GLC = gas-liquid chromatography; PAGE = polyacrylamide gel electrophoresis; PC = paper chromatography; TLL = thin-layer chromatography

Table 10 (continued)

Species	Pistil type		Tissue analyzed	Carbohydrate analysis
	Stigma[a]	Style[b]		Method
G. *Yucca aloifolia*	N	O	Isolated exudate	ion exchange and PC; GLC; quantitative analysis
Dicotyledons				
H. *Petunia hybrida*	C	C	Isolated exudate	80% ethanol extract, TLC, GF
I. *Ipomoea batatas* (18 cv of sweet potato and 14 spp. of *Ipomoea*)	N	ND	Whole extract	PC
J. *Forsythia intermedia*	N	C	Homogenate	ND
K. *Brassica oleracea*	N	C	Whole extract and homogenate	PAGE
L. *B. campestris*	N	C	Homogenate	GF and AC with lectins. Density gradient iso-electric focussing SDS-PAGE
M. *Oenothera drummondii*	C	ND	Whole extract	Molisch, Fehling and DNS tests
N. *Citrullus lanatus*	C	C	Exudate and whole extract	70% ethanol. PC and GLC
O. *Citrus aurantium*	C	C	Whole extract	Molisch, Fehling and DNS tests, TLC
P. *Aptenia cordifolia*	C	ND	Isolated exudate	Ethanol extract; auto analysis system based on affinity of borate complexes for anion exchange resin Durrum DAX 4–20.
Gymnosperms				
Q. *Taxus baccata*			Micropylar fluid	PC
R. *Ephedra helvetica*			Micropylar fluid	PC
S. *Pinus elliotti* *P. nigra*, *P. griffithii*			Micropylar fluid	PC

Carbohydrate analysis	Amino acid and protein analysis	
Result	Method[c]	Result
1.9% free sugars D-Glc, D-Fru, Sucrose		
Free sugars: sucrose, Glu and Fru (trace of Gal). 48 h later Xyl also present and exudate mucilaginous. Style secretions low MW acidic carbohydrates containing GlcA, Galp in ratio 3:2 and other acidic oligo saccharides	Acetone extraction and PAGE for proteins 70% ethanol and amino acid analysis	Minute traces of free amino acids esp. norleucine aspartic and, threonine, serine, alanine
Glc present after acid hydrolysis Sucrose and Rha present in some ethanol or acetone extracts		
	ND	
S-Glycoprotein abundant in receptive stigmas, absent at younger stages; binds ConA		
Carbohydrate: Protein ratio 1.2 (w/w) S_7 specific glycoprotein	Amino acid analysis after acid hydrolysis	High serine glutamine and glycine
Glc, Fru as free sugars	ND	
Free monosaccharides Glc, Fru, Sucrose		
No free sugars, reducing sugars present in hydrolysate	ND	
68% dry weight as sugars, 67% free sugars, 63% Glc, 36% Fru, Traces of Gal, Rib, Xyl; 33% Polysaccharide fraction, 36% GalpA, 23% Glc, 23% GlpA, 11% Fru, 4% Gal, 30% Ara Tr. of Man, Xyl, Rha	Lowry, Biuret SDS-PAGE	8% dry weight is protein, 5 major components, mw. 18,000–60,000
10% free sugars; Sucrose, D-Glc, D-Fru. Sucrose:Hexose 1:1.5 → 1:3 in different years	PC	10 free amino acids, 6 in peptides
25% free sugars; sucrose	PC	8 free amino acids, 12 in peptides
1.25% free sugars; D-Fru (40 mM), D-Glc (33 mM) and Sucrose (3 mM)	Not detected	

Table 10 (continued)

Species	Lipids				
	Method	No. of fatty acids after esterification	No. of C atoms	Major fatty acids	
A.	Ether extracts, some KOH precipitates, GLC	14	8–18	16:0 oleic acid	
B.	Ether extracts GLC	12	6–18	16:0 17:0 18:2 18:3 } capric acid	
C.	Ether extracts GLC	11	10–20	16:0 18:2 18:3 } palmitic and oleic acids	
D.	ND				
E.	ND				
F.	2:1 chloroform: methanol	0.1% lipid sterol glucoside			
G.	ND				
H.	2:1 chloroform methanol extract TLC, GLC	8	11–20	18:2 eight present esp. undecanoic, heptadecanoic and linoleic acids Iodine no. 54.25 glyco-sphingo lipids charac-terized	
I.	Ether extracts, GLC	2	10–12	10:1 capric and lauric acids	
J.	2:1 chloroform methanol extract TLC, GLC	21	7–22	16:0 palmitic acid 18:1 oleic acid 18:2 linoleic acid 18:3 linolenic acid	
K.	ND				
L.	ND				
M.	ND				
N.	Stained with Sudan dyes				
O.	ND				
P.	Lipids not detected				
Q.	ND				
R.	ND				
S.	ND				

[d] 1. MARTIN 1970, 2. PORTNOI and HOROVITZ 1977, 3. VILLERET 1974, 4. ROSENFIELD and LOEWUS 1975, 5. ASPINALL and ROSELL 1978, 6. GLEESON and CLARKE 1979, 7. KNOX et al. 1976, 8. GLEESON and CLARKE 1980, 9. CLARKE et al. 1979, 10. HOROVITZ et al. 1972, 11. KONAR and LINSKENS 1965, 1966a, b, 12. DELBART et al. 1980a, b, c,

Phenolic and other compounds/special properties	Reference[d]
95% ethanol extract, UV absorption spectroscopy. TLC. Exudate slightly soluble in H_2O but readily soluble in organic solvents. Glucose and often esters of p-coumaric acid principal UV-absorbing component	1
Acetone extracts separated by paper chromatography. Glycosides and esters of hydroxy-cinnamic acids, containing p-coumaric and ferulic acids after acid hydrolysis, Anthocyanin pigments and glycosides present (cyanidin and pelargonidin)	1, 2
ND	3
ND	2
ND	3, 4, 5, 6
ND	7, 8, 9
ND	10
Exudate – "liquid cuticle" of lipid. 80 µg exudate released in 2 h after anthesis; temperature-dependent; claimed to be KCN and DNP – independent release. Density of exudate 1.14 at 23 °C. Viscosity ~rel 1.43. Surface tension 89.42 dynes cm^{-1}; contains no water. Sterols absent from stigmas. Several phenolics, phenolic esters and glycosides present	12, 13
Ethanol and acetone extracts; TLC. No water soluble components; glycosides and esters of caffeic acid	14
Phenolic compounds osmiophilic, chelated by $AlCl_3$; argentaffin reaction, oxidized by H_2O_2 or $H O_3$; flavonoid aglycons. Sterols absent from stigmas, but present in style. Flavone glycosides	15, 16
95:1 ethanol. The extracts contain 7 phenolic compounds, esp. quercetin and 3 flavonoid glycosides	17, 18, 19
10,000 stigmas – dry wt. 0.39 g	20
ND	2
ND	21
ND	2
Exudate 74% water content. Phenolics not detected	22
ND	23
Invertase activity detected in nucellus	23
ND	24

13. MARTIN 1969, 14. KROH 1973, 15. MARTIN and TELEK 1971, 16. DUMAS 1974a, b, 1975, 1977, and DUMAS et al. 1978, 17. MOEWUS 1950, 18. SEDGLEY 1975, 19. ROBERTS et al. 1979a, b, 20. NISHIO and HINATA 1979, 21. SEDGLEY et al. 1983, 22. KRISTEN et al. 1979, 23. ZIEGLER 1959, 24. McWILLIAM 1958

24.2.4.1.1). Quantitative differences in the polar:neutral lipid ratio were detected, the values decreasing during stigma maturation. Sterols are absent from the stigma, but present in style and ovary and represent the major difference in lipid composition detected by DUMAS (1977).

Lipids and phenolic compounds are characteristic components of the stigma exudate of the sweet potato, *Ipomoea batatas* and other *Ipomoea* species (Table 10). Free sugars, including sucrose and rhamnose, are present in some extracts. Glucose is the sugar residue in the phenolic glycosides present, tentatively identified as glucosides of caffeic acid (MARTIN and TELEK 1971).

Extremely viscous exudates occur at the stigma surface of some dicotyledons, for example BAUM (1950) found that the exudate of *Koelreuteria paniculata* was soluble in acetic acid, but not in water, alcohol, or ether, and concluded it contained a terpenoid resin as major component.

The remaining seven dicotyledonous stigma types whose exudates have been analyzed (Table 10) have proved to be hydrophilic in nature, containing proteins and glycoproteins as major components. Exudates of stigmas of these genera contained free sugars, especially glucose and fructose, while sucrose is also present in watermelon *Citrullus* (Table 10, HAWKER et al. 1983).

The best-studied example is the stigma exudate of *Aptenia cordifolia* (Mesembryanthemaceae). Here, sugars represented more than two-thirds of the dry weight of the exudate (KRISTEN et al. 1979), which had a water content of 74%. No phenolics or lipids were detected. Proteins amounted to 8% of the dry weight, and included five components detectable by SDS-PAGE ranging in molecular weight from 18,000 to 60,000. In this respect, the stigma exudate resembles that of *Gladiolus* reviewed above. Most of the saccharides were present as free sugars, with glucose and fructose (2:1) predominating with traces of galactose, ribose and xylose. Polysaccharide fractions high in galacturonic and glucuronic acids, glucose and fructose were detected (Table 10). Galactose and arabinose were also present in lower concentrations (3% to 4%). In the legume *Trifolium*, the stylar canal fluid has been analyzed by J. HESLOP-HARRISON and Y. HESLOP-HARRISON (1982a). In addition to several glycoproteins, one of which is specific to the style, the fluid contains sucrose, glucose and traces of galactose and arabinose.

We can conclude that dicot stigmas have surface secretions containing a similar diversity of components as monocots. However, the extracts are quantitatively quite different. For example, the carbohydrate analyses for *Lilium* and *Aptenia* (Table 10) are qualitatively similar, but differ in the proportions of acidic and neutral monosaccharides.

Micropylar Fluid of Gymnosperms

Wind-borne pollen of gymnosperms such as pine, *Pinus* spp. or yew, *Taxus baccata*, arrive at the "stigmatic surface" of the micropyle, within the female cone. The pollen is then transported within a fluid the pollination drop (see Sect. 24.2.3). Fructose, glucose and sucrose have been detected as free sugars within the pollination drop of *Pinus* spp. (MCWILLIAM 1958) and *Taxus baccata* (ZIEGLER 1959), and their relative concentrations are given in Table 9. In *Ephedra*, ZIEGLER (1959) could detect only sucrose, which amounted to 25% of the exudate. No free amino acids were detected in *Pinus*, but were present in *Taxus* and *Ephedra* (Table 10). There is a close parallel between the composition of these secretions of the pollen chamber, and those from floral nectaries and guttation glands of vegetative organs of angiosperms.

24.4.2.2 Function of Exudate

The exudate and secretions of the stigma and style may fulfil a number of roles in the pollen–stigma interaction:

1. In the control of pollen adhesion, hydration and germination; secretion of exudates with differing osmolarities in different genera might provide a con-

venient means for discrimination of foreign pollen by preventing its germination through osmotic imbalance (HESLOP-HARRISON 1975b).
2. To protect the pistil with its female gametes from attack by animal predators or microbial infection (MARTIN and BREWBAKER 1971, LINSKENS 1976a); also as a liquid cuticle to prevent stigma dehydration (KONAR and LINSKENS 1966).
3. In flower-visitor nutrition during pollination; the sugar exudates of many genera correspond closely in concentration to nectar secreted from flowers (HOROVITZ et al. 1972).
4. In nutrition of the pollen tube during its growth through stigma and style to the ovary (LOEWUS and LABARCA 1973).

The ability of the stigmatic surface to trap the pollen grains efficiently is undoubtedly conferred by the presence of macromolecules that are adhesive, or are components of an adhesive. The role of the arabinogalactan protein and other components of *Gladiolus* stigmas in pollen adhesion is considered in Section 24.5.

The stigma surface secretion has been implicated in the control of pollen hydration and germination by several investigators. The role of the exudate in stigmas with a copious secretion has been reviewed by MARTIN and BREWBAKER (1971) while that for the dry stigmas with only a thin membranous secretion has been discussed by MATTSSON et al. (1974) and HESLOP-HARRISON et al. (1975a, b).

Enzymes

The association between esterase activity and the receptive cell surface is significant as the pollen capturing sites may actually be identified by the presence of the reaction product. The receptive site can be differentiated from brush hairs and other adaptations for insect pollination and can be precisely delimited in pistils where there are only a few receptive stigma cells; for example, the Spanish chestnut *Castanea* (Y. HESLOP-HARRISON and SHIVANNA 1977), or the mangrove *Avicennia*. Esterase activity is also present in the mucilaginous surface secretion of stigmas of seagrasses, which flower while totally submerged in the sea (DUCKER and KNOX 1976, PETTITT 1980, PETTITT et al. 1981). Here the pellicle appears to be insoluble in sea water.

Many isozymes of esterase may be present in stigma and style extracts; for example, BREWBAKER (1971) demonstrated 15 esterase isozymes in styles of *Hemerocallis* by starch gel electrophoresis. The stigma pellicle proteins are freely diffusible in detergent solutions, diffusates of *Hibiscus* stigmas giving seven bands by SDS-PAGE (HESLOP-HARRISON et al. 1975b). Two of these bands showed esterase activity. In *Raphanus* stigmas, the pellicle esterase shows an almost exponential increase in activity during stigma development. In pistils of *Trifolium pratense*, six esterase isozymes are present in the stylar canal fluid, while three of these are also in the stigma exudate (J. HESLOP-HARRISON and Y. HESLOP-HARRISON 1982). The presence of esterase in some dry stigma types is associated with the ability of the pollen tubes to penetrate the stigma cuticle (SHIVANNA and SASTRI 1981, BERNHARDT et al. 1980).

Acid phosphatase has also been detected cytochemically in the stigmatic exudate of several species, including (HERRERO and DICKINSON 1979), *Linum* (GHOSH 1981) and in the cell walls of receptive stigmas of the seagrass *Thalassia* (PETTITT 1980). Presumably this enzyme belongs to the family of esterase enzymes, but no information is available on its substrate specificity. In addition, glucose-6-phosphate dehydrogenase and an isozyme of peroxidase have been shown by gel electrophoresis to be specific to the transmitting tissue of styles of *Nicotiana alata* (BREDEMEIJER 1979), while amylase has been detected in *Petunia* (SCHLÖSSER 1961). While no specific function has been ascribed to these enzymes, surface-located enzymes presumably play a role in pollen reception or germination. Extracellular enzymes within the style could be involved in stigma autolysis, following compatible pollination.

Antigens

Several groups have examined the nature of the stigmatic surface by immunological techniques, with the object of detecting differences in components which may be products of the S-gene which controls self-incompatibility. Usually, antisera to whole surface exudates have been used and the conclusions have been based on relatively simple tests, for example, immunodiffusion and immunoelectrophoesis often after immunoabsorption (see review by KNOX and CLARKE 1980).

Evidence for the presence of an antigen with S-allele specificity, within the stigma of the cabbage, *Brassica oleracea* var. *capitata* has been obtained by NASRALLAH, WALLACE and co-workers at Cornell University. Antisera were raised against homogenates of pistils with S_2S_2 genotype. Three bands were obtained by immunodiffusion (NASRALLAH and WALLACE 1967a, b), two of which were common to pistils with different S-genotypes while the third and most intense band was considered to be S-genotype-specific (NASRALLAH 1979). The transfer of the putative S-antigens to both F_1 and F_2 generations after the plant has been crossed with another genotype has been followed immunologically (NASRALLAH et al. 1970) implicating the antigen as an S-gene product. Antigens which correlate with the S-genotype have also been detected in stigmas of other varieties of *Brassica oleracea,* kale, and brussels sprout (LANDOVA and LANDA 1973, SEDGLEY 1974a, b, KUCERA and POLAK 1975) and *B. campestris* (HINATA and NISHIO 1978, NISHIO and HINATA 1977, 1978). In stigmas of *B. campestris,* the S-associated glycoprotein binds the lectin ConA (NISHIO and HINATA 1980).

In the wet type stigmas of sweet cherry, *Prunus avium,* five major antigens have been described (RAFF et al. 1979, 1981) one of which is correlated with the S-allele group (Fig. 18). An especially interesting feature is that an antigen immunologically identical with the stigma S-antigen is present in callus cells cultured from leaf explants. The antigen is secreted into the medium of liquid suspension cultures in significant concentrations (RAFF and CLARKE 1981).

S-antigens of stigmas of two *Brassica* spp. and *Prunus* styles have recently been isolated and partially characterized (Table 11). All three are glycoporteins, with rather similar properties. The principal amino acids in all three cases are

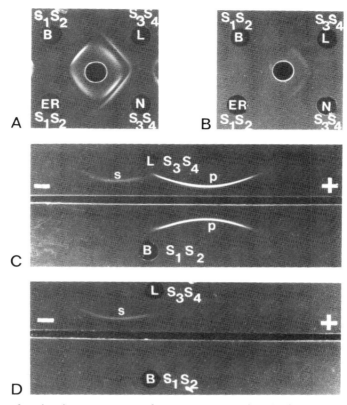

Fig. 18 A–D. Detection of antigenic components of *Prunus avium* styles. Antigens were prepared from extracts of mature styles of different *P. avium* cultivars: *B* Bedford (S_1S_2); *ER* Early Rivers (S_1S_2); *L* Lambert (S_3S_4); *N* Napoleon (S_3S_4). Antiserum was raised to extracts of mature styles of cv. Lambert (S_3S_4). **A** Immunodiffusion showing inner band (*P*-antigen) common to all cultivars tested; a strong outer band (*S*-antigen) was produced with antigen preparations from the same *S*-allele groups as the eliciting antigen for antiserum production. A faint, but detectable outer band was produced using antigen preparations from cultivars of different *S*-allele groups. **B** As for **A**, except that the antiserum was preadsorbed with antigen preparation from style extracts of var Bedford (S_1S_2). The common *P*-antigen is not present, and the *S*-antigen is only detected in antigen preparations from cultivars of *S*-allele group S_3S_4. **C** Immunoelectrophoresis at pH 8.8 showing separation of the *P*- and *S*-antigens. The *P*-antigen common to both S_1S_2-(*lower well*) and S_3S_4-(*upper well*) antigen preparations is negatively charged while the *S*-antigen detected in the S_3S_4-antigen preparations is positively charged. **D** Immunoelectrophoresis at pH 8.8, with antiserum preadsorbed with style extract from cv. Early Rivers (S_1S_2). The *P*-antigen is absent, and the S-antigen is only detected using antigen preparations from the cultivars having the same *S*-genotype as the eliciting antigen, S_3S_4. (RAFF et al. 1981)

serine, glutamic acid and glycine. The *Brassica* stigma antigens are freely diffusible from whole stigmas (NASRALLAH and WALLACE 1967a, b). The S-specific antigen is concentrated in the stigma rather than the style (NISHIO and HINATA 1977) and does not appear until 3 days before flower opening, at which time

Table 11. Properties of three purified pistil S-glycoproteins (i.e., glycoproteins which are associated with expression of particular S-alleles)

Species	S-al-leles	Tissue of origin	Apparent MW	pI	Carbo-hydrate content	Mono-saccha-rides	Reference
Brassica campestris	S_7S_7	Stigma	57,000	5.7	1.2[a]	NT	Nishio and Hinata 1979
B. oleracea	S_2S_2	Stigma	54,500	High	1.3[a]	Ara, Gal, Glc, Man, Gal, Fuc	Ferrari et al. 1981
Prunus avium	S_3S_4	Style	37,000 to 39,000	High (>8.8)	16.3%	Glc, Man, Gal, Fuc[b]	Clarke et al. 1982; Mau et al. 1982

[a] Protein: carbohydrate ratio (w/w).
[b] Indicated by lectin-binding assays only

self-incompatibility is expressed. The isolation of these glycoproteins which may be active in pollen–pistil recognition is an important development. In the 1980's, progress can be expected in elucidating the physiological, genetical and biochemical aspects of their apparent role in S-gene expression.

Lectin-like activity, in terms of the presence of agglutinins, has been detected in the pistils of *Primula* (Golynskaya et al. 1976) and ovules of *Ginkgo* (Pettitt 1977). These macromolecules are assumed to function in pollen or pollen tube interactions.

24.5 Control of Pollen–Pistil Interactions

The pollen–pistil interaction is a sequential series of events, leading to the acceptance of compatible pollen for fertilization on the one hand, and the rejection of incompatible self-pollen, and of foreign pollen, on the other hand. The evidence available suggests that it is a multi-step process (Heslop-Harrison 1975, Knox et al. 1976, Clarke and Knox 1978, Knox and Clarke 1980, Dumas and Gaude 1981) involving (1) pollen contact; (2) adhesion and attachment to the stigma surface, (3) pollen hydration, (4) germination, (5) pollen tube growth through stigma and style leading to (6) fertilization. The role of macromolecules in initiating these events will now be considered, especially in the exchange of signals between the two systems.

24.5.1 Adhesion

Adhesion is a fundamental property of all cells, and usually initiates cellular recognition (Frazier and Glaser 1979). Pollen adhesion on the stigma surface

Table 12. Physicochemical approach to pollen-stigma adhesion, suggested by the composition of an ideal adhesive (REYNOLDS 1971) and the quantitative analysis of pollen and stigma surfaces of *Gladiolus*. (CLARKE et al. 1979. Adapted from DUMAS and GAUDE 1982)

Component	Characteristic	Candidate		Reference
		Pollen surface	Stigma surface	
Adhesive base	Branched polymer of high molecular weight	Arabino-galactan	Arabino-galactan	CLARKE et al. 1979 GLEESON and CLARKE 1979
Plasticizer	Prevent adhesive becoming brittle	Free mono-saccharide	Diffusible mono-saccharide	CLARKE et al. 1979
Thickener	Increase viscosity of adhesive	Glyco-proteins	Glyco-proteins	KNOX et al. 1976; DUMAS and GAUDE 1981; NISHIO and HINATA 1978
Tackifier	Resin	Pigments of pollen coat	?	HESLOP-HARRISON 1968; CLARKE et al. 1979
Detergent	Wetting agent	Glycolipids	Glycolipids	CLARKE et al. 1979; DUMAS and GAUDE 1981

involves both impaction and the subsequent formation of attachment bonds (see DUMAS and GAUDE 1981).

Several physical parameters have been assessed in a series of experiments by WOITTIEZ and WILLEMSE (1979), these include: surface tension, wind speed, electrostatic force, gravity, inertial and electrodynamic forces. The possibilities offered by a physicochemical approach have been presented by CLARKE et al. (1979) in relation to a quantitative analysis of pollen and stigma components in *Gladiolus* (Table 12). The groups of components identified on the *Gladiolus* pollen and stigma surfaces are considered in terms of the classes of compounds that make up an ideal synthetic adhesive. In most cases, the molecules identified could provide a basis for efficient adhesion. On initial contact, the adhesive components could be contributed by either or both partners, to enhance mutual adhesion. The specificity needed for foreign pollen discrimination might be provided by the adhesive base or the thickening agents – both high molecular weight components (Table 11). Efficient glueing is known to require roughened surfaces, and smooth surfaces do not accept adhesives efficiently (REYNOLDS 1971). The pollen grain surface is apparently adapted for adhesion because even relatively smooth windborne pollens such as that of the grasses have an exine with surface patterning, providing a roughened surface for adhesion.

The "sticky" nature of the receptive stigma surface has been demonstrated experimentally be the general binding of self- and non-self-pollen preparations to *Gladiolus* stigmas (CLARKE et al. 1979). Proteins and glycoproteins of animal origin also bind to variable extents (KNOX et al. 1976). While the stigma surface is apparently generally sticky, there is nevertheless a defined capacity for adhesion demonstrated by decreased binding of the ^{125}I-labelled pollen preparations after pretreatment of the stigma with unlabelled pollen preparation. Con A binds specifically to *Gladiolus* stigma surface acceptors (KNOX et al. 1976) and this binding has been shown to decrease the adhesive capacity for pollen protein; the capacity is restored when Con A is applied in the presence of its complementary ligand, suggesting that specific binding of Con A alters the topography of the receptive surface in a way that results in a non-ideal contact surface for general adhesion (CLARKE et al. 1979). This implies that the stigma surface must be an efficient adhesive as well as a carrier of pollen receptors.

The experimental demonstration of adhesion in pollen–stigma interactions has been carried out by both observational and quantitative methods. Light and scanning electron microscopy have been employed to detect adhesion of *Brassica* pollen (ROGGEN 1972, STEAD et al. 1979). An assay for adhesion of pollen on stigmas in which the pellicle had been removed by proteolytic enzyme treatment was developed by STEAD et al. (1980). this assay has been useful also in determining the half-life or turn-over time of the stigma pellicle adhesive components (see review by ROBERTS et al. 1980). These approaches have demonstrated that self-incompatible *Brassica* pollen shows only weak, non-specific adhesion, while compatible pollen shows strong, specific adhesion. *Brassica* pollen mutants defective in expression of the enzyme β-galactosidase, have recently been shown to have reduced efficiency of adhesion to the stigma surface (SINGH and KNOX 1983).

24.5.2 Hydration

Rehydration of mature dry pollen grains is an essential prelude to germination (see Sect. 24.2.1.3, and reviews by HESLOP-HARRISON 1979a, b, 1980, DUMAS and GAUDE 1982b). Since the water potential of the dry grain is lower than that of the stigma surface, it is readily apparent that the water of hydration will move in the direction of decreasing water potential from the stigma cell to the pollen grain. In dry stigmas, the pellicle has been proposed as a regulator (MATTSSON et al. 1974) so that hydration may not occur in many foreign pollinations (KNOX et al. 1976). High relative humidity permits hydration which may be essential to restore viability of stored pollen (SHIVANNA and HESLOP-HARRISON 1981) and is known to overcome self-incompatibility (see Sect. 24.2.4.1.2), or to increase the rate of pollen germination in incompatible matings (SHIVANNA et al. 1981).

24.5.3 Models of Recognition and Receptors

The transmission and receipt of signals between cells is generally a cell-surface phenomenon, and requires the presence of receptors, which have been identified and characterized in animal cell systems (BARONDES and ROSEN 1980). In plant cell recognition, the presence of receptors has been assumed, although none have been identified (see CLARKE and GLEESON 1981, DUMAS and GAUDE 1981, 1982b). The ideal receptor will be a bifunctional molecule, with one or more binding sites, together with additional sites for specific molecular interactions. It will thus resemble lectins in its properties. What progress has been made towards a model for self-incompatibility that is experimentally testable?

First of all, any model must account for the actual recognition event/s, and indicate the mechanism by which its consequences are generated. The consequences include pollen acceptance or rejection, and metabolic changes in the pistil. Few have doubted the validity of the control of self-incompatibility by the S supergene, with its variety of sites of expression. Likewise, there has been an acceptance of the concept that the S-gene may have an S-allele-specific product distinct for each allele. The early models proposed by LEWIS (see LEWIS 1965) are based on the interaction of these putative S-gene products in pollen and pistil. According to his dimer hypothesis, identical pollen and pistil gene products complexed to form the inhibitor, while a second hypothesis developed the approach for the sterically complementary antigen/antibody or key-in-lock type of products. The second hypothesis has been extended by BURNET (1971) and SAMPSON (1977) who have proposed that the products might have common and specific regions, rather like mammalian antibodies.

Evidence supporting either of these hypotheses has not yet been obtained. LEWIS (1952) succeeded in demonstrating the existence of a stable S-gene-specific antigen in the pollen of *Oenothera*. In subsequent work, this product was shown to be readily diffusible, and to constitute up to 40% of the total pollen protein (LEWIS et al. 1967). Unfortunately, the existence of a stigma or style counterpart has not been demonstrated in *Oenothera*. In *Petunia*, LINSKENS (1953) demonstrated the existence of new proteins following self- or cross-pollination. Two proteins were specific to selfed styles, and one to crossed styles. In later work, radiolabelling experiments showed that these stigma proteins were glycosylated, with glucose originating from the pollen (LINSKENS 1960). These data were originally taken as evidence that the S-gene product of the pistil is synthesized de novo following pollination (LINSKENS 1973b, 1976a, b). This is hardly surprising since these proteins showed high antigenic specificity for the S-alleles involved. In subsequent studies with *Brassica*, NASRALLAH et al. (1967a, b) detected an S-specific antigen present in mature unpollinated stigmas. The stigma S-gene product in this system is certainly present prior to pollination, a fact that has been confirmed by other workers (NISHIO and HINATA 1978, 1979, SEDGLEY 1974a, b, ROBERTS and DICKINSON 1982). It is perhaps simplistic to expect all self-incompatibility systems to have the same type of S-gene product in the same cells, in view of the wide range of tissues in which the S-gene apparently is expressed (DE NETTANCOURT 1977).

Recently, CLARKE et al. (1977, 1979) have demonstrated the presence of antigens common to pollen and other somatic tissues of *Gladiolus gandavensis* and *Prunus avium*. Pistil antigens of *Brassica* showed no immunological reactions of identity with pollen antigens, as detected by immunodiffusion (NASRALLAH et al. 1967a, b, KNOX et al. 1975). At present, S-specific glycoproteins have only been isolated and purified from pistil tissues and not pollen grains (see Sect. 24.3.2.4).

In sporophytic self-incompatibility systems such as *Brassica*, further experimental models have been proposed. The hypothesis of FERRARI and WALLACE

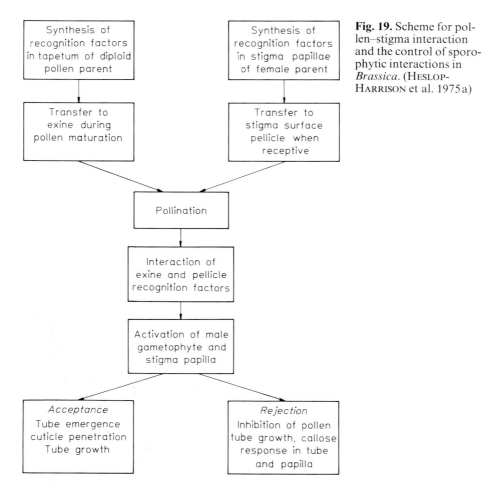

Fig. 19. Scheme for pollen–stigma interaction and the control of sporophytic interactions in *Brassica*. (Heslop-Harrison et al. 1975a)

(1977a) is based on pollen germination, and the control mechanisms of tube growth; recognizing more than eight stages (see reviews by Shivanna et al. 1979, Dumas and Gaude 1981, 1982a, b). Heslop-Harrison et al. (1975a) proposed what is today regarded as a classical cellular model (see review by Dumas and Gaude 1982b). The incompatibility reaction consists of a series of interacting events (Fig. 19), in which stigma discrimination may lead to pollen acceptance or rejection. The *callose rejection response* provides a convenient monitor of the reaction in both pollen and stigma (Dickinson and Lewis 1973a, b, Heslop-Harrison et al. 1974, 1975a, and review by Dumas and Knox 1983). The model of Heslop-Harrison et al. (1975) is the only one that takes account of recent developments in our understanding of the initial reactions of pollen–stigma interactions, namely adhesion and hydration.

Another recent model has linked possible changes in phenolic acid metabolism leading to the production of phytoalexins inhibitory to pollen tube growth

Fig. 20 A, B. Scheme für pollen-stigma interactions (DUMAS and GAUDE 1982b). **A** Self-pollination: the adhesion is slight and permits only a weak hydration of the pollen grain (reversible phenomenon). **B** Cross-pollination: the firm adhesion ensures a better hydration of the pollen grain which poromotes germination and pollen tube growth (irreversible phenomenon). During the self-pollination, the bonds of attachment are achieved by a specific interaction between S-proteins (hypothesis). S-proteins and membrane receptors of both cell surfaces interact with each other (see DUMAS and GAUDE 1982a). Υ Υ: S-proteins; Υ Ŷ: "membrane" receptors; 1 pollen contact; 2 pollen hydration; *arrows* water flow

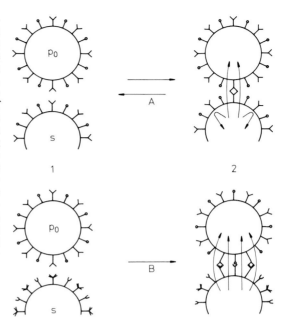

with the callose rejection response in Cruciferae. LEWIS (1980) has proposed that the 1,3-β glucans formed could act as elicitors of the phytoalexins in incompatible pollinations. In such situations, there could be extensive borate-complexing to the stigma callose, as the stigmas are known to contain high concentrations of borate. Phytoalexins, such as rishitin, are certainly able to inhibit *Solanum* pollen germination in vitro (HODGKIN and LYON 1979). Such suggestions point to the possible role of cell-wall oligosaccharides in eliciting the incompatibility interaction.

Remarkable parallels between plant host–pathogen interactions and pollen–stigma interactions have been drawn by PEGG (1977) and CLARKE and GLEESON (1981). Both depend on initial adhesion and subsequent interactions initiate growth of the invasive pollen tube or fungal hypha. The possible molecular basis for the adhesive stage of the host–pathogen interactions has recently been considered, and it appears that both specific or lectin-mediated adhesion and nonspecific glycoprotein–glycoprotein interactions may be involved (see KOSUGE Chap. 25, Vol. 13 B, this Series). The same situation may be true in the pollen–stigma interaction, with some surface components participating in pollen grain capture and adhesion and others being involved in the information exchange which initiates the cascade of pollination events.

ROBERTS and DICKINSON (1982) found evidence from isoelectric focussing studies of self- and cross-pollination in *Brassica* that the self-incompatibility response depends on the binding of pollen and stigma proteins. The initial recognition, occurring within minutes of pollen arrival in the stigma, appears to involve an interaction between (1) one or more low molecular weight proteins of the sporophytically determined pollen-grain coating, and (2) a glycoprotein from the stigma pellicle. Incompatible grains fail to adhere, and cannot take up water, so that the prevention of water flow by the stigma could regulate the type of self-incompatibility. These authors postulate a model involving an alteration in the molecular architecture of the stigma pellicle rendering it impervious to water. The callose response is seen as a secondary feature, perhaps providing a "back-up" if the initial control fails to block pollen germination.

DUMAS and GAUDE (1982a, b) have presented a model of self-incompatibility that provides for adhesion and hydration, and involves membrane receptors and S-specific glycoproteins in the interaction (Fig. 20). Self-pollination is viewed as a reversible phenomenon, giving rise to a weak interaction; compatible cross-pollinations provide for an adhesion promoting irreversible processes – pollen germination and tube growth. A feature of this model is that the pollen interaction involves the outermost layer of the exine, which in *Brassica* is distinct from the underlying areas in its electron microscopic cytochemistry, and becomes attached to the stigma pellicle.

It is a pre-requisite to explain pollen–stigma interactions that a great deal more work is carried out on the chemical nature of the pollen and stigma surfaces. The processes of adhesion and hydration certainly hold some of the keys to the recognition process.

24.5.4 Response

In order to carry out bioassays of the effects of pollen components on the stigma, characteristic stigma responses need to be detected. Conversely, the effects of stigma components on pollen germination can only be determined by monitoring pollen behaviour on the stigma or in vitro. Are responses detectable in the style and ovary? The available data on mutual responses of pollen and pistil will now be presented, with the focus on their potential use as bioassays.

24.5.4.1 In-Vivo Responses

Active rejection responses were first observed by CHARLES DARWIN (1882) during pollination of orchids (Table 13). Following self-pollination, the pollen masses failed to separate and germinate, developing a dark brown necrotic appearance, while the flowers withered prematurely. Subsequently, browning responses of the stigmas of *Orchis* were reported (Table 13), presumably the result of dramatic metabolic changes in the cells affected. Other parts of the orchid flowers show post-pollination changes: in *Cycnoches,* scent production ceases; in *Vanda,* petal colours fade; in *Cymbidium,* flowers redden, column swells and the stigma closes, while the petals show hyponasty, and there may be initiation or cessation of nectar production (ARDITTI 1979).

During pollination of grasses changes in stigma cell permeability (Table 13) indicate the changed water relations of the pollinated stigma cells essential for water transport to permit pollen hydration. This localized change in permeability around the germinating pollen was detected in staining tests by KATO

Table 13. Specific Responses to Pollination. The following selected list summarizes investigations whose objective has been to define a particular response of the germinating pollen or the receptive stigma or style to pollination

Species	Treatment	Stigma/style response	Pollen response	Reference
Notylia, Oncidium	Self-pollination	–	Pollinia turn brown and necrotic	DARWIN 1882
Orchis masculata and *O. fusca*	Interspecific pollination with *O. morio* pollen	Papillae turn brown in contact with pollinia or pollen tubes	–	STRASBURGER 1886; FITTING 1909
Secale cereale	Self- or intergeneric pollination (within Poaceae)	Papillose cells in contact with grains become more permeable to basic dyes	–	KATO (1953); KATO and WATANABE (1957)
	Self-pollination	–	Callose within arrested pollen tubes and grains	HESLOP-HARRISON (1976); VITHANAGE et al. (1980)
Cosmos bipinnatus	Self-pollination	Callose lenticules in papillae	Pollen and germinating tubes occluded with callose	KNOX (1973); HOWLETT et al. (1975)
Helianthus annuus	Self- and intergeneric pollination within Compositae	As above	As above	VITHANAGE and KNOX (1977)
Iberis spp., *Brassica oleracea, Raphanus sativus*	Self-pollination	Callose lenticules in inner wall of papillae contacting pollen	Pollen and germinated tubes occluded with callose	HESLOP-HARRISON et al. (1974, 1975a) DICKINSON and LEWIS (1973a)
Raphanus sativus	Tryphine (surface extractable material) from self-pollen	Callose induced in unpollinated stigmas	–	DICKINSON and LEWIS (1973b)
Iberis sempervirens	Self-pollen diffusate in agarose gel	Callose induced in unpollinated stigmas	–	HESLOP-HARRISON et al. (1974)
Iberis spp. *Brassica oleracea*	Self-pollen diffusates and partly purified fractions	Callose induced in unpollinated stigmas	–	HESLOP-HARRISON et al. (1975a)

Table 13 (continued)

Species	Treatment	Stigma/style response	Pollen response	Reference
Persea americana	–	Callose in pistils of aged flowers (male phase of dichogamy) when not receptive to pollen	–	SEDGLEY (1977)
Gossypium hirsutum	Self-pollination	Transmitting cells of style adjacent to pollen tube produce callose in cell walls	–	JENSEN and FISHER (1969)
Populus deltoides	Interspecific pollination with *P. alba*		–	KNOX et al. (1972 and unpublished)

and WATANABE (1957). They termed it the stigma reaction, an effect elicited specifically by self-species pollen, and pollen of other grasses. HESLOP-HARRISON and HESLOP-HARRISON (1981a) found that the permeability changes are correlated with a progressive dissolution of the plasma membrane of the stigma cells in contact with the pollen. The pit fields in the cell walls become occluded with callose, and the cells become necrotic, with their vacuolar contents filling the lumen of the cell. The lumen is not entered by the pollen tube, which is confined to the apoplast.

Specific rejection responses have recently been demonstrated during self-incompatibility in several systems, when callose, detected by its fluorescence with decolorized aniline blue (LINSKENS and ESSER 1957) is deposited in the rejected pollen grains and pollen tubes, and in some systems even in the adjacent stigmatic papillae (Table 13 and review by DUMAS and KNOX 1983). In compatible pollinations in these cases, little or no callose staining is usually detected. Responses on the male side, in the pollen grain and pollen-tube, are characteristic of gametophytic self-incompatibility systems; for example, in grasses (SHIVANNA and J. HESLOP-HARRISON 1978).

Callose responses in *both* interacting partners, the stigma cells and the pollen grains and tubes are characteristic of sporophytic self-incompatibility systems of the Cruciferae (DICKINSON and LEWIS 1973b, HESLOP-HARRISON et al. 1974) and Compositae (KNOX 1973). In these systems, the callose deposition in the stigma papillae can be experimentally induced by the application of self pollen-wall preparations (DICKINSON and LEWIS 1973a, b, HESLOP-HARRISON et al. 1974, 1975).

The specificity of the response has been confirmed by KERHOAS et al. (1983), who have also shown that it is blocked by pretreatment of self stigmas by the lectin Con A or the detergent Triton X-100. These results suggest that the stigma pellicle contains the receptors for pollen information, and initiates the stigma read-out (see review by DUMAS and KNOX 1983).

The aniline blue-stained pollen wall and tube component found after self-pollination in rye, *Secale cereale* has been isolated and analyzed by a combination of chemical and enzymic methods. It proved to be a mixed linked glucan, with both $(1 \rightarrow 4)$-β-glucan and $(1 \rightarrow 3)$-β-glucan linkages in the proportion of $83:9$ (VITHANAGE et al. 1980, and review by CLARKE Chap. 23 Vol. 13 B, this Series).

Responses are not confined to initial intercellular contact. In a recent study of interspecific incompatibility in *Rhododendron,* WILLIAMS et al. (1982a) have demonstrated striking callose responses apparently within the egg cell of the embryo sac, using an aniline blue fluorescence technique. In addition, an array of pollen tube tip error syndromes characterized certain interspecific crosses. Further, different errors were induced in pollen of one species when pollinated on pistils of other species (WILLIAMS et al. 1982b).

Callose also provides an indicator of the effective pollination period in pistils, since its appearance may coincide with loss of receptivity of stigmas, and loss of viability of ovules (see review by DUMAS and KNOX 1983).

24.5.4.2 In-Vitro Responses

Effects somewhat comparable with those of known inhibitors of pollen-tube growth (Sect. 24.2) have been obtained with stigma and style extracts. In *Brassica oleracea,* which has a sporophytic self-incompatibility system, where inhibition of self-pollen tubes occurs at the stigma surface, FERRARI and WALLACE (1977b) and FERRARI et al. (1981) have found that stigma fractions from specific S-genotypes inhibit pollen tube growth of the same S-genotypes to a greater extent than other S-genotypes. However, they have found that a low molecular weight stigma fraction is a non-specific inhibitor of pollen tube growth. SHIVANNA et al. (1981) have tested stigma extracts in the heteromorphic self incompatibility system in *Primula vulgaris,* and found that thrum stigma extracts are more inhibitory to thrum pollen than to pin pollen, and the converse (Fig. 21). Purified stylar components, associated with particular S-alleles have been tested in an in vitro pollen tube bioassay in the gametophytically inherited S system of *Prunus avium* and have been found to be inhibitory (WILLIAMS et al. 1982b). The technique holds considerable promise as a bioassay for pistil molecules active in regulating pollen germination and tube growth.

A problem with all investigations of the biological activity of molecules extracted from plants is the low yield of purified material. This means that experiments consume an inordinate amount of the extract, or assays adapted to a micro-scale must be developed. In vitro experiments with stigmas and styles have thus far been confined to their use as isolated systems in Petri plates of agar, when the effects of pollen extracts have been tested (see Sect. 24.3.3.1). HESLOP-HARRISON (1976) has demonstrated that the style transmitting cells in *Prunus* and *Malus* can be separated into filaments of cells, offering the possibility of short-term culture, and detection of specific responses to extracts from pollen grains or tubes.

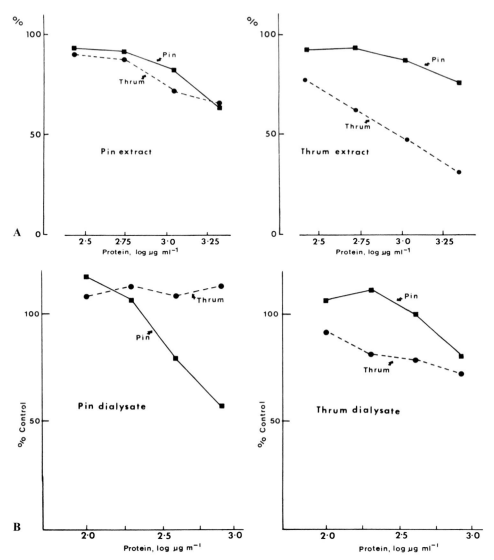

Fig. 21. Physiological effects of pin and thrum stylar extracts on **A** pollen germination and **B** pollen tube growth of *Primula vulgaris,* which has a heteromorphic self-incompatibility system (SHIVANNA et al. 1981). A crude extract was added to the culture medium in **A**, while in **B**, the extracts were dialyzed to remove low molecular weight (< 10,000) components

24.5.4.3 Evidence for Signalling

Evidence for the existence of signalling between the mating partners has been provided by the stimulation of protein synthesis within ovules of *Petunia hybrida* following pollination at the stigma (DEURENBERG 1976a, b). During compatible pollinations, the pollen tubes growth through the transmitting tissue to the ovary within about 30 h. At 3–4 h, pollen tubes enter the style, and when the tubes are about one-third of the way through the style, at 9 h, there is a sudden increase in ribosomal protein content of the ovules that is not detectable in unpollinated control ovules. By 18–30 h, there is a marked increase in ribosomal protein content of ovules from cross-pollinated pistils compared with those of self-pollinated pistils. Ribosomal protein content is detected by incorporation of ^{14}C-leucine in an in vitro protein-synthesising system. These data are interpreted as evidence for two different kinds of signals at pollination:

1. A primary signal indicating arrival of pollen tubes in the style;
2. A secondary signal indicating the kind of pollen whether compatible or not.

In this case, there is little doubt that the message is initiated by the arrival of the pollen tube in the style. GILISSEN (1977) has shown that wilting of the flower in a compatible mating is initiated when pollen tubes enter the style; furthermore, removal of the pollen-wall components (including proteins and glycoproteins amounting to 18% of the pollen dry weight) did not affect the ability of the washed pollen to effect compatible pollination (GILISSEN and BRANTJES 1978).

These data suggest that the intercellular communications between pollen and stigma are not based on a single recognition system, but involve a progressive series of mutual interactions. An interesting question that now needs an answer is which cell or tissue of the female gametophyte is the receptor of the signals. Is it the egg cell? One way in which this question may be approached is to explore the physiology of the embryo sac, and JENSEN (1983) has found that plant hormones such as gibberellic acid and auxin may mimic some of the characteristic changes that occur in cotton embryo sacs during the progamic phase of pollination. The central role of the egg cell in gametic recognition is suggested by recent observations of pollen tube/embryo sac interactions during interspecific matings of *Rhododendron* (WILLIAMS et al. 1982a, 1983). Following certain crosses, foreign pollen tubes entered the micropyle and apparently the embryo sac in some cases; in others, such foreign pollinations provoked the formation of a callose wall around the egg cell.

24.5.4.4 Pollen-Tube Growth and Gamete Competition

The guiding tissues of the style provide a track along which selection of pollen tubes may occur, especially the discrimination of self from foreign and self from pollen tubes of other individuals within a species. For gametophytic responses, the style provides the testing ground. The style also supports much of the growth of the pollen tubes through processes of heterotrophic nutrition, which amount to physiological complementation between the interacting partners (LINSKENS and PFAHLER 1977). The growth rates of different types of pollen tubes are regulated by the stylar environment, and may vary in different styles. In maize, the pollen-tube growth rates of pollen from five inbred lines were estimated in the styles of five other lines in pollen mixtures against a standard pollen source which carried a marker for aleurone colour in the kernel (OTTAVIANO et al. 1980). The rate of tube growth in the tubes carrying the marker gene could be estimated indirectly by scoring kernel colour, and F_1 kernel characters expressed as regression coefficients. Results suggest that the most rapidly growing tubes (which were able to effect fertilization) gave rise to seedlings which had a significantly increased dry weight and other sporophytic characters (OTTAVIANO et al. 1980).

These data need to be confirmed by experiments with other species. The prospects for gametophytic selection operating to modify the quality of the resulting $F1$ generation of sporophytes is exciting for two reasons:
1. Pollen tube growth rate is readily monitored cytologically in vitro and in vivo (e. g., in *Prunus* styles, RAFF 1981).
2. In dioecious crops, e. g., the date, *Phoenix,* the type of pollen may have striking effects on fruit size and quality (metaxenia, SHAFAAT and SHABANA 1980).

Growth of the pollen tube from stigma surface to ovule has seldom been followed in its entirety, although the methods are now available to do this, e. g., in *Rhododendron* (WILLIAMS et al. 1982 and references therein). Cytological monitoring of pollen-tube growth has been carried out for cherry, *Prunus* and apple, *Malus* pollinations in vivo (STÖSSER and ANVARI 1981, ANVARI and STÖSSER 1981). The growth rates of pollen tubes in the style of cherry was more rapid than in the ovary: the style was penetrated within 2 to 3 days, while 6 to 8 days were required for tubes to reach the ovule (STÖSSER and ANVARI 1981). Both the effective pollination period and the longevity of the ovules are important factors in determining fruit set.

24.5.4.5 Co-Adaptations of Pollen, Stigma and Biotic Pollinators

In the last section of this review, we turn to one of the most challenging areas of all. We consider the question of how the various adaptations or evolutionary trends detected in pollen and pistil may have arisen. Could they result from the selective pressures brought to bear by interactions with wind, water or animal pollinators?

Some of the most remarkable interactions are associated with insect pollination. The use of floral scents which resemble the insect pheromones has been reported as an attractant for pollination in certain orchids (FAEGRI and VAN DER PIJL 1979). The mechanism by which such intimate interactions arose remains obscure. Likewise, the incredible ingenuity of the translator organ which effects pollination in certain asclepiads. Here, pollinia form the means of mass sperm transfer; one pollinium contains enough pollen to help ensure a full seed set. In asclepiads, the anthers and pistil unite to form a gynostegium, which secretes the translator organ, composed of lipids and mucilage (SCHILL and JÄCKEL 1978, SCHNEPF et al. 1979). This organ attaches a pair of pollinia from adjacent anthers to visiting insects. These insects are attracted to the flowers by nectar, secreted from nectaries located within the stigmatic chamber. Sugar secretion occurs through epithelial cells with transfer-cell-like wall ingrowths (SCHNEPF and CHRIST 1980). Once attracted to the flower, the pollinia stick to the insect in a specific region of its body by the translator organ. The two caudiculae of the translator attach separately to the sporopollenin envelope of the pollinia, which may be coated with adhesives derived from the anther tapetum. The corpus is then accessible for attachment to the visiting insect. Highly coordinated secretory processes are involved in translator production, and it is a remarkably ingenious morphological adaptation for pollination.

The pollinium is an adaptation for insect pollination, providing both a convenient pollen load for the visiting insect, and often ensuring a full seed set in the receptor flower. It is a mechanism that has been exploited in pollination of certain orchids and asclepiads (FAEGRI and VAN DER PIJL 1979). In some 15% of families, composite pollen grains or polyads provide similar possibilities for insect pollination. Especially remarkable are pollen and stigma adaptations in the Leguminosae, subfamily Mimosoideae. In the tribe Mimoseae, there are genera with either polyads or free monads. The stigma morphology varies from funnel-shaped, tubular to narrow-porate. The tubular stigma of *Entada* and *Elephantorrhiza* is associated with monads, while the narrow porate type is associated with polyads in *Adenanthera* and *Mimosa* (LEWIS and ELIAS 1981). The latter genus has the smallest polyad found in the angiosperms, GUINET (1981) recording its diameter as only 6.5 μm. It is apparent that the size and morphology of the stigma is directly related to the type of pollen.

The possession of polyads might be expected to confer a selective advantage in reproduction. In *Acacia*, there is a polymorphism for polyad grain number. In Australian species, some 90% of species have grain numbers of 16, a single polyad occurring in each loculus, the product of 4 pollen mother cells. Most of the other species have 8 grain polyads, and there are tetrads in one species. In African species, the polymorphism operates in the opposite direction, with most species having 16 grain polyads, but several with 32 and even 64 grain polyads. KENRICK and KNOX (1982) have shown that polyad grain number is related to pod seed number, in that the maximum seed number does not exceed the grain number in species so far examined. There appears to be a more complex relationship with ovule number. KENRICK and KNOX (1982) believe that this apparently simplistic relationship is based on the principle that the polyad is, like the pollinium of asclepiads and orchids, a mechanism helping to ensure a full set of seeds following a single pollination event. These structural relationships of pollen and stigma are important adaptations to insect pollination. However, more data need to be collected before we can begin to explore the physiology of such functional relationships, and their possible control by either the S-supergene or closely linked genes.

24.6 Conclusions

An initial conclusion from this review is that plants, in adapting to either their diverse environments or types of pollination system, have adopted or explored the usefulness of almost every conceivable variation in pollen or pistil structure. During the 1970's, the emphasis of most physiological and biochemical investigations of plant reproduction was on the nature of the individual components of the pollen grain and pollen tube, and the stigma and style. Only at the ultrastructural level did studies approach the analysis of their interactions. In macromolecular terms, this aspect has hardly been investigated except for a few pioneering studies.

The state of play for the pollen grain is as follows:

1. The remarkable diversity in pollen-grain structure, especially of the outer exine wall has been correlated with pollination of the flowers. In animal-vectored systems, the pollen grains, >300 µm in diameter, often have a massively thickened exine, heavily adcrusted with ornamentation and bear a thick coating of yellow, blue, brown, or black pollen coat materials. Wind-borne grains are spherical 30 to 50 µm in diameter, and have a thin exine, with a different type of surface coating present in the exine arcades and in the anther cavity, that is usually light yellow in colour. Water-borne pollens usually have a reduced exine and thick intine and may be covered by mucilage; in the extreme case of filiform grains the exine is absent and grains may be >5 mm in length.

2. A key characteristic of mature pollen, its viability, may be estimated by both destructive and non-destructive methods. Mature pollen is dehydrated, with a low water content. Decrease in cell water is associated with loss of viability. In certain types of pollen with a low water content, e.g. grasses, the plasma membrane at maturity may be porous and ineffective.

3. Carbohydrates, proteins, lipids and pigments diffuse from the surface of moistened pollen grains. Quantitative analyses show that the water-soluble components comprise equal amounts of carbohydrates and proteins. The carbohydrates include both free monosaccharides and polysaccharides. Many of the proteins (and glycoproteins) have enzymatic activity; some are antigenic in rabbits, a few are potent allergens. The presence of antigens correlated with the S-(self-incompatibility) genotype is evidence that pollen grains may carry the putative S-gene product.

4. These proteins and carbohydrates are stored in three sites in the mature pollen grains: in the cytoplasm; within the polysaccharide matrix of the intine; within the arcades of the exine and on its surface. This has been determined by cytochemical observations of enzymatic activity and by immunocytochemistry using both light and electron microscopy; and by short- and long-term extraction in aqueous solvents.

5. The origins of these proteins and carbohydrates are remarkably diverse. The intine proteins are gametophytic, being synthesized by the pollen proto-plast. The exine proteins of three families are sporophytic, being synthesized by diploid tapetal cells and transferred to the exine late in pollen develop-ment. Exceptions to these rules are being discovered. The possible role of the protoplast in macromolecule elimination from mature pollen is sug-gested by observations of the dehydrated state of mature grains and tempo-rary absence of an intact plasma membrane. The occurrence of transfer cell-like wall ingrowths in the intine of some species and vesicles containing glycoproteins, suggests that the pollen wall, and particularly the intine, is adapted for secretion or elimination of molecules and macromolecules to the external surface.

6. The pollen grain exists to act as a vehicle for transmission of the nuclear and cytoplasmic elements of heredity. Two groups have shown that in certain tricellular types of pollen, the sperm cells may be linked together, and one of the sperm cells may be associated by a connection with the nucleus of the vegetative cell. This association has led to the new concept of the male germ line unit, which has important implications for fertilization.

   The pollen grain being essentially an isolated single "cell" has been an attractive subject for chemical analysis. In contrast, the pistil, with its complex morphology and differentiation into several distinct tissues has been more difficult to approach. Nevertheless, considerable progress has been made and may be summarized as follows.

7. In vascular plants, there is always a receptive surface to read-out pollen information. In gymnosperms, this may take the form of a "pollination drop" above the megaspore, comprising a secretion of carbohydrates, with traces of free amino acids. In angiosperms, the stigma cells may be covered by a thin adhesive layer on the surface of dry stigmas or by a copious liquid exudate which is secreted around the receptive cells or covers them completely (wet stigmas).

8. In dry type stigmas, the secretion consists of proteins and carbohydrates in nearly equal amounts. The carbohydrates include free monosaccharides and high molecular weight, polysaccharides, proteoglycans, and glycoproteins. In wet type stigmas, two kinds of secretion have been detected – a hydrophilic exudate consisting of carbohydrates and proteins and a hydrophobic exudate composed of lipids usually with very little or no water. The type of stigma secretion is sometimes eccrine[1], and sometimes granulocrine[1]; while a holocrine[1] secretion follows autolysis of the cells in some systems. Mixed types of secretion occur, and each different component may be secreted sequentially. The stigmatic carbohydrate secretions may have a secondary role as a reward for animal pollen vectors.

9. Interactions with the pollen tube may occur in the style, or within the ovary. A hydrophilic secretion is found in the styles, forming a mucilaginous exudate in the stylar canal, or in the apoplast in intercellular spaces of the transmitting tissue. The stylar exudate has a partial role in pollen-tube nutrition. Little is known concerning secretions within the ovary.

10. Pollen–pistil interactions may be readily monitored by modern high resolution microscopic and cytochemical techniques. The numbers of pollen tubes penetrating the style, together with the distance and rate of growth, may be estimated with precision. Gamete competition has been demonstrated, which may have applications for gametophytic selection in plant breeding.

11. Finally, what is the nature of the recognition reactions between pollen and pistil? In systems with dry type stigmas, the interactions form a well-defined sequence: pollen contact, attachment and adhesion to the stigma, pollen hydration; pollen germination and tube emergence; pollen-tube growth and penetration of stigma; tube growth through style and ovary; fertilization. There is now little doubt that the major route of biocommunication which

---

1 For definitions see LÜTTGE U, PITMAN MG, Chapter 5.1 in Vol. 2B of this Encyclopedia

initiates these interactions involves the informational molecules present at the surface of the pollen grain, and their read-out by receptors present at the stigma surface.

12. The availability of specific mutants defective in one of the steps of the interaction e.g. the multi-allelic S-gene, provide genetic tools for the analysis of specific macromolecules. In several systems, the expression of specific S-alleles has been correlated with the presence of particular glyco-proteins in the pistil secretions. Pistil extracts, and in one system, an isolated S-glycoprotein have been shown to inhibit pollen tube growth in vitro. The receptor macromolecules in the opposite pollen partner have yet to be identified. The S-locus is a complex one in many instances closely linked to and possibly controlling genes for flower structure, colour and development. The challenge for the 1980's lies in the elucidation of specific interactions between surface macromolecules of the pollen grain or tube, and the stigma or style.

Acknowledgements. I thank Dr. ADRIENNE CLARKE, Dr. CHRISTIAN DUMAS, Professor JOHN HESLOP-HARRISON and Dr. YOLANDE HESLOP-HARRISON for stimulating discussion of some of the topics in this review; Dr. HUGH DICKINSON and Dr. MARGARET SEDGLEY for kindly sending preprints of their unpublished work and these colleagues together with Dr. SOPHIE DUCKER, Dr. BARBARA HOWLETT and Dr. TERRY O'BRIEN for helpful comments on the manuscript; Miss J. KENRICK and Mrs. G. BERESFORD for bibliographic assistance and Ms. JENNIFER GILBERT for valued secretarial assistance.

References

Abadie M (1979) Différentiation pariétale et active phosphatasique acide des tuniques extra-corticales chez une bactérie géante: *Bacillus camptaspora.* Ann Sci Nat Bot Biol Veg. 13ᵉ Ser 1:87–95

Anvari SF, Stösser R (1981) Über das Pollenschlauchwachstum beim Apfel. Mitt Klosterneuberg 31:24–30

Arasu NN (1968) Self-incompatibility in angiosperms: a review. Genetica 39:1–24

Arditti J (1979) Aspects of the physiology of orchids. Adv Bot Res 7:421–655

Ascher PD, Drewlow LW (1970) Stigmatic exudate: a potential carrier of exogenous substances into the style of *Lilium longiflorum* Thunb. J Am Soc Hortic Sci 95:706–708

Ashford AE, Knox RB (1980) Characteristics of pollen diffusates and pollen-wall cytochemistry in poplars. J Cell Sci 44:1–18

Aspinall GO, Rosell K-G (1978) Polysaccharide component in the stigmatic exudate from *Lilium longiflorum.* Phytochemistry 17:919–922

Augustin R (1959a) Grass pollen allergens I. Paper chromatography and membrane diffusion studies. Immunology 2:1–18

Baker HG, Baker I, Opler P (1974) Stigmatic exudates and pollination. In: Brantjes NBM (ed) Pollination and dispersal. Univ Nijmegen, Netherlands, pp 47–60

Bailey IW (1960) Some useful techniques in the study and interpretation of pollen morphology. J Arnold Arbor Harv Univ 41:141–148

Barendse GWM, Rodrigues Pereira AS, Berkers PA, Driessen FM, Eyden-Emons A van, Linskens HF (1970) Growth hormones in pollen, styles and ovaries of *Petunia hybrida* and of *Lilium* species. Acta Bot Neerl 19:175–186

Barondes SH, Rosen SD (1980) Cell surface carbohydrate-binding proteins: role in cell recognition. In: Barondes S (ed) Neuronal recognition. Plenum, New York London, pp 331–356

Bar-Shalom D, Mattsson O (1977) Mode of hydration as an important factor in the germination of trinucleate pollen grains. Bot Tiddskr 71:245–251

Barthlott W (1981) Epidermal and seed surface characters of plants: systematic applicability and some evolutionary aspects. Nordic J Bot 1:345–355

Bateman AJ (1954) Self-incompatibility systems in angiosperms. II *Iberis amara*. Heredity 8:305–332

Baum H (1950) Das Narbensekret von *Koelreuteria paniculata*. Oesterr Bot Z 97:517–519

Beattie AJ (1982) Ants and gene dispersal in flowering plants. In: Armstrong JA, Powell JM, Richards AJ (eds) Pollination plants and evolution. Royal Bot Gardens, Sydney, pp 1–8

Belin L, Rowley JR (1971) Demonstration of birch pollen allergen from isolated pollen grains using immunofluorescence and a single radial immunodiffusion technique. Int Arch Allergy Appl Immunol 40:754–769

Bell J, Hicks G (1976) Transmitting tissue in the pistil of tobacco. Light and electron microscopic observations. Planta 131:187–200

Bernhardt P, Knox RB, Calder DM (1980) Floral biology and self-incompatibility in some Australian mistletoes of the genus *Amyema* (Loranthaceae). Aust J Bot 28:437–451

Bhandac IS, Malik CP (1980) Total and polar lipid biosynthesis during *Crotalaria juncea* L. pollen tube growth: effect of gibberellic acid, indole acetic acid and (2-chlorethyl-) phosphoric acid. J Exp Bot 31:931–935

Boney AD (1981) Mucilage: the ubiquitous algal attribute. Br Phycol J 16:115–132

Bouveng HO (1965) Polysaccharide in pollen. II The xylogalacturonan from Mountain Pine (*Pinus mugo* Turra) pollen. Acta Chem Scand 19:953–963

Bouveng HO, Lundström H (1965) Polysaccharides in pollen. III. The acidic arabinogalactan in Mountain Pine pollen. Acta Chem Scand 19:1004–1005

Bredemeijer GMM (1979) The distribution of peroxidase iso-enzymes, chlorogenic acid oxidase and glucose-6-phosphate dehydrogenese in transmitting tissue and cortex of *Nicotiana alata* styles. Acta Bot Neerl 28:197–205

Brewbaker JL (1957) Pollen cytology and incompatibility systems in plants. J Hered 48:271–277

Brewbaker JL (1971) Pollen enzymes and iso-enzymes. In: Heslop-Harrison J (ed) Pollen, development and physiology. Butterworths, London, pp 156–170

Brewbaker JL, Kwack BH (1964) The calcium ion and substances influencing pollen growth. In: Linskens HF (ed) Pollen physiology and fertilization. North Holland, Amsterdam, pp 143–151

Brooks J, Shaw G (1971) Recent developments in the chemistry, biochemistry, geochemistry and post-tetrad ontogeny of sporopollenins derived from pollen and spore exines. In: Heslop-Harrison J (ed) Pollen, development and physiology. Butterworths, London, pp 99–114

Brooks J, Shaw G (1978) Sporopollenin: a review of its chemistry, palaeochemistry and geochemistry. Grana 17:91–97

Brummitt RK, Ferguson IK, Poole MM (1980) A unique and extraordinary pollen type in the genus *Crossandra* (Acanthaceae). Pollen Spores 22:11–16

Bubar JS (1958) An association between variability in ovule development within ovaries and self-incompatibility in *Lotus* (Leguminosae). Can J Bot 36:65–72

Burgess J (1970) Cell shape and mitotic spindle formation in the generative cell of *Endymion non-scriptus* L. Planta 95:72–83

Burnet FM (1971) "Self-recognition" in colonial marine forms and flowering plants in relation to the evolution of immunity. Nature 232:230–235

Buttrose MS, Grant WJR, Lott JNA (1977) Reversible curvative of style branches of *Hibiscus trionum* L. a pollination mechanism. Aust J Bot 25:567–570

Capkova-Balatkova V, Hrabetova E, Tupy J (1980) Effects of cycloheximide on pollen of *Nicotiana tabacum* in culture. Biochem Physiol Pflanz 175:412–420

Carter AL, Williams ST, McNeilly T (1975) Scanning electron microscope studies of pollen behaviour on immature and mature Brussels sprouts *Brassica oleracea* var. *gemmifera* stigmas. Euphytica 24(1):133–142

Carpita N, Sabularse D, Montezinos D, Delmer DP (1979) Determination of the pore size of cell walls of living plant cells. Science 205:1144–1147

Chandra S, Bhatnagar SP (1974) Reproductive biology of *Triticum*. II Pollen germination, pollen tube growth and its entry into the ovule. Phytomorphology 24:211–217

Chao C (1972) A periodic acid-Schiff substance related to the directional growth of pollen tube into embryo sac in *Paspalum* ovules. Am J Bot 58:649–654

Chao C (1977) Light microscopic detection of PAS-positive substances with thiosemicarbazide in freeze-substituted ovaries of *Paspalum longifolium* before pollination. Histochemistry 54:159–168

Christ B (1959) Entwicklungsgeschichtliche und physiologische Untersuchungen über die Selbststerilität von *Cardamine pratensis* L. Z Bot 47:88–112

Clarke AE, Hoggart RM (1982) The use of lectins in the study of glycoproteins. In: Marchalonis JJ, Warr GD (eds) Antibody as a tool: applications of immunochemistry. Wiley, Chichester, pp 347–401

Clarke AE (1981) Defined components involved in pollination. In: Loewus F, Tanner W (eds) Extracellular carbohydrates. Encyclopaedia of Plant Physiology New Ser Vol 13 B. Springer, Berlin Heidelberg New York, pp 577–582

Clarke AE, Gleeson PA (1981) Molecular aspects of recognition and response in the pollen–stigma interaction. In: Loewus F, Ryan CA (eds) The phytochemistry of cell recognition and cell surface interactions. Rec Adv Phytochem 15:161–211

Clarke AE, Knox RB (1978) Cell recognition in flowering plants. Quart Rev Biol 53:3–28

Clarke AE, Knox RB (1980) Plants and immunity. Dev Comp Immunol 3:571–589

Clarke AE, Considine JA, Ward R, Knox RB (1977) Mechanism of pollination in *Gladiolus*: roles of the stigma and pollen tube guide. Ann Bot 41:15–20

Clarke AE, Gleeson P, Harrison S, Knox RB (1979) Pollen–stigma interactions: identification and characterization of surface components with recognition potential. Proc Natl Acad Sci USA 76:3358–3362

Clarke AE, Abbot A, Mandel TE, Pettitt JM (1980) Organization of the wall layers of the stigmatic papillae of *Gladiolus gandavensis*: a freeze-fracture study. J Ultrastruct Res 73:269–281

Clarke AE, Raff J, Mau S-L (1982) Nature of some components of *Prunus avium* L. styles including two antigenic glycoproteins. Phytomorphology (in press)

Condeelis JJ (1974) The identification of F-actin in the pollen tube and protoplast of *Amaryllis belladonna*. Exp Cell Res 88:435–439

Considine JA, Knox RB (1979) Development and histochemistry of the pistil of the grape, *Vitis vinifera*. Ann Bot 43:11–22

Cope RW (1962) The mechanism of pollen incompatibility in *Theobroma cacao* L. Heredity 17:157–182

Crane MB, Brown AG (1937) Incompatibility and sterility in the sweet cherry *Prunus avium* L. J Pomol 15:86–116

Crang RE, Miles GB (1969) An electron microscope study of germinating *Lychnis alba* pollen. Am J Bot 56:309–405

Cresti M, Went JL van (1976) Callose deposition and plug formation in *Petunia* pollen tubes in situ. Planta 133:35–40

Cresti M, Went JL van, Pacini E, Willemse MTM (1976a) Ultrastructure of transmitting tissue of *Lycopersicum peruvianum*. Style development and histochemistry. Planta 132:305–312

Cresti M, Went JL van, Willemse MTM, Pacini E (1976b) Fibrous masses and cell and nucleus movement in the pollen tube of *Petunia hybrida*. Acta Bot Neerl 25:381–383

Cresti M, Pacini E, Ciampolini F, Sarfatti G (1977) Germination and early tube development in vitro of *Lycopersicum peruvianum* pollen: ultrastructural features. Planta 136:239–247

Cresti M, Ciampolini F, Pacini E, Sree Ramulu K, Devreux M (1978) Gamma irradiation of *Prunus avium* L. flower buds: affects on stylar development – an ultrastructural study. Acta Bot Neerl 27:97–106

Crowe LK (1954) Incompatibility in *Cosmos bipinnatus*. Heredity 8:1–11

Crowe LK (1971) The polygenic control of outbreeding in *Borago officinalis*. Heredity 27:111–118

Darwin C (1882) The variation of animals and plants under domestication. Murray, London

Dashek WV, Harwood HI, Rosen WG (1970) The significance of a wall-bound, hydroxy-proline-containing glycopeptide in lily pollen tube elongation. In: Heslop-Harrison J (ed) Pollen, development and physiology. Butterworths, London, pp 194–200

Delbart C, Bris B, Linskens HF, Linder R, Coustaut D (1980a) Analysis of glycosphingo-lipids of *Petunia hybrida,* a self-incompatible species. II. Evolution of the fatty acids composition after cross- and self-pollination. Proc K Ned Akad Wet Ser C 83:241–254

Delbart C, Bris B, Linskens HF, Coustaut D (1980b) Analysis of glycosphingolipids of *Petunia hybrida,* a self-incompatible species. III. Evolution of the long chained buds after self- and cross-pollination. Proc K Ned Akad Wet Ser C 83:255–269

Delbart C, Linskens HF, Bris B, Moschetto Y, Coustaut D (1980c) Analysis of glyco-sphingolipids of *Petunia hybrida,* a self-incompatible species. I. Composition in fatty acids and in long-chained bases of pollen and unpollinated style. Proc K Ned Akad Wet Ser C, 83:229–239

Denborough MA (1964) Blood groups and disease. Br Med J 2:190–191

Deurenberg JJM (1976a) Activation of protein synthesis in ovaries from *Petunia hybrida* after compatible and incompatible pollination. Acta Bot Neerl 25:221–226

Deurenberg JJM (1976b) In vitro protein synthesis with polysomes from unpollinated, cross- and self-pollinated *Petunia* ovaries. Plants 128:29–33

Dexheimer J (1970) Recherches cytophysiologiques sur les grains de pollen. Rev Cytol Biol Veg 33:169–234

Dhaliwal AS, Malik CP, Singh MB (1981) Overcoming incompatibility in *Brassica cam-pestris* L. by carbon dioxide, and dark fixation of the gas by self- and cross-pollinated pistils. Ann Bot 48:227–234

Dickinson HG, Lawson J (1975a) Pollen tube growth in the stigma of *Oenothera organen-sis* following compatible and incompatible intraspecific pollination. Proc R Soc Lond B 188:327–444

Dickinson HG, Lewis D (1973a) Cytochemical and ultrastructural differences between intraspecific compatible and incompatible pollinations in *Raphanus.* Proc R Soc Lond B 183:21–35

Dickinson HG, Lewis D (1973b) The formation of the 'tryphine' coating the pollen grains of *Raphanus* and its properties relating to self-incompatibility. Proc R Soc Lond B 184:149–165

Dickinson HG, Lewis D (1974) Changes in the pollen grain wall of *Linum grandiflorum* following compatible and incompatible intraspecific pollinations. Ann Bot 38:23–29

Ducker SC, Knox RB (1976) Submarine pollination in seagrasses. Nature 263:705–706

Ducker SC, Pettitt JM, Knox RB (1978) Biology of Australian Seagrasses: pollen develop-ment and submarine pollination in *Amphibolis antarctica* and *Thalassodendron ciliatum* (Cymodoceaceae). Aust J Bot 26:265–285

Dulberger R (1974) Structural dimorphism of stigmatic papillae in distylous *Linum* spe-cies. Am J Bot 61:238–243

Dumas C (1973a) Contribution à l'étude cytophysiologique du stigmate. I. Les étapes observées durant les processus glandulaires chez une Oleaceae: *Forsythia intermedia* Zabel. Z Pflanzenphysiol 69:35–54

Dumas C (1973b) Contribution à l'étude cytophysiologique du stigmate. VII. Les va-cuoles lipidiques et les associations réticulum endoplasmique-vacuole chez *Forsythia intermedia* Z. Botaniste 56:59–80

Dumas C (1974a) Contribution à l'étude cytophysiologique du stigmate. V. Mise en évidence histochimique de la nature essentiellement lipidique de l'exsudat de *Forsythia intermedia* Z. en microscopie photonique et electronique. Acta Histochem 48:115–123

Dumas C (1974b) Some aspects of stigmatic sécrétion in *Forsythia* sp. In: Linskens HF (ed) Fertilization in higher plants. Elsevier, North Holland/ Oxford New York, pp 119–126

Dumas C (1975) Le stigmate et la sécrétion stigmatique. Ph D Thesis Univ Lyons

Dumas C (1977) Lipochemistry of the progamic stage of a self-incompatible species: neutral lipids and fatty acids of the secretory stigma during its glandular activity, and the solid style, the ovary and the anther in *Forsythia intermedia* Zabel (heterostylic species). Planta 137:177–184

Dumas C (1979) Nature et rôle de l'interface pistil–pollen au cours de la phase progamique. Coll SFME Biol Cell 35:27a

Dumas C, Gaude T (1981) Stigma–pollen recognition: a new look. In: Int Symp Adv Plant Cytoembryology. Acta Soc Bot Pol 50:235–247

Dumas C, Gaude T (1982) Sécrétions et biologie florale. II. Leurs rôles dans l'adhésion et la reconnaissance pollen-stigmate. Données récentes, hypothèses et notion d'immunité végétale. Actual Bot (in press)

Dumas C, Gaude T (1983a) Stigma-pollen recognition and pollen hydration. Phytomorphology 31:191–201

Dumas C, Gaude T (1983b) Advances in structure, composition and function of stigma in angiosperms. Int Rev Plant Sci (in press)

Dumas C, Knox RB (1983) Callose and determination of pistil viability and incompatibility. Theoret Appl Genet (in press)

Dumas C, Rougier M, Zandonella P, Ciampolini F, Cresti M, Pacini E (1978) The secretory stigma in *Lycopersicum peruvianum* Mill.: ontogenesis and glandular activity. Protoplasma 96:173–187

Dumas C, Duplan J-C, Said C, Soulier JP (1983) ^{1}H Nuclear magnetic resonance to correlate water content and pollen viability. In: Mulcahy DL, Ottaviano E (eds) Pollen biology and implications for plant breeding. Elsevier Biomedical, New York Amsterdam, Oxford, pp 15–20

Dumas C, Knox RB, Gaude T (1983) The mature viable tricellular pollen grain of *Brassica*: germ line characteristics. Protoplasma (in press)

Echlin P (1971) The role of the tapetum during microsporogenesis of angiosperms. In: Heslop-Harrison J (ed) Pollen, physiology and development. Butterworths, London, pp 41–61

Elsik WC (1971) Microbiological degradation of sporopollenin. In: Brooks J, Grant PR, Muir M, Gijsel P van, Shaw C (eds) Sporopollenin. Academic Press, London, New York, pp 480–511

Emerson SH (1940) Growth of incompatible tubes in *Oenothera organensis*. Bot Gaz 101:890–899

Engels FM (1974a) Function of Golgi vesicles in relation to cell wall synthesis in germinating *Petunia* pollen II. Chemical composition of Golgi vesicles and pollen tube wall. Acta Bot Neerl 23:81–89

Engels FM (1974b) Function of Golgi vesicles in relation to cell wall synthesis in germinating *Petunia* pollen. IV. Identification of cellulose in pollen tube walls and Golgi vesicles by X-ray diffraction. Acta Bot Neerl 23:209–216

Erdelska O (1974) Contribution to the study of fertilization in the living embryo sac. In: Linskens HF (ed) Fertilization in higher plants. North Holland, Amsterdam, pp 191–195

Erdtman G (1952) Pollen morphology and plant taxonomy. I. Angiosperms. Almquist and Wiksell, Stockholm

Erdtman G (1966) Sporoderm morphology and morphogenesis. Grana Palynol 6:317–323

Faegri K, Van Der Pijl L (1979) The principles of pollination ecology, 3rd edn. Pergamon, London New York

Faegri L, Iversen J (1975) Textbook of pollen analysis, 3rd edn. Hafner, New York

Ferrari TE, Wallace DH (1977a) A model of self-recognition and regulation of the incompatibility response of pollen. Theoret Appl Genet 50:211–225

Ferrari TE, Wallace DH (1977b) Incompatibility on *Brassica* stigmas is overcome by treating pollen with cycloheximide. Science 196:436–438

Ferrari TE, Bruns D, Wallace DH (1981) Isolation of a plant glycoprotein involved with control of intercellular recognition. Plant Physiol 67:270–277

Fett WF, Paxton JD, Dickinson DB (1976) Studies on the self-incompatibility response of *Lilium longiflorum*. Am J Bot 63:1104–1108

Fitting H (1909) Die Beeinflussung der Orchideenblüten durch die Bestäubung und durch andere Umstände. Z Bot 1:1–39

Franke WW, Herth W, Van der Woude WJ, Morre DJ (1972) Tubular and filamentous structures in pollen tubes: possible involvement as guide elements in protoplasmic streaming and vectorial migration of secretory vesicles. Planta 105:317–341

Fränkel R, Galun E (1977) Pollination mechanisms, reproduction and plant breeding. Springer, Berlin Heidelberg New York

Frankis R, Mascarenhas JP (1980) Messenger RNA in the ungerminated pollen grain: a direct demonstration of its presence. Ann Bot 45:595–599

Frazier W, Glaser L (1979) Surface components and cell recognition. Ann Rev Biochem 48:491–523

Gaude T (1982) Adhesion et reconnaissance pollen-stigmate: étude des surfaces cellulaires chez *Brassica*. PhD Thesis (3me Cycle), Univ Lyon 1, Villeurbanne, France

Gherardini GL, Healey PL (1969) Dissolution of outer wall of pollen grain during pollination. Nature 224:218–219

Gosh S (1981) Studies on pollen–pistil interaction in *Linum* and *Zephyranthes*. PhD Thesis, Univ Delhi

Ghosh S, Shivanna KR (1980a) Pollen–pistil interaction in *Linum grandiflorum*: Scanning electron microscopic observations and proteins of stigma surface. Planta 149:257–261

Ghosh S, Shivanna KR (1980b) Pollen–pistil interaction in *Linum grandiflorum*: Stigma surface proteins and stigma receptivity. Proc Indian Natl Sci Acad B 46:177–183

Gilissen LJW (1977) The influence of relative humidity on the swelling of pollen grains in vitro. Planta 137:299–301

Gilissen LJW, Brantjes NBM (1978) Function of the pollen coat in the different stages of the fertilization process. Acta Bot Neerl 27:205–209

Gladyshev BN (1962) (in Russian) Biokimiya 27:240–247

Gleeson PA, Clarke AE (1979) Structural studies on the major component of *Gladiolus* style mucilage, an arabinogalactan protein. Biochem J 181:607–621

Golynskaya EL, Bashkirova NV, Tomchuk NN (1976) Phytohemagglutinins of the pistil in *Primula* as possible proteins of generative incompatibility. Soviet Plant Physiol 23:69–76

Green JR (1984) On the germination of the pollen grain and nutrition of the pollen tube. Ann Bot 8:225–228

Guinet Ph (1981) Mimosoideae: the characters of the pollen grains. In: Polhill RM, Raven PH (eds) Advances in legume systematics Part 1. RB Gard Kew pp 835–857

Gunning BES, Pate JS (1969) Transfer cells: plant cells with wall ingrowths, specialized in relation to short distance transport of solutes – their occurrence, structure and development. Protoplasma 68:107–113

Guttenberg HV, Reiff B (1958) Der Mechanismus der Narbenbewegung von *Mimulus* sp. in seiner Abhängigkeit. Planta 50:498–503

Habura EC (1957) Parasterilität bei Sonnenblumen. Z Pflanzenzuecht 37:280–298

Hara A, Yoshihara K, Watanabe T (1970) Studies on the nucleases from Japanese cycad, *Cycas revoluta* Thunb. III. Ribonuclease and phosphomonoesterase eluted by salt solutions from cycad pollen cells. Nippon Nogei Kagaku Kaishi 44:385–392

Hara A, Yamamoto M, Horita Y, Watanabe T (1972) Invertase of cell walls from cycad pollen. Mem Fac Agric Kagoshima Univ 8:27–37

Hara A, Yamashita H, Kobayashi A (1977) Isolation of a polysaccharide from the inner cellular intine of pollen of *Cryptomeria japonica*. Plant Cell Physiol 18:381–386

Hawker JS, Sedgley M, Loveys BR (1983) The composition of stigmatic exudate nectar and pistil of watermelon [*Citrullus lanatus* (Thumb.) Matsum and Nakai] before and after pollination. Aust J Plant Physiol 10:257–264

Heitz B, Jean R, Prensier G (1971) Observation de la surface du stigmate et du grain de pollen de *Linum austriacum* L. heterostyle. CR Acad Sci Paris 273:2493–2495

Helsper JPFG, Veerkemp JH, Sassen MMA (1977) β-Glucan synthetase activity in Golgi vesicles of *Petunia hybrida*. Planta 133:303–308

Herd YR, Beadle DJ (1980) The site of the self-incompatibility mechanism in *Tradescantia pallida*. Ann Bot 45:251–256

Herdt E, Sutfeld R, Wiermann R (1978) The occurrence of enzymes involved in phenyl propanoid metabolism in the tapetum fraction of anthers. Cytobiologie 17:433–441

Herpen MMA van, Linskens HF (1981) Effect of season, plant age and temperature during plant growth on compatible and incompatible pollen tube growth in *Petunia hybrida*. Acta Bot Neerl 30:209–218

Herrero M, Dickinson HG (1979) Pollen–pistil incompatibility in *Petunia hybrida*: changes in the pistil following compatible and incompatible intraspecific crosses. J Cell Sci 36:1–18

Herrero M, Dickinson HG (1980) Pollen-tube development in *Petunia hybrida* following compatible and incompatible intraspecific matings. J Cell Sci 47:365–383

Heslop-Harrison J (1968) The pollen grain wall. Science 161:230–237

Heslop-Harrison J (1969) Scanning electron microscopic observations on the wall of the pollen grain in *Cosmos bipinnatus*. Proc Engis Stereoscan Colloq 89–96 Chicago Ill

Heslop-Harrison J (1972) Sexuality of angiosperms. In: Steward F (ed) Plant Physiology VI c. Academic Press, London New York, pp 133–291

Heslop-Harrison J (1975a) Physiology of the pollen-grain surface. Proc R Soc Lond B 190:275–300

Heslop-Harrison J (1975b) Incompatibility and the pollen–stigma interaction. Annu Rev Plant Physiol 26:403–425

Heslop-Harrison J (1976) A new look at pollination. Rep E Malling Res Stn 1975:141–157

Heslop-Harrison J (1978a) Cellular recognition systems in plants. Studies in Biol Vol 100. Arnold, London

Heslop-Harrison J (1978b) Recognition and response in the pollen–stigma interaction. Symp Soc Exp Biol 32:121–138

Heslop-Harrison J (1979a) Aspects of the structure, cytochemistry and germination of the pollen of rye (*Secale cereale* L.). Ann Bot (Suppl) 1:1–47

Heslop-Harrison J (1979b) Hydrodynamics of the grass pollen grain. Am J Bot 66:737–743

Heslop-Harrison J (1979c) Pollen walls as adaptive systems. Ann Missouri Bot Gard 66:813–829

Heslop-Harrison J (1979d) The forgotten generation: some thoughts on the genetics and physiology of angiosperm gametophytes. Proc 4th John Innes Symposium, John Innes Institute, Norwick, pp 1–14

Heslop-Harrison J (1982) Pollen–stigma interaction and cross-incompatibility in the grasses. Science 215:1358–1364

Heslop-Harrison J, Heslop-Harrison Y (1970) Evaluation of pollen viability by enzymatically induced fluorescence: intracellular hydrolysis of fluorescein diacetate. Stain Technol 45:115–120

Heslop-Harrison J, Heslop-Harrison Y (1975a) Enzymic removal of the proteinaceous pellicle of the stigma papilla prevents pollen tube entry in the Caryophyllaceae. Ann Bot (Lond) 39(159):163–165

Heslop-Harrison J, Heslop-Harrison Y (1975b) Fine structure of the stigmatic papillae of *Crocus*. Micron 6:45–52

Heslop-Harrison J, Heslop-Harrison Y (1979) Pollen–stigma interaction in other families. Rep Welsh Plant Breed Stn 1978:149–151

Heslop-Harrison J, Heslop-Harrison Y (1980a) Cytochemistry and function of the Zwischenkörper in grass pollens. Pollen et spores 22:5–10

Heslop-Harrison J, Heslop-Harrison Y (1980b) The pollen–stigma interaction in the grasses. I. Acta Bot Neerl 29:261–276

Heslop-Harrison J, Heslop-Harrison Y (1981a) The pollen–stigma interaction in the grasses. II. Pollen-tube penetration and the stigma response in *Secale*. Acta Bot Neerl 30:289–307

Heslop-Harrison J, Heslop-Harrison Y (1981b) The specialized cuticles of stigma papillae. In: Cutler DF (ed) The plant cuticle. Suppl Bot J Linn Soc 99–119

Heslop-Harrison J, Heslop-Harrison Y (1982a) Pollen–stigma interaction in the Leguminosae: constituents of the stylar fluid and stigma secretion of *Trifolium pratense* L. Ann Bot 49:729–735

Heslop-Harrison J, Heslop-Harrison Y (1982b) The growth of the grass pollen tube: 1. Characteristics of the polysaccharide particles ("P-particles") associated with apical growth. Protoplasma 112:71–80

Heslop-Harrison J, Heslop-Harrison Y, Knox RB, Howlett B (1973) Pollen–wall proteins: 'gametophytic' and 'sporophytic' fractions in the pollen walls of Malvaceae. Ann Bot 37:402–412

Heslop-Harrison J, Knox RB, Heslop-Harrison Y (1974) Pollen–wall proteins: exine-held fraction associated with the incompatibility response in Cruciferae. Theor Appl Genet 44:133–137

Heslop-Harrison J, Knox RB, Heslop-Harrison Y, Mattsson O (1975a) Pollen–wall proteins. In: Duckett JG, Racey PA (eds) Biology of the Male Gamete. Academic Press, London, pp 189–202

Heslop-Harrison J, Heslop-Harrison Y, Barber J (1975b) The stigma surface in incompatibility responses. Proc R Soc Lond B 188:287–297

Heslop-Harrison Y (1976) Localization of Concanavalin A-binding sites on the stigma surface of a grass species. Micron 7:33–36

Heslop-Harrison Y (1977) The pollen–stigma interaction: pollen tube penetration in Crocus. Ann Bot 41:913–922

Heslop-Harrison Y (1981) Stigma characteristics and angiosperm taxonomy. Nord J Bot 1:401–420

Heslop-Harrison Y, Heslop-Harrison J (1982a) Pollen–stigma interaction in the Leguminosae: the secretory system of the style in *Trifolium pratense* L. Ann Bot 50:637–649

Heslop-Harrison Y, Heslop-Harrison J (1982b) The microfibrillar component of the pollen intine: some structural features. Ann Bot 50:831–842

Heslop-Harrison Y, Shivanna KR (1977) The receptive surface of the angiosperm stigma. Ann Bot 41:1233–1258

Hideux MJ, Ferguson IK (1976) The stereostructure of the exine and its evolutionary significance in Saxifragaceae s.l. In: Ferguson IK, Muller J (eds) The evolutionary significance of the exine. Academic Press, London New York, pp 327–378

Hinata T, Nishio K (1978) S-allele specificity of stigma proteins in *Brassica oleracea* and *B. campestris*. Heredity 41:93–100

Hodgkin T, Lyon GD (1979) Inhibition of *Solanum* pollen germination in vitro by the phytoalexin rishitin. Ann Bot 44:253–255

Hoekstra FA (1979) Mitochondrial development and activity of binucleate and trinucleate pollen during germination in vitro. Planta 145:25–36

Hoekstra FA, Bruinsma J (1979) Protein synthesis of binucleate and trinucleate pollen and its relationship to tube emergence and growth. Planta 146:559–566

Hogenboom NG (1975) Incompatibility and incongruity: two different mechanisms for the non-functioning of intimate partner relationships. Proc R Soc Lond B 188:361–375

Horovitz A, Galil J, Portnoy L (1972) Soluble sugars in the stigmatic exudate of *Yucca aloifolia* L. Phyton (Buenos Aires) 29:43–46

Howlett BJ, Knox RB, Heslop-Harrison J (1973) Pollen–wall proteins: release of the allergen Antigen E from intine and exine sites in the pollen grains of ragweed and *Cosmos*. J Cell Sci 13:603–619

Howlett BJ, Knox RB, Paxton JD, Heslop-Harrison J (1975) Pollen–wall proteins: characterization and role in self-incompatibility in *Cosmos bipinnatus*. Proc R Soc Lond B 188:167–182

Howlett BJ, Vithanage HIMV, Knox RB (1979) Pollen antigens, allergens and enzymes. Curr Adv Plant Sci 35:1–17

Howlett BJ, Vithanage HIMV, Knox RB (1981) Immunofluorescence localization of two water-soluble glycoproteins, including the major allergen, from pollen of ryegrass, *Lolium perenne*. Histochem J 13:461–480

Iwanami Y, Iwadare T (1978) Inhibiting effects of myrmicacin on pollen growth and pollen tube mitosis. Bot Gaz 139:42–45

Iwanami Y, Iwadare T (1979) Myrmic acids: a group of new inhibitors analogous to myrmicacin (β-hydroxydecanoic acid). Bot Gaz 140:1–4

Iwanami Y, Okada I, Iwamatsu M, Iwadare T (1979) Inhibitory effects of royal jelly acid, myrmicacin, and their analogous compounds on pollen germination, pollen tube elongation and pollen tube mitosis. Cell Struc Func 4:135–143

Iwanami Y, Nakamura S, Miki-Hirosige H, Iwadare T (1981) Effect of myrmicacin (β-hydroxydecanoic acid) on protoplasmic movement and ultrastructure of *Camellia japonica* pollen. Protoplasma 104:341–348

Jensen WA (1962) Botanical histochemistry. Freeman, San Francisco

Jensen WA (1974) Reproduction in flowering plants. In: Roberts AW (ed) Dynamic aspects of plant ultrastructure. Academic Press, New York London, pp 481–503

Jensen WA, Ashton ME, Beasley CA (1983) Pollen-tube-embryo sac interactions in cotton. In: Mulcahy DL, Ottaviano E (eds) Pollen: Biology and implications for plant breeding. Elsevier Biomedical, Amsterdam, pp 67–72

Jensen WA, Fisher DB (1969) Cotton embryogenesis: the tissues of the stigma and style and their relation to the pollen tube. Planta 84:97–121

Jensen WA, Fisher DB, Ashton M (1968) Cotton embryogenesis: the pollen cytoplasm. Planta 81:206–228

Jermyn MA, Yeow YM (1975) A class of lectins present in the tissues of seed plants. Aust J Plant Physiol 2:501–531

Jobson S, Knox RB, Kenrick J, Dumas C (1983) Plastid development and ferritin content of stigmas of the legumes *Acacia, Lotus* and *Trifolium*. Protoplasma 116:213–218

Johnson LEB, Wilcoxson RD, Frosheiser FI (1975) Transfer cells in tissues of the reproductive system of alfalfa. Can J Bot 53:952–956

Johri BM (1981) Do "transfer cells" occur in reproductive tissues of angiosperms? Proc Nat Symp Biol Reproduct Plants. Dept Bot, Univ Delhi, pp 95–98

Johri BM, Shivanna KR (1974) Experimental studies on pollen and fertilization. In: Linskens HF (ed) Fertilization in Higher Plants. North Holland Amsterdam, pp 179–190

Kato K (1953) A phenomenon in the early stage of pollination in *Secale cereale*: the stigma reaction. Mem Coll Sci Univ Kyoto Ser B 20:203–206

Kato K, Watanabe K (1957) The stigma reaction. II. Bot Mag 70:96–101

Kenrick J, Knox RB (1981a) Post-pollination exudate from stigmas of *Acacia* (Mimosaceae). Ann Bot 48:103–106

Kenrick J, Knox RB (1981b) Structure and histochemistry of the stigma and style of some Australian species of *Acacia*. Aust J Bot 29:733–745

Kenrick J, Knox RB (1982) Function of the polyad in reproduction of *Acacia*. Ann Bot 50:721–727

Kerhoas C, Knox RB, Dumas C (1983) Specificity of the callose response in stigmas of *Brassica*. Ann Bot 52:597–602

Knight AH, Crooke WM, Shepherd H (1972) Chemical composition of pollen with particular reference to cation exchange capacity and uronic acid content. J Sci Food Agric 23:263–274

Knox RB (1971) Pollen–wall proteins: localization, enzymic and antigenic activity during development in *Gladiolus* (Iridaceae). J Cell Sci 9:209–237

Knox RB (1973) Pollen–wall proteins: cytochemical observations of pollen–stigma interactions in ragweed and *Cosmos* (Compositae). J Cell Sci 12:421–443

Knox RB (1976) Cell recognition and pattern formation in plants. In: Graham CF, Wareing PF (eds) The developmental biology of plants and animals. Saunders, Philadelphia Toronto. Blackwell Oxford

Knox RB (1979) Pollen and allergy. Studies in biology Vol 107. Arnold, London

Knox RB (1983) The pollen grain. In: Johri BM (ed) Embryology of angiosperms. Springer, Berlin Heidelberg New York (in press)

Knox RB, Clarke AE (1979) Localization of proteins and glycoproteins by binding to labelled antibodies and lectins. In: Hall JL (ed) Electron microscopy and cytochemistry of plant cells. Elsevier-North Holland Biomedical Press, Amsterdam New York, pp 149–185

Knox RB, Clarke AE (1980) Discrimination of self and non-self in plants. In: Marchalonis JJ, Cohen N (eds) Contemp Topics Immunobiol. Plenum, New York London, pp 9–36

Knox RB, Heslop-Harrison J (1969) Cytochemical localization of enzymes in the wall of the pollen grain. Nature 223:92–94

Knox RB, Heslop-Harrison J (1970) Pollen–wall proteins: Localisation and enzymic activity. J Cell Sci 6:1–27

Knox RB, Heslop-Harrison J (1971a) Pollen–wall proteins: electron microscopic localization of acid phosphatase in the intine of *Crocus vernus*. J Cell Sci 8:727–733

Knox RB, Heslop-Harrison J (1971b) Pollen–wall proteins: fate of intine-held antigens in compatible and incompatible pollinations of *Phalaris tuberosa* L. J Cell Sci 9:239–251

Knox RB, Willing R, Ashford AE (1972) Role of pollen–wall proteins as recognition substances in inter-specific hybridization in poplars. Nature 237:381–383

Knox RB, Heslop-Harrison J, Heslop-Harrison Y (1975) Pollen–wall proteins. In: Duckett JG, Racey PA (eds) The biology of the male gamete. Academic Press, London New York, pp 177–187

Knox RB, Clarke AE, Harrison S, Smith P, Marchalonis JJ (1976) Cell recognition in plants: determinants of the stigma surface and their pollen interactions. Proc Natl Acad Sci USA 73:2788–2792

Knox RB, Howlett BJ, Vithanage HIMV (1980) Botanical immunocytochemistry, a review with special reference to pollen antigens and allergens. Histochem J 12:247–272

Konar RN, Linskens HF (1965) Some observations on the stigmatic exudate of *Petunia hybrida*. Naturwissenschaften 52:625

Konar RN, Linskens HF (1966a) Physiology and biochemistry of the stigmatic fluid of *Petunia hybrida*. Planta 71:372–387

Konar RN, Linskens HF (1966b) The morphology and anatomy of the stigma of *Petunia hybrida*. Planta 71:356–371

Kress WJ, Stone DE (1982) Nature of the sporoderm in monocotyledons with special reference to the pollen grains of *Canna* and *Heliconia*. Grana 21:129–148

Kress WJ, Stone DE, Sellers SC (1978) Ultrastructure of exine-less pollen: *Heliconia* (Heliconiaceae). Am J Bot 65:1064–1076

Kristen V (1977) Granulocrine extrusion of stigmatic secretions via vesicles of endoplasmic reticulum in *Aptenia cordifolia*. Protoplasma 92:243–252

Kristen V, Biedermann M, Liebezeit G, Dawson R, Böhm L (1979) The composition of stigmatic exudate and the ultrastructural of the stigma papillae in *Aptenia cordifolia*. Europ J Cell Biol 19:281–287

Kroh M (1967) Bildung und Transport des Narbensekrets von *Petunia hybrida*. Planta 77:250–260

Kroh M (1973) Nature of the intercellular substances of stylar transmitting tissue. In: Loewus F (ed) Biogenesis of plant cell wall polysaccharides. Academic Press, New York New York, pp 195–205

Kucera V, Polak J (1975) The serological specificity of S-alleles of homozygous incompatible lines of the marrow-stem kale, *Brassica oleracea* var. *acephala*. Biol Plant 17:50–59

Kwan SC, Hamson AR, Campbell WF (1969) The effects of different chemicals on pollen germination and tube growth in *Allium cepa* L. J Am Soc Hortic Sci 94:561–562

Landova B, Landa Z (1973) Immunochemicka detekee autoinkompatibilnich genotype *Brassica oleracea*. L Genet Sel 9:249–256

Larson DA (1965) Fine structural changes in the cytoplasm of germinating pollen. Am J Bot 52:139–154

Lea DJ, Sehon AH (1962) The chemical characterization of non-dialyzable constituents of ragweed pollen. Int Arch Allergy Appl Immunol 20:203–214

Lecocq M, Dumas C (1975) Histophysiologie des stigmates normaux et des formations stigmatoides chez *Begonia tuberhybrida*. I. Observations prélliminaires. Can J Bot 53:1252–1258

Lewis D (1942) The physiology of incompatibility in plants. I Effect of temperature. Proc Roy Soc Lond B 131:13–26

Lewis D (1952) Serological reactions of pollen incompatibility substances. Proc R Soc B 140:127–135

Lewis D (1956) Incompatibility and plant breeding. Brookhaven Symp Biol 9:89–100

Lewis D (1965) A protein dimer hypothesis on incompatibility. In: Geerts, SJ (ed) Proc 11th Int Congr Genet Vol 3:657–663

Lewis D (1976) Incompatibility in flowering plants. In: Cuatrecasas P, Greaves MF (eds) Receptors and recognition. Ser A Vol 2, Chapman and Hall, London, pp 167–198

Lewis D, Crowe LK (1958) Unilateral incompatibility in flowering plants. Heredity 12:233–256

Lewis D, Burrage S, Walls D (1967) Immunological reactions of single pollen grains, electrophoresis and enzymology of pollen protein exudates. J Exp Bot 18:371–378

Lewis DH (1980) Are there inter-relations between the metabolic role of boron, synthesis of phenolic phytoalexins and the germination of pollen. New Phytol 84:261–270

Lewis GP, Elias TS (1981) Tribe 3. Mimoseae Bronn. (1822) Part 1. In: Polhill RM, Raven PH (eds) Advances in legume systematics. Bot Gard Kew, pp 155–168

Linskens HF (1953) Physiologische und chemische Unterschiede zwischen selbst- und fremdbestäubten *Petunia*-Griffeln. Naturwissenschaften 40:28–29

Linskens HF (1960) Zur Frage der Abwehrkörper der Inkompatibilitätsreaktion von *Petunia*. III. Z Bot 48:16–135

Linskens HF (1973a) Accumulation in anthers. 3rd Symp Accumulation and Translocation of nutrients and regulators in plant organisms. Proc Res Inst of Pomology Skiernewice Poland Ser E:91–106

Linskens HF (1973b) The reaction of inhibition during incompatible pollination and its elimination. Sov Plant Physiol 20:156–166

Linskens HF (1974) Translocation phenomena in the *Petunia* flower after cross- and self-pollination. In: Linskens HF (ed) Fertilization in higher plants. North-Holland, Amsterdam, pp 285–292

Linskens HF (1976a) Specific interactions in higher plants. In: Wood RKS, Graniti A (eds) Specificity in plant diseases. Plenum, New York London, p 311

Linskens HF (1976b) Stigmatic responses. In: Vardar Y, Sheikh KH (eds) Proceedings 3rd MPP Meeting, 1975. Ege Univ, Izmir Turkey, pp 1–12

Linskens HF (1977) Incompatibility reactions during the flowering period of several *Petunia* clones. Acta Bot Neerl 26:411–415

Linskens HF (1980) Physiology of fertilization and fertilization barriers in higher plants. In: Subtelny S, Wessels NK (eds) The cell surface: mediator of developmental processes. 38th Symp Soc Devel Biol, Academic Press, New York London, pp 113–126

Linskens HF, Esser K (1957) Über eine spezifische Anfärbung der Pollenschläuche und die Zahl von Kallosepfropfen nach Selbstung und Fremdung. Naturwissenschaften 44:16

Linskens HF, Esser K (1959) Stoffaufnahme der Pollenschläuche aus dem Leitgewebe des Griffels. Proc K Ned Akad Wet C 62:150–154

Linskens HF, Heinen W (1962) Cutinase-Nachweis in Pollen. Z Bot 50:338–347

Linskens HF, Kroh M (1970) Regulation of pollen-tube growth. Curr Topics in Devel Biol Academic Press, New York London 5:89–113

Linskens HF, Pfahler PL (1977) Genotypic effects on the amino-acid relationships in maize (*Zea mays* L) pollen and style. Theor Appl Genet 50:173–177

Linskens HF, Spanjers AW (1973) Changes of the electrical potential in the transmitting tissue of *Petunia* styles after cross- and self-pollination. Incompatibility News 3:81–85

Linskens HF, Havez R, Linder R, Salden M, Randoux A, Laniez D, Custaut D (1969) Etude des glycanohydrolyses en cours de la croissance du pollen chez *Petunia hybrida* auto-incompatible. CR Hebd Seanc Acad Sci Paris B 269:1855–1857

Lloyd FE (1910) Development and nutrition of the embryo, seed and kernel in the date, *Phoenix dactylifera* L. A Rep Miss Bot Gard 21:103–164

Loewus F, Labarca C (1973) Pistil secretion product and pollen tube wall formation. In: Loewus F (ed) Biogenesis of plant cell wall polysaccharides. Academic Press, New York London, pp 175–194

Loewus M, Loewus FA (1982) Myo-inositol-l-phosphatase from the pollen of *Lilium longiflorum* Thunb. Plant Physiol. 70:765–770

Loewus F, Chen M-S, Loewus MW (1973) The myo-inositol oxidation pathway to cell wall polysaccharides. In: Loewus F (ed) Biogenesis of plant cell wall polysaccharides. Academic Press, London New York, pp 1–15

Lord EM, Webster BD (1979) The stigmatic exudate of *Phaseolus vulgaris* L. Bot Gaz 140:266–271

McWilliam JR (1958) The role of the micropyle in the pollination of *Pinus*. Bot Gaz 120:109–117

Maiti L, Kolattukudy PE, Shayk M (1979) Purification and characterization of a novel cutinase from nasturtium, *Tropaeolum majus,* pollen. Arch Biochem Biophys 196:412–423

Mäkinen Y, Lewis D (1962) Immunological analysis of incompatibility proteins and of cross-reacting material in a self-compatible mutant of *Oenothera organensis*. Genet Res 3:352–363

Malik CP, Singh MB (1980) Plant enzymology and histoenzymology – a text manual. Kalyani, New Delhi

Malik CP, Mehan M, Vermani S (1975) Distribution of alkaline phosphatase in pollen grains and pollen tubes of *Calotropis procera*. Biochem Physiol Pflanzen 167:601–603

Martin FW (1969) Compounds from the stigma of ten species. Am J Bot 56:1023–1027

Martin FW (1970) The ultraviolet absorption profile of stigmatic extracts. New Phytol 69:425–430

Martin FW, Brewbaker JL (1971) The nature of the stigmatic exudate and its role in pollen germination. In: Heslop-Harrison J (ed) Pollen, development and physiology. Butterworths, London, pp 262–266

Martin FW, Telek L (1971) The stigmatic secretion of the sweet potato. Am J Bot 58:317–322

Mascarenhas JP (1975) The biochemistry of angiosperm pollen development. Bot Rev 41:259–314

Mascarenhas JP (1978) Ribonucleic acids and proteins in pollen germination. IVth Int Polynological Conf, Lucknow 1:400–406

Mattson O, Knox RB, Heslop-Harrison J, Heslop-Harrison Y (1974) Protein pellicle of stigmatic papillae as a probable recognition site in incompatibility reactions. Nature 247:298–300

Mau S-L, Raff J, Clarke AE (1982) Isolation and partial characterization of components of *Prunus avium* L. styles, including an S-allele-associated antigenic glycoprotein. Planta 156:505–516

Mehan M, Malik CP (1976) Studies on the physiology of pollen tube growth. I *Pinus roxburghii*. Acta Histochem 56:80–85

Meyer B, Stubbe W (1974) Das Zahlenverhältnis von mütterlichen und väterlichen Plastiden in den Zygoten von *Oenothera erythrosepala* Barbas (syn. *Oe. lamarckiana*). Ber Dtsch Bot Ges 87:29–38

Miki-Hirosige H, Nakamura S (1981) The metabolic incorporation of label from myoinositol-2-^{3}H by the growing young anther of *Lilium longiflorum*. Acta Bot Soc Polon 50:77–82

Morre DJ, Mollenhauer HH, Bracker CE (1971) Origin and continuity of Golgi apparatus. In: Beermann W, Reinert H, Ursprung H (eds) Origin and continuity of cell organelles. Springer, Berlin Heidelberg New York, pp 82–118

Mulcahy DL, Mulcahy GB (1983) Pollen selection: an overview. In: Mulcahy DL, Ottaviano E (ed) Pollen: biology and implications for plant breeding. Elsevier Biomedical, Amsterdam, Oxford, pp 29–34

Nasrallah ME (1979) Self-incompatibility antigens and S-gene expression in *Brassica*. Heredity 43:347–353

Nasrallah ME, Wallace D (1967a) Immunogenetics of self-incompatibility in *Brassica oleracea* L. Heredity 22:519–527

Nasrallah ME, Wallace D (1967b) Immunochemical detection of antigens in self-incompatibility genotypes of cabbage. Nature 213:700–701

Nasrallah ME, Barber J, Wallace DH (1970) Self-incompatibility proteins in plants: Detection, genetics and possible mode of action. Heredity 25:23–27

Nettancourt D de (1977) Incompatibility in angiosperms. Springer, Berlin Heidelberg
    New York
Nettancourt D de, Devreux M, Bozzini A, Cresti M, Pacini E, Sarfatti G (1973) Ultra-
    structural aspects of the self-incompatibility mechanism in *Lycopersicum peruvianum*
    Mill. J Cell Sci 12:403–419
Nishio T, Hinata K (1977) Analysis of S-specific proteins in stigma of *Brassica oleracea* L.
    by isoelectric focussing. Heredity 38:391–397
Nishio T, Hinata K (1978) Stigma proteins in self-incompatible *Brassica campestris* and
    self-compatible relatives with special reference to S-allele specificity. Jpn J Genet
    53:27–35
Nishio T, Hinata K (1979) Purification of an S-specific glycoprotein in self-incompatible
    Brassica campestris L. Jpn J Genet 54:307–311
Nishio T, Hinata K (1980) Rapid detection of S-glycoproteins of self-incompatible cru-
    cifers using Con A-reaction. Euphytica 29:217–221
Ockendon DG (1968) Biosystematic studies in the *Linum perenne* group. New Phytol
    67:787–813
Ohtani M (1955) Salk Gabuj Hokuku Nogaku 7:21 (1955) (cited by Stanley and Linskens,
    1974)
Ottaviano E, Sari-Gorla M, Mulcahy D (1980) Pollen tube growth rates in *Zea mays*:
    implications for genetic improvements of crops. Science 210:437–438
Owens SJ, Horsfield NJ (1982) A light and electron microscopic study of stigmas in
    Aneilema and *Commelina* spp. (Commelinaceae). Protoplasma (in press)
Pacini E, Sarfatti G (1976) The reproductive calendar of *Lycopersicum peruvianum* Mill.
    Bull Soc Bot Fr 125:295–299
Pankow H (1957) Über den Pollenkitt bei *Galanthus nivalis*. Flora 146:240–253
Pegg GF (1977) Glucanohydrolases of higher plants: A possible defence mechanism
    against parasitic fungi. In: Solheim B, Raa J (eds) Cell wall biochemistry related
    to specificity in host-plant pathogen interactions. Universitatsforlaget Tromso, Nor-
    way
Pettitt JM (1977) Detection in primitive gymnosperms of proteins and glycoproteins
    of possible significance in reproduction. Nature 266:530
Pettitt JM (1980) Reproduction in seagrasses: nature of the pollen and receptive surface
    of the stigma in the Hydrocharitaceae. Ann Bot 45:257–271
Pettitt JM, McConchie CA, Ducker SC, Knox RB (1980) Unique adaptations for subma-
    rine pollination in seagrasses. Nature 286:487–489
Pettitt JM, Ducker SC, Knox RB (1981) Submarine pollination. Sci Am 244:134–141
Picton JM, Steer M (1981) Determination of secretory vesicle production rates by dictyo-
    somes in pollen tubes of *Tradescantia* using cytochalasin. J Cell Sci 49:261–272
Pluijm JE van der, Linskens HF (1966) Feinstruktur der Pollenschläuche im Griffel
    von *Petunia*. Zuechter 36:220–224
Portnoi L, Horovitz A (1977) Sugars in natural and artificial pollen germination sub-
    strates. Ann Bot 41:21–27
Preil W, Keyser D (1975) Pollen germination on exposed ovules in *Begonia semperflorens*.
    In: Mulcahy DL (ed) Gamete competition in plants and animals. North-Holland,
    Amsterdam, pp 219–225
Raff JW (1981) Cell recognition in *Prunus avium* L. PhD Thesis, Univ Melbourne
Raff JW, Clarke AE (1981) Tissue specific antigens secreted by suspension cultured
    callus cells of *Prunus avium* L. Planta (in press)
Raff JW, Hutchinson JF, Knox RB, Clarke AE (1979) Cell recognition: Antigenic deter-
    minants of plant organs and their cultured callus cells. Differentiation 12:179–186
Raff JW, Knox RB, Clarke AE (1981) Style and S-allele-associated antigens of *Prunus
    avium*. Planta 153:124–129
Raff JW, Pettitt JM, Knox RB (1982) Cytochemistry of pollen tube growth in the stigma
    and style of *Prunus avium*. Phytomorphology (in press)
Rawitscher-Kunkel E, Machlis L (1961) The hormonal regulation of sexual reproduction
    in *Oedogonium*. Am J Bot 49:177–183
Reynolds GET (1971) In: Alner DJ (ed) Aspects of adhesion. Univ of London Press,
    London, pp 96–111

Roberts IN, Dickinson HG (1982) Intraspecific incompatibility on the stigma of *Brassica*. Phytomorphology (in press)

Roberts IN, Stead AD, Dickinson HG (1979a) No fundamental changes in lipids of the pollen grain coating of *Brassica oleracea* following either self- or cross-pollination. Incompatibility Newsl 11:77–83

Roberts IN, Stead AD, Ockendon DJ, Dickinson HG (1979b) A glycoprotein associated with the acquisition of the self-incompatibility system by maturing stigmas of *Brassica oleracea*. Planta 146:179–183

Roberts IN, Stead AD, Ockendon DJ, Dickinson HG (1980) Pollen stigma interactions in *Brassica oleracea*. Theor Appl Genet 58:241–246

Roggen HPJR (1972) Scanning electron microscopical observations on compatible and incompatible pollen–stigma interactions in *Brassica*. Euphytica 21:1–10

Roland F (1971) Characterization and extraction of the polysaccharides of the intine and of the generative cell wall in the pollen grains of some Ranunculaceae. Grana 11:101–106

Rosen WG (1964) Chemotropism and fine structure of pollen tubes. In: Linskens HF (ed) Pollen Physiology and Fertilization. North-Holland, Amsterdam, pp 159–169

Rosen WG (1971) Pistil–pollen interactions in *Lilium*. In: Heslop-Harrison J (ed) Pollen, physiology and development. Butterworths, London, pp 239–254

Rosen WG, Thomas HR (1970) Secretory cells of lily pistils. I. Fine structure and function. Am J Bot 57(9):1108–1114

Rosen WG, Gawlick SR, Dashek WV, Siegesmund KA (1964) Fine structure and cytochemistry of *Lilium* pollen tubes. Am J Bot 51:61–71

Rosenfield C-L, Loewus F (1975) Carbohydrate interconversions in pollen–pistil interactions of the lily. In: Mulcahy DL (ed) Gamete competition in plants and animals. North-Holland, Amsterdam, pp 151–160

Rosenfield C-L, Matile Ph (1979) Glycosidases in pear pollen tube development. Plant Cell Physiol 20:605–613

Rowley JR (1981) Pollen wall characters with emphasis upon applicability. Nord J Bot 1:357–380

Rowley JR, Dahl AO (1977) Pollen development in *Artemisia vulgaris* with special reference to glycocalyx material. Pollen Spores 19:169–284

Rowley JR, Flynn JJ (1966) Single-stage carbon replicas of microspores. Stain Technol 41:287–290

Rowley JR, Dahl AO, Rowley JS (1980) Coiled construction of exinous units in pollen of *Artemisia*. In: Bailey GW (ed) 38th Ann Proc Electron Microscopy Soc Am Claitors, Baton Rouge, La, pp 252–253

Rowley JR, Dahl AO, Sengupta S, Rowley JS (1981a) A model of exine substructure based on dissection of pollen and spore exines. Palynology 5:107–152

Rowley JR, Dahl AO, Rowley JS (1981b) Substructure in exines of *Artemisia vulgaris* (Asteraceae). Rev Palaeobot Palynol 35:1–38

Russell SD (1979) Fine structure of megagametophyte development in *Zea mays*. Can J Bot 57:1093–1110

Russell SD (1980) Participation of male cytoplasm during gamete fusion in an angiosperm, *Plumbago zeylanica*. Science 210:200–201

Russell SD, Cass DD (1981a) Ultrastructure of the sperms of *Plumbago zeylanica*. I Cytology and association with the vegetative nucleus. Protoplasma 107:85–107

Russell SD, Cass DD (1981b) Ultrastructure of fertilization in *Plumbago zeylanica*. Acta Soc Bot Polon 50:185–190

Sampson DR (1977) Dynamics of unstable sporophytic self-incompatibility systems. Can J Genet Cytol 19(1):153–165

Sanger JM, Jackson WT (1971) Fine-structure study of pollen development in *Haemanthus katherinae* Baker. J Cell Sci 8:303–315

Sassen MMA (1974) The stylar transmitting tissue. Acta Bot Neerl 23:99–108

Schill R, Jäckel U (1978) Beitrag zur Kenntnis der Asclepiadaceen-Pollinarien. Akad Wiss Lit Mainz, Trop Subtrop Pflanzenwelt 22:53–170

Schlösser K (1961) Cytologische und cytochemische Untersuchungen über das Pollenschlauchwachstum selbststeriler Petunien. Z Bot 49:266–288

Schnepf E, Christ P (1980) Unusual transfer cells in the epithelium of the nectaries of *Asclepias curassavica* L. Protoplasma 105:135–148

Schnepf E, Witzig F, Schill R (1979) Über Bildung und Feinstruktur des Translators der Pollinarien von *Asclepias curassavica* und *Gomphocarpus fruticosus* (Asclepiadaceae). Akad Wiss Lit Mainz, Trop Subtrop Pflanzenwelt 25:1–39

Schrauwen J, Linskens HF (1967) Technique for germinating *Petunia* pollen. Acta Bot Neerl 16:177

Sears M (1937) Cytological phenomena connected with self-sterility in the flowering plants. Genetics 22:130–181

Sedgley M (1974a) Assessment of serological techniques for S-allele identification in *Brassica oleracea*. Euphytica 23:543–549

Sedgley M (1974b) The concentration of S-protein in stigmas of *Brassica oleracea* plants homozygous and heterozygous for a given S-allele. Heredity 33:412–419

Sedgley M (1975) Flavonoids in pollen and stigma of *Brassica oleracea* and their effect on pollen germination in vitro. Ann Bot 39:1091–1095

Sedgley M (1977) Reduced pollen tube growth and the presence of callose in the pistil of the male floral stage of the avocado. Sci Hort 7:27–36

Sedgley M (1979) Structural changes in the pollinated and unpollinated avocado stigma and style. J Cell Sci 38:49–60

Sedgley M (1981) Ultrastructure and histochemistry of the watermelon stigma. J Cell Sci 48:137–146

Sedgley M (1982) Anatomy of the unpollinated and pollinated watermelon stigma. J Cell Sci 54:341–355

Sedgley M, Buttrose M S (1978) Structure of the stigma and style of the avocado. Aust J Bot 26:663–682

Sedgley M, Scholefield PB (1980) Stigma secretion in the watermelon before and after pollination. Bot Gaz 141:428–434

Shafaat M, Shaban R (1980) Metaxenic effects in date palm fruit. Beitr Trop Landwirtsch Veterinarmed 18:117–123

Sharma S, Singh MB, Malik CP (1981) Relation of glycosidases to *Amaryllis vittata* pollen tube growth. Plant Cell Physiol 22:81–86

Shaw G (1971) The chemistry of sporopollenin. in: Brooks J, Grant PR, Muir M, Gijsel P van, Shaw G (eds) Sporopollenin. Academic Press, London New York, pp 305–350

Shayk M, Kolattukudy PE (1977) Production of a novel extracellular cutinase composition and ultrastructure of the stigma cuticle of *Nasturtium*. Plant Physiol 60:907–915

Shchori Y (1969) Self-incompatibility studies in sunflower. Res Rep Hebrew Univ Jerusalem Sci Agric 1:549 (abstr)

Shiraishi M, Matsumoto K, Shigemasu A (1976) Morphological studies on fertilization and development of citrus fruit. I Ultrastructural examination of papillary cells of the stigma of satsuma mandarin (*Citrus unshui* Marc.) J Jpn Soc Hortic Sci 45:231–237

Shivanna KR (1982) Pollen–pistil interaction and control of fertilization. In: Johri BM (ed) Experimental embryology of vascular plants. Springer, Berlin Heidelberg New York, pp 131–174

Shivanna KR, Heslop-Harrison J (1978) Inhibition of the pollen tube in the self-incompatibility response of grasses. Incompatibility Newsl 10:5–7

Shivanna KR, Heslop-Harrison J (1981) Membrane state and pollen viability. Ann Bot 47:759–770

Shivanna KR, Sastri DC (1981) Stigma surface esterase activity and stigma receptivity in some taxa. Ann Bot 47:53–64

Shivanna KR, Jaiswal VS, Mohan Ram HY (1974) Inhibition of gamete formation by cycloheximide in pollen tubes of *Impatiens balsamina*. Planta 117:173–177

Shivanna KR, Johri BM, Sastri DC (1979) Development and physiology of angiosperm pollen. Today and Tomorrow, New Delhi

Shivanna KR, Heslop-Harrison J, Heslop-Harrison Y (1981) Heterostyly in *Primula*. 2 Sites of pollen inhibition, effects of pistil constituents on compatible and incompatible pollen tube growth. Protoplasma 107:319–337

Shukla SN, Tewari MN (1974) Antagonism between plant growth regulators in pollen tube elongation of *Calotropis procera*. Experientia 30:495

Singh MB, Marginson R, Knox RB (1983) β-galactosidase-deficient pollen of *Brassica*: evidence for gametophytic transcription of the enzyme and role in stigma interactions. Proc Nat Acad Sci USA (in press)

Sitte P (1953) Untersuchungen zur submikroskopischen Morphologie der Pollen und Sporenmembranen. Mikroskopie 8:290–291

Sondheimer E, Linskens HF (1974) Control of in vitro germination and tube extension of *Petunia hybrida* pollen. Proc K Ned Akad Wet 77 C:116–124

Southworth D (1973) Cytochemical reactivity of pollen walls. J Histochem Cytochem 21:73

Soutworth D (1974) Solubility of pollen exines. Am J Bot 61:36–44

Southworth D (1975) Lectins stimulate pollen germination. Nature 258:600–602

Stanley RG, Linskens HF (1964) Enzyme activation in germinating *Petunia* pollen. Nature 203:542–544

Stanley RG, Linskens HF (1974) Pollen. Springer, Berlin Heidelberg New York

Stead AD, Roberts IN, Dickinson HG (1979) Pollen–pistil interactions in *Brassica oleracea*: events prior to pollen germination. Planta 146:211–216

Stead AD, Roberts IN, Dickinson HG (1980) Pollen–stigma interaction in *Brassica oleracea*: the role of stigmatic proteins in pollen grain adhesion. J Cell Sci 42:417–423

Stösser R, Anvari SF (1981) Das Wachstum der Pollenschläuche im Fruchtknotengewebe von Kirschen. Gartenbauwissenschaft 46:154–158

Stout AB, Chandler C (1933) Pollen-tube behaviour in *Hemerocallis* with special reference to incompatibilities. Torrey Bot Club 60:397–416

Strasburger E (1886) Über fremdartige Bestäubung. Jahrb Wiss Bot 17:50–95

Straub J (1946) Zur Entwicklungsphysiologie der Selbststerilität von *Petunia*. Z Naturforsch 1:287–291

Thompson MM (1979) Growth and development of the pistillate flower and nut in "Barcelona" Filbert. J Am Soc Hortic Sci 104:427–432

Tilton VR, Horner HT (1980) Stigma, style and obturator of *Ornithogalum caudatum* (Liliaceae) and their function in the reproductive process. Am J Bot 67:1113–1131

Tsinger VN, Petrovskaya-Baranova TP (1961) The pollen grain wall – a living physiologically active structure. Dokl Akad Nauk SSSR 138:466–469

Uwate WJ, Lin J (1981a) Development of the stigmatic surface of *Prunus avium* L., sweet cherry. Am J Bot 68:1165–1176

Uwate WJ, Lin J (1981b) Tissue development in the stigma of *Prunus avium* L. Ann Bot 47:41–51

Vasiliev AE (1970) Ultrastructure of stigmatoid cells in *Lilium*. Fisiolog Rastenii 17:1240–1248

Villeret S (1974) Sur l'évolution des acides gras majeurs des fleurs d'*Iris pseudacorus* L. CR Acad Sci (Paris) 278:1213–1216

Vithanage HIMV, Knox RB (1976) Pollen-wall proteins: quantitative cytochemistry of the origins of intine and exine enzymes in *Brassica oleracea*. J Cell Sci 21:423–435

Vithanage HIMV, Knox RB (1977) Development and cytochemistry of stigma surface and response to self- and foreign pollination in *Helianthus annuus*. Phytomorphology 27:168–179

Vithanage HIMV, Knox RB (1979) Pollen development and quantitative cytochemistry of exine and intine enzymes in sunflower, *Helianthus annuus* L. Ann Bot 44:95–106

Vithanage HIMV, Gleeson PA, Clarke AE (1980) The nature of callose produced during self-pollination in *Secale cereale*. Planta 148:498–509

Vithanage HIMV, Howlett BJ, Jobson S, Knox RB (1982) Immunocytochemical localization of water soluble glycoproteins, including Group 1 allergen, in pollen of ryegrass, *Lolium perenne* using ferritin labelled antibody. Histochem J 14:949–966

Vuilleumier BS (1966) The origin and evolutionary development of heterostyly in angiosperms. Evolution 21:210–226

Walker J (1974) Aperture evolution in the pollen of primitive angiosperms. Am J Bot 61:1112–1137

Watanabe T, Motomura Y, Aso K (1961) (in Japanese) Tohoku J Agr Res 12:173–185
Webster BD, Tucker CL, Lynch SP (1977) A morphological study of the development
    of reproductive structures of *Phaseolus vulgaris* L. J Am Soc Hortic Sci 102:640–643
Wiermann R (1979) Stage-specific phenylpropanoid metabolism during pollen develop-
    ment. In: Luckner M, Schreiber K (eds) Regulation of secondary product and plant
    hormone metabolism. Pergamon, Oxford New York, pp 231–239
Wilkinson HP (1979) The plant surface. In: Metcalfe CR, Chalk L (eds) Anatomy of
    the dicotyledons, 2nd edn Clarendon, Oxford, pp 97–165
Williams EG, Knox RB (1981) Quantification of the growth of compatible and incompati-
    ble pollen tubes in the style of *Lycopersicum peruvianum*. Austr Plant Breed Genet
    Newsl
Williams EG, Knox RB, Rouse JL (1982a) Pollination sub-systems distinguished by
    pollen tube arrest after incompatible interspecific crosses in *Rhododendron* (Ericaceae).
    J Cell Sci 53:255–277
Williams EG, Ramm-Anderson S, Dumas C, Mau S-L, Clarke AE (1982b) The effect
    of isolated components of *Prunus avium* L. styles on in vitro growth of pollen tubes.
    Planta 156:517–519
Williams EG, Knox RB, Rouse JL (1983) Pollen–pistil interactions and the control of
    pollination. Phytomorphology (in press)
Wottiez RD, Willemse MJM (1979) Sticking of pollen on stigmas, the factors and a
    model. Phytomorphology 29:57–63
Woude WJ van der, Morre DJ (1967) Endoplasmic reticulum-dictyosome-secretory vesicle
    associations in pollen tubes of *Lilium longiflorum* Thunb Proc Indiana Acad Sci
    77:164–170
Woude WJ van der, Morre DJ, Bracker CE (1971) Isolation and characterization of
    secretory vesicles in germinated pollen of *Lilium longiflorum*. J Cell Sci 8:331–351
Yamada Y (1965) Studies on the histological and cytological changes in the tissues of
    the pistil after pollination. Jpn J Bot 19:69–82
Yamashita K (1978) Studies on self-incompatibility of hyuganatsu, *Citrus tamurana* Hort.
    ex Tanaka. I. Pollen behaviour on stigmas and pollen tube growth in styles observed
    under a scanning electron microscope and a fluorescent microscope. J Jpn Soc Hortic
    Sci 47:188–194
Zandonella P, Dumas C, Gaude T (1981) Sécrétions et biologie florale. I Nature, origine
    et role des sécrétions dans la pollinisation et la fécondation: revue des données ré-
    centes. Apidologie 12:383–396
Zeisler M (1938) Über die Abgrenzung der eigentlichen Narbenfläche mit Hilfe von
    Reaktionen. Beih Bot Zbl A 58:308–318
Zeitz SJ, McClure BS, Van Arsdel PP (1971) Specific response of human lymphocytes
    to pollen antigen in tissue culture. J Allergy 36:197–206
Zenkteler M (1970) Test-tube fertilization on ovules in *Melandrium album* Mill. with
    pollen grains of *Datura stramonium* L. Experientia 26:661–662
Ziegler H (1959) Über die Zusammensetzung des „Bestäubungstropfens" und den Me-
    chanismus seiner Sekretion. Planta 52:587–599
Zolvitsch G, Secenska M (1963) (in Bulgarian) CR Acad Bulg Sci 16:105–111

25 Mentor Effects in Pollen Interactions[1]

R. F. STETTLER and A. A. AGER

25.1 Introduction

In his extensive hybridization program involving intergeneric, interspecific, and interracial crosses, the Russian plant breeder MICHURIN (1855–1935) made use of several techniques to improve breeding success. One of these was to add to the foreign pollen a small amount of pollen of the maternal species. The maternal pollen "very likely has the capacity to more easily stimulate the pistil for the act of fertilization and, one would think, to introduce with itself the alien pollen" (MICHURIN 1948, p. 122). Subsequently referred to as mentor effect, it took several decades before critical evidence was available to substantiate the phenomenon. This delay was, in part, due to the ideological rift which for more than 30 years had set Soviet genetics apart from the scientific developments in the western world. In the meantime, the use of pollen mixes formed part of the methodological repertoire of Russian plant breeders, notably those involved in remote hybridization (TSITSIN 1962, KARPOV 1966).

The past 20 years have seen a steady flow of studies concerned with mentor pollen. They have followed a dual pathway; on the one hand, the exploration by geneticists and plant breeders of the applicability of the concept to an ever-broadening array of plant taxa; and, on the other hand, the identification by plant physiologists of underlying mechanisms and their relationship to incompatibility and incongruity mechanisms.

This review will reflect the duality of approaches to the study of mentor effects. Its focus is on phenomena associated with the use of mentor pollen; i.e. treated or untreated pollen, or substances derived from it, intended to facilitate normally difficult-to-achieve matings. The subject has been discussed in two recent reviews by DE NETTANCOURT (1977) and by CLARKE and KNOX (1978). The reader is further referred to other sections of this volume, notably the chapters dealing with pollen–stigma interactions (KNOX, Chap. 24), incompatibility (DE NETTANCOURT, Chap. 26) and incongruity (HOOGENBOOM, Chap. 27).

25.2 Survey of Cases Reported

Table 1 lists the various experiments reported in the literature dealing with pollen mixtures or substitution treatments of pollen, but not including stigmatic

---

1 This review was concluded in March 1982

Table 1. Synopsis of studies reporting mentor effects[a]

Attempted mating	Pollen treatment	Outcome	Reference
Intergeneric			
Festuca spp. × *Poa* spp.	Pre-pollination with killed (UV or long storage) pollen	Unspecified barrier(s) not overcome	MATZK et al. 1980
Lolium spp. × *Poa* spp. Lolium	Pre-pollination with killed (UV or long storage) pollen	Unspecified barrier(s) not overcome	MATZK et al. 1980
Festuca arundinacea × *Dactylis glomerata*	Pre-pollination with killed (UV or long storage) pollen	Successful hybridization not attributable to pollen treatment	MATZK 1981
Malus sp. × *Pyrus* sp. (several cultivars)	PMX or "pioneer pollen" with IRR pollen	Stimulated parthenocarpic fruit, no seed set	VISSER 1981
	PMX or "pioneer pollen" with MeOH pollen	No response	VISSER 1981
Interspecific			
Populus trichocarpa × *P.* spp.	PMX with IRR pollen	Stigmatic and stylar barriers overcome, verified hybrids	STETTLER 1968
Populus spp.	PMX with IRR pollen	Stigmatic and other barriers overcome, verified hybrids	KNOX et al. 1972b
	PMX with FT pollen	Stigmatic and other barriers overcome, verified hybrids	KNOX et al. 1972b
Populus deltoides × *P. alba*	PMX with IRR pollen	Stigmatic barrier overcome, verified hybrids	KNOX et al. 1972a
	PMX with MeOH pollen	Stigmatic barrier overcome, verified hybrids	KNOX et al. 1972a
	PMX with FT pollen	Stigmatic barrier overcome, verified hybrids	KNOX et al. 1972a
	Incompatible pollen mixed with protein extracts from pollen of maternal sp.	Stigmatic barrier overcome, verified hybrids	KNOX et al. 1972a
Populus spp.	PMX with IRR pollen	Stigmatic and stylar barriers overcome, verified hybrids	WILLING and PRYOR 1976

	Incompatible pollen coated with solvent-extracted tryphine from pollen of maternal sp.	Stigmatic and stylar barriers overcome, verified hybrids	WILLING and PRYOR 1976
	Incompatible pollen washed with anhydrous non-polar solvents	Stigmatic and stylar barriers overcome, verified hybrids	WILLING and PRYOR 1976
Populus trichocarpa × P. spp.	PMX with IRR pollen	Stigmatic and stylar barriers overcome, verified hybrids	STETTLER and GURIES 1976
	PMX with MeOH or IRR pollen	Stigmatic and stylar barriers not overcome	STETTLER and GURIES 1976
Populus spp.	PMX with IRR pollen	Stigmatic and stylar barriers overcome, verified hybrids	STETTLER et al. 1980
	Incompatible pollen washed with n-hexane	Stigmatic and stylar barriers not overcome	STETTLER et al. 1980
Nicotiana spp.	PMX with IRR pollen	Stylar barriers overcome, verified hybrids	PANDEY 1977
	PMX with IRR incompatible pollen of other Nicotiana spp.	Stylar barriers overcome, verified hybrids	PANDEY 1977
	Unmixed partially viable FT pollen	Stylar barriers overcome, verified hybrids	PANDEY 1977
Passiflora spp.	PMX with UT pollen	Fruit set stimulated, sporadic seed set, verified hybrids	PAYÀN and MARTIN 1975
Sesamum indicum × S. mulayanum	PMX with MeOH pollen	Germination barrier overcome, subsequent stylar barrier prevented fertilization	SASTRI and SHIVANNA 1976a

a Abbreviations: PMX = pollen mix containing pollen of the desired cross and compatible mentor pollen from the maternal species. Instances where mentor is incompatible and/or from another species are specifically noted

            SPMX = self-pollen mix containing self pollen and compatible mentor pollen from the maternal species. Other instances are noted as for PMX

            IRR = pollen subjected to gamma irradiation

            FT = pollen subjected to repeated freezing and thawing

            MeOH = pollen washed in anhydrous methanol

            UT = untreated pollen

Table 1 (continued)

Attempted mating	Pollen treatment	Outcome	Reference
Pinus sylvestris × *P. nigra* and reciprocal	PMX with IRR pollen	Nucellar barrier overcome, hybrids obtained	VIDAKOVIĆ 1977
	PMX with IRR incompatible pollen	Nucellar barrier overcome, hybrids obtained	VIDAKOVIĆ 1977
Ipomoea spp.	PMX with IRR pollen	Stigmatic-stylar barrier overcome in one cross only, subsequent barriers prevented fertilization	GURIES 1978
Cucumis spp.	PMX with IRR pollen	Stylar barrier overcome, fruit set stimulated, embryos aborted	DEN NIJS and OOST 1980
Cucumis africanus × *C. metuliferus* and reciprocal	PMX with IRR pollen	Stylar barrier overcome, hybrid seedlings (*C. metuliferus* × *C. africanus*) from explanted embryos	DEN NIJS et al. 1980
Cucumis spp.	PMX with IRR pollen	Unaided crosses successful, but higher fruit set and lower seed set with PMX	CUSTERS et al. 1981
Cucumis spp.	Compatible pollen of maternal species preceded by IRR compatible self-pollen	Percent empty seed increased with delayed application of untreated compatible pollen	DEN NIJS 1981
Trifolium spp.	PMX with MeOH pollen	Unaided crosses successful, fewer seeds and hybrids obtained with PMX	TAYLOR et al. 1980
Intraspecific Sporophytic self-incompatibility			
Theobroma cacao	SPMX with UT compatible pollen	Ovular barrier overcome, putative selfs	POSNETTE 1940
	SPMX with UT compatible pollen	Ovular barrier overcome, putative selfs	VAN DER KNAAP 1955
	SPMX with UT compatible pollen	Ovular barrier overcome, verified selfs	GLENDINNING 1960
Chrysanthemum cinerariaefolium	SPMX with "dead" pollen, treatment unspecified	Pollen germination barrier not overcome	BREWER and HENSTRA 1974
Cosmos bipinnatus	SPMX with IRR pollen, or pollen-wall diffusate	Incomplete stigmatic barrier overcome, verified selfs	HOWLETT et al. 1975
	SPMX with IRR incompatible pollen, or pollen-wall diffusate	No response, seed set same as self-pollination	HOWLETT et al. 1975

Species	Treatment	Result	Reference
Brassica oleracea	SPMX of UT compatible pollen and IRR incompatible pollen of maternal sp.	No response, seed set same as self-pollination	Howlett et al. 1975
	Selfing preceded by stigma treatment with ether extracts from pollen of *Brassicus napus*	Pollen germination barrier overcome, self-seed set	Roggen 1975
Gametophytic self-incompatibility			
Lotus tenuis	Unemasculated flowers pollinated with incompatible pollen of another *Lotus* sp.	Seed set enhanced over that of self-pollination	Grant et al. 1962
Lotus corniculatus	Self-pollination followed by mix of self and cross-pollen	Seed set enhanced over that of self-pollination	Grant et al. 1962
Nicotiana forgetiana and *N. alata*	SPMX with IRR pollen of maternal or other *Nicotiana* spp.	Parthenogenetic progeny containing marker genes from treated pollen	Pandey 1975
Nicotiana spp.	SPMX with IRR pollen of maternal or other *Nicotiana* spp.	Variable results, verified selfs in spp. with stylar barriers, no success with species having barrier in stigma or upper style	Pandey 1977
Nicotiana alata	SPMX with IRR pollen	Stylar barrier overcome, verified selfs	Ramulu et al. 1977
	SPMX with IRR incompatible pollen of *N. forgetiana*	Stigmatic and stylar barriers overcome, flowers not brought to seed	Pandey 1978
	SPMX with IRR pollen	Stylar barrier overcome, self-seed set	Ramulu et al. 1979
	SPMX with IRR incompatible pollen	Stylar barrier not overcome	Ramulu et al. 1979
Petunia hybrida	SPMX with UT or IRR incompatible pollen from other families	Self- and cross-pollen tubes inhibited by pollen tubes of other families	Gilissen and Linskens 1975
	Petunia hybrida pollen mixed with UT incompatible pollen of *Nicotiana alata*	Self- and cross-pollen tubes inhibited by pollen tubes of other families	Gilissen and Linskens 1975
	SPMX with "recognition" pollen (treatment unspecified)	Incomplete stylar barrier overcome, self-seed set	Sastri and Shivanna 1976b
Malus sp. (McIntosh)	SPMX with MeOH pollen of Richared Delicious	Stylar barrier overcome, self-seed set	Dayton 1974
Malus sp. (Crawley Beauty)	SPMX with MeOH compatible pollen of other varieties	Stylar barrier not overcome	Williams and Church 1975

Table 1 (continued)

Attempted mating	Pollen treatment	Outcome	Reference
Malus sp. (several varieties)	SPMX with IRR or MeOH pollen of Golden Delicious	Incomplete stylar barrier not overcome	SANSAVINI et al. 1979
Malus sp. (several varieties)	SPMX or "pioneer pollen" with IRR pollen of other varieties	Variable results: either stylar barrier overcome and seed set, or parthenocarpic fruit set stimulated	VISSER 1981
	SPMX or "pioneer pollen" with MeOH pollen of other varieties	Variable results, either no response, or parthenocarpic fruit set stimulated	VISSER 1981
Lilium longiflorum (Ace, Nellie White)	SPMX with IRR pollen of another variety	Stylar barrier not overcome	FETT et al. 1976
	Incompatible pollen mixed with protein extracts from compatible pollen	Stylar barrier not overcome	FETT et al. 1976
Lycopersicum hybrid (*L. esculentum* × *L. peruvianum*)	SPMX with IRR pollen of *L. peruvianum*	Barriers not overcome	RAMULU et al. 1977
	SPMX with IRR incompatible pollen of another genus, species, or variety	Barriers not overcome	RAMULU et al. 1977
Oenothera organensis	SPMX with IRR pollen	Stylar barrier not overcome	RAMULU et al. 1979
	SPMX with IRR incompatible pollen	Stylar barrier not overcome	RAMULU et al. 1979
Pyrus sp. (several varieties)	SPMX or "pioneer pollen" with IRR pollen of another variety	Parthenocarpic fruit set stimulated, self-seed set	VISSER 1981
	SPMX or "pioneer pollen" with MeOH pollen of another variety	Variable results, either no response or self-seed set	VISSER 1981
Pyrus sp. (Doyenné du Comice)	SPMX with IRR pollen of Conference	Stylar barrier overcome, flowers not brought to seed	VISSER and OOST 1981
	SPMX with UT pollen of Conference	Stylar barrier overcome, flowers not brought to seed	VISSER and OOST 1981
	SPMX with MeOH pollen of Conference	No response	VISSER and OOST 1981
	SPMX with MeOH "pioneer pollen" of other varieties	No response	VISSER and OOST 1981
	SPMX with IRR "pioneer pollen" of other varieties	Stylar barriers overcome, tube growth enhanced over that of simultaneous pollination with PMX	VISSER and OOST 1981

treatments (covered by KNOX in Chap. 24, this Vol.). The experiments are grouped by genus and according to the type of mating and self-incompatibility system.

If wide hybridization, as practiced by the Michurin school of plant breeders, was the original motive for the use of mentor pollen, increasing the yield of cocoa plantations provided a further impetus. By the 1930's it had become obvious that clones of *Theobroma cacao* differed significantly in their fruit set and that much of this variation was attributable to the self-incompatibility of certain clones. In an effort to identify the factors responsible for the difference, POSNETTE (1940) conducted selfings and crosses on a number of clones. After having earlier found incompatible and compatible pollen to germinate equally well on the stigma, he attempted a microscopic study of pollen-tube growth, but had difficulties in staining the tubes properly. He then devised an indirect approach by conducting double pollinations, in which he applied compatible pollen with incompatible pollen either simultaneously, or, alternatively, 2 h earlier or later. He found that pods could be set on an incompatible tree by this method, but that the addition of incompatible pollen reduced the set even when applied 2 h later. He suspected an inhibitory influence of the incompatible upon the compatible pollen, and suggested a critical experiment to verify selfing, making use of genetic markers. Such experiments were later conducted by VAN DER KNAAP (1955) and GLENDINNING (1960), and their results gave more direct evidence for selfing having, in fact, occurred, although pseudogamy could not be ruled out as an alternative explanation in either case.

A similar case of self-facilitation was described in gametophytically self-incompatible *Lotus* in which attempted interspecific crosses of unemasculated flowers had largely yielded selfs (GRANT et al. 1962). It was suggested that the foreign pollen had in some manner stimulated fertilization by contaminant pollen of the maternal parent. This was later verified in a more critical test by MIRI and BUBAR (1966), using several clones of *Lotus corniculatus* with known degrees of self-sterility and carrying genetic markers. High proportions of self seeds were obtained after plants had first been selfed and subsequently pollinated with a 1:1 mixture of self- and cross-pollen. The experiments were to simulate bee pollinations in the field where multiple pollinations may be common. It was concluded that concurrent or delayed application of compatible pollen had stimulated tube growth of self-pollen and reduced pod abortion.

A drawback inherent in the combined use of incompatible with live compatible pollen was the competitive advantage of the latter in the occupation of ovule sites and in embryogeny. A treatment devised to circumvent this "genetic contamination" was the exposure of mentor pollen to high levels of ionizing radiation, which had been found to impair pollen chromatin selectively without significantly affecting pollen-tube growth (reviewed by BREWBAKER and EMERY 1961). This method was first applied in wide hybridization trials in *Populus* (STETTLER 1968) and resulted in the production of hybrids among several species from different sections of the genus. These experiments were successfully repeated with other species of poplars by ZUFA (1971), KNOX et al. (1972b), WILLING and PRYOR (1976), HAMILTON (1976), and further elaborated upon by GURIES and STETTLER (1976), STETTLER and GURIES (1976), and STETTLER et al. (1980).

KNOX et al. (1972b) also showed that compatible pollen killed[2] with methanol, or by repeated freezing and thawing, was as effective as irradiated pollen in providing mentor functions in poplar. More importantly, these workers extracted rapidly leachable proteins from the pollen walls of compatible pollen, mixed them with incompatible pollen, and were able to obtain hybrid seedlings. This evidence suggested that the wall-held proteins served as "recognition substances" and that foreign pollen could "borrow" them and thus gain access to the ovules. Additional indirect evidence for a recognition event at the pollen–stigma interface was recently obtained in chemical analyses of pollen diffusates from *Populus deltoides* and *P. alba,* yielding wall-held glycoproteins (ASHFORD and KNOX 1980), a group of proteins having been implicated widely in recognition reactions in plants (CLARKE and KNOX 1978).

Studies by WILLING and PRYOR (1976) involving the washing of pollen or stigma with ethylacetate or anhydrous non-polar solvents directed attention to another group of compounds serving the mentor function, namely lipoidal materials. The coating of incompatible pollen with solvent-extracted "tryphine" from compatible pollen resulted in successful crosses. HAMILTON (1976) verified that n-hexane treatment of stigmas or pollen permitted incompatible pollen to engage in fertilization. He further showed lipoidal materials to reside in "waxy mounds" on the stigmatic papillae, and to differ chemically among breeding groups of poplar species. Finally he found different esterase isozymes in pollen diffusates of different species, verified later by ASHFORD and KNOX (1980), suggesting that compatibility involved a certain affinity between pollen enzymes and the stigmatic surface lipids they break down.

While all these investigations had focussed on surface barriers, several other studies showed that most incompatible poplar crosses displayed normal stigmatic penetration by the pollen tube but a slower tube growth in the style than compatible crosses (STETTLER and GURIES 1976, GURIES and STETTLER 1976, STETTLER et al. 1980). Furthermore, in one of the few combinations with a seeming surface barrier, irradiated mentor pollen grains did not seem to facilitate stigma penetration of neighbouring foreign pollen grains. In many combinations this mentor pollen stimulated ovule and pistil growth, as well as inflorescence retention, even in the absence of embryo development, whereas foreign pollen failed to do so.

The integration of all these results from research on poplars is rendered difficult by the differences in materials and methods among the various researchers and the constraints on controls inherent in experimentation with trees. The available evidence in poplars points to mentor effects being mediated both via pollen–stigma and pollen–pollen interactions as well as having distal impact on ovules and pistils preceding the arrival of pollen tubes.

The recognition phenomenon involved in mentor effects first described for interspecific relationships in poplars by KNOX et al. (1972b), was soon found to be also operative in the case of sporophytic self-incompatibility in *Cosmos bipinnatus* (HOWLETT et al. 1975). Mixing either killed compatible pollen or wall diffusates from compatible pollen with live incompatible pollen was shown

---

2 We distinguish *inviable* pollen (capable of tube growth, but genetically inert) from *killed* pollen (unable to germinate)

to partially overcome self-incompatibility. Loosely held pollen-exine proteins that are released immediately after pollen alights on the stigmatic surface were demonstrated as recognition agents. Extending this concept to a gametophytic self-incompatibility system FETT et al. (1976) examined the role of these loosely bound wall substances in *Lilium longiflorum* cv. Ace. They found that neither the removal of these compounds in incompatible pollen nor the mixing of killed compatible pollen or its extract with incompatible pollen improved tube growth of the latter. Only heat treatment of the style was able to break the incompatibility response. As is well known, recognition in the lily system does not occur at the stigma surface but part way down the style (DE NETTANCOURT 1979).

However, it had been found earlier that mentor pollen was effective in several other cases of gametophytic self-incompatibility such as in *Lotus* (GRANT et al. 1962) and *Malus* (DAYTON 1974). To shed more light on this system, PANDEY conducted a series of studies making use of several species of *Nicotiana* in which the underlying genetics was well understood (1975, 1977, 1978). He found that both self- and interspecific incompatibility could be overcome by the use of radiation-killed compatible mentor pollen, and attributed it to a hypothetical pollen-growth-promoting substance (PGS) which was transferable from treated compatible to untreated incompatible pollen. An additional and somewhat unexpected finding was that a mentor effect was exerted even by radiation-killed *in*compatible pollen of *N. langsdorffii* upon admixed untreated pollen of the same species in the styles of *N. forgetiana,* resulting in normally unobtainable hybrids between the two species (PANDEY 1977). In subsequent efforts to elucidate the quantitative variation in incompatibility responses, PANDEY (1978) found that the strength of the mentor effect was unique to the donor/recipient combination and not necessarily correlated with the length of style of the donor species. From these results he hypothesized that the donor pollen releases a regulator substance which is non-specific and can be picked up by the recipient pollen where it triggers the production of PGS, and thereby renders it compatible (1979). More recent experiments by RAMULU et al. (1979) confirmed the mentor effect of irradiated compatible pollen, but not that of irradiated incompatible pollen, in intraspecific crosses of *N. alata.*

Another puzzling result in tobacco after use of irradiated pollen (100 krad) was the appearance of some maternal-looking progenies carrying, however, a few characters derived from the irradiated pollen donor (PANDEY 1975). Initially found in *N. forgetiana* (with mentor pollen of *N. alata*) and ascribed to gametic transformation, the phenomenon has later been observed also in *N. rustica* by VIRK et al. (1977) and in several other species by PANDEY (1979). Fragments of the irradiated genome were seemingly transferred to a stimulated egg cell and permanently incorporated, as indicated by subsequent progeny tests (see PANDEY 1979, for detailed discussion).

In the meantime, efforts had been made to test the usefulness of the mentor-pollen principle in a practical context; namely, to overcome (gametophytic) self-incompatibility in apple. In a first study, DAYTON (1974) used a mix of methanol-killed pollen of Richared Delicious and untreated self-pollen and was thus able to induce successful selfing in the McIntosh variety. Similar, if somewhat lower, seed sets were obtained in Golden Delicious. However, the same technique applied to the Crawley Beauty cultivar by WILLIAMS and CHURCH (1975) was unsuccessful. Nor did either methanol-treated or heavily irradiated (100, 150, 300 krad) pollen of Golden Delicious significantly aid selfing success in Blushing Golden or Starcrimson (SANSAVINI et al. 1979). In a series of experiments with a broad array of apple and pear cultivars, VISSER (1981) found inconsistencies between trials in different years and attributed them to variation in temperature at the time of pollination, as earlier suggested by DAYTON (1974). The overall results showed consistent induction of parthenocarpy with irradiated mentor pollen without, however, concomitant seed set in either selfings or intergeneric crosses. The stimulatory effect on fruit growth was believed to have been mediated via the pollen tubes which grow well in apple even after exposure to 100 krad. By contrast, methanol-treated pollen did not germinate or formed a few short tubes only and resulted only in sporadic fruit and seed set (VISSER 1981). Another finding of interest had come from studies with repeated pollinations of the same apple blossoms at an interval of 1–2 days, using two compatible pollen lots with different markers (VISSER and VERHAEGH 1980). Judging from the significant excess of seeds resulting from the second pollinations, the first pollen appeared to have "prepared the way" for the subsequent one and was accordingly termed pioneer pollen (VISSER and VERHAEGH 1980), to be distinguished from the simultaneously functioning mentor pollen. A subsequent comparison of pioneer and mentor pollen in incompatible matings, however, showed no difference in effectiveness between the two techniques, as applied to apple (VISSER 1980), whereas a somewhat increased promotion of pollen-tube growth was induced by the former in pear (VISSER and OOST in press).

Another case of testing the mentor-pollen method in an applied breeding programme is that of Cucumis, where resistance genes to green mottle mosaic virus were to be introduced into the cultivated cucumber from two African species (DEN NIJS and OOST 1980). In unaided crosses, pollen-tube growth was arrested halfway down the style in all but one combination. By contrast, irradiated mentor pollen (100, 200 krad) grew to the ovules and, in mixture with foreign pollen, was able to increase fruit set, ovule growth, and in a few cases embryogenesis. However, because of early embryo degeneration and unsuccessful attempts at raising the embryos in vitro, the hybrid nature of the progeny could not be ascertained. The question arose whether the low number of embryos was due to the competitive effects of the mentor pollen, as earlier suggested for Populus by STETTLER and GURIES (1976). This question was addressed in a subsequent study (DEN NIJS 1981) in which the fertilizing ability of self-pollen, irradiated with 100 and 500 krad, was compared with that of fresh compatible pollen added at various time intervals. In three genotypes the percentage of filled seeds decreased with increased delay of the second pollination, whereas in two others it remained constant. Control pollinations with irradiated pollen alone consistently yielded fruits with enlarged but empty seeds; pollen irradiated with 500 krad gave a lower fruit set than 100-krad pollen. These results indicate that irradiated mentor pollen may be capable of "sterilizing" a certain proportion of ovules even in competition with fully viable compatible pollen, and that this may be even more pronounced when mixed with slower-functioning incompatible pollen (DEN NIJS 1981).

The only case in which irradiated mentor pollen was successfully used in gymnosperms was described by VIDAKOVIĆ (1977).

Finally, several additional researchers, listed in Table 1, have reported only partial success or negative results from applying the mentor-pollen method.

25.3 Hypothesized Mechanisms for Mentor Effects

Several mechanisms have been implicated in playing a role in the mentor-pollen phenomenon. They may be categorized according to the site at which the mentor effects are manifested (Table 2).

Mechanisms A and B have been proposed to explain events at the stigmatic surface. Both concern the initial "dialogue" between pollen and stigma; they differ primarily by emphasizing different substances to be critically involved. In the case of Mechanism A, exine- or intine-held recognition proteins from compatible pollen permit incompatible pollen to germinate and penetrate the stigma (KNOX et al. 1972a). Mechanism B de-emphasizes the role of protein recognition and considers n-hexane soluble (presumably mainly lipoidal) compounds from the exine more important (WILLING and PRYOR 1976). Both hypotheses are based on experiments demonstrating the effectiveness of mixing incompatible pollen with extracts of the respective substances derived from compatible pollen. Further details bearing on these mechanisms, including data from additional studies (HAMILTON 1976, ASHFORD and KNOX 1980), are reviewed by KNOX, Chapter 24, this Volume.

Mechanism C was formulated by PANDEY (1977, 1978) to account for mentor effects manifested in the style, notably in gametophytic self-incompatibility systems. The "recognition pollen" hypothesis, PANDEY felt, was inconsistent with

Table 2. Proposed mechanisms to account for mentor pollen effects

Site of mentor effects	Proposed mechanism		Reference
At stigmatic surface	A	Mentor pollen provides recognition proteins which permit incompatible pollen to germinate	KNOX et al. (1972a, b)
	B	Mentor pollen provides P-factor which interacts with S-factor of stigma to render it accessible to incompatible pollen	WILLING and PRYOR (1976)
In style	C	Mentor pollen provides a pollen-growth-promoting (PGS) or regulating substance which permits incompatible pollen to sustain tube growth	PANDEY (1977, 1978)
In distal tissues of pistil	D	Mentor pollen, after tube penetration, provides substances critical for sustained growth of ovules, ovary, and other fruit tissues	STETTLER and GURIES (1976) VISSER (1981)

the autonomous action of each pollen grain in accordance with its own genotype; nor did it account for mentor functions delivered by incompatible pollen.
Accordingly, he postulated that non-specific pollen growth substances (PGS)
or regulatory substances are released from the intine or exine of the mentor
pollen, a process that is aided by various treatments (e.g. ionizing radiation,
freezing/thawing etc.), and that these substances provide extra growth potential
for incompatible pollen tubes.

Mechanism D pertains to a group of effects manifest in distal tissues of
the pistil. Here, mentor pollen is suggested to act indirectly in facilitating both
a fertilization event and successful seed maturation. The hypothesized mechanism involves stimulatory signals to be delivered by developing mentor pollen
tubes and triggering such effects as fruit retention (i.e. inhibition of abscission),
ovule enlargement, and stimulatory parthenocarpy (STETTLER and GURIES 1976,
DEN NIJS et al. 1980, VISSER 1981, CUSTERS et al. 1981). Judging from the ineffec-
tiveness of killed mentor pollen in bringing about these reactions, pollen-tube
growth seems to be a prerequisite for this mechanism.

The effectiveness of, or need for, specific pollen treatments is closely tied
to the mechanisms postulated. While Mechanism C benefits from treatments
such as ionizing radiation, freezing and thawing, methanol, or ageing, the other
mechanisms do not require them. On the other hand, any treatment that kills
mentor pollen is detrimental in conjunction with Mechanism D. The overall
high success rate of irradiated pollen in providing mentor functions may be
attributed to the fact that this treatment is compatible with all the four hypothesized mechanisms for mentor effects while it minimizes genetic contamination.
However, pollen treatments that kill pollen may give better results in certain
instances where tubes from irradiated pollen compete with those of the admixed
pollen for available ovules.

25.4 Overall Interpretation and Conclusion

Mentor functions delivered by pollen have been described in a broad variety
of plant families, representing diverse genetic systems, breeding systems, and
incompatibility systems. Some of these plants have been domesticated for some
time and reflect selection pressures imposed by the breeder; others are still
in a more or less wild state. Given this diversity, it is not surprising that several
different processes have been shown to be associated with mentor phenomena
in different plants; and it is safe to say that additional ones will be discovered
as more species are studied. Tempting as it is to consider the evolutionary
implications of certain mentor effects, most comparative interpretations are
likely to be meaningful only within a genus. Thus, plausible arguments can
be advanced on the one hand for the benefits derived by a plant that can
take advantage of heterogeneous pollen mixtures, which often obtain under
natural conditions; and on the other hand, for mechanisms ensuring the integrity
of the species via specific pollination requirements that must be met. Recent

accounts of pollen allelopathy (GILISSEN and LINSKENS 1975, SUKHADA and JAYACHANDRA 1980) add another dimension to this picture and we might well view mentor effects as merely an arbitrary zone in the continuum of pollen interactions, from positive to negative.

Successful fertilization resulting in a normally proceeding embryogeny involves a multitude of steps, many of which may be critically affected by mentor pollen in any given mating. In the simplest case of selfing this may merely be a cascading sequence of secondary effects triggered by an initial signal. In typical crossings, particularly between species and genera, primary mentor effects are likely to be multiple and involving several mechanisms. In this case they interact with other effects generated by such important – but often ignored – variables as the genetic background, cytoplasmic and maternal effects, as well as the environment in which pollination and all the subsequent processes take place. Thus, mentor effects will often be quantitative and confounded by unique interactions of two biotypes in a specific environment (HERMSEN 1979, MATZK et al. 1980). Breeders have long recognized the role of the mother plant and made use of specific cultivars, even "foster mothers", to attempt difficult crosses. In a similar way it may be appropriate to identify successful "foster fathers" with particularly potent mentor functions, or high barrier penetration capacity (e.g. HERMSEN and SAWICKA 1979).

Byproducts of experiments involving mentor pollen have included such phenomena as haploid parthenogenesis (STETTLER 1968, PANDEY 1977) and gametic transformation (PANDEY 1975), as well as progenies suspected to have arisen from pseudogamy (STETTLER et al. 1980). Apomictic seeds can be a real source of error in the interpretation of experiments, either directly or indirectly via their competitive effects (PANDEY 1977). The use of genetic markers and subsequent progeny tests will minimize this problem.

As a practical tool to the plant geneticist, mentor pollen provides a biological system of conveniently packaged and critically timed stimuli that may facilitate normally unobtainable matings. To the plant physiologist, mentor pollen is a convenient probe for analysis to identify critical processes in the course of events leading from pollination to a viable seed. Further research to elucidate mentor phenomena will greatly benefit from the judicious use by plant physiologists of specific genetic material available from breeding programmes; and from the broader testing by plant geneticists of pollen treatments, as well as other stimuli designed to substitute for mentor pollen.

References

Ashford AE, Knox RB (1980) Characteristics of pollen diffusates and pollen-wall cytochemistry in poplars. Cell Sci 44:1–17
Brewbaker JL, Emery GC (1961) Pollen radiobotany. Radiat Bot 1:101–154
Brewer JG, Henstra S (1974) Pollen of pyrethrum (*Chrysanthemum cinerariaefolium* Vis.). Fine structure and recognition reaction. Euphytica 23:657–663
Clarke AE, Knox RB (1978) Cell recognition in flowering plants. Q Rev Biol 53:3–28
Custers JBM, Nijs APM den, Riepma AW (1981) Reciprocal crosses between *Cucumis africanus* L. f and *C. metuliferus* Naud. III. Effects of pollination aids, physiological

condition and genetic constitution of the maternal parent on crossability. Cucurbit Genet Coop Rep 4:50–53

Dayton DF (1974) Overcoming self-incompatibility in apple with killed compatible pollen. J Am Soc Hortic Sci 99:190–192

Fett WF, Paxton JD, Dickinson DB (1976) Studies on the self-incompatibility response of *Lilium longiflorum*. Am J Bot 63:1104–1108

Gilissen LJWJ, Linskens HF (1975) Pollen tube growth in styles of self-incompatible *Petunia* pollinated with radiated pollen and with foreign pollen mixtures. In: Mulcahy DL (ed) Gamete competition in plants and animals. North-Holland, Amsterdam, pp 201–205

Glendinning DR (1960) Selfing a self-incompatible cocoa. Nature 187:170

Grant WF, Bullen MR, Nettancourt D de (1962) The cytogenetics of *Lotus*. I. (Embryo-cultured) interspecific diploid hybrids closely related to *L. corniculatus* L. Can J Genet Cytol 4:105–128

Guries RP (1978) A test of the mentor pollen technique in the genus *Ipomoea*. Euphytica 27:825–830

Guries RP, Stettler RF (1976) Pre-fertilization barriers to hybridization in the poplars. Silvae Genet 25:37–44

Hamilton D (1976) Intersectional incompatibility in *Populus*. PhD Thesis, Australian National Univ, Canberra

Hermsen JGTh (1979) Factors controlling interspecific crossability and their bearing on the strategy for breaking barriers to intercrossing of tuber-bearing *Solanum* species. In: Proc conf broadening genet base crops, Wageningen, 1978. Pudoc, Wageningen, pp 311–318

Hermsen JGTh, Sawicka E (1979) Incompabitility and incongruity in tuber-bearing *Solanum* Species. In: Hawkes JG et al. (eds) The biology and taxonomy of the Solanaceae. Linn Soc Symp Series 7:445–454

Howlett BJ, Knox RB, Paxton JB, Heslop-Harrison J (1975) Pollen-wall proteins: Physiochemical characterization and role in self-incompatibility in *Cosmos bipinnatus*. J Proc R Soc B 188:167–182

Knaap WP van der (1955) Observations on the pollination of cacaoflowers. Rep 14th Intern Hort Cong 2:1287–1293

Knox RB, Willing RR, Ashford AE (1972a) Role of pollen-wall proteins as recognition substances in interspecific incompatibility in poplars. Nature 237:381–383

Knox RB, Willing RR, Pryor LD (1972b) Interspecific hybridization in poplars using recognition pollen. Silvae Genet 21:65–69

Karpov GK (1966) Remote hybridization in the work of the Michurin Central Genetical Laboratory. Genetika 2:165–170

Matzk F (1981) Successful crosses between *Festuca arundinacea* Schreb. and *Dactylis glomerata* L. Theoret Appl Gen 60:119–122

Matzk F, Gröber K, Zacharias M (1980) Ergebnisse von Art- und Gattungskreuzungen mit Gramineen im Zusammenhang mit den natürlichen Isolationsmechanismen zwischen den Arten. Kulturpflanze 28:257–284

Michurin IV (1948) Sochineniia: Vol I, Printsipy i metody. Ogiz Moscow

Miri RK, Bubar JS (1966) Self-incompatibility as an outcrossing mechanism in birdsfoot trefoil (*Lotus corniculatus*). Can J Plant Sci 46:411–418

Nettancourt D de (1977) Incompatibility in angiosperms. Springer, Berlin Heidelberg New York

Nijs APM den (1981) Competition for ovules between irradiated and fresh pollen in *Cucumis sativus* L. Incomp Newsl 13:147–150

Nijs APM den, Oost EH (1980) Effect of mentor pollen on pistil–pollen incongruities among species of *Cucumis* L. Euphytica 29:267–271

Nijs APM den, Custers JBM, Kooistra AJ (1980) Reciprocal crosses between *Cucumis africanus* L. f. and *C. metuliferus* Naud. I. Overcoming barriers to fertilization by mentor pollen and AVG. Cucurbit Gen Coop Rep 3:60–61

Payàn FR, Martin FW (1975) Barriers to the hybridization of *Passiflora* species. Euphytica 24:709–716

Pandey KK (1975) Sexual transfer of specific genes without gametic fusion. Nature 256:310–313

Pandey KK (1977) Mentor pollen: Possible role of wall-held pollen-growth-promoting substances in overcoming intra- and interspecific incompatibility. Genetica 47:219–229

Pandey KK (1978) Proposed causal mechanisms of the "mentor pollen effect". Incomp Newsl 10:87–93

Pandey KK (1979) Overcoming incompatibility and promoting genetic recombination in flowering plants. NZJ Bot 17:645–663

Posnette AF (1940) Self-incompatibility in cocoa (*Theobroma* spp). Trop Agric 17:67–71

Ramulu KS, Bredemeijer GMM, Gastel AJG van (1977) Influence of mentor pollen on gametophytic intraspecific incompatibility in *Nicotiana*. Incomp Newsl 8:87–90

Ramulu KS, Bredemeijer GMM, Dijkhuis P (1979) Mentor pollen effects on gametophytic incompatibility in *Nicotiana, Oenothera* and *Lycopersicum*. Theoret Appl Gen 54:215–218

Roggen H (1975) Stigma application of an extract from rape pollen (*Brassica napus* L.) affects self-incompatibility in Brussels sprouts (*Brassica oleracea* L. var *gemmifera* DC). Incomp Newsl 6:80–86

Sansavini S, Ragazzini D, Chresti M, Ciampolini C (1979) Studi sull' applicazione del mentor pollen per il superamento dell' autoincompatibilità nel melo. Overcoming self-incompatibility by mentor pollen in apple. Rusta Ortoflorofrutticolt. It 63(5):399–410

Sastri DC, Shivanna KR (1976a) Attempts to overcome interspecific incompatibility in *Sesamum* by using recognition pollen. Ann Bot 40:891–893

Sastri DC, Shivanna KR (1976b) Recognition pollen alters intraspecific incompatibility in *Petunia*. Incomp Newsl 7:22–24

Stettler RF (1968) Irradiated mentor pollen: Its use in remote hybridization of black cottonwood. Nature 219:746–747

Stettler RF, Guries RP (1976) The mentor pollen phenomenon in black cottonwood. Can J Bot 54:820–830

Stettler RF, Koster R, Steenackers V (1980) Interspecific crossability studies in poplars. Theoret Appl Gen 58:273–282

Sukhada DK, Jayachandra (1980) Pollen allelopathy – a new phenomenon. New Phytol 84:739–746

Taylor NL, Quarles RF, Anderson MK (1980) Methods of overcoming interspecific barriers in *Trifolium*. Euphytica 29:441–450

Tsitsin NV (ed) (1962) Wide hybridization of plants. Israel Prog Sci Transl, Jerusalem

Vidaković M (1977) Savladivanje inkompatibilnosti pri križanju nekih borova. (Overcoming the incompatibility in crossing some pine species.) Genetika 9:51–63

Virk DS, Dhahi SJ, Brumpton RJ (1977) Matromorphy in *Nicotiana rustica*. Heredity 39:287–295

Visser T (1981) Pollen and pollination experiments IV. "Mentor pollen" and "pioneer pollen" techniques regarding incompatibility and incongruity in apple and pear. Euphytica 30:363–369

Visser T, Oost EH (1981) Pollen and pollination experiments V. An empirical basis for a mentor pollen effect observed on the growth of incompatible pollen tubes in pear. (unpubl manuscript)

Visser T, Verhaegh JJ (1980) Pollen and pollination experiments. II. The influence of the first pollination on the effectiveness of the second one in apple. Euphytica 29:385–390

Williams RR, Church RM (1975) The effect of killed compatible pollen on self-compatibility in apple. J Hortic Sci 50:457–461

Willing RR, Pryor LD (1976) Interspecific hybridization in poplar. Theoret Appl Gen 47:141–151

Zufa L (1971) Summary report on poplar and pine breeding in 1968 and 1969. In: Morgenstern EK (ed) Proc 12th Mtng Comm For Tree Breed Can, Part 2, pp 53–63

26 Incompatibility [1]

D. DE NETTANCOURT

26.1 Introduction, Definitions and General References

Plant tissues are able to reject cells intruding into their territory. This capacity, expressed in many host–parasite reactions (see TEASDALE et al. 1974 and DAY, Chap. 8, this Vol.) and during certain phases of graft-rejection, is manifested through the various relationships of incompatibility which occur in flowering plants between pollen grains and specific parts of the pistil.

The primary function of such incompatibility relationships in the plant species which display them is to establish and maintain allogamy, that is to say the constant exchange of genetic information within the population. The mechanism involved is called self-incompatibility and can be defined as the inability of a fertile hermaphrodite seed-plant to produce zygotes after self-pollination; it is based upon the inherited capacity of the flowers of a self-incompatible plant to reject their own pollen and to accept pollen from other self-incompatible plants in the population which are genetically different for the specific factors involved in the recognition of self-pollen.

A second manifestation of incompatibility relationships in higher plants seems to appear in the post-pollination processes which prevent the formation of hybrid zygotes combining the genomes of two fertile species. This form of incompatibility, known as interspecific incompatibility, usually occurs unilaterally and prevents self-incompatible species from accepting the pollen of species which are able to reproduce through selfing (self-compatible species). Whether or not the genetic determinants of the self-incompatibility character consistently play a role in the manifestation of interspecific incompatibility constitutes a controversial matter (see HOGENBOOM, Chap. 27, this Vol.).

The widespread occurrence of self-incompatibility and of interspecific incompatibility, the unique features of these barriers, their evolutionary significance, as well as their importance for the study of intracellular interactions or the control of breeding systems, have stimulated a considerable amount of research since Darwinian times. Yet, although much is now known on the breeding systems of higher plants and on the genetics of self-incompatibility, only few major discoveries have been made on the intimate nature of the mechanisms which govern incompatibility relationships.

It is the purpose of the present review to summarize the essential features of self- and interspecific incompatibility and to outline briefly some of the most recent achievements in the field. Detailed descriptions of incompatibility in high-

---

1 This publication is contribution No. 1839 of the Biology, Radiation Protection and Medical Research Direction of the European Communities

er plants and accounts of earlier work are to be found in LINSKENS and KROH (1967), HESLOP-HARRISON (1975), LEWIS (1976) and DE NETTANCOURT (1977). LEWIS (1954) compared incompatibility in angiosperms and fungi. ESSER and BLAICH (1973) compiled and analyzed the data from the literature dealing with heterogenic incompatibility in fungi, higher plants and animals; they established correlations between heterogenic incompatibility and histoincompatibility.

Specific references to the events occurring at the onset of pollination are provided in this volume in the chapters by KNOX (pollen–stigma interactions) and STETTLER and AGAR (pollen–pollen interactions and mentor effects).

26.2 Self-Incompatibility

26.2.1 Classification of Self-Incompatibility Systems

Self-incompatibility is an inherited trait which results from the recognition of specific gene products or phenotypic characters in pistil and pollen. The genetic mechanism may be monofactorial (governed by a single gene, usually referred to as the *S*-gene) or polyfactorial (where other genes also participate).

Two criteria can be used for the classification of the numerous forms of monofactorial and polyfactorial self-incompatibility which are known to operate in at least 71 families and more than 250 genera of angiosperms (BREWBAKER 1959). These are the site of formation of *S*-gene products and the presence or absence of floral polymorphism.

26.2.1.1 Site of Formation of *S*-Gene Products in the Pistil

In the pistil, the incompatibility phenotype is determined in the stigma (stigmatic incompatibility), the style (stylar incompatibility) or, more rarely, in the ovary (ovarian incompatibility) at a stage which usually coincides with the opening of the flower. This means, in the case of an ordinary diploid plant, that the two alleles of each of the self-incompatibility loci present in the somatic tissues have implemented, independently or through interactions, a barrier to self-pollen which is effective when the flower reaches the receptive stage for cross-pollination.

Immature pistils, as demonstrated in several species by means of bud pollinations (YASUDA 1934, ATTIA 1950, LINSKENS 1964, SHIVANNA and RANGASWAMY 1969, SHIVANNA et al. 1978) usually do not express an incompatibility phenotype and were found to lack, in *Brassica,* a specific glycoprotein (ROBERTS et al. 1979). HERRERO and DICKINSON (1980a) consider that the self-incompatibility response involves a pollination-stimulated release of certain polypeptides which does not take place in young pistils. This view is supported by the observation (CRESTI et al. 1978, HERRERO and DICKINSON 1980a) that the proteinaceous fraction of the intercellular matrix in *Lycopersicum, Prunus* and *Petunia* is formed late during stylar development.

26.2.1.2 Site of Formation of S-Gene Products in the Stamen

The stage at which the male gametophyte receives the information necessary for the determination of its incompatibility phenotype may vary greatly from one family of plants to the next. All incompatibility systems fall, on this basis, into two distinct categories:

1. *Sporophytic incompatibility,* in which the incompatibility phenotype of the pollen is governed by the genotype of the pollen producing plant. This means, in the case of a sporophytic system governed by a single locus, *S*, with two different allelic states, *S*-dominant and *s*-recessive, that the heterozygous plant will produce two classes of pollen grains, *S* and *s*, which will both express *S* in phenotype. In the Compositae and in the Cruciferae, the determination of the pollen phenotype by the *S*-alleles present in the diploid cells of the sporophyte is accomplished through the transfer from the tapetum of substances which accumulate in the cavities of the pollen exine (HESLOP-HARRISON et al. 1974). DICKINSON and LEWIS (1973a, b) extracted from *Raphanus* pollen the tryphine emitted by tapetal cells around pollen grains and showed that it stimulated, when deposited upon the stigma of the same flower, the production below the stigmatic papillae of a callosic body typical of the incompatibility reaction. HESLOP-HARRISON and co-workers (1974) and KNOX et al. (1975) induced a similar rejection reaction in two other different ways, with agar or agarose gels into which pollen-wall materials had been allowed to diffuse, and with isolated fragments of the tapetum taken from anthers of incompatible plants. These results do not imply that the pollen phenotype is determined by the tapetum in all species expressing sporophytic incompatibility. In certain cases, such as in *Linum grandiflorum,* where differences in turgor pressures between pollen and style adequately explain the occurrence of self-incompatibility (LEWIS 1943), it is even possible that the recognition of self and the manifestation of incompatibility excludes a chemical identification of *S*-gene products from pollen and pistil.

2. *Gametophytic incompatibility* where the genotype of the individual microspore determines the phenotype of the pollen grain. The exact timing of events is unknown and occurs, according to BREWBAKER (1957, 1959), in the microspores of species with binucleate pollen and in the pollen tubes for species with trinucleate pollen. The *S*-gene-specified proteins detected in *Petunia* pollen (LINSKENS 1960) and in *Oenothera* pollen (LEWIS et al. 1967) by means of immunological tests with rabbit antisera to pollen of known *S*-genotypes are probably intine-located (HESLOP-HARRISON 1975).

26.2.1.3 Floral Polymorphism

There exist, morphologically, two basic types of self-incompatibility: heteromorphic and homomorphic.

1. *In heteromorphic incompatibility,* always sporophytic and represented in 23 families and 134 genera (VUILLEUMIER 1967), cross-compatible groups of plants

within a population display noticeable differences in floral architecture and pollen cytology, which contribute to reinforcing the efficiency of the outbreeding mechanism promoted by self-incompatibility. BAKER (1948) has shown that incompatibility probably predated polymorphism in heteromorphic species.

Heteromorphic species may be distylic, with two types of floral architecture segregating in the population, or tristylic, with three groups of plants distinct in their morphology. A self-incompatibility system operates in such species and prevents each plant from setting seed upon selfing or after cross-pollination with a member of the group to which it belongs. The incompatibility genes (S in distylic species; S and M in tristylic) are complex and consist of tightly linked elements controlling incompatibility and floral morphology (LEWIS 1949). DULBERGER (1975) established the significance of the heteromorphic characters for the physiology of incompatibility.

2. Homomorphic self-incompatibility is more widespread than heteromorphy and has been detected in about half the families of flowering plants (LEWIS 1976). All species with a gametophytic system and many species with a sporophytic system are homomorphic. In such species no systematic morphological differences can be found, in principle, between plants which are cross-compatible; mating types cannot be distinguished unless pollinations or certain tests in vitro (see DE NETTANCOURT 1977) are carried out which reveal the typical reaction of self-incompatibility. Observations by CLARKE et al. (1977) of antigens common to pollen and to other tissues of the same plant in *Gladiolus* do suggest, however, that incompatibility substances in homomorphic species are part of a wider system of recognition factors which may, one day, become detectable directly in leaves, stems or other organs of the plants. Yet, to date, all attempts to demonstrate a clear association between incompatibility and other properties (see DE NETTANCOURT 1977) such as graft affinity or concentration of specific peroxidase in the leaf have failed.

26.2.2 Site of Inhibition

The correlations established by BREWBAKER (1957, 1959) between the site of inhibition and pollen cytology reveal that gametophytic incompatibility is usually manifested in the style, while sporophytic homomorphic incompatibility, which tends to characterize species with tri-nucleated pollen, is stigmatic. These correlations do not apply in heteromorphic species and cannot be extended to the Gramineae or the Onagraceae where self-incompatibility is gametophytic and stigmatic.

HESLOP-HARRISON (1975) noted that the site of inhibition in homomorphic sporophytic systems is on the stigma because the pollen grains cannot carry any further the wall-held message which determines their incompatibility phenotype. In such cases, and in contrast with the situation in gametophytic systems usually characterized by abundant stigmatic exudates, the stigma is sufficiently dry to avoid the diffusion to adjacent pollen grains of the incompatibility substances carried in the exine cavities. Methods for inducing the physiological breakdown of self-incompatibility have been reviewed by DE NETTANCOURT (1977) and RAMULU et al. (1979).

26.2.3 Genetic Control

Incompatibility systems can be classified, on the basis of their genetic features, into two distinct groups:
- with polyallelic series at one, two or, more rarely, several loci which are typical of all gametophytic systems and of certain homomorphic sporophytic species
- with two alleles per locus and one or several loci; typical of sporophytic control and always found in heterostyled species.

Relationships between different alleles of a same locus can be established only in sporophytic systems or in the pistil of gametophytic species. They may be of independence (as in the stigmas and styles of plants with gametophytic incompatibility), dominance or mutual interactions.

Relationships between different loci occur in systems governed by more than one locus. They may be of independence, complementarity or epistasis. Detailed evidence (LUNDQUIST et al. 1973) demonstrates that the involvement of large number of self-incompatibility loci is frequent in certain species of Ranunculaceae and Chenopodiaceae. At least four complementary loci have been detected in the sugar beet, *Beta vulgaris,* and it is probable that such cases of complex genetic control, considered as ancestral by LUNDQUIST and coworkers, are commonly spread in several families of plants. An extreme case of polyfactorial self-incompatibility has been reported by CROWE (1971) in *Borago officinalis,* where barriers to selfing are post-zygotic, polygenic, facultative and versatile. In this species, self-incompatibility is thus expressed as a quantitative trait with wide and continuous variations between generations and between families.

The number of different alleles segregating in gametophytic systems and in certain sporophytic species is extremely high (cf. LEWIS 1949, LUNDQUIST 1964) and perhaps surprisingly so, since MAYO and HAYMAN (1973) have shown that the proportion N of incompatible pollinations within a population remains very low as soon as more than a few alleles are in segregation. For an equal number, n, of gametophytic alleles at all segregating loci and equal frequencies of alleles, $N = \dfrac{2}{n}$ for a single locus, $\dfrac{4}{n(n+3)}$ for two loci and $\dfrac{8(n^3-1)}{n^3(n+1)^3-8}$ for a trifactorial system.

26.2.4 Mutability and Structure of Self-Incompatibility Loci

The mechanism of monofactorial gametophytic self-incompatibility is well adapted for mutation studies and large-scale screening tests, because any fertile pollen grain bearing a mutation for a loss of function at the *S*-locus or for

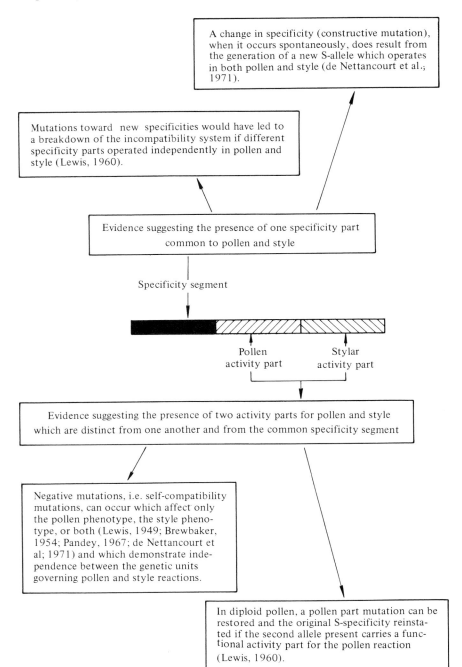

Fig. 1. Argumentation in favour of LEWIS' hypothesis (1960) on the tripartite structure of the *S*-locus. (DE NETTANCOURT 1977)

a change of specificity will pass the pistil barrier upon self-pollination and will transmit the mutated character to the resulting offspring.

26.2.4.1 Loss of Function in Pollen and Pistil

Exposure of plants with a gametophytic monofactorial system of self-incompatibility to mutagens (ionizing rays in most cases) leads to the formation of self-compatible mutants where either the pollen or the pistil expresses self-compatibility. The occurrence of such pollen and pistil mutants, together with the results of complementation analyses in diploid pollen grains and considerations on the requirements for the establishment of polyallelic series, led LEWIS (1960) to suggest a tripartite structure for the S-locus (Fig. 1). The hypothesis of LEWIS is consistent with all known facts and stipulates that the S-locus in gametophytic systems comprises a "specificity" segment which is activated in the pollen by a "pollen activity part" and in the style (or stigma) by a "stylar (or stigmatic) activity part". It is the specificity segment which individualizes the S-allele and which carries the specific genetic information necessary for preventing, through an absence of stimulation or an inhibition of growth, the penetration of the pollen tube in all pistils where one of the two specificity segments present in the diploid cells of the stigmate or style is also present in the haploid pollen grain. This model can easily be transposed to sporophytic self-incompatibility systems by substituting the notion of a "pollen activity part" to that of a "tapetum activity part" which switches on the specificity segment in the tapetal cells at the stage where substances are being formed and transferred to the cavities of the pollen exine.

26.2.4.2 The Generation of New S-Alleles

Experimental mutagens have never been observed to induce constructive mutations at the S-locus, that is to say the transformation of one self-incompatibility allele (S_x for instance) into another functional self-incompatibility allele (S_y for instance). It is therefore possible that the mutagens (essentially ionizing radiations) used in the past lack the capacity to induce point mutations at the S-locus or that the structure of the specificity segment is complex and involves redundant information distributed over the entire genome. The mutability of S-alleles cannot, at any rate, be questioned because several reports (DENWARD 1963, DE NETTANCOURT and ECHOCHARD 1969, PANDEY 1970, DE NETTANCOURT et al. 1971, HOGENBOOM 1972, ANDERSON et al. 1974) demonstrate that new S-alleles appear spontaneously and at high frequencies in the offspring of inbred plants obtained by means of the various techniques (bud pollinations, heat stocks, applications of hormones) which allow the illegitimate formation of seed after self-pollination.

26.2.5 Physiology and Biochemistry of Self-Incompatibility

Very numerous studies, initiated by the pioneering research of LINSKENS and of his co-workers (LINSKENS and ESSER 1957, LINSKENS and KROH 1967, LINS-

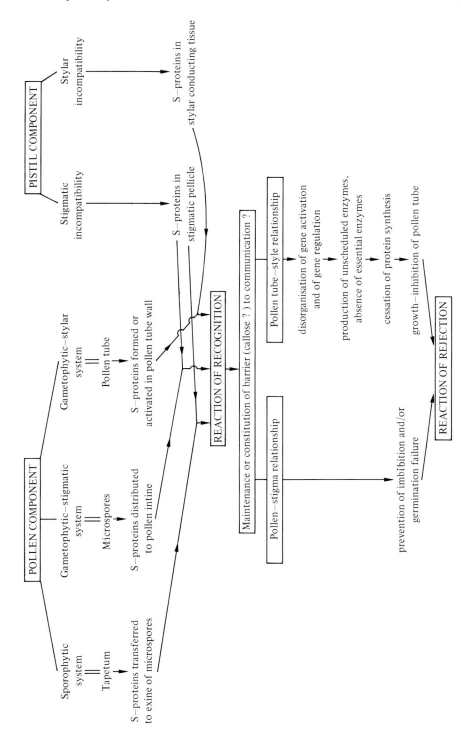

Fig. 2. Attempt to conciliate in a general scheme the various features of different self-incompatibility systems. (DE NETTANCOURT 1977)

KENS 1965, LINSKENS et al. 1960) have been made on the physiology, ultrastructure and biochemistry of self-incompatibility. These studies, reviewed by HES-
LOP-HARRISON et al. (1974, 1975), HESLOP-HARRISON (1975), LEWIS (1976), DE
NETTANCOURT (1977), FERRARI and WALLACE (1977) and LINSKENS (1980), revealed in particular:
- the role of the tapetum in the control of sporophytic incompatibility (see
  Sect. 2.1.2 above)
- the presence of S-allele specific proteins in the pollen or pistil of several
  species with homomorphic incompatibility (see MÄKINEN and LEWIS 1962,
  LEWIS et al. 1967, NASRALLAH and WALLACE 1967, NISHIO and HINATA 1977,
  1979, NASRALLAH 1979, ROBERTS et al. 1979, BREDEMEIJER and BLAAS 1981)
- the probably involvement of such specific substances in a highly localized
  recognition reaction which immediately follows the emission of pollen-wall
  proteins after pollination of the entry of the pollen tube in the stylar tissue
- as a consequence of the recognition of incompatible pollen by the pistil,
  the cessation of pollen germination or pollen-tube growth through a rejection
  process which in stigmatic incompatibility results from the deposition of callose in the stigmatic papilla adjacent to the point of contact with the pollen
  grain (HESLOP-HARRISON et al. 1975). Although they do not deny the role
  of callose in the expression of self-incompatibility, ROBERTS et al. (1980) proposed that in *Brassica* the recognition of pollen proteins by molecules in
  the stigmatic pellicule leads to the formation of an active complex which
  inhibits water supply to the incompatible grain. All other manifestations of
  self-incompatibility constitute, in the opinion of these authors, the consequences of this initial shortage of water. In spite of detailed descriptions
  of the ultrastructure of pollen-tube development after compatible and incompatible matings (see VAN DER PLUIJM and LINSKENS 1966, DE NETTANCOURT
  et al. 1973, 1974, CRESTI and VAN WENT 1976, CRESTI et al. 1979, HERRERO
  and DICKINSON 1979, 1980b, 1981), the basic principle of the rejection mechanism has not yet been understood in systems where the incompatibility reaction occurs in the style.
    An attempt to summarize and to conciliate in a general theory of the self-
incompatibility mechanism some of the information presently available on the
physiology and biochemistry of different self-incompatibility system is presented
in Fig. 2.

26.2.6 Evolution and Distribution of Self-Incompatibility

It is generally considered (GRANT 1949, WHITEHOUSE 1950, STEBBINS 1950) that
self-incompatibility contributed heavily to the expansion of angiosperms during
the Cretaceous period and several authors (WHITEHOUSE 1950, BREWBAKER 1957,
1959, CROWE 1964, PANDEY 1958, 1960) attributed the rapidity and success
of this expansion to the high outbreeding efficiency of gametophytic polyallelic
incompatibility. This means, if one accepts the hypothesis, substantiated by
WHITEHOUSE (1950) and challenged by BATEMAN (1952), of a single ancestor
to all self-incompatible species of angiosperms, that sporophytic self-incompati-

bility, heteromorphism and self-compatibility are in fact derived conditions of multi-allelic gametophytic incompatibility. Should such be the case it becomes difficult, at first sight, to explain why so many plant species have lost their self-incompatibility characters and reverted to self-compatibility. However, several reasons (for a review, see DE NETTANCOURT 1977) and, among these, the obvious superiority of self-fertilizers to colonize a new habitat from a single dispersed seed, have been found for the expansion of self-incompatibility in modern times. Since most of the families which include self-incompatible forms contain self-compatible species it appears furthermore that any discussion of the adaptive value conferred by self-compatibility must take into account the function of self-compatibility in the establishment of interspecific incompatibility (see Sect. 3, below).

Self-incompatibility is widespread throughout many orders and families of angiosperms and has been discovered in gymnosperms by GUSTAFSSON and in the pteridophytes by WILKIE (see BATEMAN 1952). EAST (1940) and VUILLEUMIER (1967) reviewed the distribution of homomorphic and heteromorphic incompatibility among the angiosperms. There is generally a strong tendency for the incompatibility system to be the same within any given family. An analysis of the distribution of self-incompatibility in families of high agricultural value reveals that the gametophytic monofactorial system is common in the Leguminosae and Solanaceae and that the Gramineae often display a bifactorial gametophytic control. A sporophytic-homomorphic system occurs frequently in the Cruciferae and the Compositae. A list of the self-incompatibility systems recorded in genera which play an important role in plant-breeding sciences is to be found in DE NETTANCOURT (1977).

26.3 Interspecific Incompatibility

Interspecific incompatibility is a reproductive barrier defined by DE NETTANCOURT (1977) as any of the post-pollination processes which prevent, through an absence of pollen germination or an abnormal behaviour of pollen tubes, the formation of hybrid zygotes combining the genomes of two fertile species or fertile ecotypes.

26.3.1 Relationships to Self-Incompatibility

Three main lines of evidence indicate that interspecific incompatibility is often closely related to self-incompatibility.

26.3.1.1 Morphological Similarities

Although only few studies have been made to determine the site of expression of interspecific incompatibility, it appears to be stylar in genera containing self-incompatible species with a stylar system of rejection and stigmatic in genera where self-incompatibility is expressed on the stigma. Yet the processes of rejection are not always identical and several authors observed that interspecific incompatibility is earlier in action that the inhibition of selfing (see LEWIS and CROWE 1958) or anatomically different (see HOGENBOOM 1975 and Chap. 27, this Vol.). A comparison under the electron microscope of the ultrastructure of incompatible pollen tubes in *Lycopersicum* (DE NETTANCOURT et al. 1973, 1974) revealed similarities between self-incompatibility and interspecific incompatibility during the early phases of the rejection processes (apparition in the pollen tube cytoplasm of a concentric endoplasmic reticulum and progressive disappearance of the callose-rich inner wall of the tube associated to an accumulation of bi-partite particles in the cytoplasm). At later stages, however, differences in the morphology and evolution of the tube apices (for a review, see DE NETTANCOURT 1977) distinguish interspecific incompatibility from self-incompatibility. Comparative studies of this kind with scanning electron microscopy have also been carried out by STETTLER et al. (1980) in *Populus,* and have emphasized the necessity to take into account variations at the intraspecific level before making comparisons between intraspecific and interspecific crosses. These authors consider, on the basis of their observations, that the multiple crossing barriers among poplar species should be viewed as a case of incongruity (see Sect. 26.3.2 below) rather than incompatibility.

26.3.1.2 The *SI* × *SC* Rule

Interspecific incompatibility usually occurs unilaterally as a barrier which prevents self-incompatible species (*SI*) from accepting the pollen or the pollen tubes of self-compatible species (*SC*). This relationship, first defined by HARRISON and DARBY (1955), was clearly established by LEWIS and CROWE (1958) who surveyed, experimentally and through an analysis of data reported in the literature, the outcome of interspecific pollinations involving all possible cross-combinations (*SC* × *SC*, *SI* × *SI*, *SC* × *SI*, *SI* × *SC*) and all known systems of self-incompatibility in several families of angiosperms. LEWIS and CROWE observed that the main exception to the general rule that interspecific incompatibility is unilateral and restricted to *SI* × *SC* combinations was to be found in crosses involving self-compatible species believed to have recently lost the self-incompatibility character.

26.3.1.3 The Dual Function of the Self-Incompatibility Locus

The considerations presented above suggest a direct involvement of the *S*-locus in the control of interspecific incompatibility. Experimental evidence in support of this hypothesis is limited, however, and stems essentially from research in *Petunia* (MATHER 1943, BATEMAN 1943) and in *Nicotiana* (ANDERSON and DE WINTON 1931, PANDEY 1964) where an allele of the *S*-locus, S_F, was detected and observed to establish unilateral incompatibility between species which were, till then, reciprocally cross-compatible. Other arguments in favour of the dual function of the *S*-locus and a review of the various models (LEWIS and CROWE

1958, SAMPSON 1962, PANDEY 1962b, MARTIN 1964, 1968, ABDALLA and HERM-
SEN 1972) presented in the past for explaining unilateral interspecific incompatibility are to be found in DE NETTANCOURT (1977).

26.3.2 Incongruity

HOGENBOOM (1972, 1973, 1975) is of the opinion, however, that interspecific
incompatibility, which he designates as incongruity, is a process distinct from
self-incompatibility and independent from any participation of the S-locus. His
theory (see HOOGENBOOM, Chap. 26, this Vol. and STETTLER et al. 1980), backed
by numerous arguments, represents interspecific incompatibility as a consequence of evolutionary divergence rather than as a cause for speciation. These
two concepts do not exclude one another, and it is possible that both interspecific
incompatibility (particularly between species and genera) and incongruity (between families) exist and operate in nature.

26.3.3 Distribution of Interspecific Incompatibility

It can be concluded, if one assumes the universality of the $SI \times SC$ rule, that
incompatibility acts unilaterally as a reproductive barrier between species within
all genera which consist of SI and SC species. The number of unilateral cross-
incompatible relations in such genera will approximate $N \times M$, where N is
the number of SI species and M the number of SC species. This means, in
the case of a genus composed of N SI species and M SC species, that
$\left(\dfrac{NM}{(N+M)^2 - 2(N+M)} \right)$ interspecific relations will be incompatible.

26.3.4 Methods for Overcoming Interspecific Incompatibility

Several different techniques (for a review see DE NETTANCOURT 1977) have been
devised for by-passing interspecific incompatibility. The efficiency of these techniques, which generally involve the selection of inbred acceptors of SC pollen,
the use of radiations or the establishment of mentor pollen effects, is, however,
restricted to certain specific combinations of crosses and no universal method
has yet been found which permits to by-passing the $SI \times SC$ barrier consistently.
The best and most sophisticated examples of success are to be found in the
experiments of KNOX et al. (1972) and of STETTLER (1980) (see STETTLER and
KNOX, Chap. 25, this Vol.), who produced hybrids in black cottonwood and
in poplars through the use of irradiated compatible mentor pollen. KNOX et al.
(1972) demonstrated, in *Populus,* that interspecific hybrids between *P. deltoides*
and *P. nigra* could also be obtained by depositing on the stigma surface of
P. deltoides pollen of *P. nigra* to which had been added proteins extracted
from the walls of compatible *deltoides* pollen (for a recent discussion of the
mechanisms involved see STETTLER et al. 1980). Another spectacular method

for transferring germplasm between different species in vivo has been proposed
by Pandey (1975), who considers that in *Nicotiana* foreign pollen exposed to
very heavy radiation doses maintains a certain capacity to reach the ovary
and to transmit small fragments of genetic material to unfertilized ovules. Jinks
et al. (1981) have presented data which appear to corroborate the results obtained by Pandey.

References

Abdalla MMF, Hermsen JGT (1972) Unilateral incompatibility: hypotheses, debate and
    its implications for plant breeding. Euphytica 21:32–47
Anderson E, Winton D de (1931) The genetic analysis of an unusual relationship between
    self-sterility and self-fertility in *Nicotiana*. Ann MO Bot Gard 18:97–116
Anderson MK, Taylor NL, Duncan JF (1974) Self-incompatibility, genotype identifica-
    tion and stability as influenced by inbreeding in red clover (*Trifolium pratense* L.).
    Euphytica 23:140–148
Attia MS (1950) The nature of incompatibility in cabbage. Proc Am Soc Hortic Sci
    56:369–371
Baker HG (1948) Dimorphism and monomorphism in the Plumbaginaceae. I.A. survey
    of the family. Ann Bot 12:207–219
Bateman AJ (1943) Specific differences in *Petunia*. II. Pollen growth. J Genet XLV:
    236–242
Bateman AJ (1952) Self-incompatibility systems in angiosperms. I. Theory. Heredity
    6:285–310
Bredemeijer GMM, Blaas J (1981) S-specific proteins in styles of self-incompatible *Nico-
    tiana alata*. Theor Appl Genet 59:185–190
Brewbaker JL (1954) Incompatibility in autotetraploid *Trifolium repens* I Competition
    and self-compatibility. Genetics 39:307–316
Brewbaker JL (1957) Pollen cytology and incompatibility systems in plants. J Hered
    48:217–277
Brewbaker JL (1959) Biology of the angiosperm pollen grain. Indian J Genet Plant
    Breed 19:121–133
Clarke AE, Harrison S, Knox RB, Raff J, Smith P, Marchalonis JJ (1977) Common
    antigens and male-female recognition in plants. Nature 265:161–162
Cresti M, Went JL van (1976) Callose deposition and plug formation in *Petunia* pollen
    tubes in situ. Planta 133:35–40
Cresti M, Ciampolini F, Pacini E, Ramulu KS, Devreux M (1978) Gamma irradiation
    of *Prunus avium* L flower bud: Effects on stylar development, an ultrastructural
    study. Acta Bot Neerl 27:97–106
Cresti M, Ciampolini F, Pacini E, Sarfatti G, Van Went JL (1979) Ultrastructural differ-
    ences between compatible and incompatible pollen tubes in the stylar transmitting
    tissue of *Petunia hybrida*. J Submicrosc Cytol 11:202–219
Crowe LK (1964) The evolution of outbreeding in plants. I. The angiosperms. Heredity
    19:435–457
Crowe LK (1971) The polygenic control of outbreeding in *Borago officinalis*. Heredity
    27 Part 1:111–118
Denward T (1963) The function of the incompatibility alleles in red clover (*Trifolium
    pratense* L.) Hereditas 49:189–334
Dickinson HG, Lewis D (1973a) The formation of the tryphine coating the pollen grains
    of *Raphanus*, and its properties relating to the self-incompatibility system. Proc R
    Soc Lond Ser B 184:148–165
Dickinson HG, Lewis D (1973b) Cytochemical and ultrastructural differences between
    intraspecific compatible and incompatible pollinations in *Raphanus*. Proc R Soc Lond
    Ser B 183:21–28

Dulberger R (1975) S-gene action and the significance of characters in the heterostylous syndrome. Heredity 35:407–415

East EM (1940) The distribution of self-sterility in flowering plants. Proc Am Phil Soc 82:449–518

Esser K, Blaich R (1973) Heterogenic incompatibility in plants and animals. Adv Gen 17:107–152

Ferrari TE, Wallace DH (1977) A model for self-recognition and regulation of the incompatibility response of pollen. Theor Appl Genet 50:211–225

Grant V (1949) Pollination systems as isolating mechanisms in angiosperms. Evolution 3:82–97

Harrison BJ, Draby L (1955) Unilateral hybridization. Nature 176:982

Herrero M, Dickinson HG (1979) Pollen–pistil incompatibility in *Petunia hybrida*. Changes in the pistil following compatible and incompatible intraspecific crosses. J Cell Sci 36:1–18

Herrero M, Dickinson HG (1980a) Ultrastructural and physiological differences between buds and mature flowers of *Petunia hybrida* prior to and following pollination. Planta 148:138–145

Herrero M, Dickinson HG (1980b) Pollen tube growth following compatible and incompatible intraspecific pollinations. Planta 148:217–221

Herrero M, Dickinson HG (1981) Pollen tube development in *Petunia hybrida* following compatible and incompatible intraspecific matings. J Cell Sci 47:365–383

Heslop-Harrison J (1975) Incompatibility and the pollen stigma interaction. Annu Rev Plant Physiol 26:403–425

Heslop-Harrison J, Knox RB, Heslop-Harrison Y (1974) Pollen-wall proteins: exine-helf fractions associated with the incompatibility response in Cruciferae. Theor Appl Genet 44:133–137

Heslop-Harrison J, Knox RB, Heslop-Harrison Y, Mattsson O (1975) Pollen-wall proteins: emission and role in incompatibility responses. In: Duckett JG, Racey PA (eds) The biology of the male gamete. Biol J Linn Soc Suppl 1, Vol VII, pp 189–202

Hogenboom NG (1972) Breaking breeding barriers in *Lycopersicon* 3. Inheritance of self-compatibility in *L. peruvianum* (L.) Mill. Euphytica 21:244–256

Hogenboom NG (1973) A model for incongruity in intimate partner relationships. Euphytica 22:219–233

Hogenboom NG (1975) Incompatibility and incongruity: two different mechanisms for the non-functioning of intimate partner relationships. Proc R Soc Lond Ser B 188:361–375

Jinks JL, Caligari PDS, Ingram NR (1981) Gene transfer in *Nicotiana rustica* using irradiated pollen. Nature 291:586–588

Knox RB, Willing R, Ashford AE (1972) Role of pollen-wall proteins as recognition substances in interspecific incompatibility in poplars. Nature 237:381–383

Knox RB, Heslop-Harrison J, Heslop-Harrison Y (1975) Pollen-wall proteins: localization and characterization of gametophytic and sporophytic fractions. In: Duckett JG, Racey PA (eds) The biology of the male gamete. Biol J Linn Soc Suppl 1, Vol VII, pp 177–187

Lewis D (1943) The physiology of incompatibility in plants. II. *Linum grandiflorum*. Ann Bot II 7:115–122

Lewis D (1949) Structure of the incompatibility gene. II. Induced mutation rate. Heredity 3:339–355

Lewis D (1954) Comparative incompatibility in angiosperms and fungi. Adv Genet 6:235–285

Lewis D (1960) Genetic control of specificity and activity of the S antigen in plants. Proc R Soc Lond Ser B 151:468–477

Lewis D (1976) Incompatibility in flowering plants. In: Cuatrecasas P, Greaves MF (eds) Receptors and recognition. Chapman and Hall, London Ser A, V 2, pp 167–197

Lewis D, Crowe LK (1958) Unilateral incompatibility in flowering plants. Heredity 12:233–256

Lewis D, Burrace S, Walls D (1967) Immunological reactions of single pollen grains, electrophoresis and enzymology of pollen protein exudates. J Exp Bot 18:371–378

Linskens HF (1960) Zur Frage der Entstehung der Abwehr-Körper bei der Inkompatibilitätsreaktion von *Petunia*. III. Mitteilung: Serologische Teste mit Leitgewebs- und Pollen-Extrakten. Z. Bot 48:126–135

Linskens HF (1964) The influence of castration on pollen-tube growth after self-pollination. In: Linskens HF (ed) Pollen physiology and fertilization. North-Holland, Amsterdam, pp 230–236

Linskens HF (1965) Biochemistry of incompatibility. In: Geerts SJ (ed) Proc 11th Intern Congr Gent, The Hague. Genetics Today 3, pp 621–636

Linskens HF (1980) Befruchtungs-Barrieren bei höheren Pflanzen. Naturwiss Rundsch 33:11–20

Linskens HF, Esser KL (1957) Über eine spezifische Anfärbung der Pollenschläuche im Griffel und die Zahl der Kallosepfropfen nach Selbstbefruchtung und Fremdbefruchtung. Naturwissenschaften 44:1–2

Linskens HF, Kroh M (1967) Inkompatibilität der Phanerogamen. In: Ruhland W (ed) Encyclopedia of plant physiology. Springer, Berlin Heidelberg New York, Vol XVIII, pp 506–530

Linskens HF, Schrauwen JAM, Van der Donk M (1960) Überwindung der Selbstinkompatibilität durch Röntgenbestrahlung des Griffels. Naturwissenschaften 46:547

Lundqvist A (1964) The nature of the two-loci incompatibility system in grasses. IV Interaction between the loci in relation to pseudo-compatibility in *Festuca pratensis* Huds. Hereditas 52:221–234

Lundqvist A, Østerbye U, Larsen K, Linde-Laursen I (1973) Complex self-incompatibility systems in *Ranunculus acris L.* and *Beta vulgaris L.* Hereditas 74:161–168

Mäkinen YLA, Lewis D (1962) Immunological analysis of incompatibility (S) proteins and of cross-reacting material in a self-compatible mutant of *Oenothera organensis*. Genet Res 3:352–363

Martin FW (1964) The inheritance of unilateral incompatibility in *Lycopersicon hirsutum*. Genetics 50:459–469

Martin FW (1968) The behavior of *Lycopersicon* incompatibility alleles in an alien genetic milieu. Genetics 60:101–109

Mather K (1943) Specific differences in *Petunia*. I. Incompatibility. J Genet 45:215–235

Mayo O, Hayman DL (1973) The stability of systems of gametophytically determined self-incompatibility. Incomp Newslett Assoc EURATOM-ITAL Wageningen 2:15–18

Nasrallah ME (1979) Self-incompatibility antigens and S gene expression in *Brassica*. Heredity 43:259–263

Nasrallah ME, Wallace DH (1967) Immunogenetics of self-incompatibility in *Brassica oleracea L.* Heredity 22:519–527

Nettancourt D de (1977) Incompatibility in angiosperms. Springer, Berlin Heidelberg New York

Nettancourt D de, Ecochard R (1969) New incompatibility specificities in the M_3 progeny of a clonal population of *L. peruvianum*. Tomato Genet Coop Rep 19:16–17

Nettancourt D de, Ecochard R, Perquin MDG, Drift T van der, Westerhof M (1971) The generation of new S-alleles at the incompatibility locus of *L. peruvianum* Mill. Theoret Appl Genet 41:120–129

Nettancourt D de, Devreux M, Bozzini A, Cresti M, Pacine E, Sarfatti G (1973) Ultrastructural aspects of self-incompatibility mechanism in *Lycopersicum peruvianum*. Mill. J Cell Sci 12:403–419

Nettancourt D de, Devreux M, Laneri U, Cresti M, Pacini E, Sarfatti G (1974) Genetical and ultrastructural aspects of self- and cross-incompatibility in interspecific hybrids between self-compatible *Lycopersicum esculentum* and self-incompatible *L. peruvianum*. Theoret Appl Genet 44:278–288

Nishio T, Hinata K (1977) Analysis of S-specific proteins in stigma of *Brassica oleracea L.* by isoelectric focusing. Heredity 38:391–396

Nishio T, Hinata K (1979) Purification of an S-specific glycoprotein in self-incompatible *Brassica campestris L.* Jpn J Genet 54:307–311

Pandey KK (1958) Time of S-allele action. Nature 181:1220–1221

Pandey KK (1960) Incompatibility in *Abitulon hybridum*. Am J Bot 47:877–883

Pandey KK (1962b) A theory of S-gene structure. Nature 196:236–238

Pandey KK (1964) Elements of the S-gene complex. I. The S_{F1} alleles in *Nicotiana*. Genet Res 2 Camb:397–409

Pandey KK (1970) Elements of the S-gene complex. VI. Mutations of the self-incompatibility gene, pseudo-compatibility and origin of new incompatibility alleles. Genetica 41:477–516

Pandey KK (1975) Sexual transfer of specific genes without gametic fusion. Nature 256:311–312

Pluijm J van der, Linskens FH (1966) Feinstruktur der Pollenschläuche im Griffel von Petunia. Züchter 36:220–224

Ramulu KS, Bredemeijer GMM, Dijkhuis P, Nettancourt D de, Schibilla H (1979) Mentor pollen effects on gametophytic incompatibility in *Nicotiana, Oenothera* and *Lycopersicum*. Theor Appl Genet 54:215–218

Roberts IN, Stead AD, Ockendon DJ, Dickinson HG (1979) A glycoprotein associated with the acquisition of the self-incompatibility system by maturing stigmas of *Brassica oleracea*. Planta 146:179–183

Roberts IN, Stead AD, Ockendon DJ, Dickinson HG (1980) Pollen stigma interactions in *Brassica oleracea*. Theor Appl Genet 58:241–246

Sampson DR (1962) Intergeneric pollen-stigma incompatibility in the *Cruciferae*. Can J Genet Cytol 4:38–49

Shivanna KR, Rangaswamy NS (1969) Overcoming self-incompatibility in *Petunia axillaris*. I. Delayed pollination, pollination with stored pollen, and bud pollination. Phytomorphology 19:372–380

Shivanna KR, Heslop-Harrison Y, Heslop-Harrison J (1978) The pollen–stigma interaction: bud pollination in the Cruciferae. Acta Bot Neerl 27:107–119

Stebbins GL (1950) Variation and evolution in plants. Columbia Univ Press, New York

Stettler RF, Kaster R, Steenackers V (1980) Interspecific crossability studies in poplars. Theor Appl Genet 58:273–282

Teasdale, J, Daniels D, Davis WC, Eddy R, Hadwiger LA (1974) Physiological and cytological similarities between disease resistance and cellular incompatibility responses. Plant Physiol 54:690–695

Vuilleumier BS (1967) The origin and evolutionary development of heterostyly in the angiosperms. Evolution 21:210–226

Whitehouse HLK (1950) Multiple-allelomorph incompatibility of pollen and style in the evolution of the angiosperms. Ann Bot New Ser 14:198–216

Yasuda S (1934) Physiological research on self-incompatibility in *Petunia violacea*. Bull Coll Agric Morioka 20:1

27 Incongruity: Non-Functioning of Intercellular and Intracellular Partner Relationships Through Non-Matching Information

N. G. HOGENBOOM

27.1 Introduction

Intimate relationships between partners exist in a multiplicity of forms, within and between organisms. Their functioning is based on well-regulated interactions.

Fertilization and zygote development in higher plants present challenging fields of research on interactions between and within cells (LINSKENS 1980). Fertilization is the outcome of a complex series of physiological and biochemical events and mutual influences, taking place after the arrival of the pollen on the stigma. It starts with pollen germination and pollen tube formation, followed by penetration of the pollen tube into the stigma and stylar tissues. The pollen tube grows into the ovary and finds its way into the ovule and embryo sac, where syngamy and fusion of the gametic nuclei take place. Up to that moment the events deal with intercellular interactions of the partners.

After double fertilization and zygote formation, embryo and endosperm development starts, leading to full-grown seeds. After germination of these seeds a plant develops and, when mature, forms flowers and gametes. In this part of the life cycle of the plant, the diplophase, two partners – the parental genomes – interact within the nucleus of the cell. During all the life cycle of the plant the nuclear content interacts with the cytoplasm and its organelles.

Each of the above developmental activities results from a series of events based on a chain of processes in both partners. For a successful development it is necessary that the processes in one partner are accurately coordinated in their interactions with those in the other partner. This means, that the genetic information of the partners with regard to their joint functioning must be fully matching (HOGENBOOM 1973, 1975).

How is this achieved? This paper reviews the development and regulation of intimate partner relationships, and discusses causes and consequences of non-functioning of these relationships due to incongruity, from a plant breeder's point of view. For illustration, the text is concentrated mainly on the pistil–pollen relationship in higher plants, but the approach can equally well be applied to other intimate partner relationships, such as between symbionts or between host and parasite.

27.2 Coevolution

Complexity can exist on the basis of interaction and coordination. Evolution of the cell includes evolution of regulation of the interaction between cell constit-

uents. Evolution of multicellular organisms requires regulation of cellular interactions. From the moment of aggregation, the units become an essential element
of the environment of other units, so that mutual influence and mutual dependence of constituents arise or become stronger. Evolutionary changes in one
constituent may put genes in another under selective pressure and thus direct
its evolution. If this mutual influence and mutually dependent development
between partners is lasting, we may speak of coevolution.

By the evolution of sexual reproduction, recombination of genetic material
was highly accelerated. Products of recombination have to cope with each other
after fusion for a normal zygotic development, and for the production of the
next generation of gametes. Between these intracellular partners accurate coordination is therefore necessary. This is well demonstrated in, for example, the
coordination of replication processes and of processes during the meiotic stages,
when chromosomes find each other and form pairs, take part in crossing over
for exchange of material and separate again to form two recombined genomes
(DOVER and RILEY 1977, HOLLIDAY 1977, RILEY and FLAVELL 1977, STERN
and HOTTA 1978). The partners in fusion set each other's limits for changes
relevant to joint action. They are mutually selective. A change in one partner
will only be accepted if the other can adequately adapt to it; otherwise it will
fail.

The differentiation of sexes in higher plants could only succeed through
stepwise co-evolution of the sexes. As the physiological barriers enveloping the
female gamete increased, more information was necessary in the pollen for
adequate functioning as a penetrator. Each change of pistil characters relevant
to fertilization could only be realized if by selective pressure the pollen information could be matched to it. The initiative in the evolution of the pistil–pollen
relationship could, of course, just as well lie on the side of the pollen (HOGEN-
BOOM 1973, 1975).

The consequence of this principle of interdependent development is that
within groups of individuals sharing a common gene pool partners stay closely
adapted to each other. More and more processes in one of them are accurately
coordinated to those in the other. But each separate group of individuals may
build its own, increasingly specialized, set of characters and relationships, differing from those of related groups, in a process eventually blocking gene exchange
between groups.

27.3 Genetics of an Intimate Partner Relationship

As a result of co-evolution in an intimate relationship, the elements of
the interaction between the partners are well coordinated. The different
activities and reactions occur at the right moment and in the right quality
and degree.

As an illustration we can draw the following picture of the pistil–pollen
relationship. The pistil is a physical barrier between the female gamete and

the outside world. It is a complex physiological system, a multitude of tissues and capacities. Some of these may act as barriers to pollen tubes, others as promotors. Each of the pistil characters and processes relevant to fertilization is based on the action of one or more genes.

To be operational the pollen must be provided with all genetic information necessary to act and react adequately in all stages of the progamic phase and in fertilization. Each capacity of the pollen and pollen tube relevant to fertilization is governed by one or more genes. Thus, as a counterpart of each relevant character and process in the pistil, the pollen must contain the potential for corresponding and adequate action and reaction. This requires matching genic systems in pistil and pollen.

If, for brevity, we call the pistil a *barrier* and the genes governing its characters *barrier genes,* and the pollen a *penetrator* and the genes governing its characters *penetration genes,* we can state that corresponding to each barrier gene or gene complex acting in the pistil, there must be a penetration gene or gene complex active in the pollen. Each pair of these corresponding units governs a link of the chain of processes and interactions in the progamic phase and in fertilization. This implies that the pistil–pollen relationship is based on a series of gene-for-gene, genes-for-gene, gene-for-genes, and genes-for-genes correspondences.

For intracellular relationships the picture of coordination will be different. The extreme intimacy of the relationship between genomes requires that they are near-identical for all developmental processes of the plant. Here, gene activities will generally be mutually complementary.

The regulatory principles of relationships between partners are scarcely known. The field requires much molecular, physiological and genetic research.

27.4 Incongruity

Incongruity is the non-functioning of an intimate partner relationship resulting from a lack of genetic information in one partner about some relevant character of the other. It may concern absence of information as well as absence of the appropriate regulation, i.e. it may concern the timing of the regulation of gene activities as well as the quality and quantity of the product.

Incongruity is the consequence of non-matching of the partners for the genetic information regulating interaction and coordination. If, for instance, the pollen lacks information on one or more of the many pistil characters which are relevant to fertilization, at some moment in the progamic phase the pollen tube will fail and the pistil–pollen interaction will be unsuccessful. If, in another example, two gametes fuse and the genomes differ in the time schedule for essential aspects of mitotic division, the zygote will fail rapidly.

Incongruity has only recently been defined as a separate mechanism for the non-functioning of intimate partner relationships (Hogenboom 1973, 1975). This has made reinterpretation of earlier research on non-functioning relationships necessary. Since then, application of this concept has been gaining ground (HERMSEN et al. 1974, 1978, FRÄNKEL and GALUN 1977, HERMSEN 1978, BUSHNELL 1979, HERMSEN and SAWICKA 1979, HOGENBOOM 1979a, KHO et al. 1980, LINSKENS 1980, STETTLER et al. 1980, BUSHNELL and ROWELL 1981). The concept may have consequences for the better understanding of functioning and non-functioning of intimate relationships in different fields of research, as will be considered further Section 27.11. The distinction of incongruity and incompatibility is considered in Section 27.8.

27.5 Evolution of Incongruity

Incongruity results from evolutionary divergence. Within a population of freely interbreeding individuals, partners in a relationship carry all the necessary information for joint functioning. By encountering a different environment, or as a consequence of a different way of adaptation to the same habitat (STEBBINS 1950), or by accident, a sub-population may be formed in which, through natural selection, a change arises in one of the partners. It may, for example, concern the appearance of an extra character in the pistil. This may result from selection pressure for this character itself, but also from pleiotropic action of, or linkage with, other genes under selective pressure.

If the new character forms an extra barrier in the pistil–pollen relationship, it can only survive if in the pollen population genetic information is available which corresponds to the new pistil character. If not, this new pistil character would disappear as a result of incongruity, as no pollen would be capable of fertilizing the plant(s) carrying the character. If the information is available in the pollen population, the selective pressure on the genes concerned will restore full congruity in the subpopulation.

Such a process of differentiation may continue and step by step form a subpopulation in which the pistil–pollen relationship is essentially different from that of the original population. Thus, an isolating mechanism between the original population and the sub-population has arisen: the pollen of one population no longer carries full information about the pistil characters of the other. This incongruity between the populations may start as a slight lack of crossability, but if the divergence proceeds it will gradually increase. In general, the degree of incongruity will depend on the extent to which the populations are related (SANZ 1945, BELLARTZ 1956, LEWIS and CROWE 1958, GRUN 1961, SMITH 1968).

The direction of the evolution of the pistil–pollen relationship has important consequences for the crossability relationships between populations. One possibility is that evolution involves the appearance of extra characters in the pistil, accompanied by selection for extra information in the pollen. The result will be a unilateral incongruity with the original population: the pollen of the new population carries full information concerning the pistil characters of the original population, but the pollen of the original population lacks information on the extra pistil characters in the new population.

A second possibility is that evolution favours the loss of certain pistil characters. In that case the corresponding part of the pollen information is no longer
under selective pressure and may, in the long run, erode. Such a subpopulation
would thus evolve from bilateral congruity with the original population to unilateral incongruity, as its pollen no longer carries information about certain
of the original pistil characters. Similar patterns of evolution may occur when
the initiative is with the pollen partner.

It is probable that in many cases evolutionary divergence will involve a
combination of the two possibilities: the loss of some and appearance of other
pistil and pollen characters. In the long run this leads to bilateral incongruity,
and this is indeed most frequently found.

The above reasoning on the evolution of the pistil–pollen relationship can,
with some modifications, also be applied to the intracellular interactions, such
as between genomes. Evolutionary divergence may bring about a multitude
of differences in genomic regulation. These may concern gene location, whether
or not in combination with gene interaction, chromosomal positioning, the rate
of processes in transcription, translation and processing or replication, etc. Deviations with a selective advantage will attain a high frequency in the subpopulation. Lasting divergence may result in the formation of genomes which can
no longer interact successfully with those of the original population. If they
were brought together by crossing, certain processes after fertilization may fail
or have disadvantageous effects. Consequences may be expressed as inviability
of the fusion product, embryo abortion, chromosome elimination, hybrid weak-
ness, lethality or sterility, etc. In this relationship we shall, in general, not expect
unilateral incongruity, unless the female cytoplasm could compensate for a certain amount of disharmony. Non-matching genomes will mostly cause bilateral
incongruity.

Incongruity between populations may concern only one developmental aspect and this could be broken down easily. But frequently, as evolutionary
divergence proceeds, more complex combinations of different expressions of
incongruity will occur.

27.6 Genetics of Incongruity

Although the pistil possesses a set of promoting and barring capacities and
the pollen a corresponding set of capacities for reaction and penetration, we
may, for convenience, call the total of the pistil characters relevant to fertilization the *barrier capacity* of a population, and the total of information in the
pollen corresponding to it the *penetration capacity* of the population. If, starting
from an original population, subpopulations evolve in which stepwise either
extra pistil characters and the corresponding information in the pollen are incorporated, or pistil characters and the corresponding information in the pollen
are lost, these populations will show the simplest form of a complex crossability
pattern, as follows (+ = congruity; − = incongruity):

	♂ Population			
	1	2	3	4
♀ Population				
1	+	+	+	+
2	−	+	+	+
3	−	−	+	+
4	−	−	−	+

This stepwise crossability pattern between related populations will occur if the barrier capacity and the corresponding penetration capacity increased from population 1 to population 4, or decreased from population 4 to population 1. The latter may concern the evolution of self-compatible species from self-incompatible species (cf. HOGENBOOM 1973).

Such crossability patterns have been found between several closely related taxa (STOUT 1952, McGUIRE and RICK 1954, MARTIN 1961, CHMIELEWSKI 1968, PANDEY 1969). Here incongruity is unilateral. If crosses between the populations can be made without difficulty and without disturbance of segregation, such material would be suitable for an analysis of the genetics of incongruity.

If genetic variation for incongruity, needed for the analysis of its inheritance, is not readily available within a group of intercrossing individuals, it may be sought or deliberately developed. HOGENBOOM (1972a, b, 1973) showed that inbreeding in a population with a high degree of heterozygosity reveals genetic variation for barrier capacity.

As yet the genetics of incongruity has been studied only in a restricted number of cases and with closely related material. Conclusions from these studies concerned the number of barrier genes on which the incongruity was based, i.e. the number of barrier genes for which the corresponding information was lacking in the pollen. In general two or more of such genes were found. They were dominant and acted independently (GRUN and AUBERTIN 1966, HOGENBOOM 1972a, b, HERMSEN et al. 1974, 1978).

In order to analyze the genetics of the pistil–pollen relationship in detail, genetic variation is required for barrier capacity as well as penetration capacity. To our knowledge, the only report on such an analysis until now was given by HERMSEN et al. (1978). Although the data gave rise to several questions that could not yet be answered, the authors proposed a model suggesting a gene-for-gene relationship between dominant barrier genes of *Solanum tuberosum* and recessive penetration genes of *S. verrucosum*.

In total four possibilities for genetic correspondence exist for each pair of barrier and penetration characters:
– 1 barrier gene corresponds with 1 penetration gene
– 1 barrier gene corresponds with more than 1 penetration genes
– more than 1 barrier genes correspond with 1 penetration gene
– more than 1 barrier genes correspond with more than 1 penetration genes.

Qualitative characters may be inherited in a simple or a complex way. But quantitative characters, to be expressed in degrees, are generally polygenically inherited. Both types will play a role in the coordination between pistil and pollen. Simple 1 gene-for-1 gene correspondences may therefore very well prove to be exceptional and more complicated situations to be the rule.

Research on the genetics of the intracellular relationships, such as between genomes, may concern the regulations or disturbances of coordination, some of which have been well characterized. One may think of the genetics of recombination, desynapsis, asynapsis and chiasma frequency (HOLLIDAY 1977, LINDSLEY and SANDLER 1977, RILEY and FLAVELL 1977, STERN and HOTTA 1978). Such genes condition exchange processes of partners. It is questionable whether this can be regarded as an example of mutual regulation between partners or, for example, simply as an influence on the physiological environment.

27.7 Nature of Incongruity

The intimate relationship between partners includes several characters, activities and reactions. Some of these may be preformed, others can occur as a reaction to contact. In the pistil–pollen relationship we can think of signal and sensor functions for recognition and mutual activation, production of the right substrate and enzymatic activities, physical and chemical signals for directed growth of the pollen tube, structures for the exchange of material and information, instruments for close contact of cell surfaces like selective receptor systems, etc.

The set of information in the pollen enables it to function in the environment to which the information is adapted. Each essential deviation from this environment causes failure of the relationship. This means that incongruity can be expressed in as many forms as there are essential elements in the interaction between pistil and pollen. This is exactly what the literature on interspecific crosses shows.

Two important consequences arise from this. One is that the cause of a certain incongruity, in the sense of the physiological or biochemical mechanism, may be difficult to discover. The other is that a study of incongruity could teach us much about the relationship between the parnters, as it is a study of the revealed missing links. By analyzing these missing links the picture of a relationship can be filled in step by step.

The knowledge of the physiology and biochemistry of the pistil–pollen relationship at the molecular level is growing, but still scarce (HESLOP-HARRISON et al. 1975, LINSKENS 1975, 1980, HESLOP-HARRISON 1978a, b). Questions such as to what extent reactions and materials are preformed and to what extent they are the result of mutual or unilateral influences through gene regulation of the other partner, are only partly answered (GILISSEN 1978). The same holds for the connected field of research on signal transport between partners and the cell-to-cell communication (GILISSEN 1978, 1981, GILULA 1980, HADWIGER and LOSCHKE 1981).

One reason for the scanty knowledge is the site of the reactions. It is not surprising that in the pistil–pollen relationship stigma-bound barriers are the best analyzed. From experiments with mentor pollen (STETTLER 1968, KNOX et al. 1972, STETTLER et al. 1980), or from chemical influences on the stigma surface (WILLING and PRYOR 1976), compensating for a lack of information in the pollen of the male parent species, and from research on recognition mechanisms (CLARKE et al. 1977, HESLOP-HARRISON 1978a, b) some first contact situations are now better understood. Results of these studies should encourage research on processes that are less easy to handle.

27.8 Revision of Visions: Incongruity and Incompatibility

Until some years ago non-functioning of the pistil–pollen relationship within, as well as between, populations was generally thought to be governed by incompatibility genes or S-genes. Few authors suggested other mechanisms for non-functioning in crosses between populations (STOUT 1952, BELLARTZ 1956, SWAMINATHAN and MURTY 1957, SAMPSON 1962, MARTIN 1963, GRUN and AUBERTIN 1966, HOGENBOOM 1972a, b).

To maintain the hypothesis of S-gene-based pollen tube inhibition in crosses between populations, many additional hypotheses on the nature and structure of S-genes were necessary. They concerned the dual role of the S-gene, mutation sequences in the S-gene, S-gene polymorphism and the primary and secondary specificities of the S-gene (LEWIS and CROWE 1958, PANDEY 1968, 1969, 1979).

It is now more and more accepted that non-functioning of interpopulational relationships can only be explained adequately with the concept of incongruity, a separate and independent principle (HOGENBOOM 1973, 1975).

The distinction between incongruity and incompatibility includes the following essential aspects:

Incompatibility frustrates the functioning of the pistil–pollen relationship, though the potential for functioning of the partners is complete. Incongruity is due to incompleteness of the relationship, the partners do not fit together.

Incompatibility is an outbreeding mechanism. It is widespread in plants and is mostly based on multiple alleles at one locus or two loci. Its inhibiting principle results from an interaction between S-specific proteins in or on pistil and pollen (LEWIS 1965, LINSKENS 1968, HESLOP-HARRISON et al. 1975, HESLOP-HARRISON 1978a, b, FERRARI et al. 1981). This principle of inhibition is the same in a large number of species. Incongruity may be expressed in processes of a very variable nature. Each part of the pistil–pollen relationship and of the relation between genomes may be the starting point of incongruity.

Incompatibility acts in the progamic phase on intercellular interactions; incongruity may act before and after fertilization, i.e. intercellular and intracellular.

As the genetics of incompatibility is based on S-alleles, it is generally simple, monogenic or oligogenic; the genetics of incongruity may range from simple to complicated, inter alia depending on the degree of relatedness of the partners. In general, incongruity will be genetically more complex than incompatibility.

Incompatibility is an evolutionary solution to the problem of inbreeding. Incongruity is a by-product of evolutionary divergence (DOBZHANSKY 1947, STEBBINS 1950, GRANT 1971). In certain cases, if populations shared a habitat and crosses between them resulted in unfit progeny, it might be strengthened by selective pressure for isolation.

In meetings of partners in the plant kingdom as a whole, incompatibility is an exception and compatibility the rule, whereas incongruity will be the rule and congruity an exception.

An extensive test of the hypothesis on incongruity with a variety of results from studies on interspecific crosses has demonstrated its general applicability (HOGENBOOM 1973, 1975). Not only phenomena such as unilateral and bilateral (in)crossability, complex crossability patterns between taxa, the genetics of (in)crossability and effects of the assumed S-gene polymorphism proved to fit into the model for incongruity, but also the puzzling behaviour of self-compatible populations in interpopulational crosses in relation to their evolutionary development. The additional hypotheses on the nature and structure of S-genes could therefore be abandoned.

27.9 Incongruity and the Prospects of Gene Manipulation in Higher Plants

New techniques of cell biology and molecular genetics for recombination of genetic material have been developed in recent years, also for application in higher plants (cf. CHALEFF 1981). The aim of these methods is the formation of plants with new and reproduceable combinations of genetic material through addition, without the use of the sexual cycle. Complete cells, or isolated nuclei, chromosomes or chromosome fragments, organelles or pure DNA are transferred to other cells.

These techniques are sometimes wrongly presented as a panacea for all crossability problems. Somatic hybridization, for example, may be used for the elimination of crossability problems resulting from incongruity in the progamic phase and in the interaction between embryo and endosperm. But many other problems, resulting from non-matching information – between genomes or between nucleus and cytoplasm – which may lead to deviating or nonfunctioning fusion products, will be the same as in crosses between plants. This limits the prospects of such a technique for recombination.

The same holds for techniques in which subcellular units are transferred to an alien physiological and genetical environment to which they are not adapted. Non-matching information may then impede the acceptance, the incorporation, or the expression of transferred material.

These possible consequences of incongruity in intracellular interactions should be kept in mind in the evaluation of prospects of the new techniques, especially when characters of agricultural importance are concerned. Many of these characters are governed by a large number of genes with practically unknown location and function. This has a great impact on the usefulness of this approach of recombination. From the many transfers of simply inherited characters from one species to another, through interspecific crosses, it is known that single genes can very well be expressed in an alien environment. This may indicate the most promising fields for application of new techniques.

In view of the incorporation and expression of genes, the chance for success of a gene transfer will generally depend on the degree of relatedness of supplier and recipient. If, for example, symbiosis of plants with nitrogen-fixing bacteria is to be exploited, transfer of genes for nitrogen fixation from a bacterium to higher plants is very probably a less promising approach than the transfer of certain genes regulating symbiosis or fixation from one bacterium to another.

27.10 Exploitation of Incongruity

Until now two possibilities for the exploitation of incongruity have been identified and applied or worked out in detail. One is the use of the phenomenon for the regulation of fertilization (HOGENBOOM 1979b). In hermaphroditic plants the prevention of self-fertilization is of interest to plant breeders in relation to hybrid seed production. To this end self-incompatibility, male sterility and femaleness are used. Incongruity offers an extra mechanism, and its exploitation functions according to the following simple principle: from a related species an extra barrier gene is introduced into lines of the crop species to be used as female parents of hybrid cultivars (the extra barrier prevents self-fertilization); at the same time the corresponding penetration gene(s) is (are) introduced into the male parent lines.

Models for the introgression of barrier and penetration genes showed the feasibility of this method (HOGENBOOM 1979b). As incongruity is frequently found between a crop species and its relatives it may be widely applicable. This will mainly depend on the selection of a barrier gene with a high level of expression in different genetic backgrounds.

A second field for the application of incongruity is the eradication of populations of noxious organisms. In unisexuals, incongruent gametes may be brought together. If incongruity is expressed just before, during or just after fertilization, it results in a waste of gametes and no progeny is produced. This approach proved to be useful in insect control (LAVEN 1967). The same techniques may be developed for the eradication of plant parasites, such as fungi and bacteria, and would offer an interesting aspect of integrated control.

27.11 Related Fields of Research; with Special Reference to the Host–Parasite Relationship

For a fruitful study of the interactions between pistil and pollen in the progamic phase and between genomes after fertilization, connections should be made with related fields of biological research. One may think of comparable principles for interaction and non-functioning in intimate relationships in fungi (ESSER 1967, ESSER and BLAICH 1973). An equally obvious parallel is the field of host–parasite relationships and the related area of symbiosis. As previously stated (HOGENBOOM 1973), it is very important that these fields should be more closely connected and that analogies and homologies be fully explored.

In genome relationships within cells we deal with two partners sharing a common physiological environment. This is one of the most intimate relationships between partners in nature. The same holds for the relationship between the nuclear content and the cytoplasm with its organelles. The mutual dependence for functioning is maximal and direct. With regard to the degree of intimacy relationships between hosts and viruses may be compared to that of genomes.

In pistil–pollen relationships the partners interact less intimately, but mutual dependence for functioning exists, as a result of co-evolution, and the partners are samples from a common gene pool. Pistil characters relevant to fertilization may result from environmental selective stress, directly or indirectly through pleiotropy. The realization also depends on genetic reserves in the pollen populations.

In symbiotic and host–parasite relationships the partners are not part of a common gene pool, and dependence is not necessarily mutual, but the interaction of the partners may very well be compared to that of the pistil–pollen relationship. Their way of coevolving is the same; in both, selection based on unilateral or mutual influence shapes the interaction. The results of co-evolution in the two types of relationship have much in common.

Host characters relevant to the interaction with the parasite will result from two categories of selective forces. One is selection based on environmental stress, without any influence of the parasite. The other is selection based on stress resulting from damage by the parasite, i.e. selection for defense mechanisms. These selective forces may reinforce or counteract each other. Characters resulting from these two categories of selective forces may influence the host–parasite relationship in the same direction or in opposite directions. Characters of the first category may be favourable to the parasite, neutral, or unfavourable, those of the second category will be unfavourable to the parasite. Both groups of characters will play a role in the evolution of the parasite, together with other environmental factors.

Knowledge concerning host–parasite relationships is extensive and the discussion on resistance mechanisms is very fruitful (VANDERPLANK 1963, 1968, 1978, DAY 1973, EENINK 1976, BROWNING 1979, ELLINGBOE 1979, 1981, PARLEVLIET 1981). It seems that in much of the discussion on the genetics and evolution of host–parasite interactions the same has taken place as in that on pistil–pollen relationships. Models for the explanation of resistance principles are not fully lucid, due to a lack of distinction between two basically different mechanisms for non-functioning of intimate partner relationships: incongruity and incompatibility.

It is justified to conclude that, on the analogy of the pistil–pollen relationship, in the host–parasite relationship two distinct mechanisms for non-functioning exist. One is incompatibility, based on interaction between specific substances in host and parasite. This prevents or disturbs the functioning of the relationship, though the potential for functioning of both partners is complete. The other mechanism is incongruity, resulting from a lack of information in the parasite on some relevant host character. The latter is non-functioning through incompleteness of the relationship.

Application of the insight that there is resistance based on incongruity (i.e. lack of genetic information) and resistance based on incompatibility (i.e. a specificity interaction) may have consequences for thinking on genetics, evolution and durability of resistances to diseases and pests. A further discussion on this matter will be given elsewhere.

27.12 Conclusions

The regulation of interaction and coordination between partners in intimate relationships results from coevolution of the partners and is based on matching genic systems in the partners.

Incongruity is the non-functioning of an intimate partner relationship due to the non-matching of partners for the genetic information regulating the interaction.

Incongruity results from evolutionary divergence. The genetics of incongruity may range from simple monogenic to very complex. Its nature is very diverse.

Exploitation of incongruity is feasible through the manipulation of the fertilization.

As incongruity occurs in all types of intimate partner relationships, some of which show great similarity and are studied separately, closer connections should be made between related research fields for the full exploration of the available knowledge.

References

Bellartz S (1956) Das Pollenschlauchwachstum nach arteigener und artfremder Bestäubung einiger Solanaceen und die Inhaltsstoffe ihres Pollens und ihrer Griffel. Planta 47:588–612

Browning JA (1979) Genetic protective mechanisms of plant–pathogen populations: their coevolution and use in breeding for resistance. In: Harris MK (ed) Biology and breeding for resistance to arthropods and pathogens in agricultural plants. Texas A & M Univ, pp 52–75

Bushnell WR (1979) The nature of basic compatibility: comparisons between pistil–pollen and host–parasite interaction. In: Daly JM, Uritani I (eds) Recognition and specificity in plant host–parasite interactions. Univ Park Press, Baltimore, pp 221–227

Bushnell WR, Rowell JB (1981) Suppressors of defense reactions: a model for roles in specificity. Phytopathology 71:1012–1014

Chaleff RS (1981) Genetics of higher plants. Applications of cell culture. Cambridge Univ Press, Cambridge London New York Melbourne

Chmielewski T (1968) Cytogenetical and taxonomical studies on a new tomato form, Part II. Genet Pol 9:97–124

Clarke AE, Harrison S, Knox RB, Raff J, Smith P, Marchalonis JJ (1977) Common antigens and male–female recognition in plants. Nature 265:161–163

Day PR (1973) Genetics of host–parasite interaction, Freeman, San Francisco

Dobzhansky T (1947) Genetics and the origin of species, 2nd edn. Columbia Univ Press, New York

Dover GA, Riley R (1977) Inferences from genetical evidence on the course of meiotic chromosome pairing in plants. Phil Trans R Soc Lond B 277:313–326

Eenink AH (1976) Genetics of host–parasite relationships and uniform and differential resistance. Neth J Pl Path 82:133–145

Ellingboe AH (1979) Inheritance of specificity: the gene-for-gene hypothesis. In: Daly JM, Uritani I (eds) Recognition and specificity in plant host–parasite interactions. Univ Park Press, Baltimore, pp 3–17

Ellingboe AH (1981) Changing concepts in host–pathogen genetics. Annu Rev Phytopathol 19:125–143

Esser K (1967) Pilze als Objekte genetischer Forschung. Ber Dtsch Bot Ges 80:453–469

Esser K, Blaich R (1973) Heterogenic incompatibility in plants and animals. Adv Genet 17:107–152

Ferrari TE, Lee SS, Wallace DH (1981) Biochemistry and physiology of recognition in pollen–stigma interactions. Phytopathology 71:752–755

Fränkel R, Galun E (eds) (1977) Pollination mechanisms, reproduction and plant breeding. Monographs on Theor Appl Genet Vol 2. Springer, Berlin Heidelberg New York

Gilissen LJWJ (1978) Bevruchtingsbiologische aspecten van zelf-incompatibele planten van *Petunia hybrida* L. PHD Thesis, Nijmegen, pp 68

Gilissen LJWJ (1981) Van cel tot cel. Vakbl Biol 15:328–334

Gilula NB (1980) Cell-to-cell communication and development. In: Subtelny S, Wessells NK (eds) The cell surface: mediator of developmental processes. Academic Press, London New York, pp 23–41

Grant V (1971) Plant speciation. Columbia Univ Press, New York London

Grun P (1961) Early stages in the formation of internal barriers to gene exchange between diploid species of *Solanum*. Am J Bot 48:79–89

Grun P, Aubertin M (1966) The inheritance and expression of unilateral incompatibility in *Solanum*. Heredity 21:131–138

Hadwiger LA, Loschke DC (1981) Molecular communication in host–parasite interactions: hexosamine polymers (chitosan) as regulator compounds in race-specific and other interactions. Phytopathology 71:756–762

Hermsen JGTh (1978) General considerations on interspecific hybridization. In: Sanchez-Monge E, Garcia-Olmedo F (eds) Interspecific hybridization in plant breeding. Proc 8th Congress Eucarpia, Madrid, pp 299–304

Hermsen JGTh, Sawicka E (1979) Incompatibility and incongruity in tuber-bearing *Solanum* species. In: Hawkes JG, Lester RN, Skelding AD (eds) The biology and taxonomy of the Solanaceae, Academic Press, London New York, pp 445–453

Hermsen JGTh, Olsder J, Jansen P, Hoving E (1974) Acceptance of self-compatible pollen from *Solanum verrucosum* in dihaploids from *S. tuberosum*. In: Linskens HF (ed) Fertilization in higher plants. North Holland, Amsterdam, pp 37–40

Hermsen JGTh, Govaert I, Hoekastra S, van Loon C, Neefjes C (1978) Analysis of the effect of parental genotypes on crossability of diploid *Solanum tuberosum* with *S. verrucosum*. A gene-for-gene relationship? In: Sanchez-Monge E, Garcia-Olmedo F (eds) Interspecific hybridization in plant breeding. Proc 8th Congress Eucarpia, Madrid, pp 305–312

Heslop-Harrison J (1978a) Recognition and response in the pollen–stigma interaction. In: Curtis A (ed) Cell-cell recognition. Soc Exp Biol Symp 32, Cambridge University Press, Cambridge, London New York Melbourne, pp 121–138

Heslop-Harrison J (1978b) Cellular recognition systems in plants. Studies in Biology vol 100. Arnold, London

Heslop-Harrison J, Heslop-Harrison Y, Barber J (1975) The stigma surface in incompatibility responses. Proc R Soc Lond B 188:287–297

Hogenboom NG (1972a) Breaking breeding barriers in *Lycopersicon*. 4. Breakdown of unilateral incompatibility between *L. peruvianum* (L.) Mill. and *L. esculentum* Mill. Euphytica 21:397–404

Hogenboom NG (1972b) Breaking breeding barriers in *Lycopersicon*. 5. The inheritance of the unilateral incompatibility between *L. peruvianum* (L.) Mill. and *L. esculentum* Mill. and the genetics of its breakdown. Euphytica 21:405–414

Hogenboom NG (1973) A model for incongruity in intimate partner relationships. Euphytica 22:219–233

Hogenboom NG (1975) Incompatibility and incongruity: two different mechanisms for the non-functioning of intimate partner relationships. Proc R Soc Lond B 188:361–375

Hogenboom NG (1979a) Incompatibility and incongruity in *Lycopersicon*. In: Hawkes JG, Lester RN, Skelding AD (eds) The biology and taxonomy of the Solanaceae, Academic Press, London New York, pp 435–444

Hogenboom NG (1979b) Exploitation of incongruity, a new tool for hybrid seed production. In: Zeven AC, van Harten AM (eds) Proc Conf Broadening Genet Base Crops. PUDOC Wageningen, pp 299–309

Holliday R (1977) Recombination and meiosis. Phil Trans R Soc Lond B 277:359–370

Kho YO, den Nijs APM, Franken J (1980) Interspecific hybridization in *Cucumis* L. II. The crossability of species, an investigation of in vivo pollen tube growth and seed set. Euphytica 29:661–672

Knox RB, Willing RR, Ashford AE (1972) Pollen-wall proteins: role as recognition substances in interspecific incompatibility in poplars. Nature 237:381–383

Laven H (1967) Eradication of *Culex pipiens fatigans* through cytoplasmic incompatibility. Nature 216:383–384

Lewis D (1965) A protein dimer hypothesis on incompatibility. Genet Today 3:657–663

Lewis D, Crowe LK (1958) Unilateral interspecific incompatibility in flowering plants. Heredity 12:233–256

Lindsley DL, Sandler L (1977) The genetic analysis of meiosis in female *Drosophila melanogaster*. Phil Trans R Soc Lond B 277:295–312

Linskens HF (1968) Egg–sperm interactions in higher plants. Accademia Nazionale Dei Lincei, Quaderno 104:47–56

Linskens HF (1975) Incompatibility in *Petunia*. Proc R Soc Lond B 188:299–311

Linskens HF (1980) Physiology of fertilization and fertilization barriers in higher plants. In: Subtelny S, Wessells NK (eds) The cell surface: mediator of developmental processes. Academic Press, London New York, pp 113–126

Martin FW (1961) Complex unilateral hybridization in *Lycopersicon hirsutum*. Proc Natl Acad Sci USA 47:855–857

Martin FW (1963) Distribution and interrelationships of incompatibility barriers in the *Lycopersicon hirsutum* Humb et Bonpl complex. Evolution 17:519–528

McGuire DC, Rick CM (1954) Self-incompatibility in species of *Lycopersicon* sect. *Eriopersicon* and hybrids with *L. esculentum*. Hilgardia 23:101–124

Pandey KK (1968) Compatibility relationships in flowering plants: role of the S-gene complex. Am Nat 102:475–489

Pandey KK (1969) Elements of the S-gene complex. V. Interspecific cross-compatibility relationships and theory of the evolution of the S complex. Genetica 40:447–474

Pandey KK (1979) The genus *Nicotiana*: evolution of incompatibility in flowering plants. In: Hawkes JG, Lester RN, Skelding AD (eds) The biology and taxonomy of the Solanaceae. Academic Press, London New York, pp 421–434

Parlevliet JE (1981) Disease resistance in plants and its consequences for plant breeding. In: Frey KJ (ed) Plant Breeding II. The Iowa State Univ Press, Ames, pp 309–364

Riley R, Flavell RB (1977) A first view of the meiotic process. Phil Trans R Soc Lond B 277:191–199

Sampson DR (1962) Intergeneric pollen–stigma incompatibility in the Cruciferae. Can J Genet 4:38–49

Sanz C (1945) Pollen-tube growth in intergeneric pollinations on *Datura stramonium*. Proc Natl Acad Sci USA 31:361–367

Smith EB (1968) Pollen competition and relatedness in *Haplopappus* section *Isopappus*. Bot Gaz 129:371–373

Stebbins GL (1950) Variation and evolution in plants, Columbia Univ Press, New York

Stern H, Hotta Y (1978) Regulatory mechanisms in meiotic crossing-over. Annu Rev Plant Physiol 29:415–436

Stettler RF (1968) Irradiated mentor pollen: its use in remote hybridization of black cottonwood. Nature 219:746–747

Stettler RF, Koster R, Steenackers V (1980) Interspecific crossability studies in poplars. Theor Appl Genet 58:273–282

Stout AB (1952) Reproduction in *Petunia*. Mem Torrey Bot Club 20:1–202

Swaminathan MS, Murty BR (1957) One-way incompatibility in some species crosses in the genus *Nicotiana*. Indian J Genet Pl Breed 17:23–26

Vanderplank JE (1963) Plant diseases: epidemics and control. Academic Press, London New York

Vanderplank JE (1968) Disease resistance in plants. Academic Press, London New York

Vanderplank JE (1978) Genetic and molecular basis of plant pathogenesis. Springer, Berlin Heidelberg New York

Willing RR, Pryor LD (1976) Interspecific hybridization in poplar. Theor Appl Genet 47:141–151

28 Allergic Interactions

B. J. HOWLETT and R. B. KNOX

28.1 Introduction

Throughout history, man has used plants for food, medicine, clothing and protection. By experience, he has learned to recognize and avoid species of plants
which cause harmful effects such as food poisoning, skin irritations, hayfever
or seasonal asthma. Some of these human responses to plants are induced by
mechanical injury or by toxic or irritant compounds in the plants, and occur
in all people who come into contact with the plant. In other cases, the response
is restricted to a proportion of the population who react specifically to the
plant or its components. The particular substances which provoke these latter
responses are known as allergens. Several allergens have been isolated and characterized, and their role in initiating the allergic response established. However,
few attempts have been made to determine the role of the allergens in the
plant. In this review we consider the occurrence and properties of natural plant
compounds which may cause allergic reactions in man. We also discuss the
mechanisms of the allergic response and possible roles of the allergens within
the plant.

28.2 Allergic Responses Caused by Plants

The term allergy was assigned by VON PIRQUET in 1906 to describe the condition
following exposure to certain molecules where patients acquire a "specific, altered capacity to react" to a subsequent exposure. Plant compounds are important in triggering two major types of allergic response. Firstly, atopic allergy
or immediate-type hypersensitivity, is mediated by antibodies, in particular,
Immunoglobulin E (IgE) (Table 1) and is triggered by pollen grains or fungal
spores inhaled from the atmosphere, or seeds consumed as food. Secondly,
delayed-type hypersensitivity is mediated by cells, in particular, by lymphocytes
(Table 1). It is initiated by contact with plant surfaces, for example, leaves
and stems.

Some plants are capable of producing both types of responses, for example
ragweed pollen, *Penicillium* spores and sawdust of western red cedar (for review
see MITCHELL and ROOK 1979). However, each type of response is almost always
provoked by different plant components.

Table 1. Features of immediate and delayed type hypersensitivity reactions caused by plants

	Allergy type	
	Immediate	Delayed
Transferred by	Serum (IgE)	Cells (lymphocytes)
Time after first exposure for sensitization to occur	Up to 21 days	Up to 28 days
Time after subsequent exposure for clinical symptoms to occur	1 min	24–48 h
Vasoactive chemicals	Histamine, serotonin	Lymphokines
Clinical diseases	Allergic rhinitis, atopic eczema	Allergic contact dermatitis
Detection assays	Skin test, radioallergosorbent test	Patch test
Causative agents	Pollen of grasses, rag-weeds, etc. seeds of soy-bean, peanut, castor beans	Leaves of poison ivy, primula, etc.

28.2.1 Atopic Allergy

When certain genetically predisposed persons (~15% of the population, BRIDGES-WEBB 1974) are exposed to atopic allergens, specific IgE is formed and binds to membranes of basophils and mast cells, present in mucosal and epithelial tissue. Upon subsequent exposure, the allergen binds to pairs of adjacent IgE molecules on the cell surface, resulting, within a matter of minutes, in degranulation of the cells and release of pharmacologically active chemicals including histamine (for review see ISHIZAKA 1976). These cause vasoconstriction, contraction of smooth muscle of the bronchioles, itching and inflammation (Fig. 1 A). In extreme cases, these responses may be fatal, especially if the allergen has been administered systemically, for example, in penicillin allergy. The IgE receptor from mast cell plasma membranes has been isolated (KANELLOPOULOS et al. 1980) and characterized as a glycoprotein (MW 50,000, 30% carbohydrate).

There are four biological assays for atopic allergens. Two detect IgE bound to cellular receptors, and the others measure levels of allergen-specific IgE in the serum, which are extremely low (10^{-4} times those of IgG). The assays are (1) direct skin testing of patients: a small sample of the suspected allergen (μg) is pricked intradermally on the forearm, giving a swollen red weal in allergic patients. This can be compared with a histamine control to give an indication of the allergenicity of the sample; (2) histamine release: leukocytes of allergic individuals are incubated in the presence of the suspected allergen and the amount of histamine released is measured (OSLER et al. 1968); (3) radioallergosorbent test (RAST): suspected allergens are covalently conjugated to paper

Fig. 1. A Diagram showing sequence of events during atopic allergic response. (Modified from KNOX 1979.) **B** Diagram showing sequence of events during allergic contact dermatitis

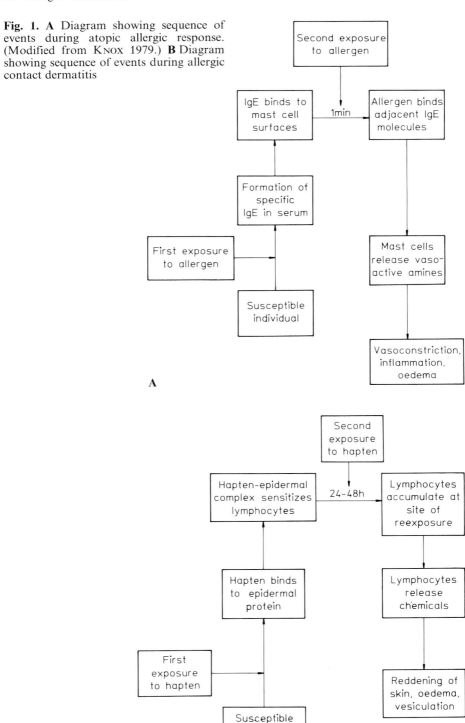

discs, and incubated in allergic patient's serum. After washing, the discs are exposed to ^{125}I-labelled antiserum raised against human IgE; counts bound to the disc indicate the amount of allergen-specific IgE, which is related to the allergenicity of the pollen extract (WIDE et al. 1967); (4) RAST-inhibition: a soluble allergen is added to compete with that on the disc for IgE in the allergic serum. By using a series of increasing concentrations of soluble allergen, the reduction in binding to the disc can be measured. This gives an indication of the allergenicity of the soluble allergen with reference to the allergen insolubilized in the disc (GLEICH et al. 1975).

28.2.2 Delayed-Type Hypersensitivity

The most common delayed-type hypersensitivity response caused by plants is allergic contact dermatitis. This response is initiated by low molecular weight compounds (haptens) which contact the skin and bind to proteins in the epidermis. These hapten–epidermal protein complexes are able to sensitize lymphocytes, and a population of lymphocytes with receptors complementary to the hapten–protein complex proliferates. On subsequent exposure to the hapten, circulating lymphocytes accumulate at the site of contact and release proteins, known as lymphokines, which, in turn, attract more lymphocytes to the site, increase vascular permeability and damage the tissue over a period of 24–48 h. This results in inflammation, oedema and vesiculation (small blisters within the epidermis) (Fig. 1B). Very little is known about the epidermal proteins which bind the haptens, and the nature of the chemicals released by the sensitized lymphocytes is an active area of current research (for review see POLAK 1980). Allergic contact dermatitis can be detected by a patch test, whereby a suspected allergen is applied to the skin in a non-irritant dose and the skin examined for symptoms of dermatitis after 48 h (FISHER 1973).

Plants are responsible for several other types of contact dermatitis, such as primary irritation, phototoxicity, and mechanical irritation (for review see EVANS and SCHMIDT 1980). These are non-allergic responses affecting all individuals if exposed to high enough concentrations of the irritant chemicals. These compounds will not be discussed in this review.

28.2.3 Other Factors Influencing Allergic Responses

The extent and capacity of individuals to react with allergens depends on many factors including genetic factors, and the intensity, duration and frequency of exposure to the allergen. In man, the existence of an Ir (immune response) gene controlling IgE responses to particular ragweed allergens and linked to the HLA-B7 histocompatibility complex has been proposed by MARSH et al. (1973, 1979).

28.3 Occurrence and Nature of Allergens

28.3.1 Atopic Allergens

The higher plant families responsible for atopic allergic disease include Gramineae, Compositae, Plantaginaceae, Chenopodiaceae, Amaranthaceae, Betulaceae and Fagaceae and certain gymnosperms. The most important allergen source in plants of these families, in terms of incidence of allergic diseases, is the pollen grain. All the allergenic pollens are from wind-pollinated species, which produce much more pollen than insect-pollinated species do, and release it in considerable amounts into the atmosphere during the flowering period of the plant (SMART and KNOX 1979). The properties of atopic pollen allergens will be considered later in this section.

Fungal spores occur in the atmosphere in fivefold excess over pollen grains (MARSH 1975, SMART and KNOX 1979) but do not appear to cause as many cases of allergic disease as does pollen. They are, however, responsible for some clinically important diseases such as penicillin allergy, bronchial asthma and farmers' lung. These latter diseases are caused by spores of *Alternaria* spp., *Helminthosporum* spp., *Penicillium* spp. and *Aspergillis* spp. (for review see RIP- PON 1974). Very little is known about the nature of atopic fungal allergens and apart from penicillin, a pure preparation has not been isolated as yet.

Seeds are another source of atopic allergens in plants. Allergens have been detected in aqueous extracts of castor bean seeds, *Ricinus communis* (YOULE and HUANG 1978), soybean seeds, *Glycine max* (SHIBASAKI et al. 1980), sesame seeds, *Sesamum indicum* (MALISH et al. 1981), and in several cereal flours such as rye, wheat and barley (BALDO and WRIGLEY 1978). The only atopic seed allergen which has been isolated is one from peanut, *Arachis hypogea* (SACHS et al. 1981, see Table 2).

Several allergens have been purified from pollen of herbaceous plants (ragweed, *Ambrosia* spp., rye grass, *Lolium perenne* and timothy, *Phleum pratense*) and of trees (birch, *Betula verrucosa,* and mountain cedar) (for reviews see MARSH 1975, KING 1976, HOWLETT et al. 1979). Their properties are summarized in Table 2. They are all either proteins or glycoproteins, ranging in molecular weight from 5000 to 60,000, and many are isolated in multiple electrophoretic forms (Table 2). In the case of the Group 1 allergens from rye grass pollen, the charge heterogeneity is accounted for in terms of differing glutamate to glutamine ratios (JOHNSON and MARSH 1966a).

Some pollen allergens are more potent than others; for instance ragweed Antigen E elicits skin reactions in more individuals than does any other ragweed pollen allergen. Antigen E is the most abundant allergen, accounting for 7% of the total protein in ragweed pollen (KING 1976), and thus susceptible persons are probably exposed to high amounts of it. The relative abundances of allergens in pollen cannot, however, account in all cases for differences in allergenicity of pollen components. For instance Glycoproteins 1 and 2 are present in comparable amounts in rye grass pollen; Glycoprotein 1 is a very potent allergen, whilst Glycoprotein 2 is only weakly allergenic (HOWLETT and CLARKE 1981a).

Table 2. Some properties of purified atopic allergens from plants

Plant source	Allergen	Protein (%)	Carbo-hydrate (%)	Mono-saccharide composition	Molecular weight (charge)	Reference
Pollen						
Ambrosia elatior (ragweed)	Antigen E	99	<0.6	Ara	38,000 (acidic)	King and Norman (1962), Paull et al. (1979)
	Antigen K [a]	99	0.6	Ara	38,000 (acidic)	King et al. (1964, 1967)
	Antigen Ra3	92	8		11,000 (basic)	Underdown and Goodfriend (1969), Klapper et al. (1980)
	Antigen Ra4 (BPA-R)	ND	+		23,000 (basic)	Griffiths and Brunet (1971)
	Antigen Ra5	100	0		5,000 (basic)	Lapkoff and Goodfriend (1974), Mole et al. (1975)
Lolium perenne (ryegrass)	Group 1 [a]	95	5	Gal, Man	27,000 (acidic)	Johnson and Marsh (1965)
	Glycoprotein 1 (Group 1)	95	5	Glc:Gal:Man:Ara:GlcNAC 3 : 3 : 1 : 2 : 1	32,000 (acidic)	Howlett and Clarke (1981 a)
	Group II [a]	ND			11,000 (acidic)	Johnson and Marsh (1965)
	Group III	ND	ND		11,000 (basic)	Marsh (1974)
	Glycoprotein 2 [b] (Antigen A)	88	12	Glc:Gal:Man:Fuc:Xyl:Ara 4 : 7 : 13 : 5 : 8 : 6	58,000 (basic)	Howlett and Clarke (1981 a)

	Antigen			MW		Reference
Phleum pratense (timothy)	Antigen 3	ND	ND	10,000 (acidic)		MALLEY and HARRIS (1967), LOWENSTEIN (1978a, b)
	Antigen 25 (AgB)	ND	13	10,000 (acidic)		
	Antigen 30	ND	ND	15,000 (basic)		
Poa pratensis (Kentucky blue grass)	C-1-2-6[a]	ND	+	10,000 (acidic)		EKRAMODDOULLAH et al. (1980)
Betula verrucosa (birch)	Antigen 2[a] (BV4)	ND	+	29,000 (acidic)		BELIN (1972)
	Antigen 7-8	ND	ND	10,000–17,000 (acidic)		VIANDER et al. (1979), APOLD et al. (1981)
Juniperus mexicana (Mountain cedar) ND		ND	ND	50,000 ND		GROSS et al. (1978)
Seed						
Arachis hypogea (peanut)	Peanut 1	11% nitrogen	8.7%	? (acidic)		SACHS et al. (1981)

[a] Allergen is present in multiple electrophoretic forms
[b] Weakly allergenic

In some pollens there is a high degree of antigenic cross-reactivity, for example between Antigens E and K; E and Ra_4 of ragweed, (King 1976) and between birch pollen allergens (Viander et al. 1979) detected using rabbit antisera in immunodiffusion studies. Using radioimmunoassay, a more sensitive technique than immunodiffusion, we have detected antigenic cross-reactivity between two quite different glycoproteins from rye grass pollen (Howlett and Clarke 1981 b). This cross-reactivity is due, at least in part, to small amounts of common saccharides in each glycoprotein.

A problem in assessing the literature on pollen allergens is the variety of names assigned by different authors to the same allergen. For example, timothy pollen Antigen 25 (Lowenstein 1978 b), corresponds to Antigen B (Malley and Harris 1967), and rye grass Group 1 allergen (Johnson and Marsh 1965) probably corresponds to Glycoprotein 1 (Howlett and Clarke 1981 a, b). Furthermore, at present there is no single method used for the standardization of allergen preparations (see review by Baldo et al. 1981). These are important deficiencies in view of the widespread use of allergens in diagnostic, therapeutic and research investigations.

Atopic allergens are present in house dust mites, animal danders, fish, eggs and milk, as well as in plants (for review see King 1976). In all these cases they are either proteins or glycoproteins with molecular weights between 5000 and 60,000. There do not appear to be any common properties of these molecules which may confer allergenicity (the ability to elicit IgE synthesis) upon them. The sequences of two pollen allergens are available (ragweed Antigens Ra3, Klapper et al. 1980, Ra5, Mole et al. 1975), and these do not appear to reveal any unusual features. In the case of codfish allergen, native protein configuration is not crucial, since this allergen remains active even after treatment with 9M urea (Aas 1978).

Most of the allergens have only small amount of carbohydrate ($<10\%$ w/w) (Table 2). The importance of this moiety in determining allergenic activity is not known. In a preliminary study, Johnson and Marsh (1966 b) used enzymes, including cellulase and β-galactosidase, to remove part of the carbohydrate of Group 1 allergen of rye grass pollen. The treated allergen showed no loss of skin-sensitizing activity, indicating that the complete carbohydrate moiety is probably not essential for allergenicity. There is a need for further experiments of this kind to be performed to determine which parts of the allergen molecule are necessary for allergenic activity.

28.3.2 Contact Allergens

Contact allergens are present in a diverse range of higher plants, fungi and algae. These compounds are much smaller than the atopic allergens. Their structures have been more readily determined, and molecular formulae of many contact allergens are established. They can be divided into several closely related groups and have been extensively reviewed by Evans and Schmidt (1980) and Mitchell and Rook (1979). An outline of the properties of allergens of the three major groups will now be presented.

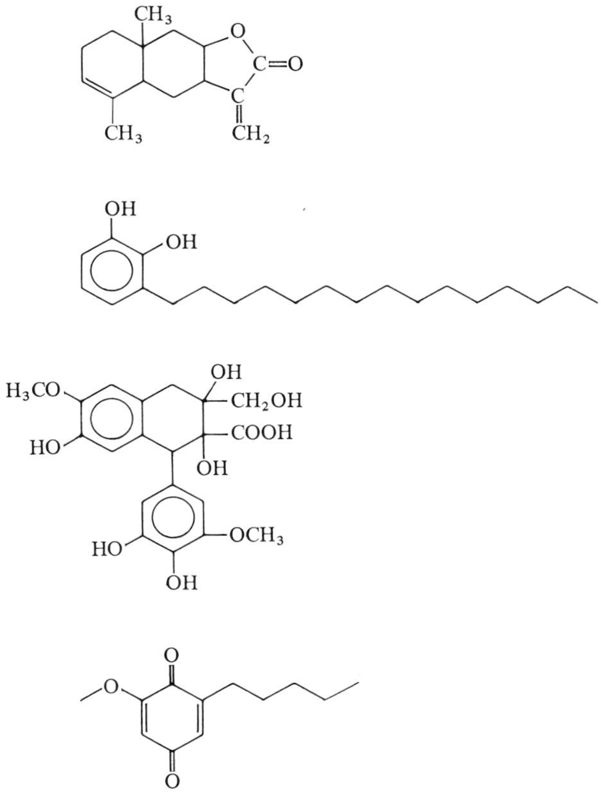

Fig. 2 A–D. Chemical structures of some contact allergens. **A** Alantolactone, a sesquiter-pene lactone from *Inula* spp. **B** 3n-pentadecylcatechol, an urushiol from *Toxicodendron* spp. **C** Plicatic acid from Western red cedar (*Thuja plicata*). **D** Primin, a quinone from *Primula* spp.

28.3.2.1 Sesquiterpene Lactones

These are 15-carbon atom compounds which are derived from isoprene units. They are usually volatile, and are extracted by steam or ether from various parts of the plant to give essential oils which are widely used as perfumes, food flavouring and sometimes medicines. The structure of a typical allergenic sesquiterpene lactone is shown in Fig. 2 A.

Allergenic compounds of this nature are present in most genera of the large Compositae family, as well as in Umbelliferae, Magnoliaceae, Lauraceae and Aristolochiaceae. Contact dermatitis caused by parthenin, a sesquiterpene lac-tone from *Parthenium hysterophorus,* affects many people in India where the weed is widely spread (SOHI et al. 1979). Lower plants such as the liverwort, *Frullania* spp., also contain these compounds (MITCHELL et al. 1970, MITCHELL and DUPOIS 1971). Structural studies show that a γ-lactone with an exocyclic α-methylene group is a partial requirement for activity; if the α-methylene group is reduced then the molecule is not allergenic (Mitchell and Dupois 1971).

28.3.2.2 Catechols

These are found principally in the family Anacardiaceae whose 60 genera probably cause more contact allergy than all other plant families together (Mitchell and Rook 1979). The family includes species important for their edible nuts: cashew (*Anacardium occidentale*), pistachio (*Pistacia vera*); fruit: mango (*Mangifera indica*); resins, oils and lacquers obtained from the varnish tree (*Toxicodendron vernicifera*) and garden ornamentals including the smoke tree (*Cotinus coggygria*) and sumac (*Rhus* spp.).

Allergenic catechols are responsible for *Rhus* or poison ivy/oak dermatitis, which is the most important cause of plant dermatitis in North America. Poison ivy (*Toxicodendron radicans*) contains four allergenic components; all are catechols with a 15-carbon atom side chain in the C3 position of the ring (see Fig. 2B). These compounds are collectively known as urushiol, and vary in the degree of unsaturation and the position of the double bonds in the side chain. The more unsaturated or the longer the side chain, the more allergenically reactive is the compound (Johnson et al. 1972). Urushiol and related compounds are remarkably stable and can persist on the skin or clothing for long periods, and thus can repeatedly initiate the allergic response.

Other allergenic catechols include plicatic acid (Fig. 2 C) which is unique to the western red cedar (*Thuja plicata*). Plicatic acid is a major component of the heart wood, constituting up to 40% (w/w) of the non-volatile components (Chan-Yeung et al. 1974). As well as provoking contact dermatitis in wood workers, it also causes immediate-type respiratory responses. Plicatic acid, however, is not responsible for "cedar poisoning" which occurs in forest workers. This disease is caused by allergenic sesquiterpene lactones present in lichens and liverworts (*Frullania* spp.) growing on the cedar trees (Mitchell and Chan-Yeung 1974).

28.3.2.3 Quinones

In Europe, the most common plants causing allergic contact dermatitis are *Primula* spp., which contain the quinone, primin (Fig. 2 D), in their leaves, stems and petals (Hausen 1978). Quinones are also found in the heartwood of many mature trees (for example, *Acacia melanoxylon,* Hausen and Schmalle 1981) and can cause allergic respiratory symptoms in timber workers.

28.3.2.4 Other Contact Allergens

Other sources of contact allergens in higher plants include *Tulipa* spp. (tulip) which contains tulipalin A (α-methylene γ-butyrolactone), a phytoalexin, in the fleshy outer layers of its bulbs, as well as in the flower, stems and leaves (Slob 1973). This compound does not occur naturally but is formed upon enzymic hydrolysis of the glycoside tuliposide A, when the plant or bulb is damaged. The condition evoked by tulipalin in humans is known as "tulip fingers".

The scents of many common garden plants including members of the Oleaceae, privet (*Ligustrum* spp.), lilac (*Syringa vulgaris*) and jasmine (*Jasminum* spp.), cause allergic diseases in some individuals, but allergens have not been isolated from these plants as yet (for review see MITCHELL and ROOK 1979).

Fungi are responsible for several contact allergic diseases. Spores of *Alternaria* spp., *Aspergillus* spp. and *Penicillium* spp., as well as causing atopic responses, can also cause allergic dermatitis. Several species of rusts and smuts are present on cereal grains and produce skin irritations in workers who harvest and mill the grains (for reviews see FINEGOLD 1976, and MITCHELL and ROOK 1979).

Marine algae can cause skin irritations to swimmers who inadvertently touch them. Some species of blue-green algae contain the pigment phycocyanin, which has been implicated as an allergen (for review see MITCHELL and ROOK 1979). Some species of lichens contain an allergen, usnic acid, which, like the sesquiterpene lactones of *Frullania* spp. (see Sect. 3.2.1) can cause allergic disease in lumbermen (MITCHELL et al. 1972).

28.3.2.5 Mechanism of Allergenicity

The exact mechanism by which contact allergens induce their response is not fully understood. It has been proposed by BYCK and DAWSON (1968) that allergenic catechols are oxidized within the skin to quinones which can then react by nucleophilic addition, with either amino or sulphydryl groups on cellular proteins to form the allergenic hapten–skin protein complexes. Such complexes have now been prepared in vitro and tested for their ability to stimulate lymphocytes (DUPOIS 1979). Urushiol–skin protein complexes and urushiol–serum albumin complexes were prepared. Both complexes were able to stimulate isolated lymphocytes of patients sensitive to poison ivy; however, neither was able to stimulate lymphocytes of control subjects. These exciting preliminary results indicate that this approach may be useful in studying the mechanism of contact allergy.

28.3.3 Allergenic Cross-Reactivity

Within both groups of allergens there is a considerable degree of allergenic cross-reactivity. Some patients are sensitive to a whole range of allergens, to which they have not been previously exposed.

For atopic allergens the usual methods of detection of allergenic cross-reactivity are skin-testing and RAST inhibition (see Sect. 2.1). Generally the level of cross-reactivity between pollens is a reflection of their taxonomic relationships, the more related species showing the highest degree of cross-reactivity (GLEICH et al. 1975, BERNSTEIN et al. 1976). More recently we have shown a high level of allergenic cross-reactivity between quite unrelated pollens including the legume *Acacia* and the monocots rye grass and *Gladiolus* (HOWLETT et al.

1982). Indeed allergenic cross-reactivity has been shown between such diverse tissues as apple fruit pulp and birch pollen (LAHTI et al. 1980). These data indicate that the allergens must possess common determinants that can be recognized by previously induced Immunoglobulin E antibody.

Since closely related contact allergens are present throughout the plant kingdom from algae to angiosperms, a high degree of allergenic cross-reactivity may be expected. This is seen, for example, where patients sensitive to the sesquiterpenes lactones in the liverwort, *Frullania* spp. will also respond to similar compounds in the leaves of higher plants, for example *Chrysanthemum* spp. (MITCHELL and ROOK 1979).

28.4 Sites of Allergens Within the Plant

The occurrence of allergens in higher plants, fungi and algae, and the organs in which they are present, have been discussed in the previous section. At the cellular level, the only allergens that have been localized are from pollen. Cellular sites of Group 1 allergen, the principal allergen of rye grass pollen, have been determined by immunocytochemistry using both light and transmission electron microscopy (HOWLETT et al. 1981, VITHANAGE et al. 1980, 1982, Fig. 3 A, B, C). The allergen is present in arcades of the exine wall layer and in pollenkitt on the surface of the grains, as well as in vesicles within the intine layer and cytoplasm. An allergen from another pollen, Antigen E of ragweed, is also present in exine and intine pollen wall sites as revealed by light microscopy (KNOX et al. 1970, KNOX and HESLOP-HARRISON 1971, HOWLETT et al. 1973), whilst birch pollen allergens have been localized at the surface of the grain (BELIN and ROWLEY 1971).

The pollen allergens diffuse rapidly from the grain upon moistening. This has been demonstrated in several ways: (1) their presence in short-term extracts; (2) their appearance external to grains in rapidly freeze-sectioned pollen (KNOX and HESLOP-HARRISON 1970); (3) pollen prints on agarose films (HOWLETT et al. 1973). The latter technique is particularly convincing as the pollen can be exposed to the agarose film for varying periods, and the emitted antigens detected close to their sites of release by immunofluorescence. With ragweed pollen, the release of Antigen E from the exine micropores has been detected in five second pollen prints, while after 30 s, allergen is released from the intine via the three apertures (Fig. 4 A, B, C).

The precise sites of contact allergens within the plant have not been established. The allergen, primin, from *Primula* spp. is found in tiny glandular hairs on the leaves and stems. Two types of glandular hairs, distinguishable by the appearance of their secretions in the transmission electron microscope, are present, but it is not known which type contains primin (WOLLENWEBER and SCHNEPF 1970). One type secretes a white crystalline "farina", mainly composed of flavonoids, while the other type secretes an oil, which contains flavonoids and forms a film on the gland surface. In both types, secretion takes place through

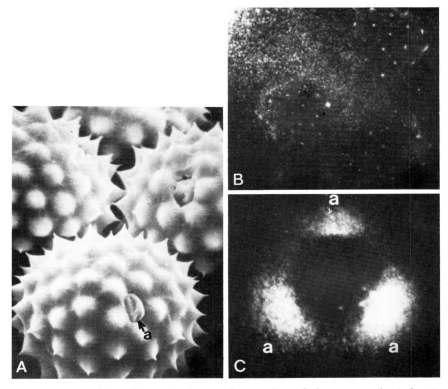

Fig. 3 A–C. Release of the allergen Antigen E from pollen of giant ragweed, *Ambrosia trifida*. **A** scanning electron micrograph showing appearance of pollen grains; *a* aperture × 5000. **B**, **C** Pollen prints showing release of Antigen E into an agarose film after exposure for **B** 5 s, **C** 30 s. The pollen was removed, the agarose film rapidly air dried, and Antigen E localized by immunofluorescence. In **B** the fluorescence is associated with the spinules of the exine; in **C** with the apertures. (HOWLETT et al. 1973)

the plasma membrane, i.e. it is of the merocrine type (WOLLENWEBER and SCHNEPF 1970), and it is possible that primin is secreted onto the leaf hair surface in this way.

Allergens are present in leaf hairs of the Anacardiaceae, for example urushiol compounds from poison ivy and oak, and are released when the leaf is touched. In other Anacardiaceae, the allergens are found within resin canals of the leaf which must be injured in order to release them (MITCHELL and ROOK 1979).

28.5 Role of Allergens in Plants

In terms of structure-property relationships, it is apparent that the two types of allergens, the immediate type, and the delayed type, might be expected to

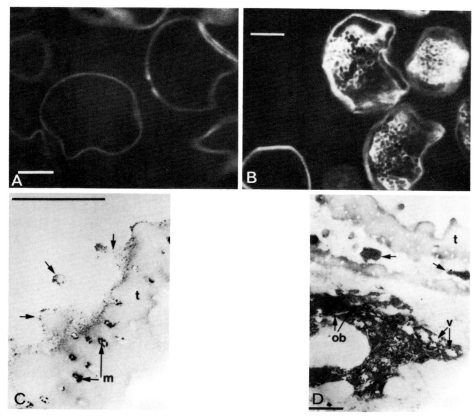

Fig. 4 A–D. Localization of Group 1 allergen in pollen of ryegrass, *Lolium perenne,* by immunocytochemistry. **A, B** Immunofluorescence localization. Freeze-dried anthers were embedded in plastic resin and sections treated for localization of Group 1 allergen by the indirect method. **A** Control section treated with pre-immune IgG. **B** Test section treated with anti-Group 1 allergen IgG. The allergen is located in the pollen wall and cytoplasmic sites. (Howlett et al. 1981). **C, D** Localization by immuno-electron microscopy using a double-embedding method. Sections were treated with anti-allergen IgG, then anti-antibody labelled with ferritin. **C** Ferritin labelling on surface of exine and in microspores. **D** Ferritin labelling in exine arcades and in vesicles at interface of intine and cytoplasm. (Vithanage et al. 1981)

have quite different functions in the plant. As the classes of molecule involved are quite different, they will be considered separately.

28.5.1 Atopic Allergens

There is no direct evidence for a role of atopic allergens in the plant. Very few studies have been performed for several reasons. (1) Very few atopic allergens have been purified and only small amounts of these are available. (2) Most workers involved in allergen research are in medical fields, and do not have the interest or expertise to investigate biological activity of the allergen in the plant. (3) It is obviously difficult to design experiments seeking a role for allergens without first assuming a role, and very few clues about the function of plant allergens are available. Hence, a firm basis for research of this kind is lacking at present.

The clues as to the role of allergens that are available have been obtained from experiments in pollen biology. From this work, several speculations have been made for the presence of allergens in pollen grains. They may be cell recognition factors, enzymes involved in pollen germination, or reserve storage proteins for pollen-tube growth (HOWLETT et al. 1979). Circumstantial evidence for pollen allergens as recognition factors in incompatibility responses is provided from a study on self-incompatibility of *Cosmos bipinnatus*. Pollen diffusates from compatible (mentor) pollen were able to partially overcome the self-incompatibility reaction on the stigma; fractions from mentor pollen enriched in antigens which immunologically cross-reacted with Antigen E of ragweed were effective in this response (HOWLETT et al. 1975). As yet this is the only study which implicates allergens as having a role in incompatibility responses.

An alternative role for pollen allergens might be as enzymes involved in pollen germination and tube growth or penetration of the stigma. A range of hydrolytic enzymes is present in pollen diffusates, and in the media of germinating pollen (see review by HOWLETT et al. 1979); several have been detected in crude extracts of ragweed pollen allergens (KING and NORMAN 1962), and acid phosphatase activity has been claimed to be associated with a timothy pollen allergen (WEEKE et al. 1974). However, there is no reported instance of enzymic activity being associated with a purified pollen allergen. Clearly there is little evidence either for or against a role of pollen allergens as enzymes or incompatibility factors and more research is necessary to gain insight into their biological role.

28.5.2 Contact Allergens

Contact allergens, like atopic allergens, appear to have no obvious functions in the plant. They may be important in preventing attack by fungal parasites, or insect or ruminant foraging. The fact that the catechols of the Anacardiaceae are present primarily in the resin ducts and are only released on injury suggests that they may be involved in the wound response in sealing off the injured tissue. Since the latex from the Japanese lacquer tree contains more than 50% urushiol compounds (TOYAMA 1923, cited by MITCHELL and ROOK 1979) and is a most effective sealing agent in furniture manufacture, it is likely that the urushiol compounds may function in sealing off the leaf after injury.

The sesquiterpene lactone, parthenin, from *Parthenium hysterophorus* (carrotweed) has the remarkable property of blocking pollen germination and thus seed-setting of a wide range of other plants (CHAR 1977). This is potentially an adaptation of great evolutionary advantage for *Parthenium* in its rapid spread as a weed.

28.6 Conclusions

Several atopic allergens and contact allergens from a diverse range of plant species have now been chemically defined. These two types of allergen are quite distinct in terms of their chemical structures and in the manner in which they provoke responses in humans. As yet, however, they cannot be assigned a role in the plant. Whether they are just end products of metabolism or whether they have some fundamental significance offers an exciting challenge for the future.

Acknowledgements. We are grateful to Drs. A.E. Clarke and G.J. Howlett for critical appraisal of the manuscript and to the Australian Research Grants Committee for financial support.

References

Aas K (1978) What makes an allergen, an allergen? Allergy 33:3–14

Apold J, Florvaag E, Elsayed S (1981) Comparative studies on tree pollen allergens. 1. Isolation and partial characterization of a major allergen from birch pollen (*Betula verrucosa*). Int Arch Allergy Appl Immunol 64:439–448

Baldo BA, Wrigley CW (1978) IgE antibodies to wheat flour components. Clin Allergy 8:109–124

Baldo BA, Krilis S, Basten A (1981) Selective approaches to the isolation and standardization of allergens. Contemporary topics in molecular immunology 8:41–88

Belin L (1972) Separation and characterization of birch pollen antigens with special reference to the allergenic components. Int Arch Allergy Appl Immunol 42:329–342

Belin L, Rowley JR (1971) Demonstration of birch pollen allergen from isolated pollen grains using immunofluorescence and a single radial immunodiffusion technique. Int Arch Allergy Appl Immunol 40:754–769

Bernstein IL, Perara M, Gallagher J, Michael JG, Johannson SGO (1976) *In vitro* cross-allergenicity of major aeroallergenic pollens by the radioallergosorbent technique. J Allergy Clin Immunol 57:141–152

Bridges-Webb CC (1974) The Traralgon health and illness survey III. Int J Epidemol 3:233–246

Byck JS, Dawson CR (1968) Assay of protein-quinone coupling involving compounds structurally related to the active principle of poison ivy. Anal Biochem 25:123–135

Chan-Yeung M, Barton GM, MacLean L, Grzybowski S (1974) Occupational asthma and rhinitis due to western red cedar (*Thuja plicata*). Am Rev Resp Dis 108:1094–1099

Char MBS (1977) Pollen allelopathy. Naturwissenschaften 64:489–494

Dupois G (1979) Studies on poison ivy. In vitro lymphocyte transformation by urushiol–protein conjugates. Br J Dermatol 101:617–624

Ekramoddoullah AK, Chakrabarty S, Kisil FT, Sehon AH (1980) Isolation of a purified allergen from Kentucky blue grass pollen. Int Arch Allergy Appl Immunol 63:220–235

Evans FJ, Schmidt RJ (1980) Plants and plant products that induce contact dermatitis (review). Planta Med 38:289–316

Feinberg SM (1944) Penicillin allergy; on the probability of allergic reactions in fungus sensitive individuals. J Allergy 15:271–273

Finegold I (1976) Allergic reactions to moulds. Cutis 17:1080–1089

Fisher AA (1973) Contact dermatitis. Lea and Febiger, Philadelphia

Gleich GJ, Larson JB, Jones RT, Baer H (1975) Measurement of the potency of allergy extracts by their inhibitory capacities in the radioallergosorbent test. J Allergy Clin Immunol 55:334–340

Griffiths GW, Brunet R (1971) Isolation of a basic protein antigen of low ragweed pollen. Can J Biochem 49:396–400

Gross GN, Zimburean JM, Capra JD (1978) Isolation and partial characterization of the allergen in mountain cedar pollen. Scand J Immunol 8:437–441

Hausen BM (1978) On the occurrence of the contact allergen primin and other quinoid compounds in species of the family of Primulaceae. Arch Derm Res 261:311–316

Hausen BM, Schmalle H (1981) Quinonoid constituents as contact sensitizers in Australian blackwood (*Acacia melanoxylon*). Br J Ind Med 38:105–109

Howlett BJ, Clarke AE (1981a) Isolation and partial characterization of two antigenic glycoproteins from rye grass pollen. Biochem J 197:695–706

Howlett BJ, Clarke AE (1981b) Role of carbohydrate as an antigenic determinant of a glycoprotein from rye grass (*Lolium perenne*) pollen. Biochem J 197:707–714

Howlett BJ, Knox RB, Heslop-Harrison J (1973) Pollen-wall proteins: release of the allergen Antigen E from intine and exine sites in the pollen grains of ragweed and *Cosmos*. J Cell Sci 13:603–619

Howlett BJ, Knox RB, Paxton JD, Heslop-Harrison J (1975) Pollen-wall proteins: physicochemical characterization and role in self-incompatibility in *Cosmos bipinnatus*. Proc R Soc Lond B 188:166–182

Howlett BJ, Vithanage HIMV, Knox RB (1979) Pollen antigens, allergens and enzymes. Curr Adv Plant Sci 35:1–17

Howlett BJ, Vithanage HIMV, Knox RB (1981b) Immunofluorescent localization of two water-soluble glycoproteins including the major allergen from pollen of rye grass (*Lolium perenne*). Histochem J 13:461–480

Howlett BJ, Hill DJ, Knox RB (1982) Cross-reactivity between *Acacia* (wattle) and rye grass pollen allergens. Clin Allergy 12:259–268

Ishizaka K (1976) Cellular events in the IgE antibody response. Adv Immunol 23:1–76

Johnson P, Marsh DG (1965) The isolation and characterization of allergens from the pollen of rye grass. (*Lolium perenne*). Europ Polymer J 1:63–77

Johnson P, Marsh DG (1966a) Allergens from common rye grass pollen. I. Chemical composition and structure. Immunochemistry 3:91–100

Johnson P, Marsh DG (1966b) Allergens from common rye grass pollen. II. The allergenic determinants and carbohydrate moiety. Immunochemistry 3:101–110

Johnson RA, Baer H, Kirkpatrick CH, Dawson CR, Khurana RG (1972) Comparison of the contact allergenicity of the four pentadecylcatechols derived from poison ivy urushiol in humans. J Allergy Clin Immunol 49:27–35

Kanellopoulos JM, Liu TY, Poy G, Metzger H (1980) Composition and subunit structure of the cell receptor for Immunoglobulin E. J Biol Chem 255:9060–9066

King TP (1976) Chemical and biological properties of some atopic allergens. Adv Immunol 23:77–105

King TP, Norman PS (1962) Isolation studies of allergens from ragweed pollen. Biochemistry 1:709–720

King TP, Norman PS, Connell JT (1964) Isolation and characterization of allergens from ragweed pollen. Biochemistry 3:458–468

King TP, Norman PS, Lichtenstein LM (1967) Properties of a ragweed allergen, Antigen K. Ann Allergy 10:541–544

Klapper DG, Goodfriend L, Capra JD (1980) Amino acid sequence of ragweed allergen Ra3. Biochemistry 19:5729–5734

Knox RB (1979) Pollen and allergy. Studies in Biology 107:54. Arnold. London

Knox RB, Heslop-Harrison J (1970) Pollen-wall proteins: localization and enzymic activity. J Cell Sci 6:1–27

Knox RB, Heslop-Harrison J (1971) Pollen-wall proteins: localization of antigenic and allergenic proteins in the pollen grain walls of *Ambrosia* spp. (ragweeds). Cytobios 4:49–54

Knox RB, Heslop-Harrison J, Reed CE (1970) Localization of antigens associated with the pollen wall by immunofluorescence. Nature 225:1066–1068

Lahti A, Bjorksten F, Hannuksela M (1980) Allergy to birch pollen and apple and cross-reactivity of the allergens studied with RAST. Allergy 35:297–300

Lapkoff CB, Goodfriend L (1974) Isolation of a low molecular weight ragweed pollen allergen, Ra5. Int Arch Allergy Appl Immunol 46:215–229

Lowenstein H (1978a) Quantitative immunoelectrophoretic methods as a tool for the analysis and isolation of allergens. Prog Allergy 25:1–62

Lowenstein H (1978b) Isolation and partial characterization of three allergens of timothy pollen. Allergy 33:30–41

Malish D, Glousky MM, Hoffman DR, Ghekiere L, Hawkins JM (1981) Anaphylaxis after sesame seed ingestion. J Allergy Clin Immunol 67:35–38

Malley A, Harris RL (1967) Biological properties of a non-precipitating antigen from timothy pollen extracts. J Immunol 99:825–830

Marsh DG (1974) Purification of pollen allergens. Use in genetic studies of immune responsiveness in man. 8th Int Congr Allergol Proc Amsterdam Excerpta Med, pp 381–393

Marsh DG (1975) Allergens and the genetics of allergy. In: Sela M (ed) The antigens. Academic Press, London New York, pp 271–359

Marsh DG, Bias WB, Hsu SH, Goodfriend L (1973) Association of the HLA 7 cross-reacting group with a specific reaginic antibody response in allergic man. Science 179:691–693

Marsh DG, Bias WB, Hsu SH, Goodfriend L (1979) Association of HLA antigens and total serum IgE level with allergic response and failure to respond to ragweed allergen, Ra3. Proc Natl Acad Sci USA 76:2903–2907

Mitchell JC, Chan-Yeung M (1974) Contact allergy from *Frullania* and respiratory allergy from *Thuja*. Can Med Assoc J 110:653–657

Mitchell JC, Dupois G (1971) Allergic contact dermatitis from sesquiterpenoids of the Compositae family of plants. Br J Dermatol 84:139–150

Mitchell J, Rook A (1979) Botanical dermatology. Plants and plant products injurious to the skin. Greenglass, Vancouver

Mitchell JC, Fritig B, Singh B, Towers GHN (1970) Allergic contact dermatitis from *Frullania* and Compositae. The role of sesquiterpene lactones. J Invest Dermatol 54:233–239

Mitchell JC, Dupois G, Geissman TA (1972) Allergic contact dermatitis from sesquiterpenoids of plants. Br J Dermatol 87:235–240

Mole LE, Goodfriend L, Lapkoff CB, Kehoe JM, Capra JD (1975) The amino acid sequence of ragweed pollen allergen Ra5. Biochemistry 14:1216–1220

Osler AG, Lichenstein LM, Levy DA (1968) *In vitro* studies of human reaginic activity. Adv Immunology 8:183–231

Paull BR, Gleich GJ, Atassi MZ (1979) Structure and activity of ragweed Antigen E. II. Allergenic cross-reactivity of the subunits. J Allergy Clin Immunol 64:539–545

Pirquet C von (1906) Allergie. Muench Med Wochenschr 53:1457 (cited by Stanley and Linskens 1974)

Polak L (1980) Immunological aspects of contact sensitivity. In: Dukor P, Kallós P, Trnka Z, Waksman BH, de Weck AL (eds) Monographs in allergy, vol 15. Karger, Basel

Rippon JW (ed) (1974) Allergic diseases. Medical mycology. Saunders, Philadelphia London Toronto, pp 502–507

Sachs MI, Jones RT, Yuninger JW (1981) Isolation and partial characterization of a major peanut allergen. J Allergy Clin Immunol 67:27–34

Shibasaki M, Suzuki S, Tajima S, Nemoto H, Kuroume T (1980) Allergenicity of major component proteins of soybean. Int Arch Allergy Appl Immunol 61:441–448

Slob A (1973) Tulip allergens in *Alstroemeria* and some other Liliiflorae. Phytochemistry 12:811–815

Smart IJ, Knox RB (1979) Aerobiology of grass pollen in the city atmosphere of Melbourne: Quantitative analysis of seasonal and diurnal changes. Aust J Bot 27:317–331

Sohi AS, Tiwari VD, Lonkar A, Nagasampagi BA (1979) Allergenic nature of *Parthenium hysterophorus*. Contact Dermatitis 5:133–136

Toyama I (1923) Further observations on lacquer dermatitis. Trop Disease Bull 22:231–234

Underdown GJ, Goodfriend L (1969) Isolation and characterization of a protein allergen from ragweed pollen. Biochemistry 8:980–989

Viander M, Fraki J, Djupsund BM, Laine S (1979) Antigens and allergens in birch pollen extract. Allergy 34:289–302

Vithanage HIMV, Howlett BJ, Knox RB (1980) Localization of grass pollen allergen by immunocytochemistry. Micron 11:411–412

Vithanage HIMV, Howlett BJ, Jobson S, Knox RB (1982) Immunocytochemical localization of water-soluble glycoproteins, including Group 1 allergen, in pollen of rye grass using ferritin-labelled antibody. Histochem J 14:949–966

Weeke B, Lowenstein H, Neilsen L (1974) Allergens in timothy pollen identified by crossed radioimmunoelectrophoresis. Acta Allerg 29:409–417

Wide L, Bennich H, Johansson SGO (1967) Diagnosis of allergy by an *in vitro* test for allergen antibodies. Lancet 2:1105

Wollenweber E, Schnepf E (1970) Vergleichende Untersuchungen über die flavonoiden Exkrete von Mehl- und Öl-Drüsen bei Primeln und die Feinstruktur der Drüsenzellen. Z Pflanzenphysiol Bd 62:216–227

Youle RJ, Huang AHC (1978) Evidence that castor bean allergens are the albumin storage proteins in the protein bodies of castor bean. Plant Physiol 61:1040–1042

Author Index

Page numbers in *italics* refer to the references

Index of Plant Genera

Subject Index

Encyclopedia of Plant Physiology

New Series

Editors: A. Pirson, M. H. Zimmermann

Volume 1
Transport in Plants I
Phloem Transport
Editors: M. H. Zimmermann, J. A. Milburn
With contributions by numerous experts
1975. 93 figures. XIX, 535 pages
ISBN 3-540-07314-0

Volume 2 A
Transport in Plants II
Cells
Editors: U. Lüttge, M. G. Pitman
With contributions by numerous experts
1976. 97 figures, 64 tables. XVI, 419 pages
ISBN 3-540-07452-X

Volume 2 B
Transport in Plants II
Tissues and Organs
Editors: U. Lüttge, M. G. Pitman
With contributions by numerous experts
1976. 129 figures, 45 tables. XII, 475 pages
ISBN 3-540-07453-8

Volume 3
Transport in Plants III
Intracellular Interactions and Transport Processes
Editors: C. R. Stocking, U. Heber
With contributions by numerous experts
1976. 123 figures. XXII, 517 pages
ISBN 3-540-07818-5

Volume 4
Physiological Plant Pathology
Editors: R. Heitefuss, P. H. Williams
With contributions by numerous experts
1976. 92 figures. XX, 890 pages
ISBN 3-540-07557-7

Volume 6
Photosynthesis II
Photosynthetic Carbon Metabolism and Related Processes
Editors: M. Gibbs, E. Latzko
With contributions by numerous experts
1979. 75 figures, 27 tables. XX, 578 pages
ISBN 3-540-09288-9

Volume 7
Physiology of Movements
Editors: W. Haupt, M. E. Feinleib
With contributions by numerous experts
1979. 185 figures, 19 tables. XVII, 731 pages
ISBN 3-540-08776-1

Volume 8
Secondary Plant Products
Editors: E. A. Bell, B. V. Charlwood
With contributions by numerous experts
1980. 176 figures, 44 tables and numerous schemes and formulas. XVI, 674 pages
ISBN 3-540-09461-X

Volume 9
Hormonal Regulation of Development I
Molecular Aspects of Plant Hormones
Editor: J. MacMillan
With contributions by numerous experts
1980. 126 figures. XVII, 681 pages
ISBN 3-540-10161-6

Volume 10
Hormonal Regulation of Development II
From the Cell to the Whole Plant
Editor: T. K. Scott
With contributions by numerous experts
ISBN 3-540-10196-9
In preparation

Springer-Verlag
Berlin
Heidelberg
New York
Tokyo

Springer-Verlag
Berlin
Heidelberg
New York
Tokyo